Integrated Mathematics

COURSE III   Second Edition

Edward P. Keenan

Curriculum Associate, Mathematics
East Williston Union Free School District
East Williston, New York

AND

Ann Xavier Gantert

Department of Mathematics
Nazareth Academy
Rochester, New York

Dedicated to serving

AMSCO

our nation's youth

When ordering this book, please specify: *either* **R 532 P** *or*
INTEGRATED MATHEMATICS: COURSE III, SECOND EDITION, PAPERBACK

AMSCO SCHOOL PUBLICATIONS, INC.
315 Hudson Street   New York, N.Y. 10013

ISBN 0-87720-277-X

Preface

INTEGRATED MATHEMATICS: COURSE III, *Second Edition* is a revision of the textbook that has been a leader in presenting high school mathematics in a contemporary, integrated manner. Over the past decade, this integrated approach has undergone further changes and refinements. Amsco's Second Edition reflects these developments.

The Amsco book parallels the integrated approach to the teaching of high school mathematics that is being promoted by the National Council of Teachers of Mathematics (NCTM) in its STANDARDS FOR SCHOOL MATHEMATICS. Moreover, the Amsco book implements many of the suggestions set forth in the NCTM Standards, which are the acknowledged guidelines for achieving a higher level of excellence in the study of mathematics.

In this Second Edition of INTEGRATED MATHEMATICS: COURSE III, which fully satisfies the requirements of the revised New York State Syllabus:

● **Problem solving** has been expanded by (1) adding nonroutine problems for selected topics and to Chapter Reviews, and (2) providing, in the Teacher's Manual, Bonus questions for each chapter.

● **Integration** of Geometry, Algebra, and Trigonometry, for which the First Edition was well known, has been continued.

● **Algebraic skills** from Courses I and II are maintained and strengthened.

● **Enrichment** has been extended by (1) increasing the number of optional topics, (2) increasing the number of challenging exercises, and (3) adding to the Teacher's Manual more thought-provoking aspects of topics in the text.

● **Hands-on activities** have been included in the Teacher's Manual to promote understanding through discovery.

The First Edition of this series was written to provide effective teaching materials for a unified three-year program. In the Second Edition of this series, adjustments have been made in the emphasis and placement of certain topics.

For example, the solution of quadratic equations by formula, formerly considered to be an 11th-year topic, has been introduced earlier, as optional in Course I and required in Course II. Now, in Course III, the quadratic formula is applied to equations with roots that are complex numbers, and to the solution of linear-quadratic systems. This tri-level presentation better serves to maintain and extend algebraic skills. In addition, transformations, now introduced intuitively in Course I, and presented in greater depth in their relationship to the coordinate plane in Course II, are used throughout Course III in the development of functions, trigonometry, and logarithms.

Both Amsco editions of Course III use the trigonometry of the right triangle as an introduction to the general definitions of the sine, cosine, and tangent functions. Teachers who prefer not to use this approach, however, may omit the introductory sections and present the definitions of the trigonometric functions in terms of the coordinates of points on a unit circle. For those who use this approach, the trigonometric ratios of the right triangle are derived as an application of the general trigonometic functions.

Logarithmic functions are introduced as the inverses of the exponential functions. Although the use of logarithms to solve computational problems is no longer important in our world of calculators and computers, such problems, by providing concrete applications of the laws of logarithms, can lead to an understanding of this important class of functions.

Statistics, introduced in Course I, is reviewed and extended to include several types of deviation from the mean, particularly standard deviation and its application to a normal distribution. The chapter on probability builds on the students' earlier work with permutations and combinations. The probability of success in an experiment with two outcomes is related to the expansion of a binomial.

This Second Edition is offered so that teachers may effectively continue to help students comprehend, master, and enjoy mathematics from an integrated point of view.

This book is dedicated to Anna Gantert, whose encouragement throughout its writing was a continuation of the loving support she has given to her daughter all her life. The book is also dedicated to Mary, David, Jennifer, and Joanna Keenan, who are a constant source of pride to their father.

Edward P. Keenan

Ann Xavier Gantert

Contents

CHAPTER 4   The Irrational Numbers and the Real Numbers

CHAPTER 5   Transformation Geometry and Coordinates

CHAPTER 6   Relations and Functions

CHAPTER 7   Functions and Transformation Geometry

CHAPTER 8   Trigonometric Functions

Chapter 1

The Rational Numbers

1-1 THE NUMBER LINE AND SETS OF NUMBERS

The numbers stamped onto a ruler are the first numbers we learned as children, namely, 1, 2, 3, and so on. This set of numbers, called the *counting numbers* or the *natural numbers*, is written in set notation as {1, 2, 3, 4, 5, 6, 7, 8, 9, 10, 11, 12, . . .}. The three dots indicate that the numbers continue in the same pattern without end.

By combining zero with the counting numbers, we form the set of *whole numbers*, represented as {0, 1, 2, 3, 4, 5, . . .}.

The ruler is a model of a *number line*. Once we have assigned to any two distinct points on a straight line the values "0" and "1," we have determined a segment whose length is the *unit measure*. By continuing in the direction of 1 from 0, it is possible to mark off equally spaced points and to assign to these points the numbers 2, 3, 4, and so on, as seen on the ruler.

Since a line extends infinitely in both directions, we can begin at zero to mark off even more equally spaced points in the direction opposite to that used to assign the whole numbers. We assign to these points the numbers -1, -2, -3, and so on. The set of numbers assigned to equally spaced points on the number line is the set of *integers*, written symbolically as {. . . , -4, -3, -2, -1, 0, 1, 2, 3, 4, . . .}.

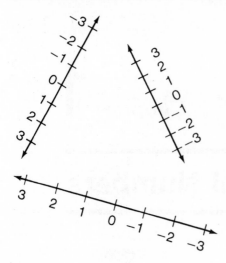

A number line can go in any direction and can have any length as its unit measure. However, once "0" and "1" have been assigned to points on this line, we cannot change the direction and we cannot change the unit measure or the scale for the remainder of the line. Arrowheads indicate that a line has no beginning and no end, just as the set of integers has no beginning and no end.

Of all the possible directions that can be used for number lines, we have come to use two directions most often in our daily lives. A *vertical* number line is used in thermometers to show us temperatures and in rulers to measure height. In a vertical line, we agree that numbers increase as we move up the line and decrease as we move down the line.

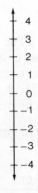

A *horizontal* line, like the ruler first shown, is drawn so that numbers increase as we move to the right and decrease as we move to the left. Thus, we agree that the *positive integers* 1, 2, 3, ... are placed to the right of zero and the *negative integers* -1, -2, -3, ... lie to the left of zero. In the horizontal line that follows, as in the ruler first shown, the unit measure is 1 centimeter.

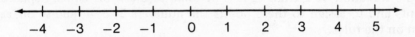

The Rational Numbers

Once the integers have been assigned to points on a number line, it is possible to divide the segments whose endpoints represent the integers into halves, thirds, quarters, tenths, and so on. In this way, we can assign fractions, decimals, and mixed numbers to specific points on the number line.

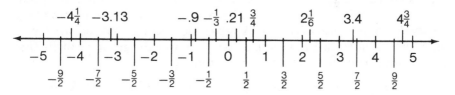

All numbers indicated on the line just displayed, including the integers, are examples of rational numbers.

■ **DEFINITION.** A number is a *rational number* if and only if it can be expressed in the form $\frac{a}{b}$ where a and b are integers and $b \neq 0$.

Notice that every integer x is a rational number because the integer can be written in the form $\frac{x}{1}$. Using examples from the number line just displayed, notice how decimals and mixed numbers are written in the required form:

$$.21 = \tfrac{21}{100} \qquad 2\tfrac{1}{6} = \tfrac{13}{6} \qquad\qquad 3.4 = 3\tfrac{4}{10} = \tfrac{34}{10}$$

$$-.9 = \frac{-9}{10} \qquad -4\tfrac{1}{4} = \frac{-17}{4} \qquad -3.13 = -3\frac{13}{100} = \frac{-313}{100}$$

In other words, every rational number can be expressed in the form of a fraction. To express a rational number (named as a fraction) in decimal form, divide the numerator by the denominator.

MODEL PROBLEM

Express as a decimal: **a.** $\tfrac{3}{8}$ **b.** $\tfrac{1}{3}$ **c.** $\tfrac{5}{11}$ **d.** $2\tfrac{1}{6}$

Solution

a. $\tfrac{3}{8} = 8\overline{)3.00000}^{\,.37500}$ **b.** $\tfrac{1}{3} = 3\overline{)1.00000}^{\,.33333\ldots}$

c. $\tfrac{5}{11} = 11\overline{)5.000000}^{\,.454545\ldots}$ **d.** $2\tfrac{1}{6} = \tfrac{13}{6} = 6\overline{)13.00000}^{\,2.16666\ldots}$

In each of the examples just given, we see that once the division is performed, a group of digits appears in the quotient so that the digits repeat continuously in the same order. Decimals that, from some point onward, repeat a sequence of digits without end are called *repeating decimals*, or *periodic decimals*. A repeating decimal may be written in an abbreviated form by placing a bar ($^{\frown}$) over the group of digits that are to be continuously repeated. For example:

$$.33333\ldots = .\overline{3} \qquad .454545\ldots = .\overline{45} \qquad 2.16666\ldots = 2.1\overline{6}$$

In some conversions, such as $\frac{3}{8} = .375$, the decimal obtained is called a *terminating decimal* because a point is reached where the division appears to end or to be completed. After this point is reached, however, it can be seen that zeros will repeat endlessly in the quotient of every terminating decimal. For this reason, every terminating decimal can be expressed as a repeating decimal. For example:

$$.375 = .375000\ldots = .375\overline{0}$$

Therefore, we make the following important observation:

■ **Every rational number can be expressed as a repeating decimal.**

In *Course I*, we learned a procedure by which a repeating decimal could be converted into a rational number in fractional form.

| MODEL PROBLEM |

Find a fraction that names the same rational number as: a. .13333 ... b. .606060 ...

How to Proceed	*Solution*

a. 1. Let N = .1333 ...

   2. Multiply both members of the equation in step 1 by 10, because a one-digit repetition appears in the number. $10N = 1.33333\ldots$

   3. Subtract the equation in step 1 from the equation in step 2. $\underline{-N = -.13333\ldots}$
                                          $9N = 1.2$

   4. Solve the resulting equation for N. $N = \dfrac{1.2}{9.0} = \dfrac{12}{90} = \dfrac{2}{15}$

   Answer: .13333 ... = $\frac{2}{15}$

b. 1. Let $N = .606060\ldots$

2. Multiply both members of the equation in step 1 by 100, because a two-digit repetition appears in the number. $\quad 100N = 60.606060\ldots$

3. Subtract the equation in step 1 from the equation in step 2. $\quad \underline{-N = -.606060\ldots}$
$$99N = 60$$

4. Solve the resulting equation for N. $\qquad N = \frac{60}{99} = \frac{20}{33}$

Answer: $.606060\ldots = \frac{20}{33}$

These examples illustrate the truth of the following statement, which is the *converse* of our last observation:

■ Every repeating decimal represents a rational number.

Since both the original observation and its converse are true, we know from our study of logic that a *biconditional* statement can be made:

■ A number is a rational number if and only if it can be represented as a repeating decimal.

EXERCISES

In 1–8, tell whether the statement is true or false.

1. Every counting number is a rational number.
2. Every whole number is an integer.
3. Every integer is a whole number.
4. The smallest whole number is 0.
5. The smallest natural number is 0.
6. Every rational number can be expressed as a repeating decimal.
7. Every repeating decimal is a rational number.

8. If a and b are integers, then $\frac{a}{b}$ is a rational number.

In 9–20, express each rational number as a repeating decimal. (Note that zeros repeat in all terminating decimals.)

9. $\frac{4}{9}$ 10. $\frac{4}{11}$ 11. $-\frac{2}{3}$ 12. $\frac{4}{5}$ 13. $1\frac{1}{3}$ 14. $-1\frac{1}{6}$

15. $-2\frac{3}{4}$ 16. $\frac{7}{60}$ 17. $-\frac{4}{15}$ 18. $\frac{1}{90}$ 19. $1\frac{1}{20}$ 20. $3\frac{1}{7}$

In 21–40, express each rational number as a fraction.

21. 5   **22.** −3   **23.** $3\frac{1}{2}$ **24.** 0   **25.** $-2\frac{5}{7}$

26. .7   **27.** .23   **28.** −.8   **29.** 4.1   **30.** .666 . . .

31. $.\overline{7}$ **32.** $.\overline{23}$ **33.** $.\overline{2}$ **34.** $.\overline{02}$ **35.** $.\overline{27}$

36. $8.\overline{3}$ **37.** $.8\overline{3}$ **38.** $.08\overline{3}$ **39.** $.3\overline{6}$ **40.** $.\overline{142857}$

41. True or false: .999 . . . = 1. Explain the reason for your answer.

1-2 QUANTIFIERS

Consider these sentences and their related truth values.

p: Integers are rational numbers.   (True)
q: Negative numbers are greater than zero. (False)
r: Integers are primes.   (Uncertain)

We know that p is true because *all* integers are rational numbers. We know that q is false because *no* negative number is greater than zero. We are uncertain of the truth value of r because *some* integers are primes and *some* integers are not primes. The words we have just used, namely, *all*, *no*, and *some*, are called *quantifiers*.

By using the proper quantifier in a given sentence, we can form a quantified statement that is always true. Thus:

All integers are rational numbers.   (True)
No negative number is greater than zero. (True)
Some integers are primes.   (True)

■ DEFINITION. A *domain* is a set of all possible replacements for a given variable. For example, in statements p and r, the domain is the set of integers. In statement q, the domain is the set of negative numbers. Thus, a given sentence that involves a variable may be written using a quantifier.

■ DEFINITION. A *quantifier* is a phrase that describes in general terms the part of the domain for which the sentence is true.

The Universal Quantifier

Whenever a sentence has the same truth value for all replacements from the domain, we can write this sentence as a *universally quantified statement*.

The word *all* is the commonly used expression for the *universal quantifier*, represented symbolically as $\forall_x$. This symbol is formed by writing a capital letter A upside down and by including a subscript x to indicate the variable. The symbol, $\forall_x$, is read in many ways as shown in the box to the right.

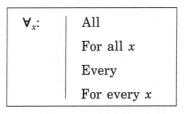

$\forall_x$:	All
	For all x
	Every
	For every x

Universal Quantifier

Let us consider two types of sentences in which the universal quantifier is used.

1. *Sentences that are always true.*

Let b represent the sentence "$x + x = 2x$."

When the domain is the set of rational numbers, every replacement for the variable x results in a true statement: $5 + 5 = 2(5)$, $3.1 + 3.1 = 2(3.1)$, and so on. Thus, we can attach the universal quantifier $\forall_x$ to the sentence and form a true statement. Notice that the statement is expressed symbolically in only one way. However, the statement may be expressed in words in many ways.

$\forall_x$: b | For *all* values of x, $x + x = 2x$.
| For *every* x, $x + x = 2x$.

In the same way, consider the sentence studied earlier in this section, namely, p: Integers are rational numbers. Since this statement is always true, we can use the universal quantifier *all* as follows:

$\forall_x$: p | *All* integers are rational numbers.
| *Every* integer is a rational number.
| For *all* integers, integers are rational numbers.

2. *Sentences that are always false.*

Let d represent the sentence "$x \cdot x = -1$."

When the domain is the set of rational numbers, there is no number that when multiplied by itself results in the product -1. Since the sentence d: $x \cdot x = -1$, is always *false*, it follows that its negation $\sim d$: $x \cdot x \neq -1$, is always *true*. Therefore, we quantify a false statement by attaching the universal quantifier, $\forall_x$, to the negation of the statement. Notice the various ways by which the words *no* or *not* enter the sentences.

$\forall_x\colon\ \sim d$ | For all x, $x \cdot x \neq -1$.

For all x, it is *not* true that $x \cdot x = -1$.

There are *no* values of x for which $x \cdot x = -1$.

For every x, $x \cdot x$ is *not* equal to -1.

In the same way, consider the sentence studied earlier in this section, namely q: Negative numbers are greater than zero. This statement is always false. To form a true statement, we negate the given sentence and attach a universal quantifier. This can be expressed by using the quantifier *no*.

$\forall_x\colon\ \sim q$ | *No* negative numbers are greater than zero.

Every negative number is *not* greater than zero.

All negative numbers are *not* greater than zero.

Therefore we may conclude:

■ **DEFINITION.** A ***universal quantifier*** is a phrase that describes the statement as being true for *all* elements of the domain or for *no* element of the domain.

The Existential Quantifier

Whenever a sentence is true for at least one replacement from the domain, write this sentence as an *existentially quantified statement*.

The words *there exists* or the word *some* are commonly-used expressions for the *existential quantifier*, represented symbolically as $\exists_x$. This symbol is formed by writing a capital letter E backwards and by including a subscript x to indicate the variable. The symbol $\exists_x$ is read in many ways, as shown in the box at the right.

$\exists_x\colon$	There exists
	At least one
	For some
	There is one
	Some

Existential Quantifier

Let h represent the sentence "$x + 1 = 5$."

When the domain is the set of rational numbers, there is at least one replacement for the variable x that results in a true statement. When $x = 4$, it is true that $x + 1 = 5$. Thus, we can attach the existential quantifier, $\exists_x$, to the sentence and form a true statement.

$\exists_x\colon h$ | *There exists* a value for x such that $x + 1 = 5$.

For *some* value of x, $x + 1 = 5$.

There is *at least one* value of x where $x + 1 = 5$.

In the same way consider the sentence studied earlier in this section, namely, r: Integers are primes. We know that some integers are primes. Therefore, we know at least one integer exists that is a prime. We attach the existential quantifier to form the following true statement:

$\exists_x: r$ | There *exists* an integer that is a prime.

There *is at least one* integer that is a prime.

Some integers are primes.

Therefore, we may conclude:

■ **DEFINITION.** An *existential quantifier* is a phrase that describes the statement as being true for *at least one* replacement from the domain.

It should be noted that an existential quantifier may be used for a sentence that is universally true. Thus, while all replacements of the domain are true, it is *not false* to say that *at least one* replacement results in a true statement. The following statements are described more accurately by using a universal quantifier, but they are still *true* statements.

$\exists_x: \quad b$ | There is an x such that $x + x = 2x$.

$\exists_x: \sim d$ | There exists at least one x such that $x \cdot x \neq -1$.

$\exists_x: \quad p$ | Some integers are rational numbers.

$\exists_x: \sim q$ | There is at least one negative number that is not greater than zero.

| **MODEL PROBLEMS** |

In 1–4, describe the sentence as being universally quantified, existentially quantified, or not quantified.

Solutions

1. Some books are written in Greek.

1. The word *some* indicates existentially quantified. *Ans.*

2. This book is Greek to me.

2. Not quantified. *Ans.*

3. Every Greek has a rich cultural background.

3. The word *every* indicates universally quantified. *Ans.*

4. All Greeks are not wealthy.

4. The word *all* indicates universally quantified. *Ans.*

5. Which statement is true for the set of real numbers?
   (1) $\exists_x \ x > x + 8$ (2) $\forall_x \ x > x + 8$ (3) $\exists_x \ x = 8$ (4) $\forall_x \ x = 8$

Solution: The open sentence $x > x + 8$ in choices (1) and (2) is never true. The open sentence $x = 8$ in choices (3) and (4) is true for only one value, namely, 8. Thus, the sentence $x = 8$ can be existentially quantified.   *Answer:* (3) $\exists_x \ x = 8$

| EXERCISES |

In 1–8: **a.** Describe the sentence as being universally quantified, existentially quantified, or not quantified. **b.** If the sentence is quantified, select the word or words that act as the quantifiers.

1. All whole numbers are rational numbers.
2. Some angles are obtuse angles.
3. No one in his right mind brings a camel into the subway.
4. The probability that you are correct is $\frac{1}{3}$.
5. There are certain whole numbers that are divisible by 5.
6. This is a difficult decision for me to make.
7. Every important decision requires careful thought.
8. At least one person is a true friend to me.

In 9–12, write the given sentence in words. Let $x \in \{cars\}$.

9. $\forall_x$ cars have wheels.   **10.** $\forall_x$ cars do not fly.
11. $\exists_x$ cars are red.   **12.** $\exists_x$ cars need tune-ups.

In 13–17, write the expression in symbolic form, using $\forall_x$ or $\exists_x$.

13. For every integer x, x is less than $x + 5$.
14. For some integer x, x is less than 5.
15. For no integer x, x is equal to $x + 5$.
16. For every rational value x, it is not true that $x + 5$ equals 8.
17. There is a rational value x such that $x + 5$ equals 8 is true.

In 18–22, use the domain of rational numbers to tell which one of the three given statements is true.

18. (1) $\forall_x: x^2 = 0$ (2) $\forall_x: x^2 > 0$ (3) $\forall_x: x^2 \geq 0$
19. (1) $\exists_x: x^2 = 0$ (2) $\exists_x: x^2 < 0$ (3) $\forall_x: x^2 < 0$
20. (1) $\exists_x: 2x + 1 = 5$ (2) $\forall_x: 2x + 1 = 5$ (3) $\forall_x: 2x + 1 \neq 5$
21. (1) $\forall_x: x - 3 \geq 7$ (2) $\exists_x: x - 3 \geq 7$ (3) $\exists_x: x - 3 \geq x$
22. (1) $\forall_x: x + x \neq x$ (2) $\forall_x: x + x > x$ (3) $\exists_x: x + x < x$

In 23–25, select the quantified statement that is true.

23. (1) All people are tall.
(2) No people are tall.
(3) Some people are tall.

24. (1) All rational numbers are integers.
(2) Some rational numbers are integers.
(3) No rational numbers are integers.

25. (1) All rectangles are parallelograms.
(2) Some rectangles are not parallelograms.
(3) Every rectangle is not a parallelogram.

1-3 NEGATIONS OF QUANTIFIED STATEMENTS

Negating a Universally Quantified Statement

"Every month has 31 days." We know that this statement is false. To prove that it is false, it is not necessary to consider every month. We only need to show one month that does not have 31 days. For example, since September has 30 days, the statement is not true for every month. The one month, September, that we used to show that the universally quantified statement was false is called a ***counterexample***. The counterexample shows that there is at least one month that does not have 31 days.

We can express these statements in symbols and in words as follows:

Let $x \in$ {months of the year}.
Let p represent "x has 31 days."

$\forall_x p$: Every month has 31 days. (False)

$\exists_x \sim p$: Some months do not have 31 days. (True)

Let us study another example, starting with a universally quantified statement that is true. We know that all segments have two endpoints. Thus, a statement that "at least one segment exists that does not have two endpoints" would be false. In symbols and in words, we say:

Let $x \in$ {line segments}.
Let q represent "x has two endpoints."

$\forall_x q$: *All* segments have two endpoints. (True)

$\exists_x \sim q$: *Some* segments do *not* have two endpoints. (False)

OR

There is *at least one* segment that does *not* have two endpoints. (False)

We may conclude:

■ The *negation of a universally quantified statement p* is an existentially quantified statement of the negation of *p*.

In symbolic form:

The negation of $\forall_x p$ is given as $\exists_x \sim p$.

Negating an Existentially Quantified Statement

"Some months have 32 days." We know that this statement is false. To prove that it is false, it is necessary to consider every month. After we have shown that every month has less than 32 days, we know that no month exists that has 32 days.

We can express these statements in symbols and in words as follows:

Let $x \in$ {months of the year}.
Let *p* represent "*x* has 32 days."

$\exists_x p$: Some months have 32 days.   (False)
$\forall_x \sim p$: All months do not have 32 days. (True)

OR

No month has 32 days.

In the following example, we start with an existentially quantified statement that is true.

Let $x \in$ {polygons}.
Let *t* represent "*x* has five sides."

$\exists_x t$: *Some* polygons have five sides. (True)
$\forall_x \sim t$: *No* polygon has five sides.   (False)

We may conclude:

■ The *negation of an existentially quantified statement p* is a universally quantified statement of the negation of *p*.

In symbolic form:

The negation of $\exists_x p$ is given as $\forall_x \sim p$.

The Difference Between *All Are Not* and *Not All Are*

The negation of a quantified statement is sometimes indicated by negating the quantifier. Let us note the differences that occur because of the placement of a negation. The following examples illustrate the difference between the phrases:

all are not ($\forall_x \sim p$, where the negation is on *p*), and
not all are ($\sim \forall_x p$, where the negation is on $\forall$).

Let $x \in \{$persons$\}$.   Let p represent "x is honest."

We use *all are not* when the statement p, which is universally quantified, is being negated:

$\forall_x\ p$: All persons are honest.
$\forall_x\ {\sim}p$: All persons are not honest. OR No persons are honest.

We use *not all are* when the universal quantifier $\forall$ is being negated:

${\sim}\forall_x\ p$: Not all persons are honest.

Note that since the negation of a universally quantified statement is an existentially quantified statement, we have:

${\sim}\forall_x\ p \leftrightarrow \exists_x\ {\sim}p$: Not all persons are honest.
OR Some persons are not honest.
OR At least one person is not honest.

MODEL PROBLEMS

In 1–4, write the negation of the quantified statement.

1. $\forall_x$: $x = 2$ **2.** $\exists_x$: $x > 7$ **3.** $\forall_x\ {\sim}k$ **4.** $\exists_x\ {\sim}m$

Solutions

1. $\exists_x$: $x \neq 2$ **2.** $\forall_x$: $x \leq 7$ **3.** $\exists_x\ k$ **4.** $\forall_x\ m$

Note. In problem 2 above, the expression $x \leq 7$ is equivalent to the expression $x \not> 7$.

5. Which is the negation of the statement "Some rectangles are squares"?

(1) Some rectangles are not squares.
(2) All rectangles are squares.
(3) All rectangles are not squares.
(4) All squares are rectangles.

Solution

Examine the original statement: "Some rectangles are squares."

Write the statement in symbolic form: $\exists_x\ p$

Negate this quantified statement: $\forall_x\ {\sim}p$

Translate the negation: "All rectangles are not squares."

This is choice (3) All rectangles are not squares. *Ans.*

| EXERCISES |

In 1–22, write the negation of the quantified statement.

1. $\forall_x: x = 3$ **2.** $\exists_x: x \neq 5$ **3.** $\exists_x: x > 4$ **4.** $\forall_x: x < 9$

5. $\exists_x: x \geq 2$ **6.** $\forall_x: x = x$ **7.** $\forall_x b$ **8.** $\exists_x d$

9. All men are human.   **10.** Some men are handsome.
11. All women are beautiful.   **12.** Some women are not rich.
13. Some animals can fly.   **14.** All roses are not red.
15. Some chairs are not soft. **16.** No chairs are tables.
17. All people are not fat.   **18.** Every frog can not sing.
19. All segments have a midpoint.
20. No segments are lines.
21. Every segment does not have two midpoints.
22. Some segments do not have a length of 5 meters.

In 23–30, select the numeral preceding the statement that best completes the sentence or answers the question.

23. Which is the negation of the statement, "All math is fun"?
   (1) No math is fun.   (2) All math is not fun.
   (3) Some math is fun.   (4) Some math is not fun.

24. Which is the negation of the statement, "Some numbers are odd"?
   (1) All numbers are odd.
   (2) All numbers are not odd.
   (3) Some numbers are even.
   (4) Some numbers are not odd.

25. Which is the negation of $\forall_x \; x > 5$?
   (1) $\exists_x \; x > 5$ (2) $\forall_x \; x \leq 5$ (3) $\exists_x \; x \leq 5$ (4) $\exists_x \; x < 5$

26. The negation of "Some angles are acute" is
   (1) Some angles are not acute.   (2) No angles are acute.
   (3) All angles are acute.   (4) No angles are not acute.

27. Which is the negation of "Every triangle has three sides"?
   (1) No triangle has three sides.
   (2) Some triangle does not have three sides.
   (3) All triangles have three sides.
   (4) Some triangles have three sides.

28. Which is the negation of "No flowers are pink"?
   (1) Some flowers are pink.   (2) All flowers are pink.
   (3) Some flowers are not pink.   (4) All flowers are not pink.

29. The negation of "No cowboys are Indians" is
(1) All cowboys are Indians.
(2) All cowboys are not Indians.
(3) Some cowboys are Indians.
(4) Some cowboys are not Indians.

30. Which is the negation of "There is an n such that $n^2 = n$"?
(1) For all n, $n^2 = n$.   (2) For some n, $n^2 \neq n$.
(3) For some n, $n^2 = n$.   (4) For all n, $n^2 \neq n$.

1-4 MATHEMATICAL SYSTEMS AND GROUPS

In geometry, you learned that a *postulational system* is one in which definitions and postulates are used to prove theorems; and then definitions, postulates, and proven theorems are used to *deduce* other theorems. This type of *deductive reasoning* is not limited to geometry but is also used to develop systems that are numerical or algebraic in nature.

Binary Operations

When we operate on two elements from a set and the result is an element from that set, we are performing a *binary operation.* Addition, subtraction, multiplication, and division are examples of binary operations on the set of rational numbers. Of course we sometimes add columns consisting of more than two numbers, but to do so we add only two numbers at a time.

■ **DEFINITION.** A *binary operation* $*$ in a set S is a way of assigning to every ordered pair of elements from S a unique response from S.

In symbolic form, we write:

$$\forall a, b \in S: \quad a * b = c \text{ and } c \in S$$

If S is a finite set, a given binary operation may be defined by showing each result in a table.

□ EXAMPLE: Let $S = \{0, 1, 2, 3, 4\}$. Let $\oplus$ be the symbol for an operation defined by the table shown at the right. Note that to every pair of elements from S, there is assigned a unique member of S.

$\oplus$	0	1	2	3	4
0	0	1	2	3	4
1	1	2	3	4	0
2	2	3	4	0	1
3	3	4	0	1	2
4	4	0	1	2	3

We can illustrate these results by using a cyclic arrangement of numbers on a dial. For example:

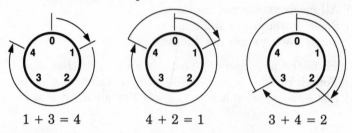

$$1 + 3 = 4 \qquad 4 + 2 = 1 \qquad 3 + 4 = 2$$

The operation ⊕ could also have been defined by describing each result in terms of ordinary addition and subtraction.

$$a \oplus b = \begin{cases} a + b \text{ if } a + b < 5 \\ a + b - 5 \text{ if } a + b \geq 5 \end{cases}$$

The operation ⊕ is often called *clock addition* or *modular addition.* The set S together with the operation ⊕ is a *mathematical system.*

■ **DEFINITION.** A *mathematical system* consists of:
1. A known set of elements.
2. One or more operations defined on this set of elements.

In general, the system also includes:
3. Definitions and postulates concerning operations on the set of elements.
4. Theorems that can be deduced from the given definitions and postulates.

The mathematical system that consists of the set of five elements {0, 1, 2, 3, 4} and the operation modular addition ⊕ is written in symbols as (Clock 5, ⊕).

Closure

Once a binary operation is defined on a set, we know that the set is *closed* under the operation.

■ **DEFINITION.** The set S is *closed* under operation ∗ if and only if:

$$\forall a, b \in S: \ a * b = c \text{ and } c \in S$$

☐ EXAMPLE 1: Addition is a binary operation on the set of integers. The set of integers is closed under addition.

☐ EXAMPLE 2: Clock addition is a binary operation on {0, 1, 2, 3, 4}. The set {0, 1, 2, 3, 4} is closed under the operation clock addition.

☐ EXAMPLE 3: The set of integers is not closed under division. For example, $5 \div 4$ is not an integer.

Associativity

■ **DEFINITION.** The operation $*$ is *associative* on set S if and only if:

$$\forall a, b, c \in S: \ (a * b) * c = a * (b * c)$$

☐ EXAMPLE 1: Addition is associative on the set of integers.

$$(-3 + 2) + 5 \stackrel{?}{=} -3 + (2 + 5)$$
$$-1 \ + 5 \stackrel{?}{=} -3 + \ 7$$
$$4 \ = \ 4$$

☐ EXAMPLE 2: Modular addition is associative on clock 5. We would need to verify $5 \cdot 5 \cdot 5$ or 125 arrangements in order to prove this. Two such arrangements are shown.

$$(2 \oplus 2) \oplus 4 \stackrel{?}{=} 2 \oplus (2 \oplus 4) \qquad\qquad (1 \oplus 3) \oplus 2 \stackrel{?}{=} 1 \oplus (3 \oplus 2)$$
$$4 \quad \oplus 4 \stackrel{?}{=} 2 \oplus \ 1 \qquad\qquad\quad 4 \quad \oplus 2 \stackrel{?}{=} 1 \oplus \ 0$$
$$3 \ = \ 3 \qquad\qquad\qquad\qquad 1 \ = \ 1$$

☐ EXAMPLE 3: Subtraction and division are not associative on the set of integers.

$$(5 - 8) - 3 \stackrel{?}{=} 5 - (8 - 3) \qquad\qquad (12 \div 6) \div 2 \stackrel{?}{=} 12 \div (6 \div 2)$$
$$-3 \ - 3 \stackrel{?}{=} 5 - \ 5 \qquad\qquad\quad 2 \div 2 \stackrel{?}{=} 12 \div \ 3$$
$$-6 \ \neq \ 0 \qquad\qquad\qquad\qquad 1 \ \neq \ 4$$

Commutativity

■ **DEFINITION.** The operation $*$ is *commutative* on set S if and only if:

$$\forall a, b \in S: \ a * b = b * a$$

☐ EXAMPLE 1: The operations addition and multiplication are commutative on the set of integers.

☐ EXAMPLE 2: The operation $\oplus$ defined on the set $\{0, 1, 2, 3, 4\}$ is commutative. There are $5 \cdot 4$ or 20 pairs to be verified. For example:

$$2 \oplus 3 \stackrel{?}{=} 3 \oplus 2 \qquad\quad 1 \oplus 2 \stackrel{?}{=} 2 \oplus 1 \qquad\quad 3 \oplus 4 \stackrel{?}{=} 4 \oplus 3$$
$$0 \ = \ 0 \qquad\qquad\quad 3 \ = \ 3 \qquad\qquad\quad 2 \ = \ 2$$

☐ EXAMPLE 3: The operations subtraction and division are not commutative on the set of integers. For example, $5 - 3 \neq 3 - 5$ and $2 \div 1 \neq 1 \div 2$.

The Identity Element

■ **DEFINITION.** For set S and operation $*$, e is the ***identity element*** if and only if:

$$\exists e \in S, \forall a \in S: \; a * e = e * a = a$$

☐ EXAMPLE 1: In the set of integers, the identity element for addition is 0.

$$9 + 0 = 0 + 9 = 9$$

☐ EXAMPLE 2: The identity element for (Clock 5, $\oplus$) is 0.

$$0 \oplus 0 = 0 \qquad 1 \oplus 0 = 0 \oplus 1 = 1 \qquad 2 \oplus 0 = 0 \oplus 2 = 2$$
$$3 \oplus 0 = 0 \oplus 3 = 3 \qquad 4 \oplus 0 = 0 \oplus 4 = 4$$

☐ EXAMPLE 3: In the set of integers, there is no identity element for subtraction and no identity element for division.

$$12 - 0 = 12 \text{ but } 0 - 12 \neq 12 \quad | \quad 12 \div 1 = 12 \text{ but } 1 \div 12 \neq 12$$

Inverse Elements

■ **DEFINITION.** An element a of set S has a *unique **inverse** a^{-1}* under the operation $*$ if and only if:

there exists an identity e for $*$ in S, and
$$\forall a \in S, \exists a^{-1} \in S: \; a * a^{-1} = a^{-1} * a = e$$

☐ EXAMPLE 1: Every integer a has an inverse $-a$ for the operation addition since $a + (-a) = -a + a = 0$.

☐ EXAMPLE 2: In (Clock 5, $\oplus$), every element has an inverse.

0 is the inverse of 0: $0 \oplus 0 = 0$
4 is the inverse of 1: $1 \oplus 4 = 4 \oplus 1 = 0$
3 is the inverse of 2: $2 \oplus 3 = 3 \oplus 2 = 0$
2 is the inverse of 3: $3 \oplus 2 = 2 \oplus 3 = 0$
1 is the inverse of 4: $4 \oplus 1 = 1 \oplus 4 = 0$

Groups

The properties that we have been using enable us to define an important structure in mathematics called a *group*.

■ **DEFINITION.** A ***group*** is a mathematical system consisting of a set G and an operation $*$ that satisfies four properties:
1. The set G is *closed* under $*$.
2. The operation $*$ is *associative* on the elements of G.
3. There is an element e of G that is the *identity* for $*$.
4. Every element a in G has, for $*$, an *inverse* a^{-1} in G.

This definition can be written in symbolic form.

■ **DEFINITION.** $(G, *)$ is a group if and only if:
1. $\forall a, b \in G$: $a * b = c$, $c \in G$. (*Closure*)
2. $\forall a, b, c \in G$: $(a * b) * c = a * (b * c)$. (*Associativity*)
3. $\exists e \in G$, $\forall a \in G$: $a * e = e * a = a$. (*Identity*)
4. $\forall a \in G$, $\exists a^{-1} \in G$: $a * a^{-1} = a^{-1} * a = e$. (*Inverses*)

The integers form a group under addition and the clock 5 numbers form a group under clock addition.

Groups were first defined in the eighteenth century. Examples of groups can be found in algebra, geometry, art, and nature. The properties of groups play an important role in science. For example, the symmetries of crystals of minerals form groups, and the symmetries of particles and of fields of force form groups.

Commutativity is not required for a mathematical system to be a group. However, many groups have operations that are commutative. When this occurs, the group is called a commutative group, or an Abelian group. (Integers, $+$) and (Clock 5, $\oplus$) are commutative groups.

Clock multiplication or *modular multiplication,* often indicated by the symbol $\odot$, can be described in terms of ordinary multiplication. In the clock 5 system shown in the following model problems:

$$a \odot b = \begin{cases} a \cdot b \text{ if } a \cdot b < 5 \\ a \cdot b - 5k \text{ if } a \cdot b \geq 5 \end{cases}$$

k represents an integer such that the value of $a \cdot b - 5k$ is in the set $\{0, 1, 2, 3, 4,\}$.

MODEL PROBLEMS

The system (Clock 5, $\odot$) consists of the set $\{0, 1, 2, 3, 4\}$ and the operation clock multiplication defined by the table.

1. Prove that (Clock 5, $\odot$) is not a group.

Solution:

$\odot$	0	1	2	3	4
0	0	0	0	0	0
1	0	1	2	3	4
2	0	2	4	1	3
3	0	3	1	4	2
4	0	4	3	2	1

(Clock 5, $\odot$)

1. *Closure:* Every pair of numbers from clock 5 has a product in clock 5 as seen in the table. Therefore, (Clock 5, $\odot$) is closed.

2. *Associativity:* We could demonstrate that the associative property holds for all numbers in clock 5. We will agree that all clock additions and multiplications are associative.

3. *Identity:* The identity is 1 because, for all clock 5 numbers x:
   $x \odot 1 = 1 \odot x = x$.

4. *Inverses:* This condition fails.
   The inverse of 1 is 1 because $1 \odot 1 = 1$.
   The inverse of 2 is 3 because $2 \odot 3 = 3 \odot 2 = 1$.
   The inverse of 3 is 2 because $3 \odot 2 = 2 \odot 3 = 1$.
   The inverse of 4 is 4 because $4 \odot 4 = 1$.
   However, there is no inverse for 0.

2. Let **clock 5/{0}** be clock 5 after eliminating 0 from the set. Prove that (Clock 5/{0}, $\odot$) is a group.

$\odot$	1	2	3	4
1	1	2	3	4
2	2	4	1	3
3	3	1	4	2
4	4	3	2	1

Solution:

The table shows (Clock 5/{0}, $\odot$).   **(Clock 5/{0}, $\odot$)**

1. *Closure:* Every pair of numbers from clock 5/{0} has a product of 1, 2, 3, or 4 as seen from the table. Therefore, (Clock 5/{0}, $\odot$) is closed.

2. *Associativity:* We agreed to accept the associativity of clock multiplication.

3. *Identity:* The identity is 1 because, for all clock 5 numbers x:
   $x \odot 1 = 1 \odot x = x$.

4. *Inverses:* Every element has an inverse.
   The inverse of 1 is 1 because $1 \odot 1 = 1$.
   The inverse of 2 is 3 because $2 \odot 3 = 3 \odot 2 = 1$.
   The inverse of 3 is 2 because $3 \odot 2 = 2 \odot 3 = 1$.
   The inverse of 4 is 4 because $4 \odot 4 = 1$.

Therefore, (Clock 5/{0}, $\odot$) is a group.

3. Tell why each of the following is not a group.
   a. The whole numbers under addition.
   b. The even integers under multiplication.
   c. The odd integers under addition.

Answer:

a. The whole numbers have no additive inverses.
b. The even numbers have no multiplicative identity, and, therefore, it is meaningless to discuss inverses.
c. The odd integers are not closed under addition, no identity element exists, and it is meaningless to discuss inverses.

EXERCISES

In 1–10, answer the following questions for the set and operation shown in the accompanying table.

a. Is the set closed under the given operation?
b. Is the operation associative on the given set?
c. Name the identity element for the system.
d. For every element having an inverse, name the element and its inverse.
e. Is the system a group?

1. Set S = {pos, neg}, to represent positive and negative integers. The operation is multiplication.

2. Set S = {−1, 0, 1}. The operation is multiplication.

·	pos	neg
pos	pos	neg
neg	neg	pos

Ex. 1

·	−1	0	1
−1	1	0	−1
0	0	0	0
1	−1	0	1

Ex. 2

3. Set S = clock 3 = {0, 1, 2}. The operation is multiplication.

4. Set S = clock 3/{0} = {1, 2}. The operation is multiplication.

⊙	0	1	2
0	0	0	0
1	0	1	2
2	0	2	1

Ex. 3

⊙	1	2
1	1	2
2	2	1

Ex. 4

5. Set S = {odd, even}, to represent odd and even integers. The operation is multiplication.

6. Set S = {1, 2, 3}. The operation is *min*, finding the minimum, or smaller, of two numbers.

·	odd	even
odd	odd	even
even	even	even

Ex. 5

min	1	2	3
1	1	1	1
2	1	2	2
3	1	2	3

Ex. 6

7. Set S = {a, b, c, d}. The operation * is defined by the table.

8. Set S = {w, x, y, z}. The operation # is defined by the table.

*	a	b	c	d
a	c	d	a	b
b	d	a	b	c
c	a	b	c	d
d	b	c	d	a

Ex. 7

#	w	x	y	z
w	x	w	z	y
x	w	x	y	z
y	z	y	x	w
z	y	z	w	x

Ex. 8

9. Set $S = \{e, f, g\}$. The operation @ is defined by the table.

@	e	f	g
e	e	f	g
f	f	e	g
g	g	f	g

Ex. 9

10. Set S = clock 8/$\{0\}$ = $\{1, 2, 3, 4, 5, 6, 7\}$. The operation is multiplication.

⊙	1	2	3	4	5	6	7
1	1	2	3	4	5	6	7
2	2	4	6	0	2	4	6
3	3	6	1	4	7	2	5
4	4	0	4	0	4	0	4
5	5	2	7	4	1	6	3
6	6	4	2	0	6	4	2
7	7	6	5	4	3	2	1

Ex. 10

11. **a.** Construct the table for (Clock 6, ⊕).
    b. Is (Clock 6, ⊕) a group?
12. **a.** Construct the table for (Clock 6, ⊙).
    b. Is (Clock 6, ⊙) a group?
13. **a.** Construct the table for (Clock 6/$\{0\}$, ⊙).
    b. Is (Clock 6/$\{0\}$, ⊙) a group?
14. True/false: For any counting number n, (Clock n, ⊕) is a group.
15. True/false: For any counting number n, (Clock $n/\{0\}$, ⊙) is a group.
16. True/false: For any prime number n, (Clock $n/\{0\}$, ⊙) is a group.
17. **a.** If $S = \{0\}$, is $(S, +)$ a group?
    b. If $S = \{1\}$, is $(S, \cdot)$ a group?
    c. If S contains a single element x, and $x * x = x$, is $(S, *)$ a group?

+	0
0	0

a.

·	1
1	1

b.

*	x
x	x

c.

Exercises 18–22 refer to *digital multiplication* with $S = \{1, 3, 5, 7, 9\}$.

18. In digital multiplication, answers are single digits, obtained by writing only the units digit from a product in standard multiplication. Compare the given examples.

⊙	1	3	5	7	9
1					
3			5	1	7
5					
7					
9					

Ex. 18 to 22

Standard:
$3 \cdot 5 = 15$ $3 \cdot 7 = 21$ $3 \cdot 9 = 27$

Digital:
$3 \odot 5 = 5$ $3 \odot 7 = 1$ $3 \odot 9 = 7$

Copy and complete the table shown above for digital multiplication with the set $\{1, 3, 5, 7, 9\}$.

19. **a.** Compute $(3 \odot 9) \odot 7$.   **b.** Compute $3 \odot (9 \odot 7)$.
    c. Is $(3 \odot 9) \odot 7 = 3 \odot (9 \odot 7)$?
    d. Is digital multiplication associative? Explain why.
20. **a.** Name the identity element for $(\{1, 3, 5, 7, 9\}, \odot)$.
    b. For every element having an inverse, name the element and its inverse.

21. Is the set closed under the operation of digital multiplication?
22. Is the set {1, 3, 5, 7, 9} under digital multiplication a group? Why?
23. **a.** Construct a table for the set {1, 3, 7, 9} under digital multiplication.
    b. Is the set {1, 3, 7, 9} under digital multiplication a group? Why?
    c. How does the table for part **a** compare to the table constructed in exercise 18?

24. **a.** Using set $S = \{0, 2, 4\}$ and the operation of average, symbolized by avg, copy and complete the table shown at the right.
    b. Give three reasons why this system is *not* a group.

avg	0 2 4
0	
2	
4	

1-5 FIELDS

In previous courses, you studied the distributive property of multiplication over addition.

$$a(b + c) = ab + ac \text{ and } ab + ac = a(b + c)$$

The importance of the distributive property is that it links two operations defined on a set of elements. The distributive property is not limited to multiplication over addition, but sometimes holds for other pairs of operations as well.

☐ EXAMPLE 1: Is multiplication distributive over *max*?

$$a(b \text{ max } c) \overset{?}{=} ab \text{ max } ac$$
$$-3(5 \text{ max } 7) \overset{?}{=} -3(5) \text{ max } - 3(7)$$
$$-3(7) \overset{?}{=} -15 \text{ max } -21$$
$$-21 \neq -15$$

Multiplication is not distributive over *max*.

☐ EXAMPLE 2: Is squaring distributive over multiplication?

$$(ab)^2 \overset{?}{=} a^2b^2$$

By using the definition of a square and the associative and commutative properties of multiplication we see that:

$$(ab)^2 = (ab)(ab)$$
$$= (aa)(bb)$$
$$= a^2b^2$$

Squaring is distributive over multiplication.

We have defined clock addition and clock multiplication on the set of clock 5 numbers {0, 1, 2, 3, 4} as shown in the tables. (Clock 5, $\oplus$, $\odot$) is a mathematical system.

$\oplus$	0 1 2 3 4
0	0 1 2 3 4
1	1 2 3 4 0
2	2 3 4 0 1
3	3 4 0 1 2
4	4 0 1 2 3

$\odot$	0 1 2 3 4
0	0 0 0 0 0
1	0 1 2 3 4
2	0 2 4 1 3
3	0 3 1 4 2
4	0 4 3 2 1

☐ EXAMPLE 3: Is multiplication distributive over addition in the system (Clock 5, $\oplus$, $\odot$)?

$$3 \odot (2 \oplus 4) \stackrel{?}{=} 3 \odot 2 \oplus 3 \odot 4$$
$$3 \odot 1 \quad \stackrel{?}{=} \quad 1 \quad \oplus \quad 2$$
$$3 \quad = \quad 3$$

This is just one case for which the distributive property holds in the given system. By considering all possible sets of numbers from clock 5, we could show that multiplication is distributive over addition in the system (Clock 5, $\oplus$, $\odot$).

Although it is possible to write a more general definition of the distributive property of one operation * over another operation #, as $a * (b \# c) = (a * b) \# (a * c)$, we will limit our definition to the more familiar operations.

■ **DEFINITION.** In set S, multiplication is *distributive* over addition if and only if

$$\forall a, b, c \in S: a(b + c) = ab + ac \text{ and } ab + ac = a(b + c)$$

A Field

Just as the distributive property links together two operations, there is a mathematical system called a *field* that contains two operations. Since most fields consist of the operations of addition and multiplication, we will use these operations in our definition.

■ **DEFINITION.** A *field* is a mathematical system consisting of a set F and two operations, normally addition and multiplication, which satisfies eleven properties:

1. The set F is a commutative group under the operation of addition, satisfying five properties: closure, associativity, the existence of an identity for addition (usually 0), the existence of inverses under addition, and commutativity.

2. The set F without the additive identity (usually $F/\{0\}$) is a commutative group under the operation of multiplication, satisfying five properties: closure, associativity, the existence of an identity for multiplication (usually 1), the existence of inverses under multiplication, and commutativity.
3. The second operation, multiplication, is distributive over the first operation, addition.

Thus, if we know the properties of a group, it becomes relatively easy to remember the definition of a field. The definition is now rewritten in symbolic form. Notice that *two* operations are included with the set F by writing $(F, +, \cdot)$.

> ■ **DEFINITION.** $(F, +, \cdot)$ is a *field* if and only if
>
> 1. $(F, +)$ is a commutative group.
> 2. $(F/\{0\}, \cdot)$ is a commutative group.
> 3. Multiplication distributes over addition.

| **MODEL PROBLEMS** |

1. Prove that (Clock 3, $\oplus$, $\odot$) is a field.

Solution:
First, establish that (Clock 3, $\oplus$) is a commutative group.
1. (Clock 3, $\oplus$) is closed.
2. (Clock 3, $\oplus$) is associative.
3. In (Clock 3, $\oplus$), the identity is 0.
4. In (Clock 3, $\oplus$):
   The additive inverse of 0 is 0.
   The additive inverse of 1 is 2.
   The additive inverse of 2 is 1.
5. (Clock 3, $\oplus$) is commutative.

$\oplus$	0 1 2
0	0 1 2
1	1 2 0
2	2 0 1

(Clock 3, $\oplus$)

Next, establish that (Clock 3/$\{0\}$, $\odot$) is a commutative group.
6. (Clock 3/$\{0\}$, $\odot$) is closed.
7. (Clock 3/$\{0\}$, $\odot$) is associative.
8. In (Clock 3/$\{0\}$, $\odot$), the identity is 1.
9. In (Clock 3/$\{0\}$, $\odot$):
   The multiplicative inverse of 1 is 1.
   The multiplicative inverse of 2 is 2.
10. (Clock 3/$\{0\}$, $\odot$) is commutative.

$\odot$	1 2
1	1 2
2	2 1

(Clock 3/$\{0\}$, $\odot$)

Finally, establish that multiplication distributes over addition.
11. In (Clock 3, $\oplus$, $\odot$), $a \odot (b \oplus c) = a \odot b \oplus a \odot c$.

Let us list some examples of the distributive property that can be shown in (Clock 3, $\oplus$, $\odot$).

$1(0 \oplus 2) \overset{?}{=} 1 \odot 0 \oplus 1 \odot 2$
     $1 \odot 2 \overset{?}{=} 0 \oplus 2$
        $2 = 2$

$2(2 \oplus 1) \overset{?}{=} 2 \odot 2 \oplus 2 \odot 1$
     $2 \odot 0 \overset{?}{=} 1 \oplus 2$
        $0 = 0$

$2(1 \oplus 1) \overset{?}{=} 2 \odot 1 \oplus 2 \odot 1$
     $2 \odot 2 \overset{?}{=} 2 \oplus 2$
        $1 = 1$

It can be shown that all possible arrangements of clock 3 numbers in the rule $a \odot (b \oplus c) = a \odot b \oplus a \odot c$ will result in true statements.

2. Prove that (Integers, $+$, $\cdot$) is not a field.

Solution:

Of the eleven field properties, one fails to be satisfied. With the exception of 1 and -1, integers do *not* have *multiplicative inverses*.

EXERCISES

In 1–6: **a.** Is the system a field? **b.** If the system is not a field, name one field property that is not satisfied.

1. (Whole numbers, $+$, $\cdot$)
2. (Positive numbers, $+$, $\cdot$)
3. (Even integers, $+$, $\cdot$)
4. (Odd integers, $+$, $\cdot$)
5. (Clock 3, $\oplus$, $\odot$)
6. (Clock 4, $\oplus$, $\odot$)

7. Give a reason why (Rational numbers, $+$) is not a field.
8. If S = $\{0\}$, give a reason why (S, $+$, $\cdot$) is not a field.

Exercises 9–16 refer to the field (Clock 5, $\oplus$, $\odot$). The tables for these operations are found on page 24.

9. What element does not have an inverse under multiplication?
10. What element is its own inverse under multiplication?
11. What is the additive inverse of 4?
12. Evaluate $3 \oplus 3 \oplus 3$.
13. Evaluate $3 \odot 3 \odot 3$.

In 14–16, evaluate parts a and b; then answer part c.

14. **a.** $4 \odot (2 \oplus 4)$ **b.** $4 \odot 2 \oplus 4 \odot 4$
    c. Is $4 \odot (2 \oplus 4) = 4 \odot 2 \oplus 4 \odot 4$?

15. **a.** $2 \odot (3 \oplus 2)$ **b.** $2 \odot 3 \oplus 2 \odot 2$
    c. Is $2 \odot (3 \oplus 2) = 2 \odot 3 \oplus 2 \odot 2$?

16. a. $3 \odot 4 \oplus 3 \odot 2$ **b.** $3 \odot (4 \oplus 2)$
   c. Is $3 \odot 4 \oplus 3 \odot 2 = 3 \odot (4 \oplus 2)$?

17. What is the field property being tested in exercises 14–16?

Exercises 18–26 refer to set $S = \{0, 2, 4, 6, 8\}$ under the operations of digital addition and digital multiplication, shown in the tables.

⊕	0	2	4	6	8
0	0	2	4	6	8
2	2	4	6	8	0
4	4	6	8	0	2
6	6	8	0	2	4
8	8	0	2	4	6

⊙	0	2	4	6	8
0	0	0	0	0	0
2	0	4	8	2	6
4	0	8	6	4	2
6	0	2	4	6	8
8	0	6	2	8	4

18. Is $(S, \oplus)$: **a.** closed? **b.** associative? **c.** commutative?

19. Name the identity for $(S, \oplus)$.

20. For every element in $(S, \oplus)$ having an inverse, name the element and its inverse.

21. Is $(S, \oplus)$ a commutative group?

22. Is $(S/\{0\}, \odot)$: **a.** closed? **b.** associative? **c.** commutative?

23. Name the identity for $(S/\{0\}, \odot)$.

24. For every element in $(S/\{0\}, \odot)$ having an inverse, name the element and its inverse.

25. Is $(S/\{0\}, \odot)$ a commutative group?
   a. Is $4 \odot (2 \oplus 6) = (4 \odot 2) \oplus (4 \odot 6)$?
   b. Is $8 \odot (2 \oplus 4) = (8 \odot 2) \oplus (8 \odot 4)$?
   c. Does the operation $\odot$ distribute over the operation $\oplus$?

26. Is $(S, \oplus, \odot)$ a field?

Exercises 27–31 refer to the set $S = \{a, b, c\}$ under the operations of $\triangle$ and $*$, as shown in the tables.

△	a	b	c
a	b	c	a
b	c	a	b
c	a	b	c

*	a	b	c
a	b	a	c
b	a	b	c
c	c	c	c

27. In $(S, \triangle)$, the operation $\triangle$ is associative and the identity element is c. Is $(S, \triangle)$ a commutative group?

28. In $(S, *)$, which element does not have an inverse?

29. By removing c (the identity element under $\triangle$) from the set S, the set $S/\{c\}$ is formed. Is $(S/\{c\}, *)$ a commutative group?

30. One of the operations distributes over the other operation.
   a. Is $b \triangle (a * c) = (b \triangle a) * (b \triangle c)$?
   b. Is $b * (a \triangle c) = (b * a) \triangle (b * c)$?
   c. Which operation is distributive over the other?

31. Is $(S, \triangle, *)$ a field?

1-6 PROPERTIES OF THE RATIONAL NUMBERS

Field Properties of the Rational Numbers

The set of rational numbers under the operations of addition and multiplication forms a field.

■ (Rational numbers, +, ·) is a field satisfying eleven properties:

1. (Rational numbers, +) is closed.

   $\forall_{a,b} \in$ Rationals: $a + b = c$ where $c \in$ Rationals.

2. (Rational numbers, +) is associative.

   $\forall_{a,b,c} \in$ Rationals: $(a + b) + c = a + (b + c)$.

3. A unique identity element (zero) exists for addition.

   $\exists_0 \in$ Rationals, $\forall_x \in$ Rationals: $x + 0 = x$, and $0 + x = x$.

4. Every element has an inverse $(-x)$ under addition.

   $\forall_x \in$ Rationals, $\exists_{(-x)} \in$ Rationals: $x + (-x) = 0$, and
   $(-x) + x = 0$.

5. (Rational numbers, +) is commutative.

   $\forall_{a,b} \in$ Rationals: $a + b = b + a$.

6. (Rational numbers/{0}, ·) is closed.

   $\forall_{a,b} \in$ Rationals/{0}: $ab = c$ where $c \in$ Rationals/{0}.

7. (Rational numbers/{0}, ·) is associative.

   $\forall_{a,b,c} \in$ Rationals/{0}: $(ab)c = a(bc)$.

8. A unique identity element (one) exists for multiplication.

   $\exists_1 \in$ Rationals/{0}, $\forall_x \in$ Rationals/{0}: $x \cdot 1 = x$, and
   $1 \cdot x = x$.

9. Every element in Rationals/{0} has an inverse $\left(\dfrac{1}{x}\right)$ under multiplication.

   $\forall_x \in$ Rationals/{0}, $\exists_{\frac{1}{x}} \in$ Rationals/{0}: $x \cdot \dfrac{1}{x} = 1$, and

   $\dfrac{1}{x} \cdot x = 1$.

10. (Rational numbers/$\{0\}$, $\cdot$) is commutative.

$$\forall_{a,b} \in \text{Rationals}/\{0\}: \quad ab = ba.$$

11. Multiplication is distributive over addition.

$$\forall_{a,b,c} \in \text{Rationals}: \quad a(b + c) = ab + ac, \text{ and } ab + ac = a(b + c).$$

We will use these eleven field properties of the rational numbers to solve equations and to perform computations throughout this chapter.

Properties of Order

Two numbers a and b are equal, that is, $a = b$, when they name the same number. For example, the integer 3 is expressed in rational form as $\frac{3}{1}$, and $3 = \frac{3}{1}$.

On a standard horizontal number line, the point assigned to the number 3 is to the right of the point assigned to the number 2. Certainly $3 \neq 2$, read as "3 is not equal to 2." However, it is possible to describe the *order* of these numbers in more specific terms:

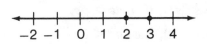

$3 > 2$, read as "3 is greater than 2."

OR

$2 < 3$, read as "2 is less than 3."

Using this conventional number line, we may state the following generalization: If a and b are two rational numbers, then $a > b$ (or $b < a$) if and only if the graph of a is to the right of the graph of b.

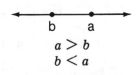

$$a > b$$
$$b < a$$

These examples help us to understand the following properties of order, each of which is true for the set of rational numbers.

1. **The Trichotomy Property**

   Given the numbers a and b, then one and only one of the following sentences is true:

   $$a > b, \text{ or } a = b, \text{ or } a < b$$

2. **The Transitive Property of Inequalities**

   For all quantities a, b, and c: If $a > b$ and $b > c$, then $a > c$.

   AND

   If $c < b$ and $b < a$, then $c < a$.

Using the rational numbers, we observe:

☐ EXAMPLE: If $5 > 3\frac{1}{2}$ and $3\frac{1}{2} > 3$, then $5 > 3$.

3. The Addition Properties of Inequalities

For all quantities a, b, and c: If $a > b$, then $a + c > b + c$.

AND

If $a < b$, then $a + c < b + c$.

Using the rational numbers, we observe:

☐ EXAMPLE 1:   If $3\frac{1}{2} > 2$, then $3\frac{1}{2} + 7 > 2 + 7$, or $10\frac{1}{2} > 9$.

☐ EXAMPLE 2:   If $4 < 6$,   then $4 + \frac{1}{2} < 6 + \frac{1}{2}$, or $4\frac{1}{2} < 6\frac{1}{2}$.

4. The Multiplication Properties of Inequalities

For all quantities a and b:

If $a > b$ and $c > 0$ (c is positive), then $ac > bc$.

If $a < b$ and $c > 0$ (c is positive), then $ac < bc$.

If $a > b$ and $c < 0$ (c is negative), then $ac < bc$.

If $a < b$ and $c < 0$ (c is negative), then $ac > bc$.

Using the rational numbers, we observe:

☐ EXAMPLE 1:   If $8 > 6$ and $\frac{1}{2} > 0$, then $8(\frac{1}{2}) > 6(\frac{1}{2})$, or $4 > 3$.

☐ EXAMPLE 2:   If $1\frac{1}{2} < 5$ and $2 > 0$, then $1\frac{1}{2}(2) < 5(2)$, or $3 < 10$.

☐ EXAMPLE 3:   If $8 > 6$ and $-\frac{1}{2} < 0$,   then $8(-\frac{1}{2}) < 6(-\frac{1}{2})$, or $-4 < -3$.

☐ EXAMPLE 4:   If $1\frac{1}{2} < 5$ and $-2 < 0$, then $1\frac{1}{2}(-2) > 5(-2)$, or $-3 > -10$.

■ **DEFINITION.** An *ordered field* $(F, +, \cdot, >)$ is a mathematical system in which a set F under the usual operations of addition and multiplication satisfies the eleven properties of a field and the four order properties of trichotomy, transitivity, addition of inequalities, and multiplication of inequalities.

Not every field is an ordered field. In Chapter 14 of this book, we will study a set of numbers that forms a field but lacks the order properties necessary to be an ordered field. For now, let us observe:

■ **The set of rational numbers is an ordered field.**

The Property of Density

■ **DEFINITION.** A set is *dense* if and only if there is at least one element of the set between any two given elements of the set.

Between any two given rational numbers, it is always possible to find at least one more rational number by finding the *average* (avg) of the two given numbers. Thus, the set of rational numbers is a *dense set*, or a set having the property of *density*.

□ EXAMPLE 1:   Find a rational number between 1 and $1\frac{1}{2}$.

Solution: $1 \text{ avg } 1\frac{1}{2} = \frac{1 + 1\frac{1}{2}}{2} = \frac{2\frac{1}{2}}{2} = \frac{2.50}{2} = 1.25, \text{ or } 1\frac{1}{4}$ *Ans.*

□ EXAMPLE 2:   Find a rational number between $\frac{1}{7}$ and $\frac{2}{7}$.

Solution: $\frac{1}{7} \text{ avg } \frac{2}{7} = \frac{\frac{1}{7} + \frac{2}{7}}{2} = \frac{\frac{3}{7}}{2} = \frac{3}{7} \div \frac{2}{1} = \frac{3}{7} \cdot \frac{1}{2} = \frac{3}{14}$ *Ans.*

In fact, choose any two rational numbers that might be graphed on a number line, such as 0 and $\frac{1}{2}$. Think of one-place decimals be-

tween these two numbers, that is, .1, .2, .3, and .4. Now think of two-place decimals, namely .01, .02, .03, . . . , .49, and three-place decimals, .001, .002, .003, . . . , .499. This process of increasing the number of decimal places leads us to make the following observation, not only for 0 and $\frac{1}{2}$, but for any two rational numbers:

■ **Between any two rational numbers, there is an infinite number of rational numbers.**

We have seen that every rational number is associated with a point on the number line, but the converse of this statement is not true. For example, between the points associated with 1 and $1\frac{1}{2}$ on a number line, there is an infinite number of points associated with other rational numbers, but there is also a point associated with $\sqrt{2}$, a number that is not rational. We will study irrational numbers like $\sqrt{2}$ in Chapter 4 of this book. For now, realize that no matter how densely packed the points associated with rational numbers are, there are still "holes" in the rational number line.

MODEL PROBLEMS

1. Write the negation of the expression $x > 5$ in two ways.

Solution

By the Trichotomy Property, $x > 5$, $x = 5$ or $x < 5$. If we know that $x \not> 5$, then $x = 5$ or $x < 5$, that is, $x \leq 5$.

Answer: $x \not> 5$ or $x \leq 5$

2. Arrange the numbers $\frac{3}{5}$, $\frac{4}{7}$, and $\frac{1}{2}$ in proper order, using the symbol $<$.

Solution

1. Change the fractions to decimal form.

$$\frac{3}{5} = 5\overline{)3.00}^{.60} \qquad \frac{4}{7} = 7\overline{)4.000000}^{.571428\ldots} \qquad \frac{1}{2} = 2\overline{)1.00}^{.50}$$

2. Since $.50 < .\overline{571428}$ and $.\overline{571428} < .60$, we can now see that $\frac{1}{2} < \frac{4}{7}$ and $\frac{4}{7} < \frac{3}{5}$. A more compact form of this inequality is $\frac{1}{2} < \frac{4}{7} < \frac{3}{5}$.

Answer: $\frac{1}{2} < \frac{4}{7} < \frac{3}{5}$

EXERCISES

In 1–8, name the property illustrated for the set of rational numbers.

1. $3\frac{1}{2} + 7 = 7 + 3\frac{1}{2}$

2. $(17 \cdot \frac{1}{2}) \cdot 4 = 17 \cdot (\frac{1}{2} \cdot 4)$

3. $\frac{3}{2} + (-\frac{3}{2}) = 0$

4. $\frac{3}{2} \cdot \frac{2}{3} = 1$

5. $6(3 + \frac{1}{3}) = 6 \cdot 3 + 6 \cdot \frac{1}{3}$

6. $x > 0$ or $x = 0$ or $x < 0$

7. $\frac{2}{5} + 0 = \frac{2}{5}$

8. $4.321(1) = 4.321$

In 9-18, tell whether the statement is true or false.

9. $5 < 8$ 10. $8 < 7$ 11. $-3 > 0$ 12. $12 \geq 2$

13. $-6 \leq -1$ 14. $\frac{3}{2} > -\frac{5}{2}$ 15. $\frac{5}{1} \geq 5$ 16. $\frac{1}{3} > \frac{1}{4}$

17. $-\frac{1}{3} > -\frac{1}{4}$ 18. $-1.6 \leq -2$

In 19-26, replace the question mark with one of the symbols $>$ or $<$ to make the sentence true.

19. If $x < 9$, then $x + 2$? $9 + 2$.
20. If $x > 5$, then $3x$? 15.
21. If $x + 3 > -2$, then x ? -5.
22. If $-2 < x$, then -3 ? $x - 1$.
23. If $7 > 6$ and $x > 0$, then $7x$? $6x$.
24. If $x < 6$ and $6 < 8$, then x ? 8.
25. If $7 > 6$ and $6 > x$, then 7 ? x.
26. If $6 < 8$ and $x < 0$, then $6x$? $8x$.

In 27-31, write the negation of the given expression in two ways.

27. $x > 2$ 28. $x < 3$ 29. $x \leq 12$ 30. $x \geq -8$ 31. $x = 4$

In 32-39, arrange the given numbers in proper order, using the symbol $<$.

32. $-2, 5, 1$ 33. $-4, -2, -7$ 34. $.2, .12, .21$

35. $\frac{1}{5}, \frac{1}{10}, \frac{3}{20}$ 36. $-\frac{3}{2}, -1, -1.6$ 37. $\frac{2}{3}, \frac{3}{5}, \frac{5}{8}$

38. $\frac{8}{9}, \frac{7}{8}, \frac{9}{11}$ 39. $1\frac{4}{5}, 1\frac{3}{10}, 1\frac{3}{4}$

In 40-47: a. Is the system a group? b. If the answer to part a is "No," name all group properties that fail.

40. (Counting numbers, +) 41. (Integers, +)
42. (Integers/{0}, ·) 43. (Rational numbers, ·)
44. (Rational numbers, +) 45. (Rational numbers/{0}, ·)
46. (Positive rational numbers, +) 47. (Positive rational numbers, ·)

In 48-57, find the *average* of the two given rational numbers to find a rational number that lies between these given numbers.

48. $7, 8$ 49. $0, -3$ 50. $-8, -7$ 51. $-5, 2$ 52. $3, 3.3$

53. $2, 2\frac{1}{2}$ 54. $\frac{2}{5}, \frac{3}{5}$ 55. $1.9, 2$ 56. $\frac{2}{3}, \frac{5}{6}$ 57. $\frac{3}{4}, \frac{7}{8}$

58. Name three rational numbers that lie between 17.1 and 17.

1-7 FIRST-DEGREE EQUATIONS AND INEQUALITIES

In working with the field of rational numbers, as well as other ordered fields, we deal with quantities, operations, and relationships. The addition and multiplication properties for inequalities were stated in the last section. When two quantities are equal, other postulates are needed.

Postulates of Equality

A *postulate* or an *axiom* is a statement that is accepted as being true without proof. Let us recall some postulates of equality, studied in earlier courses, that are used when dealing with equalities and solving equations.

1. **The Reflexive Property** $\forall_a$: $a = a$.

2. **The Symmetric Property** $\forall_{a,b}$: If $a = b$, then $b = a$.

3. **The Transitive Property** $\forall_{a,b,c}$: If $a = b$ and $b = c$, then $a = c$.

4. **The Substitution Property** $\forall_{a,b}$: If $a = b$, then a may be replaced by b, or b may be replaced by a, in any expression.

5. **The Addition Property** $\forall_{a,b,c}$: If $a = b$, then $a + c = b + c$.

6. **The Multiplication Property** $\forall_{a,b,c}$: If $a = b$, then $ac = bc$.

Equations and Inequalities

An *equation* is a sentence that uses the symbol $=$ to state that two quantities are equal, such as $\frac{2}{4} = \frac{1}{2}$. An *inequality* is a sentence that uses one of the symbols of order, namely, $>$, $<$, $\geq$, or $\leq$, to show the order relation of two quantities.

In an *open sentence*, one or more of the quantities in the relationship contains a variable. For example:

Equation: $x + 7 = 3$ Inequality: $x + 7 < 3$

In each of these open sentences, the left-hand member ($x + 7$) contains the variable x. The right-hand member (3) as well as the term 7 are called *constants*. A *variable* is a placeholder that represents a member or an element of a given set. Such a set is called the *domain*, or the *replacement set*, of the variable. The open sentence is not true or false until we replace the variable with elements of the domain.

Equation: $x + 7 = 3$	Inequality: $x + 7 < 3$
If $x = -6$, then $-6 + 7 = 3$	If $x = -6$, then $-6 + 7 < 3$
or $1 = 3$ (False)	or $1 < 3$ (True)
If $x = -4$, then $-4 + 7 = 3$	If $x = -4$, then $-4 + 7 < 3$
or $3 = 3$ (True)	or $3 < 3$ (False)

When a sentence can be judged to be true or false, it is called a *statement*, or a *closed sentence*. In a statement there are no variables.

A *solution set* is a subset of the domain consisting of those elements or members of the domain that make the open sentence true. A solution to an equation is sometimes called a *root* of the equation.

☐ EXAMPLE 1:

Domain = Integers
Solve for x: $x + 7 = 3$

Solution: Add the inverse of 7 to both members and simplify.

$$x + 7 = 3$$
$$x + 7 + (-7) = 3 + (-7)$$
$$x = -4$$

Answer: $x = -4$, or the solution
set = $\{-4\}$

☐ EXAMPLE 2:

Domain = Integers
Solve for x: $x + 7 < 3$

Solution: Add the inverse of 7 to both members and simplify.

$$x + 7 < 3$$
$$x + 7 + (-7) < 3 + (-7)$$
$$x < -4$$

Answer: $x < -4$, or the solution
set = $\{-5, -6, -7, \ldots\}$

In solving the open sentences in examples 1 and 2, we used many of the field properties, the properties of equality, and the properties of order. These included: the addition property of equality; the addition property of inequality; associativity under addition; additive inverses; the additive identity; and closure under addition. Rather than use a lengthy procedure to write the application of each property as a different step, let us agree to use abbreviated procedures to solve all open sentences. (*See these procedures in the model problems that follow.*)

Types of Solution Sets

We have learned that sets may be finite, infinite, or empty. This statement is also true for solution sets, as we will now see.

A *finite set* is a set whose elements can be counted. For example:

1. Given the domain of integers and $x + 7 = 3$, the solution set is $\{-4\}$.
2. Given the domain of whole numbers and $x < 4$, the solution set is $\{0, 1, 2, 3\}$.
3. Given the domain of counting numbers and $x \leq 300$, the solution set is $\{1, 2, 3, \ldots, 300\}$.

An *infinite set* is a set whose elements cannot be counted. Here, the counting process does not come to an end. For example:

1. Given the domain of integers and $x + 7 < 3$, the solution set is $\{-5, -6, -7, \ldots\}$.
2. Given the domain of rational numbers and $x + 7 < 3$, it is *not* possible to list a pattern of numbers as the solution set. Rather, the solution set is written in *set-builder notation* as follows:

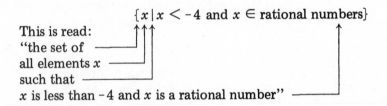

$$\{x \mid x < -4 \text{ and } x \in \text{rational numbers}\}$$

This is read:
"the set of
all elements x
such that
x is less than -4 and x is a rational number"

Note that set-builder notation may be used to indicate any set. However, if it is possible to list the elements of a set in a simpler form, we should do so. For example, $\{x \mid x = 4 \text{ and } x \in \text{integers}\}$ can be listed simply as $\{4\}$.

The *empty set*, or *null set*, is the set that has no elements.

1. Given the domain of integers and $x + \frac{1}{2} = 2$, the solution set is the empty set, written as $\{\ \}$ or $\varnothing$.

2. Given the domain of whole numbers and $x < -4$, the solution set is again $\{\ \}$ or $\varnothing$.

MODEL PROBLEMS

Unless otherwise noted, assume that the domain for all equations and inequalities in this chapter is the set of rational numbers.

1. Solve and check: $5(x - 1) - 3 = 2x - (3 - x)$

Procedure for Solving a First-degree Equation	*Solution*
1. Write the equation.	$5(x - 1) - 3 = 2x - (3 - x)$
2. Clear parentheses by using the distributive property. [*Note:* $-(3 - x)$ means $-1(3 - x)$.]	$5x - 5 - 3 = 2x - 3 + x$
3. Combine like terms on each side of the equation.	$5x - 8 = 3x - 3$
4. Use additive inverses to form an equivalent equation with only variable terms on one side and only constant terms on the other.	$5x - \cancel{8} + (-3x + \cancel{8}) = \cancel{3x} - 3 + (-\cancel{3x} + 8)$ $5x + (-3x) = -3 + (8)$
5. Combine like terms.	$2x = 5$
6. Use a multiplicative inverse (also called a reciprocal) to isolate the variable.	$\frac{1}{2}(2x) = \frac{1}{2}(5)$ $x = \frac{5}{2}$, or $2\frac{1}{2}$, or 2.5

Procedure for Checking an Equation	*Check*
1. Check the solution (or the root) only in the orginal equation.	$5(x - 1) - 3 = 2x - (3 - x)$ $5(2\frac{1}{2} - 1) - 3 \stackrel{?}{=} 2(2\frac{1}{2}) - (3 - 2\frac{1}{2})$
2. Simplify terms within parentheses.	$5(1\frac{1}{2}) - 3 \stackrel{?}{=} 2(2\frac{1}{2}) - (\frac{1}{2})$

3. Multiply (and divide). $7\frac{1}{2} - 3 \overset{?}{=} 5 - \frac{1}{2}$

4. Add (and subtract). $4\frac{1}{2} = 4\frac{1}{2}$ (True)

Answer: $x = \frac{5}{2}$, or $2\frac{1}{2}$, or 2.5; OR solution set = $\{2\frac{1}{2}\}$

2. Solve within the set of rational numbers: $2(5 - x) > 3 + 1$

How to Proceed	*Solution*
1. Write the inequality.	$2(5 - x) > 3 + 1$
2. Clear parentheses.	$10 - 2x > 3 + 1$
3. Combine like terms on each side of the inequality.	$10 - 2x > 4$
4. Use additive inverses to form an equivalent inequality with only variable terms on one side and only constant terms on the other.	$\cancel{10} - 2x + (-\cancel{10}) > 4 + (-10)$ $-2x > 4 + (-10)$
5. Combine like terms.	$-2x > -6$
6. Use a multiplicative inverse (reciprocal) to isolate the variable. (*Note:* When multiplying by a negative number, the order is reversed.)	$-\frac{1}{2}(-2x) < -\frac{1}{2}(-6)$ $x < 3$

Answer: $\{x \mid x < 3 \text{ and } x \in \text{rational numbers}\}$

Since this solution set is infinite, it is not possible to check all values that make the inequality true. However, you may select one or more rational numbers less than 3 and check these values in $2(5 - x) > 3 + 1$, the original open sentence.

| EXERCISES |

In 1–7, list the elements of the solution set, or indicate that the solution is the empty set.

1. $\{x \mid x + 5 = 16 \text{ and } x \in \text{whole numbers}\}$
2. $\{x \mid x - 3 < 1 \text{ and } x \in \text{counting numbers}\}$
3. $\{y \mid 2y + 5 = 8 \text{ and } y \in \text{rational numbers}\}$
4. $\{y \mid 3 - 4y = 2y \text{ and } y \in \text{natural numbers}\}$

5. $\{x \mid 3(4 + x) \leq 27 \text{ and } x \in \text{whole numbers}\}$
6. $\{x \mid x + 21 = 3 - 2x \text{ and } x \in \text{integers}\}$
7. $\{y \mid 5 - 2y = 4(y - 7) \text{ and } y \in \text{integers}\}$

In 8–36, solve and check the equation. Use the domain of rational numbers.

8. $x - 7 = 10$	9. $y + 8 = 3$	10. $4z = 1$
11. $-3 = 2w$	12. $.8 = a + .5$	13. $2b = \frac{1}{4}$
14. $.4c = 6$	15. $\frac{5}{3}d = \frac{15}{9}$	16. $-2\frac{1}{3} = -3 + p$
17. $\frac{q}{3} = -2\frac{1}{3}$	18. $r - .35 = .2$	19. $s + .1 = .21$
20. $2x + 7 = 1$	21. $4y - 3 = 4$	22. $3z = 20 + z$
23. $\frac{3}{2}m = 12 - \frac{5}{2}m$	24. $2.5k = 11 + .3k$	25. $4(t + 3) = 8$
26. $2(x - 1) = -2$	27. $4 = 8(y + 1)$	28. $4(b + .2) = 4$
29. $6x - 6 = 3x - 2$	30. $3y - 5 + 2y = 8 - 3$	
31. $b - .24 = .4 - 3b$	32. $2k - 3 = .4k + 3.4$	
33. $3x - 2(x - 3) = 15$	34. $2y - (y + 10) = 5(y + 6)$	
35. $2y - 3(2y - 3) = y + 29$	36. $4 - (x + .2) = 2x - .1$	

In 37–52, solve within the set of rational numbers. Use set-builder notation to write the solution set.

37. $x + 1\frac{1}{2} < 4$	38. $3x + 1 > 13$	39. $-3x < -18$
40. $15 - 2x < 1$	41. $x \geq 7x - 6$	42. $25 < 9x - 2$
43. $\frac{3}{4}y \geq \frac{3}{8}$	44. $\frac{y}{2} \leq .15$	45. $3 - x < 11$
46. $4x < 3(x + 4)$	47. $3(x + 2) \leq 2$	48. $2(3 - x) \leq 10$
49. $8 + 3y > 10 - 2y$		50. $14 + y \leq 4 + 2y$
51. $x - 2(x - 2) \leq 2$		52. $x - 3 \leq 3(2x - 1)$

In 53–55, select the numeral preceding the expression that best completes the sentence or answers the question.

53. The solution set of $x + 8 = 7$ is { } when the domain is the set of:
    (1) integers   (2) whole numbers.
    (3) negative numbers   (4) rational numbers
54. For which domain will the solution set of $4x + 1 < 13$ be $\{1, 2\}$?
    (1) integers   (2) whole numbers
    (3) natural numbers   (4) rational numbers
55. The solution set of $3x - 2 = 5$ is *not* empty when the domain is the set of:
    (1) integers   (2) whole numbers
    (3) natural numbers   (4) rational numbers

1-8   RATIO AND PROPORTION

Take the first five letters of the word "rational" to form a closely associated mathematical term, "ratio."

■ **DEFINITION.** The *ratio* of two numbers a and b, where b is not zero, is the number $\frac{a}{b}$.

For example, the ratio of 2 and 4 in the order given is the number $\frac{2}{4}$. Recall that any ratio of a to b may also be written in the form $a:b$. Thus, the ratio of 2 to 4 may be written as $\frac{2}{4}$ or as $2:4$.

Either form, $\frac{a}{b}$ or $a:b$, may be used when comparing two numbers.

In our study of geometry, we used ratios to compare lengths of segments and measures of angles. Keep in mind that lengths and angle measures are *numbers*, not the physical objects themselves. When comparing three or more numbers in a *continued ratio*, we use the form $a:b$. For example, the ratio of the angle measures of a $30°$-$60°$-$90°$ triangle is expressed as $30:60:90$. (See Fig. 1.)

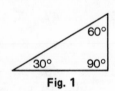
Fig. 1

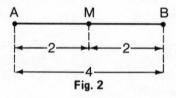
Fig. 2

If M is a midpoint of segment $\overline{AB}$, and AB has a length of 4 centimeters, then $AM = 2$ centimeters and $MB = 2$ centimeters. (See Fig. 2.)

In this example, the ratio $\frac{AM}{AB}$ is the number $\frac{2}{4}$, or $2:4$. Since the ratio $\frac{2}{4}$ is equal to the ratio $\frac{1}{2}$, we may write $\frac{2}{4} = \frac{1}{2}$. We may also write $2:4 = 1:2$, which is read as "2 is to 4 as 1 is to 2." Each of these equations is an example of a proportion.

■ **DEFINITION.** A *proportion* is an equation that states that two ratios are equal.

In general terms, a proportion is written as $\frac{a}{b} = \frac{c}{d}$, or as $a:b = c:d$, where $b \neq 0$ and $d \neq 0$. The first and fourth terms, a and d, are called the *extremes* of the proportion. The second and third terms, b and c, are called the *means* of the proportion.

1. Given any proportion:

$$\frac{a}{b} = \frac{c}{d}$$

2. Apply the multiplication property of equality.

$$\frac{a}{b}(bd) = \frac{c}{d}(bd)$$

3. Simplify.

$$ad = bc$$

Conversely, by starting with $ad = bc$, and multiplying each side by $\frac{1}{bd}$, we will obtain $\frac{a}{b} = \frac{c}{d}$. Thus, we observe:

■ **A proportion exists if and only if the product of the means is equal to the product of the extremes.**

Thus, given $b \neq 0$ and $d \neq 0$, we can state:

$$\frac{a}{b} = \frac{c}{d} \text{ if and only if } ad = bc.$$

For example, given $\frac{6}{15} = \frac{8}{20}$, then $6 \cdot 20 = 15 \cdot 8$, or $120 = 120$.

Using the multiplication property for inequalities where $b > 0$ and $d > 0$, we can prove that similar relations are true for fractions that are not equal, or terms not in a proportion:

$\frac{a}{b} > \frac{c}{d}$ if and only if $ad > bc$	$\frac{a}{b} < \frac{c}{d}$ if and only if $ad < bc$
For example:	For example:
If $\frac{2}{3} > \frac{4}{11}$, then $2 \cdot 11 > 3 \cdot 4$	If $\frac{4}{7} < \frac{8}{9}$, then $4 \cdot 9 < 7 \cdot 8$
or $22 > 12$.	or $36 < 56$.

Equivalent Fractions

Two fractions, such as $\frac{2}{4}$ and $\frac{1}{2}$, that are different expressions for the same rational number are called *equivalent fractions*. By stating that equivalent fractions are equal to each other, as in $\frac{2}{4} = \frac{1}{2}$, we form a proportion. Conversely, when a proportion is stated, the left-hand and right-hand members of the equation represent or name equivalent fractions.

There are two common methods to find fractions that are equivalent to any given rational number $\frac{a}{b}$:

1. EXTENSION. Multiply both numerator and denominator by the same number x, where $x \neq 0$. Note that $\frac{x}{x}$ is equal to $\frac{1}{1}$, or 1, the identity element under multiplication.

$$\frac{a}{b} = \frac{a \cdot x}{b \cdot x} = \frac{ax}{bx}$$

For example, $\frac{3}{4} = \frac{3 \cdot 5}{4 \cdot 5} = \frac{15}{20}$, and $\frac{3}{4} = \frac{3(-2)}{4(-2)} = \frac{-6}{-8}$. Thus, $\frac{3}{4}, \frac{15}{20}$, and $\frac{-6}{-8}$ are all equivalent fractions.

2. CANCELLATION. When possible, divide both numerator and denominator by the same number x, where $x \neq 0$. To determine a number for this common division, find a factor common to both parts of the fraction.

For example, $\frac{8}{12} = \frac{4 \cdot 2}{4 \cdot 3}$. Thus, $\frac{8}{12} = \frac{8 \div 4}{12 \div 4} = \frac{2}{3}$. This is sometimes written in the form of a cancellation: $\frac{8}{12} = \frac{\overset{2}{\cancel{8}}}{\underset{3}{\cancel{12}}} = \frac{2}{3}$. Thus, $\frac{8}{12}$ and $\frac{2}{3}$ are equivalent fractions.

Proportions in Geometry

Let us recall some applications of proportions from geometry.

■ **DEFINITION.** *Two polygons are similar* if and only if there is a one-to-one correspondence between their vertices such that the corresponding angles are congruent, and the ratios of the lengths of corresponding sides are equal.

For example, if $\triangle ABC \sim \triangle A'B'C'$, it is understood that A corresponds to A', B corresponds to B', and C corresponds to C'. Since the ratios of lengths of corresponding sides are equal, we may write:

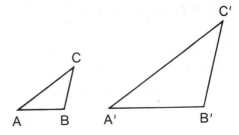

$$\frac{AB}{A'B'} = \frac{BC}{B'C'} \qquad \frac{AB}{A'B'} = \frac{CA}{C'A'} \qquad \frac{BC}{B'C'} = \frac{CA}{C'A'}$$

Note: When the ratios of the lengths of the corresponding sides of two polygons are equal, as just shown, we say that *the corresponding sides of the two polygons are in proportion.*

It is interesting to note that $\dfrac{AB}{A'B'}$, $\dfrac{BC}{B'C'}$, and $\dfrac{CA}{C'A'}$ all name the same rational number and are therefore equivalent fractions.

Some familiar examples involving similar triangles appear in the exercises to follow. For now, let us recall three specific situations.

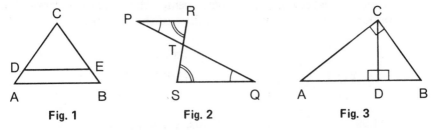

Fig. 1   Fig. 2   Fig. 3

1. In Fig. 1, if $\overleftrightarrow{DE} \parallel \overleftrightarrow{AB}$, then $\triangle DEC \sim \triangle ABC$. Many proportions can be formed, including: $\dfrac{CD}{DA} = \dfrac{CE}{EB}$; $\dfrac{CD}{CA} = \dfrac{CE}{CB}$; $\dfrac{CD}{DE} = \dfrac{CA}{AB}$

2. In Fig. 2, if $\overleftrightarrow{PR} \parallel \overleftrightarrow{SQ}$, then $\angle P \cong \angle Q$ and $\angle R \cong \angle S$. Since $\triangle PRT \sim \triangle QST$ by a.a. $\cong$ a.a., proportions include: $\dfrac{PR}{QS} = \dfrac{RT}{ST}$; $\dfrac{PR}{QS} = \dfrac{PT}{QT}$

3. In Fig. 3, if altitude $\overline{CD}$ is drawn to hypotenuse $\overline{AB}$ in right $\triangle ABC$, three similar triangles are formed, $\triangle ABC \sim \triangle ACD \sim \triangle CBD$. Among the proportions: $\dfrac{DA}{DC} = \dfrac{DC}{DB}$; $\dfrac{AD}{AC} = \dfrac{AC}{AB}$; $\dfrac{BD}{BC} = \dfrac{BC}{BA}$

MODEL PROBLEMS	

1. Replace the question mark "?" between each pair of rational numbers with $>$, $<$, or $=$ to make the statement true.

   a. $\frac{4}{7}$? $\frac{12}{21}$ b. $\frac{8}{15}$? $\frac{6}{10}$

   Solution

   Method 1: Change each fraction to an equivalent fraction so that both rational numbers have the same denominator.

   a. $\frac{4}{7}$? $\frac{12}{21}$ b. $\frac{8}{15}$? $\frac{6}{10}$

   $\frac{4 \cdot 3}{7 \cdot 3}$? $\frac{12}{21}$ $\frac{8 \cdot 2}{15 \cdot 2}$? $\frac{6 \cdot 3}{10 \cdot 3}$

   $\frac{12}{21} = \frac{12}{21}$ $\frac{16}{30} < \frac{18}{30}$

   Thus, $\frac{4}{7} = \frac{12}{21}$ Thus, $\frac{8}{15} < \frac{6}{10}$

   Method 2: Determine the relationship that exists between the product of the means and the product of the extremes.

   a. $\frac{4}{7}$? $\frac{12}{21}$ b. $\frac{8}{15}$? $\frac{6}{10}$

   $4(21)$? $7(12)$ $8(10)$? $15(6)$

   $84 = 84$ $80 < 90$

   Thus, $\frac{4}{7} = \frac{12}{21}$ Thus, $\frac{8}{15} < \frac{6}{10}$

   Answer: a. $\frac{4}{7} = \frac{12}{21}$ b. $\frac{8}{15} < \frac{6}{10}$

2. Given: $\overline{AB}$ intersects $\overline{CD}$ at E, $\angle A \cong \angle D$, $\angle C \cong \angle B$, $AC = 36$, $CE = 15$, $EA = 30$, and $DB = 24$.
   a. Explain why $\triangle ACE \sim \triangle DBE$.
   b. Find the lengths of $\overline{DE}$ and $\overline{BE}$.

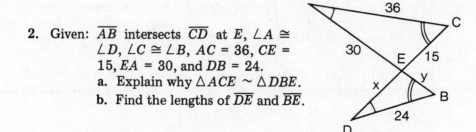

Solution

a. Since $\angle A \cong \angle D$ and $\angle C \cong \angle B$, it follows that $\triangle ACE \sim \triangle DBE$ by a.a. $\cong$ a.a.

b. 1. Identify the variables.

Let $x = DE$. | Let $y = BE$.

2. The corresponding sides of similar triangles are in proportion.

$\dfrac{DE}{AE} = \dfrac{DB}{AC}$ | $\dfrac{BE}{CE} = \dfrac{BD}{CA}$

3. Substitute known values.

$\dfrac{x}{30} = \dfrac{24}{36}$ | $\dfrac{y}{15} = \dfrac{24}{36}$

4. The product of the means is equal to the product of the extremes.

$36x = 30(24)$ | $36y = 15(24)$

$36x = 720$ | $36y = 360$

5. Solve the equations.

$x = 20$ | $y = 10$

Answer: **a.** $DE = 20$ **b.** $BE = 10$

$\left(\text{Note that } \dfrac{DE}{AE} = \dfrac{20}{30} = \dfrac{2}{3}, \dfrac{BE}{CE} = \dfrac{10}{15} = \dfrac{2}{3}, \text{ and } \dfrac{DB}{AC} = \dfrac{24}{36} = \dfrac{2}{3}. \text{ The}\right.$
ratios are equivalent fractions because they name the same rational
number, $\left. \frac{2}{3}. \right)$

EXERCISES

1. If $AB = 6$ and $BC = 9$, state each ratio in simplest form (or reduced form).

   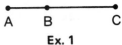

   A   B   C

   Ex. 1

   a. $\dfrac{AB}{BC}$ **b.** $\dfrac{BC}{AB}$ **c.** $\dfrac{AB}{AC}$ **d.** $\dfrac{AC}{BC}$

2. If M is the midpoint of segment $\overline{DE}$, find the rational number represented by each ratio.

   a. $\dfrac{EM}{MD}$ **b.** $\dfrac{DE}{EM}$ **c.** $\dfrac{DM}{DE}$

3. Tell whether or not each statement represents a true proportion.

   a. $\frac{2}{3} = \frac{8}{12}$ **b.** $\frac{9}{4} = \frac{36}{14}$ **c.** $\frac{16}{10} = \frac{24}{15}$

   d. $4:18 = 6:27$ **e.** $6:18 = 4:27$

4. Replace the question mark "?" between each pair of rational numbers with $>$, $<$, or $=$ to make the statement true.

   a. $\frac{9}{20}$? $\frac{4}{10}$ b. $\frac{3}{6}$? $\frac{6}{12}$ c. $\frac{20}{30}$? $\frac{12}{18}$ d. $\frac{14}{16}$? $\frac{18}{20}$

   e. $\frac{4}{12}$? $\frac{5}{15}$ f. $\frac{10}{8}$? $\frac{9}{7}$ g. $\frac{15}{2}$? $\frac{-30}{4}$ h. $5\frac{1}{3}$? $\frac{32}{6}$

In 5–10, $\triangle ABC \sim \triangle A'B'C'$.

5. If $AB = 10$, $BC = 6$, and $A'B' = 15$, find $B'C'$.

6. If $AC = 18$, $AB = 12$, and $A'C' = 30$, find $A'B'$.

7. If $A'B' = 30$, $B'C' = 20$, and $BC = 14$, find AB.

8. If $A'B' = 40$, $B'C' = 30$, and $AB = 26$, find BC.

9. If $AB = 8$, $BC = 6$, $AC = 10$, and $A'B' = 12$, find $B'C'$ and $A'C'$.

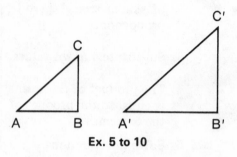

Ex. 5 to 10

10. If $A'B' = 35$, $B'C' = 25$, $C'A' = 40$, and $BC = 15$, find AB and CA.

In 11–16, D is a point on $\overline{AC}$ and E is a point on $\overline{BC}$ such that $\overleftrightarrow{DE} \parallel \overleftrightarrow{AB}$.

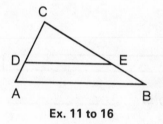

Ex. 11 to 16

11. If $AD = 6$, $DC = 8$, and $BE = 12$, find EC.

12. If $DE = 12$, $AB = 18$, and $CA = 12$, find CD.

13. If $CE = 5$, $ED = 5$, and $EB = 3$, find BA.

14. If $CD = 6$, $DE = 12$, and $DA = 5$, find AB.

15. If $CE = 6$, $ED = 10$, and $AB = 15$, find EB.

16. If $AD = 5$, $DC = 10$, and $BC = 24$, find BE and EC.

In 17–20, $\overline{AB}$ and $\overline{CD}$ meet at E, $\angle A \cong \angle B$, and $\angle C \cong \angle D$.

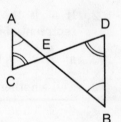

17. If $AC = 12$, $CE = 9$, and $ED = 12$, find BD.

18. If $AE = 8$, $EB = 12$, and $DE = 9$, find EC.

19. If $AE = 3$, $EB = 6$, and $DE = 10$, find EC.

20. If $AC = 4$, $CE = 5$, $EA = 6$, and $BD = 6$, find BE and ED.

Ex. 17 to 20

In 21-27, $\overline{CD}$ is the altitude to hypotenuse $\overline{AB}$ in right $\triangle ABC$.

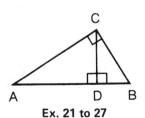
Ex. 21 to 27

21. If $DB = 3$ and $DC = 6$, find DA.
22. If $DB = 4$ and $BC = 6$, find BA.
23. If $DB = 8$ and $DC = 10$, find DA.
24. If $AB = 10$ and $BC = 4$, find DB.
25. If $AD = 20$ and $DB = 5$, find CD.
26. If $CB = 6$ and $BA = 18$, find BD and DA.
27. If $AD = 16$ and $DB = 9$, find DC, BC, and AC.

In 28–33, $\overline{AB}$ intersects $\overline{CD}$ at $E, \angle A \cong \angle C$, and $\angle D \cong \angle B$.

28. If $DE = 12, EA = 8$, and $BE = 9$, find EC.
29. If $DE = 14, BE = 10$, and $EA = 7$, find EC.
30. If $BE = 5, EA = 3$, and $CE = 2$, find ED.
31. If $DA = 16, AE = 6$, and $EC = 3$, find CB.
32. If $BE = 9, EA = 2$, and the ratio of $DE:EC$ is $2:1$, find DE and EC.
33. If $DE = 16, EC = 3$, and the ratio of $BE:EA$ is $3:1$, find BE and EA.

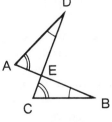
Ex. 28 to 33

34. If a photograph 8 cm by 5 cm is enlarged so that its longer side measures 20 cm, what is the measure of its shorter side?
35. There are 75 calories in a $2\frac{1}{2}$-ounce serving of cottage cheese. How many calories are in an 8-ounce container of cottage cheese?
36. Find three numbers in the ratio of $1:2:4$ whose sum is 35.
37. Doris and Danny play checkers. The ratio of the number of games won by Doris to the number of games won by Danny is $5:3$. If Doris won 6 more games than Danny, find the total number of games of checkers that they played.

1-9 OPERATIONS WITH POLYNOMIALS

You have learned that a rational number can be expressed in the form of a fraction, such as $\frac{2}{5}, \frac{8}{1}$, and $\frac{-17}{3}$. An **algebraic fraction**, sometimes called a **rational expression**, is a quotient of two polynomials. Examples of algebraic fractions include $\frac{3}{x}, \frac{x+5}{x-2}$, and $\frac{-4x}{x^2 - 3x + 2}$.

To prepare for operations with algebraic fractions, we will review operations with polynomial expressions.

Algebraic Terms

A *term* is a number, a variable, or the indicated product or quotient of numbers and variables. Examples of terms include: 8, x, $-5ab$, $\dfrac{x}{y}$

In a product, any factor is the *coefficient* of the remaining factors. Thus, for the product $2ax$, we say that $2a$ is the coefficient of x, and $2x$ is the coefficient of a. If a numeral only is the coefficient, we call this factor the *numerical coefficient*. Given $2ax$, we say that 2 is the numerical coefficient of ax.

Let us agree that the word "coefficient," unless otherwise stated will mean numerical coefficient. Therefore, the coefficient of $7ab$ is 7, and the coefficient of y is 1, since $y = 1y$.

In a term such as x^3, the *base* (x) is the quantity that is used as a factor two or more times; the *exponent* (3) is the number that tells us how many times the base is used as a factor; and the *power* (x^3) is the product. While x^3 is read as x to the third power, note that the power is $x \cdot x \cdot x$, or simply x^3.

Numerical Example		
$5^2 = 25$ $\begin{cases}\\ \\ \\ \end{cases}$	Base	$= 5$
	Exponent	$= 2$
	Power	$= 5^2$ or 25

Algebraic Term		
x^3 $\begin{cases}\\ \\ \\ \end{cases}$	Base	$= x$
	Exponent	$= 3$
	Power	$= x^3$

Like terms are two or more terms that contain the same variables, with corresponding variables having the same exponents. Terms that are not like are called *unlike terms*. Notice the differences.

LIKE TERMS: $5x$ and $6x$ t^2 and $-2t^2$ $3ab^2$ and ab^2

UNLIKE TERMS: $5x$ and $6y$ t^2 and $-2t$ $3ab^2$ and a^2b

Polynomials

An algebraic expression consisting of one term that is a constant, a variable, or the product of constants and variables is a *monomial*. Thus, we say:

■ **DEFINITION.** A *polynomial* is a monomial or any sum of monomials.

Since *poly* means many, we often think of a polynomial as an expression with many terms. You have also learned to identify types of polynomials by special names. For example:

1. MONOMIALS (mono means "one"), such as 8, x, and $-5ab$.
2. BINOMIALS (bi means "two"), such as $3a - 2b$ and $x + 7$.
3. TRINOMIALS (tri means "three"), such as $a + b + c$ and $x^2 - x - 6$.

The *degree of a monomial* is the sum of the exponents of the variables. For example, the degree of $8a^3bx^2$ is $3 + 1 + 2$, or 6. The *degree of a polynomial* is the highest degree in the expression. Thus, the degree of $5a^2x^2 + x$ is 4; the degree of $x^2 - 5x + 8$ is 2; and the degree of $10x - 7$ is 1.

By arranging the terms according to their exponents, we may write a polynomial containing only one variable in *ascending order* ($5 - 2x + x^2$) or in *descending order* ($x^2 - 2x + 5$). The *standard form* of a polynomial in one variable uses descending order. Thus, $2x^2 + 8x - 5$ and $5x - 9$ are polynomials in standard form. In general terms, these standard forms are written as $ax^2 + bx + c$ where $a \neq 0$, and $ax + b$ where $a \neq 0$.

Addition

We learned to add like terms, or like monomials, by using the distributive property:

$$8y + 6y = (8 + 6)y = 14y$$

Eventually, we eliminated the middle step and simply wrote $8y + 6y = 14y$. Although we can *combine like terms* under addition, the sum of two unlike terms will always be a binomial. For example, the sum of $4a$ and $3b$ is $4a + 3b$. Unlike terms cannot be combined under addition.

■ **PROCEDURE.** To add polynomials, combine like terms.

☐ EXAMPLE 1:

Add: $3x^2 - 7x - 5$
$\phantom{\text{Add: }}2x^2 + 5x - 3$
$\phantom{\text{Add: }}\underline{-x^2 + 2x + 1}$
Answer: $4x^2 - 7$

☐ EXAMPLE 2:

$(5a - 3b) + (-4a - 2b)$
$= 5a - 3b - 4a - 2b$
$= a - 5b$ *Ans.*

Subtraction

In $9x - 7x$, the minuend is $9x$ and the subtrahend is $7x$. The opposite, or additive inverse, of $7x$ is $-7x$. Just as $9x - 7x = 2x$, so too does $9x + (-7x) = 2x$. Note that like terms are combined under subtraction.

■ **PROCEDURE.** To subtract polynomials, add the opposite (additive inverse) of the subtrahend to the minuend.

☐ EXAMPLE 1:

Subtract:

$$\begin{array}{r} 3x^2 - 7x \\ \underline{-2x^2 \quad\quad + 5} \end{array}$$

Answer: $5x^2 - 7x - 5$

☐ EXAMPLE 2:

$(5a - 3b) - (4a - 2b)$

$= (5a - 3b) + (-4a + 2b)$

$= 5a - 3b - 4a + 2b$

$= a - b$ Ans.

Multiplication

Recall the procedures learned in multiplying monomials:

1. In multiplying terms with the same base, the product contains the same base, but its exponent is the sum of the exponents of the terms. For example, $x^3 \cdot x^2 = (x \cdot x \cdot x) \cdot (x \cdot x) = x^5$. Also, $y \cdot y^6 = y^7$. In general, if a and b are positive integers:

$$x^a \cdot x^b = x^{a+b}$$

2. In repeated multiplication of the same term, sometimes called the power of a power, the product can be found by addition or multiplication of exponents. For example $(x^3)^4 = x^3 \cdot x^3 \cdot x^3 \cdot x^3 = x^{12}$. Here, $3 + 3 + 3 + 3 = 12$, and $3 \cdot 4 = 12$. In general, if a and c are positive integers:

$$(x^a)^c = x^{ac}$$

3. To multiply monomials, we can express the product by a series of factors: first, we multiply the numerical coefficients; then, in alphabetical order, we multiply variable factors that are powers having the same base. For example:

$(3x)(6x) = (3 \cdot 6)(x \cdot x) = 18x^2;\ (2y)(7y^2) = (2 \cdot 7)(y \cdot y^2) = 14y^3$
$(5a)(4b) = (5 \cdot 4)(a \cdot b) = 20ab;\ (-8a^3b)(ab^2x) = -8a^4b^3x$

■ **PROCEDURE.** To multiply a polynomial by a monomial, use the distributive property to multiply each term of the polynomial by the monomial; then add the resulting products.

☐ EXAMPLE: $2x(x^2 - 3x + 4) = 2x(x^2) + 2x(-3x) + 2x(4)$

$= (2x^3) + (-6x^2) + (8x)$

$= 2x^3 - 6x^2 + 8x$ Ans.

■ **PROCEDURE.** To multiply a polynomial by a polynomial, use the distributive property to multiply each term of one polynomial (the multiplicand) by each term of the other polynomial (the multiplier); then combine like terms.

It is advisable to write each polynomial in standard form before multiplying whenever possible.

☐ EXAMPLE 1:
Multiply:

$$x^2 + 8x + 9$$
$$\underline{\hspace{1.5cm} x - 4}$$

$x(x^2 + 8x + 9) = \overline{x^3 + 8x^2 + 9x}$
$-4(x^2 + 8x + 9) = \underline{\hspace{1cm} -4x^2 - 32x - 36}$

Answer: $\quad x^3 + 4x^2 - 23x - 36$

☐ EXAMPLE 2:
$(x + 2)(3x - 5)$
$= x(3x - 5) + 2(3x - 5)$
$= 3x^2 - 5x + 6x - 10$
$= 3x^2 + x - 10 \quad$ *Ans.*

Let us redo example 2, using the process to multiply binominals mentally:

1. Multiply the first terms of the binomials.

2. Multiply the first term of each binomial by the last term of the other binomial and add these products. (Here, $-5x + 6x = x$.)

3. Multiply the last terms of the binomials.

4. Add the results from steps 1, 2, and 3.

Division

Recall the procedures learned in dividing monomials:

1. In dividing terms with the same base, the quotient contains the same base but its exponent is the difference of the exponents of the terms in the order given. For example:

$$x^5 \div x^3 = \frac{\overset{1}{\cancel{x}} \cdot \overset{1}{\cancel{x}} \cdot \overset{1}{\cancel{x}} \cdot x \cdot x}{\underset{1}{\cancel{x}} \cdot \underset{1}{\cancel{x}} \cdot \underset{1}{\cancel{x}}} = x^2$$

Also, $y^7 \div y = y^7 \div y^1 = y^6$. In general, if a and b are integers, $a > b$, and $x \neq 0$:

$$x^a \div x^b = x^{a-b}$$

2. To divide monomials, we can express the quotient as a series of factors: First, we divide the numerical coefficients; then, in alphabetical order, we divide variable factors that are powers of the same base. For example:

$$(2x \div 3x) = \frac{2x}{3x} = \frac{2}{3} \qquad 4a^4y^2 \div 8a^3y^2 = \frac{4a^4y^2}{8a^3y^2} = \frac{a}{2}$$

■ **PROCEDURE.** To divide a polynomial by a monomial, divide each term of the polynomial by the monomial.

☐ EXAMPLE: $(8x^3 + 6x^2) \div (-2x) = \dfrac{8x^3}{-2x} + \dfrac{6x^2}{-2x} = -4x^2 - 3x$ *Ans.*

■ **PROCEDURE.** To divide a polynomial by a polynomial, follow the process of long division from arithmetic until reaching a remainder of 0 or a remainder whose degree is less than the degree of the divisor.

Before starting the division process, write each polynomial in standard form. This process is shown with two examples, one from arithmetic and one from algebra, to allow for a comparison.

	☐ EXAMPLE 1:	☐ EXAMPLE 2:
	$1079 \div 43$	$\dfrac{x^2 + 8x + 19}{x + 5}$
How to Proceed	*Solution:*	*Solution:*
1. Write the usual division form.	$43\overline{)1079}$	$x + 5\overline{)x^2 + 8x + 19}$
2. Divide the first term of the dividend by the first term of the divisor.	$\dfrac{2}{43\overline{)1079}}$	$\dfrac{x}{x + 5\overline{)x^2 + 8x + 19}}$
3. Multiply the entire divisor by the first term of the quotient.	$\dfrac{2}{43\overline{)1079}}$ 86	$\dfrac{x}{x + 5\overline{)x^2 + 8x + 19}}$ $x^2 + 5x$
4. Subtract this product from the dividend. Bring down the next term to obtain the new dividend.	$\dfrac{2}{43\overline{)1079}}$ $\underline{86}$ 219	$\dfrac{x}{x + 5\overline{)x^2 + 8x + 19}}$ $\underline{x^2 + 5x}$ $3x + 19$

5. Repeat steps 2 to 4 with the next term of the quotient, and so on, until the remainder is 0 or the degree of the remainder is less than the degree of the divisor.

$$\begin{array}{r} 25 \\ 43\overline{)1079} \\ \underline{86} \\ 219 \\ \underline{215} \\ 4 \end{array}$$

$$\begin{array}{r} x + 3 \\ x + 5\overline{)x^2 + 8x + 19} \\ \underline{x^2 + 5x} \\ 3x + 19 \\ \underline{3x + 15} \\ 4 \end{array}$$

(In example 2, the remainder 4 has a degree of 0, while the divisor $x + 5$ has a degree of 1.)

Answer:

$$25\frac{4}{43}$$

Answer:

$$x + 3 + \frac{4}{x + 5}$$

Check

To check a division problem, multiply the quotient by the divisor. When the remainder is added to this product, the result should be the dividend.

Check

$$\begin{array}{r} 25 \\ \times 43 \\ \hline 75 \\ 100 \\ \hline 1075 \\ +4 \\ \hline 1079 \end{array}$$

Check

$x\ \ + 3$	Quotient
$x\ \ + 5$	Divisor
$x^2 + 3x$	
$\ \ \ \ + 5x + 15$	
$x^2 + 8x + 15$	
$\ \ \ \ \ \ \ \ + \ \ 4$	Remainder
$x^2 + 8x + 19$	Dividend

MODEL PROBLEMS

1. Simplify: $3x + 2(3x + 2) + (x - 1)^2$

How to Proceed	*Solution*
1. Write the expression.	$3x + 2(3x + 2) + (x - 1)^2$
2. Multiply (or divide) from left to right.	$= 3x + 2(3x + 2) + (x - 1)(x - 1)$ $= 3x + 6x + 4 + x^2 - 2x + 1$
3. Add (or subtract) by combining like terms.	$= x^2 + (3x + 6x - 2x) + (4 + 1)$ $= x^2 + 7x + 5$ *Ans.*

2. Simplify: $4y^2 - [3y + 4y(y - 3)]$

How to Proceed	Solution

If an expression contains one grouping within another, such as parentheses within brackets, work from the innermost grouping first. Clear grouping symbols by using the distributive property.

$4y^2 - [3y + 4y(y - 3)]$
$= 4y^2 - [3y + 4y^2 - 12y]$
$= 4y^2 - 3y - 4y^2 + 12y$
$= 9y$ *Ans.*

| EXERCISES |

In 1-5, add.

1. $3ab$
       ab
      $-2ab$

2. $4x^2 -\ \ x - 3$
       $2x^2 - 3x + 1$
       $x^2 + 4x + 8$

3. $(2a - 3b + c) + (5b - 6c)$
4. $(8m - 7) + (5 - m) + (2 - 7m)$
5. $(3b - a) + (2a - 3c) + (c - b)$

In 6-12, subtract the lower polynomial from the upper polynomial.

6. $18ax$
       $12ax$

7. $5by^2$
      $-4by^2$

8. $-3k + 2$
       $-8k + 3$

9. $4c - 3d$
       $c + 2d$

10. $x^2 - 5x - 6$
        $2x^2 - 3x - 6$

11. $4 - 3x - x^2$
        $3 - 4x - 2x^2$

12. $8a\ \ \ \ \ \ \ - 4c$
        $-7a + 3b + 6c$

13.   Subtract $x^2 - 8x + 2$ from $3x^2 - 8x + 1$.
14.   From the sum of $2x + 8$ and $x - 13$, subtract $4x - 3$.
15.   Subtract $3y - y^2$ from the sum of $y^2 - 2$ and $8y - 8$.
16.   By how much does $4k - 7$ exceed $3k + 2$?
17.   How much greater than $a^2 + a - 3$ is $3a^2 + a + 5$?
18.   The sum of two polynomials is $x^2 - 8x + 1$. If one of the polynomials is $2x^2 - x + 1$, what is the other polynomial?
19.   a. What polynomial when added to $3a - b$ produces a sum of 0?
        b. What polynomial when subtracted from $3a - b$ produces a difference of 0?
        c. How are the answers to parts a and b related?

In 20-46, multiply.

20. $3x^3 \cdot 2x^2$
21. $(-m)(8m)$
22. $y^3 \cdot y \cdot y^5$
23. $9ab(2ab^3)$
24. $x^a \cdot x^{3a}$
25. $y^b \cdot y \cdot y$
26. $(-8x^a)(-x^a)$
27. $y^{a+5} \cdot y^{3-a}$

28. $-4(2a - b)$ 29. $5x(x + 3)$ 30. $-c^2(8c - 3c^2)$
31. $xy(2x^2 - y)$ 32. $ab^2(ab - a^2)$ 33. $(x + 5)(x + 4)$
34. $(y - 8)(y + 8)$ 35. $(k - 1)(k - 6)$ 36. $(t - 5)(t - 5)$
37. $(5x + 2)(3x - 1)$ 38. $(2y - 3)(2y + 3)$ 39. $(2x - 1)(x - 5)$
40. $(3a + b)(a + b)$ 41. $(2x - y)(2x - y)$ 42. $(x + 3b)(x - 7b)$
43. $(x^2 - 2x + 1)(x - 1)$ 44. $(y^2 + y + 1)(y - 1)$
45. $(d^2 + 3d - 4)(2d - 6)$ 46. $(2x^2 - x + 3)(x + 3)$

In 47–64, simplify the expression.

47. $5(x - 2) + 10$ 48. $6k - (7 - 6k)$ 49. $8x - 2(4x - 1)$
50. $-[x - 2(x - 1)]$ 51. $3 - [2 - (1 - x)]$ 52. $(y^2)^3 + (y^3)^2$
53. $(2k^2)^3 - k^6$ 54. $[(6x - x) + 3]^2$ 55. $6x - (x + 3)^2$
56. $(y + 2)^2 - y^2$ 57. $3(x - 3)^2 - 27$ 58. $[3(x - 2) + 5]^2$
59. $ab(3a - 4c) - bc(2b - 4a)$ 60. $x(x - 8) - 2x(x - 3)$
61. $3y[4y - 3(y - 2) - 5]$ 62. $k - 3[k - 3(k - 3)]$
63. $(2x + 3)^2 - (2x - 3)^2$ 64. $(x^a + 6)(x^a - 6) - (x^2)^a$

In 65–79, divide and check.

65. $(18a - 6b) \div 6$ 66. $(3x^2 + 8x) \div x$ 67. $(4y^2 - 2ay) \div 2y$
68. $\dfrac{8x^2 - 12x + 4}{4}$ 69. $\dfrac{28t^3 - 21t^4}{7t^3}$ 70. $\dfrac{6a^2b - 3ab^2 + 3ab}{3ab}$
71. $(y^2 + 9y + 14) \div (y + 7)$ 72. $(x^2 - 2x - 15) \div (x + 3)$
73. $(2x^2 + 7x - 4) \div (x + 4)$ 74. $(6y^2 + 7y + 2) \div (2y + 1)$
75. $(k^2 - 11k + 30) \div (k - 3)$ 76. $(4x^2 + 8x - 19) \div (2x - 3)$
77. $\dfrac{x^2 - 16}{x - 4}$ 78. $\dfrac{y^2 + 12y + 36}{y + 6}$ 79. $\dfrac{x^2 - 4x - 6}{x - 1}$

80. If the length and width of a rectangle measure $2y + 7$ and $3y - 10$, respectively, express the area of the rectangle as a polynomial.
81. A side of a square measures $7x - 3$. Express its area as a trinomial.
82. If one factor of $x^2 - 8x - 9$ is $x - 9$, find the other factor.
83. In a parallelogram whose area is $5k^2 + 13k - 6$, a base measures $k + 3$. Find the measure of the altitude to that base in terms of k.

84. The measures of the length and width of a rectangle are $3x + 4$ and $2x + 2$, respectively. Copy the diagram of the rectangle shown here.

   a. Find the area of each of the four regions into which the

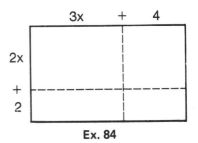

Ex. 84

rectangle is partitioned. Then add these values to find the area of the rectangle.

b. Show that the area of the rectangle is found by multiplying $(3x + 4)(2x + 2)$.

c. Express the perimeter of the rectangle as a binomial.

1-10 FACTORING POLYNOMIALS

To *factor a number*, we find two or more numbers whose product is the given number. Although $20 = \frac{1}{2} \cdot 40$, we restrict factors of integers to numbers that are integers only, as in $20 = 4 \cdot 5$. It is sometimes useful to find all possible pairs of factors for a number. However, a number is *factored completely* only when it is expressed as a product of prime factors. Recall that a *prime number*, such as 2 and 5, is a positive integer greater than 1 whose only factors are 1 and the number itself.

Pairs of Factors	Prime Factors
$20 = 1 \cdot 20$	
$20 = 2 \cdot 10$	$20 = 2 \cdot 2 \cdot 5$
$20 = 4 \cdot 5$	

To *factor a polynomial*, we find two or more algebraic expressions whose product is the given polynomial. In saying that $\frac{1}{2}a + \frac{1}{2}b = \frac{1}{2}(a + b)$, we are factoring this polynomial over the set of rational numbers because $\frac{1}{2}$ is a rational number. It is generally understood, however, that polynomials with integral coefficients are factored with respect to the set of integers. In other words, all coefficients of polynomial factors are integers.

Consider this example: $6x^3 - 12x^2 = 6x^2(x - 2)$. Here, all coefficients of the factors $6x^2$ and $(x - 2)$ are integers. Let us examine these factors carefully:

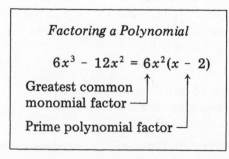

Factoring a Polynomial

$$6x^3 - 12x^2 = 6x^2(x - 2)$$

Greatest common monomial factor

Prime polynomial factor

1. The factor $6x^2$ is called the *greatest common monomial factor* of the polynomial because it is the greatest monomial term that is a factor of each term of the polynomial. In writing this greatest common monomial factor, we do not factor its numerical coefficient or any powers of variables, that is, $6x^2$ is not written as $2 \cdot 3 \cdot x \cdot x$.

2. The second factor $(x - 2)$ is called a *prime polynomial* because it has no factors other than 1 and the polynomial itself, with respect to the set of integers.

■ A *polynomial is factored completely* when it is expressed as a product that may include only the greatest common monomial factor and prime polynomial factors.

Binomial Factors

After the greatest common monomial has been factored from a polynomial expression, or if no common term exists, we next look for binomial factors. In earlier courses, we learned to recognize two polynomial forms having special types of factors:

1. The Difference of Two Squares	2. A Perfect Square Trinomial
$x^2 - a^2 = (x + a)(x - a)$	$x^2 + 2bx + b^2 = (x + b)(x + b)$

□ EXAMPLE:
Factor $25y^2 - 4k^2$.

Solution

Since $25y^2 - 4k^2 = (5y)^2 - (2k)^2$, the difference of two squares, we factor the expression to fit the form:

$25y^2 - 4k^2$
$= (5y + 2k)(5y - 2k)$ *Ans.*

□ EXAMPLE:
Factor $4y^2 + 12y + 9$.

Solution

Since the first and last terms are perfect squares, $(2y)^2$ and $(3)^2$, and the middle term $12y = 2(3)(2y)$, we factor the expression to fit the form:

$4y^2 + 12y + 9$
$= (2y + 3)(2y + 3)$ *Ans.*

■ **PROCEDURE.** To factor any trinomial of the form $ax^2 + bx + c$ where $a \neq 0$, we find two binomial factors such that:

1. The product of the first terms of the binomials is ax^2.

2. The product of the last terms of the binomials is c.

3. When the first term of each binomial is multiplied by the last term of the other, the sum of the two products is bx.

☐ EXAMPLE: Factor $2x^2 - x - 6$.

Solution

1. The product of the first terms of the binomials must be $2x^2$. Thus, we write: $2x^2 - x - 6 = (2x \quad)(x \quad)$.
2. The product of the last terms of the binomials must be -6. Pairs of factors of -6 are $(1)(-6)$, $(2)(-3)$, $(3)(-2)$, and $(6)(-1)$. Since the order of the placement for these pairs is important, possible factors include:

$$(2x + 1)(x - 6) \qquad (2x - 6)(x + 1)$$
$$(2x + 2)(x - 3) \qquad (2x - 3)(x + 2)$$
$$(2x + 3)(x - 2) \qquad (2x - 2)(x + 3)$$
$$(2x + 6)(x - 1) \qquad (2x - 1)(x + 6)$$

3. When the first term of each binomial is multiplied by the last term of the other, only one pair of binomial factors has a middle term of $-x$. Note that the sum of the products $+3x$ and $-4x$ is $-x$.

$$(2x + 3) \, (x - 2)$$
$$+3x$$
$$-4x$$

Answer: $2x^2 - x - 6 = (2x + 3)(x - 2)$

Note: A shortcut could have been taken in step 2 of the example just given. Since $2x^2 - x - 6$ has no common monomial factors other than 1, we know that the binomial factors of this expression must be prime. Therefore, it was not necessary to include the following possible factors:

$$(2x + 2)(x - 3) \qquad (2x - 6)(x + 1)$$
$$(2x + 6)(x - 1) \qquad (2x - 2)(x + 3)$$

In each case, the first binomial is not prime because it has a common factor of 2.

■ **PROCEDURE.** To factor a polynomial completely:

1. First find the greatest common monomial factor, if one exists.

2. Then factor the remaining expression into binomials or other polynomials until all such factors are prime.

You have learned how to factor a polynomial by finding the greatest common monomial factor. It is sometimes possible to factor a polynomial by finding a common binomial factor. Compare the following:

$$3xy - 5y \qquad = y(3x - 5)$$
$$3x(x + 2) - 5(x + 2) = (x + 2)(3x - 5)$$

In the first example, y is a common monomial factor. In the second example, $(x + 2)$ is a common binomial factor.

Often it is necessary to factor *pairs* of terms first in order to identify a common binomial factor.

☐ EXAMPLE: Factor $x^3 - 5x^2 + 2x - 10$.

Find a common monomial factor for each pair of terms.

$$x^3 - 5x^2 + 2x - 10$$
$$= x^2(x - 5) + 2(x - 5)$$

Factor out the common binomial.

$$= (x - 5)(x^2 + 2)$$

Note: A polynomial with four terms can be factored into two binomials when the product of the first and last terms equals the product of the two middle terms.

| MODEL PROBLEMS |

In 1–4, factor completely.

1. Factor: $12a^3b + 3a^2b^2 - 6a^2b^3$

Solution

The greatest common monomial factor is $3a^2b$. Once this term has been factored out, the trinomial factor that remains is prime.

$$12a^3b + 3a^2b^2 - 6a^2b^3 = 3a^2b(4a + b - 2b^2) \ \ Ans.$$

2. Factor: $2x^3 + 14x^2 - 60x$

How to Proceed

Solution

1. Find the greatest common monomial factor.

$$2x^3 + 14x^2 - 60x$$
$$= 2x(x^2 + 7x - 30)$$

2. Factor the trinomial into two prime binomials.

$$= 2x(x + 10)(x - 3) \ \ Ans.$$

3. Factor: $y^4 - 81$

Solution

$$y^4 - 81$$
$$= (y^2 + 9)(y^2 - 9)$$
$$= (y^2 + 9)(y + 3)(y - 3) \ \ Ans.$$

4. Factor: $x^{b+2} - 16x^b$

Solution

$$x^{b+2} - 16x^b$$
$$= x^b(x^2 - 16)$$
$$= x^b(x + 4)(x - 4) \ \ Ans.$$

5. Factor: $ax^2 - bx - ax + b$

<table>
<tr><td>How to Proceed</td><td>Solution</td></tr>
<tr><td>1. Find the greatest common monomial factor for each pair of terms. Note that 1 or -1 may be a factor.</td><td>$ax^2 - bx - ax + b$
$= x(ax - b) - 1(ax - b)$</td></tr>
<tr><td>2. Factor out the common binomial.</td><td>$= (ax - b)(x - 1)$ Ans.</td></tr>
</table>

EXERCISES

In 1–54, factor completely.

1. $3x^2 - 12x$	**2.** $18 - 6y - 12x$	**3.** $4ab^2 - 12a^2b$
4. $9y^3 + 3y^2$	**5.** $x^{b+1} - x^b$	**6.** $y^{a+2} + 2y^a$
7. $k^2 - 49$	**8.** $100 - x^2$	**9.** $a^2b^2 - 144$
10. $x^2 + 10x + 25$	**11.** $y^2 - 12y + 36$	**12.** $9x^2 + 6x + 1$
13. $y^2 + 10y + 9$	**14.** $x^2 - 12x + 27$	**15.** $k^2 + 5k - 14$
16. $x^2 + 2x - 24$	**17.** $3y^2 + 4y + 1$	**18.** $2x^2 + 13x + 6$
19. $2k^2 - 7k + 6$	**20.** $2x^2 + 7x - 4$	**21.** $6y^2 - 13y - 5$
22. $3y^2 - 12$	**23.** $80 - 5d^2$	**24.** $100 - 4x^2$
25. $7a^2 - 7b^2$	**26.** $x^3 - 121x$	**27.** $3y^3 - 192y$
28. $4c^3 - 36cx^2$	**29.** $y^4 - 1$	**30.** $x^4 - 625$
31. $x^3 + 3x^2 - 10x$	**32.** $4x^2 - 20x + 24$	**33.** $ax^3 - 9ax$
34. $2y^3 + 50y$	**35.** $36c^2 - 100d^2$	**36.** $y^{2a} - 1$
37. $x^{2k} - 16$	**38.** $x^{2+k} - 4x^k$	**39.** $x^4 - 5x^2 + 4$
40. $y^4 - 7y^2 - 18$	**41.** $2x^3 + 6x^2 + 2x$	**42.** $4x^2 + 40x + 64$

43. $5x^2 - 5ax - 60a^2$	**44.** $2ax^2 - 8ax - 12a$
45. $8y^3 - 60y^2 - 32y$	**46.** $9ay^2 - 21a^2y + 6a^3$
47. $3bx^2 + b^2x^2 - 2b^3x$	**48.** $x^{c+2} - 14x^{c+1} + 24x^c$
49. $x^2 - 2x - 3x + 6$	**50.** $3x^3 - 6x^2 + 2x - 4$
51. $4x^5 - 8x^3 - 3x^2 + 6$	**52.** $b^2y^2 + by^2 + 3b^2 + 3b$
53. $c^2x^2 + x^2 - c^2 - 1$	**54.** $4x^2 - 4 - x^2y^2 + y^2$

55. If $3x^2 + 20x + 12$ represents the area of a rectangle, find two binomials that can represent its dimensions.

56. If $25x^2 + 20d^2x + 4d^4$ represents the area of a square, express the measure of a side of the square as a binomial.

1-11 REVIEW EXERCISES

In 1–5, express each rational number as a repeating decimal.

1. $\frac{2}{9}$ 2. $\frac{5}{6}$ 3. $\frac{6}{11}$ 4. $3\frac{2}{5}$ 5. $-\frac{7}{30}$

In 6–10, express each rational number as a fraction.

6. $3\frac{1}{4}$ 7. -2.8 8. $.05$ 9. $.\overline{5}$ 10. $.\overline{63}$

In 11–14, name the property illustrated for the set of rational numbers.

11. $\frac{3}{7} \cdot \frac{7}{3} = 1$
12. $\frac{1}{2} \cdot 4 + \frac{1}{2} \cdot 12 = \frac{1}{2}(4 + 12)$
13. $\frac{3}{7} + 0 = \frac{3}{7}$
14. $(\frac{1}{2} \cdot 4) \cdot 12 = \frac{1}{2} \cdot (4 \cdot 12)$

In 15–20, solve and check. Use the domain of rational numbers.

15. $x + .2 = 6$ 16. $3y - 2 = 20$ 17. $k + .4 = 3k$
18. $6(b - 2) = 3$ 19. $5c + 4 = 3c - 7$ 20. $x - 2(x - 2) = -2$

In 21–25, using a domain of rational numbers, write the solution set in set-builder notation.

21. $\frac{y}{4} > 3\frac{1}{2}$ 22. $k \leq 4k - 6$ 23. $2(5 - w) > 8$

24. $5(x - 2) \leq 3(x + 1)$ 25. $y + 3(y + 3) < 2(y + 3) + 3$

In 26–29, replace the question mark between each pair of rational numbers with $>$, $<$, or $=$ to make the sentence true.

26. $\frac{4}{5}$? $\frac{7}{10}$ 27. $\frac{11}{4}$? $\frac{8}{3}$ 28. $\frac{8}{12}$? $\frac{12}{18}$ 29. $2\frac{2}{3}$? $\frac{16}{6}$

In 30–32, $\overline{AB}$ intersects $\overline{CD}$ at E, $\angle A \cong \angle D$, and $\angle C \cong \angle B$.

30. If $AC = 10$, $CE = 4$, and $EB = 6$, find DB.
31. If $CE = 4$, $AE = 12$, and $BE = 5$, find DE.
32. If $AE = 16$, $BE = 9$, and the ratio of $CE : ED$ is $1 : 4$, find CE and DE.

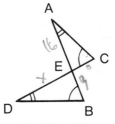

Ex. 30 to 32

In 33–38, perform the indicated operations and simplify.

33. $(x + 2b) + (x - 3b)$ 34. $(3x^2 - 2x + 7) - (x^2 - 2x - 7)$
35. $(x + 2b)(x - 3b)$ 36. $(2y^2 - 7y - 15) \div (y - 5)$
37. $[3x - 2(x - 3) + 2]^2$ 38. $(2k + 1)(2k - 1) - (2k - 1)^2$

In 39–46, factor completely.

39. $x^2 + 14x + 49$ **40.** $6b^2 + 9b$ **41.** $4x^2 - 25$
42. $3y^2 + 7y - 6$ **43.** $4x^2 - 12x + 8$ **44.** $3y^3 - 48y$
45. $x^3 - 2x^2 - 9x + 18$ **46.** $ac + 2c - 3ab - 6b$

In 47–50, write the given sentence in words.
Let $x \in$ {rational numbers}.

47. $\forall_x : x^2 \geq 0$ **48.** $\exists_x : x$ is prime
49. $\exists_x : x$ is not even   **50.** $\sim \exists_x : x + 1 = 1$

In 51–54, write the negation of the given sentence.

51. All rhombuses are squares.
52. There exists an integer that is divisible by 5.
53. Some integers do not have additive inverses.
54. No rational numbers are even.

In 55–69, $S = \{a, b, c\}$ and $*$ and $\#$ are operations defined by the tables.

55. Is S closed under $*$?   **56.** Is $*$ commutative?
57. Give two examples to show that $*$ is associative.
58. What is the identity element for $*$?
59. For each element of S, name the inverse under $*$.
60. Is $(S, *)$ a commutative group?
61. Is S closed under $\#$?   **62.** Is $\#$ commutative?
63. Give two examples to show that $\#$ is associative.
64. What is the identity element for $\#$?
65. For each element of S, name the inverse under $\#$, if it exists.
66. Is $S/\{c\}$ closed under $\#$?
67. Is $(S/\{c\}, \#)$ a commutative group?
68. Give two examples to show that $\#$ is distributive over $*$.
69. Is $(S, *, \#)$ a field?

$*$	a	b	c
a	b	c	a
b	c	a	b
c	a	b	c

$\#$	a	b	c
a	a	b	c
b	b	a	c
c	c	c	c

70. Find the error in the following proof that $4 = 0$.

$$\text{Given: } a = 4$$
$$a^2 = 4a$$
$$a^2 - 16 = 4a - 16$$
$$(a + 4)(a - 4) = 4(a - 4)$$
$$a + 4 = 4$$
$$a = 0$$
$$\text{But, } a = 4 \text{ is given.}$$
$$\text{Therefore, } 4 = 0.$$

Chapter 2

Rational Expressions

2-1 REDUCING RATIONAL EXPRESSIONS

In this chapter, we will study algebraic fractions, more commonly called rational expressions.

■ **DEFINITION.** A *rational expression* is the quotient of two polynomials. Examples of rational expressions include $\dfrac{3}{5}$, $\dfrac{-x}{8}$, $\dfrac{2}{x}$, $\dfrac{6}{x-7}$, $\dfrac{3x-6}{x^2-2x-8}$.

Rational expressions follow the same rules and possess the same properties as those of rational numbers. Compare the two forms that follow:

In a *rational number* $\dfrac{a}{b}$:	In a *rational expression* $\dfrac{A}{B}$:
a and b are integers, and $b \neq 0$.	A and B are polynomials, and $B \neq 0$.

Since division by zero is not possible, we know that $\frac{3}{0}$ has no meaning. In the same way, any rational expression whose denominator equals zero is *meaningless* or *not defined*. By setting each denominator equal to zero, we see that $\frac{2}{x}$ has no meaning when $x = 0$, and $\frac{6}{x-7}$ has no meaning when $x = 7$.

Reducing to Lowest Terms

A rational expression is *reduced to lowest terms*, or stated in *simplest form*, when its numerator and denominator have no common factors other than 1 and -1.

63

Start with a fraction that is not in lowest terms. We learned to divide its numerator and its denominator by the same nonzero number to find an equivalent fraction in lowest terms. The same procedure can be applied to rational expressions.

□ EXAMPLE 1:

$$\frac{5}{10} = \frac{5 \div 5}{10 \div 5} = \frac{1}{2}$$

□ EXAMPLE 2:

$$\frac{3x}{x^2 + 2x} = \frac{3x \div x}{(x^2 + 2x) \div x} = \frac{3}{x + 2}$$

In example 2 above, it is easy to see that both the numerator and the denominator can be divided by x. In general, if we *completely factor* the numerator and denominator of a rational expression, we can discover all common factors before reducing. This process, indicated as a *cancellation*, is simply an alternate form of dividing the numerator and denominator by the same nonzero number.

□ EXAMPLE 1:

$$\frac{5}{10} = \frac{\overset{1}{\cancel{5}}}{\cancel{5} \cdot 2} = \frac{1}{2}$$

□ EXAMPLE 2:

$$\frac{3x}{x^2 + 2x} = \frac{3 \cdot \overset{1}{\cancel{x}}}{\underset{1}{\cancel{x}}(x + 2)} = \frac{3}{x + 2}$$

In addition to noting the importance of complete factorization in the next example, we will use example 3 to make some observations.

□ EXAMPLE 3:

$$\frac{4y^2 - 4y}{4y^2 - 12y + 8} = \frac{4y(y - 1)}{4(y^2 - 3y + 2)} = \frac{\overset{1}{\cancel{4}}y\overset{1}{\cancel{(y - 1)}}}{\underset{1}{\cancel{4}}(y - 2)\underset{1}{\cancel{(y - 1)}}} = \frac{y}{y - 2}$$

We observe:

1. The denominator in factored form is $4(y - 2)(y - 1)$. Let this equal zero to see that $\dfrac{4y^2 - 4y}{4y^2 - 12y + 8}$ has no meaning when $y = 2$ or $y = 1$.

2. To give a numerical demonstration that $\dfrac{4y^2 - 4y}{4y^2 - 12y + 8} = \dfrac{y}{y - 2}$, substitute any number for y in both expressions except $y = 2$ and $y = 1$. This may serve as a check of your work.

For example, let $y = 3$:

$$\frac{4y^2 - 4y}{4y^2 - 12y + 8} = \frac{4(3)^2 - 4(3)}{4(3)^2 - 12(3) + 8} = \frac{4(9) - 4(3)}{4(9) - 12(3) + 8}$$

$$= \frac{36 - 12}{36 - 36 + 8} = \frac{24}{8} = 3$$

Also, $\dfrac{y}{y - 2} = \dfrac{(3)}{(3) - 2} = \dfrac{3}{1} = 3$.

3. The rational expression $\dfrac{y}{y - 2}$ is reduced to simplest form because both numerator and denominator are prime polynomials. It is incorrect to cancel y from both numerator and denominator because y is *not a factor* in the denominator. Rather, y is a term in the binomial $(y - 2)$ in the denominator.

Reducing to -1

In some rational expressions, the numerator and the denominator (or factors thereof) are *additive inverses*. In these cases, the rational expressions (or the factors thereof) reduce to -1.

☐ EXAMPLE 1:

$$\frac{-9}{9} = \frac{-1 \cdot \overset{1}{\cancel{9}}}{\cancel{9}} = -1$$

☐ EXAMPLE 2:

$$\frac{3 - k}{k - 3} = \frac{-1(-3 + k)}{k - 3} = \frac{-1\overset{1}{\cancel{(k - 3)}}}{\cancel{(k - 3)}} = -1$$

☐ EXAMPLE 3:

$$\frac{12 - 6x}{x^2 - 4} = \frac{6(2 - x)}{(x + 2)(x - 2)} = \frac{6(-1)(-2 + x)}{(x + 2)(x - 2)} = \frac{6(-1)\overset{1}{\cancel{(x - 2)}}}{(x + 2)\cancel{(x - 2)}} = \frac{-6}{x + 2}$$

Note: In example 2, if $k = 3$, then the original expression $\dfrac{3 - k}{k - 3}$ is undefined. Thus, $\dfrac{3 - k}{k - 3} = -1$ if and only if $k \neq 3$.

Similarly, in example 3, $\dfrac{12 - 6x}{x^2 - 4}$ is undefined when $x^2 - 4 = 0$, or $x^2 = 4$, or $x = \pm 2$. Thus, $\dfrac{12 - 6x}{x^2 - 4} = \dfrac{-6}{x + 2}$, if and only if $x \neq \pm 2$.

■ In general, whenever rational expressions are reduced, it is understood that the equivalent form of the expression equals the original expression only for those values whereby the original fraction is defined, or the original fraction has meaning.

| **MODEL PROBLEMS** |

1. For what values of x does $\dfrac{4x + 8}{x^2 - 2x - 8}$ have no meaning?

How to Proceed	*Solution*
1. Set the denominator equal to zero.	Let $x^2 - 2x - 8 = 0$
2. Factor the trinomial.	$(x - 4)(x + 2) = 0$
3. Let each factor equal zero.	$x - 4 = 0 \quad\mid\quad x + 2 = 0$
4. Solve the resulting equations.	$x = 4 \quad\mid\quad x = -2$

Answer: $\dfrac{4x + 8}{x^2 - 2x - 8}$ is meaningless, or not defined, when $x = 4$ or $x = -2$.

2. Reduce the rational expression $\dfrac{4x + 8}{x^2 - 2x - 8}$ to lowest terms.

How to Proceed	*Solution*
1. Factor both numerator and denominator completely.	$\dfrac{4x + 8}{x^2 - 2x - 8} = \dfrac{4(x + 2)}{(x - 4)(x + 2)}$
2. Cancel, or divide numerator and denominator by all common factors.	$= \dfrac{4(\cancel{x + 2})^{1}}{(x - 4)(\cancel{x + 2})_{1}}$
	$= \dfrac{4}{x - 4} \quad Ans.$

3. Reduce to simplest form: $\dfrac{a^2x - ax^2}{a^2x^2 - a^3x}$

$$Solution: \quad \dfrac{a^2x - ax^2}{a^2x^2 - a^3x} = \dfrac{ax(a - x)}{a^2x(x - a)} = \dfrac{\overset{1}{\cancel{ax}}(-1)\overset{1}{\cancel{(x - a)}}}{\underset{a}{\cancel{a^2x}}\underset{1}{\cancel{(x - a)}}} = \dfrac{-1}{a} \quad Ans.$$

| EXERCISES |

In 1-8, for what value(s) of x does the expression have no meaning?

1. $\dfrac{5}{x}$ 2. $\dfrac{5}{x^2}$ 3. $\dfrac{8x}{x - 6}$ 4. $\dfrac{x - 6}{8x}$

5. $\dfrac{14 - x}{7 + x}$ 6. $\dfrac{6x}{3x - 1}$ 7. $\dfrac{x + 2}{x^2 - 9}$ 8. $\dfrac{2x - 5}{x^2 + x - 20}$

In 9-12, find the value(s) of the variable for which the rational expression is not defined.

9. $\dfrac{9y}{(y + 3)^2}$ 10. $\dfrac{4b}{b^2 + 5b}$ 11. $\dfrac{z^2 + 10z}{z^2 - 7z + 10}$ 12. $\dfrac{2x - 8}{x^3 - 16x}$

In 13-36, reduce the rational expression to lowest terms.

13. $\dfrac{6x^3}{8x}$ 14. $\dfrac{8a^2y^2}{4ay^6}$ 15. $\dfrac{3bx^2}{(3bx)^2}$ 16. $\dfrac{15(x - 4)}{20(x - 4)}$

17. $\dfrac{x - 25}{25 - x}$ 18. $\dfrac{2y + 4}{3y + 6}$ 19. $\dfrac{2z}{2z^2 + 6z}$ 20. $\dfrac{16x^2}{2x^2 - 4x}$

21. $\dfrac{bx^3 - b^2x^2}{bx^3 - 2b^2x^2}$ 22. $\dfrac{k^2 - 25}{k + 5}$ 23. $\dfrac{y^2 - 81}{(y - 9)^2}$ 24. $\dfrac{(x - 1)^2}{2 - 2x}$

25. $\dfrac{3a + 3}{a^2 - 1}$ 26. $\dfrac{2p - 8}{(4 - p)^2}$ 27. $\dfrac{(x - a)^2}{x^2 - a^2}$

28. $\dfrac{x^3y - x^2y^2}{xy^3 - x^2y^2}$ 29. $\dfrac{1 - 9x^2}{3x^2 - x}$ 30. $\dfrac{5a^2 - 45}{5a - 15}$

31. $\dfrac{x^2 - 81}{81 - x^2}$ 32. $\dfrac{y^2 - 2y}{y^2 - y - 2}$ 33. $\dfrac{2y^3 - 8y}{4y^2 - 8y}$

34. $\dfrac{x^2 + x - 2}{x^2 + 3x - 4}$ 35. $\dfrac{3y^2 - 12y}{y^3 - 16y}$ 36. $\dfrac{16 - 6x - x^2}{x^2 - 64}$

37. Find a fraction equivalent to $\dfrac{9}{x - 2}$ whose denominator is $2 - x$.

In 38–42, select the numeral preceding the expression that best completes the sentence or answers the question.

38. The expression $\dfrac{y}{y-5}$ equals:

   (1) $\dfrac{y}{5-y}$ (2) $\dfrac{-y}{y-5}$ (3) $\dfrac{-y}{5-y}$ (4) $\dfrac{y}{y+5}$

39. Which rational expression is in simplest form?

   (1) $\dfrac{x^2+2x}{x^2+2x}$ (2) $\dfrac{x^2+2x}{x^2+4x}$ (3) $\dfrac{x^2+2x}{x+2}$ (4) $\dfrac{x^2+2x}{x^2+4}$

40. Which expression is defined for every rational number?

   (1) $\dfrac{x^2+5}{x^2+4}$ (2) $\dfrac{x^2+5x}{x^2+4x}$ (3) $\dfrac{x^2-5}{x^2-4}$ (4) $\dfrac{4x^2-1}{4x-1}$

41. If $x \neq a$, which is a true statement?

   (1) $\dfrac{x-a}{a-x}=1$ (2) $\dfrac{x-a}{a-x}=-1$

   (3) $\dfrac{x^2-a^2}{x-a}=1$ (4) $\dfrac{x^2-a^2}{x-a}=-1$

42. The expression $\dfrac{x-3}{x^2-9}=\dfrac{1}{x+3}$ for which of the following domains?

   (1) all rational numbers   (2) all rational numbers except $x=9$
   (3) all rational numbers except $x=3$ (4) all rational numbers
   except $x=3$ and $x=-3$

2-2 MULTIPLYING RATIONAL EXPRESSIONS

The *product of two fractions* is a fraction whose numerator is the product of the given numerators and whose denominator is the product of the given denominators. In general, for $\dfrac{a}{b}$ and $\dfrac{x}{y}$ where $b \neq 0$ and $y \neq 0$:

$$\frac{a}{b} \cdot \frac{x}{y} = \frac{ax}{by}$$

In the following example, two procedures are used. In method 1, we multiply terms and reduce the answer to simplest form. In method 2, we divide the numerator and the denominator by all common factors before finding the final product. Notice that this second procedure, called *cancellation*, reduces the computation required.

| *Method 1* | *Method 2* |

$$\frac{5}{12} \cdot \frac{8}{5} = \frac{5 \cdot 8}{12 \cdot 5} = \frac{40}{60} = \frac{2 \cdot \overset{1}{\cancel{20}}}{3 \cdot \underset{1}{\cancel{20}}} = \frac{2}{3} \qquad \Bigg| \qquad \frac{5}{12} \cdot \frac{8}{5} = \frac{\overset{1}{\cancel{5}}}{\underset{3}{\cancel{12}}} \cdot \frac{\overset{2}{\cancel{8}}}{\underset{1}{\cancel{5}}} = \frac{2}{3}$$

The *product of two rational expressions* is found by the same procedures as those used for fractions in arithmetic. For example:

Method 1

$$\frac{2x^2}{5y} \cdot \frac{10y^3}{5x^2} = \frac{2x^2 \cdot 10y^3}{5y \cdot 5x^2} = \frac{20x^2y^3}{25x^2y} = \frac{4y^2 \cdot \overset{1}{\cancel{5x^2y}}}{5 \cdot \underset{1}{\cancel{5x^2y}}} = \frac{4y^2}{5}$$

Method 2

$$\frac{2x^2}{5y} \cdot \frac{10y^3}{5x^2} = \frac{\overset{2}{\cancel{2x^2}}}{\underset{1}{\cancel{5y}}} \cdot \frac{\overset{2y^2}{\cancel{10y^3}}}{\underset{5}{\cancel{5x^2}}} = \frac{4y^2}{5}$$

If rational expressions contain polynomials of two or more terms, these polynomials should be *factored* before the cancellation procedure is used. Study the model problem that follows.

| MODEL PROBLEM |

Multiply and express the product in simplest form: $\dfrac{xy + 3y}{6x} \cdot \dfrac{2x^2 - 6x}{x^2 - 9}$

How to Proceed	*Solution*

1. Factor all numerators and denominators, except those that are monomials.

$$\frac{xy + 3y}{6x} \cdot \frac{2x^2 - 6x}{x^2 - 9}$$

$$= \frac{y(x + 3)}{6x} \cdot \frac{2x(x - 3)}{(x + 3)(x - 3)}$$

2. Cancel, that is, divide the numerators and denominators by all common factors.

$$= \frac{y\overset{1}{\cancel{(x + 3)}}}{\underset{3}{\cancel{6x}}} \cdot \frac{\overset{1}{\cancel{2x}}(\overset{1}{\cancel{x - 3}})}{(\underset{1}{\cancel{x + 3}})(\underset{1}{\cancel{x - 3}})}$$

3. Multiply remaining factors in the numerator; multiply remaining factors in the denominator.

$$= \frac{y}{3} \quad Ans.$$

| EXERCISES |

In 1–21, multiply and express the product in its simplest form.

1. $\dfrac{8}{x} \cdot \dfrac{x}{9}$

2. $\dfrac{3b}{4b} \cdot \dfrac{4x}{6}$

3. $\dfrac{18a^2x}{5b^3} \cdot \dfrac{3b}{27ax}$

4. $\dfrac{x+7}{2bc} \cdot \dfrac{4bc^2}{x+7}$

5. $\dfrac{x-a}{12a^3} \cdot \dfrac{8a^4}{x-a}$

6. $\dfrac{x-y}{5z} \cdot \dfrac{5z}{y-x}$

7. $\dfrac{y-2}{3m} \cdot \dfrac{m^3}{2-y}$

8. $\dfrac{4ax^2}{(3y)^3} \cdot \dfrac{3y^3}{(2ax)^2}$

9. $\dfrac{2x+6}{x^2} \cdot \dfrac{3x^2}{6x+18}$

10. $\dfrac{xy}{x^2-4x} \cdot \dfrac{x^2-16}{4y}$

11. $\dfrac{b+8}{5b^2} \cdot \dfrac{3b^2-24b}{b^2-64}$

12. $\dfrac{4x-20}{4x+20} \cdot \dfrac{3x^2+30x}{3x^2-15x}$

13. $\dfrac{a^3-a^2b}{a^2-b^2} \cdot \dfrac{ab^2+b^3}{a^2b^3}$

14. $\dfrac{y^2+y}{y+2} \cdot \dfrac{y^2-4}{y^2+3y}$

15. $\dfrac{(x+1)^2}{x^3-x} \cdot \dfrac{(x-1)^2}{x}$

16. $\dfrac{2x^2-10x}{x^2+2x} \cdot \dfrac{x^2+5x+6}{x^2-2x-15}$

17. $\dfrac{y^2+2y-8}{y^2+3y-4} \cdot \dfrac{3y^2+3y}{3y-6}$

18. $\dfrac{6a^2+2a}{9a^2+6a+1} \cdot \dfrac{9a^2-1}{6a^2}$

19. $\dfrac{2b^2+3b-2}{4b^2+8b} \cdot \dfrac{8b^3-8b^2}{2b^2-b}$

20. $\dfrac{2x-12}{3x-6} \cdot \dfrac{x^2-4}{x^2-36} \cdot \dfrac{3x+18}{4x+8}$

21. $\dfrac{x^2-12x+27}{x^2-81} \cdot \dfrac{x+9}{3-x}$

In 22–25, select the numeral preceding the expression that best completes the sentence.

22. The product of $\dfrac{5-y}{(3y)^2}$ and $\dfrac{9y^2}{y-5}$ is:

(1) 1   (2) -1 (3) 3   (4) -3

23. If the side of a square measures $\dfrac{x+1}{x+2}$, the area of the square in

terms of x is:

(1) $\dfrac{x+1}{x+2}$ (2) $\dfrac{x^2+1}{x^2+4}$ (3) $\dfrac{x^2+2x+1}{x^2+4x+4}$ (4) $\dfrac{1}{4}$

24. The product $\dfrac{x^2-100}{x-2} \cdot \dfrac{x-2}{5x-50} = \dfrac{x+10}{5}$ for all rational values

of x where:

(1) $x \neq 2$ (2) $x \neq 10$ (3) $x \neq 2$, and $x \neq 10$

(4) $x \neq 2$, $x \neq 10$, and $x \neq 5$

25. If $x = 12$, the value of the product $\dfrac{3x^2 + 21x}{12x^2} \cdot \dfrac{4x^2 - 28x}{x^2 - 49}$ is:

(1) 1   (2) 0   (3) 12   (4) $\frac{1}{12}$

2-3 DIVIDING RATIONAL EXPRESSIONS

There are many ways to indicate a problem in division, such as "10 divided by 2 equals 5." In each of the following formats, 10 is the *dividend*, 2 is the *divisor*, and 5 is the *quotient:*

$$2\overline{)10}^{\,5} \qquad 10 \div 2 = 5 \qquad \tfrac{10}{2} = 5$$

Since $10 \div 2 = \frac{10}{2} = 10 \cdot \frac{1}{2}$, it follows that a division problem can be answered by performing a related problem in multiplication. This is also true when dividing any fraction by a nonzero fraction.

> *Division:* $10 \div 2 = 5$
> *Multiplication:* $10 \cdot \frac{1}{2} = 5$

The *quotient of two fractions*, that is, the dividend divided by a non-zero divisor, is found by multiplying the dividend by the *reciprocal* (or multiplicative inverse) of the divisor.

$$\frac{9}{8} \div \frac{3}{2} = \frac{9}{8} \cdot \frac{2}{3} = \frac{\overset{3}{\cancel{9}}}{\underset{4}{\cancel{8}}} \cdot \frac{\overset{1}{\cancel{2}}}{\underset{1}{\cancel{3}}} = \frac{3}{4}$$

In general, for $\frac{a}{b}$ and $\frac{x}{y}$, where $b \neq 0$, $x \neq 0$, and $y \neq 0$:

$$\frac{a}{b} \div \frac{x}{y} = \frac{a}{b} \cdot \frac{y}{x} = \frac{ay}{bx}$$

The *quotient of two rational expressions* is found by the same procedure as that used for fractions in arithmetic. Since the division of rational numbers is restated as a multiplication, polynomials of two or more terms should be *factored* before the cancellation method is used. For example:

$$\frac{x^2 - 9}{4x} \div \frac{3x + 9}{2x} = \frac{x^2 - 9}{4x} \cdot \frac{2x}{3x + 9}$$

$$= \frac{(\cancel{x + 3})(x - 3)}{\underset{2}{\cancel{4x}}} \cdot \frac{\overset{1}{\cancel{2x}}}{3(\cancel{x + 3})} = \frac{x - 3}{6}$$

Note: The quotient $\dfrac{x^2 - 9}{4x} \div \dfrac{3x + 9}{2x} = \dfrac{x - 3}{6}$ if and only if $x \neq 0$ and $x \neq -3$ because these values would have produced fractions with denominators of zero, which are meaningless.

MODEL PROBLEM

Perform the division and express the quotient in simplest form:

$$\frac{a^2 - ay - 2y^2}{3a^3} \div \frac{a^2 - 4y^2}{3a^3 + 6a^2y}$$

How to Proceed	*Solution*
1. Rewrite the problem, indicating that the dividend is to be multiplied by the reciprocal of the divisor.	$\dfrac{a^2 - ay - 2y^2}{3a^3} \div \dfrac{a^2 - 4y^2}{3a^3 + 6a^2y}$ $= \dfrac{a^2 - ay - 2y^2}{3a^3} \cdot \dfrac{3a^3 + 6a^2y}{a^2 - 4y^2}$
2. Factor all numerators and denominators, except those that are monomials.	$= \dfrac{(a + y)(a - 2y)}{3a^3} \cdot \dfrac{3a^2(a + 2y)}{(a - 2y)(a + 2y)}$
3. Cancel, that is, divide the numerators and denominators by all common factors.	$= \dfrac{(a + y)\cancel{(a - 2y)}}{\cancel{3a^3}\,a} \cdot \dfrac{\cancel{3a^2}(a + 2y)}{\cancel{(a - 2y)}\cancel{(a + 2y)}}$
4. Multiply remaining factors.	$= \dfrac{a + y}{a}$ *Ans.*

EXERCISES

In 1–19, divide and express the quotient in its simplest form.

1. $\dfrac{y}{2} \div \dfrac{y}{3}$ 2. $\dfrac{3k}{7} \div \dfrac{6k^2}{14}$ 3. $\dfrac{1}{ab} \div \dfrac{1}{ab^2}$

4. $\dfrac{x + 5}{3a^2} \div \dfrac{x + 5}{2a}$ 5. $\dfrac{y - b}{b^3y} \div \dfrac{y - b}{by^3}$ 6. $\dfrac{x - 2}{4m} \div \dfrac{2 - x}{4m}$

7. $\dfrac{2y - 3}{y^3} \div \dfrac{3 - 2y}{3y}$ 8. $\dfrac{(2x)^2}{(3y)^3} \div \dfrac{2x^2}{3y^3}$ 9. $\dfrac{3k - 3}{k} \div \dfrac{12k - 12}{4k^2}$

10. $\dfrac{x^2 - 4}{2y} \div \dfrac{x^2 - 2x}{xy}$ 11. $\dfrac{(y - 3)^2}{y^2 - 9} \div \dfrac{3y - 9}{y + 3}$

12. $\dfrac{x^2 + 1}{x^2 - 1} \div \dfrac{1}{x - 1}$

13. $\dfrac{m + 3}{3m} \div \dfrac{6m}{m + 3}$

14. $\dfrac{2x^2 - 8x}{x^2 - 16} \div \dfrac{8x^2}{(x + 4)^2}$

15. $\dfrac{c^3 - c^2y}{c^2 - y^2} \div \dfrac{c^3y}{cy^2 + y^3}$

16. $\dfrac{x^2 - 25}{x^2 + 7x} \div \dfrac{x^2 + 7x + 10}{x^2 + 9x + 14}$

17. $\dfrac{3r^3t - 3r^2t^2}{3r^2t + 3rt^2} \div \dfrac{(r - t)^2}{r^2 - t^2}$

18. $\dfrac{2y^2 + 11y + 5}{4y^2 + 4y + 1} \div \dfrac{2y^3 + 10y^2}{4y^3}$

19. $\dfrac{x^3 - 36x}{x^2 + 7x + 6} \div \dfrac{6x^2 - x^3}{x^2 + x}$

In 20 and 21, select the numeral preceding the expression that best completes the sentence.

20. $\dfrac{x - 2}{x - 1} \div \dfrac{x - 2}{x - 1} = 1$ for all rational values of x where:

(1) $x \neq 1$ (2) $x \neq 2$

(3) $x \neq 1$ and $x \neq 2$ (4) $x \neq 1$, $x \neq 2$, and $x \neq 0$

21. The quotient $\dfrac{x - 3}{x} \div \dfrac{x - 2}{x} = \dfrac{x - 3}{x - 2}$ for all rational values of x

where:

(1) $x \neq 0$ (2) $x \neq 0$, $x \neq 2$

(3) $x \neq 0$, $x \neq 3$ (4) $x \neq 0$, $x \neq 2$, $x \neq 3$

22. If $\dfrac{x^2 - 49}{2x + 6}$ represents the area of a rectangle and $\dfrac{x + 7}{x + 3}$ represents

the measure of its length, find the rational expression that represents the measure of its width.

23. Evaluate $\dfrac{9 - 2x}{x - 7} \div \dfrac{2x^2 - 9x}{x^2 - 7x}$ when:

a. $x = 1$ **b.** $x = 10$ **c.** $x = 78$

2-4 ADDING OR SUBTRACTING RATIONAL EXPRESSIONS THAT HAVE THE SAME DENOMINATOR

The *sum (or difference) of two fractions that have the same denominator* is a fraction whose numerator is the sum (or difference) of the given numerators and whose denominator is the common denominator of the given fractions. We can show that this is true by writing each fraction as the product of the numerator and the reciprocal of the denominator, and then using the distributive property, as follows.

☐ EXAMPLE 1:

$$\frac{3}{10} + \frac{1}{10} = 3 \cdot \frac{1}{10} + 1 \cdot \frac{1}{10} = (3 + 1) \cdot \frac{1}{10} = 4 \cdot \frac{1}{10} = \frac{4}{10} = \frac{2}{5}$$

☐ EXAMPLE 2:

$$\frac{7}{12} - \frac{5}{12} = 7 \cdot \frac{1}{12} - 5 \cdot \frac{1}{12} = (7 - 5) \cdot \frac{1}{12} = 2 \cdot \frac{1}{12} = \frac{2}{12} = \frac{1}{6}$$

Once we understand the mathematical principles upon which the addition (or subtraction) of fractions depends, we eliminate the middle steps and write:

☐ EXAMPLE 1:

$$\frac{3}{10} + \frac{1}{10} = \frac{4}{10} = \frac{2}{5}$$

☐ EXAMPLE 2:

$$\frac{7}{12} - \frac{5}{12} = \frac{2}{12} = \frac{1}{6}$$

In general, for $\frac{a}{b}$ and $\frac{c}{b}$, where $b \neq 0$:

$$\frac{a}{b} + \frac{c}{b} = \frac{a + c}{b} \quad \text{AND} \quad \frac{a}{b} - \frac{c}{b} = \frac{a - c}{b}$$

The *sum (or difference) of two rational expressions that have the same denominator* is found by the same procedure as that used for fractions in arithmetic. Keep in mind that final answers are reduced to lowest terms, as seen in the following example:

$$\frac{3a^2}{8x} + \frac{7a^2}{8x} - \frac{6a^2}{8x} = \frac{3a^2 + 7a^2 - 6a^2}{8x} = \frac{4a^2}{8x} = \frac{\overset{1}{\cancel{4}} \cdot a^2}{\underset{1}{\cancel{4}} \cdot 2x} = \frac{a^2}{2x}$$

■ **PROCEDURE.** To add (or subtract) rational expressions that have the same denominator:

Step 1: Write a rational expression whose numerator is the sum (or difference) of the given numerators and whose denominator is the common denominator. Since a fraction bar is a symbol of grouping, place all numerators of two or more terms in parentheses (as seen in the model problems that follow).

Step 2: In the resulting expression, factor the numerator and factor the denominator. Then, cancel all common factors.

| MODEL PROBLEMS |

In 1 and 2, perform the operation. Reduce answers to lowest terms.

$$1. \quad \frac{x^2 - 5}{2x^2} + \frac{5 - 4x}{2x^2} \qquad 2. \quad \frac{7y - 3}{y^2 - 9} - \frac{y + 15}{y^2 - 9}$$

How to Proceed	*Solution*	*Solution*
Step 1: Add or subtract.	$\dfrac{x^2 - 5}{2x^2} + \dfrac{5 - 4x}{2x^2}$ $= \dfrac{(x^2 - 5) + (5 - 4x)}{2x^2}$ $= \dfrac{x^2 - 5 + 5 - 4x}{2x^2}$ $= \dfrac{x^2 - 4x}{2x^2}$	$\dfrac{7y - 3}{y^2 - 9} - \dfrac{y + 15}{y^2 - 9}$ $= \dfrac{(7y - 3) - (y + 15)}{y^2 - 9}$ $= \dfrac{7y - 3 - y - 15}{y^2 - 9}$ $= \dfrac{6y - 18}{y^2 - 9}$
Step 2: Factor and reduce to lowest terms.	$= \dfrac{\overset{1}{\cancel{x}}(x - 4)}{\underset{2x}{\cancel{2x^2}}}$ $= \dfrac{x - 4}{2x} \quad Ans.$	$= \dfrac{6(\cancel{y - 3})^{1}}{(y + 3)(\cancel{y - 3})_{1}}$ $= \dfrac{6}{y + 3} \quad Ans.$

| EXERCISES |

In 1–21, perform the operation. Reduce answers to lowest terms.

1. $\dfrac{3x}{5} + \dfrac{4x}{5} + \dfrac{3x}{5}$

2. $\dfrac{7}{4y} + \dfrac{3}{4y} - \dfrac{2}{4y}$

3. $\dfrac{9a}{7b} - \dfrac{a}{7b} + \dfrac{3a}{7b}$

4. $\dfrac{x}{2c} + \dfrac{y}{2c} + \dfrac{z}{2c}$

5. $\dfrac{2b}{2b + 3} + \dfrac{3}{2b + 3}$

6. $\dfrac{4}{x - 4} - \dfrac{x}{x - 4}$

7. $\dfrac{5x - 3}{4x - 4} - \dfrac{2x}{4x - 4}$

8. $\dfrac{y}{y^2 - 36} - \dfrac{6}{y^2 - 36}$

9. $\dfrac{5z + 9}{z^2 - 25} + \dfrac{16}{z^2 - 25}$

10. $\dfrac{3a}{(a - 2)^2} - \dfrac{6}{(a - 2)^2}$

11. $\dfrac{4b}{5b - 10} - \dfrac{b + 6}{5b - 10}$

12. $\dfrac{2k^2}{k^2 + 3k} - \dfrac{k^2 + 9}{k^2 + 3k}$

13. $\dfrac{c^2d + 1}{c^2 - d^2} - \dfrac{cd^2 + 1}{c^2 - d^2}$

14. $\dfrac{x^2y + 2}{2xy} - \dfrac{xy^2 + 2}{2xy}$

15. $\dfrac{3p + 8}{p^2 - 4} + \dfrac{2p + 7}{p^2 - 4}$

16. $\dfrac{x^2 + x}{4x^2 - 1} + \dfrac{x^2}{4x^2 - 1}$

17. $\dfrac{5y}{y^2 - 5y} - \dfrac{y^2}{y^2 - 5y}$

18. $\dfrac{x^2 + 1}{x^2 - 16} + \dfrac{5x + 3}{x^2 - 16}$

19. $\dfrac{y^2 + 6}{y^2 - 2y} - \dfrac{5y}{y^2 - 2y}$

20. $\dfrac{x^2 + 16}{4 - x} - \dfrac{8x}{4 - x}$

21. $\dfrac{d^2 + 8}{d^3 - d} - \dfrac{8 - d}{d^3 - d}$

In 22–26, copy and complete the table by adding, subtracting, multiplying, and dividing the expressions that represent A and B. Express all answers in simplest form.

	A	B	$A + B$	$A - B$	AB	$A \div B$
22.	$\dfrac{x^2}{a}$	$\dfrac{x^2}{a}$				
23.	$\dfrac{12b}{y}$	$\dfrac{4b}{y}$				
24.	$\dfrac{5a}{2c}$	$\dfrac{3b}{2c}$				
25.	$\dfrac{2x}{x + 3}$	$\dfrac{6}{x + 3}$				
26.	$\dfrac{x}{x - y}$	$\dfrac{y}{x - y}$				

In 27, select the numeral preceding the expression that best completes the sentence.

27. The sum $\dfrac{x - 4}{x - 3} + \dfrac{1}{x - 3} = 1$ for all rational values of x where:

(1) $x \neq 3$

(2) $x \neq 3, x \neq 4$

(3) $\dot{x} \neq 3, x \neq 0$

(4) $x \neq 0, x \neq 3, x \neq 4$

2-5 ADDING OR SUBTRACTING RATIONAL EXPRESSIONS THAT HAVE DIFFERENT DENOMINATORS

We have learned that the multiplicative identity for the set of rational numbers is 1. Recall that any rational number of the form $\frac{x}{x}$, where $x \neq 0$, is equivalent to 1. By combining these ideas, we can add or subtract any fractions.

$$\frac{a}{b} = \frac{a}{b} \cdot 1$$

$$\frac{a}{b} = \frac{a}{b} \cdot \frac{x}{x} = \frac{ax}{bx}$$

The *sum (or difference) of two fractions that have different denominators* is found by transforming all values to equivalent fractions having the same denominator before adding or subtracting the fractions.

☐ EXAMPLE: Add $\frac{5}{8}$ and $\frac{1}{6}$.

Step 1: Find the *lowest common denominator* (L.C.D.), which is the product of the highest power of each of the prime factors of the denominators, 8 and 6.

Finding the L.C.D.

$$8 = 2 \cdot 2 \cdot 2 = 2^3$$
$$6 = 2 \cdot 3 \quad\ \ = 2 \cdot 3$$
$$\overline{\text{L.C.D.} \quad\quad = 2^3 \cdot 3 = 24}$$

Step 2: Transform $\frac{5}{8}$ and $\frac{1}{6}$ to equivalent fractions having the common denominator of 24 by multiplying each fraction by an appropriate form of the identity element 1.

$$\frac{5}{8} = \frac{5}{8} \cdot \frac{3}{3} = \frac{15}{24} \quad \text{AND} \quad \frac{1}{6} = \frac{1}{6} \cdot \frac{4}{4} = \frac{4}{24}$$

Step 3: Using these equivalent fractions, add the numerators and maintain the common denominator. All steps in this solution can be seen in the following line:

$$\frac{5}{8} + \frac{1}{6} = \frac{5}{8} \cdot \frac{3}{3} + \frac{1}{6} \cdot \frac{4}{4} = \frac{15}{24} + \frac{4}{24} = \frac{15 + 4}{24} = \frac{19}{24} \quad Ans.$$

Note: A geometric model of this example is now shown. By comparing the areas of the shaded regions, first using solid lines and then using dotted lines, we can see why $\frac{5}{8} = \frac{15}{24}$ and why $\frac{1}{6} = \frac{4}{24}$. Then, by adding the shaded boxes in the first two rectangles, we can see the addition of the numerators. (See the three figures on top of the next page.)

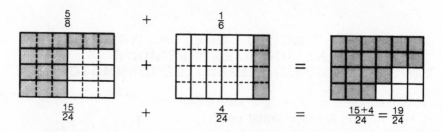

The *sum (or difference) of two rational expressions that have differ-ent denominators* is found by the same procedures as those used for fractions in arithmetic, as seen in the model problems that follow.

| MODEL PROBLEMS |

1. Subtract and express the answer in simplest form: $\dfrac{x + 2}{x^2 - x} - \dfrac{6}{x^2 - 1}$

How to Proceed

1. Factor the de-nominators to find the lowest common de-nominator. No-tice that $(x - 1)$ is a factor in both denomina-tors. Thus, the L.C.D. is: $x(x - 1)(x + 1)$.

Solution

$$x^2 - x = x(x - 1)$$
$$x^2 - 1 = (x - 1)(x + 1)$$
$$\text{L.C.D.} = x(x - 1)(x + 1)$$

2. Transform the ra-tional expressions to equivalent ex-pressions having an L.C.D. of $x(x - 1)(x + 1)$. To do this, multiply each expression by an appropriate form of the identity ele-ment 1.

$$\frac{x + 2}{x^2 - x} \qquad - \qquad \frac{6}{x^2 - 1}$$

$$= \frac{x + 2}{x(x - 1)} \qquad - \qquad \frac{6}{(x - 1)(x + 1)}$$

$$= \frac{x + 2}{x(x - 1)} \cdot \frac{(x + 1)}{(x + 1)} - \frac{(x)}{(x)} \cdot \frac{6}{(x - 1)(x + 1)}$$

$$= \frac{(x + 2)(x + 1)}{x(x - 1)(x + 1)} \qquad - \qquad \frac{6(x)}{x(x - 1)(x + 1)}$$

$$= \frac{x^2 + 3x + 2}{x(x - 1)(x + 1)} \qquad - \qquad \frac{6x}{x(x - 1)(x + 1)}$$

3. Subtract the numerators and maintain the common denominator.

$$= \frac{x^2 + 3x + 2 - 6x}{x(x - 1)(x + 1)}$$

$$= \frac{x^2 - 3x + 2}{x(x - 1)(x + 1)}$$

4. Factor the numerator and, if possible, reduce the resulting expression to simplest form by cancellation.

$$= \frac{\overset{1}{(\cancel{x - 1})}(x - 2)}{x\overset{}{(\cancel{x - 1})}(x + 1)}$$

$$= \frac{x - 2}{x(x + 1)} \quad \text{OR} \quad \frac{x - 2}{x^2 + x} \quad Ans.$$

2. Add and express the sum in simplest form: $\dfrac{3y}{2y - 6} + \dfrac{9}{6 - 2y}$

How to Proceed   *Solution*

1. Since one denominator is the additive inverse of the other, multiply one of the rational expressions by $\dfrac{(-1)}{(-1)}$ to form common denominators.

$$\frac{3y}{2y - 6} + \frac{9}{6 - 2y}$$

$$= \frac{3y}{2y - 6} + \frac{(-1)}{(-1)} \cdot \frac{9}{(6 - 2y)}$$

$$= \frac{3y}{2y - 6} + \frac{-9}{2y - 6}$$

2. Add the numerators and maintain the common denominator.

$$= \frac{3y + (-9)}{2y - 6}$$

$$= \frac{3y - 9}{2y - 6}$$

3. Factor and reduce to simplest form.

$$= \frac{3\overset{1}{(\cancel{y - 3})}}{2\underset{1}{(\cancel{y - 3})}} = \frac{3}{2} \quad Ans.$$

Mixed Expressions

A mixed number such as $3\frac{1}{4}$ is the sum of an integer and a fraction, that is, $3 + \frac{1}{4}$. A mixed number can be transformed into a fraction by following the procedures just seen.

$$3\frac{1}{4} = 3 + \frac{1}{4} = \frac{3}{1} \cdot \frac{4}{4} + \frac{1}{4} = \frac{12}{4} + \frac{1}{4} = \frac{12 + 1}{4} = \frac{13}{4}$$

A *mixed expression* is the sum (or difference) of a polynomial and a rational expression. Once again, the same procedures just learned can be used to transform a mixed expression into a rational expression. For example:

$$x + \frac{3}{x + 5} = \frac{x}{1} \cdot \frac{(x + 5)}{(x + 5)} + \frac{3}{x + 5} = \frac{x(x + 5)}{x + 5} + \frac{3}{x + 5}$$

$$= \frac{x^2 + 5x}{x + 5} + \frac{3}{x + 5} = \frac{x^2 + 5x + 3}{x + 5} \quad Ans.$$

MODEL PROBLEMS

1. Transform $y - 5 + \dfrac{3}{y + 2}$ into a rational expression.

How to Proceed

1. Multiply all terms in the polynomial $y - 5$ by $\dfrac{(y + 2)}{(y + 2)}$, a form of the identity 1.

2. Add the numerators and maintain the common denominator. The expression cannot be reduced.

Solution

$$y - 5 + \frac{3}{y + 2}$$

$$= \frac{(y - 5)}{1} \cdot \frac{(y + 2)}{(y + 2)} + \frac{3}{y + 2}$$

$$= \frac{y^2 - 3y - 10}{y + 2} + \frac{3}{y + 2}$$

$$= \frac{y^2 - 3y - 10 + 3}{y + 2}$$

$$= \frac{y^2 - 3y - 7}{y + 2} \quad Ans.$$

2. Simplify: $\left(x - \dfrac{16}{x}\right)\left(1 + \dfrac{4}{x - 4}\right)$

How to Proceed

1. Transform each mixed expression into a rational expression.

Solution

$$\left(x - \frac{16}{x}\right)\left(1 + \frac{4}{x - 4}\right)$$

$$= \left(\frac{x}{1} \cdot \frac{x}{x} - \frac{16}{x}\right)\left(\frac{1}{1} \cdot \frac{x - 4}{x - 4} + \frac{4}{x - 4}\right)$$

$$= \left(\frac{x^2}{x} - \frac{16}{x}\right)\left(\frac{x - 4}{x - 4} + \frac{4}{x - 4}\right)$$

$$= \left(\frac{x^2 - 16}{x}\right)\left(\frac{x}{x - 4}\right)$$

2. Perform the indicated multiplication after factoring and cancelling all common factors.

$$= \frac{(x + 4)\overset{1}{\cancel{(x - 4)}}}{\underset{1}{\cancel{x}}} \cdot \frac{\overset{1}{\cancel{x}}}{\underset{1}{\cancel{(x - 4)}}}$$

$$= x + 4 \quad Ans.$$

EXERCISES

In 1–31, perform the operation. Express answers in simplest form.

1. $\dfrac{3x}{8} + \dfrac{x}{4}$

2. $\dfrac{7b}{10} - \dfrac{2b}{5}$

3. $\dfrac{3}{4x} - \dfrac{3}{8x}$

4. $\dfrac{y}{2} + \dfrac{y}{3} - \dfrac{y}{6}$

5. $\dfrac{3}{4a} - \dfrac{2}{3a} + \dfrac{1}{6a}$

6. $\dfrac{c + 2}{2} + \dfrac{c - 3}{3}$

7. $\dfrac{1}{a} - \dfrac{1}{b}$

8. $\dfrac{4x + 3}{3x} + \dfrac{x - 1}{x}$

9. $\dfrac{4y + 3}{3y} - \dfrac{y + 2}{y}$

10. $\dfrac{x + 5}{x} - \dfrac{8}{x^2}$

11. $\dfrac{3a - 2}{5a} + \dfrac{2a - 3}{4a}$

12. $\dfrac{x - y}{xy^2} + \dfrac{x + y}{x^2 y}$

13. $\dfrac{2}{x} + \dfrac{3}{x - 5}$

14. $\dfrac{1}{y - 2} + \dfrac{5}{y - 3}$

15. $\dfrac{1}{c - d} - \dfrac{1}{c + d}$

16. $\dfrac{2y}{y - 5} + \dfrac{10}{5 - y}$

17. $\dfrac{2m}{6m - 3} + \dfrac{1}{3 - 6m}$

18. $\dfrac{a^2}{a - b} + \dfrac{b^2}{b - a}$

19. $\dfrac{x}{x + 2} - \dfrac{8}{x^2 - 4}$

20. $\dfrac{y}{y - 3} - \dfrac{18}{y^2 - 9}$

21. $\dfrac{z + 3}{z - 2} - \dfrac{10}{z^2 - 2z}$

22. $\dfrac{3y + 1}{y^2 - 16} + \dfrac{y - 2}{2y + 8}$

23. $\dfrac{b}{(b - 7)^3} - \dfrac{1}{(b - 7)^2}$

24. $\dfrac{x + y}{x - y} + \dfrac{x - y}{x + y}$

25. $\dfrac{2}{x^2 - 36} - \dfrac{1}{x^2 + 6x}$

26. $\dfrac{4}{y^2 - 9} - \dfrac{2}{y^2 - 3y}$

27. $\dfrac{10x}{x^2 - 25} + \dfrac{5}{5 - x}$

28. $\dfrac{1}{x^2 + 4x + 3} + \dfrac{1}{x^2 - 1}$

29. $\dfrac{7}{y^2 - 49} - \dfrac{6}{y^2 - 2y - 35}$

30. $\dfrac{9}{x^2 + 7x + 10} + \dfrac{3}{x + 5} - \dfrac{1}{x + 2}$

31. $\dfrac{2}{2y - 1} - \dfrac{1}{2y + 1} - \dfrac{2}{4y^2 - 1}$

In 32–43, transform the mixed expression into a rational expression.

32. $a + \dfrac{b}{3}$

33. $x + \dfrac{1}{x}$

34. $1 - \dfrac{x}{x + y}$

35. $x - 3 + \dfrac{2}{x}$

36. $y - 1 + \dfrac{y - 1}{y}$

37. $z + \dfrac{z}{z - 1}$

38. $a - \dfrac{ab}{a + b}$ **39.** $2x + \dfrac{ax}{x - a}$ **40.** $y - 1 + \dfrac{1}{y + 1}$

41. $x + 4 + \dfrac{12}{x - 4}$ **42.** $k + 2 - \dfrac{4k}{k + 2}$ **43.** $y + 3 - \dfrac{y - 6}{y - 2}$

In 44–55, perform the indicated operations and express the answer in simplest form.

44. $\left(4 - \dfrac{1}{x}\right)\left(\dfrac{2x}{4x - 1}\right)$ **45.** $\left(1 - \dfrac{5}{y}\right)\left(\dfrac{y}{5 - y}\right)$

46. $\left(1 + \dfrac{y}{x}\right)\left(\dfrac{x}{x^2 - y^2}\right)$ **47.** $\left(\dfrac{3b}{b - 3}\right)\left(2 - \dfrac{6}{b}\right)$

48. $\left(k + 2 + \dfrac{1}{k}\right)\left(\dfrac{1}{k + 1}\right)$ **49.** $\left(y + \dfrac{4}{y - 4}\right)\left(\dfrac{y - 4}{2y - 4}\right)$

50. $\left(1 - \dfrac{8}{x}\right)\left(1 + \dfrac{8}{x - 8}\right)$ **51.** $\left(2 + \dfrac{6}{d}\right)\left(d - \dfrac{3d}{d + 3}\right)$

52. $\left(y - \dfrac{25}{y}\right) \div \left(1 + \dfrac{5}{y}\right)$ **53.** $\left(x - \dfrac{36}{x}\right) \div \left(x - 8 + \dfrac{12}{x}\right)$

54. $\left(x - 5 + \dfrac{6}{x}\right) \div \left(3 - \dfrac{6}{x}\right)$ **55.** $\left(y - 2 + \dfrac{3}{y + 2}\right) \div \left(y + \dfrac{1}{y + 2}\right)$

In 56–59, copy and complete the table by adding, subtracting, multiplying, and dividing the expressions that represent A and B. Express all answers in simplest form.

	A	B	$A + B$	$A - B$	AB	$A \div B$
56.	$\dfrac{1}{y}$	$\dfrac{1}{x}$				
57.	$\dfrac{c}{d}$	$\dfrac{d}{c}$				
58.	$\dfrac{x}{x - 2}$	$\dfrac{x}{2 - x}$				
59.	$4y$	$\dfrac{1}{y}$				

In 60 and 61, select the numeral preceding the expression that best completes the sentence.

60. The difference $\dfrac{x - 1}{x - 2} - \dfrac{2x - 1}{x^2 - 4} = \dfrac{x^2 - x - 1}{x^2 - 4}$ for all rational values of x where:

(1) $x \neq 1, x \neq 2$ (2) $x \neq 2$

(3) $x \neq \pm 2$ (4) $x \neq \pm 1, x \neq \pm 2$

61. The expression $\left(1 - \dfrac{3}{x}\right)\left(2 + \dfrac{6}{x-3}\right) = 2$ for all rational values of

 x where:
 (1) $x \neq 0$ (2) $x \neq 3$ (3) $x \neq 3, x \neq 0$ (4) $x \neq 3, x \neq -3$

62. a. When a number is divided by 4, its quotient is 7 and its remainder is 3, written $7\frac{3}{4}$. Find the number that is the dividend.

 b. When a polynomial is divided by $x + 3$, its quotient is $x - 4$
 and its remainder is 2, written $x - 4 + \dfrac{2}{x + 3}$. Find the polynomial that is the dividend.

2-6 SIMPLIFYING COMPLEX FRACTIONS AND COMPLEX RATIONAL EXPRESSIONS

A *complex fraction* contains one or more fractions in its numerator, its denominator, or both. There are two procedures by which a complex fraction is transformed into a simple fraction:

Complex Fractions		
$\dfrac{\frac{1}{5}}{\frac{9}{10}}$	$\dfrac{2\frac{1}{3}}{7}$	$\dfrac{\frac{3}{8}}{3\frac{1}{4}}$

Method 1

Multiply the complex fraction by $\dfrac{k}{k}$, a form of the identity element 1, where k is the L.C.D. of all fractions that are found in the complex fraction.

☐ EXAMPLE 1:
 (For $\frac{1}{5}$ and $\frac{9}{10}$, the L.C.D. = 10.)

$$\frac{\frac{1}{5}}{\frac{9}{10}} = \frac{\frac{1}{5}}{\frac{9}{10}} \cdot \frac{10}{10} = \frac{\frac{1}{5} \cdot \overset{2}{\cancel{10}}}{\frac{9}{10} \cdot \cancel{10}} = \frac{2}{9}$$

Method 2

Change the numerator to a single fraction; change the denominator to a single fraction; and then divide the numerator by the denominator.

☐ EXAMPLE 1:

$$\frac{\frac{1}{5}}{\frac{9}{10}} = \frac{1}{5} \div \frac{9}{10} = \frac{1}{\cancel{5}} \cdot \frac{\overset{2}{\cancel{10}}}{9} = \frac{2}{9}$$

□ EXAMPLE 2:

(For $2\frac{1}{3}$ and 7, the L.C.D. = 3.)

$$\frac{2\frac{1}{3}}{7} = \frac{\frac{7}{3}}{\frac{7}{1}} \cdot \frac{3}{3} = \frac{\frac{7}{\cancel{3}} \cdot \cancel{3}}{\frac{7}{1} \cdot 3} = \frac{7}{21} = \frac{1}{3}$$

□ EXAMPLE 2:

$$\frac{2\frac{1}{3}}{7} = \frac{\frac{7}{3}}{\frac{7}{1}} = \frac{7}{3} \div \frac{7}{1} = \frac{\cancel{7}}{3} \cdot \frac{1}{\cancel{7}} = \frac{1}{3}$$

A *complex rational expression* contains one or more rational expressions in its numerator, its denominator, or both. Complex expressions are simplified by the same procedures used to simplify complex fractions, as seen in the model problems that follow.

Complex Rational Expressions

$$\frac{\frac{3}{ax}}{\frac{6}{bx}} \qquad \frac{\frac{x^2}{16} - 1}{\frac{x}{8} - \frac{1}{2}}$$

MODEL PROBLEMS

1. Express in simplest form: $\dfrac{\dfrac{3}{ax}}{\dfrac{6}{bx}}$

Method 1

The L.C.D. of $\frac{3}{ax}$ and $\frac{6}{bx}$ is abx.

Multiply by $\frac{abx}{abx}$ and reduce.

$$\frac{\frac{3}{ax}}{\frac{6}{bx}} = \frac{abx}{abx} \cdot \frac{\frac{3}{ax}}{\frac{6}{bx}} = \frac{\cancel{abx} \cdot \frac{3}{\cancel{ax}}^{b}}{\cancel{abx} \cdot \frac{6}{\cancel{bx}}_{a}}$$

$$= \frac{3b}{6a} = \frac{b}{2a} \quad Ans.$$

Method 2

Divide the numerator by the denominator and simplify.

$$\frac{\frac{3}{ax}}{\frac{6}{bx}} = \frac{3}{ax} \div \frac{6}{bx}$$

$$= \frac{\overset{1}{\cancel{3}}}{\underset{a}{\cancel{ax}}} \cdot \frac{\overset{b}{\cancel{bx}}}{\underset{2}{\cancel{6}}} = \frac{b}{2a} \quad Ans.$$

2. Express in simplest form: $\dfrac{\dfrac{x^2}{16} - 1}{\dfrac{x}{8} - \dfrac{1}{2}}$

Method 1	*Method 2*

Method 1

The L.C.D. of the expressions is 16. Multiply by $\frac{16}{16}$ and reduce.

$$\dfrac{\dfrac{x^2}{16} - 1}{\dfrac{x}{8} - \dfrac{1}{2}} = \dfrac{16}{16} \cdot \dfrac{\left(\dfrac{x^2}{16} - 1\right)}{\left(\dfrac{x}{8} - \dfrac{1}{2}\right)}$$

$$= \dfrac{\overset{1}{\cancel{16}} \cdot \dfrac{x^2}{\cancel{16}} - 16 \cdot 1}{\underset{}{\overset{2}{\cancel{16}} \cdot \dfrac{x}{\cancel{8}} - \overset{8}{\cancel{16}} \cdot \dfrac{1}{\cancel{2}}}}$$

$$= \dfrac{x^2 - 16}{2x - 8}$$

$$= \dfrac{(x + 4)\overset{1}{\cancel{(x - 4)}}}{2\underset{1}{\cancel{(x - 4)}}}$$

$$= \dfrac{x + 4}{2} \quad Ans.$$

Method 2

Change both numerator and denominator to single fractions. Then, divide and simplify.

$$\dfrac{\dfrac{x^2}{16} - 1}{\dfrac{x}{8} - \dfrac{1}{2}} = \dfrac{\dfrac{x^2}{16} - 1\left(\dfrac{16}{16}\right)}{\dfrac{x}{8} - \dfrac{1}{2}\left(\dfrac{4}{4}\right)}$$

$$= \dfrac{\dfrac{x^2}{16} - \dfrac{16}{16}}{\dfrac{x}{8} - \dfrac{4}{8}} = \dfrac{\dfrac{x^2 - 16}{16}}{\dfrac{x - 4}{8}}$$

$$= \dfrac{x^2 - 16}{16} \div \dfrac{x - 4}{8}$$

$$= \dfrac{x^2 - 16}{16} \cdot \dfrac{8}{x - 4}$$

$$= \dfrac{(x + 4)\overset{1}{\cancel{(x - 4)}}}{\underset{2}{\cancel{16}}} \cdot \dfrac{\overset{1}{\cancel{8}}}{\underset{1}{\cancel{(x - 4)}}}$$

$$= \dfrac{x + 4}{2} \quad Ans.$$

EXERCISES

In 1–45, express each complex fraction or rational expression in simplest form.

1. $\dfrac{\dfrac{3}{7}}{\dfrac{4}{7}}$ **2.** $\dfrac{\dfrac{x}{5}}{\dfrac{2x}{5}}$ **3.** $\dfrac{\dfrac{3}{8}}{\dfrac{3}{4}}$ **4.** $\dfrac{\dfrac{5}{3x}}{\dfrac{1}{2x}}$

5. $\dfrac{2\frac{1}{2}}{3}$

6. $\dfrac{x + \dfrac{1}{x}}{6}$

7. $\dfrac{7}{8\frac{3}{4}}$

8. $\dfrac{y - 1}{y - \dfrac{1}{y}}$

9. $\dfrac{\dfrac{a^2}{b}}{\dfrac{a}{b^2}}$

10. $\dfrac{\dfrac{2}{k}}{1 + \dfrac{2}{k}}$

11. $\dfrac{\dfrac{a + b}{2a}}{\dfrac{a + b}{3a}}$

12. $\dfrac{\dfrac{24}{x - 3}}{\dfrac{36}{x - 3}}$

13. $\dfrac{\dfrac{1}{d}}{\dfrac{1}{d} - 1}$

14. $\dfrac{y - \dfrac{1}{2}}{y + \dfrac{1}{2}}$

15. $\dfrac{1 - \dfrac{2}{x}}{1 - \dfrac{4}{x^2}}$

16. $\dfrac{z + \dfrac{1}{5}}{z^2 - \dfrac{1}{25}}$

17. $\dfrac{\dfrac{1}{7} + \dfrac{1}{b}}{\dfrac{1}{b}}$

18. $\dfrac{\dfrac{1}{r} + \dfrac{1}{m}}{\dfrac{1}{r} - \dfrac{1}{m}}$

19. $\dfrac{x - \dfrac{1}{x}}{\dfrac{1 - x^2}{x}}$

20. $\dfrac{\dfrac{a^2 - b^2}{a}}{1 - \dfrac{b}{a}}$

21. $\dfrac{\dfrac{x - 5}{x}}{\dfrac{x}{5} - 1}$

22. $\dfrac{\dfrac{b}{a} - 1}{\dfrac{1}{a} - \dfrac{1}{b}}$

23. $\dfrac{\dfrac{y}{3} + \dfrac{3}{y}}{\dfrac{1}{3} + \dfrac{1}{y}}$

24. $\dfrac{\dfrac{x + y}{x}}{\dfrac{1}{x} + \dfrac{1}{y}}$

25. $\dfrac{\dfrac{b^2}{4} - 1}{\dfrac{b}{4} - \dfrac{1}{2}}$

26. $\dfrac{\dfrac{2}{x^2} + \dfrac{2}{y^2}}{\dfrac{4}{xy}}$

27. $\dfrac{\dfrac{4}{x} - \dfrac{8}{x^2}}{1 - \dfrac{2}{x}}$

28. $\dfrac{\dfrac{k}{2} - \dfrac{k}{6}}{\dfrac{k}{2} + \dfrac{k}{3}}$

29. $\dfrac{\dfrac{1}{7} - \dfrac{1}{x}}{\dfrac{x}{7} - \dfrac{7}{x}}$

30. $\dfrac{1 - \dfrac{y}{8}}{\dfrac{1}{8} - \dfrac{1}{y}}$

31. $\dfrac{\dfrac{9}{2} + \dfrac{3}{2x}}{\dfrac{9x}{2} - \dfrac{1}{2x}}$

32. $\dfrac{\dfrac{a}{b} - \dfrac{b}{a}}{1 - \dfrac{b}{a}}$

33. $\dfrac{6 + \dfrac{12}{t}}{3t - \dfrac{12}{t}}$

34. $\dfrac{\dfrac{x}{2} - \dfrac{8}{x}}{\dfrac{1}{4} - \dfrac{1}{x}}$

35. $\dfrac{\dfrac{3}{a^2} + \dfrac{5}{a^3}}{\dfrac{10}{a} + 6}$

36. $\dfrac{\dfrac{1}{n} - 3}{3n - 1}$

37. $\dfrac{1 + \dfrac{4}{x} + \dfrac{3}{x^2}}{1 - \dfrac{9}{x^2}}$

38. $\dfrac{1 + \dfrac{2}{y} - \dfrac{24}{y^2}}{1 + \dfrac{4}{y} - \dfrac{12}{y^2}}$

39. $\dfrac{\dfrac{1}{k} - \dfrac{3}{k^2} + \dfrac{2}{k^3}}{\dfrac{1}{k} - \dfrac{4}{k^2} + \dfrac{4}{k^3}}$

40. $\dfrac{1 + \dfrac{7}{y - 2}}{1 + \dfrac{3}{y + 2}}$

41. $\dfrac{1 + \dfrac{4}{x + 1}}{x - 1 - \dfrac{24}{x + 1}}$

42. $\dfrac{\dfrac{3}{x - 2} - \dfrac{3}{x + 2}}{\dfrac{12}{x^2 - 4}}$

43. $\dfrac{\dfrac{3}{b} - 1}{1 - \dfrac{6}{b} + \dfrac{9}{b^2}}$

44. $\dfrac{\dfrac{5}{a + b} - \dfrac{5}{a - b}}{\dfrac{10}{a^2 - b^2}}$

45. $1 - \dfrac{1}{1 + \dfrac{1}{x}}$

In 46, select the numeral preceding the expression that best answers the question.

46. For what rational values of x will $\dfrac{\dfrac{x^2}{5} - 5}{\dfrac{x}{5} - 1} = x + 5$?

(1) all rational numbers   (2) all rational numbers where $x \neq 0$ and $x \neq 5$ (3) all rational numbers where $x \neq 5$ (4) all rational numbers where $x \neq \pm 5$

2-7 SOLVING FRACTIONAL EQUATIONS

To solve an equation containing a numerical coefficient that is a fraction, we may use one of two procedures. In each method, the equation is transformed into a series of simpler equivalent equations.

□ EXAMPLE: Solve for x: $\frac{1}{5} x + 2 = 6$

Method 1

Use the standard procedure to simplify a first-degree equation.

$$\frac{1}{5} x + 2 = 6$$
$$\frac{1}{5} x + 2 - 2 = 6 - 2$$
$$\frac{1}{5} x = 4$$
$$5 \cdot \frac{1}{5} x = 5 \cdot 4$$
$$x = 20 \quad \textit{Ans.}$$

Method 2

First, clear the equation of all fractions. To do this, multiply both members by the L.C.D. of all fractions in the equation. Then, use standard procedures.

$$\frac{1}{5} x + 2 = 6$$
$$5(\tfrac{1}{5} x + 2) = 5(6)$$
$$5 \cdot \tfrac{1}{5} x + 5 \cdot 2 = 5(6)$$
$$x + 10 = 30$$
$$x = 20 \quad \textit{Ans.}$$

Fractional Equations

An equation is called a *fractional equation* when a variable appears in the *denominator* of one or more of its terms. Thus, $\frac{1}{5}x + 2 = 6$ is not a true fractional equation but simply an equation with rational coefficients. Examples of fractional equations include:

$$\frac{1}{12} + \frac{1}{y} = \frac{1}{4} \qquad \frac{x}{x-1} = \frac{2}{x} + \frac{1}{x-1}$$

To solve a fractional equation, we use the procedure stated in method 2 above, namely:

■ Clear the equation of all fractions by multiplying both of its members by the L.C.D. of all fractions and rational expressions in the equation.

| MODEL PROBLEM |

Solve for y and check: $\frac{1}{12} + \frac{1}{y} = \frac{1}{4}$

How to Proceed	*Solution*	*Check*
1. Write the equation.	$\frac{1}{12} + \frac{1}{y} = \frac{1}{4}$	$\frac{1}{12} + \frac{1}{y} = \frac{1}{4}$
2. Multiply by the L.C.D., which is $12y$.	$12y\left(\frac{1}{12} + \frac{1}{y}\right) = 12y\left(\frac{1}{4}\right)$	$\frac{1}{12} + \frac{1}{6} \stackrel{?}{=} \frac{1}{4}$
3. Apply the distributive property, and cancel wherever possible.	$\overset{y}{\cancel{12y}} \cdot \frac{1}{\cancel{12}} + \overset{12}{\cancel{12y}} \cdot \frac{1}{\cancel{y}} = \overset{3y}{\cancel{12y}}\left(\frac{1}{\cancel{4}}\right)$	$\frac{1}{12} + \frac{2}{12} \stackrel{?}{=} \frac{1}{4}$
4. Solve the resulting equation.	$y + 12 = 3y$	$\frac{3}{12} \stackrel{?}{=} \frac{1}{4}$
	$12 = 2y$	
	$6 = y$	$\frac{1}{4} = \frac{1}{4}$
		(True)

Answer: $y = 6$, or solution set = $\{6\}$.

Fractional Equations With Extraneous Roots

If we multiply both members of an equation by the L.C.D. of all denominators in the equation, we do *not necessarily* form an equivalent equation. Let us study a situation where this happens.

☐ EXAMPLE: Solve for x and check: $\dfrac{x}{x-1} = \dfrac{2}{x} + \dfrac{1}{x-1}$

How to Proceed	*Solution*

1. Multiply by the L.C.D., which is $x(x-1)$.

$$x(x-1)\left(\frac{x}{x-1}\right) = x(x-1)\left(\frac{2}{x} + \frac{1}{x-1}\right)$$

2. Apply the distributive property, and simplify.

$$x(x-1) \cdot \frac{x}{(x-1)} = x(x-1) \cdot \frac{2}{x} + x(x-1) \cdot \frac{1}{(x-1)}$$

3. Combine like terms and write an equivalent quadratic equation with one side equal to zero.

$$x \cdot x = 2(x-1) + x$$
$$x^2 = 2x - 2 + x$$
$$x^2 = 3x - 2$$
$$x^2 - 3x + 2 = 0$$

4. Factor, and let each factor equal zero.

$$(x-2)(x-1) = 0$$
$$x - 2 = 0 \mid x - 1 = 0$$
$$x = 2 \mid x = 1$$

5. *Check* the roots of the quadratic equation, namely, $x = 2$ and $x = 1$, in the original fractional equation *before* writing an answer.

Check for $x = 2$	*Check for $x = 1$*
$\dfrac{x}{x-1} = \dfrac{2}{x} + \dfrac{1}{x-1}$	$\dfrac{x}{x-1} = \dfrac{2}{x} + \dfrac{1}{x-1}$
$\dfrac{2}{2-1} \overset{?}{=} \dfrac{2}{2} + \dfrac{1}{2-1}$	$\dfrac{1}{1-1} \overset{?}{=} \dfrac{2}{1} + \dfrac{1}{1-1}$
$\dfrac{2}{1} \overset{?}{=} \dfrac{2}{2} + \dfrac{1}{1}$	$\dfrac{1}{0} \overset{?}{=} 2 + \dfrac{1}{0}$
$2 \overset{?}{=} 1 + 1$	Division by 0 is not defined. Thus, the statement here is meaningless, and 1 is not a root of the original equation.
$2 = 2$ (True)	

6. Write the answer.

Answer: $x = 2$, or solution set $= \{2\}$.

In the example just given, $x = 1$ is called an *extraneous root*, or an "extra" root, because it is a root of the *derived* equation ($x^2 = 3x - 2$), but it is not a root of the *original* equation $\left(\dfrac{x}{x-1} = \dfrac{2}{x} + \dfrac{1}{x-1}\right)$. How is this possible? Notice that the derived equation was formed when we multiplied the members of the original equation by the L.C.D. of $x(x - 1)$.

But, wait. If $x = 1$, then the L.C.D. $= x(x - 1) = 1(1 - 1) = 1(0) = 0$. Just as $x = 1$ is meaningless for the original equation, so too is it meaningless to say that an L.C.D. equals 0, and to multiply both members of the equation by 0.

Recall that the **multiplication property of zero** states that the product of zero and any number is zero. In general terms:

For any number a: $a \cdot 0 = 0$ AND $0 \cdot a = 0$

While this statement is true for all numbers, multiplying members of an equation by zero may pose problems, as we have just seen. If the members of an equation are multiplied by a polynomial expression that might represent zero, we observe:

■ Since the derived equation is not necessarily equivalent to the original equation, each root of the derived equation should be checked only in the original equation to see if it is a member of the solution set.

MODEL PROBLEMS

1. Solve for x: $2 + \dfrac{4}{x-4} = \dfrac{x}{x-4}$

How to Proceed

Solution

1. Multiply by the L.C.D., $(x - 4)$.

$$\left(2 + \frac{4}{x-4}\right) \cdot (x - 4) = \left(\frac{x}{x-4}\right) \cdot (x - 4)$$

2. Apply the distributive property, and simplify the resulting equation.

$$2 \cdot (x - 4) + \frac{4}{(x-4)} \cdot (x-4) = \left(\frac{x}{x-4}\right) \cdot (x-4)$$

$$2(x - 4) + 4 = x$$
$$2x - 8 + 4 = x$$
$$2x - 4 = x$$
$$2x = x + 4$$
$$x = 4$$

3. *Check* the only possible root, $x = 4$, in the original equation. Since the statement formed is not defined, 4 is not a root of the equation. Thus, the solution set is empty, or no root exists.

Answer: $\emptyset$, or $\{\ \}$

Check for x = 4

$$2 + \frac{4}{x - 4} = \frac{x}{x - 4}$$

$$2 + \frac{4}{4 - 4} \stackrel{?}{=} \frac{4}{4 - 4}$$

$$2 + \frac{4}{0} \stackrel{?}{=} \frac{4}{0} \quad \text{(Not defined)}$$

2. Solve for y: $\dfrac{y - 2}{y} = \dfrac{4}{y^2 - 2y}$

How to Proceed	*Solution*

1. Factor all denominators to find the L.C.D., $y(y - 2)$. Then, multiply by this L.C.D.

$$\frac{y - 2}{y} = \frac{4}{y(y - 2)}$$

$$y(y - 2) \cdot \frac{y - 2}{y} = y(y - 2) \cdot \frac{4}{y(y - 2)}$$

2. Cancel wherever possible.

$$\overset{1}{\cancel{y}}(y - 2) \cdot \frac{y - 2}{\underset{1}{\cancel{y}}} = \overset{1}{\cancel{y}}\overset{1}{(\cancel{y - 2})} \cdot \frac{4}{\underset{1}{\cancel{y}}\underset{1}{(\cancel{y - 2})}}$$

3. Solve the resulting quadratic equation.

$$(y - 2)(y - 2) = 4$$
$$y^2 - 4y + 4 = 4$$
$$y^2 - 4y = 0$$
$$y(y - 4) = 0$$
$$y = 0 \ \bigg| \ \begin{array}{l} y - 4 = 0 \\ y = 4 \end{array}$$

4. *Check* the possible roots, $y = 0$ and $y = 4$, in the original equation.

Check for y = 0

$$\frac{y - 2}{y} = \frac{4}{y^2 - 2y}$$

$$\frac{0 - 2}{0} \stackrel{?}{=} \frac{4}{0 - 0}$$

With denominators of 0, the statement is meaningless. Thus, $y = 0$ is an extraneous root, and 0 is not part of the solution.

Answer: $y = 4$, or solution set = $\{4\}$.

Check for y = 4

$$\frac{y - 2}{y} = \frac{4}{y^2 - 2y}$$

$$\frac{4 - 2}{4} \stackrel{?}{=} \frac{4}{16 - 8}$$

$$\frac{2}{4} \stackrel{?}{=} \frac{4}{8}$$

$$\frac{1}{2} = \frac{1}{2} \quad \text{(True)}$$

| EXERCISES |

In 1–47, solve and check.

1. $\dfrac{4}{3x} + \dfrac{1}{3} = 1$

2. $\dfrac{y + 3}{2y} = \dfrac{2}{3}$

3. $\dfrac{3}{2z} + \dfrac{1}{z} = \dfrac{1}{2}$

4. $\dfrac{5a}{a + 4} = \dfrac{5}{2}$

5. $\dfrac{9}{2b + 7} = \dfrac{3}{b}$

6. $\dfrac{1}{c} + \dfrac{1}{3c} = \dfrac{2}{3}$

7. $\dfrac{3x + 12}{x + 4} = \dfrac{5}{3}$

8. $\dfrac{4y - 1}{5y} = \dfrac{3}{y}$

9. $\dfrac{k}{6} = \dfrac{5}{6} - \dfrac{1}{k}$

10. $\dfrac{1}{6} + \dfrac{1}{12} = \dfrac{1}{m}$

11. $\dfrac{4}{3w + 7} = \dfrac{1}{2}$

12. $\dfrac{t}{t - 3} = \dfrac{3}{4}$

13. $\dfrac{x}{x + 3} = \dfrac{8}{x + 6}$

14. $\dfrac{12}{y} = \dfrac{9}{y - 3}$

15. $\dfrac{2z}{z - 4} = \dfrac{2z - 4}{z - 5}$

16. $\dfrac{b - 2}{2} = \dfrac{5}{b - 5}$

17. $\dfrac{c - 5}{c - 5} = \dfrac{1}{c}$

18. $\dfrac{3 - 2d}{3 + 2d} = \dfrac{1}{2}$

19. $\dfrac{x + 3}{2x + 5} = \dfrac{1}{x + 3}$

20. $\dfrac{1}{15} + \dfrac{1}{y} = \dfrac{1}{6}$

21. $\dfrac{h + 2}{h + 6} = \dfrac{h}{h + 2}$

22. $\dfrac{y + 1}{y - 1} = \dfrac{y + 4}{y + 5}$

23. $\dfrac{2p - 1}{2p + 5} = \dfrac{p - 1}{p + 3}$

24. $\dfrac{x}{3} = \dfrac{4}{x + 4}$

25. $\dfrac{1}{k - 2} = \dfrac{6}{k^2 - 2k}$

26. $\dfrac{3x - 6}{2 - x} = \dfrac{3}{2}$

27. $\dfrac{y^2 - 2}{y^2 - 16} = \dfrac{y - 2}{y - 4}$

28. $\dfrac{x + 2}{x - 2} - \dfrac{4}{3} = \dfrac{5}{x - 2}$

29. $\dfrac{y + 3}{y + 2} = \dfrac{2}{y} + \dfrac{1}{y + 2}$

30. $\dfrac{b}{b + 4} - \dfrac{1}{b} = \dfrac{2}{b + 4}$

31. $\dfrac{2}{w - 3} + 4 = \dfrac{2}{w - 3}$

32. $\dfrac{x}{x - 3} - \dfrac{4}{x} = \dfrac{3}{x - 3}$

33. $\dfrac{1}{m + 10} + \dfrac{1}{5} = \dfrac{3}{m + 10}$

34. $\dfrac{3y}{y - 7} - \dfrac{3}{2} = \dfrac{21}{y - 7}$

35. $\dfrac{x}{x - 2} - \dfrac{5}{x} = \dfrac{2}{x - 2}$

36. $\dfrac{1}{2} + \dfrac{1}{d - 2} = \dfrac{d}{16}$

37. $\dfrac{1}{h + 1} + \dfrac{1}{h - 1} = \dfrac{6}{h^2 - 1}$

38. $\dfrac{x}{x + 8} + \dfrac{16}{x^2 - 64} = \dfrac{1}{x - 8}$

39. $\dfrac{t}{t + 6} + \dfrac{16}{t^2 - 36} = \dfrac{1}{t - 6}$

40. $\dfrac{4}{y + 2} + \dfrac{1}{y^2 - 4} = \dfrac{1}{y - 2}$

41. $\dfrac{x + 2}{x + 5} + \dfrac{x - 1}{x^2 - 25} = 1$

42. $\dfrac{x - 1}{x - 5} - \dfrac{1}{x} = \dfrac{20}{x^2 - 5x}$

43. $\dfrac{2y + 1}{3y - 18} - \dfrac{5}{y - 6} = \dfrac{1}{3}$

44. $\dfrac{1}{2b + 6} + \dfrac{1}{2b - 6} = \dfrac{4}{b^2 - 9}$

45. $\dfrac{x}{2x + 8} + \dfrac{1}{x - 4} = \dfrac{16}{x^2 - 16}$

46. $\dfrac{x+1}{x} + \dfrac{x-1}{x} = \dfrac{x+6}{x+1}$

47. $\dfrac{y}{y+3} + \dfrac{y}{y-3} = \dfrac{18}{y^2-9}$

48. Let $\dfrac{x}{x+4}$ represent a fraction. If 1 is subtracted from the numerator of the fraction, then the new fraction formed, $\dfrac{x-1}{x+4}$, is equal to $\frac{1}{2}$. Find the original fraction.

49. In a fraction, the denominator is 3 more than the numerator. If 1 is added to the numerator and 1 is added to the denominator, the new fraction formed equals $\frac{3}{4}$. Find the original fraction.

50. Let $\dfrac{3k}{5k}$ represent a fraction that is equivalent to $\dfrac{3}{5}$. If 4 is subtracted from the numerator and 4 is subtracted from the denominator, the new fraction equals $\frac{1}{2}$. Find the original fraction.

51. A fraction is equivalent to $\frac{2}{3}$. If 5 is added to its numerator and 5 is added to its denominator, the new fraction is equal to $\frac{3}{4}$. Find the original fraction.

52. Pipe A can fill an industrial tank in 8 hours, and pipe B can fill the same tank in 4 hours. Let x represent the number of hours needed to fill the tank when both pipes are in operation. In one hour, pipe A fills $\frac{1}{8}$ of the tank, pipe B fills $\frac{1}{4}$ of the tank, and, with both pipes working, $\dfrac{1}{x}$ of the tank is filled. Use the equation $\dfrac{1}{8} + \dfrac{1}{4} = \dfrac{1}{x}$ to find the time needed to fill the tank, with both pipes working together.

53. If it takes Felix 12 hours to paint an average room and it takes Oscar 8 hours to paint an average room, how many hours will it take them working together to paint an average room?

54. If it takes 20 minutes for one clerk to sort the mail and 30 minutes for a second clerk to do the same job, how many minutes will it take both clerks working together to sort the mail?

2-8 REVIEW EXERCISES

In 1–3, find the value(s) of x for which the rational expression is not defined.

1. $\dfrac{x-2}{3x-12}$

2. $\dfrac{5x}{x^2+5x}$

3. $\dfrac{x-6}{x^2+5x-24}$

In 4–14, perform the operation. Express the answer in simplest form.

4. $\dfrac{a - b}{16b^3} \cdot \dfrac{12b^4}{a - b}$

5. $\dfrac{y^2 - 4y}{y^2 + 3y} \cdot \dfrac{y^2 - 9}{y - 4}$

6. $\dfrac{2x - 9}{x + 1} \div \dfrac{9 - 2x}{3x + 3}$

7. $\dfrac{(y + 1)^2}{y^2 + y} \div \dfrac{y^2 - 1}{y^2}$

8. $\dfrac{9x}{3x + 5} + \dfrac{15}{3x + 5}$

9. $\dfrac{2y - 5}{y^2 - 25} - \dfrac{5}{y^2 - 25}$

10. $\dfrac{x + 1}{x} + \dfrac{x - 3}{3x}$

11. $\dfrac{2y}{y - 7} + \dfrac{14}{7 - y}$

12. $\dfrac{k}{k - 2} - \dfrac{8}{k^2 - 4}$

13. $\dfrac{1}{y + 6} + \dfrac{2}{y - 6} + \dfrac{12}{y^2 - 36}$

14. $\dfrac{x^2 - 4x - 32}{x^2 + 12x + 32} \div \dfrac{(x - 8)^2}{x^2 - 64}$

In 15–18, change the expression to its simplest form.

15. $\dfrac{1 - \dfrac{8}{x}}{\dfrac{8}{x}}$

16. $\dfrac{\dfrac{x - y}{x}}{\dfrac{x}{y} - 1}$

17. $\dfrac{\dfrac{a^2}{20} - 5}{\dfrac{a}{20} - \dfrac{1}{2}}$

18. $\dfrac{\dfrac{1 + 2t}{4t}}{t - \dfrac{1}{4t}}$

19. The Sullivans went to a ski lodge for the weekend. They traveled for 70 miles at a constant rate of speed. Then they encountered snow-covered roads, and had to reduce their speed by 30 mph for the last 12 miles. The trip took 2 hours. Find the rate of speed for each part of the trip.

   Hint: Express the time for each part of the trip as $\frac{\text{distance}}{\text{rate}}$.

20. The product $\dfrac{x + 4}{x - 4} \cdot \dfrac{x + 12}{x + 4} = \dfrac{x + 12}{x - 4}$ for all rational values of x where:

   (1) $x \neq 4$
   (2) $x \neq -4$
   (3) $x \neq 4, x \neq -4$
   (4) $x \neq 4, x \neq -4, x \neq -12$

In 21–25, solve and check.

21. $\dfrac{1}{k} + \dfrac{1}{4} = \dfrac{9}{4k}$

22. $\dfrac{x}{6} = \dfrac{1}{6} + \dfrac{2}{x}$

23. $\dfrac{w - 3}{w - 3} = \dfrac{1}{w}$

24. $\dfrac{2}{x + 6} = \dfrac{3x + 4}{x^2 + 6x}$

25. $\dfrac{y}{y + 7} + \dfrac{42}{y^2 - 49} = \dfrac{3}{y - 7}$

Chapter **3**

Geometry of the Circle

3-1 ARCS AND ANGLES

In our work in geometry in *Course II*, we showed that the locus of points equidistant from a given point is a circle. In this chapter, we will prove some important relationships of the measures of angles, arcs, and line segments of circles.

■ **DEFINITION.** A *circle* is a set of points in a plane such that the points are equidistant from a fixed point called the center of the circle.

If the center of a circle is point O, the circle is called circle O, written in symbols as $\odot O$.

A *radius* of a circle (plural, *radii*) is a line segment from the center of the circle to any point of the circle. In Fig. 1, points A, B, and C are points of circle O, and $\overline{OA}$, $\overline{OB}$, and $\overline{OC}$ are radii of the circle. Since every point of the circle is equidistant from its center, $OA = OB = OC$. Thus, $\overline{OA} \cong \overline{OB} \cong \overline{OC}$, illustrating the truth of the following statement:

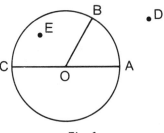

Fig. 1

■ **All radii of the same circle are congruent.**

A circle separates a plane into three sets of points. In Fig. 1, let the length of the radius of circle O be r. Then:

Point C is on the circle if $OC = r$.
Point D is outside the circle if $OD > r$.
Point E is inside the circle if $OE < r$.

The *interior of a circle* is the set of all points whose distance from the center of the circle is less than the length of the radius of the circle.

The *exterior of a circle* is the set of all points whose distance from the center of the circle is greater than the length of the radius of the circle.

Central Angles

Recall that an *angle* is the union of two rays having a common endpoint. The common endpoint is called the *vertex* of the angle.

■ **DEFINITION.** A *central angle* of a circle is an angle whose vertex is the center of the circle.

For example, in Fig. 1, $\angle AOB$ and $\angle BOC$ are central angles because the vertex of each angle is point O, the center of the circle.

Types of Arcs

An *arc* of a circle is any part of the circle. In Fig. 2, A, B, C, and D are points on circle O; and $\angle AOB$ intersects the circle at two distinct points, A and B, separating the circle into two arcs.

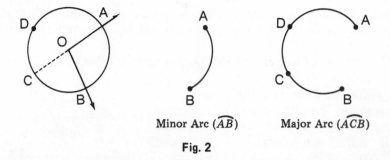

Minor Arc $(\overarc{AB})$ Major Arc $(\overarc{ACB})$

Fig. 2

1. If $m\angle AOB < 180$, points A and B and the points of the circle in the interior of $\angle AOB$ make up *minor arc AB*, written $\overarc{AB}$.
2. Points A and B and the points of the circle not in the interior of $\angle AOB$ make up *major arc AB*. A major arc is usually named by three points: the two endpoints and any other point on the major arc. Thus, the major arc with endpoints A and B is written $\overarc{ACB}$ or $\overarc{ADB}$.

Note: If $m\angle AOC = 180$, points A and C separate circle O into two equal parts, each of which is called a *semicircle*. While $\overarc{AC}$ names a semicircle in Fig. 2, notice that $\overarc{ADC}$ and $\overarc{ABC}$ clearly identify the two semicircles.

An arc of a circle is *intercepted* by an angle if each endpoint of the arc is on a different ray of the angle and the other points of the arc are in the interior of the angle.

A *quadrant* is an arc that is one-fourth of a circle.

Degree Measure of an Arc

■ **DEFINITION.** The *degree measure of an arc* is equal to the measure of the central angle that intercepts the arc.

In circle O (see Fig. 3), if $m\angle FOG = 80$, then the degree measure of arc FG is also 80, written as $m\overset{\frown}{FG} = 80$. Using this same figure, let us make some additional observations:

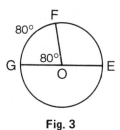

Fig. 3

1. The *degree measure of a major arc* is equal to 360 minus the degree measure of the minor arc having the same endpoints. Thus:

$$m\overset{\frown}{FEG} = 360 - m\overset{\frown}{FG} = 360 - 80 = 280.$$

2. The *degree measure of a semicircle* is 180. Here, $m\overset{\frown}{EFG} = 180$.

Caution: Do not confuse the degree measure of an arc with the length of an arc, which will be discussed in a later chapter. For example, if the circumference of circle O, shown in Fig. 3, is 9 cm, then the length of $\overset{\frown}{FG}$ is $\frac{80}{360}(9 \text{ cm}) = \frac{720}{360} \text{ cm} = 2 \text{ cm}$. Notice that the degree measure of the arc (80°) is not the same as the length of the arc (2 cm).

Congruent Circles, Congruent Arcs, and Arc Addition

Congruent circles are circles with congruent radii. In Fig. 4, if $\overline{O'A'} \cong \overline{OA}$, then circles O' and O are congruent.

Congruent arcs are arcs of the same or congruent circles that are equal in measure. In Fig. 4, if $m\overset{\frown}{AB} = m\overset{\frown}{BC}$, then $\overset{\frown}{AB} \cong \overset{\frown}{BC}$. Furthermore, if $\overline{OA} \cong \overline{O'A'}$ and $m\overset{\frown}{AB} = m\overset{\frown}{A'B'}$, then $\overset{\frown}{AB} \cong \overset{\frown}{A'B'}$.

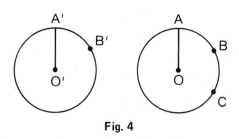

Fig. 4

■ *Postulate 1.* Arc Addition Postulate. If $\overset{\frown}{AB}$ and $\overset{\frown}{BC}$ are arcs of the same circle having a common endpoint and no other points in common, then $\overset{\frown}{AB} + \overset{\frown}{BC} = \overset{\frown}{ABC}$ and $m\overset{\frown}{AB} + m\overset{\frown}{BC} = m\overset{\frown}{ABC}$.

The arc that is the sum of two arcs may be a minor arc, a major arc, or a semicircle.

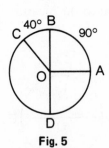

Fig. 5

For example, in Fig. 5, A, B, C, and D are points of circle O, $m\overarc{AB}$ = 90, $m\overarc{BC}$ = 40, and $\overrightarrow{OB}$ and $\overrightarrow{OD}$ are opposite rays.

1. *Minor arc:* $m\overarc{AC} = m\overarc{AB} + m\overarc{BC}$ = 90 + 40 = 130.
   Also, $\overarc{AB} + \overarc{BC} = \overarc{AC}$, a minor arc.

2. *Semicircle:* Since $\overrightarrow{OB}$ and $\overrightarrow{OD}$ are opposite rays, $\angle BOD$ is a straight angle. Thus, $\overarc{BC} + \overarc{CD} = \overarc{BCD}$, a semicircle.
   Also, $m\overarc{BC} + m\overarc{CD} = m\overarc{BCD}$ = 180.

3. *Major arc:* $m\overarc{ABD} = m\overarc{AB} + m\overarc{BCD}$ = 90 + 180 = 270.
   Also, $\overarc{AB} + \overarc{BCD} = \overarc{ABD}$, a major arc.

■ *Theorem 1.* **In a circle or in congruent circles, congruent central angles intercept congruent arcs.**

Given: Circle $O \cong$ circle O'.
           $\angle AOB \cong \angle COD$.
           $\angle AOB \cong \angle A'O'B'$.

To prove: $\overarc{AB} \cong \overarc{CD}$.
           $\overarc{AB} \cong \overarc{A'B'}$.

Plan: We will use the definition of the measure of an arc to prove that the arcs are congruent.

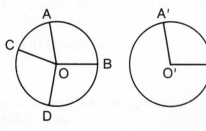

Proof: *Statements*	*Reasons*
1. Circle $O \cong$ circle O'.	1. Given.
2. $\angle AOB \cong \angle COD$. $\angle AOB \cong \angle A'O'B'$.	2. Given.
3. $m\angle AOB = m\angle COD$. $m\angle AOB = m\angle A'O'B'$.	3. Congruent angles are angles that have the same measure.
4. $m\angle AOB = m\overarc{AB}$. $m\angle COD = m\overarc{CD}$. $m\angle A'O'B' = m\overarc{A'B'}$.	4. The degree measure of an arc is equal to the measure of the central angle that intercepts the arc.
5. $m\overarc{AB} = m\overarc{CD}$. $m\overarc{AB} = m\overarc{A'B'}$.	5. Transitive property of equality.
6. $\overarc{AB} \cong \overarc{CD}$. $\overarc{AB} \cong \overarc{A'B'}$.	6. Congruent arcs are arcs of the same or congruent circles that are equal in measure.

The converse of this theorem can be proved using these same definitions and postulates.

■ **Theorem 2.** In a circle or in congruent circles, congruent arcs are intercepted by congruent central angles.

Use the diagram from theorem 1. Given that circle $O \cong$ circle O', $\overarc{AB} \cong \overarc{CD}$, and $\overarc{AB} \cong \overarc{A'B'}$, it can be proved that $\angle AOB \cong \angle COD$ and $\angle AOB \cong \angle A'O'B'$.

[The proof is left to the student.]

In theorems 1 and 2, the condition that the angles and arcs be in the same circle or in congruent circles is important. Let us examine some angles and arcs that have the same measure.

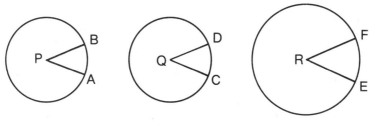

Fig. 6

In Fig. 6, $\overline{PA} \cong \overline{QC}$, but $\overline{PA} \not\cong \overline{RE}$. If m$\angle APB$ = m$\angle CQD$ = m$\angle ERF$, then $\angle APB \cong \angle CQD \cong \angle ERF$ and m$\overarc{AB}$ = m$\overarc{CD}$ = m$\overarc{EF}$. In congruent circles, arcs that have equal measures are congruent. Thus, $\overarc{AB} \cong \overarc{CD}$. In circles that are not congruent, arcs that have equal measures are not congruent. Thus, $\overarc{AB} \not\cong \overarc{EF}$.

MODEL PROBLEM

In circle O, $\overrightarrow{OA}$ and $\overrightarrow{OB}$ are opposite rays and m$\angle BOC$ = 15.

Find:

a. m$\angle AOC$ b. m$\overarc{AC}$ c. m$\overarc{BC}$

d. m$\overarc{AB}$ e. m$\overarc{ABC}$

Solution:

a. m$\angle AOC$ = m$\angle AOB$ - m$\angle BOC$
      = 180 - 15 = 165   *Ans.*

b. $m\overset{\frown}{AC} = m\angle AOC = 165$ *Ans.*

c. $m\overset{\frown}{BC} = m\angle BOC = 15$ *Ans.*

d. $m\overset{\frown}{AB} = m\overset{\frown}{ACB} = m\overset{\frown}{AC} + m\overset{\frown}{CB}$
$$= 165 + 15 = 180 \quad Ans.$$

e. $m\overset{\frown}{ABC} = m\overset{\frown}{AB} + m\overset{\frown}{BC}$
$$= 180 + 15 = 195 \quad Ans.$$

| EXERCISES |
---

1. Find the measure of a central angle that intercepts an arc whose measure is:
      a. 70   **b.** 140   **c.** 23   **d.** 178   **e.** *r*

2. Find the measure of the arc intercepted by a central angle whose measure is:
      a. 30   **b.** 65   **c.** 117   **d.** 145   **e.** *r*

3. In circle O, $m\angle AOB = 87$, $m\angle BOC = 93$, and $m\angle COD = 35$. Find the measure of each of the following:

      a. $\angle DOA$ **b.** $\overset{\frown}{AB}$ **c.** $\overset{\frown}{BC}$
      d. $\overset{\frown}{ABC}$ **e.** $\overset{\frown}{DC}$ **f.** $\overset{\frown}{AD}$
      g. $\overset{\frown}{BCD}$ **h.** $\overset{\frown}{CDB}$ **i.** $\overset{\frown}{DBC}$

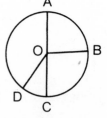

Ex. 3

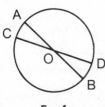

Ex. 4

4. Lines $\overset{\leftrightarrow}{AB}$ and $\overset{\leftrightarrow}{CD}$ intersect at O, the center of the circle, and $m\angle AOC = 25$. Find the measure of each of the following:

      a. $\angle COB$ **b.** $\angle BOD$ **c.** $\angle DOA$
      d. $\overset{\frown}{AC}$ **e.** $\overset{\frown}{BC}$ **f.** $\overset{\frown}{BD}$
      g. $\overset{\frown}{AB}$ **h.** $\overset{\frown}{ACD}$ **i.** $\overset{\frown}{CBA}$

5. In circle O, $m\angle POQ = 100$, $m\angle ROS = 40$, and $\angle POR \cong \angle QOS$. Find the measure of each of the following:

      a. $\overset{\frown}{PQ}$ **b.** $\overset{\frown}{RS}$ **c.** $\angle QOS$ **d.** $\overset{\frown}{SQ}$ **e.** $\overset{\frown}{PR}$
      f. $\overset{\frown}{QPS}$ **g.** $\angle QOR$ **h.** $\overset{\frown}{QR}$ **i.** $\overset{\frown}{QPR}$

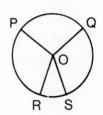

Ex. 5

6. In circle O, $\angle AOC$ and $\angle COB$ are supplementary. If $m\angle AOC = 2x$, $m\angle COB = x + 90$, and $m\angle AOD = 3x + 10$, find:

a. x b. $m\angle AOC$ c. $m\angle COB$

d. $m\angle AOD$ e. $m\angle DOB$ f. $m\widehat{AC}$

g. $m\widehat{BC}$ h. $m\widehat{AB}$ i. $m\widehat{AD}$

j. $m\widehat{DB}$ k. $m\widehat{ADC}$ l. $m\widehat{BCD}$

Ex. 6

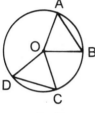  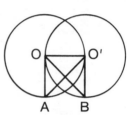

Ex. 7   **Ex. 8**   **Ex. 9**

7. *Given:*   Circle O with $\widehat{AB} \cong \widehat{CD}$.
      To prove: $\triangle ABO \cong \triangle CDO$.

8. *Given:* $\overleftrightarrow{AB}$ intersects $\overleftrightarrow{CD}$ at O, the center of the circle.
      To prove: $\overline{AC} \cong \overline{BD}$.

9. *Given:* O' on circle O and O on circle O'.
             $\overleftrightarrow{OA} \perp \overline{OO'}$ and $\overleftrightarrow{O'B} \perp \overline{OO'}$.
      To prove: $\overline{AO'} \cong \overline{BO}$.

3-2 ARCS AND CHORDS

■ **DEFINITION.** A *chord* of a circle is a line segment whose endpoints are points of the circle.

■ **DEFINITION.** A *diameter* of a circle is a chord that has as one of its points the center of the circle.

In the diagram, $\overline{AB}$ and $\overline{AC}$ are chords of circle O, and $\overline{AC}$ is a diameter of circle O. Recall that the midpoint of a line segment is the point that separates the segment into two congruent parts. Since the radii $\overline{OA}$ and $\overline{OC}$ are two congruent parts of $\overline{AC}$, O is the midpoint of diameter $\overline{AC}$.

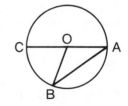

If the length of a radius of a circle is r and the length of a diameter of that circle is d, then: $d = 2r$

A chord, like a central angle, determines two points on the circle and, therefore, a major arc and a minor arc. In the diagram, chord $\overline{AB}$, central ∠AOB, minor arc $\overset{\frown}{AB}$, and major arc $\overset{\frown}{ACB}$ are determined by points A and B of circle O. When we refer to the arc of a chord that is not a diameter, we mean the minor arc that has the same endpoints as the chord.

■ *Theorem 3.* In a circle or in congruent circles, congruent central angles have congruent chords.

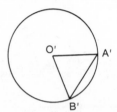

Given: ⊙O ≅ ⊙O'.
          ∠AOB ≅ ∠COD.
          ∠AOB ≅ ∠A'O'B'.

To prove: $\overline{AB} ≅ \overline{CD}$.
           $\overline{AB} ≅ \overline{A'B'}$.

Plan: Prove triangles AOB, COD, and A'O'B' congruent using s.a.s. ≅ s.a.s.

■ *Theorem 4.* In a circle or in congruent circles, congruent arcs have congruent chords.

Given: ⊙O ≅ ⊙O'.
          $\overset{\frown}{AB} ≅ \overset{\frown}{CD}$.
          $\overset{\frown}{AB} ≅ \overset{\frown}{A'B'}$.

To prove: $\overline{AB} ≅ \overline{CD}$.
           $\overline{AB} ≅ \overline{A'B'}$.

Plan: Draw $\overline{OA}$, $\overline{OB}$, $\overline{OC}$, $\overline{OD}$, $\overline{O'A'}$, $\overline{O'B'}$. Use theorem 2 to show that the central angles are congruent. Then use theorem 3.

The converse of theorem 3 and the converse of theorem 4 can be proved in a similar way.

■ *Theorem 5.* In a circle or in congruent circles, congruent chords have congruent central angles.

Using circles O and O' in the diagram for theorem 4, if $\overline{AB} ≅ \overline{CD}$ and $\overline{AB} ≅ \overline{A'B'}$, then ∠AOB ≅ ∠COD and ∠AOB ≅ A'O'B'.

■ *Theorem 6.* In a circle or in congruent circles, congruent chords have congruent arcs.

Using circles O and O' in the diagram for theorem 4, if $\overline{AB} ≅ \overline{CD}$ and $\overline{AB} ≅ \overline{A'B'}$, then $\overset{\frown}{AB} ≅ \overset{\frown}{CD}$ and $\overset{\frown}{AB} ≅ \overset{\frown}{A'B'}$.

Chords Equidistant From the Center of a Circle

In *Course II*, we defined the distance from a point to a line as the length of the perpendicular from the point to the line. The perpendicular is the shortest line segment that can be drawn from a point to a line.

■ *Theorem 7.* A diameter perpendicular to a chord bisects the chord and its arcs.

Given: Circle O, diameter $\overline{CD}$, chord $\overline{AB}$,
$\overleftrightarrow{CD} \perp \overline{AB}$.

To prove: $\overline{AE} \cong \overline{BE}$.
$\overparen{AC} \cong \overparen{BC}$.
$\overparen{AD} \cong \overparen{BD}$.

Plan: Draw radii $\overline{OA}$ and $\overline{OB}$. Prove right triangles $\cong$ by hy. leg $\cong$ hy. leg.

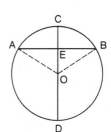

Proof: *Statements*	*Reasons*
1. Draw radii $\overline{OA}$ and $\overline{OB}$.	1. Two points determine a line.
2. $\overleftrightarrow{CD} \perp \overline{AB}$.	2. Given.
3. $\angle AEO$ and $\angle BEO$ are right angles.	3. Perpendicular lines intersect to form right angles.
4. $\overline{OA} \cong \overline{OB}$. (hy. $\cong$ hy.)	4. All radii of the same circle are congruent.
5. $\overline{OE} \cong \overline{OE}$. (leg $\cong$ leg)	5. Reflexive property of congruence.
6. Rt. $\triangle AEO \cong$ rt. $\triangle BEO$.	6. Hy. leg $\cong$ hy. leg.
7. $\overline{AE} \cong \overline{BE}$.	7. Corresponding parts of congruent triangles are congruent.
8. $\angle AOE \cong \angle BOE$.	8. Reason 7.
9. $\overparen{AC} \cong \overparen{BC}$.	9. In a circle, congruent central angles have congruent chords.
10. $\angle AOD$ is supplementary to $\angle AOE$. $\angle BOD$ is suppl. to $\angle BOE$.	10. If two angles form a linear pair, then they are supplementary.
11. $\angle AOD \cong \angle BOD$.	11. Supplements of congruent angles are congruent.
12. $\overparen{AD} \cong \overparen{BD}$.	12. Reason 9.

☐ COROLLARY *7-1.* **The perpendicular bisector of a chord of a circle passes through the center of the circle.**

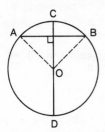

In the diagram, $\overleftrightarrow{CD}$ is the perpendicular bisector of chord $\overline{AB}$ in circle O. In *Course II*, we proved that the perpendicular bisector of a line segment is the locus of points, or set of all points, equidistant from the endpoints of the segment. Therefore, a point is on $\overleftrightarrow{CD}$ if and only if it is equidistant from A and B. Since $OA = OB$, point O, the center of the circle, is on $\overleftrightarrow{CD}$.

■ *Theorem 8.* **If two chords of a circle are congruent, they are equidistant from the center of the circle.**

Given: Circle O with $\overline{AB} \cong \overline{CD}$.

$\overline{OE} \perp \overline{AB}$.

$\overline{OF} \perp \overline{CD}$.

To prove: $\overline{OE} \cong \overline{OF}$.

Plan: Draw $\overline{OB}$ and $\overline{OD}$. Show that right triangles BOE and DOF are congruent by hy. leg $\cong$ hy. leg.

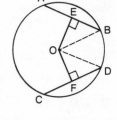

Proof: *Statements*

Statements	Reasons
1. Draw radii $\overline{OB}$ and $\overline{OD}$.	1. Two points determine a line.
2. $\overline{OE} \perp \overline{AB}$; $\overline{OF} \perp \overline{CD}$.	2. Given.
3. $\angle OEB$ and $\angle OFD$ are right angles.	3. Perpendicular lines intersect to form right angles.
4. $\overline{OE}$ bisects $\overline{AB}$. $\overline{OF}$ bisects $\overline{CD}$.	4. A diameter perpendicular to a chord bisects the chord.
5. $\overline{AB} \cong \overline{CD}$.	5. Given.
6. $\overline{BE} \cong \overline{DF}$. (leg $\cong$ leg)	6. Halves of congruent segments are congruent.
7. $\overline{OB} \cong \overline{OD}$. (hy. $\cong$ hy.)	7. All radii of the same circle are congruent.
8. Rt. $\triangle BOE \cong$ rt. $\triangle DOF$.	8. Hy. leg $\cong$ hy. leg.
9. $\overline{OE} \cong \overline{OF}$.	9. Corresponding parts of congruent triangles are congruent.

■ *Theorem 9.* If two chords of a circle are equidistant from its center, the chords are congruent.

Using the diagram of theorem 8, if $\overline{OE} \perp \overline{AB}$, $\overline{OF} \perp \overline{CD}$, and $OE = OF$ in circle O, then $\overline{AB} \cong \overline{CD}$. Notice that theorem 9 is the converse of theorem 8. The proof of this theorem uses a procedure similar to that used to prove theorem 8, and is left to the student as an exercise.

A Polygon Inscribed in a Circle

If a polygon is inscribed in a circle, the vertices of the polygon are points of the circle and the sides of the polygon are chords of the circle. We can also say that the circle is circumscribed about the polygon. For example, in the accompanying diagram, A, B, C, and D are points of circle O. Therefore:

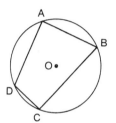

1. Polygon $ABCD$ is *inscribed* in circle O.
2. Circle O is *circumscribed* about polygon $ABCD$.

To circumscribe a circle about $\triangle ABC$:

1. Construct $\overleftrightarrow{PQ}$, the $\perp$ bisector of $\overline{AB}$.
2. Construct $\overleftrightarrow{RS}$, the $\perp$ bisector of $\overline{BC}$.
3. Lines $\overleftrightarrow{PQ}$ and $\overleftrightarrow{RS}$ intersect at O.
   Since O is on $\overleftrightarrow{PQ}$, $\overline{OA} \cong \overline{OB}$.
   Since O is on $\overleftrightarrow{RS}$, $\overline{OB} \cong \overline{OC}$.
   By the transitive property, $\overline{OA} \cong \overline{OC}$.
   Therefore O is on the perpendicular bisector of $\overline{AC}$.
4. Using $\overline{OA}$ as a radius, draw circle O. Triangle ABC is inscribed in $\odot O$.

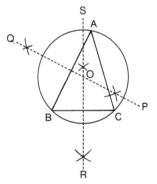

The construction to circumscribe a circle about a triangle, or to inscribe a triangle in a circle, indicates that the following statements can be proved:

■ Any three non-collinear points determine a circle.

■ The perpendicular bisectors of the sides of a triangle are concurrent, that is, the three perpendicular bisectors meet at one point.

| MODEL PROBLEMS |

1. Find the length of a chord 3 cm from the center of a circle whose radius measures 5 cm.

Solution

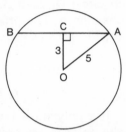

Since $\overleftrightarrow{OC} \perp \overline{BA}$, $\triangle AOC$ is a right triangle. By the Pythagorean Theorem:

$$(CA)^2 + 3^2 = 5^2$$
$$(CA)^2 + 9 = 25$$
$$(CA)^2 = 16$$
$$CA = 4 \quad \text{(A length is positive.)}$$

Since $\overline{OC}$ bisects $\overline{BA}$, $BC = CA$. Thus, $BA = BC + CA = 4 + 4 = 8$.

Answer: The length of the chord is 8 cm.

2. In a circle of radius 10, $m\widehat{AB} = 90$. **a.** Find the length of chord $\overline{AB}$. **b.** Find the distance of $\overline{AB}$ from the center of the circle.

Solution

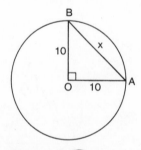

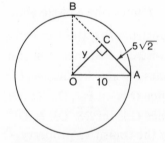

a. Since $m\widehat{AB} = 90$, therefore $m\angle AOB = 90$ and $\triangle AOB$ is a right triangle. Let $x = AB$.

$$(AB)^2 = (OB)^2 + (OA)^2$$
$$x^2 = 10^2 + 10^2$$
$$x^2 = 100 + 100$$
$$x^2 = 200$$
$$x = \pm\sqrt{200}$$
$$x = \pm10\sqrt{2}$$

(A length is positive.)

$$AB = 10\sqrt{2} \quad Ans.$$

b. Since the distance, OC, is the length of the perpendicular to $\overline{AB}$, C is the midpoint of $\overline{AB}$ and $\triangle AOC$ is a right triangle.

$$AC = \tfrac{1}{2}(10\sqrt{2}) = 5\sqrt{2}$$

Let $y = OC$.
$$(OC)^2 + (AC)^2 = (OA)^2$$
$$y^2 + (5\sqrt{2})^2 = 10^2$$
$$y^2 + 50 = 100$$
$$y^2 = 50$$
$$y = \pm\sqrt{50}$$
$$y = \pm5\sqrt{2}$$
$$OC = 5\sqrt{2} \quad Ans.$$

| EXERCISES |

1. Find the length of the radius of a circle whose diameter measures:
   a. 10 in.   b. 12 m   c. 9 ft.   d. 2.6 cm   e. d
2. Find the length of the diameter of a circle whose radius measures:
   a. 7 in.   b. 12 m   c. 8 ft.   d. 9.2 cm   e. r
3. In the diagram, O is the center of the circle, and A, B, C, and D are points of the circle. Name:
   a. 4 radii   b. 2 diameters
   c. 4 chords   d. 4 central angles
4. In circle O, $\overline{AB}$ is a diameter, $AO = 3x - 1$, and $AB = 5x$. Find the length of a radius of the circle.
5. Points A, C, and D are points of circle O such that $\angle DAC$ is a right angle, $DA = 6$, and $AC = 8$.
   a. Find the length of $\overline{CD}$, a diameter of the circle.   b. Find OA.

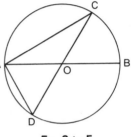

Ex. 3 to 5

In 6–19, $\overline{DE}$ is a diameter of circle O, and $\overline{DE} \perp$ chord $\overline{AB}$ at point C.

6. If $AB = 6$ and $OC = 4$, find OB.
7. If $AB = 14$ and $OC = 24$, find OB.
8. If $AB = 30$ and $OB = 17$, find OC.
9. If $AB = 32$ and $OB = 20$, find OC.
10. If $OB = 13$ and $OC = 5$, find AB.
11. If $OB = 15$ and $OC = 12$, find AB.
12. If $m\angle AOB = 90$, find: a. $m\widehat{AB}$
    b. $m\widehat{AD}$ c. $m\widehat{AEB}$ d. $m\widehat{AE}$
13. If $m\angle AOE = 140$, find:
    a. $m\angle AOC$ b. $m\angle AOB$
    c. $m\widehat{AB}$ d. $m\widehat{BD}$ e. $m\widehat{AEB}$
14. If $m\angle AOB = 90$ and $OC = 3$, find AB.
15. If $m\angle AOB = 90$ and $OA = \sqrt{8}$, find: a. OB b. AB c. AC
    d. OC
16. If $m\angle AOB = 60$ and $OA = 12$, find: a. $m\angle OAB$ b. $m\angle OBA$
    c. AB d. AC e. OC
17. If $m\angle AOB = 60$ and $BC = 2$, find: a. AC b. AB c. OB
    d. OA e. OC
18. If $OD = 41$ and $CD = 32$, find: a. OA b. OC c. AC d. AB
19. Let $AB = 24$, $CD = 8$, and $OC = x$.   a. Represent the length of $\overline{OB}$ in terms of x.   b. Using OB, OC, and CB, write an equation that can be used to find x.   c. Find OC.   d. Find OB.

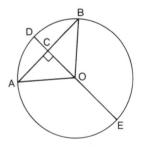

Ex. 6 to 19

20. **a.** Are angles that have the same measure always congruent?
   b. Are arcs that have the same degree measure always congruent?

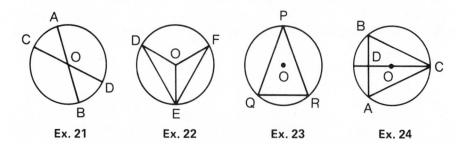

Ex. 21   Ex. 22   Ex. 23   Ex. 24

21. *Given:*   In circle O, chords $\overline{AB}$ and $\overline{CD}$ intersect at O.
   To prove: $m\widehat{AC} = m\widehat{BD}$.
22. *Given:*   In circle O, $\overline{DE} \cong \overline{FE}$.
   To prove: $\triangle DOE \cong \triangle FOE$.
23. *Given:* $\triangle PQR$ is inscribed in circle O, $\overline{PQ} \cong \overline{PR}$.
   To prove: **a.** $\widehat{PQ} \cong \widehat{PR}$.   **b.** $\widehat{PQR} \cong \widehat{PRQ}$.
24. *Given:*   In circle O, $\overleftrightarrow{DOC} \perp$ chord $\overline{AB}$.
   To prove: **a.** $\triangle CDB \cong \triangle CDA$.   **b.** $\triangle ABC$ is isosceles.

25. Draw an obtuse triangle. Construct circle O so that the triangle is inscribed in the circle.
26. Tell which two of the given quadrilaterals can *always* be inscribed in a circle: square, parallelogram, rhombus, rectangle, trapezoid.

3-3 INSCRIBED ANGLES AND THEIR MEASURES

■ **DEFINITION.** An *inscribed angle* of a circle is an angle whose vertex is on the circle and whose sides contain chords of the circle.

In the diagram, $\angle ABC$ is an inscribed angle that intercepts $\widehat{AC}$. Notice that $\overrightarrow{BA}$ and $\overrightarrow{BC}$, the rays that form the angle, contain two chords of the circle, namely, $\overline{BA}$ and $\overline{BC}$.

In order to determine how the measure of an inscribed angle is related to the measure of its intercepted arc, we must consider three cases, as shown in the following diagram:

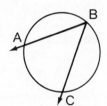

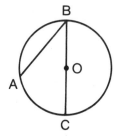 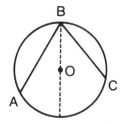 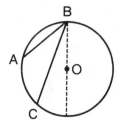

Case I. The center of the circle is contained in one ray of ∠ABC.

Case II. The center of the circle is in the interior of ∠ABC.

Case III. The center of the circle is in the exterior of ∠ABC.

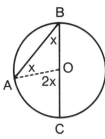

Case I

In Case I, draw $\overline{OA}$ to form isosceles $\triangle AOB$ in which $m\angle OAB = m\angle OBA = x$. Since the measure of an exterior angle of a triangle is equal to the sum of the measures of the two remote interior angles, $m\angle AOC = x + x = 2x$. Thus, $m\widehat{AC} = 2x$, since the measure of the central angle, $\angle AOC$, is equal to the measure of its intercepted arc, $\widehat{AC}$. Therefore:

$$m\angle ABC = x = \tfrac{1}{2}(2x) = \tfrac{1}{2}m\widehat{AC}$$

The measure of the inscribed angle is one-half the measure of its intercepted arc. Let us now present a formal proof of the theorem for all cases.

■ *Theorem 10.* The measure of an inscribed angle of a circle is equal to one-half the measure of its intercepted arc.

Case I. The center of the circle is a point on one ray of the inscribed angle.

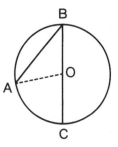

Given: ∠ABC inscribed in circle O, with O on $\overrightarrow{BC}$.

To prove: $m\angle ABC = \tfrac{1}{2}m\widehat{AC}$.

Plan: Draw $\overline{AO}$ and show that the measure of exterior $\angle AOC$ is twice the measure of ∠ABC.

Proof: *Statements*	*Reasons*
1. $\angle ABC$ is inscribed in circle O; O is a point on $\overrightarrow{BC}$.	1. Given.
2. Draw $\overline{AO}$.	2. Two points determine a line.
3. $\overline{AO} \cong \overline{OB}$.	3. All radii of the same circle are congruent.
4. $\angle A \cong \angle B$.	4. If two sides of a triangle are congruent, the angles opposite these sides are congruent.
5. $m\angle A = m\angle B$.	5. Congruent angles are equal in measure.
6. $m\angle AOC = m\angle A + m\angle B$.	6. The measure of an exterior angle of a triangle is equal to the sum of the measures of the two remote interior angles.
7. $m\angle AOC = m\angle B + m\angle B$ $= 2m\angle B$.	7. Substitution.
8. $m\angle AOC = m\widehat{AC}$.	8. The measure of an arc of a circle is equal to the measure of the central angle that intercepts the arc.
9. $2m\angle B = m\widehat{AC}$.	9. Transitive property of equality.
10. $m\angle B = \frac{1}{2} m\widehat{AC}$.	10. Multiplication property of equality.

Case II. The center of the circle is a point in the interior of the inscribed angle.

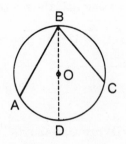

Given: $\angle ABC$ is inscribed in circle O, with O in the interior of $\angle ABC$.

To prove: $m\angle ABC = \frac{1}{2} m\widehat{AC}$.

Plan: Draw $\overleftrightarrow{BO}$ and use Case I.

Proof: *Statements*	*Reasons*
1. $\angle ABC$ is inscribed in circle O; O is a point in the interior of $\angle ABC$.	1. Given.

2. Draw $\overleftrightarrow{BO}$ intersecting the circle at D.

2. Two points determine a line.

3. $m\angle ABC = m\angle ABD + m\angle DBC$.

3. The whole is equal to the sum of its parts.

4. $m\angle ABD = \frac{1}{2} m\overarc{AD}$.
   $m\angle DBC = \frac{1}{2} m\overarc{DC}$.

4. Case I.

5. $m\angle ABC = \frac{1}{2} m\overarc{AD} + \frac{1}{2} m\overarc{DC}$.

5. Substitution.

6. $m\angle ABC = \frac{1}{2} (m\overarc{AD} + m\overarc{DC})$.

6. Distributive property.

7. $m\angle ABC = \frac{1}{2} m\overarc{AC}$.

7. Arc addition postulate.

Case III. The center of the circle is a point in the exterior of the inscribed angle.

Given: $\angle ABC$ inscribed in circle O, with O in the exterior of $\angle ABC$.

To prove: $m\angle ABC = \frac{1}{2} m\overarc{AC}$.

Plan: Draw $\overleftrightarrow{BO}$ and use Case I.

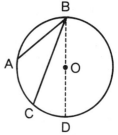

Proof: *Statements*

Reasons

1. $\angle ABC$ inscribed in circle O; O is a point in the exterior of $\angle ABC$.

1. Given.

2. Draw $\overleftrightarrow{OB}$ intersecting the circle at D.

2. Two points determine a line.

3. $m\angle ABC + m\angle CBD = m\angle ABD$.

3. The whole is equal to the sum of its parts.

4. $m\angle ABC = m\angle ABD - m\angle CBD$.

4. Addition property of equality.

5. $m\angle ABD = \frac{1}{2} m\overarc{AD}$.
   $m\angle CBD = \frac{1}{2} m\overarc{CD}$.

5. Case I.

6. $m\angle ABC = \frac{1}{2} m\overarc{AD} - \frac{1}{2} m\overarc{CD}$.

6. Substitution.

7. $m\angle ABC = \frac{1}{2} (m\overarc{AD} - m\overarc{CD})$.

7. Distributive property.

8. $m\angle ABC = \frac{1}{2} m\overarc{AC}$.

8. Substitution.

☐ CoROLLARY *10-1.*   An angle inscribed in a semicircle is a right angle.

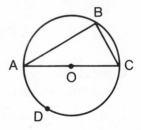

In the diagram, $\angle ABC$ is inscribed in $\widehat{ABC}$, a semicircle of circle O. The arc intercepted by $\angle ABC$, $\widehat{ADC}$, is also a semicircle whose degree measure is 180. Thus:

$$m\angle ABC = \tfrac{1}{2}\,m\widehat{ADC} = \tfrac{1}{2}(180) = 90.$$

☐ CoROLLARY *10-2.*   Two inscribed angles of a circle that intercept the same arc are congruent.

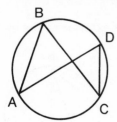

In the diagram, $\angle ABC$ and $\angle ADC$ are inscribed angles, each of which intercepts arc $\widehat{AC}$. Since $m\angle ABC = \tfrac{1}{2}\,m\widehat{AC}$, and $m\angle ADC = \tfrac{1}{2}\,m\widehat{AC}$, then $m\angle ABC = m\angle ADC$ and $\angle ABC \cong \angle ADC$.

Constructing a Right Triangle

Corollary 10-1 suggests a way of constructing a right triangle with a given line segment as the hypotenuse. Imagine that you are given only the line segment $\overline{AB}$. If $\overline{AB}$ is to be the hypotenuse of a right triangle, bisect $\overline{AB}$. Let the midpoint of $\overline{AB}$ be O. With O as the center and $\overline{AO}$ as a radius, draw a circle. Any point on the circle can be the vertex of a right angle be-

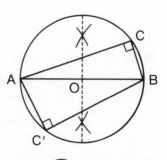

cause the angle will be inscribed in the semicircle, $\widehat{AB}$. Two possible triangles, $\triangle ABC$ and $\triangle ABC'$, are shown in the diagram.

MODEL PROBLEM

Given: $\triangle ABC$ is inscribed in a circle, $m\widehat{BC} = 100$, and $m\angle B = 70$.

Find: **a.** $m\angle A$ **b.** $m\widehat{AC}$

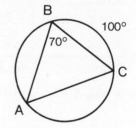

Solution

The measure of an inscribed angle of a circle equals one-half the degree measure of its intercepted arc. Therefore:

a. $m\angle A = \frac{1}{2} m\widehat{BC}$

$m\angle A = \frac{1}{2}(100)$

$m\angle A = 50$ *Ans.*

b. $m\angle B = \frac{1}{2} m\widehat{AC}$

$70 = \frac{1}{2} m\widehat{AC}$

$140 = m\widehat{AC}$ *Ans.*

EXERCISES

1. Find the measure of an inscribed angle that intercepts an arc whose degree measure is:
   a. 60   b. 140   c. 200   d. 75   e. *r*
2. Find the degree measure of an arc intercepted by an inscribed angle whose measure is:
   a. 60   b. 120   c. 15   d. 90   e. *r*

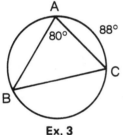

Ex. 3

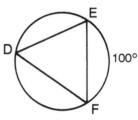

Ex. 4

3. Triangle *ABC* is inscribed in a circle, $m\angle A = 80$, and $m\widehat{AC} = 88$. Find: a. $m\widehat{BC}$ b. $m\angle B$ c. $m\angle C$ d. $m\widehat{AB}$
4. Triangle *DEF* is inscribed in a circle, $\overline{DE} \cong \overline{EF}$, and $m\widehat{EF} = 100$. Find: a. $m\angle D$ b. $m\widehat{DE}$ c. $m\angle F$

In 5–7, chords $\overline{AC}$ and $\overline{BD}$ of a circle intersect at *E*.

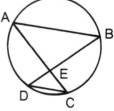

Ex. 5 to 7

5. If $m\angle B = 40$ and $m\angle AEB = 102$, find:
   a. $m\angle A$ b. $m\widehat{BC}$ c. $m\widehat{AD}$ d. $m\angle D$
   e. $m\angle C$
6. If $\overleftrightarrow{AB} \parallel \overleftrightarrow{DC}$ and $m\angle B = 36$, find:
   a. $m\angle D$ b. $m\widehat{AD}$ c. $m\widehat{BC}$ d. $m\angle A$
   e. $m\angle C$
7. If $m\widehat{AD} = 100$, $m\widehat{AB} = 110$, and $m\widehat{BC} = 90$, find:
   a. $m\widehat{DC}$ b. $m\angle A$ c. $m\angle B$ d. $m\angle C$ e. $m\angle D$

In 8 and 9, diameter $\overline{DOE} \perp$ chord $\overline{AB}$ at F, $\overline{AOC}$ is a diameter, and $\overline{BC}$ is a chord of circle O.

8. If $m\widehat{BC} = 60$, find: **a.** $m\widehat{AB}$ **b.** $m\angle A$
c. $m\angle C$ **d.** $m\widehat{AD}$ **e.** $m\angle AOD$
f. $m\widehat{CE}$

9. If $m\angle AOD = 35$, find: **a.** $m\widehat{AD}$ **b.** $m\widehat{DB}$
c. $m\widehat{BC}$ **d.** $m\angle A$ **e.** $m\angle C$ **f.** $m\angle B$
g. $m\widehat{CE}$ **h.** $m\widehat{AE}$

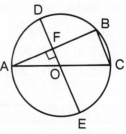

Ex. 8 and 9

10. If $\triangle ABC$ is inscribed in a circle so that $m\widehat{AB}:m\widehat{BC}:m\widehat{CA} = 2:3:4$, find: **a.** $m\widehat{AB}$ **b.** $m\widehat{BC}$ **c.** $m\widehat{CA}$ **d.** $m\angle A$ **e.** $m\angle B$
f. $m\angle C$

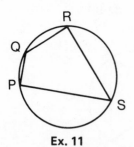

Ex. 11

11. Quadrilateral $PQRS$ is inscribed in a circle, and $m\widehat{PQ}:m\widehat{QR}:m\widehat{RS}:m\widehat{SP} = 1:2:4:5$. Find: **a.** $m\widehat{PQ}$ **b.** $m\widehat{QR}$ **c.** $m\widehat{RS}$
d. $m\widehat{SP}$ **e.** $m\angle S$ **f.** $m\angle Q$ **g.** $m\angle R$
h. $m\angle P$

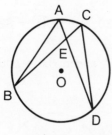

Ex. 12

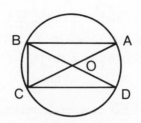

Ex. 13

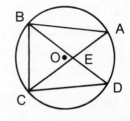

Ex. 14

12. *Given:*   Chords $\overline{AD}$ and $\overline{BC}$ of circle O intersect at E, $\overline{AB} \cong \overline{CD}$.
To prove: $\triangle ABE \cong \triangle CDE$.

13. *Given:*   Diameters $\overline{BOD}$ and $\overline{COA}$ intersect at the center of circle O.
To prove: $\triangle ABC \cong \triangle DCB$.

14. *Given:*   Chords $\overline{AC}$ and $\overline{BD}$ of circle O intersect at E, $\overline{AB} \cong \overline{CD}$.
To prove: $\triangle ABC \cong \triangle DCB$.

3-4   TANGENTS AND SECANTS

A line may have two points, one point, or no points in common with a circle, as shown in the diagram.

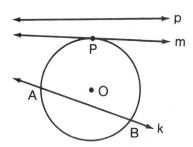

■ **DEFINITION.** A *tangent* to a circle is a line in the plane of the circle that intersects the circle in exactly one point.

In the diagram, line *m* is tangent to circle *O* because the line intersects the circle at only one point, *P*.

■ **DEFINITION.** A *secant* of a circle is a line that intersects the circle in two points.

In the diagram, line *k* is a secant of circle *O* because the line intersects the circle at two points, *A* and *B*.

■ *Postulate 2.* At a given point on a circle, there is one and only one tangent to the circle.

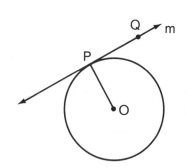

Let *P* be any point on circle *O* and $\overline{OP}$ a radius to that point. If line *m* containing points *P* and *Q* is perpendicular to $\overleftrightarrow{OP}$, it follows that $OQ > OP$ because the perpendicular is the shortest distance from a point to a line. Therefore, every point on line *m* except *P* is outside of circle *O*, and line *m* must be tangent to circle *O*. From this discussion we may prove the following theorem.

■ *Theorem 11.* If a line is perpendicular to a radius at its point of intersection with the circle, the line is tangent to the circle.

It is left to the student to write a formal proof of the theorem.
Our next theorem is the converse of theorem 11.

■ *Theorem 12.* If a line is tangent to a circle, the line is perpendicular to the radius drawn to the point of tangency.

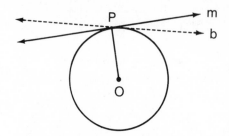

We are given that line m is tangent to circle O at point P, and we will use an indirect proof to show that $m \perp \overline{OP}$.

Either $m \perp \overline{OP}$ or m is not $\perp \overline{OP}$. If we assume that m is not $\perp \overline{OP}$, then there is some line, b, that is $\perp \overline{OP}$ at P. But, by theorem 11, if line $b \perp \overline{OP}$, then b is tangent to circle O at P. However, two lines, m and b, tangent to circle O at P, contradicts postulate 2. Therefore, our assumption that m is not $\perp \overline{OP}$ is false, and line $m \perp \overline{OP}$ must be true.

Construction of Tangents

Theorem 11 suggests a method of constructing a tangent to a circle at a given point on the circle. If P is a point on circle O at which the tangent is to be constructed, draw $\overleftrightarrow{OP}$. Construct a perpendicular to $\overleftrightarrow{OP}$ at P. Since $\overleftrightarrow{AP} \perp \overleftrightarrow{OP}$, line $\overleftrightarrow{AP}$ is tangent to circle O at P.

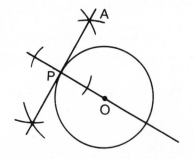

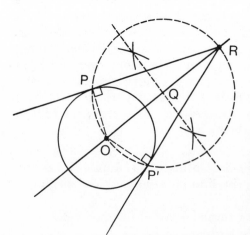

If R is any point outside of circle O, two tangents can be drawn to circle O from R. The construction is shown in the diagram. Draw $\overline{OR}$ and bisect it. If the midpoint of $\overline{OR}$ is Q, draw a circle with Q as the center and $\overline{OQ}$ as the radius. This circle will intersect circle O in two points, P and P'. Draw $\overleftrightarrow{RP}$ and $\overleftrightarrow{RP'}$. Since an angle inscribed in a semicircle is a right angle, $\angle OPR$ and $\angle OP'R$ are right angles, each of which is inscribed in a semicircle of circle Q. Thus, $\overleftrightarrow{RP} \perp \overline{OP}$ and $\overleftrightarrow{RP'} \perp \overline{OP'}$ because of these right angles. The lines $\overleftrightarrow{RP}$ and $\overleftrightarrow{RP'}$ are therefore tangent to circle O since each line is perpendicular to a radius at its point of contact with the circle.

Common Tangents

■ **DEFINITION.** A *common tangent* is a line that is tangent to each of two circles.

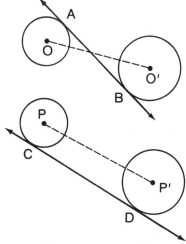

In Fig. 1, $\overleftrightarrow{AB}$ is tangent to circle O at A and to circle O' at B. Line $\overleftrightarrow{CD}$ is tangent to circle P at C and to circle P' at D. Therefore, $\overleftrightarrow{AB}$ and $\overleftrightarrow{CD}$ are common tangents.

Common *internal* tangents intersect the line segment joining the centers of the circles. Line $\overleftrightarrow{AB}$ is a common internal tangent because $\overleftrightarrow{AB}$ intersects $\overline{OO'}$. Common *external* tangents do not intersect the line segment joining the centers of the circles. Line $\overleftrightarrow{CD}$ is a common external tangent because $\overleftrightarrow{CD}$ does not intersect $\overline{PP'}$.

Fig. 1

Two circles in the same plane are said to be tangent to each other if they are tangent to the same line at the same point.

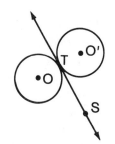

In Fig. 2, $\overleftrightarrow{ST}$ is tangent to circle O and to circle O' at T. Circle O is tangent to circle O'. Since all of the points of each circle, except the point of tangency, are exterior points of the other circle, the circles are externally tangent. Since $\overleftrightarrow{ST}$ intersects $\overline{OO'}$, $\overleftrightarrow{ST}$ is a common internal tangent.

Externally Tangent Circles
$\overleftrightarrow{ST}$ is a common internal tangent.

Fig. 2

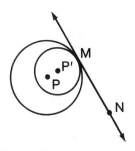

In Fig. 3, $\overleftrightarrow{MN}$ is tangent to circle P and to circle P' at M. Circle P is tangent to circle P'. Since all of the points of one circle, except the point of tangency, are in the interior of the other circle, the circles are internally tangent. Since $\overleftrightarrow{MN}$ does not intersect $\overline{PP'}$, $\overleftrightarrow{MN}$ is a common external tangent to circles P and P'.

Internally Tangent Circles
$\overleftrightarrow{MN}$ is a common external tangent.

Fig. 3

As shown in the diagram, two circles can have four, three, two, one, or no common tangents.

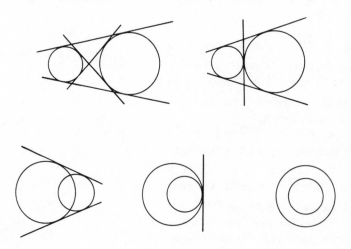

Tangent Segments

■ **DEFINITION.** A *tangent segment* is a segment of a tangent line, one of whose endpoints is the point of tangency.

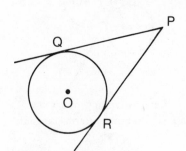

In the diagram, $\overline{PQ}$ and $\overline{PR}$ are tangent segments of the tangents $\overleftrightarrow{PQ}$ and $\overleftrightarrow{PR}$ to the circle O from P.

■ *Theorem 13.* Tangent segments drawn to a circle from an external point are congruent.

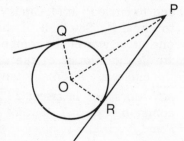

Given: $\overleftrightarrow{PQ}$ tangent to circle O at Q.
$\overleftrightarrow{PR}$ tangent to circle O at R.

To prove: $\overline{PQ} \cong \overline{PR}$.

Plan: Draw $\overline{OQ}$, $\overline{OR}$, and $\overline{OP}$. Prove
right triangles PQO and PRO
congruent by hy. leg $\cong$ hy. leg.

Proof: *Statements*	*Reasons*
1. $\overleftrightarrow{PQ}$ tangent to circle O at Q. $\overleftrightarrow{PR}$ tangent to circle O at R.	1. Given.
2. $\overleftrightarrow{PQ} \perp \overline{OQ}$. $\overleftrightarrow{PR} \perp \overline{OR}$.	2. If a line is tangent to a circle, it is perpendicular to the radius drawn to the point of tangency.
3. $\angle PQO$ and $\angle PRO$ are right angles.	3. Perpendicular lines intersect and form right angles.
4. $\overline{QO} \cong \overline{RO}$. (leg $\cong$ leg)	4. All radii of the same circle are congruent.
5. $\overline{OP} \cong \overline{OP}$. (hy. $\cong$ hy.)	5. Reflexive property of equality.
6. Rt. $\triangle PQO \cong$ rt. $\triangle PRO$.	6. Hy. leg $\cong$ hy. leg.
7. $\overline{PQ} \cong \overline{PR}$.	7. Corresponding parts of congruent triangles are congruent.

☐ COROLLARY *13-1.* **If two tangents are drawn to a circle from an external point, the line determined by that point and the center of the circle bisects the angle formed by the tangents.**

In the proof of theorem 13, $\triangle PQO$ and $\triangle PRO$ were shown to be congruent. Therefore, $\angle QPO$ and $\angle RPO$ are congruent, and $\overleftrightarrow{PO}$ bisects $\angle QPR$.

A Polygon Circumscribed About a Circle

A polygon is circumscribed about a circle if each side of the polygon is tangent to the circle. When a polygon is circumscribed about a circle, the circle is inscribed in the polygon.

For example, in the accompanying diagram $\overline{AB}$ is tangent to circle O at E, $\overline{BC}$ is tangent to circle O at F, $\overline{CD}$ is tangent to circle O at G, and $\overline{DA}$ is tangent to circle O at H. Therefore:

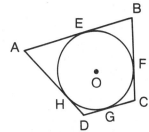

1. Polygon $ABCD$ is circumscribed about circle O.
2. Circle O is inscribed in polygon $ABCD$.

From corollary 13-1, we know that the line joining the center of a circle to an external point bisects the angle formed by the tangents to the circle from this point. Therefore, to inscribe a circle in a polygon, we must determine that the bisectors of each angle of the polygon meet in a point that is the center of the inscribed circle.

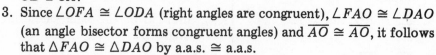

To demonstrate the truth of this statement, consider the following construction for any given triangle ABC.

1. Bisect $\angle A$ and bisect $\angle C$. The bisectors of these angles will meet at a point called O.
2. Draw $\overline{OF} \perp \overline{AB}$, $\overline{OE} \perp \overline{BC}$, and $\overline{OD} \perp \overline{CA}$.
3. Since $\angle OFA \cong \angle ODA$ (right angles are congruent), $\angle FAO \cong \angle DAO$ (an angle bisector forms congruent angles) and $\overline{AO} \cong \overline{AO}$, it follows that $\triangle FAO \cong \triangle DAO$ by a.a.s. $\cong$ a.a.s.
4. Similarly, $\triangle DCO \cong \triangle ECO$ by a.a.s. $\cong$ a.a.s.
5. Therefore, $\overline{OF} \cong \overline{OD}$, and $\overline{OD} \cong \overline{OE}$ since corresponding parts of congruent triangles are congruent.
6. Draw circle O with radii $\overline{OF}$, $\overline{OD}$, and $\overline{OE}$. Since these radii are perpendicular, respectively, to $\overline{AB}, \overline{AC}$, and $\overline{BC}$, it follows that the sides of the triangle are tangent to circle O and $\triangle ABC$ is circumscribed about circle O.

| MODEL PROBLEMS |

1. Triangle ABC is circumscribed about circle O, with D, E, and F being points of tangency for $\overleftrightarrow{AB}$, $\overleftrightarrow{BC}$, and $\overleftrightarrow{CA}$, respectively. If $AF = 6$ and $EB = 7$, find the length of $\overline{AB}$.

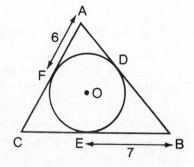

Solution: Since tangent segments drawn to a circle from an external point are congruent, it follows that $AF = AD = 6$ and $EB = DB = 7$. Thus, $AB = AD + DB = 6 + 7 = 13$.

Answer: $AB = 13$

2. Point P is 10 cm from the center of a circle whose radius measures 6 cm. Find the length of a tangent segment from P to the circle.

Solution: Let R be a point at which a line from P is tangent to circle O. Since $\angle ORP$ is a right angle, $\triangle ORP$ is a right triangle in which $OR = 6$ and $OP = 10$.

Let PR = the length of the tangent segment
from P to the circle.

By the Pythagorean Theorem:

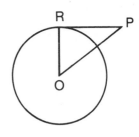

$$(OR)^2 + (PR)^2 = (OP)^2$$
$$6^2 + (PR)^2 = 10^2$$
$$36 + (PR)^2 = 100$$
$$(PR)^2 = 64$$

(A length is positive.) $PR = 8$

Answer: 8 cm

EXERCISES

In 1 and 2, $\triangle ABC$ is circumscribed about a circle, and D, E, and F
are points of tangency.

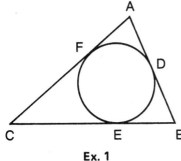

Ex. 1

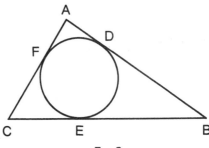

Ex. 2

1. Let AD = 5, EB = 5, and CF = 10.
   a. Find the lengths AB, BC, and CA.
   b. Show that $\triangle ABC$ is isosceles.
2. Let AF = 10, CE = 20, and BD = 30.
   a. Find the lengths AB, BC, and CA.
   b. Show that $\triangle ABC$ is a right triangle.

In 3-9, $\overleftrightarrow{PQ}$ is tangent to circle O at P, $\overleftrightarrow{SQ}$ is tangent to circle O at S,
and $\overleftrightarrow{OQ}$ intersects circle O at T and R.

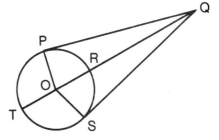

3. If OP = 15 and PQ = 20, find:
   a. OQ b. QS
4. If OQ = 25 and PQ = 24, find
   PO.
5. If OP = 10 and PQ = 10, find:
   a. QS b. OQ
6. If RT = 12 and RQ = 4, find:
   a. PO b. OQ c. PQ d. QS

7. If $OP = 9$ and $RQ = 6$, find: a. OQ b. PQ c. QS

8. If $PQ = 2x - 4$, $SQ = x + 4$, and $OP = \frac{1}{2}x + 1$, find: a. x b. PQ
   c. OP d. OQ

9. If $PQ = 4x - 1$, $SQ = x + 11$, and $OP = 2x$, find: a. x b. PQ
   c. OP d. OQ

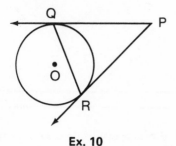

Ex. 10

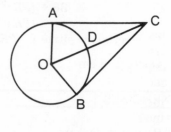

Ex. 11

10. *Given:* $\overleftrightarrow{PQ}$ and $\overleftrightarrow{PR}$ tangent to circle O at Q and R.
    To prove: $\angle PQR \cong \angle PRQ$.

11. *Given:* $\overleftrightarrow{AC}$ and $\overleftrightarrow{BC}$ tangent to circle O at A and B.
    To prove: $\overrightarrow{OC}$ bisects $\angle AOB$.
    $\overset{\frown}{AD} \cong \overset{\frown}{BD}$.

12. *Given:* $\triangle ABC$ circumscribed about circle O with points of tangency D, E, and F; $\overline{AB} \cong \overline{AC}$.
    To prove: E is the midpoint of $\overline{BC}$.

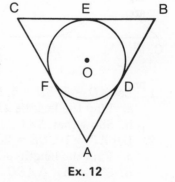

Ex. 12

3-5 ANGLES FORMED BY TANGENTS, CHORDS, AND SECANTS

We have seen how the measures of central angles and inscribed angles are related to the measures of their intercepted arcs. In this section, we will study other angles formed by tangents, chords, and secants.

■ *Theorem 14.* The measure of an angle formed by a tangent and a chord intersecting at the point of tangency is equal to one-half the measure of the intercepted arc.

Given: Tangent $\overleftrightarrow{PS}$ intersects $\overline{PR}$ at point P on circle O.

To prove: $m\angle SPR = \frac{1}{2}m\widehat{PR}$.

Plan: The following steps outline a proof of the theorem.

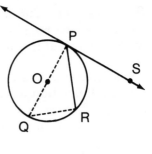

1. Draw diameter $\overline{POQ}$ and chord $\overline{QR}$, forming $\triangle PQR$ with $\angle R$ a right angle.
2. Angle RQP and angle RPQ are complementary because they are the acute angles of a right triangle.
3. Since radius $\overline{OP} \perp$ tangent $\overleftrightarrow{PS}$, $\angle SPR$ and $\angle RPQ$ are complementary.
4. Since the complements of the same angle are congruent, $\angle SPR \cong \angle RQP$.
5. Since $\angle RQP$ is an inscribed angle, $m\angle RQP = \frac{1}{2}m\widehat{PR}$.
6. By substitution: $m\angle SPR = \frac{1}{2}m\widehat{PR}$.

[The formal proof is left to the student.]

■ *Theorem 15.* The measure of an angle formed by two chords intersecting within a circle is equal to one-half the sum of the measures of the arcs intercepted by the angle and by its vertical angle.

Given: Chords $\overline{AB}$ and $\overline{CD}$ intersect at E within the circle.

To prove: $m\angle AED = \frac{1}{2}(m\widehat{AD} + m\widehat{BC})$.
$m\angle CEB = \frac{1}{2}(m\widehat{AD} + m\widehat{BC})$.

Plan: Draw chord $\overline{BD}$, forming inscribed angles B and D.
Thus, $m\angle B = \frac{1}{2}m\widehat{AD}$, and $m\angle D = \frac{1}{2}m\widehat{BC}$.
Note that $\angle AED$ is an exterior angle of $\triangle BED$.
Therefore, $m\angle AED = m\angle B + m\angle D$
$= \frac{1}{2}m\widehat{AD} + \frac{1}{2}m\widehat{BC}$
$= \frac{1}{2}(m\widehat{AD} + m\widehat{BC})$.
Since vertical angles are congruent, $m\angle CEB = m\angle AED$
Therefore, $m\angle CEB = \frac{1}{2}(m\widehat{AD} + m\widehat{BC})$.

[The formal proof is left to the student.]

Note: To show that $m\angle CEA = m\angle BED = \frac{1}{2}(m\widehat{AC} + m\widehat{BD})$, draw $\overline{AD}$.

■ *Theorem 16.* The measure of an angle formed by a tangent and a secant, or two secants, or two tangents intersecting outside a circle is equal to one-half the difference of the intercepted arcs.

Since three separate cases are involved in this theorem, we will present an illustration of each case. By supplying the reasons for the given statements in each of the following cases, the student can write a formal proof of this theorem.

Case I. An angle formed by a tangent and a secant intersecting outside the circle.

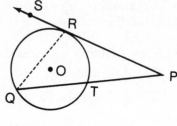

Given: Tangent $\overleftrightarrow{PRS}$ and secant $\overleftrightarrow{PTQ}$ intersect at P outside circle O.

To prove: $m\angle P = \frac{1}{2}(m\widehat{RQ} - m\widehat{RT})$.

Proof:
1. Draw chord $\overline{RQ}$.
2. $m\angle P + m\angle Q = m\angle SRQ$.
3. $\quad m\angle P = m\angle SRQ - m\angle Q$.
4. $\quad m\angle SRQ = \frac{1}{2}m\widehat{RQ}$.
5. $\quad m\angle Q = \frac{1}{2}m\widehat{RT}$.
6. $\quad m\angle P = \frac{1}{2}m\widehat{RQ} - \frac{1}{2}m\widehat{RT}$.
7. $\quad m\angle P = \frac{1}{2}(m\widehat{RQ} - m\widehat{RT})$.

Case II. An angle formed by two secants intersecting outside the circle.

Given: Secants $\overline{PTR}$ and $\overline{PQS}$ intersect at P outside circle O.

To prove: $m\angle P = \frac{1}{2}(m\widehat{RS} - m\widehat{TQ})$.

Proof:
1. Draw chord $\overline{RQ}$.
2. $m\angle P + m\angle R = m\angle RQS$.
3. $\quad m\angle P = m\angle RQS - m\angle R$.
4. $\quad m\angle RQS = \frac{1}{2}m\widehat{RS}$.
5. $\quad m\angle R = \frac{1}{2}m\widehat{TQ}$.
6. $\quad m\angle P = \frac{1}{2}m\widehat{RS} - \frac{1}{2}m\widehat{TQ}$.
7. $\quad m\angle P = \frac{1}{2}(m\widehat{RS} - m\widehat{TQ})$.

Case III. An angle formed by two tangents intersecting outside the circle. (Note that the intercepted arcs are a major arc and a minor arc having common endpoints.)

Given: Tangents $\overleftrightarrow{PRS}$ and $\overleftrightarrow{PQ}$ intersect at P outside circle O.

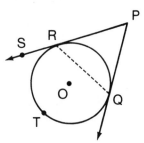

To prove: $m\angle P = \frac{1}{2}(m\widehat{RTQ} - m\widehat{RQ})$.

Proof:
1. Draw chord $\overline{RQ}$.
2. $m\angle P + m\angle PQR = m\angle SRQ$.
3. $\qquad m\angle P = m\angle SRQ - m\angle PQR$.
4. $\qquad m\angle SRQ = \frac{1}{2}m\widehat{RTQ}$.
5. $\qquad m\angle PQR = \frac{1}{2}m\widehat{RQ}$.
6. $\qquad m\angle P = \frac{1}{2}m\widehat{RTQ} - \frac{1}{2}m\widehat{RQ}$.
7. $\qquad m\angle P = \frac{1}{2}(m\widehat{RTQ} - m\widehat{RQ})$.

| **MODEL PROBLEMS** |

1. A tangent and a secant are drawn to circle O from an external point P. The tangent intersects the circle at Q and the secant at R and S.
   If $m\widehat{QR} : m\widehat{RS} : m\widehat{SQ} = 2 : 3 : 4$, find:
   a. $m\widehat{QR}$ **b.** $m\widehat{RS}$ **c.** $m\widehat{SQ}$
   d. $m\angle P$ **e.** $m\angle PQR$ **f.** $m\angle PRQ$

Solution

a. Let $m\widehat{QR} = 2x$
$m\widehat{RS} = 3x$
$m\widehat{SQ} = 4x$

$2x + 3x + 4x = 360$
$9x = 360$
$x = 40$
$m\widehat{QR} = 2x = 80$ *Ans.*

b. $m\widehat{RS} = 3x = 120$ *Ans.* | **c.** $m\widehat{SQ} = 4x = 160$ *Ans.*

d. $m\angle P = \frac{1}{2}(m\widehat{SQ} - m\widehat{QR})$

$\qquad = \frac{1}{2}(160 - 80)$

$\qquad = \frac{1}{2}(80) = 40$ *Ans.*

e. $m\angle PQR = \frac{1}{2}m\widehat{QR}$

$\qquad = \frac{1}{2}(80)$

$\qquad = 40$ *Ans.*

f. $m\angle PRQ = 180 - (m\angle P + m\angle PQR)$

$\qquad = 180 - (40 + 40)$

$\qquad = 180 - 80$

$\qquad = 100$ *Ans.*

Note: In part **f**, the measure of $\angle PRQ$, an angle formed by a secant and a chord, is *not* equal to one-half the intercepted arc.

2. Two tangents are drawn to a circle from an external point R such that $m\angle R = 70$. Find the measures of the major arc and the minor arc into which the circle is divided by the points of tangency.

Solution: Let P and Q be the points of tangency and let S be any point on the major arc.

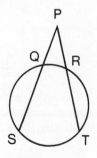

$$x = m\widehat{PQ}$$
$$360 - x = m\widehat{PSQ}$$
$$m\angle R = \frac{1}{2}(m\widehat{PSQ} - m\widehat{PQ})$$
$$70 = \frac{1}{2}(360 - x - x)$$
$$70 = \frac{1}{2}(360 - 2x)$$
$$70 = 180 - x$$
$$x = 110$$

Answer: Measure of the minor arc is 110.

Measure of the major arc is 250.

EXERCISES

In 1–6, secants $\overleftrightarrow{SQP}$ and $\overleftrightarrow{TRP}$ intersect at P.

1. If $m\widehat{ST} = 200$ and $m\widehat{QR} = 100$, find $m\angle P$.
2. If $m\widehat{ST} = 150$ and $m\widehat{QR} = 70$, find $m\angle P$.
3. If $m\widehat{ST} = 100$ and $m\widehat{QR} = 10$, find $m\angle P$.
4. If $m\widehat{ST} = 210$ and $m\angle P = 50$, find $m\widehat{QR}$.
5. If $m\widehat{ST} = 185$ and $m\angle P = 80$, find $m\widehat{QR}$.
6. If $m\angle P = 10$ and $m\widehat{QR} = 10$, find $m\widehat{ST}$.

Ex. 1 to 6

In 7–12, tangent $\overleftrightarrow{PQ}$ and secant $\overleftrightarrow{PRT}$ intersect at P.

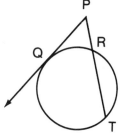

7. If $m\widehat{QT} = 90$ and $m\widehat{QR} = 30$, find $m\angle P$.
8. If $m\widehat{QT} = 112$ and $m\widehat{QR} = 75$, find $m\angle P$.
9. If $m\widehat{QT} = 88$ and $m\widehat{QR} = 10$, find $m\angle P$.
10. If $m\widehat{QT} = 200$ and $m\angle P = 80$, find $m\widehat{QR}$.
11. If $m\widehat{QT} = 150$ and $m\angle P = 25$, find $m\widehat{QR}$.
12. If $m\angle P = 45$ and $m\widehat{QR} = 90$, find $m\widehat{QT}$.

Ex. 7 to 12

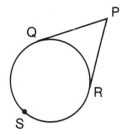

In 13–18, tangents $\overleftrightarrow{PQ}$ and $\overleftrightarrow{PR}$ intersect at P; point S is on major arc $\widehat{QR}$.

13. If $m\widehat{QSR} = 200$, find $m\angle P$.
14. If $m\widehat{QSR} = 300$, find $m\angle P$.
15. If $m\widehat{QR} = 110$, find $m\angle P$.
16. If $m\widehat{QR} = 150$, find $m\angle P$.
17. If $m\angle P = 90$, find $m\widehat{QR}$.
18. If $m\angle P = 75$, find $m\widehat{QSR}$.

Ex. 13 to 18

In 19–24, chords $\overline{AB}$ and $\overline{CD}$ of the circle intersect at E.

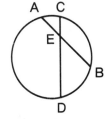

19. If $m\widehat{AC} = 20$ and $m\widehat{BD} = 100$, find $m\angle AEC$.
20. If $m\widehat{CB} = 120$ and $m\widehat{AD} = 180$, find $m\angle AED$.
21. If $m\widehat{AC} = 27$ and $m\widehat{DB} = 81$, find $m\angle DEB$.
22. If $m\widehat{AC} = 30$ and $m\angle AEC = 55$, find $m\widehat{BD}$.
23. If $m\widehat{BD} = 75$ and $m\angle DEB = 60$, find $m\widehat{AC}$.
24. If $m\angle AED = 100$ and $m\widehat{CB} : m\widehat{DA} = 3:5$ find:
    a. $m\widehat{CB}$ b. $m\widehat{DA}$

Ex. 19 to 24

25. Two tangent segments $\overline{PA}$ and $\overline{PB}$ are drawn to circle O from an external point P. If the measure of major arc $\widehat{AB}$ is 220, find $m\angle APB$.

26. Two tangent segments $\overline{PA}$ and $\overline{PB}$ are drawn to circle O from external point P. If the measure of the major arc $\widehat{AB}$ is twice the measure of the minor arc $\widehat{AB}$, find $m\angle APB$.

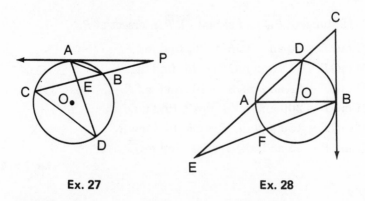

Ex. 27   Ex. 28

27. Line $\overleftrightarrow{PA}$ is tangent to circle O at A and secant $\overleftrightarrow{PBC}$ intersects circle O at B and C. Chord $\overline{AD}$ intersects chord $\overline{BC}$ at E, $\overline{AB}$ and $\overline{CD}$ are chords, $\text{m}\widehat{AC} = 70$, $\text{m}\angle P = 15$, and $\text{m}\widehat{CD}:\text{m}\widehat{BD} = 3:2$. Find:
   a. $\text{m}\widehat{AB}$ b. $\text{m}\widehat{BD}$ c. $\text{m}\widehat{CD}$ d. $\text{m}\angle PAB$
   e. $\text{m}\angle BCD$ f. $\text{m}\angle CED$ g. $\text{m}\angle ABP$ h. $\text{m}\angle PAD$

28. In the diagram, $\overline{AOB}$ is a diameter of circle O, $\overleftrightarrow{CB}$ is tangent to the circle at B, $\overleftrightarrow{EAC}$ and $\overleftrightarrow{EFB}$ are secants, $\overline{OD}$ is a radius, $\text{m}\widehat{AD} = 100$, $\text{m}\angle ABF = 20$. Find:
   a. $\text{m}\angle AOD$ b. $\text{m}\angle CAB$ c. $\text{m}\widehat{AF}$ d. $\text{m}\angle E$
   e. $\text{m}\angle EAB$ f. $\text{m}\angle C$ g. $\text{m}\angle ABC$

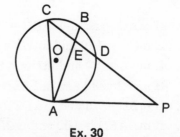

Ex. 29   Ex. 30

29. In the diagram, $\overleftrightarrow{PA}$ and $\overleftrightarrow{PB}$ are tangent to circle O at A and B, respectively. Diameter $\overline{BD}$ and chord $\overline{AC}$ intersect at E, $\text{m}\widehat{CB} = 120$, and $\text{m}\angle P = 50$. Find:
   a. $\text{m}\widehat{AB}$ b. $\text{m}\widehat{AD}$ c. $\text{m}\widehat{CD}$ d. $\text{m}\angle DEC$ e. $\text{m}\angle PAC$

30. In the diagram, $\overleftrightarrow{PA}$ is tangent to circle O at A and $\overleftrightarrow{PC}$ intersects circle O at D and C. Chord $\overline{CA}$ is drawn, chords $\overline{BA}$ and $\overline{CD}$ intersect at E, $m\widehat{BC} = 45$, $m\angle BEC = 65$, $m\widehat{BD} = x$, $m\widehat{CA} = 2x + 20$. Find:
   a. $m\widehat{DA}$ b. $m\widehat{BD}$ c. $m\angle CPA$ d. $m\angle BAC$ e. $m\angle BAP$

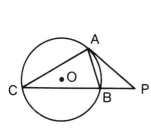

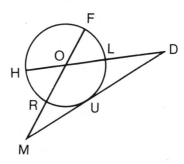

Ex. 31   Ex. 32

31. In the diagram, $\overleftrightarrow{PA}$ is tangent to circle O at A, $\overleftrightarrow{PBC}$ is a secant, $\overline{AB}$ and $\overline{CA}$ are chords, $m\angle P = 40$, and $m\widehat{AC}:m\widehat{AB} = 7:3$. Find:
   a. $m\widehat{AC}$ b. $m\widehat{BC}$ c. $m\angle ACB$ d. $m\angle ABC$ e. $m\angle PAB$
   f. $m\angle ABP$
32. In the diagram, $\overleftrightarrow{MUD}$ is tangent to circle O at U, secants $\overleftrightarrow{HOLD}$ and $\overleftrightarrow{FORM}$ intersect at O, and $m\widehat{RU} = m\widehat{UL} = 65$. Find:
   a. $m\widehat{HR}$ b. $m\widehat{FL}$ c. $m\angle FOL$ d. $m\angle LOR$ e. $m\angle FMD$
   f. $m\angle HDM$

33. $\overleftrightarrow{BUS}$ is tangent to circle O at U, secant $\overleftrightarrow{BAN}$ intersects diameter $\overline{TOE}$ at R, $m\widehat{EN} = 60$, $m\widehat{EA} = 40$, and $m\widehat{AU}:m\widehat{UT} = 4:3$. Find:
   a. $m\widehat{NT}$ b. $m\angle ERA$
   c. $m\widehat{AU}$ d. $m\widehat{UT}$
   e. $m\angle ERN$ f. $m\angle BSE$
   g. $m\angle NBS$

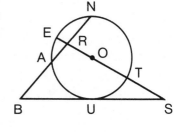

Ex. 33

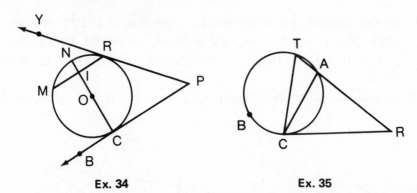

Ex. 34   **Ex. 35**

34. In the diagram, $\overrightarrow{PRY}$ and $\overrightarrow{PCB}$ are tangents to circle O at R and C, respectively. Diameter $\overline{COIN}$ and chord $\overline{RIM}$ intersect at point I, $\text{m}\widehat{NM} = 50$, and $\text{m}\angle P = 40$. Find:

   a. $\text{m}\widehat{MC}$ b. $\text{m}\widehat{RC}$ c. $\text{m}\widehat{NR}$ d. $\text{m}\angle RIC$

   e. $\text{m}\angle RIN$ f. $\text{m}\angle OCB$ g. $\text{m}\angle MRY$ h. $\text{m}\angle MRP$

35. In the diagram, tangent $\overleftrightarrow{RC}$ intersects the circle at C, secant $\overleftrightarrow{RAT}$ intersects the circle at A and T, $\overline{AC}$ and $\overline{TC}$ are chords, $\widehat{CBT}$ is a major arc, and $\text{m}\widehat{TA} : \text{m}\widehat{AC} : \text{m}\widehat{CBT} = 1:3:5$. Find:

   a. $\text{m}\widehat{TA}$ b. $\text{m}\widehat{AC}$ c. $\text{m}\widehat{CBT}$ d. $\text{m}\angle CTA$

   e. $\text{m}\angle ACR$ f. $\text{m}\angle CRT$ g. $\text{m}\angle ACT$ h. $\text{m}\angle CAR$

Miscellaneous Exercises

In 36–41, secants $\overleftrightarrow{PAB}$ and $\overleftrightarrow{PCD}$ are drawn to circle O from P. Chords $\overline{BC}$ and $\overline{AD}$ intersect at E. Line $\overleftrightarrow{GF}$ is tangent to the circle at D. Chord $\overline{BOC}$ is a diameter and $\overline{OD}$ is a radius.

36. Express each of the following in terms of the measures of intercepted arcs.

   a. $\text{m}\angle COD$ b. $\text{m}\angle ABC$

   c. $\text{m}\angle BEA$ d. $\text{m}\angle P$

   e. $\text{m}\angle CDF$

Ex. 36 to 41

37. Name an angle congruent to each of the following:

   a. $\angle ABC$ b. $\angle BAD$ c. $\angle CED$ d. $\angle CDB$ e. $\angle PAD$

38. If $\text{m}\widehat{AB} = 140$, find $\text{m}\angle ABC$.

39. If $m\overarc{BD}$ = 80 and $m\overarc{AC}$ = 30, find:
    a. $m\angle P$ b. $m\angle AEC$ c. $m\angle COD$ d. $m\angle CDF$ e. $m\angle BOD$
40. If $m\overarc{BD}$ = $m\overarc{CD}$ and $m\overarc{BA} : m\overarc{AC}$ = 4:1, find:
    a. $m\overarc{BD}$ b. $m\overarc{AC}$ c. $m\overarc{AB}$ d. $m\angle P$ e. $m\angle ADF$
    f. $m\angle AEB$ g. $m\angle ADC$
41. If $m\angle AEB$ = 130 and $m\overarc{AB}$ = 150, find:
    a. $m\overarc{CD}$ b. $m\overarc{AC}$ c. $m\overarc{BD}$ d. $m\angle COD$ e. $m\angle BCD$
    f. $m\angle P$ g. $m\angle CDA$ h. $m\angle PAD$ i. $m\angle CDF$

3-6 MEASURES OF CHORDS, TANGENT SEGMENTS, AND SECANT SEGMENTS

We have been using some theorems to establish the relationships between the measures of angles of a circle and the measures of the intercepted arcs. Now we will study the measures of line segments related to the circle. To do this, we will need to use similar triangles. Recall that two polygons are similar if there is a one-to-one correspondence between vertices such that:

1. all pairs of corresponding angles are congruent; and
2. the ratios of the lengths of all pairs of corresponding sides are equal.

As a consequence of this definition, we know that the following statement is true:

■ Corresponding sides of similar triangles are in proportion.

The most commonly used method of proving that two triangles are similar is the following theorem:

■ Two triangles are similar if two angles of one triangle are congruent to two angles of the other.

We will use these statements about similar triangles to prove the following theorem:

■ *Theorem 17.* If two chords intersect within a circle, the product of the measures of the segments of one chord equals the product of the measures of the segments of the other.

Given: Chords $\overline{AB}$ and $\overline{CD}$ intersect at E, an interior point of circle O.

To prove: $(AE)(EB) = (CE)(ED)$.

Plan: Draw $\overline{AD}$ and $\overline{BC}$. Prove $\triangle AED \sim \triangle CEB$ by a.a. $\cong$ a.a.

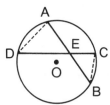

Proof: *Statements*	*Reasons*
1. Draw $\overline{AD}$ and $\overline{BC}$.	1. Two points determine a line.
2. $\angle A \cong \angle C$. (a. $\cong$ a.)	2. Inscribed angles of a circle that intercept the same arc are congruent.
3. $\angle D \cong \angle B$. (a. $\cong$ a.)	3. Same as reason 2.
4. $\triangle AED \sim \triangle CEB$.	4. a.a. $\cong$ a.a.
5. $\dfrac{AE}{CE} = \dfrac{ED}{EB}$.	5. Corresponding sides of similar triangles are in proportion.
6. $(AE)(EB) = (CE)(ED)$.	6. In a proportion, the product of the means is equal to the product of the extremes.

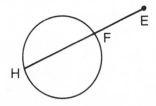

Does a similar relationship exist for two secants intersecting outside the circle? First let us identify the segments formed when a secant line intersects a circle. In the diagram, $\overleftrightarrow{EH}$ intersects the circle in points F and H, forming three segments:

1. $\overline{FH}$ is a *chord* of the circle because its endpoints are points on the circle.

2. $\overline{EF}$ is called an *external segment of the secant* because it is a line segment whose endpoints are an external point and a point on the circle nearer to the external point than any other point of chord $\overline{FH}$.

3. $\overline{EH}$ is called a *secant segment* because it is a line segment whose endpoints are an external point and a point on the circle farther from the external point than any other point of chord $\overline{FH}$. The secant segment is the sum of the chord and the external segment of the secant.

■ *Theorem 18.* If two secants intersect outside a circle, then the product of the measures of one secant segment and its external segment is equal to the product of the measures of the other secant segment and its external segment.

Given: Secants $\overline{ABC}$ and $\overline{ADE}$ intersect at A, outside a circle.

To prove: $(AC)(AB) = (AE)(AD)$.

Plan: Draw $\overline{BE}$ and $\overline{CD}$. Since $\angle C \cong \angle E$, and $\angle A \cong \angle A$, it follows by a.a. $\cong$ a.a. that $\triangle ACD \sim \triangle AEB$.

Thus, $\dfrac{AC}{AE} = \dfrac{AD}{AB}$, or

$(AC)(AB) = (AE)(AD)$.

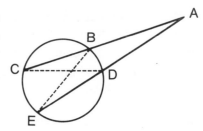

[The formal proof is left to the student.]

A similar theorem can be proved when a tangent and a secant intersect.

■ *Theorem 19.* **If a tangent and a secant are drawn to a circle from an external point, then the square of the measure of the tangent segment is equal to the product of the measures of the secant segment and its external segment.**

Given: Tangent $\overline{PA}$ and secant $\overline{PBC}$ intersect at P, outside a circle.

To prove: $(PA)^2 = (PC)(PB)$.

Plan: Draw $\overline{AC}$ and $\overline{AB}$. Prove $\triangle PAB \sim \triangle PCA$ by a.a. $\cong$ a.a.

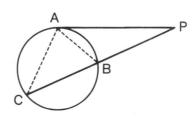

Proof:

Statements	Reasons
1. Draw $\overline{AC}$ and $\overline{AB}$.	1. Two points determine a line.
2. $m\angle PAB = \frac{1}{2}m\widehat{AB}$.	2. Theorem 14.
3. $m\angle C = \frac{1}{2}\widehat{AB}$.	3. Theorem 10.
4. $m\angle PAB = m\angle C$.	4. Transitive property of equality.
5. $\angle PAB \cong \angle C$.	5. Two angles are congruent if they have the same measure.
6. $\angle P \cong \angle P$.	6. Reflexive property of congruence.
7. $\triangle PAB \sim \triangle PCA$.	7. a.a. $\cong$ a.a.
8. $\dfrac{PB}{PA} = \dfrac{PA}{PC}$.	8. Corresponding sides of similar triangles are in proportion.
9. $(PA)^2 = (PC)(PB)$.	9. In a proportion, the product of the means equals the product of the extremes.

Note: If two means of a proportion are equal, then either mean is called the *mean proportional* between the remaining two terms of the proportion. For example, in $\frac{2}{6} = \frac{6}{18}$, and in $(6)^2 = (2)(18)$, 6 is the mean proportional between 2 and 18. Therefore, theorem 19 could be restated as follows:

■ If a tangent and a secant are drawn to a circle from an external point, then the measure of the tangent segment is the *mean proportional* between the measures of the secant segment and its external segment.

| MODEL PROBLEMS |

1. In the accompanying diagram, $\overleftrightarrow{PC}$ is tangent to the circle at C, and $\overleftrightarrow{PAB}$ is a secant with $PA = 3$ and $AB = 9$. Find PC.

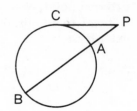

Solution

$PB = PA + AB = 3 + 9 = 12.$
Let $x = PC$.

1. Since a tangent and a secant are drawn to a circle from an external point, it follows that: $(PC)^2 = (PA)(PB)$
2. Substitute values: $(x)^2 = (3)(12)$
                                     $x^2 = 36$
3. Solve for x: $x = \pm 6$
4. Since a length is positive: $PC = 6$

Answer: $PC = 6$

2. Chords $\overline{AB}$ and $\overline{CD}$ intersect in a circle at point E. If $AE = 6$, $EB = 10$, and $ED = 12$, find CE.

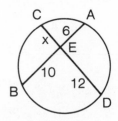

Solution

1. As an aid, draw and label the chords in a circle, as shown at the right.
Let $x = CE$.
2. Since the product of the measures of the segments of one chord equals the product of the measures of the segments of the other: $(CE)(ED) = (AE)(EB)$
3. Substitute values: $(x)(12) = (6)(10)$

4. Solve for x:

$$(x)(12) = (6)(10)$$
$$12x = 60$$
$$x = 5$$

Answer: $CE = 5$

3. Find the length of a chord 3 cm from the center of a circle whose radius measures 5 cm.

Solution

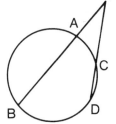

1. As an aid, draw and label the chords in a circle, as shown at the right. Here, the distance from the center of the circle to chord $\overline{AB}$ is shown by $OE = 3$.
   Thus, $CE = OC - OE = 5 - 3 = 2$
   $ED = EO + OD = 3 + 5 = 8$
2. Since a diameter $\perp$ to a chord bisects the chord, let $x = AE$, and $x = EB$.
3. Write the relationship: $(AE)(EB) = (CE)(ED)$
4. Substitute values: $(x)(x) = (2)(8)$
5. Solve for x: $x^2 = 16$
                                 $x = \pm 4$
6. Since a length is positive: $AB = AE + EB = 4 + 4 = 8$

Answer: The length of the chord is 8 cm.

Note: An alternate solution to this problem is found on page 106, model problem 1.

4. From point P, secants $\overleftrightarrow{PAB}$ and $\overleftrightarrow{PCD}$ are drawn to a circle. If $PA = 3$, $PB = 8$, and CD is 2 less than PC, find PD.

Solution

1. Let $PC = x$,
   $CD = x - 2$,
   $PD = PC + CD = 2x - 2$.
2. Since the product of the measures of one secant segment and its external segment equals the product of the measures of the other secant segment and its external segment, write the relationship: $(PD)(PC) = (PB)(PA)$
3. Substitute values: $(2x - 2)(x) = (8)(3)$
4. Multiply: $2x^2 - 2x = 24$

5. Solve the quadratic equation:

$$2x^2 - 2x - 24 = 0$$
$$2(x^2 - x - 12) = 0$$
$$2(x - 4)(x + 3) = 0$$
$$x - 4 = 0 \mid x + 3 = 0$$
$$x = 4 \mid x = -3 \quad \text{(Reject the negative value.)}$$

Answer: PD = 2x - 2 = 2(4) - 2 = 6.

EXERCISES

In 1–10, chords $\overline{AB}$ and $\overline{CD}$ intersect at E.

1. If $CE = 12, ED = 2$, and $AE = 3$, find EB.
2. If $CE = 18, ED = 2$, and $BE = 6$, find AE.
3. If $AE = 8, EB = 9$, and $CE = 12$, find ED.
4. If $CE = 18, ED = 6$, and $AE = 4$, find EB.
5. If $AB = 9, EA = 4$, and $CE = 2$, find ED.
6. If $AB = 20, BE = 8$, and $CE = 16$, find ED.
7. If $AE = 9, EB = 16$, and $CE = ED$, find CE.
8. If $EB = 8, ED = 10$, and AE is 1 more than CE, find CE.
9. If $AE = 3, CE = 5$, and ED is 4 less than EB, find EB.
10. If $CE = 4, ED = 12$, and EB is 2 more than AE, find AE.

Ex. 1 to 10

11. Chords $\overline{CD}$ and $\overline{JK}$ intersect at point E within a circle. If $CE = 10$, $ED = 6$, and $JE = 4$, find EK.
12. Chords $\overline{AB}$ and $\overline{CD}$ intersect at point E within a circle; $AE = 9$ and $EB = 1$.   **a.** If $CD = 6$ and $CE = x$, what expression represents ED in terms of x?   **b.** Find CE.   **c.** Find ED.
13. If chords $\overline{AB}$ and $\overline{RS}$ intersect at point E within a circle so that $AE = 3, EB = 16$, and $RE:ES = 3:4$, find:
    a. RE **b.** ES **c.** RS

In 14–21, $\overleftrightarrow{AF}$ is tangent to the circle at F, and secant $\overleftrightarrow{ABC}$ intersects the circle at B and C.

14. If $AF = 8$ and $AB = 4$, find AC.
15. If $AC = 12$ and $AB = 3$, find AF.
16. If $AF = 6$ and $AC = 9$, find AB.
17. If $AB = 2$ and $BC = 6$, find AF.
18. If $AB = 4$ and $BC = 21$, find AF.
19. If $AC:AB = 4:1$ and $AF = 12$, find AB.
20. If $AB:BC = 1:3$ and $AF = 4$, find AB.
21. If $AF = 10$ and $BC = 15$, find AB.

Ex. 14 to 21

In 22-28, secants $\overleftrightarrow{ABC}$ and $\overleftrightarrow{ADE}$ intersect at point A outside the circle.

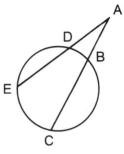

22. If $AC = 15$, $AB = 6$, and $AE = 10$, find AD.
23. If $AC = 20$, $AB = 5$, and $AD = 4$, find AE.
24. If $AB = 9$, $BC = 11$, and $AD = 10$, find AE.
25. If $AB = 3$, $BC = 7$, and $AD = 5$, find DE.
26. If $AB = 2$, $AD = 3$, and $DE = 5$, find BC.
27. If $AD = 6$, $AE = 10$, and $AC = 12$, find:
    a. AB b. BC
28. If $AC = 18$, $AB = 4$, and $ED = 1$, find AD.

Ex. 22 to 28

29. In a circle, diameter $\overline{AB}$ is extended through B to an external point P, and tangent segment $\overline{PC}$ is drawn to point C on the circle. If $BP = 4$ and $PC = 6$, find the length of $\overline{AB}$.

30. Secant $\overleftrightarrow{ABC}$ and tangent $\overleftrightarrow{AD}$ intersect at A outside the circle, while B, C, and D are points on the circle. If $AB = 16$ and $BC = 9$, find the length of the tangent segment, $\overline{AD}$.

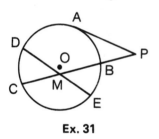

Ex. 31

31. In the diagram, tangent $\overleftrightarrow{PA}$ and secant $\overleftrightarrow{PBC}$ are drawn to circle O from point P. Chord $\overline{DE}$ bisects chord $\overline{BC}$ at M, $PA = 4$, $PB = 2$, and $DE = 10$. Find:
    a. PC b. BC c. CM
    d. DM where $DM > ME$

32. In the diagram, secants $\overleftrightarrow{PAB}$ and $\overleftrightarrow{PCD}$ are drawn to a circle from P. Chords $\overline{AD}$ and $\overline{BC}$ intersect at E, with $BE > EC$. If $PA = 14$, $AB = 10$, $PC = 16$, $AE = 4$, $ED = 4$, and $BC = 10$, find:
    a. PD b. CD c. CE d. EB

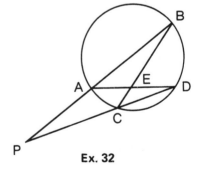

Ex. 32

3-7 REVIEW EXERCISES

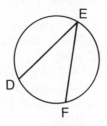

Ex. 1   Ex. 2

1. In circle O, $\overline{AOB}$ is a diameter, $\overline{OC}$ is a radius, and $m\angle AOC = 140$. Find:

   a. $m\widehat{AC}$ b. $m\widehat{BC}$ c. $m\widehat{BAC}$

2. In a circle, chord $\overline{DE} \cong$ chord $\overline{FE}$, and $m\widehat{DF} = 80$. Find:

   a. $m\angle DEF$ b. $m\widehat{DE}$ c. $m\widehat{DFE}$

3. In a circle, $\overline{AOB}$ is a diameter, $\overline{OC}$ is a radius, $AB = 3x - 1$, and $OC = 2x - 3$. Find:

   a. x b. OC c. AB d. $m\widehat{ACB}$

In 4–8, chords $\overline{AB}$ and $\overline{CD}$ of a circle intersect at E.

4. If $m\widehat{AC} = 50$ and $m\widehat{DB} = 110$, find $m\angle AEC$.

5. If $m\angle DEB = 60$ and $m\widehat{AC} = 45$, find $m\widehat{DB}$.

6. If $DE = 9$, $EC = 2$, and $AE = 3$, find EB.

7. If $DE = 8$, $EC = 2$, and $AE = EB$, find AE.

8. If $\overline{DC}$ is a diameter, $m\widehat{AC} = 40$, and $m\widehat{CB} = 80$, find:

   a. $m\widehat{AD}$ b. $m\angle AED$ c. $m\widehat{ADB}$

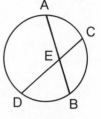

Ex. 4 to 8

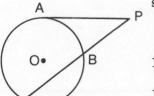

Ex. 9 to 13

In 9–13, $\overleftrightarrow{PA}$ is tangent to circle O at A, and secant $\overleftrightarrow{PBC}$ intersects circle O at B and C.

9. If $m\widehat{AC} = 170$ and $m\widehat{AB} = 110$, find $m\angle P$.

10. If $m\widehat{AC} = 160$ and $m\widehat{AB} = m\widehat{BC}$, find $m\angle P$.

11. If $PB = 15$ and $PC = 60$, find PA.

12. If $PB = 4$ and $BC = 5$, find PA.

13. If $PA = 8$ and $BC = 12$, find:

    a. PB b. PC

In 14–18, diameter $\overline{AB}$ of circle O is perpendicular to chord $\overline{CD}$ at E.

14. If $CD = 10$, find ED.
15. If $AE = 18$ and $EB = 8$, find CE.
16. If $OE = 6$ and $OB = 10$, find:
   a. EB **b.** AE **c.** CE **d.** CD
17. If $EB = 4$ and $CE = 6$, find:
   a. AE **b.** AB **c.** OA
18. If $CE = 8$ and $AB = 20$, find OE, the distance of the chord from the center of the circle.

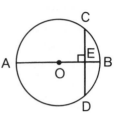

Ex. 14 to 18

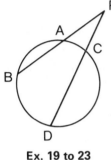

In 19–23, secants $\overleftrightarrow{PAB}$ and $\overleftrightarrow{PCD}$ meet at an external point P, and intersect the circle at points $A, B, C,$ and D.

19. If $\overset{\frown}{BD} = 130$ and $m\overset{\frown}{AC} = 40$, find $m\angle P$.
20. If $m\angle P = 40$ and $m\overset{\frown}{BD} = 150$, find $m\overset{\frown}{AC}$.
21. If $PA = 4, PB = 15,$ and $PC = 6$, find PD.
22. If $PA = 4, AB = 16,$ and $PC = 5$, find:
   a. PD **b.** CD
23. If $PC = 2, CD = 4,$ and $AB = 1$, find:
   a. PA **b.** PB

Ex. 19 to 23

24. Two tangents to circle O from point P meet the circle at points A and B.
   a. If the measure of the major arc $\overset{\frown}{AB}$ is 200, find $m\angle P$.
   b. If the ratio of the measures of major arc $\overset{\frown}{AB}$ to minor arc $\overset{\frown}{AB}$ is 13:5, find $m\angle P$.
25. Find the length of a chord that is 5 cm from the center of a circle whose radius measures 13 cm.
26. Find the length of the radius of a circle if a chord 24 cm long is 9 cm from the center of the circle.

27. In the diagram, diameter $\overline{AOB}$ is extended to point C, $\overleftrightarrow{CD}$ is tangent to the circle at D, $\overline{DOE}$ is a diameter and $m\overset{\frown}{BD}:m\overset{\frown}{AD} = 2:7$. Find:
   a. $m\overset{\frown}{BD}$ **b.** $m\overset{\frown}{AD}$ **c.** $m\angle E$
   d. $m\angle C$ **e.** $m\overset{\frown}{AE}$ **f.** $m\overset{\frown}{EB}$
   g. $m\angle CDE$ **h.** $m\angle ADC$

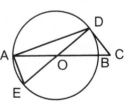

Ex. 27

28. In the diagram, $\overleftrightarrow{PA}$ and $\overleftrightarrow{PB}$ are tangent to circle O at A and B, respectively. Diameter $\overline{BOD}$ and chord $\overline{AC}$ intersect at E, $m\widehat{CB} = 160$, and $m\angle APB = 40$. Find:

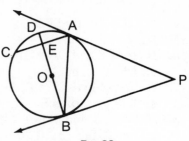

a. $m\widehat{ADCB}$ b. $m\widehat{AB}$
c. $m\widehat{CD}$ d. $m\widehat{DA}$
e. $m\angle DEC$ f. $m\angle PAC$
g. $m\angle PBD$ h. $m\angle PBA$

Ex. 28

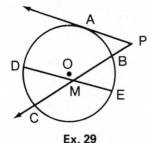

Ex. 29

29. In the diagram, $\overleftrightarrow{PA}$ is tangent to circle O at A, $\overleftrightarrow{PBC}$ intersects the circle at B and C, chord $\overline{DE}$ bisects chord $\overline{BC}$ at M, $PA = 9, PB = 3$, and $DE = 30$. Find:

a. PC b. BC
c. CM d. EM when $EM < DM$

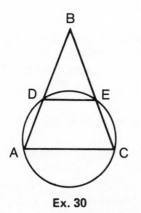

Ex. 30

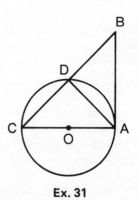

Ex. 31

30. *Given:*   Secants $\overleftrightarrow{ADB}$ and $\overleftrightarrow{CEB}$ intersect at B. $\overline{AD} \cong \overline{CE}$.
       To prove: $\triangle ABC$ is isosceles.

31. *Given:*   Tangent $\overleftrightarrow{AB}$ and secant $\overleftrightarrow{BDC}$ intersect at B outside circle O. $\overline{AOC}$ is a diameter, and $\overline{AD}$ is a chord.
       To prove: $\triangle ABC \sim \triangle DAC$.

Chapter 4

The Irrational Numbers and the Real Numbers

4-1 COMPLETING THE REAL-NUMBER LINE

In Chapter 1, we learned that every rational number can be associated with a point on the number line. Although the set of rational numbers is a dense set and an infinite number of rationals lies between every two rational numbers, the number line that contains only the graphs of all rational numbers is not complete. There are "holes" in the line reserved for numbers that are not rational. For example, $\sqrt{2}$, $\sqrt{3}$, and $\sqrt{5}$ are not rational numbers, but we can locate a point associated with each of these numbers on the number line.

In Fig. 1, a square is constructed on the number line so that each of the sides of the square measures 1. When its diagonal, which measures d, is drawn, a right triangle is formed. By the Pythagorean Theorem:

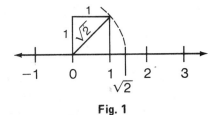

Fig. 1

$$d^2 = 1^2 + 1^2$$
$$d^2 = 1 + 1 = 2$$
$$d = \sqrt{2} \text{ (a length is positive)}$$

By placing the point of a pair of compasses on 0 on the number line and opening the compasses to the length $\sqrt{2}$, we can swing an arc so as to locate the point associated with $\sqrt{2}$ on the number line.

In Fig. 2, we can use the same process to construct a rectangle whose sides measure 1 and $\sqrt{2}$. A diagonal whose measure is d is drawn. The Pythagorean Theorem is applied:

$$d^2 = 1^2 + (\sqrt{2})^2$$
$$d^2 = 1 + 2 = 3$$
$$d = \sqrt{3}$$

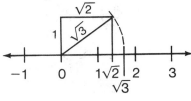

Fig. 2

Another arc is then swung to locate the point associated with $\sqrt{3}$ on the number line.

In Fig. 3, a rectangle whose sides measure 1 and 2 is used to help us locate the graph of $\sqrt{5}$ on the number line.

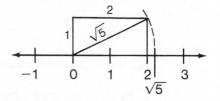

Fig. 3

The Irrational Numbers

The numbers $\sqrt{2}$, $\sqrt{3}$, and $\sqrt{5}$ are irrational numbers. To define an irrational number, let us recall a discovery made in Chapter 1; namely, a number is *rational* if and only if the number can be expressed as a repeating decimal.

■ **DEFINITION.** An *irrational number* is a non-repeating decimal (one that does not terminate or end).

There are infinitely many irrational numbers. Consider these examples:

.10110111011110111110 . . .   AND   −.1234567891011121314 . . .

In each case, the three dots indicate that the pattern continues without end. Note, however, that the same sequence of numbers does not repeat, as in $.\overline{3} = .333 \ldots = \frac{1}{3}$. Thus, there is no way to represent either number shown in the form of a rational number, that is, $\frac{a}{b}$ where a and b are integers and $b \neq 0$.

If k is a positive number but not a perfect square, then $\sqrt{k}$ is irrational. Such irrational numbers include $\sqrt{2}$, $\sqrt{3}$, $\sqrt{5}$, $\sqrt{1.4}$, and $\sqrt{\frac{1}{2}}$. *Pi*, written symbolically as π, is also an irrational number.

π = 3.14159265358979323846264338327950288841971 . . .

Here, the three dots indicate that the number continues, but we are uncertain of the next digit or we do not wish to display the digits any further. While modern computers can calculate π to thousands of decimal places, the exact value of π cannot be written as a finite decimal. For this reason, we often use *approximate rational values* when working with π and other irrational numbers. For example, $\pi \approx \frac{22}{7}$ or $\pi \approx 3.14$ or $\pi \approx 3.1416$.

The symbol "$\approx$" is read as "is approximately equal to." The table of square roots on page 746 also indicates approximate rational values for many irrational numbers, such as:

$$\sqrt{2} \approx 1.414 \qquad \sqrt{3} \approx 1.732 \qquad \sqrt{5} \approx 2.236$$

The Real Numbers

■ **DEFINITION.** The *set of real numbers* is the union of the set of rational numbers and the set of irrational numbers.

In other words, every number that can be expressed as a decimal, whether the decimal is repeating or non-repeating, is a real number. By identifying points that correspond to irrational numbers as well as points that correspond to rational numbers, we completely fill in the *real-number line*.

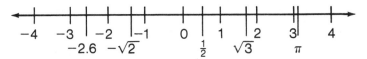

There is a *one-to-one correspondence* between the set of real numbers and the set of points on the number line. Thus:

1. For every real number, there is one and only one point on the number line, and
2. For every point on the number line, there is one and only one real number.

The Properties of the Real Numbers

The set of real numbers, symbolized here as R, is an ordered field under the operations of addition and multiplication. The properties that held for the set of rational numbers in Chapter 1 are now extended to the set of real numbers. Let us restate these properties for R, both in words and in symbols.

■ **(Real Numbers, +, ·, >) is an ordered field, satisfying the following properties:**

In Words	*In Symbols*
1. $(R, +)$ is *closed*.	1. $\forall_{a,b} \in R$: $a + b = c$ where $c \in R$.
2. $(R, +)$ is *associative*.	2. $\forall_{a,b,c} \in R$: $(a + b) + c = a + (b + c)$.
3. $(R, +)$ has a unique *identity* element (zero).	3. $\exists_0 \in R, \forall_x \in R$: $x + 0 = x$, and $0 + x = x$.
4. Every element of $(R, +)$ has an *inverse* $(-x)$.	4. $\forall_x \in R, \exists_{(-x)} \in R$: $x + (-x) = 0$, and $(-x) + x = 0$.

5. $(R, +)$ is *commutative*.

5. $\forall_{a,b} \in R:$ $a + b = b + a.$

6. $(R/\{0\}, \cdot)$ is *closed*.

6. $\forall_{a,b} \in R/\{0\}:$
$ab = c$ where $c \in R/\{0\}.$

7. $(R/\{0\}, \cdot)$ is *associative*.

7. $\forall_{a,b,c} \in R/\{0\}:$ $(ab)c = a(bc).$

8. $(R/\{0\}, \cdot)$ has a unique *identity* element (one).

8. $\exists_1 \in R/\{0\}, \forall_x \in R/\{0\}:$
$x \cdot 1 = x,$ and $1 \cdot x = x.$

9. Every element of $(R/\{0\}, \cdot)$ has an *inverse* $\left(\dfrac{1}{x}\right).$

9. $\forall_x \in R/\{0\}, \exists_{\frac{1}{x}} \in R/\{0\}:$
$x \cdot \dfrac{1}{x} = 1,$ and $\dfrac{1}{x} \cdot x = 1.$

10. $(R/\{0\}, \cdot)$ is *commutative*.

10. $\forall_{a,b} \in R/\{0\}:$ $ab = ba.$

11. Multiplication *distributes* over addition.

11. $\forall_{a,b,c} \in R:$ $a(b + c) = ab + ac,$
and $ab + ac = a(b + c).$

12. *Trichotomy Property*.

12. $\forall_{a,b} \in R,$ one and only one is true:
$a > b,$ or $a = b,$ or $a < b.$

13. *Transitive Property of Inequalities*.

13. $\forall_{a,b,c} \in R:$
If $a > b$ and $b > c,$ then $a > c.$
If $c < b$ and $b < a,$ then $c < a.$

14. *Addition Property of Inequalities*.

14. $\forall_{a,b,c} \in R:$
If $a > b,$ then $a + c > b + c.$
If $a < b,$ then $a + c < b + c.$

15. *Multiplication Property of Inequalities*.

15. $\forall_{a,b,c} \in R$ where c is *positive*:
If $a > b$ and $c > 0,$ then $ac > bc.$
If $a < b$ and $c > 0,$ then $ac < bc.$
$\forall_{a,b,c} \in R$ where c is *negative*:
If $a > b$ and $c < 0,$ then $ac < bc.$
If $a < b$ and $c < 0,$ then $ac > bc.$

In addition to the properties of an ordered field, the set of real numbers obeys the postulates of equality and the zero property of multiplication:

1. *Reflexive Property of Equality*.

1. $\forall_a \in R:$ $a = a.$

2. *Symmetric Property of Equality*.

2. $\forall_{a,b} \in R:$ If $a = b,$ then $b = a.$

3. *Transitive Property of Equality.*	3. $\forall_{a,b,c} \in R$: If $a = b$ and $b = c$, then $a = c$.
4. *Substitution Property of Equality.*	4. $\forall_{a,b} \in R$: If $a = b$, then a or b may replace each other in any expression.
5. *Addition Property of Equality.*	5. $\forall_{a,b,c} \in R$: If $a = b$, then $a + c = b + c$.
6. *Multiplication Property of Equality.*	6. $\forall_{a,b,c} \in R$: If $a = b$, then $ac = bc$.
7. *Zero Property of Multiplication.*	7. $\forall_x \in R$: $x \cdot 0 = 0$, and $0 \cdot x = 0$.

From this point on, unless otherwise stated, assume that all work is to be performed using the domain of real numbers.

EXERCISES

In 1–15, tell whether the number is rational or irrational.

1. -2 2. 0 3. $\sqrt{2}$ 4. $\sqrt{625}$ 5. π 6. $\sqrt{90}$

7. $\frac{1}{2}$ 8. 3.67 9. $\sqrt{\frac{1}{4}}$ 10. $\sqrt{\frac{1}{3}}$ 11. $\sqrt{3.6}$ 12. $\sqrt{.36}$

13. $.20202020\ldots$ 14. $.2020020002\ldots$ 15. $.248163264\ldots$

16. Is the set of irrational numbers closed under addition? (*Hint:* Add $\sqrt{3}$ and $-\sqrt{3}$.)

17. Is the set of irrational numbers closed under multiplication?

18. a. Construct a rectangle whose sides measure 1 and 3 so that one of its vertices touches 0 on a number line.

   b. Using this rectangle, locate the graph of $\sqrt{10}$ on the number line.

   c. Locate the graph of $-\sqrt{10}$ on the same number line.

In 19–26, identify the property of the real numbers that is illustrated by the given statement.

19. $\sqrt{19} + 0 = \sqrt{19}$ 20. $\pi + (-\pi) = 0$ 21. $\sqrt{10} \cdot 1 = \sqrt{10}$

22. $\sqrt{5} \cdot \frac{1}{4} = \frac{1}{4} \cdot \sqrt{5}$ 23. $\frac{\sqrt{3}}{2} \cdot \frac{2}{\sqrt{3}} = 1$ 24. $\pi + 4 = 4 + \pi$

25. $3(2 + \sqrt{7}) = 3 \cdot 2 + 3\sqrt{7}$ 26. $3 + (2 + \sqrt{7}) = (3 + 2) + \sqrt{7}$

In 27–30, tell which of the three given statements is true.

27. $.2 > .\overline{2}$ or $.2 = .\overline{2}$ or $.2 < .\overline{2}$

28. $\sqrt{2} > 1.4$ or $\sqrt{2} = 1.4$ or $\sqrt{2} < 1.4$

29. $\pi > \frac{22}{7}$ or $\pi = \frac{22}{7}$ or $\pi < \frac{22}{7}$ **30.** $.\overline{5} > \frac{5}{9}$ or $.\overline{5} = \frac{5}{9}$ or $.\overline{5} < \frac{5}{9}$

4-2 GRAPHING SOLUTION SETS INVOLVING ONE VARIABLE ON A NUMBER LINE

The *graph of a solution set* of an open sentence involving *one variable* is the set of points on the *real-number line* that are associated with elements of the solution set. In constructing or reading such graphs, we observe:

1. A darkened circle ● is drawn as a point on the number line to represent a number in the solution set.
2. A darkened segment ——— or a darkened ray ———▶ indicates that all real numbers associated with points on the segment or the ray are in the solution set.
3. A non-darkened circle ○ shown as an endpoint of a segment or a ray indicates that the number associated with that point is not found in the solution set.

| MODEL PROBLEM |

Graph the solution set of $3x + 8 < 5$.

How to Proceed

1. Simplify the given inequality. Since $3x + 8 < 5$ and $x < -1$ are equivalent, both sentences have the same solution set and, thus, the same graph.

2. The graph consists of all points to the left of -1, and a non-darkened circle shows that -1 is not included.

Solution

$$3x + 8 < 5$$
$$3x + 8 - 8 < 5 - 8$$
$$3x < -3$$
$$x < -1$$

Answer:

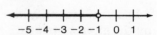
$-5\ -4\ -3\ -2\ -1\ \ \ 0\ \ \ 1$

Graphing Conjunctions and Disjunctions

The expression $1 < x < 5$ is read as "1 is less than x, and x is less than 5." In logic, we learned that a *conjunction* is a compound sentence that combines two simple sentences by using the word "and" (in symbols: $\wedge$). Thus, we observe:

$1 < x < 5$ is equivalent to the conjunction $(1 < x) \wedge (x < 5)$.
Also, $1 < x < 5$ is equivalent to the conjunction $(x > 1) \wedge (x < 5)$.

The solution set of a conjunction of two open sentences contains only those values of the variable that are true for *both* open sentences. Thus, the graph of the solution set of a conjunction is the *intersection* of the graphs of the simple open sentences.

□ EXAMPLE: Graph the solution set of $(x > 1) \wedge (x < 5)$.

How to Proceed	*Solution*
1. Think of the graph of $x > 1$ and the graph of $x < 5$ as they would appear on the same number line.	*Think:* 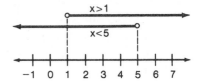
2. The graph of the solution set of the conjunction is the intersection of the two sets of points seen in step 1. Thus, any point on the graph will be true for both conditions, $(x > 1)$ *and* $(x < 5)$.	*Write:* 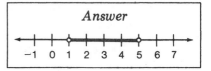

Note: The graph of $(x > 1) \wedge (x < 5)$ is also the graph of $1 < x < 5$.

A *disjunction* is a compound sentence that combines two simple sentences by using the word "or" (in symbols: $\vee$). The solution set of a disjunction of two open sentences contains all values of the variable that are true for *either* one or the other open sentence, or both. Thus, the graph of the solution set of a disjunction is the *union* of the graphs of the simple open sentences.

□ EXAMPLE: Graph the solution set of $(x \geq 2) \vee (6 < x)$.

How to Proceed	*Solution*
1. Think of the graph of $x \geq 2$ and the graph of $6 < x$ (equivalent to $x > 6$) as they would appear on the same number line.	*Think:* 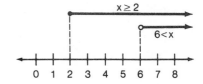

2. The graph of the solution set of the disjunction is the union of the two sets of points seen in step 1. Thus, any point on the graph will fit one or both of the conditions $(x \geq 2)$ *or* $(6 < x)$.

Write:

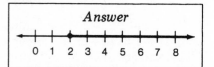

MODEL PROBLEM

Which of the following expressions is represented by the graph at the right?

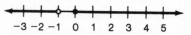

(1) $(x < -1) \wedge (x \geq 0)$ (2) $(x \leq -1) \wedge (x > 0)$
(3) $(x < -1) \vee (x \geq 0)$ (4) $(x \leq -1) \vee (x > 0)$

Solution: The graph shows the *union* of two sets of points having no elements in common. Since union indicates a *disjunction*, study only choices (3) or (4), which involve the word "or," shown by the symbol $\vee$. The graph indicates that -1 is not an element of the set, while 0 is an element. Therefore, the correct choice is $(x < -1) \vee (x \geq 0)$.

Answer: (3) $(x < -1) \vee (x \geq 0)$

EXERCISES

In 1–19, graph the solution set of the given sentence on a number line.

1. $x + 7 < 8$
2. $3x \geq 18$
3. $2 \geq x$
4. $2y - 3 \geq 7$
5. $5 + 4y > 5$
6. $6y - 5 = 7$
7. $5k < 4k - 2$
8. $5 < 8 + w$
9. $3 \leq 2 - p$
10. $2x + 6 = 9$
11. $4y - 1 \geq 1$
12. $x + 4 > 4 + \pi$
13. $\dfrac{b}{2} < -2$
14. $\dfrac{x}{3} - 1 \geq \dfrac{x}{2}$
15. $-3 \leq x < 0$
16. $(x > -2) \wedge (x \leq 4)$
17. $(x > 2) \vee (x \leq -4)$
18. $(y \leq 1) \vee (y < 5)$
19. $(y \geq -2) \vee (y > -1)$

In 20–23, graph the given set on a real-number line.

20. $\{x | 3 \leq x \leq 8\}$
21. $\{y | -2\frac{1}{2} < y < 1\frac{1}{2}\}$
22. $\{k | (k < -5) \vee (k > -3)\}$
23. $\{x | (x \leq -3) \vee (x \leq -1.3)\}$

24. Which of the following expressions is represented by the graph
at the right?

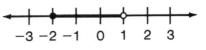

   -3 -2 -1 0 1 2 3

   (1) $-2 < x \leq 1$ (2) $(x > -2) \vee (x \leq 1)$
   (3) $-2 \leq x < 1$ (4) $(x \geq -2) \vee (x < 1)$

25. Which of the following expressions is represented by the
graph at the right?

   -5 -4 -3 -2 -1 0 1

   (1) $(x < -3) \wedge (x \geq -1)$ (2) $(x < -3) \vee (x \geq -1)$
   (3) $(x \leq -3) \wedge (x > -1)$ (4) $(x \leq -3) \vee (x > -1)$

26. Which of the following expressions is represented by the graph
at the right?

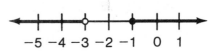

   -4 -3 -2 -1 0 1 2

   (1) $(y \geq 1) \vee (y > -2)$ (2) $(y > 1) \vee (y \geq -2)$
   (3) $(y \geq 1) \wedge (y > -2)$ (4) $(y > 1) \wedge (y \geq -2)$

27. Which expression has the entire real-number line as its graph?
   (1) $(x > 6) \wedge (x \leq 10)$ (2) $(x < 6) \wedge (x \geq 10)$
   (3) $(x > 6) \vee (x \leq 10)$ (4) $(x < 6) \vee (x \geq 10)$

4-3   ABSOLUTE-VALUE EQUATIONS

The absolute value of a real number n, written in symbols as $|n|$, can
be defined in various ways.

■ **ARITHMETIC DEFINITION.** The *absolute value* of a real number
n is the maximum of the number n and its additive inverse $(-n)$. In
symbols:

$$|n| = n \text{ max } (-n)$$

☐ EXAMPLES:

1. Since $|4| = 4$ max $(-4) = 4$, we write: $|4| = 4$
2. Since $|-3| = (-3)$ max $3 = 3$, we write: $|-3| = 3$
3. The absolute value of 0 is defined as 0 itself: $|0| = 0$
   Note that $|0| = 0$ max $0 = 0$.

■ **GEOMETRIC DEFINITION.** The *absolute value* of a real number n
is the distance between the graphs of the numbers n and 0 on the real-
number line.

☐ EXAMPLE 1:

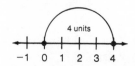

The distance between the graphs of 4 and 0 is 4. Thus:

$$|4| = 4$$

☐ EXAMPLE 2:

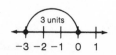

The distance between the graphs of -3 and 0 is 3. Thus:

$$|-3| = 3$$

☐ EXAMPLE 3:

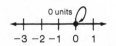

The distance from 0 to itself is 0. Thus:

$$|0| = 0$$

☐ EXAMPLE 4:

The solution set of the absolute-value equation $|x| = 2$ is $\{2, -2\}$ because the graphs of 2 and -2 are each at a distance of 2 units from the graph of 0. Thus, $|-2| = 2$ and $|2| = 2$.

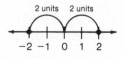

Some mathematicians write $|x| = 2$ as $|x - 0| = 2$ to show that the distance between the graphs of x and 0 is 2.

■ **ALGEBRAIC DEFINITION.** The *absolute value* of a real number n is defined as:

$$|n| = n, \text{ if } n \geq 0 \quad \text{AND} \quad |n| = -n, \text{ if } n < 0$$

In example 4 above, using a geometric approach, we saw that the solution set of $|x| = 2$ is $\{2, -2\}$. Let us redo this problem, using the algebraic definition.

☐ EXAMPLE: Solve $|x| = 2$.

If $x \geq 0$, then $|x| = x$.   If $x < 0$, then $|x| = -x$.
When $|x| = 2$,   When $|x| = 2$,
$x = 2$ $-x = 2$
                                  $x = -2$

Answer: $x = 2$ or $x = -2$ OR   solution set = $\{2, -2\}$.

Derived Equations

In the example just seen, the equations "$x = 2$ or $x = -2$" are *derived* from the absolute-value equation $|x| = 2$. In general:

If $|x| = k$ where $k > 0$, we derive the equations "$x = k$ or $x = -k$."

The procedure for solving an absolute-value equation involves the use of such derived equations. This procedure is outlined in model problems 1 and 2 that follow. Remember, the roots of all derived equations must be checked in the original equation so that extraneous roots are not included in the answer or the solution set.

| MODEL PROBLEMS |

1. Solve $|x - 4| = 3$ and graph the solution set.

How to Proceed	*Solution*
1. Write the absolute-value equation.	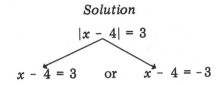
2. Using disjunction, write two derived equations.	$x - 4 = 3$ or $x - 4 = -3$
3. Solve the derived equations.	$x - 4 + 4 = 3 + 4$ $\qquad$ $x - 4 + 4 = -3 + 4$ $x = 7$ $\qquad\qquad$ $x = 1$
4. Check the roots of the derived equations in the absolute-value equation.	Check for $x = 7$. $\quad$ Check for $x = 1$. $|x - 4| = 3$ $\qquad$ $|x - 4| = 3$ $|7 - 4| \stackrel{?}{=} 3$ $\qquad$ $|1 - 4| \stackrel{?}{=} 3$ $|3| \stackrel{?}{=} 3$ $\qquad$ $|-3| \stackrel{?}{=} 3$ $3 = 3$ $\qquad\quad$ $3 = 3$ (True) $\qquad\qquad$ (True)
5. Write the answer or the solution set.	*Answer:* $x = 7$ or $x = 1$ OR solution set = $\{1, 7\}$
6. Graph the solution set.	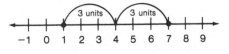

Note: In geometric terms, $|x - 4| = 3$ means that the distance between the graphs of x and 4 is 3 units. Given $|x - 4| = 3$, we 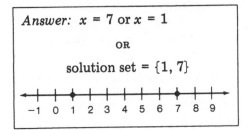 know that $x = 1$ or $x = 7$. At the right, notice how the distance between the graphs of 1 and 4 is 3, and the distance between the graphs of 7 and 4 is also 3.

2. Solve $|x| + 3 = 2x$ and graph the solution set.

How to Proceed	*Solution*

1. Write the given equation.

$$|x| + 3 = 2x$$

2. Transform the equation into an equivalent sentence in which the absolute-value expression is isolated on one side of the equation.

$$|x| + 3 - 3 = 2x - 3$$
$$|x| = 2x - 3$$

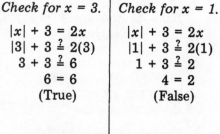

3. Using the algebraic definition, write two equations.

If $x \geq 0$, $|x| = x$.
$$|x| = 2x - 3$$
$$x = 2x - 3$$

If $x < 0$, $|x| = -x$.
$$|x| = 2x - 3$$
$$-x = 2x - 3$$

4. Solve the equations. (Notice the second equation. Since $x < 0$ and $x = 1$ are not both true, $x = 1$ cannot be a root. See the check.)

$$-x = -3$$
$$x = 3$$

$$-3x = -3$$
$$x = 1$$

5. Check the roots of the derived equations in the original absolute-value equation. (Note that $x = 1$ is an *extraneous* root because 1 is not a root of the absolute-value equation.)

Check for x = 3.
$$|x| + 3 = 2x$$
$$|3| + 3 \stackrel{?}{=} 2(3)$$
$$3 + 3 \stackrel{?}{=} 6$$
$$6 = 6$$
(True)

Check for x = 1.
$$|x| + 3 = 2x$$
$$|1| + 3 \stackrel{?}{=} 2(1)$$
$$1 + 3 \stackrel{?}{=} 2$$
$$4 = 2$$
(False)

6. Write the answer.

7. Graph the solution set.

Answer: $x = 3$

OR

solution set = $\{3\}$

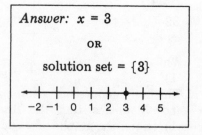

Alternate Method: In step 3, write the two derived equations $x = 2x - 3$ and $x = -(2x - 3)$. These equations have the roots $x = 3$ and $x = 1$, respectively.

3. Which is the solution set of $|n| = -5$?

(1) $\{5\}$ (2) $\{-5\}$ (3) $\{5, -5\}$ (4) $\{\ \}$

Solution

There is no number whose absolute value is a *negative* number. This can be demonstrated by checking 5 and -5 in the equation $|n| = -5$.

Check for n = 5.	*Check for n = -5.*
$|5| \overset{?}{=} -5$	$|-5| \overset{?}{=} -5$
$5 = -5$ (False)	$5 = -5$ (False)

Thus, the solution set is the empty set.

Answer: (4) $\{\ \}$

| EXERCISES |

In 1–31: **a.** Solve the absolute-value equation.   **b.** Graph the solution set if it is not empty.

1. $|x| = 4$ **2.** $|y| = \frac{1}{2}$ **3.** $|-w| = 2$ **4.** $|n| = -9$
5. $|p| = 0$ **6.** $|2k| = 7$ **7.** $|x| + 3 = 4$ **8.** $|y| - 4 = 2$
9. $5 + |m| = 1$ **10.** $|x + 3| = 4$ **11.** $|y - 4| = 2$
12. $|5 + m| = 1$ **13.** $|k + 4| = 6$ **14.** $|2 - k| = 3$
15. $|2b + 9| = 0$ **16.** $|y - 3| = 3$ **17.** $|y - 3| = -3$
18. $|2 - b| = 4$ **19.** $|4 - 3x| = x$ **20.** $|4 - x| = 3x$
21. $|x| - 4 = 3x$ **22.** $|3x| + 4 = x$ **23.** $|2k - 3| = k$
24. $|k - 3| = 2k$ **25.** $|k| - 3 = 2k$ **26.** $|2k| + 3 = k$
27. $|y - 1| = 3y$ **28.** $|2x + 5| = x + 4$ **29.** $|2x + 5| = x + 1$
30. $|y + 3| + 5 = 2y$ **31.** $|y - 4| - 3y = 6$

In 32–34, select the numeral preceding the expression that best answers the question.

32. Which of the given equations has $\{\ \}$ as its solution set?
     (1) $|x| = 7$ (2) $|-x| = 7$ (3) $|x| = -7$ (4) $|x - 7| = 0$
33. Which of the given equations has a solution set consisting of one and only one real number?
     (1) $|y| = 1$ (2) $|y| = 2$ (3) $|y| = -3$ (4) $|y| = 0$
34. Which expression correctly states the derived equations for $|x| + 3 = 5$?
     (1) $(x + 3 = 5) \lor (x + 3 = -5)$ (2) $(x = 2) \lor (x = -2)$
     (3) $(x + 3 = 5) \land (x + 3 = -5)$ (4) $(x = 2) \land (x = -2)$

4-4 ABSOLUTE-VALUE INEQUALITIES

Study the three conditions that involve absolute value and their corresponding graphs, shown here.

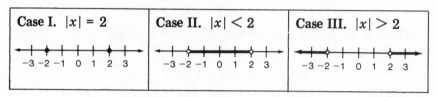

Case I. $|x| = 2$	Case II. $|x| < 2$	Case III. $|x| > 2$

Case I. If $|x| = 2$, we learned that "$x = -2$ or $x = 2$." Recall that the distance between the graphs of -2 and 0 is 2 units, and the distance between the graphs of 2 and 0 is also 2 units.

Case II. If $|x| < 2$, then the distance between the graphs of x and 0 is *less than 2 units*. Thus, x may be any real number less than 2 and greater than -2. The solution set, written as $\{x \mid -2 < x < 2\}$, is equivalent to the *conjunction* $\{x \mid x > -2 \text{ and } x < 2\}$.

Let us apply the algebraic definition of absolute value to the general case: $|x| < k$, where k is positive.
1. If $x \geq 0$, $|x| = x$. When $|x| < k$,
$$x < k. \text{ Therefore, } 0 \leq x < k.$$
2. If $x < 0$, $|x| = -x$. When $|x| < k$,
$$-x < k, \text{ or}$$
$$x > -k. \text{ Therefore, } -k < x < 0.$$
3. The solution of $|x| < k$, when $k > 0$, is $(-k < x < 0) \lor (0 \leq x < k)$. The union of these disjoint sets can be written as $-k < x < k$.

We have therefore proved:

■ If $|x| < k$, where k is positive, its solution set is $\{x \mid -k < x < k\}$.

The graph of $\{x \mid -k < x < k\}$, as shown at the right, is a segment that includes the graphs of all real numbers between $-k$ and k, but excludes the graphs of these endpoints.

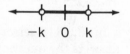

From $|x| < k$, where k is positive, we can derive the expressions:

$$\text{"} -k < x < k \text{"} \quad \text{OR} \quad \text{"} x > -k \text{ and } x < k \text{"}$$
$$\text{OR} \quad \text{"} (x > -k) \land (x < k) \text{"}$$

Case III. If $|x| > 2$, then the distance between the graphs of x and 0 is *greater than 2 units*. Thus, x may be any real number greater than 2 or less than -2. The solution set, written as $\{x \mid x < -2 \text{ or } x > 2\}$ or as $\{x \mid (x < -2) \lor (x > 2)\}$, is a *disjunction* of two sets of numbers with no elements in common.

Again, we will apply the algebraic definition of absolute value to the general case: $|x| > k$, where k is positive.

1. If $x \geq 0$, $|x| = x$. When $|x| > k$,
$$x > k. \text{ Therefore, } x > k.$$

2. If $x < 0$, $|x| = -x$. When $|x| > k$,
$$-x > k, \text{ or}$$
$$x < -k. \text{ Therefore, } x < -k.$$

3. The solution of $|x| > k$, when $k > 0$, is $(x < -k) \lor (x > k)$.

We have therefore proved:

■ If $|x| > k$, where k is positive, its solution set is:

$$\{x \mid x < -k \text{ or } x > k\}$$

The graph of $\{x \mid x < -k \text{ or } x > k\}$, as shown at the right, is the union of two rays, one to the right of k and one to the left of $-k$, excluding the graphs of the endpoints of these rays.

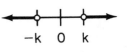
−k 0 k

From $|x| > k$, where k is positive, we can derive the expressions:

$$\text{``} x < -k \text{ or } x > k \text{''} \quad \text{OR} \quad \text{``} (x < -k) \lor (x > k) \text{''}$$

| **MODEL PROBLEMS** |

1. Solve $|2x + 3| < 7$ and graph the solution set.

How to Proceed	*Solution*
1. Write the given inequality.	$|2x + 3| < 7$
2. Write a derived expression.	$-7 < 2x + 3 < 7$
3. Simplify: add -3 to each member of the inequality, and divide each member by 2.	$-7 - 3 < 2x + 3 - 3 < 7 - 3$ $-10 < 2x < 4$ $-5 < x < 2$
4. Write the solution set.	
5. Graph the solution set.	*Solution set:* $\{x \mid -5 < x < 2\}$ 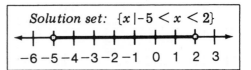 $-6\ -5\ -4\ -3\ -2\ -1\ \ 0\ \ 1\ \ 2\ \ 3$

Note: In step 2, it is possible to write "$2x + 3 > -7$ *and* $2x + 3 < 7$" as the derived expression. By simplifying this expression, we arrive at the solution set $\{x \mid x > -5 \text{ } and \text{ } x < 2\}$, which is equivalent to the solution set $\{x \mid -5 < x < 2\}$.

2. Solve $|3 + y| - 2 \geq 0$ and graph the solution set.

How to Proceed	Solution
1. Write the given inequality.	$\|3 + y\| - 2 \geq 0$
2. Transform the inequality into an equivalent sentence in which the absolute value is isolated on one side of the inequality.	$\|3 + y\| - 2 + 2 \geq 0 + 2$ $\|3 + y\| \geq 2$
3. Write a derived expression.	$3 + y \leq -2 \text{ or } 3 + y \geq 2$
4. Add -3 to each member of the inequality.	$3 + y - 3 \leq -2 - 3 \quad \mid \quad 3 + y - 3 \geq 2 - 3$ $y \leq -5 \text{ or } y \geq -1$
5. Write the solution set.	*Solution set:* $\{y \mid y \leq -5 \text{ or } y \geq -1\}$
6. Graph the solution set.	

EXERCISES

In 1–18: **a.** Solve the absolute-value inequality. **b.** Graph the solution set.

1. $|x| < 3$
2. $|y| > 3$
3. $|w| \leq 4$
4. $\left|\dfrac{p}{2}\right| \geq 4$
5. $|2x| > 5$
6. $|-3y| < 18$
7. $|x| - 1 < 4$
8. $3 + |y| \leq 5$
9. $|x| - 2 \geq 1.5$
10. $|x - 1| < 4$
11. $|3 + y| \leq 5$
12. $|x - 2| \geq 1.5$
13. $\left|\dfrac{w}{5}\right| \geq 1.2$
14. $\left|\dfrac{x + 5}{3}\right| < 2$
15. $\left|\dfrac{k}{3} + 2\right| \leq 2$
16. $|2b + 3| > 5$
17. $|2 + 3d| \geq 4$
18. $|2m - 1| > 2$

In 19–24, select the numeral preceding the expression or the diagram that best completes the sentence or answers the question.

19. Which is the solution set of $|2x - 3| < 5$?
    (1) $\{x \,|-4 < x < 1\}$ (2) $\{x \,| x < -4 \text{ or } x > 1\}$
    (3) $\{x \,|-1 < x < 4\}$ (4) $\{x \,| x < -1 \text{ or } x > 4\}$

20. Which represents the solution set for y in $|8 + y| > 3$?
    (1) $\{y \,|(y < -11) \lor (y > -5)\}$ (2) $\{y \,|-11 < y < -5\}$
    (3) $\{y \,|(y < -5) \lor (y > 5)\}$ (4) $\{y \,|-5 < y < 5\}$

21. If $|2 - x| > 8$, the solution set is:
    (1) $\{x \,|-6 < x < 10\}$ (2) $\{x \,| x > -6 \text{ or } x < 10\}$
    (3) $\{x \,|-10 < x < 6\}$ (4) $\{x \,| x < -6 \text{ or } x > 10\}$

22. Which of the given inequalities has ϕ as its solution set?
    (1) $|x| > 2$ (2) $|x| < 2$ (3) $|x| > -2$ (4) $|x| < -2$

23. Which is the graph of the solution set of $|x - 3| > 4$?

    (1)
        -1 7

    (2)
        -1 7

    (3)
        -1 7

    (4)
        -1 7

24. Which of the following inequalities has a solution set represented by the graph at the right?

-4 -2 0 2 4 6 8

    (1) $|x - 4| < 2$ (2) $|x - 2| < 4$
    (3) $|x - 4| > 2$ (4) $|x - 2| > 4$

4-5 ROOTS AND RADICALS

Square Root

To solve a quadratic equation, we often work with real numbers that have a *radical sign*, $\sqrt{}$. Consider the three examples that follow.

☐ EXAMPLE 1:

  Solve $x^2 = 25$
$$x = \pm\sqrt{25}$$
$$x = \sqrt{25} \text{ or } -\sqrt{25}$$
$$x = 5 \text{ or } -5$$
Solution set $= \{5, -5\}$

☐ EXAMPLE 2:

  Solve $y^2 = 3$
$$y = \pm\sqrt{3}$$
$$y = \sqrt{3} \text{ or } -\sqrt{3}$$
Solution set $= \{\sqrt{3}, -\sqrt{3}\}$

☐ EXAMPLE 3:

If the area of a square is 3, find the length of one side of the square.

y ☐
y

Let y = the length of one side.
Since side^2 = area:

$$y^2 = 3$$
$$y = \pm\sqrt{3}$$

(A length is positive only.)

Solution set = $\{\sqrt{3}\}$

■ **DEFINITION.** A *square root* of a number is one of two equal factors.

We say that the square root of a number k is x if and only if $x \cdot x = k$, or $x^2 = k$.

In example 1 above, 5 is a square root of 25 because $(5)(5) = 25$. In the same way, -5 is a square root of 25 because $(-5)(-5) = 25$. Thus, 25 has two square roots, written $\pm\sqrt{25}$ or ±5. These square roots are *rational* numbers.

In example 2, $\sqrt{3}$ is a square root of 3 because $(\sqrt{3})(\sqrt{3}) = 3$, and $-\sqrt{3}$ is a square root of 3 because $(-\sqrt{3})(-\sqrt{3}) = 3$. Thus, 3 has two square roots, written $\pm\sqrt{3}$ or $\sqrt{3}$ or $-\sqrt{3}$. These square roots are *irrational* numbers.

In certain situations, such as the geometric problem in example 3, we restrict the solution set to include only the positive square root. Notice, however, that we determined the solution set by solving a quadratic equation and finding both square roots of a number. All three examples shown above lead us to the following observation:

■ **Every positive real number has two square roots.**

If $k > 0$, the square roots of k are $\pm\sqrt{k}$, that is, $\sqrt{k}$ or $-\sqrt{k}$. To indicate that both square roots are to be found, we place a plus sign and a minus sign together in front of the radical sign. For example:

$$\pm\sqrt{49} = \pm7 \qquad \pm\sqrt{\tfrac{4}{9}} = \pm\tfrac{2}{3} \qquad \pm\sqrt{.16} = \pm.4$$

■ **DEFINITION.** The *principal square root* of a positive number k is its positive square root, $\sqrt{k}$.

Thus, a principal square root indicates only one root. For example:

$$\sqrt{49} = 7 \qquad \sqrt{\tfrac{4}{9}} = \tfrac{2}{3} \qquad \sqrt{.16} = .4$$

To indicate that only the *negative square root* of a number is to be found, we place a minus sign in front of the radical sign. For example:

$$-\sqrt{49} = -7 \qquad -\sqrt{\tfrac{4}{9}} = -\tfrac{2}{3} \qquad -\sqrt{.16} = -.4$$

For numbers that are not positive real numbers, we observe:

1. **The square root of zero is zero,** that is, $\sqrt{0} = 0$. This is true because 0 is the only number whose square is 0.
2. **The square roots of a negative real number are not real numbers.** If $k < 0$, then $\pm\sqrt{k}$ are not real numbers. For example, $\sqrt{-4}$ is not a real number because there is no real number that, when squared, is -4. We will study these square roots in Chapter 14, The Complex Numbers.

Cube Root

■ **DEFINITION.**   A *cube root* of a number is one of three equal factors.

The cube root of a number k is x, written $\sqrt[3]{k} = x$, if and only if $x \cdot x \cdot x = k$, or $x^3 = k$. Since $(2)(2)(2) = 8$, we can say that 2 is a cube root of 8.

Every nonzero real number has three cube roots, one of which is a *real number* called the ***principal cube root*** of the number. For example, when we write $\sqrt[3]{8} = 2$, we are indicating that 2 is the principal cube root of 8. The other two cube roots of 8 are not real numbers; these roots will be studied in Chapter 14, The Complex Numbers.

Since $(-5)(-5)(-5) = -125$, we know that $\sqrt[3]{-125} = -5$. Notice that -5 is a real number. By the definition stated earlier, -5 is the principal cube root of -125. Let us consider other examples:

$$\sqrt[3]{64} = 4 \qquad \sqrt[3]{-64} = -4 \qquad \sqrt[3]{\tfrac{1}{27}} = \tfrac{1}{3} \qquad \sqrt[3]{-\tfrac{1}{27}} = -\tfrac{1}{3}$$

We observe that the principal cube root of a positive real number is a positive real, and the principal cube root of a negative real number is a negative real. Because 0 is the only number whose cube is 0, we know that $\sqrt[3]{0} = 0$.

The *n*th Root of a Number

■ **DEFINITION.**   The *nth root* of a number (where n is any counting number) is one of n equal factors. For example:

Since $(3)(3)(3)(3) = 81$, then 3 is a fourth root of 81, written $\sqrt[4]{81} = 3$.
Since $x \cdot x \cdot x \cdot x \cdot x = x^5$, then x is a fifth root of x^5, written $\sqrt[5]{x^5} = x$.

The *principal nth root* of a number k is written as:

$$\sqrt[n]{k}$$

where: k = the *radicand*
  n = the *index*, a counting number that indicates the root to be taken
  $\sqrt[n]{k}$ = the *radical*, or principal nth root of k

For example, $\sqrt[5]{32}$ is a radical whose radicand is 32 and whose index is 5. When the radical is a square root, as in $\sqrt{81}$, it is understood that the index is 2.

To determine the value of $\sqrt[n]{k}$, the principal nth root of k, we observe:

1. **If n is an odd counting number**, then $\sqrt[n]{k}$ = the one real number r such that $r^n = k$. For example:

$$\sqrt[3]{125} = 5 \qquad \sqrt[5]{-32} = -2 \qquad \sqrt[7]{1} = 1 \qquad \sqrt[9]{-1} = -1 \qquad \sqrt[9]{0} = 0$$

2. **If n is an even counting number and k is a non-negative number**, then $\sqrt[n]{k}$ = the non-negative real number r such that $r^n = k$. For example:

$$\sqrt{100} \ (\text{or} \ \sqrt[2]{100}) = 10 \qquad \sqrt[4]{625} = 5 \qquad \sqrt[6]{1} = 1 \qquad \sqrt[8]{0} = 0$$

(*Note:* If n is an even counting number and k is a negative number, as in $\sqrt{-9}$ and $\sqrt[4]{-1}$, then $\sqrt[n]{k}$ is not defined in the set of real numbers. These roots will be defined in Chapter 14, The Complex Numbers.)

| MODEL PROBLEMS |

1. Find the square root of 144.

   Solution: There are two square roots of 144, written $\pm\sqrt{144}$. Since $(12)(12) = 144$, and $(-12)(-12) = 144$, then $\pm\sqrt{144} = \pm12$.

   Answer: +12 or −12   OR ±12

2. Evaluate the given expression by finding the indicated root. Let each variable represent a positive number.

   a. $\sqrt{144}$ b. $\sqrt{100x^4}$ c. $\sqrt[3]{-8y^3}$ d. $-\sqrt[5]{-1}$

Solution	*Answer*
a. Since $\sqrt{144}$ indicates the principal square root of 144, then $\sqrt{144} = 12$.	**a.** $\sqrt{144} = 12$
b. Since $(10x^2)(10x^2) = 100x^4$, then $\sqrt{100x^4} = 10x^2$.	**b.** $\sqrt{100x^4} = 10x^2$
c. Since $(-2y)(-2y)(-2y) = -8y^3$, then $\sqrt[3]{-8y^3} = -2y$.	**c.** $\sqrt[3]{-8y^3} = -2y$
d. Here, $-\sqrt[5]{-1} = -(\sqrt[5]{-1}) = -(-1) = 1$.	**d.** $-\sqrt[5]{-1} = 1$

3. Find the *smallest integral* value of x for which $\sqrt{2x - 9}$ represents a real number.

Solution

1. A square-root radical is a real number if and only if the radicand is greater than or equal to zero. Thus, we write the inequality: $\qquad 2x - 9 \geq 0$
2. Simplify the inequality. $\qquad\qquad\qquad\qquad 2x \geq 9$

$$x \geq 4\tfrac{1}{2}$$
$$x = 5$$

3. Since $x \geq 4\tfrac{1}{2}$ is equivalent to the inequality $2x - 9 \geq 0$, select the smallest integer x so that $x \geq 4\tfrac{1}{2}$ is true.
4. *Check* the answer $(x = 5)$ and the next smallest integer $(x = 4)$. If $x = 5$, then $\sqrt{2x - 9} = \sqrt{2(5) - 9} = \sqrt{10 - 9} = \sqrt{1}$, a real number. If $x = 4$, then $\sqrt{2x - 9} = \sqrt{2(4) - 9} = \sqrt{8 - 9} = \sqrt{-1}$, *not* a real number.

Answer: 5

EXERCISES

In 1-40, evaluate the given expression by finding the indicated root(s). Let each variable represent a positive number.

1. $\sqrt{9}$	2. $\pm\sqrt{9}$	3. $-\sqrt{9}$	4. $\sqrt{225}$
5. $-\sqrt{121}$	6. $\sqrt{n^2}$	7. $\sqrt{.25}$	8. $\pm\sqrt{.04}$
9. $\sqrt{64}$	10. $\sqrt[3]{64}$	11. $\sqrt[6]{64}$	12. $-\sqrt{400}$
13. $\sqrt{81x^2}$	14. $\sqrt{4y^4}$	15. $-\sqrt{b^6}$	16. $\sqrt{16x^{16}}$
17. $\pm\sqrt{\dfrac{4}{49}}$	18. $-\sqrt{\dfrac{25}{9}}$	19. $\pm\sqrt{\dfrac{a^2}{36}}$	20. $\sqrt{\dfrac{m^8}{100}}$
21. $\sqrt[3]{27}$	22. $\sqrt[3]{-27}$	23. $-\sqrt[3]{27}$	24. $-\sqrt[3]{-27}$

25. $\sqrt[3]{1000}$ **26.** $\sqrt[3]{k^3}$ **27.** $\sqrt[3]{-125}$ **28.** $\sqrt[3]{216}$

29. $\sqrt[3]{-8y^{12}}$ **30.** $\sqrt[4]{81}$ **31.** $-\sqrt[4]{625}$ **32.** $\sqrt[4]{0}$

33. $\sqrt[4]{16x^{16}}$ **34.** $\sqrt[5]{-243}$ **35.** $\sqrt[5]{32y^5}$ **36.** $-\sqrt[5]{x^{10}}$

37. $\sqrt[7]{-1}$ **38.** $-\sqrt[9]{-1}$ **39.** $-\sqrt[3]{.008}$ **40.** $-\sqrt[7]{-x^7}$

In 41–52, state whether the number is rational or irrational.

41. $\sqrt{25}$ **42.** $\sqrt{8}$ **43.** $\sqrt[3]{8}$ **44.** $\sqrt[3]{-1}$

45. $\sqrt{27}$ **46.** $-\sqrt{196}$ **47.** $\sqrt{1.44}$ **48.** $\sqrt{.04}$

49. $\sqrt{.4}$ **50.** $\sqrt[3]{4}$ **51.** $\sqrt[3]{-343}$ **52.** $\sqrt[3]{.64}$

In 53–56, find the *smallest* value of x for which the radical represents a real number.

53. $\sqrt{x-8}$ **54.** $\sqrt{x+3}$ **55.** $\sqrt{2x-5}$ **56.** $\sqrt{3x-10}$

In 57–64, find the *smallest integral* value of x for which the radical represents a real number.

57. $\sqrt{2x-5}$ **58.** $\sqrt{3x-10}$ **59.** $\sqrt{5x}$ **60.** $\sqrt{4x-25}$

61. $\sqrt{2x+1}$ **62.** $\sqrt{4+3x}$ **63.** $\sqrt{\dfrac{2x-9}{9}}$ **64.** $\sqrt{\dfrac{2x+7}{5}}$

65. *True* or *False:* The principal square root of a positive number is positive.

66. *True* or *False:* The principal nth root of a nonzero number is positive.

67. *True* or *False:* If n is an even counting number, then $\sqrt[n]{k^n} = |k|$.

68. a. Solve for x: $5^2 + x^2 = 13^2$
    b. In a right triangle whose hypotenuse measures 13, one leg has a measure of 5. Find the measure of the other leg.
    c. In what ways are parts a and b alike?
    d. In what ways are parts a and b different?

69. a. Solve for n: $n^2 = 7$
    b. If the area of a square is 7, find the length of one of its sides.
    c. In what ways are parts a and b different?

70. a. Solve for x: $2x^2 = 4$
    b. If the diagonal of a square has a length of 2, find the length of one side of the square.
    c. In what ways are parts a and b different?
    d. Find the area of the square described in part b.

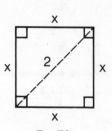

Ex. 70

71. a. Find the one real number y such that $y^3 = 1000$.
    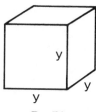
    b. If the volume of a cube is 1000 cubic centimeters (written cm^3), find the length of one edge of the cube.
    c. In what ways are parts a and b alike?
    d. If a cube whose volume is 1000 cm^3 is filled with water, the cube will hold 1 liter (or 1000 milliliters) of the liquid. How many milliliters are contained in 1 cm^3 of water?

    Ex. 71

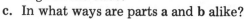

In 72–83, solve for the variable when the replacement set is the set of real numbers.

72. $x^2 = 16$ 73. $y^2 = 169$ 74. $w^2 = 5$
75. $3y^2 = 75$ 76. $k^2 = 75$ 77. $n^2 - 4 = 0$
78. $x^2 - 10 = 0$ 79. $x^2 + 1 = 0$ 80. $y^3 + 1 = 0$
81. $8b^3 = 1$ 82. $y^4 = 16$ 83. $n^4 - 81 = 0$

In 84 and 85, select the numeral preceding the expression that best answers the question.

84. Which radical is *not* a real number?
    (1) $-\sqrt{1}$ (2) $\sqrt{-1}$ (3) $-\sqrt[3]{1}$ (4) $\sqrt[3]{-1}$
85. Which of the following statements is *false*?
    (1) $-\sqrt[3]{(4)^3} = -4$ (2) $\sqrt[3]{(-4)^3} = -4$
    (3) $\sqrt[4]{(-3)^4} = -3$ (4) $\sqrt[4]{(-3)^4} = 3$

86. Let k represent any non-negative real number.
    a. For what real numbers is the sentence $k = \sqrt{k}$ true?
    b. For what real numbers is the sentence $k > \sqrt{k}$ true?
    c. For what real numbers is the sentence $k < \sqrt{k}$ true?

4-6 APPROXIMATING RATIONAL VALUES OF SQUARE ROOTS

To evaluate the square root of a number k where $k \geq 0$, we observe:

1. If k equals the square of some rational number, then $\sqrt{k}$ is a *rational* number. Square-root radicals that are rational numbers include:

$$\sqrt{25} = \sqrt{(5)^2} = 5 \qquad \sqrt{\tfrac{4}{9}} = \sqrt{(\tfrac{2}{3})^2} = \tfrac{2}{3} \qquad \sqrt{6.25} = \sqrt{(2.5)^2} = 2.5$$

2. If k is not equal to the square of some rational number and $k > 0$,

then $\sqrt{k}$ is an *irrational* number. Square-root radicals that are irrational numbers include:

$$\sqrt{2} \qquad \sqrt{40} \qquad \sqrt{\tfrac{1}{3}} \qquad \sqrt{.18}$$

In our daily lives, we often use *approximate rational values* for irrational numbers. For example, if we know mathematically that a piece of wood needed for a project is to be $\sqrt{2}$ meters long, we would probably ask the worker at the lumber yard for a piece of wood slightly longer than 1.4 meters. Although $\sqrt{2} \neq 1.4$ and $\sqrt{2} \neq 1.414$, the irrational number $\sqrt{2}$ is close enough to the rational number 1.414 so that we can state $\sqrt{2} \approx 1.414$, or $\sqrt{2}$ "is approximately equal to" 1.414.

Let us consider some methods for evaluating a square root. If the radical is an irrational number, each method will lead us to an approximate rational value. Since the equipment needed for methods 1 and 2 might not always be available, it is recommended that either method 3 or method 4 be learned in addition to the first two methods.

Method 1: The Calculator

After entering the radicand k and pressing the "square root" key in the proper sequence needed, we find the value (or  approximate rational value) of $\sqrt{k}$ by reading the number displayed on the calculator. For example, the approximate rational value of $\sqrt{5}$ is displayed on one calculator as shown above.

The number 2.236068 may be rounded off to any convenient number of decimal places less than the six places in the display, as shown below.

$\sqrt{5} \approx 2.236$ or $\sqrt{5} = 2.236$ to the nearest thousandth

$\sqrt{5} \approx 2.24$ or $\sqrt{5} = 2.24$ to the nearest hundredth

$\sqrt{5} \approx 2.2$ or $\sqrt{5} = 2.2$ to the nearest tenth

In general, to *round off* a number to n decimal places, we drop all digits to the right of that place. If, however, the digit in the $(n + 1)$ decimal place is *5 or more*, we add 1 to the digit in the n decimal place and drop all appropriate digits to obtain the rounded-off number.

Method 2: The Square-Root Table

The table on page 746 helps us to evaluate square roots.

■ *Procedure 1:* The table on page 746 lists the values of the principal square roots of integers from 1 through 150. Approximate rational values are given to the nearest thousandth. For example:

$$\sqrt{31} \approx 5.568 \quad \text{AND} \quad \sqrt{143} \approx 11.958$$

■*Procedure 2:* Study the portion of the table reproduced here at the right to see how we may determine the square roots of certain numbers greater than 150. Since $30^2 = 900$, we know that $\sqrt{900} = 30$. Thus, for every number in the "Square" column, its principal square root is found

No.	Square	Square Root
30	900	5.477
31	961	5.568
32	1,024	5.657

to the immediate left in the "No." column. Here, we also see:

$$\sqrt{961} = 31 \qquad \sqrt{1024} = 32$$

■*Procedure 3:* If a number is greater than 150 but is not a perfect square, we can find a quick approximation of the square root of the number. For example, to approximate $\sqrt{1000}$, consider the portion of the table reproduced above:

$$\text{Since } 961 < 1000 < 1024,$$
$$\text{then } \sqrt{961} < \sqrt{1000} < \sqrt{1024},$$
$$\text{or } 31 < \sqrt{1000} < 32$$

Thus, $\sqrt{1000}$ lies between 31 and 32. In methods 3 and 4, which follow, we will learn how to find more precise approximations.

Method 3: Divide and Average

■*Observation 1:* If the divisor of a number and the resulting quotient are equal, then the square root of the number is either the divisor or the quotient. For example, 144 divided by 12 is 12. Thus, $\sqrt{144} = 12$.

$$\textit{Division: } 12\overline{)144}^{\,12} \qquad \textit{Square Root: } \sqrt{144} = 12$$

■*Observation 2:* If the divisor of a number and the resulting quotient are unequal, then the square root of the number lies somewhere between the divisor and the quotient. For example:

$$\textit{Division: } 9\overline{)144}^{\,16} \text{ indicates that } \sqrt{144} \text{ lies between 9 and 16.}$$

The average of the divisor and the quotient is an approximation of the square root of the number. By repeating this process of "divide and average," where the average is used as the new divisor, we will find a better approximation of the square root. For example:

Divide and Average	*Divide and Average*
$9\overline{)144}^{\,16} \rightarrow \dfrac{16+9}{2} = \dfrac{25}{2} = 12.5$	$12.5\overline{)144.00}^{\,11.5} \rightarrow \dfrac{11.5+12.5}{2} = \dfrac{24}{2} = 12$

This process may be repeated as many times as needed. With each step, the quotient may be found to one more decimal place than there is in the divisor. Study the model problem that follows.

| MODEL PROBLEM |

By the "divide and average" method, find $\sqrt{1000}$ to the nearest tenth.

How to Proceed

1. Select a number as an approximation of $\sqrt{1000}$.

2. Divide 1000 by 31, finding the quotient to one decimal place (here, 32.2). Then, find the average of the quotient 32.2 and the divisor 31.

3. Divide 1000 by 31.6, the average just found. The quotient is now found to two decimal places (here, 31.64). Then, find the average of the quotient 31.64 and the divisor 31.6.

4. Round off the approximation to the nearest tenth.

Solution

Since $31 < \sqrt{1000} < 32$, select 31 as the approximation.

Divide   and   Average

$$31\overline{)1000.0}$$
$$\frac{32.2}{}$$
$$\underline{93}$$
$$70$$
$$\underline{62}$$
$$80$$
$$\underline{62}$$
$$18$$

$$\frac{32.2 + 31}{2}$$
$$= \frac{63.2}{2}$$
$$= 31.6$$

Divide   and   Average

$$31.6\overline{)1000.0\,00}$$
$$\frac{3\ 1.64}{}$$
$$\underline{948}$$
$$520$$
$$\underline{316}$$
$$2040$$
$$\underline{1896}$$
$$1440$$
$$\underline{1264}$$
$$176$$

$$\frac{31.64 + 31.6}{2}$$
$$= \frac{63.24}{2}$$
$$= 31.62$$

$\sqrt{1000} \approx 31.62$

$\sqrt{1000} = 31.6$ to the nearest tenth

Answer: $\sqrt{1000} = 31.6$ to the nearest tenth

Method 4: The Square-Root Algorithm

An *algorithm* is a process or a method used in calculation. An algorithm to find the square root of a number (or an approximate rational value for the square root) is described in the model problem that follows.

MODEL PROBLEM

Use the square-root algorithm to find $\sqrt{1000}$ to the nearest tenth.

How to Proceed　　　　　　　　*Solution*

1. Place a decimal point in the number and in the quotient. From the decimal point in the number, move to the left and mark off groups of two digits. For each group of two digits, there will be a single digit in the quotient. Thus, place two groups of "00" to the right of the decimal in the number so that the quotient will be in hundredths.

Step 1:

$$\sqrt{1000}$$

$$\sqrt{10\ 00.00\ 00}$$

2. Below the first group at the left, write the largest perfect square that is equal to or less than that group. Subtract the perfect square from the group. Write the square root of the perfect square above the first group in the quotient.

Step 2:

$$\begin{array}{r} 3 \\ \sqrt{10\ 00.00\ 00} \\ \underline{9} \\ 1 \end{array}$$

3. Bring down the next group of two digits and annex this group to the remainder. Form a divisor by doubling the root in the quotient found to this point and annexing a "_" to represent a digit. Disregard decimal points in writing the divisor. (Here, doubling the root 3 gives 6. Thus, "6_" is a divisor in the "sixties," and "_" below it represents a single digit.)

Step 3:

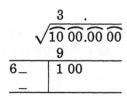

4. Replace each "_" with the same digit so that the product of the divisor and the digit is equal to or less than the remainder. Write the product under the remainder and subtract. Write the digit that replaces "_" above the next group of two digits in the quotient. (Here, 61 · 1 = 61, and 100 − 61 = 39. Note that 62 · 2 is a product larger than the remainder 100. Place the digit 1 in the quotient so that the root is now 31.)

Step 4:

$$
\begin{array}{r}
3\ \ 1. \\
\sqrt{10\ 00.00\ 00} \\
\end{array}
$$

	9
61	1 00
1	61
	39

5. Repeat Step 3. (Here, doubling the root 31 gives 62. Thus, "62_" is a three-place number.)

Step 5:

$$
\begin{array}{r}
3\ \ 1. \\
\sqrt{10\ 00.00\ 00} \\
\end{array}
$$

	9
61	1 00
1	61
62_	39 00
_	

6. Repeat Step 4. (Here, 626 · 6 = 3756, and 3900 − 3756 = 144. Place the digit 6 in the quotient; the root is now 31.6.)

Step 6:

$$
\begin{array}{r}
3\ \ 1.\ 6 \\
\sqrt{10\ 00.00\ 00} \\
\end{array}
$$

	9
61	1 00
1	61
626	39 00
6	37 56
	1 44

7. Repeat Step 3. (Here, doubling the root 316 gives 632. Thus, "632_" is a four-place number.)

Step 7:

$$
\begin{array}{r}
3\ \ 1.\ 6 \\
\sqrt{10\ 00.00\ 00} \\
\end{array}
$$

	9
61	1 00
1	61
626	39 00
6	37 56
632_	1 44 00
_	

8. Repeat Step 4. (Here, $6322 \cdot 2 =$ 12644, and $14400 - 12644 = 1756$. The digit 2 in the quotient makes the root 31.62.)

Step 8:

$$
\begin{array}{r}
3\ 1.\ 6\ 2 \\
\sqrt{10\ 00.00\ 00} \\
9 \\
\hline
\end{array}
$$

61	1 00
1	61
626	39 00
6	37 56
6322	1 44 00
2	1 26 44
	17 56

9. Round off the root to the nearest tenth.

Step 9:

$\sqrt{1000} \approx 31.62$

$\sqrt{1000} = 31.6$ to the nearest tenth

Answer: $\sqrt{1000} = 31.6$ to the nearest tenth

(*Note:* In the square-root algorithm, Steps 3 and 4 are repeated over and over again. However, if an exact square root can be found, as in $\sqrt{6.25} = 2.5$, a remainder of zero will be reached in the process.)

| EXERCISES |

In 1–10, use the table on page 746 to find the principal square root of the given number. If necessary, express the square root correct to the nearest thousandth.

1. 81	**2.** 121	**3.** 54	**4.** 130	**5.** 109
6. 196	**7.** 149	**8.** 784	**9.** 1849	**10.** 7744

In 11–20, use the table on page 746 to find the principal square root of the given number correct to the nearest tenth.

11. 11	**12.** 38	**13.** 57	**14.** 65	**15.** 13
16. 99	**17.** 102	**18.** 82	**19.** 120	**20.** 143

In 21–24, use the table on page 746 to find two consecutive positive integers such that the given radical lies between these integers.

21. $\sqrt{300}$ **22.** $\sqrt{1200}$ **23.** $\sqrt{9300}$ **24.** $\sqrt{13,706}$

In 25–44, the principal square root of the given number is an integer, or a rational number terminating in one decimal place. Find the principal square root by: (a) the square-root algorithm (b) the "divide-and-average" method.

25. 2601	**26.** 6561	**27.** 1089	**28.** 576
29. 2809	**30.** 361	**31.** 441	**32.** 1681
33. 53.29	**34.** 6.76	**35.** 46.24	**36.** 1162.81
37. 26,244	**38.** 19,881	**39.** 24,025	**40.** 372.49
41. 54,756	**42.** 112.36	**43.** 432.64	**44.** 94,249

In 45–60, find the principal square root of the given number to the nearest tenth by: (a) the square-root algorithm (b) the "divide-and-average" method.

45. .18	**46.** .32	**47.** .88	**48.** .75
49. 27	**50.** 2.65	**51.** 70	**52.** 41.56
53. 3.21	**54.** 70.5	**55.** 39.8	**56.** 2.48
57. 180	**58.** 500	**59.** 222	**60.** 345

4-7 SIMPLIFYING A RADICAL

Square-Root Radicals With Integral Radicands

Since $\sqrt{9 \cdot 16} = \sqrt{144} = 12$ and $\sqrt{9} \cdot \sqrt{16} = 3 \cdot 4 = 12$, it can be said that $\sqrt{9 \cdot 16} = \sqrt{9} \cdot \sqrt{16}$. This example illustrates the following property:

■ **The square root of a product of positive numbers is equal to the product of the square roots of the numbers.**

In general, if a and b are positive numbers:

$$\sqrt{a \cdot b} = \sqrt{a} \cdot \sqrt{b}$$

If the radicand of a square-root radical is a positive integer and if the radicand contains a factor that is a perfect square, then we can use the given rule to find one or more equivalent radicals. For example:

$$\sqrt{200} = \sqrt{4 \cdot 50} = \sqrt{4} \cdot \sqrt{50} = 2\sqrt{50}$$
$$\sqrt{200} = \sqrt{25 \cdot 8} = \sqrt{25} \cdot \sqrt{8} = 5\sqrt{8}$$
$$\sqrt{200} = \sqrt{100 \cdot 2} = \sqrt{100} \cdot \sqrt{2} = 10\sqrt{2}$$

■ **DEFINITION.** The *simplest form of a square-root radical* is a monomial of the form $k\sqrt{r}$ where k is a nonzero rational number and the radicand r is a positive integer containing no perfect-square factors other than 1.

In other words, the greatest perfect square has been factored out of the radicand. The greatest perfect-square factor of 200 is 100. Since $\sqrt{200} = \sqrt{100} \cdot \sqrt{2} = 10\sqrt{2}$ and since the radicand 2 contains no perfect-square factors other than 1, we can state:

$$\sqrt{200} \text{ in its simplest form is } 10\sqrt{2}$$

(*Note:* In $2\sqrt{50}$, the radicand 50 contains a perfect-square factor of 25. In $5\sqrt{8}$, the radicand 8 contains a perfect-square factor of 4. For these reasons, neither $2\sqrt{50}$ nor $5\sqrt{8}$ is the simplest form of $\sqrt{200}$.)

Square-Root Radicals With Fractional Radicands

Since $\sqrt{\dfrac{9}{16}} = \dfrac{3}{4}$ and $\dfrac{\sqrt{9}}{\sqrt{16}} = \dfrac{3}{4}$, then $\sqrt{\dfrac{9}{16}} = \dfrac{\sqrt{9}}{\sqrt{16}}$. This example illustrates the following property:

■ **The square root of a quotient of positive numbers is equal to the quotient of the square roots of the numbers.**

In general, if a and b are positive numbers: $\sqrt{\dfrac{a}{b}} = \dfrac{\sqrt{a}}{\sqrt{b}}$

This rule is used to simplify a square-root radical in which the radicand is a fraction. Remember that the simplest form of a square-root radical is $k\sqrt{r}$ where k is rational and the radicand r is a positive *integer*. Let us consider two cases.

Case I. If the denominator of the fractional radicand is a perfect square, apply the rule directly. For example:

$$\textit{In simplest form:} \quad \sqrt{\dfrac{5}{16}} = \dfrac{\sqrt{5}}{\sqrt{16}} = \dfrac{\sqrt{5}}{4} \quad \textit{Ans.}$$

Since $\dfrac{\sqrt{5}}{4} = \dfrac{1}{4} \cdot \sqrt{5} = \dfrac{1}{4}\sqrt{5}$, we can write the simplest form of $\sqrt{\dfrac{5}{16}}$ in two ways, namely, $\dfrac{\sqrt{5}}{4}$ or $\dfrac{1}{4}\sqrt{5}$. Compare each monomial to the general form $k\sqrt{r}$ to see that $k = \frac{1}{4}$, a rational number, and the radicand $r = 5$, a positive *integer*.

Case II. If the denominator of the fractional radicand is *not* a perfect square, multiply the fraction by some form of the identity element 1 to find an equivalent fraction whose denominator is a perfect square. Then, apply the given rule. For example:

In simplest form: $\sqrt{\dfrac{2}{3}} = \sqrt{\dfrac{2}{3} \cdot \dfrac{3}{3}} = \sqrt{\dfrac{6}{9}} = \dfrac{\sqrt{6}}{\sqrt{9}} = \dfrac{\sqrt{6}}{3}$ *Ans.*

Here, observe that $\frac{2}{3}$ is multiplied by $\frac{3}{3}$ (a form of the identity element 1) to transform $\frac{2}{3}$ to the equivalent fraction $\frac{6}{9}$. The rule is applied at this point, showing us that the simplest form of $\sqrt{\dfrac{2}{3}}$ is $\dfrac{\sqrt{6}}{3}$, or $\dfrac{1}{3}\sqrt{6}$. When this expression is compared to the general form $k\sqrt{r}$, we see that $k = \frac{1}{3}$, a rational number, and the radicand $r = 6$, a positive *integer*.

(*Note:* It is true that $\sqrt{\dfrac{2}{3}} = \dfrac{\sqrt{2}}{\sqrt{3}}$, but the fraction $\dfrac{\sqrt{2}}{\sqrt{3}}$ does not fit the general form of a radical in simplest form, namely, $k\sqrt{r}$. Later in this chapter, we will study a method for handling fractions such as $\dfrac{\sqrt{2}}{\sqrt{3}}$. For now, let us follow the procedures outlined in Case II.)

Radicals of Index *n*

The rules we have just learned for square-root radicals (or radicals of index 2) can be extended to radicals having another index.

In general, if a and b are positive numbers and the index n is a counting number, it can be shown that:

$$\sqrt[n]{a \cdot b} = \sqrt[n]{a} \cdot \sqrt[n]{b} \qquad \text{AND} \qquad \sqrt[n]{\dfrac{a}{b}} = \dfrac{\sqrt[n]{a}}{\sqrt[n]{b}}$$

Just as we found perfect-square factors to simplify radicals of index 2, we will find perfect-cube factors to simplify radicals of index 3, and so on. The *simplest form of a radical* whose index is a counting number n is now written as $k\sqrt[n]{r}$. Here, k is a nonzero rational number and the radicand r is a positive integer containing no factor other than 1 that is a perfect nth power. For example:

1. *In simplest form:* $\sqrt[3]{40} = \sqrt[3]{8 \cdot 5} = \sqrt[3]{8} \cdot \sqrt[3]{5} = 2\sqrt[3]{5}$ *Ans.*

2. *In simplest form:* $\sqrt[4]{\dfrac{9}{16}} = \dfrac{\sqrt[4]{9}}{\sqrt[4]{16}} = \dfrac{\sqrt[4]{9}}{2}$ or $\dfrac{1}{2}\sqrt[4]{9}$ *Ans.*

3. *In simplest form:* $\sqrt[3]{\dfrac{9}{16}} = \sqrt[3]{\dfrac{9}{16} \cdot \dfrac{4}{4}} = \sqrt[3]{\dfrac{36}{64}} = \dfrac{\sqrt[3]{36}}{\sqrt[3]{64}} = \dfrac{\sqrt[3]{36}}{4}$

Ans.

| MODEL PROBLEMS |

1. Simplify the radicals: **a.** $\sqrt{48}$ **b.** $\sqrt[3]{48}$

How to Proceed	*Solution*	*Solution*
1. Factor the radicand so that the perfect power is a factor.	**a.** $\sqrt{48} = \sqrt{16 \cdot 3}$	**b.** $\sqrt[3]{48} = \sqrt[3]{8 \cdot 6}$
2. Express the radical as the product of the roots of the factors.	$\sqrt{48} = \sqrt{16} \cdot \sqrt{3}$	$\sqrt[3]{48} = \sqrt[3]{8} \cdot \sqrt[3]{6}$
3. Simplify the radical containing the largest perfect power.	$\sqrt{48} = 4\sqrt{3}$ *Answer:* $4\sqrt{3}$	$\sqrt[3]{48} = 2\sqrt[3]{6}$ *Answer:* $2\sqrt[3]{6}$

2. Simplify the radical expressions. (All variables represent positive numbers.)

   a. $2\sqrt{98}$ **b.** $\frac{1}{4}\sqrt{96x^2}$ **c.** $\sqrt{45a^4b^3}$ **d.** $\sqrt[3]{n^7}$

   Solution: Use the procedure outlined in model problem 1. If necessary, simplify terms outside the radical as the fourth step.

   a. $2\sqrt{98} = 2\sqrt{49 \cdot 2} = 2\sqrt{49} \cdot \sqrt{2} = 2 \cdot 7\sqrt{2} = 14\sqrt{2}$ *Ans.*

   b. $\frac{1}{4}\sqrt{96x^2} = \frac{1}{4}\sqrt{16x^2 \cdot 6} = \frac{1}{4}\sqrt{16x^2} \cdot \sqrt{6} = \frac{1}{4} \cdot 4x\sqrt{6} = x\sqrt{6}$
   Ans.

   c. $\sqrt{45a^4b^3} = \sqrt{9a^4b^2 \cdot 5b} = \sqrt{9a^4b^2} \cdot \sqrt{5b} = 3a^2b\sqrt{5b}$ *Ans.*

   d. $\sqrt[3]{n^7} = \sqrt[3]{n^6 \cdot n} = \sqrt[3]{n^6} \cdot \sqrt[3]{n} = n^2\sqrt[3]{n}$ *Ans.*

3. Express in simplest form: a. $\sqrt{\frac{1}{12}}$ b. $\sqrt[3]{\frac{3}{4}}$

How to Proceed	*Solution*	*Solution*
1. Change the radicand to an equivalent fraction whose denominator is a perfect power.	a. $\sqrt{\frac{1}{12}} = \sqrt{\frac{1}{12} \cdot \frac{3}{3}}$ $= \sqrt{\frac{3}{36}}$	b. $\sqrt[3]{\frac{3}{4}} = \sqrt[3]{\frac{3}{4} \cdot \frac{2}{2}}$ $= \sqrt[3]{\frac{6}{8}}$
2. Express the radical as the quotient of two roots.	$= \dfrac{\sqrt{3}}{\sqrt{36}}$	$= \dfrac{\sqrt[3]{6}}{\sqrt[3]{8}}$
3. Simplify the radical in the denominator.	$= \dfrac{\sqrt{3}}{6}$	$= \dfrac{\sqrt[3]{6}}{2}$

Answer: a. $\dfrac{\sqrt{3}}{6}$ or $\dfrac{1}{6}\sqrt{3}$ b. $\dfrac{\sqrt[3]{6}}{2}$ or $\dfrac{1}{2}\sqrt[3]{6}$

4. Simplify the radicals: a. $\sqrt{\frac{8}{25}}$ b. $\sqrt[3]{\frac{8}{25}}$

Solution: Use the procedure outlined in model problem 3. However, the radical remaining in the numerator may sometimes be simplified, as seen in these problems.

a. $\sqrt{\dfrac{8}{25}} = \dfrac{\sqrt{8}}{\sqrt{25}} = \dfrac{\sqrt{4 \cdot 2}}{\sqrt{25}} = \dfrac{\sqrt{4} \cdot \sqrt{2}}{\sqrt{25}} = \dfrac{2\sqrt{2}}{5}$ or $\dfrac{2}{5}\sqrt{2}$ *Ans.*

b. $\sqrt[3]{\dfrac{8}{25}} = \sqrt[3]{\dfrac{8}{25} \cdot \dfrac{5}{5}} = \sqrt[3]{\dfrac{8 \cdot 5}{125}} = \dfrac{\sqrt[3]{8 \cdot 5}}{\sqrt[3]{125}} = \dfrac{\sqrt[3]{8} \cdot \sqrt[3]{5}}{\sqrt[3]{125}} = \dfrac{2\sqrt[3]{5}}{5}$

Ans.

EXERCISES

In 1–28, simplify the radical expressions. (All variables represent positive numbers.)

1. $\sqrt{75}$	2. $\sqrt{32}$	3. $\sqrt{45}$	4. $-\sqrt{300}$
5. $\sqrt{24}$	6. $\sqrt[3]{24}$	7. $\sqrt{54}$	8. $\sqrt[3]{54}$
9. $2\sqrt{50}$	10. $-3\sqrt{28}$	11. $5\sqrt{12}$	12. $3\sqrt[3]{56}$
13. $\frac{1}{2}\sqrt{8}$	14. $\frac{2}{5}\sqrt{500}$	15. $-\frac{2}{3}\sqrt{27}$	16. $\frac{5}{8}\sqrt{80}$
17. $\sqrt{49x}$	18. $\sqrt{a^2 b}$	19. $\sqrt{n^3}$	20. $\sqrt{4cd^2}$
21. $\sqrt{9y^9}$	22. $7\sqrt{40x^2}$	23. $\sqrt{200w}$	24. $\sqrt{18a^3 b^5}$
25. $-2\sqrt{63k^4}$	26. $\frac{3}{2}\sqrt{20y}$	27. $\sqrt[3]{16x^6}$	28. $\frac{4}{5}\sqrt[3]{125y^5}$

In 29-48, write the expression in simplest form. (All variables represent positive numbers.)

29. $\sqrt{\frac{2}{9}}$

30. $12\sqrt{\frac{5}{36}}$

31. $4\sqrt{\frac{17}{64}}$

32. $\sqrt{\frac{18}{49}}$

33. $\sqrt{\frac{1}{5}}$

34. $6\sqrt{\frac{1}{2}}$

35. $-\sqrt{\frac{2}{3}}$

36. $\sqrt{\frac{3}{50}}$

37. $8\sqrt{\frac{5}{32}}$

38. $-\sqrt{\frac{7}{27}}$

39. $\sqrt{\frac{4}{5}}$

40. $4\sqrt{\frac{9}{8}}$

41. $\sqrt[3]{\frac{4}{27}}$

42. $4\sqrt[3]{\frac{7}{64}}$

43. $\sqrt[3]{\frac{4}{9}}$

44. $\sqrt[3]{\frac{8}{25}}$

45. $\sqrt{\frac{x}{y^2}}$

46. $\sqrt{\frac{a^4}{b}}$

47. $\sqrt{\frac{w^2}{7}}$

48. $\sqrt{\frac{p}{q}}$

In 49-60: **a.** Write the expression in simplest form. **b.** Find the approximate rational value of the expression to the nearest tenth by using $\sqrt{2} \approx 1.414$, $\sqrt{3} \approx 1.732$, or $\sqrt{5} \approx 2.236$.

49. $\sqrt{98}$

50. $\sqrt{20}$

51. $\sqrt{108}$

52. $\sqrt{180}$

53. $\sqrt{147}$

54. $\frac{1}{3}\sqrt{72}$

55. $2\sqrt{125}$

56. $\sqrt{\frac{2}{49}}$

57. $\sqrt{\frac{3}{4}}$

58. $\sqrt{\frac{1}{3}}$

59. $\sqrt{\frac{1}{2}}$

60. $\sqrt{\frac{4}{5}}$

In 61-63, select the numeral preceding the expression that best completes the sentence or answers the question.

61. Which radical expression is *not* equivalent to $\sqrt{800}$?
    (1) $10\sqrt{8}$ (2) $4\sqrt{50}$ (3) $5\sqrt{16}$ (4) $20\sqrt{2}$

62. Which radical expression is *not* equivalent to $\sqrt[3]{1000}$?
    (1) 10 (2) $5\sqrt[3]{8}$ (3) $2\sqrt[3]{125}$ (4) $4\sqrt[3]{25}$

63. If the sides of a triangle measure $6\sqrt{6}$, $3\sqrt{24}$, and $2\sqrt{54}$, then the triangle is *best* described as being:
    (1) isosceles   (2) scalene
    (3) equilateral   (4) right

4-8 ADDING AND SUBTRACTING RADICALS

Like radicals are radicals that have the same index and the same radicand. For example, $3\sqrt{7}$ and $2\sqrt{7}$ are like radicals. Also, $5\sqrt[3]{2}$ and $\sqrt[3]{2}$ are like radicals.

Recall that addition and subtraction of like monomial terms is based on the distributive property. For example, $3y + 2y = (3 + 2)y = 5y$. In the same way, we use the distributive property to add and subtract like radicals. Consider the examples on the next page.

☐ Example 1:

Find the perimeter of a rectangle whose sides measure $3\sqrt{7}$ and $2\sqrt{7}$.

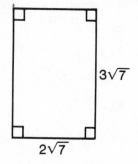

$3\sqrt{7}$

$2\sqrt{7}$

Addition:

$$3\sqrt{7} + 2\sqrt{7} + 3\sqrt{7} + 2\sqrt{7}$$
$$= (3 + 2 + 3 + 2)\sqrt{7}$$
$$= 10\sqrt{7} \quad Ans.$$

☐ Example 2:

Subtract $\sqrt[3]{2}$ from $5\sqrt[3]{2}$.

Subtraction:

$$5\sqrt[3]{2} - \sqrt[3]{2}$$
$$= 5\sqrt[3]{2} - 1\sqrt[3]{2}$$
$$= (5 - 1)\sqrt[3]{2}$$
$$= 4\sqrt[3]{2} \quad Ans.$$

Unlike radicals are radicals that do not have the same index, or the same radicand, or both. In many cases, the sum or the difference of unlike radicals cannot be stated as a single term. For example:

1. $\sqrt{5}$ and $\sqrt{3}$ are unlike radicals whose sum is $\sqrt{5} + \sqrt{3}$.

2. $\sqrt[3]{7}$ and $\sqrt{7}$ are unlike radicals whose difference is $\sqrt[3]{7} - \sqrt{7}$, in the order given.

But wait! In some cases, radicals that are unlike can be transformed into radicals that are like. When this happens, as seen in the following example, the sum or the difference of the radicals can be combined into a single term.

☐ Example 3: Simplify $\sqrt{48} + \sqrt{12} - \sqrt{3}$.

$$\sqrt{48} + \sqrt{12} - \sqrt{3} = \sqrt{16 \cdot 3} + \sqrt{4 \cdot 3} - \sqrt{3}$$
$$= \sqrt{16} \cdot \sqrt{3} + \sqrt{4} \cdot \sqrt{3} - \sqrt{3}$$
$$= 4\sqrt{3} + 2\sqrt{3} - 1\sqrt{3}$$
$$= (4 + 2 - 1)\sqrt{3}$$
$$= 5\sqrt{3} \quad Ans.$$

■ **PROCEDURE.**   To add (or subtract) radicals:

1. If necessary, *simplify* the radicals to be added or subtracted.

2. *Combine like radicals* by using the distributive property.

Since a radical sometimes represents an irrational number, let us make two general observations regarding sums (or differences) involving irrational numbers.

1. *The sum (or difference) of two irrational numbers may be either an irrational number or a rational number.*

□ EXAMPLE 1:   Add the irrational numbers $\sqrt{5}$ and $\sqrt{3}$.
The sum is $\sqrt{5} + \sqrt{3}$, an irrational number.

□ EXAMPLE 2:. Add the irrational numbers $\sqrt{5}$ and $^-\sqrt{5}$.
The sum is $\sqrt{5} + (-\sqrt{5}) = 0$, a rational number.

2. *The sum (or difference) of an irrational number and a rational number is an irrational number.*

□ EXAMPLE 3:   Add $\sqrt{3}$ (irrational) and 8 (rational).
The sum is $\sqrt{3} + 8$, or $8 + \sqrt{3}$, an irrational number.

□ EXAMPLE 4:   Subtract 8 from $\sqrt{3}$.
The difference is $\sqrt{3} - 8$, an irrational number.

| MODEL PROBLEMS |

1. Express in simplest form: $\sqrt{27} + 6\sqrt{\frac{1}{3}} + \sqrt{50} - \sqrt{8}$

How to Proceed   *Solution*

1. Simplify each radical.

$$\sqrt{27} \quad + \quad 6\sqrt{\tfrac{1}{3}} \quad + \quad \sqrt{50} \quad - \sqrt{8}$$

$$= \sqrt{9 \cdot 3} \quad + 6\sqrt{\tfrac{1}{3} \cdot \tfrac{3}{3}} + \quad \sqrt{25 \cdot 2} \ - \sqrt{4 \cdot 2}$$

$$= \sqrt{9} \cdot \sqrt{3} + 6 \cdot \frac{\sqrt{3}}{\sqrt{9}} + \sqrt{25} \cdot \sqrt{2} - \sqrt{4} \cdot \sqrt{2}$$

$$= 3\sqrt{3} \quad + \overset{2}{\cancel{6}} \cdot \frac{\sqrt{3}}{\underset{1}{\cancel{3}}} + \quad 5\sqrt{2} \quad - 2\sqrt{2}$$

2. Combine like radicals by using the distributive property.

$$= 3\sqrt{3} + 2\sqrt{3} + 5\sqrt{2} - 2\sqrt{2}$$
$$= (3 + 2)\sqrt{3} + (5 - 2)\sqrt{2}$$
$$= 5\sqrt{3} + 3\sqrt{2} \quad Ans.$$

2. **a.** Find the sum of the irrational numbers: $2\sqrt{5}$ and $(9 - \sqrt{20})$
   b. Is the sum rational or irrational?

Solution

a. $2\sqrt{5} + (9 - \sqrt{20}) = 2\sqrt{5} + 9 - \sqrt{4 \cdot 5} = 2\sqrt{5} + 9 - 2\sqrt{5} = 9.$
b. The sum, 9, is a rational number.

Answer: **a.** 9   **b.** rational

EXERCISES

In 1–15, combine the radicals. (All variables represent positive numbers.)

1. $8\sqrt{3} + 9\sqrt{3}$ 2. $6\sqrt{7} - 4\sqrt{7}$ 3. $8\sqrt{6} - \sqrt{6}$

4. $\sqrt{72} + \sqrt{18}$ 5. $\sqrt{48} + \sqrt{75}$ 6. $\sqrt{63} - \sqrt{28}$

7. $\sqrt{180} - \sqrt{80}$ 8. $2\sqrt{8} - \sqrt{32}$ 9. $\sqrt{54} + 3\sqrt{24}$

10. $\sqrt{81x} + \sqrt{9x}$ 11. $\sqrt{45y} - \sqrt{20y}$ 12. $9\sqrt{x^3} - \sqrt{9x^3}$

13. $\dfrac{3\sqrt{7}}{7} - \sqrt{\dfrac{1}{7}}$ 14. $2\sqrt{\dfrac{1}{8}} + \sqrt{\dfrac{1}{2}}$ 15. $\sqrt{125} + \sqrt{\dfrac{1}{5}}$

In 16–25, write the expression in simplest form. (All variables represent positive numbers.)

16. $\sqrt{700} + 8\sqrt{7} - 3\sqrt{28}$ 17. $\sqrt{160} - \sqrt{40} + \sqrt{90}$

18. $\sqrt{50} - \sqrt{98} + \sqrt{128}$ 19. $\sqrt{192} - \sqrt{27} - \sqrt{108}$

20. $\sqrt{20} + \sqrt{5} + \sqrt{150} - \sqrt{96}$ 21. $\sqrt{125} + \sqrt{12} - \sqrt{45} + \sqrt{75}$

22. $\sqrt[3]{64x} - \sqrt[3]{8x} + \sqrt[3]{27x}$ 23. $\sqrt[3]{250y^3} - \sqrt[3]{16y^3}$

24. $\sqrt{\dfrac{1}{10}} + \sqrt{\dfrac{2}{5}} - \dfrac{1}{10}\sqrt{90}$ 25. $\sqrt{\dfrac{4}{3}} + \sqrt{\dfrac{1}{3}} + \sqrt{44}$

In 26–35, two irrational numbers are given. **a.** Find the sum of the numbers in simplest form. **b.** Is the sum rational or irrational?

26. $\sqrt{14} + 3\sqrt{14}$ 27. $3\sqrt{17} + (-\sqrt{17})$

28. $5\sqrt{7} + (-\sqrt{175})$ 29. $12\sqrt{3} + (-3\sqrt{12})$

30. $(3 + \sqrt{2}) + (9 - \sqrt{2})$

31. $6\sqrt{15} + (2 - 3\sqrt{60})$

32. $(8 - \sqrt{5}) + (3 + \sqrt{20})$

33. $(10 - 8\sqrt{6}) + \sqrt{216}$

34. $2\sqrt{13} + (\sqrt{256} - \sqrt{52})$

35. $(\sqrt{\frac{4}{9}} - \sqrt{13}) + \frac{1}{3}\sqrt{117}$

In 36 and 37, select the numeral preceding the expression that best answers the question.

36. Which expression is equivalent to the sum of $\sqrt{600}$ and $\sqrt{400}$?

(1) $10\sqrt{10}$ (2) $10\sqrt{6} + 20$ (3) $12\sqrt{6}$ (4) $30\sqrt{6}$

37. Which expression is equivalent to $\sqrt{\frac{3}{2}} + \sqrt{\frac{2}{3}} + \sqrt{\frac{1}{6}}$?

(1) $\sqrt{6}$ (2) $\dfrac{\sqrt{3}}{2} + \dfrac{\sqrt{2}}{3} + \dfrac{\sqrt{6}}{6}$ (3) $\dfrac{\sqrt{66}}{11}$ (4) $\dfrac{\sqrt{21}}{3}$

In 38–41, express the perimeter of the figure: **(a)** in simplest radical form **(b)** as a rational number correct to the nearest tenth.

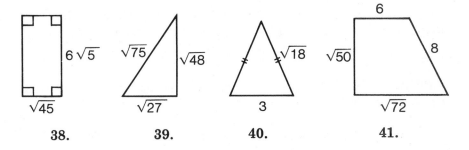

38.   39.   40.   41.

4-9 MULTIPLYING RADICALS WITH THE SAME INDEX

Monomial Terms

If a and b are positive numbers and the index n is a counting number, we have learned that $\sqrt[n]{a \cdot b} = \sqrt[n]{a} \cdot \sqrt[n]{b}$. By applying the symmetric property of equality to the given statement, we form a rule to find the product of two radicals with the same index, namely:

If a and b are positive numbers and the index n is a counting number, then:

$$\sqrt[n]{a} \cdot \sqrt[n]{b} = \sqrt[n]{a \cdot b}$$

For example, $\sqrt{2} \cdot \sqrt{7} = \sqrt{14}$, and $\sqrt[3]{5} \cdot \sqrt[3]{4} = \sqrt[3]{20}$. Let us extend this concept to the multiplication of terms containing radicals with coefficients.

□ Example : If the sides of a rectangle measure $5\sqrt{3}$ and $2\sqrt{6}$, find the area of the rectangle.

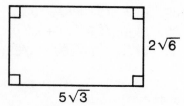

$2\sqrt{6}$

$5\sqrt{3}$

Solution: In a rectangle, area $A = lw$. Notice how the associative and commutative properties of multiplication are used to find the area.

$$A = lw = (5\sqrt{3})(2\sqrt{6}) = 5 \cdot (\sqrt{3} \cdot 2) \cdot \sqrt{6} = 5 \cdot (2 \cdot \sqrt{3}) \cdot \sqrt{6}$$
$$= (5 \cdot 2)(\sqrt{3} \cdot \sqrt{6}) = 10\sqrt{18}$$

The area contains a radical that can be simplified:

$$A = 10\sqrt{18} = 10\sqrt{9 \cdot 2} = 10\sqrt{9} \cdot \sqrt{2} = 10 \cdot 3\sqrt{2} = 30\sqrt{2}$$

Answer: $30\sqrt{2}$

From this example, we make the following generalization: If a and b are positive numbers and the index n is a counting number, then:

$$x \sqrt[n]{a} \cdot y \sqrt[n]{b} = xy \sqrt[n]{ab}$$

■ **PROCEDURE.** To multiply monomial terms containing radicals with the same index:

1. Multiply the coefficients to find the coefficient of the product.

2. Multiply the radicands to find the radicand of the product.

3. If possible, simplify the radical in the resulting product.

Polynomial Terms

By using the distributive property and the procedure outlined above, we can multiply polynomial expressions containing radical terms.

□ Example 1: *Multiplying a polynomial by a monomial.*

$$\sqrt{5}(3\sqrt{2} + \sqrt{7}) = \sqrt{5} \cdot 3\sqrt{2} + \sqrt{5} \cdot \sqrt{7}$$
$$= 3\sqrt{10} + \sqrt{35} \quad Ans.$$

□ Example 2: *Multiplying a polynomial by a polynomial.*

Three methods, similar to those learned in multiplying algebraic polynomials, are given here to find the product $(2 + \sqrt{3})(6 - \sqrt{3})$. Notice the use of the distributive property in each method.

Method 1

$$(2 + \sqrt{3})(6 - \sqrt{3})$$
$$= 2(6 - \sqrt{3}) + \sqrt{3}(6 - \sqrt{3})$$
$$= 2 \cdot 6 - 2 \cdot \sqrt{3} + \sqrt{3} \cdot 6 - \sqrt{3} \cdot \sqrt{3}$$
$$= 12 - 2\sqrt{3} + 6\sqrt{3} - 3$$
$$= 12 - 3 - 2\sqrt{3} + 6\sqrt{3}$$
$$= 9 + 4\sqrt{3} \quad Ans.$$

Method 2

$$6 - \sqrt{3}$$
$$2 + \sqrt{3}$$

$2(6 - \sqrt{3}) \rightarrow \overline{12 - 2\sqrt{3}}$

$\sqrt{3}(6 - \sqrt{3}) \rightarrow \underline{\quad +6\sqrt{3} - 3}$

$$12 + 4\sqrt{3} - 3 = 9 + 4\sqrt{3} \quad Ans.$$

Method 3

Here, we use the procedure learned to multiply binomials mentally. To state the answer in simplest terms, we must combine 12 and -3, as well as the like radicals $6\sqrt{3}$ and $-2\sqrt{3}$.

$$= 12 + 4\sqrt{3} - 3$$
$$= 9 + 4\sqrt{3} \quad Ans.$$

Since a radical sometimes represents an irrational number, let us make two general observations regarding products involving irrational numbers.

1. *The product of two irrational numbers may be either an irrational number or a rational number.*

For example, $\sqrt{3} \cdot \sqrt{5} = \sqrt{15}$ (irrational),
while $\sqrt{2} \cdot \sqrt{18} = \sqrt{36} = 6$ (rational).

2. *The product of an irrational number and any nonzero rational number is an irrational number.*

For example, the product of $\sqrt{15}$ and 6 is $6\sqrt{15}$ (irrational).

| MODEL PROBLEMS |

1. Multiply: **a.** $8\sqrt{5} \cdot \frac{1}{4}\sqrt{15}$ **b.** $4\sqrt[3]{x^2} \cdot \sqrt[3]{16x^4}$

How to Proceed	*Solution*	*Solution*
1. Multiply the coefficients, and multiply the radicands.	**a.** $8\sqrt{5} \cdot \frac{1}{4}\sqrt{15}$ $= (8 \cdot \frac{1}{4})(\sqrt{5} \cdot \sqrt{15})$ $= 2\sqrt{75}$	**b.** $4\sqrt[3]{x^2} \cdot \sqrt[3]{16x^4}$ $= (4 \cdot 1)(\sqrt[3]{x^2} \cdot \sqrt[3]{16x^4})$ $= 4\sqrt[3]{16x^6}$
2. Simplify the radical in the resulting product.	$= 2\sqrt{25 \cdot 3}$ $= 2\sqrt{25} \cdot \sqrt{3}$ $= 2 \cdot 5\sqrt{3}$ $= 10\sqrt{3}$ *Ans.*	$= 4\sqrt[3]{8x^6 \cdot 2}$ $= 4\sqrt[3]{8x^6} \cdot \sqrt[3]{2}$ $= 4 \cdot 2x^2\sqrt[3]{2}$ $= 8x^2\sqrt[3]{2}$ *Ans.*

2. Find the area of a square whose side measures $2\sqrt{5}$.

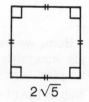

Solution

$$\text{Area} = (2\sqrt{5})^2 = 2\sqrt{5} \cdot 2\sqrt{5}$$
$$= (2 \cdot 2)(\sqrt{5} \cdot \sqrt{5})$$
$$= 4\sqrt{25}$$
$$= 4 \cdot 5$$
$$= 20 \quad Ans.$$

$2\sqrt{5}$

3. **a.** Multiply: $(4 + \sqrt{6})(4 - \sqrt{6})$
   b. Is the product rational or irrational?

Solution

a. $(4 + \sqrt{6})(4 - \sqrt{6})$
$= 4(4 - \sqrt{6}) + \sqrt{6}(4 - \sqrt{6})$
$= 4 \cdot 4 - 4 \cdot \sqrt{6} + \sqrt{6} \cdot 4 - \sqrt{6} \cdot \sqrt{6}$
$= 16 - 4\sqrt{6} + 4\sqrt{6} - 6$
$= 10 \quad Ans.$

b. The product 10 is rational. *Ans.*

| EXERCISES |

In 1-24, multiply and express the product in simplest form. (All variables represent positive numbers.)

1. $\sqrt{5} \cdot \sqrt{20}$ 2. $3\sqrt{32} \cdot \sqrt{2}$ 3. $4\sqrt{5} \cdot \sqrt{10}$

4. $\frac{1}{3}\sqrt{18} \cdot \sqrt{6}$ 5. $8\sqrt{3} \cdot \frac{1}{2}\sqrt{15}$ 6. $2\sqrt{14} \cdot 6\sqrt{7}$

7. $\sqrt{\frac{1}{2}} \cdot \sqrt{72}$ 8. $\sqrt{\frac{3}{4}} \cdot \sqrt{3}$ 9. $5\sqrt{\frac{1}{3}} \cdot \sqrt{18}$

10. $\sqrt{x} \cdot \sqrt{x^5}$ 11. $\sqrt{6y^3} \cdot \sqrt{2y}$ 12. $\sqrt{ab^3} \cdot \sqrt{ab^5}$

13. $\sqrt[3]{5} \cdot \sqrt[3]{25}$ 14. $2\sqrt[3]{2} \cdot 4\sqrt[3]{4}$ 15. $\sqrt[3]{9x^2} \cdot \sqrt[3]{6x}$

16. $\sqrt{3}(4\sqrt{3} + \sqrt{12})$ 17. $3\sqrt{2}(2\sqrt{8} - \sqrt{3})$

18. $2\sqrt{5}(3\sqrt{2} + \sqrt{45} - 4\sqrt{5})$ 19. $\sqrt{\frac{2}{3}}(4\sqrt{6} - 2\sqrt{24} + 5\sqrt{3})$

20. $(2 + \sqrt{7})(3 + \sqrt{7})$ 21. $(9 - \sqrt{2})(7 + \sqrt{2})$

22. $(5 + \sqrt{3})^2$ 23. $(10 - \sqrt{5})(10 + \sqrt{5})$

24. $(3 + 2\sqrt{3})(3 - 2\sqrt{3})$

In 25-36, raise the expression to the indicated power, and simplify the result.

25. $(\sqrt{17})^2$ 26. $(2\sqrt{10})^2$ 27. $(3\sqrt{7})^2$

28. $(\frac{1}{2}\sqrt{8})^2$ 29. $(8\sqrt{\frac{1}{2}})^2$ 30. $(4\sqrt{2})^2$

31. $(\sqrt[3]{2})^3$ 32. $(\sqrt[3]{9})^3$ 33. $(2\sqrt[3]{6})^3$

34. $(4 + \sqrt{5})^2$ 35. $(3 - \sqrt{2})^2$ 36. $(1 - \sqrt{8})^2$

In 37-46, two irrational numbers are given. a. Find the product of the numbers in simplest form. b. Is the product rational or irrational?

37. $5\sqrt{5} \cdot 3\sqrt{3}$ 　　　　　　　38. $8\sqrt{8} \cdot 2\sqrt{2}$

39. $\frac{2}{5}\sqrt{3} \cdot 10\sqrt{12}$ 　　　　　　40. $\frac{3}{8}\sqrt{10} \cdot 4\sqrt{\frac{2}{5}}$

41. $\sqrt{5}(2\sqrt{20} + \sqrt{2})$ 　　　　42. $3\sqrt{6}(2\sqrt{6} - \sqrt{24})$

43. $(2 + \sqrt{10})(5 + \sqrt{10})$ 　　　44. $(2 + \sqrt{10})(5 - \sqrt{10})$

45. $(5 + \sqrt{10})(5 - \sqrt{10})$ 　　　46. $(2 + \sqrt{8})(1 - \sqrt{2})$

In 47-50, express the area of the figure in simplest form.

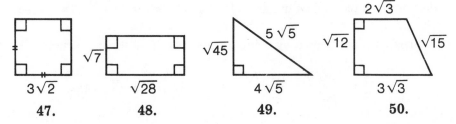

47.　　　　　48.　　　　　49.　　　　　50.

51. The base and the height of a parallelogram measure $3\sqrt{7}$ cm and $2\sqrt{14}$ cm, respectively. Find the number of square centimeters in the area of the parallelogram, expressed: (a) in simplest radical form (b) as a rational number correct to the nearest tenth.

52. Find the value of $x^2 - 4$ when:

    a. $x = 5$ **b.** $x = \sqrt{5}$ **c.** $x = -2$

    d. $x = \sqrt{3}$ **e.** $x = \sqrt{3} - 1$ **f.** $x = 3 + \sqrt{2}$

53. Find the value of $x^2 - 4x - 1$ when:

    a. $x = 3$ **b.** $x = \sqrt{3}$ **c.** $x = 2 + \sqrt{5}$

4-10 DIVIDING RADICALS WITH THE SAME INDEX

If a and b are positive numbers and the index n is a counting number, we have learned that $\sqrt[n]{\dfrac{a}{b}} = \dfrac{\sqrt[n]{a}}{\sqrt[n]{b}}$. By applying the symmetric property of equality to the given statement, we form a rule to find the quotient of two radicals with the same index, namely:

If a and b are positive numbers and the index n is a counting number, then:

$$\frac{\sqrt[n]{a}}{\sqrt[n]{b}} = \sqrt[n]{\frac{a}{b}}$$

For example, $\dfrac{\sqrt{30}}{\sqrt{6}} = \sqrt{\dfrac{30}{6}} = \sqrt{5}$ and $\dfrac{\sqrt[3]{48}}{\sqrt[3]{6}} = \sqrt[3]{\dfrac{48}{6}} = \sqrt[3]{8} = 2$. Let us extend this concept to the division of radical terms having coefficients.

□ EXAMPLE: A parallelogram whose area is $10\sqrt{6}$ has a height $h = 2\sqrt{3}$. Find the length of its base b.

Solution: In a parallelogram, area $A = bh$. Therefore, $b = \dfrac{A}{h}$. The necessary division is performed by using a property of fractions: $\dfrac{wx}{yz} = \dfrac{w}{y} \cdot \dfrac{x}{z}$.

$$b = \frac{A}{h} = \frac{10\sqrt{6}}{2\sqrt{3}} = \frac{10}{2} \cdot \frac{\sqrt{6}}{\sqrt{3}} = \frac{10}{2} \cdot \sqrt{\frac{6}{3}} = 5\sqrt{2} \quad Ans.$$

From this example, we make the following generalization: If a and b are positive numbers and the index n is a counting number, then:

$$x\sqrt[n]{a} \div y\sqrt[n]{b} = \frac{x\sqrt[n]{a}}{y\sqrt[n]{b}} = \frac{x}{y} \cdot \sqrt[n]{\frac{a}{b}}$$

■ **PROCEDURE.** To divide monomial terms containing radicals with the same index:

1. Divide the coefficients to find the coefficient of the quotient.
2. Divide the radicands to find the radicand of the quotient.
3. If possible, simplify the radical in the resulting quotient.

Since a radical sometimes represents an irrational number, let us make two general observations regarding quotients involving irrational numbers.

1. *The quotient of two irrational numbers may be either an irrational number or a rational number.*

   For example, $\sqrt{40} \div \sqrt{8} = \sqrt{5}$ (irrational),
   while $\sqrt{40} \div \sqrt{10} = \sqrt{4} = 2$ (rational).

2. *The quotient of an irrational number and any nonzero rational number, in either order, is an irrational number.*

   For example, $\sqrt{5} \div 2 = \dfrac{\sqrt{5}}{2}$ (irrational).

   Also, $2 \div \sqrt{5} = \dfrac{2}{\sqrt{5}}$ (irrational).

| MODEL PROBLEMS |

1. Divide:
   a. $48\sqrt{54} \div 12\sqrt{3}$ b. $\sqrt[3]{32x^4} \div 2\sqrt[3]{x}$

How to Proceed	*Solution*	*Solution*
1. Divide the coefficients, and divide the radicands.	a. $\dfrac{48\sqrt{54}}{12\sqrt{3}}$ $= \dfrac{48}{12} \cdot \dfrac{\sqrt{54}}{\sqrt{3}}$ $= 4\sqrt{18}$	b. $\dfrac{\sqrt[3]{32x^4}}{2\sqrt[3]{x}}$ $= \dfrac{1}{2} \cdot \dfrac{\sqrt[3]{32x^4}}{\sqrt[3]{x}}$ $= \dfrac{1}{2}\sqrt[3]{32x^3}$

2. Simplify the radical in the resulting quotient.

$$= 4\sqrt{18}$$

$$= 4\sqrt{9} \cdot \sqrt{2}$$

$$= 4 \cdot 3\sqrt{2}$$

$$= 12\sqrt{2} \quad Ans.$$

$$= \frac{1}{2}\sqrt[3]{32x^3}$$

$$= \frac{1}{2}\sqrt[3]{8x^3} \cdot \sqrt[3]{4}$$

$$= \frac{1}{2} \cdot 2x\sqrt[3]{4}$$

$$= x\sqrt[3]{4} \quad Ans.$$

2. Simplify the expression: $\dfrac{15\sqrt{8} - 5\sqrt{6}}{5\sqrt{2}}$

Solution: Divide each term in the numerator by the denominator.

$$\frac{15\sqrt{8} - 5\sqrt{6}}{5\sqrt{2}} = \frac{15\sqrt{8}}{5\sqrt{2}} - \frac{5\sqrt{6}}{5\sqrt{2}} = \frac{15}{5} \cdot \frac{\sqrt{8}}{\sqrt{2}} - \frac{5}{5} \cdot \frac{\sqrt{6}}{\sqrt{2}}$$

$$= 3\sqrt{4} - 1\sqrt{3} = 3 \cdot 2 - 1 \cdot \sqrt{3} = 6 - \sqrt{3} \quad Ans.$$

EXERCISES

In 1–18, divide and express the quotient in simplest form. (All variables represent positive numbers.)

1. $\sqrt{80} \div \sqrt{5}$

2. $12\sqrt{7} \div 2\sqrt{7}$

3. $\sqrt{170} \div \sqrt{17}$

4. $\sqrt{150} \div \sqrt{3}$

5. $10\sqrt{24} \div 2\sqrt{2}$

6. $9\sqrt{60} \div 3\sqrt{3}$

7. $\dfrac{28\sqrt{90}}{7\sqrt{2}}$

8. $\dfrac{24\sqrt{48}}{48\sqrt{6}}$

9. $\dfrac{45\sqrt{40}}{60\sqrt{10}}$

10. $\dfrac{30\sqrt[3]{128}}{6\sqrt[3]{2}}$

11. $\dfrac{6\sqrt[3]{60}}{15\sqrt[3]{10}}$

12. $\dfrac{3\sqrt[3]{96}}{12\sqrt[3]{4}}$

13. $\dfrac{\sqrt{98x^3}}{\sqrt{2x}}$

14. $\dfrac{2\sqrt{54y^5}}{6\sqrt{2y}}$

15. $\dfrac{\sqrt{20ab^5}}{2\sqrt{5ab^3}}$

16. $\dfrac{\sqrt{75} + \sqrt{48}}{\sqrt{3}}$

17. $\dfrac{2\sqrt{5} + 6\sqrt{15}}{2\sqrt{5}}$

18. $\dfrac{\sqrt{108} - \sqrt{150}}{\sqrt{6}}$

In 19–24, two irrational numbers are given. a. Find the quotient of the numbers in simplest form. b. Is the quotient rational or irrational?

19. $\sqrt{54} \div \sqrt{6}$

20. $2\sqrt{90} \div \sqrt{18}$

21. $\sqrt{192} \div 3\sqrt{3}$

22. $\dfrac{\sqrt{72} + \sqrt{32}}{\sqrt{8}}$

23. $\dfrac{8\sqrt{24} + 12\sqrt{2}}{4\sqrt{2}}$

24. $\dfrac{\sqrt{50} - \sqrt{8}}{4\sqrt{2}}$

In 25-27, express the quotient: (a) in simplest radical form (b) as the approximate rational value to the nearest tenth.

25. $\dfrac{12\sqrt{70}}{24\sqrt{7}}$ **26.** $\dfrac{\sqrt{96} + \sqrt{27}}{\sqrt{3}}$ **27.** $\dfrac{3\sqrt{20} - \sqrt{15}}{\sqrt{5}}$

28. In a rectangle, if the area $A = 12\sqrt{30}$ and the base $b = 3\sqrt{5}$, find the measure of the height h.

29. In a triangle, if the area $A = 9\sqrt{6}$ and the base $b = 3\sqrt{3}$, find the measure of the height h.

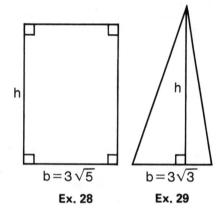

$b = 3\sqrt{5}$ $b = 3\sqrt{3}$

Ex. 28 Ex. 29

30. a. If the perimeter of a square is $\sqrt{80}$, what is the length of one side of the square? **b.** What is the area of the square?

4-11 RATIONALIZING A DENOMINATOR CONTAINING A RADICAL

Monomial Denominators

In the last section, it was stated that $2 \div \sqrt{5} = \dfrac{2}{\sqrt{5}}$, an irrational number. While the fraction $\dfrac{2}{\sqrt{5}}$ has an irrational number as its denominator, it is possible to multiply this fraction by some form of the identity element 1 to find an equivalent fraction with a rational denominator. For example:

$$\frac{2}{\sqrt{5}} = \frac{2}{\sqrt{5}} \cdot 1 = \frac{2}{\sqrt{5}} \cdot \frac{\sqrt{5}}{\sqrt{5}} = \frac{2\sqrt{5}}{5}$$

■ **DEFINITION.** To *rationalize the denominator* of a fraction (where the denominator is not a rational number) means to find an equivalent fraction in which the denominator is a rational number.

Thus, when we rationalize the denominator of $\dfrac{2}{\sqrt{5}}$, we find the equivalent fraction $\dfrac{2\sqrt{5}}{5}$ where the denominator 5 is a rational number.

The process of rationalizing a denominator allows us to simplify the computation needed to approximate the rational value of the expression. Compare the following computations:

1. *Without* rationalizing the denominator:

$$\frac{2}{\sqrt{5}} \approx \frac{2}{2.236} = \frac{2.000\,000}{2.236} = \frac{2000.000}{2236} \approx .894$$

2. *With* rationalizing the denominator:

$$\frac{2}{\sqrt{5}} = \frac{2\sqrt{5}}{5} \approx \frac{2(2.236)}{5} = \frac{4.472}{5} \approx .894$$

These examples show us that the rational approximation for $\dfrac{2}{\sqrt{5}}$ is .894 to three decimal places. Without rationalizing the denominator, we must divide by 2.236. However, once the denominator is rationalized and $\dfrac{2}{\sqrt{5}}$ becomes $\dfrac{2\sqrt{5}}{5}$, we need only divide by 5 to find the approximate rational value.

Binomial Denominators

Given the fraction $\dfrac{2}{4 + \sqrt{11}}$, how can we rationalize its denominator?

First, let us recall that the product of two binomials of the form $(a + b)$ and $(a - b)$ is a binomial that is written as the difference of two squares:

$$(a + b)(a - b) = a^2 - b^2$$

We will apply this product rule to a special set of binomials. Let a or b (or both terms) be a square-root radical that is irrational. Let any term, a or b, that is not a radical be a rational number. For example, if b is a square-root radical that is irrational, such as $\sqrt{11}$, then $b^2 = 11$, a rational number. Of course, if b is rational, then b^2 is rational. The same rules apply to a and a^2. Therefore, using this set, we find that the product $a^2 - b^2$ is the difference of two rational numbers, which is a rational number. Consider the following examples:

1. $(4 + \sqrt{11})(4 - \sqrt{11}) = 16 - 11 = 5$ The product is rational.
2. $(\sqrt{7} + 5)(\sqrt{7} - 5) = 7 - 25 = -18$ The product is rational.
3. $(\sqrt{8} - \sqrt{5})(\sqrt{8} + \sqrt{5}) = 8 - 5 = 3$ The product is rational.

Binomials such as $(4 + \sqrt{11})$ and $(4 - \sqrt{11})$ are called *conjugates* of each other. The conjugate of $(\sqrt{7} + 5)$ is $(\sqrt{7} - 5)$, and the conjugate of $(\sqrt{8} - \sqrt{5})$ is $(\sqrt{8} + \sqrt{5})$.

Therefore, to rationalize the denominator in $\dfrac{2}{4 + \sqrt{11}}$, we multiply this fraction by a form of the identity element 1 where both numerator and denominator are equal to the conjugate of the given denominator.

□ EXAMPLE: Rationalize the denominator of $\dfrac{2}{4 + \sqrt{11}}$.

Solution

$$\frac{2}{4 + \sqrt{11}} = \frac{2}{4 + \sqrt{11}} \cdot 1 = \frac{2}{(4 + \sqrt{11})} \cdot \frac{(4 - \sqrt{11})}{(4 - \sqrt{11})}$$

$$= \frac{8 - 2\sqrt{11}}{16 - 11} = \frac{8 - 2\sqrt{11}}{5} \quad Ans.$$

The procedures for rationalizing a denominator containing a radical are stated in the model problems that follow.

MODEL PROBLEMS

1. Rationalize the denominator: $\dfrac{10}{\sqrt{8}}$

How to Proceed	*Solution*
1. Multiply the fraction by a form of the identity element 1 to find an equivalent fraction with a rational denominator. Let the numerator and denominator of the identity 1 equal the least radical to yield the rational denominator.	$\dfrac{10}{\sqrt{8}} = \dfrac{10}{\sqrt{8}} \cdot 1 = \dfrac{10}{\sqrt{8}} \cdot \dfrac{\sqrt{2}}{\sqrt{2}}$ $= \dfrac{10\sqrt{2}}{\sqrt{16}}$ $= \dfrac{10\sqrt{2}}{4}$
2. Simplify the result.	$= \dfrac{5\sqrt{2}}{2} \quad Ans.$

Note: If the identity 1 does not involve the least radical, then the radical in the resulting fraction can be simplified.

$$\frac{10}{\sqrt{8}} = \frac{10}{\sqrt{8}} \cdot \frac{\sqrt{8}}{\sqrt{8}} = \frac{10\sqrt{8}}{8} = \frac{10\sqrt{4 \cdot 2}}{8} = \frac{10 \cdot 2\sqrt{2}}{8} = \frac{20\sqrt{2}}{8} = \frac{5\sqrt{2}}{2}$$

2. Rationalize the denominator: $\dfrac{\sqrt[3]{6}}{4\sqrt[3]{9}}$

Solution: Use the procedure outlined in model problem 1. Here, the resulting fraction cannot be simplified.

$$\frac{\sqrt[3]{6}}{4\sqrt[3]{9}} = \frac{\sqrt[3]{6}}{4\sqrt[3]{9}} \cdot \frac{\sqrt[3]{3}}{\sqrt[3]{3}} = \frac{\sqrt[3]{18}}{4\sqrt[3]{27}} = \frac{\sqrt[3]{18}}{4 \cdot 3} = \frac{\sqrt[3]{18}}{12} \quad Ans.$$

3. Express $\dfrac{6}{3 - \sqrt{5}}$ as an equivalent fraction with a rational denominator.

How to Proceed	*Solution*
1. Multiply the fraction by a form of the identity element 1, where both numerator and denominator of the identity 1 equal the conjugate of the denominator of the given fraction.	$\dfrac{6}{3 - \sqrt{5}} = \dfrac{6}{(3 - \sqrt{5})} \cdot \dfrac{(3 + \sqrt{5})}{(3 + \sqrt{5})}$ $= \dfrac{6(3 + \sqrt{5})}{9 - 5}$ $= \dfrac{6(3 + \sqrt{5})}{4}$
2. Simplify the result.	$= \dfrac{\overset{3}{\cancel{6}}(3 + \sqrt{5})}{\underset{2}{\cancel{4}}}$

Answer: $\dfrac{3(3 + \sqrt{5})}{2}$ or $\dfrac{9 + 3\sqrt{5}}{2}$

4. The expression $\dfrac{\sqrt{7} + 1}{\sqrt{7} - 2}$ is equivalent to:

(1) $\dfrac{9 + 3\sqrt{7}}{5}$ (2) $\dfrac{5 - \sqrt{7}}{3}$ (3) $3 + \sqrt{7}$ (4) $3 + 3\sqrt{7}$

Solution: Use the procedure outlined in model problem 3. Notice that the numerator of the resulting fraction is factored to simplify the result.

$$\frac{\sqrt{7}+1}{\sqrt{7}-2} = \frac{\sqrt{7}+1}{\sqrt{7}-2} \cdot 1 = \frac{(\sqrt{7}+1)}{(\sqrt{7}-2)} \cdot \frac{(\sqrt{7}+2)}{(\sqrt{7}+2)} = \frac{7+3\sqrt{7}+2}{7-4}$$

$$= \frac{9+3\sqrt{7}}{3} = \frac{\overset{1}{\cancel{3}}(3+\sqrt{7})}{\underset{1}{\cancel{3}}} = \frac{3+\sqrt{7}}{1} = 3+\sqrt{7}$$

Answer: (3) $3 + \sqrt{7}$

EXERCISES

In 1–44, rationalize the denominator. If possible, simplify the result.

1. $\dfrac{1}{\sqrt{7}}$ **2.** $\dfrac{9}{\sqrt{2}}$ **3.** $\dfrac{8}{\sqrt{5}}$ **4.** $\dfrac{15}{\sqrt{10}}$ **5.** $\dfrac{3}{\sqrt{6}}$

6. $\dfrac{6}{\sqrt{3}}$ **7.** $\dfrac{4}{\sqrt{18}}$ **8.** $\dfrac{6}{\sqrt{8}}$ **9.** $\dfrac{15}{\sqrt{50}}$ **10.** $\dfrac{6}{\sqrt{27}}$

11. $\dfrac{4}{\sqrt{48}}$ **12.** $\dfrac{3}{2\sqrt{2}}$ **13.** $\dfrac{3}{2\sqrt{3}}$ **14.** $\dfrac{9}{4\sqrt{6}}$ **15.** $\dfrac{10}{3\sqrt{20}}$

16. $\dfrac{5\sqrt{2}}{\sqrt{5}}$ **17.** $\dfrac{\sqrt{6}}{4\sqrt{2}}$ **18.** $\dfrac{3\sqrt{8}}{4\sqrt{18}}$ **19.** $\dfrac{2}{\sqrt[3]{16}}$ **20.** $\dfrac{4\sqrt[3]{3}}{3\sqrt[3]{2}}$

21. $\dfrac{5}{2-\sqrt{3}}$ **22.** $\dfrac{4}{3+\sqrt{2}}$ **23.** $\dfrac{1}{4+\sqrt{5}}$ **24.** $\dfrac{5}{4+\sqrt{6}}$

25. $\dfrac{6}{4-\sqrt{10}}$ **26.** $\dfrac{9}{5-\sqrt{13}}$ **27.** $\dfrac{6}{\sqrt{7}+2}$ **28.** $\dfrac{4}{\sqrt{15}-3}$

29. $\dfrac{4}{\sqrt{5}-3}$ **30.** $\dfrac{11}{\sqrt{3}-5}$ **31.** $\dfrac{12}{\sqrt{17}+5}$ **32.** $\dfrac{\sqrt{7}}{3-\sqrt{7}}$

33. $\dfrac{\sqrt{3}}{\sqrt{3}+1}$ **34.** $\dfrac{2\sqrt{5}}{\sqrt{5}-1}$ **35.** $\dfrac{\sqrt{2}}{2-\sqrt{2}}$ **36.** $\dfrac{2+\sqrt{3}}{4-\sqrt{3}}$

37. $\dfrac{6 - \sqrt{7}}{5 - \sqrt{7}}$ 38. $\dfrac{1 + \sqrt{11}}{4 - \sqrt{11}}$ 39. $\dfrac{1 + \sqrt{5}}{3 - \sqrt{5}}$ 40. $\dfrac{1 + \sqrt{3}}{3 - \sqrt{3}}$

41. $\dfrac{1 + \sqrt{5}}{5 + \sqrt{5}}$ 42. $\dfrac{\sqrt{10} - 3}{\sqrt{10} - 2}$ 43. $\dfrac{\sqrt{7} + \sqrt{3}}{\sqrt{7} - \sqrt{3}}$ 44. $\dfrac{5\sqrt{2} + 1}{2\sqrt{2} - 1}$

In 45–52, using $\sqrt{2}$ = 1.414 and $\sqrt{3}$ = 1.732, approximate the value of the fraction rounded off to *two decimal places*. (*Hint:* Rationalize the denominator before finding the approximate rational value.)

45. $\dfrac{1}{\sqrt{2}}$ 46. $\dfrac{1}{\sqrt{3}}$ 47. $\dfrac{3}{2\sqrt{2}}$ 48. $\dfrac{1 + \sqrt{2}}{\sqrt{2}}$

49. $\dfrac{\sqrt{3} - 1}{\sqrt{3}}$ 50. $\dfrac{1}{3 - \sqrt{2}}$ 51. $\dfrac{13}{4 + \sqrt{3}}$ 52. $\dfrac{2 - \sqrt{3}}{2 + \sqrt{3}}$

In 53–56, change the given expression to an equivalent fraction that does not have a radical in its denominator. (All variables represent positive numbers.)

53. $\dfrac{1}{\sqrt{k}}$ 54. $\dfrac{y}{\sqrt{y}}$ 55. $\dfrac{\sqrt{a}}{\sqrt{b}}$ 56. $\dfrac{6}{\sqrt{6x}}$

In 57–60, select the numeral preceding the expression that best completes the sentence.

57. The expression $\dfrac{6}{4 - \sqrt{7}}$ is equivalent to:

    (1) $\dfrac{4 + \sqrt{7}}{3}$ (2) $\dfrac{2(4 + \sqrt{7})}{3}$ (3) $4 + \sqrt{7}$ (4) $6(4 + \sqrt{7})$

58. An expression equivalent to $\dfrac{2}{\sqrt{15} - 4}$ is:

    (1) $2\sqrt{15} + 4$ (2) $2\sqrt{15} + 8$ (3) $-2\sqrt{15} - 4$ (4) $-2\sqrt{15} - 8$

59. The expression $\dfrac{\sqrt{2} + 2}{\sqrt{2} + 1}$ is equivalent to:

    (1) $\sqrt{2}$ (2) 2 (3) $\frac{3}{2}$ (4) $4 + \sqrt{2}$

60. The expression $\dfrac{2 - \sqrt{6}}{3 - \sqrt{6}}$ is equivalent to:

    (1) $\dfrac{\sqrt{6}}{3}$ (2) $\dfrac{-\sqrt{6}}{3}$ (3) $\sqrt{2}$ (4) $-\sqrt{2}$

4-12 THE QUADRATIC FORMULA AND QUADRATIC EQUATIONS WITH REAL ROOTS

We have learned that a *quadratic equation* is an equation that can be written in the form $ax^2 + bx + c = 0$, where a, b, and c are real numbers and $a \neq 0$.

If a quadratic equation has *rational roots*, it is possible to solve the equation by factoring $ax^2 + bx + c$, and setting each factor equal to zero. An example of a quadratic equation with rational roots is shown at the right; the check is left to the student.

Solve: $x^2 + 3x - 10 = 0$

$$(x - 2)(x + 5) = 0$$

$$x - 2 = 0 \quad | \quad x + 5 = 0$$
$$x = 2 \quad | \quad x = -5$$

Answer: $x = 2$ or $x = -5$

OR

solution set $= \{2, -5\}$

Not every quadratic equation has rational roots. Therefore, some quadratic equations cannot be solved by factoring. Recall that we learned another procedure to find the roots of a quadratic equation $ax^2 + bx + c = 0$, namely, the use of the *quadratic formula:*

$$x = \frac{-b \pm \sqrt{b^2 - 4ac}}{2a}$$

In *Course II*, we learned two derivations of the quadratic formula, each based upon the completion of a square. One of these derivations is restated here.

Derivation of the Quadratic Formula

How to Proceed

Solution

Given the general quadratic equation, where a, b, and c are real numbers and $a \neq 0$:

$$ax^2 + bx + c = 0 \quad (a \neq 0)$$

1. Multiply both sides of the equation by $4a$.

$$4a(ax^2 + bx + c) = 4a \cdot 0$$
$$4a^2x^2 + 4abx + 4ac = 0$$

2. Transform the equation, keeping all terms containing x on the left-hand side.

$$4a^2x^2 + 4abx \qquad = -4ac$$

3. Form a perfect square. Since $(2ax + b)^2 = 4a^2x^2 + 4abx + b^2$, add b^2 to both sides.

$$4a^2x^2 + 4abx + b^2 = b^2 - 4ac$$

4. Show the square of a binomial on the left.

$$(2ax + b)^2 = b^2 - 4ac$$

5. Take the square root of both sides. Show two roots at the right by writing ± before the radical.

$$2ax + b = \pm\sqrt{b^2 - 4ac}$$

6. Add $-b$ to both sides.

$$2ax = -b \pm \sqrt{b^2 - 4ac}$$

7. Divide both sides by $2a$ to obtain the *quadratic formula*.

$$x = \frac{-b \pm \sqrt{b^2 - 4ac}}{2a}$$

Thus, by the quadratic formula, the two roots of the equation are:

$$x_1 = \frac{-b + \sqrt{b^2 - 4ac}}{2a} \quad \text{AND} \quad x_2 = \frac{-b - \sqrt{b^2 - 4ac}}{2a}$$

Recall that the *discriminant* $b^2 - 4ac$ (the expression under the radical sign) indicates the nature of the roots of the quadratic equation:

1. If the discriminant $b^2 - 4ac$ is a positive number that is a perfect square or if $b^2 - 4ac = 0$, the quadratic equation has *rational roots*. (See model problem 1 that follows.)
2. If the discriminant $b^2 - 4ac$ is a positive number that is not a perfect square, the quadratic equation has *irrational roots*. (See model problem 2 that follows.)
3. If the discriminant $b^2 - 4ac$ is a negative number, the roots of the quadratic equation are not real. We will study such equations in Chapter 14 of this book.

For now, let us limit the study of quadratic equations to those equations having real roots, that is, roots that are rational or irrational.

MODEL PROBLEMS

1. Using the quadratic formula, find the roots of $2x^2 + 5x = 12$.

How to Proceed	*Solution*
1. Transform the equation so that one side is 0.	$2x^2 + 5x = 12$ $2x^2 + 5x - 12 = 0$
2. Compare the equation to $ax^2 + bx + c = 0$ to determine a, b, and c.	$a = 2, b = 5, c = -12$

3. Substitute the values of a, b, and c in the quadratic formula, and simplify.

$$x = \frac{-b \pm \sqrt{b^2 - 4ac}}{2a}$$

$$x = \frac{-(5) \pm \sqrt{(5)^2 - 4(2)(-12)}}{2(2)}$$

$$x = \frac{-5 \pm \sqrt{25 + 96}}{4}$$

$$x = \frac{-5 \pm \sqrt{121}}{4} = \frac{-5 \pm 11}{4}$$

4. (The check of each root in the original equation is left to the student.)

$$x_1 = \frac{-5 + 11}{4} \qquad x_2 = \frac{-5 - 11}{4}$$

$$= \frac{6}{4} = \frac{3}{2} \qquad\qquad = \frac{-16}{4} = -4$$

Answer: $x = \frac{3}{2}$ or $x = -4$ OR   solution set $= \{\frac{3}{2}, -4\}$

Note: The discriminant $b^2 - 4ac = 121$, a positive number that is a perfect square. Since the roots of $2x^2 + 5x - 12 = 0$ are *rational*, it is possible to solve this equation by factoring: $(2x - 3)(x + 4) = 0$. When each factor is set equal to zero, the roots are $x = \frac{3}{2}$ and $x = -4$.

2. a. Solve $x^2 - 10x + 13 = 0$ for values of x expressed in simplest radical form, and check the roots.
   b. Solve $x^2 - 10x + 13 = 0$ for values of x to the nearest tenth.

Solution

a. 1. Compare $x^2 - 10x + 13 = 0$ with $ax^2 + bx + c = 0$ to determine that $a = 1$, $b = -10$, and $c = 13$.

2. Substitute these values in the quadratic formula, and simplify.

$$x = \frac{-b \pm \sqrt{b^2 - 4ac}}{2a} = \frac{-(-10) \pm \sqrt{(-10)^2 - 4(1)(13)}}{2(1)}$$

$$= \frac{10 \pm \sqrt{100 - 52}}{2} = \frac{10 \pm \sqrt{48}}{2}$$

$$= \frac{10 \pm \sqrt{16 \cdot 3}}{2} \quad \text{(\textit{Note:} Simplify the radical.)}$$

$$= \frac{10 \pm 4\sqrt{3}}{2} = \frac{10}{2} \pm \frac{4\sqrt{3}}{2} = 5 \pm 2\sqrt{3} \quad \textit{Ans.}$$

3. Check each root in the original equation.

Check for $x = 5 + 2\sqrt{3}$

$$x^2 \qquad\quad - 10x \qquad\qquad + 13 = 0$$
$$(5 + 2\sqrt{3})(5 + 2\sqrt{3}) - 10(5 + 2\sqrt{3}) + 13 \overset{?}{=} 0$$
$$25 + 20\sqrt{3} + 12 - 50 - 20\sqrt{3} + 13 \overset{?}{=} 0$$
$$25 + \cancel{20\sqrt{3}} + 12 - 50 - \cancel{20\sqrt{3}} + 13 \overset{?}{=} 0$$
$$50 - 50 \overset{?}{=} 0$$
$$0 = 0 \quad \text{(True)}$$

Check for $x = 5 - 2\sqrt{3}$

$$x^2 \qquad\quad - 10x \qquad\qquad + 13 = 0$$
$$(5 - 2\sqrt{3})(5 - 2\sqrt{3}) - 10(5 - 2\sqrt{3}) + 13 \overset{?}{=} 0$$
$$25 - 20\sqrt{3} + 12 - 50 + 20\sqrt{3} + 13 \overset{?}{=} 0$$
$$25 - \cancel{20\sqrt{3}} + 12 - 50 + \cancel{20\sqrt{3}} + 13 \overset{?}{=} 0$$
$$50 - 50 \overset{?}{=} 0$$
$$0 = 0 \quad \text{(True)}$$

b. 1. Since the roots of $x^2 - 10x + 13 = 0$ are $5 \pm 2\sqrt{3}$, use any acceptable method (such as the algorithm shown at the right) to find a rational approximation for $\sqrt{3}$. This approximation is found to two decimal places, one more than that required in the answer.

```
      1. 7  3
   √3.00 00
     1
 27 | 200
  7 | 189
343 | 1100
  3 | 1029
        71
```

2. Substitute the approximation 1.73 for $\sqrt{3}$, and compute each root. Only in the final step is the root rounded off to the nearest tenth.

$$x_1 = 5 + 2\sqrt{3} \approx 5 + 2(1.73) = 5 + 3.46$$
$$= 8.46 = 8.5 \text{ (nearest tenth)}$$
$$x_2 = 5 - 2\sqrt{3} \approx 5 - 2(1.73) = 5 - 3.46$$
$$= 1.54 = 1.5 \text{ (nearest tenth)}$$

Answer: **a.** $x = 5 + 2\sqrt{3}$ or $x = 5 - 2\sqrt{3}$

(sometimes written $x = 5 \pm 2\sqrt{3}$)

OR

solution set $= \{5 + 2\sqrt{3}, 5 - 2\sqrt{3}\}$

b. $x = 8.5$ or $x = 1.5$ OR solution set $= \{8.5, 1.5\}$

Note: The discriminant $b^2 - 4ac = 48$, a positive number that is not a perfect square. Since the roots of $x^2 - 10x + 13 = 0$ are *irrational*, the equation cannot be solved by factoring.

| EXERCISES |

In 1–9, find the roots of the equation by using the quadratic formula, and check. Express irrational roots in simplest radical form.

1. $x^2 + 2x - 24 = 0$ **2.** $3x^2 + 7x + 2 = 0$ **3.** $2x^2 + 5 = 11x$
4. $x^2 - 6x + 7 = 0$ **5.** $x^2 - 6x + 9 = 0$ **6.** $x^2 = 6x + 1$
7. $2x^2 - 9 = 0$ **8.** $x^2 = 6x + 31$ **9.** $x(x + 4) = 1$

In 10–15, solve the quadratic equation for values of x expressed in simplest radical form, and check.

10. $x^2 - 8x + 13 = 0$ **11.** $x^2 - 10x + 18 = 0$ **12.** $x^2 = 2x + 5$
13. $x^2 = 6x + 11$ **14.** $3x^2 - 5 = 0$ **15.** $x(x + 8) = 34$

In 16–21, solve the quadratic equation for values of x: **(a)** expressed in simplest radical form **(b)** to the nearest tenth.

16. $x^2 - 12x + 29 = 0$ **17.** $x^2 + 10x + 1 = 0$ **18.** $x^2 + 3 = 7x$
19. $2x^2 + 1 = 4x$ **20.** $3x^2 = 2(x + 2)$ **21.** $4x^2 = 4x + 39$

In 22–30, find the roots of the equation to the nearest tenth.

22. $x^2 - 3x - 5 = 0$ **23.** $2x^2 + x = 7$ **24.** $3x(x - 2) = 1$
25. $x^2 + 8 = 10x$ **26.** $5x^2 = 3 - x$ **27.** $x^2 = 4(3x - 2)$
28. $\dfrac{x + 2}{2} + \dfrac{3}{x} = 5$ **29.** $2 - \dfrac{3}{x} = \dfrac{4}{x^2}$ **30.** $\dfrac{x - 2}{x + 2} = \dfrac{1}{x - 3}$

31. The length of a rectangle is 4 more than its width.
    a. Find the dimensions of the rectangle if its area is 12. Check.
    b. Find the dimensions of the rectangle if its area is 8. Check.

4-13 SOLVING RADICAL EQUATIONS

A *radical equation* in one variable, such as $\sqrt{x - 2} = 5$, is an equation in which the variable is contained in a radical. To solve a radical equation, we find a derived equation that does not contain radicals, as outlined in the procedure and model problems that follow.

■ **PROCEDURE.** To solve a radical equation containing only one radical:

1. Isolate the radical so that it is the only term on one side of the equation.

2. If the radical is a square root, square both sides of the equation. If the radical is a cube root, cube both sides, and so forth.

3. Solve the derived equation that is formed.

4. Since the derived equation may not be equivalent to the original radical equation, check all roots in the radical equation and reject any roots that are extraneous.

| MODEL PROBLEMS |

1. Solve and check: $\sqrt{x-2} = 5$

Solution   *Check*

1. Write the equation. $\sqrt{x-2} = 5$ $\sqrt{x-2} = 5$

2. Square both sides. $(\sqrt{x-2})^2 = (5)^2$ $\sqrt{27-2} \overset{?}{=} 5$

3. Solve the derived equation. $x - 2 = 25$ $\sqrt{25} \overset{?}{=} 5$
$x = 27$

4. Check the root in the original equation (shown at the right). $5 = 5$
(True)

Answer: $x = 27$ OR   solution set = {27}

2. Solve and check: $\sqrt{2y-1} + 7 = 4$

Solution   *Check*

1. Write the equation. $\sqrt{2y-1} + 7 = 4$ $\sqrt{2y-1} + 7 = 4$

2. Isolate the radical. $\sqrt{2y-1} = -3$ $\sqrt{2(5)-1} + 7 \overset{?}{=} 4$

3. Square both sides. $(\sqrt{2y-1})^2 = (-3)^2$ $\sqrt{9} + 7 \overset{?}{=} 4$

4. Solve the derived equation. $2y - 1 = 9$ $3 + 7 \overset{?}{=} 4$
$2y = 10$ $10 = 4$
$y = 5$

5. Check the root in the original equation (shown at the right).   (False)

Since $y = 5$ is an *extraneous root* of the radical equation, it is rejected. Therefore, the equation has no root, and the solution set is empty.

Answer: { } or ∅

3. Solve and check: $x = 1 + \sqrt{x + 5}$

Solution

1. Write the equation.

$$x = 1 + \sqrt{x + 5}$$

2. Isolate the radical.

$$x - 1 = \sqrt{x + 5}$$

3. Square both sides.

$$(x - 1)^2 = (\sqrt{x + 5})^2$$

4. Solve the derived quadratic equation.

$$x^2 - 2x + 1 = x + 5$$
$$x^2 - 3x - 4 = 0$$
$$(x - 4)(x + 1) = 0$$

$$x - 4 = 0 \quad | \quad x + 1 = 0$$
$$x = 4 \quad | \quad x = -1$$

5. Check each root in the original equation.

Check for $x = 4$ | *Check for $x = -1$*

$$x = 1 + \sqrt{x + 5}$$
$$4 \overset{?}{=} 1 + \sqrt{4 + 5}$$
$$4 \overset{?}{=} 1 + \sqrt{9}$$

$$4 \overset{?}{=} 1 + 3$$
$$4 = 4 \quad \text{(True)}$$

$$x = 1 + \sqrt{x + 5}$$
$$-1 \overset{?}{=} 1 + \sqrt{-1 + 5}$$
$$-1 \overset{?}{=} 1 + \sqrt{4}$$

$$-1 \overset{?}{=} 1 + 2$$
$$-1 = 3 \quad \text{(False)}$$
(Reject the root $x = -1$.)

Answer: $x = 4$ OR   solution set $= \{4\}$

4. Solve and check: $3\sqrt{x - 2} - 2\sqrt{x + 8} = 0$

Note: If a radical equation contains *two radicals*, there is no definite procedure to use in eliminating the radicals. However, as shown in the solution that follows, it is often helpful to transform the equation so that one radical is isolated on one side of the equation. In this case, we can then eliminate the radicals by squaring both sides.

Solution

$$3\sqrt{x - 2} - 2\sqrt{x + 8} = 0$$
$$3\sqrt{x - 2} = 2\sqrt{x + 8}$$
$$(3\sqrt{x - 2})^2 = (2\sqrt{x + 8})^2$$
$$9(x - 2) = 4(x + 8)$$
$$9x - 18 = 4x + 32$$
$$5x = 50$$
$$x = 10$$

Check

$$3\sqrt{x - 2} - 2\sqrt{x + 8} = 0$$
$$3\sqrt{10 - 2} - 2\sqrt{10 + 8} \overset{?}{=} 0$$
$$3\sqrt{8} - 2\sqrt{18} \overset{?}{=} 0$$
$$3\sqrt{4 \cdot 2} - 2\sqrt{9 \cdot 2} \overset{?}{=} 0$$
$$3 \cdot 2\sqrt{2} - 2 \cdot 3\sqrt{2} \overset{?}{=} 0$$
$$6\sqrt{2} - 6\sqrt{2} \overset{?}{=} 0$$
$$0 = 0$$
(True)

Answer: $x = 10$ OR   solution set $= \{10\}$

| EXERCISES |

In 1–40, solve the radical equation and check.

1. $\sqrt{x} = 3$ 2. $\sqrt{3x} = 6$ 3. $3\sqrt{x} = 6$

4. $\sqrt{6x} = 3$ 5. $\sqrt[3]{y} = 4$ 6. $\sqrt[3]{2x} = -2$

7. $\sqrt{2x} = -2$ 8. $7 + \sqrt{x} = 13$ 9. $5 + \sqrt{y} = 3$

10. $\sqrt{y - 6} = 2$ 11. $\sqrt{y + 8} = 4$ 12. $\sqrt{5 + y} = 3$

13. $\sqrt{4 - x} = 3$ 14. $\sqrt{2x + 3} = 7$ 15. $\sqrt{9 - 2k} = 5$

16. $4 - \sqrt{x} = 7$ 17. $\sqrt{2y + 5} = -3$ 18. $x = \sqrt{6x + 7}$

19. $y = \sqrt{6y + 16}$ 20. $x = \sqrt{x}$ 21. $y = 2\sqrt{2y - 3}$

22. $x = 2\sqrt{3 - x}$ 23. $y - 2 = \sqrt{y}$ 24. $x - 3 = 2\sqrt{x}$

25. $\sqrt{x^2 + 13} = x + 1$ 26. $\sqrt{x^2 - 12} = x - 2$

27. $y - 3 = \sqrt{y^2 + 3y}$ 28. $5 + \sqrt{4y - 3} = 2$

29. $x - 1 = \sqrt{5x - 9}$ 30. $x + 2 = \sqrt{3x + 16}$

31. $\sqrt{2 - 2y} = y + 3$ 32. $y = 3 + \sqrt{30 - 2y}$

33. $x = 4 + \sqrt{2x - 8}$ 34. $3x = 2\sqrt{3x - 1}$

35. $\sqrt{5k - 3} = \sqrt{k + 13}$ 36. $\sqrt{k^2 - 1} = \sqrt{k + 5}$

37. $3\sqrt{y - 3} = \sqrt{3y + 3}$ 38. $2\sqrt{y + 7} - \sqrt{y + 25} = 0$

39. $\sqrt{x^2 - 9} = \sqrt{x + 3}$ 40. $\sqrt{x^2 + 4} = 2\sqrt{x + 4}$

In 41–48, select the numeral preceding the expression that best completes the sentence or answers the question.

41. The solution set of the equation $x = \sqrt{5x + 14}$ is:
    (1) $\{7, -2\}$ (2) $\{-2\}$ (3) $\{7\}$ (4) $\{\ \}$

42. The equation $\sqrt{2 - x} = 2 - x$ has for its roots:
    (1) 1, only   (2) 2, only   (3) 1 and 2   (4) neither 1 nor 2

43. The equation $\sqrt{y - 2} = 2 - y$ has for its roots:
    (1) 2 and 3   (2) 2, only   (3) 3, only   (4) neither 2 nor 3

44. The solution set of $5 - \sqrt{2x} = 9$ is:
    (1) $\{8\}$ (2) $\{2\}$ (3) $\{2, -2\}$ (4) $\{\ \}$

45. The solution set of $\sqrt{x^2 + 8} = 2\sqrt{2x - 1}$ is:
    (1) $\{2, 6\}$ (2) $\{2\}$ (3) $\{6\}$ (4) $\emptyset$

46. Given the equation $y = 3\sqrt{y}$, its roots are:
    (1) 0, only   (2) 0 and 3   (3) 3, only   (4) 0 and 9

47. Which equation has 4 as a root?
    (1) $x = \sqrt{20 - x}$ (2) $\sqrt{x} = 2 - x$
    (3) $\sqrt{x + 5} = 1 - x$ (4) $\sqrt{2x - 7} = -1$

48. Which equation has roots of 1 and 3?

(1) $y = \sqrt{2y - 1}$ (2) $y = \sqrt{4y - 3}$

(3) $y = \sqrt{6y - 9}$ (4) $y - 3 = \sqrt{6 - 2y}$

4-14 REVIEW EXERCISES

In 1–5, identify the property of the real numbers that is illustrated by the given statement.

1. $\sqrt{7} \cdot 1 = \sqrt{7}$ **2.** $\sqrt{5} + (-\sqrt{5}) = 0$

3. $\sqrt{2} + (\sqrt{2} + \sqrt{3}) = (\sqrt{2} + \sqrt{2}) + \sqrt{3}$

4. $\sqrt{2}(\sqrt{2} + \sqrt{3}) = \sqrt{2} \cdot \sqrt{2} + \sqrt{2} \cdot \sqrt{3}$

5. $\sqrt{3} > 2$ or $\sqrt{3} = 2$ or $\sqrt{3} < 2$

In 6–11, solve the absolute-value equation.

6. $|x| - 3 = 5$ **7.** $|x - 3| = 5$ **8.** $|2x - 3| = 5$

9. $|y - 2| = -2$ **10.** $|3y - 4| = y$ **11.** $|y| - 4 = 3y$

In 12–17, graph the solution set of the given sentence on a real-number line.

12. $4x - 3 \geq 7$ **13.** $1 < 3 + y$ **14.** $-2 \leq x < 3$

15. $(x < 2) \vee (x \geq 3)$ **16.** $|x - 3| > 4$ **17.** $4 + |y| \leq 6$

In 18–21, select the numeral preceding the expression or the diagram that best completes the sentence or answers the question.

18. The solution set of $|3x + 1| = 2$ is:

(1) $\{\frac{1}{3}\}$ (2) $\{\frac{1}{3}, -\frac{1}{3}\}$ (3) $\{\frac{1}{3}, -1\}$ (4) $\{-1, 1\}$

19. Which represents the solution set for x in the inequality $|3 - x| > 6$?

(1) $\{x \mid x < -9$ or $x > 3\}$ (2) $\{x \mid x < -3$ or $x > 9\}$

(3) $\{x \mid x < -9$ or $x > -3\}$ (4) $\{x \mid -3 < x < 9\}$

20. Which is the graph of the solution set of $|x + 4| < 2$?

(1)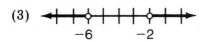

(2)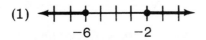

(3)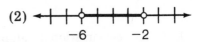

(4)

21. Which is the graph of the solution set of $|6 + y| \geq 1$?

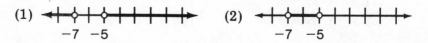

(1) -7 -5 (2) -7 -5

(3) -7 -5 (4) -7 -5

In 22–24, find the smallest integral value of x for which the radical represents a real number.

22. $\sqrt{x + 9}$ 23. $\sqrt{3x - 14}$ 24. $\sqrt{7 + 2x}$

In 25–32, evaluate the expression by finding the indicated root(s).

25. $\sqrt{81}$ 26. $-\sqrt{625}$ 27. $\pm\sqrt{196}$ 28. $\sqrt{.64}$
29. $\sqrt[3]{0}$ 30. $\sqrt[3]{-64}$ 31. $-\sqrt[3]{-8}$ 32. $\sqrt[5]{-1}$

In 33–36, write the expression in simplest radical form.

33. $\sqrt{28}$ 34. $\sqrt{8b^2}$ 35. $\sqrt{\frac{3}{5}}$ 36. $6\sqrt{\frac{4}{3}}$

In 37–46, perform the indicated operation, and write the answer in simplest form.

37. $\sqrt{90} + \sqrt{40}$ 38. $\sqrt{98} - 2\sqrt{18}$ 39. $(3\sqrt{5})^2$

40. $\sqrt{\frac{8}{3}} + \sqrt{\frac{2}{3}}$ 41. $2\sqrt{5} \cdot \sqrt{15}$ 42. $\frac{6\sqrt{60}}{24\sqrt{3}}$

43. $\sqrt{300} + \sqrt{50} - \sqrt{72} + \sqrt{3}$ 44. $\sqrt{3}(2\sqrt{27} - \sqrt{6})$

45. $(2 + \sqrt{5})(3 - \sqrt{5})$ 46. $\frac{3\sqrt{7} + 12\sqrt{21}}{3\sqrt{7}}$

In 47–50, rationalize the denominator. If possible, simplify the result.

47. $\frac{30}{\sqrt{20}}$ 48. $\frac{4}{3 + \sqrt{5}}$ 49. $\frac{3}{\sqrt{10} - 2}$ 50. $\frac{4 + \sqrt{2}}{3 - \sqrt{2}}$

In 51–54, two irrational numbers are given. **a.** Perform the indicated operation. **b.** Is the result rational or irrational?

51. $(2 + 3\sqrt{6}) + (5 - \sqrt{54})$ 52. $(8 + \sqrt{3})(5 - \sqrt{3})$
53. $3\sqrt{50} - 5\sqrt{18}$ 54. $(7\sqrt{2} + 10)(7\sqrt{2} - 10)$

In 55–57, solve the quadratic equation for values of x: (a) expressed in simplest radical form (b) to the nearest tenth.

55. $x^2 - 8x + 4 = 0$ **56.** $x^2 + 4x = 6$ **57.** $3x^2 = 2(x + 1)$

In 58–60, solve the radical equation and check.

58. $\sqrt{4x + 1} = 5$ **59.** $x = \sqrt{3x + 28}$ **60.** $2 - \sqrt{x} = 9$

In 61–63, select the numeral preceding the expression that best completes the sentence or answers the question.

61. The expression $\dfrac{7}{4 + \sqrt{2}}$ is equivalent to:

     (1) $\dfrac{4 - \sqrt{2}}{2}$ (2) $\dfrac{4 + \sqrt{2}}{2}$ (3) $2 - \sqrt{2}$ (4) $2 + \sqrt{2}$

62. The equation $x + 2 = \sqrt{34 - 3x}$ has for its roots:
     (1) -10 and 3   (2) -10, only
     (3) 3, only   (4) neither -10 nor 3

63. Which of the following equations has a solution set that is empty?
     (1) $\sqrt{x} + 1 = 2$ (2) $\sqrt{x + 1} = 2$
     (3) $\sqrt{x} + 2 = 1$ (4) $\sqrt{x + 2} = 1$

64. The park department wants to place a distance marker at F, the beginning of two straight paths through a rectangular park, $ABCD$. One path extends to an adjacent side and the other to an opposite corner of the park, as shown in the diagram. The length of the park, CD, is 12 km and the total length of the two paths, $FE + FD$, is 18 km. If F is equidistant from E and C, what are the distances FE and FD to be listed on the marker?

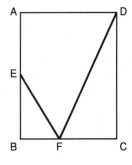

65. Two congruent circles drawn in the interior of a square are each tangent to two adjacent sides of the square and to each other, as shown in the diagram. If the measure of a side of the square is 16, find the radius of each circle.

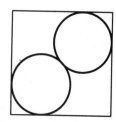

Chapter 5

Transformation Geometry and Coordinates

5-1 WHAT IS A TRANSFORMATION?

Imagine that a string of beads contains one large bead, two smaller dark beads, and thirty white beads of equal size. The string breaks! Beads bounce on the floor. Now imagine that someone gets a new cord and starts to restring the beads. The single large bead will be in the same position on the new string. The two smaller dark beads may be in the same position, or they may exchange places. The thirty beads of equal size, in all probability, will take up new positions. However, when completed, the new string of beads will look exactly like the old one.

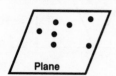

Plane

Now think of any plane or flat surface. The points on the plane are like beads. Under a *transformation of the plane*, points will move about: some points may remain fixed and other points will change position. However, after the transformation, the plane appears once again, full and complete, with no missing points.

■ **DEFINITION.** A *transformation of the plane* is a one-to-one correspondence between the points in a plane that demonstrates a change in position or a fixed position for each point in the plane.

An infinite number of transformations can take place in a plane. In this chapter, we will review a few special transformations, each of which follows a definite pattern or rule. These coordinate rules have been studied in *Course II*.

5-2 LINE REFLECTIONS AND LINE SYMMETRY

Line Reflections

At the right, $\triangle ABC \cong \triangle A'B'C$. One triangle will "fit exactly" on top of the other if we fold this page along line k, the *line of reflection*. Thus, points A and A' correspond to each other, and points B and B' correspond to each other. Point C is called a *fixed point* because C is found on the line of reflection.

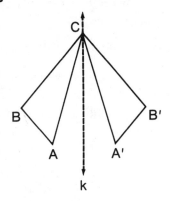

The term *image* is used to describe the relationship of these points, just as we might think of one point as being the mirror image of another.

In Symbols	*In Words*
$A \rightarrow A'$, and $A' \rightarrow A$.	The image of A is A', and the image of A' is A.
$B \rightarrow B'$, and $B' \rightarrow B$.	The image of B is B', and the image of B' is B.
$C \rightarrow C$.	The image of C is C.

If the image of A is A', we may also say that the *preimage* of A' is A.

A reflection in the line k is indicated in symbols as r_k. Thus, to show that the images are formed under a reflection in line k, we can write:

$r_k(A) = A'$.	Under a reflection in line k, the image of A is A'.
$r_k(B) = B'$.	Under a reflection in line k, the image of B is B'.
$r_k(C) = C$.	Under a reflection in line k, the image of C is C.

In the diagram, $\overline{AA'}$ and $\overline{BB'}$ are drawn to connect two points and their images. Notice how line k, the line of reflection, is related to these segments:

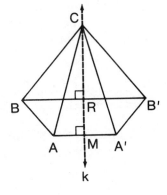

1. Line k is perpendicular to each segment $\overline{AA'}$ and $\overline{BB'}$.

2. Line k bisects each of these segments. If M is the intersection of $\overline{AA'}$ and line k, then $AM = MA'$. Similarly, if R is the intersection of $\overline{BB'}$ and line k, then $BR = RB'$.

■ **DEFINITION.** A *reflection in a line k* is a transformation of the plane such that:

1. If point P is not on line k, the image of P is P', where line k is the perpendicular bisector of $\overline{PP'}$.

2. If point P is on line k, the image of P is P.

Under a line reflection, the image of a segment is another segment, as in $\overline{AB} \rightarrow \overline{A'B'}$, or $r_k(\overline{AB}) = \overline{A'B'}$. Also, the image of an angle is another angle, as in $\angle ABC \rightarrow \angle A'B'C$, or $r_k(\angle ABC) = \angle A'B'C$.

Line Symmetry

If every point in a figure moves to its image through a reflection in a line, and the figure appears to be unchanged, this line of reflection is called an *axis of symmetry*, and the figure is said to have *line symmetry*.

As drawn here, the butterfly has *vertical symmetry* and the word BIKE has *horizontal symmetry*.

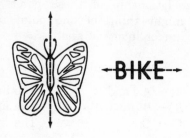

■ **DEFINITION.** *Line symmetry* occurs in a figure when the figure is its own image under a reflection in a line. Such a line is called the *axis of symmetry*.

It is possible for a figure to have more than one axis of symmetry, or reflection line. Rectangle *EFGH* has two axes of symmetry, namely, line m and line p. Thus, rectangle *EFGH* has both *vertical symmetry* and *horizontal symmetry*.

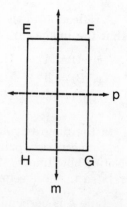

EXERCISES

1. Tell which of the following words have line symmetry and, if such symmetry exists, draw a line of reflection through the word.

   a. BOXED   **b.** COOKIE   **c.** AXIOM   **d.** CHECKERS

   e. MOUTH   **f.** TOOT   **g.** MYTH   **h.** WITHOUT

2. Using the words from exercise 1 that do *not* have line symmetry, print the letters of each word in a vertical column. Which of these words now has line symmetry?

In 3–9, which refer to the accompanying diagram, the reflection of $\triangle ABC$ in line k is $\triangle DEC$.

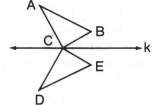

3. What is the image of point A under the line reflection?
4. $r_k(B) = ?$ 5. $r_k(C) = ?$ 6. $r_k(D) = ?$
7. What is the preimage of point B under the line reflection?
8. $r_k(\angle ABC) = ?$ 9. $r_k(\overline{DE}) = ?$

In 10–19: **a.** On your paper, copy the given figure or sketch the geometric figure named. **b.** Tell the number of lines of symmetry each figure has, if any, and sketch them on your drawing.

10.	11.	12. rectangle	13. square

10. parabola   11. star

12. rectangle   13. square
14. rhombus   15. parallelogram
16. circle   17. equilateral triangle
18. ellipse   19. isosceles triangle

5-3 LINE REFLECTIONS IN COORDINATE GEOMETRY

Many transformations in the coordinate plane can be described by rules involving the variables x and y. These rules are often discovered by inductive reasoning.

☐ EXAMPLE 1: At the right, the vertices of $\triangle ABC$ are $A(1, 6)$, $B(4, 3)$, and $C(2, -1)$. These vertices are reflected in the y-axis; and their images, when connected, form $\triangle A'B'C'$. We observe:

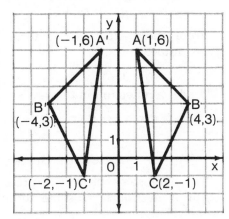

$$A(1, 6) \rightarrow A'(-1, 6)$$
$$B(4, 3) \rightarrow B'(-4, 3)$$
$$C(2, -1) \rightarrow C'(-2, -1)$$

From these examples we form a general rule.

■ Under a reflection in the y-axis:

$$P(x, y) \rightarrow P'(-x, y) \qquad \text{OR} \qquad r_{y\text{-axis}}(x, y) = (-x, y)$$

□ EXAMPLE 2: In the diagram, the endpoints of $\overline{AB}$ are $A(3, 0)$ and $B(4, 3)$. These endpoints are reflected in the line whose equation is $y = x$, and $\overline{A'B'}$ is formed. We observe:

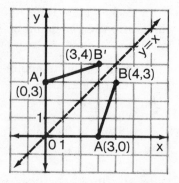

$$A(3, 0) \rightarrow A'(0, 3)$$
$$B(4, 3) \rightarrow B'(3, 4)$$

From these examples we form a general rule.

■ Under a reflection in the line $y = x$:

$$P(x, y) \rightarrow P'(y, x) \qquad \text{OR} \qquad r_{y=x}(x, y) = (y, x)$$

□ EXAMPLE 3: At the right, the vertices of quadrilateral $ABCD$ are $A(1, 3), B(2, 1), C(7, 1),$ and $D(5, 5)$. These vertices are reflected in the x-axis; and their images, when connected, form quadrilateral $A'B'C'D'$. We observe:

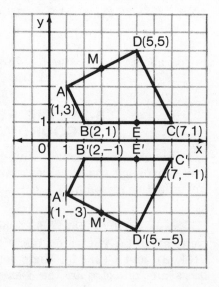

$$A(1, 3) \rightarrow A'(1, -3)$$
$$B(2, 1) \rightarrow B'(2, -1)$$
$$C(7, 1) \rightarrow C'(7, -1)$$
$$D(5, 5) \rightarrow D'(5, -5)$$

From these examples we form a general rule.

■ Under a reflection in the x-axis:

$$P(x, y) \rightarrow P'(x, -y) \qquad \text{OR} \qquad r_{x\text{-axis}}(x, y) = (x, -y)$$

Properties Under a Line Reflection

To study some properties preserved under a line reflection, let us recall from *Course II* some formulas used in coordinate geometry, namely:

Distance. The distance between two points whose coordinates are (x_1, y_1) and (x_2, y_2) is given by the formula:

$$d = \sqrt{(x_1 - x_2)^2 + (y_1 - y_2)^2}$$

Slope. If the coordinates of two points on a line are (x_1, y_1) and (x_2, y_2) where $x_1 \neq x_2$, the slope of the line is:

$$m = \frac{y_2 - y_1}{x_2 - x_1}$$

Midpoint. If the endpoints of a line segment have the coordinates (x_1, y_1) and (x_2, y_2), the coordinates of the midpoint are:

$$\left(\frac{x_1 + x_2}{2}, \frac{y_1 + y_2}{2} \right)$$

Now, let us use the quadrilaterals $ABCD$ and $A'B'C'D'$ from example 3 above to observe some properties preserved under a line reflection.

1. *Distance is preserved*; that is, each segment and its image are equal in length.

    In example 3, $\overline{AB} \rightarrow \overline{A'B'}$.

    For A (1, 3) and B (2, 1):

    $$AB = \sqrt{(1 - 2)^2 + (3 - 1)^2}$$
    $$= \sqrt{(-1)^2 + 2^2}$$
    $$= \sqrt{1 + 4}$$
    $$= \sqrt{5}$$

    For A' (1, −3) and B' (2, −1):

    $$A'B' = \sqrt{(1 - 2)^2 + [-3 - (-1)]^2}$$
    $$= \sqrt{(-1)^2 + (-2)^2}$$
    $$= \sqrt{1 + 4}$$
    $$= \sqrt{5}$$

    Therefore, when the image of $\overline{AB}$ is $\overline{A'B'}$, the lengths are equal; $AB = A'B' = \sqrt{5}$. Similarly, $\overline{BC} \rightarrow \overline{B'C'}$, and $BC = B'C' = 5$.

2. *Angle measure is preserved*; that is, each angle and its image are equal in measure.

    In example 3, $\angle DAB \rightarrow \angle D'A'B'$.

    For $A(1, 3)$, $D(5, 5)$ and $B(2, 1)$:

    Slope of $\overline{DA} = \dfrac{5 - 3}{5 - 1} = \dfrac{2}{4} = \dfrac{1}{2}$

    Slope of $\overline{AB} = \dfrac{3 - 1}{1 - 2} = \dfrac{2}{-1} = -2$

    $\overline{DA} \perp \overline{AB}$ and m $\angle DAB = 90°$.

    For $A'(1, -3)$, $D'(5, -5)$ and $B'(2, -1)$:

    Slope of $\overline{D'A'}$

    $= \dfrac{-5 - (-3)}{5 - 1} = \dfrac{-2}{4} = -\dfrac{1}{2}$

    Slope of $\overline{A'B'} = \dfrac{-3 - (-1)}{1 - 2} = \dfrac{-2}{-1} = 2$

    $\overline{D'A'} \perp \overline{A'B'}$ and m $\angle D'A'B' = 90°$.

    Thus, $\angle DAB \rightarrow \angle D'A'B'$ and m$\angle DAB$ = m$\angle D'A'B' = 90°$.

3. *Parallelism is preserved*; that is, if two lines are parallel, then their images will be parallel lines.

In example 3, the slope of $\overline{AB}$ is -2 and the slope of $\overline{DC}$ is -2. Since the *slopes are equal*, then $\overline{AB} \parallel \overline{DC}$. Examine their images: $\overline{AB} \rightarrow \overline{A'B'}$, and $\overline{DC} \rightarrow \overline{D'C'}$. Since the slope of $\overline{A'B'}$ is $+2$ and the slope of $\overline{D'C'}$ is $+2$, then $\overline{A'B'} \parallel \overline{D'C'}$.

4. *Collinearity is preserved*; that is, if three or more points lie on a straight line, their images will also lie on a straight line.

In example 3, points B, E, and C are collinear points that lie on the line whose equation is $y = 1$. Their images B', E', and C' are also collinear since these points lie on the line whose equation is $y = -1$.

5. *A midpoint is preserved*; that is, given three points such that one is the midpoint of the line segment whose endpoints are the other two points, then their images will be related in the same way.

In example 3, let $A(1, 3) = (x_1, y_1)$ and $D(5, 5) = (x_2, y_2)$. The midpoint of $\overline{AD}$ is M, and the coordinates of M are:

$$\left(\frac{x_1 + x_2}{2}, \frac{y_1 + y_2}{2} \right) = \left(\frac{1 + 5}{2}, \frac{3 + 5}{2} \right) = \left(\frac{6}{2}, \frac{8}{2} \right) = (3, 4)$$

The images of these points are $A'(1, -3)$, $D'(5, -5)$, and $M'(3, -4)$. Using $A'(1, -3)$ and $D'(5, -5)$, we see that the coordinates of the midpoint of $\overline{A'D'}$ are:

$$\left(\frac{1 + 5}{2}, \frac{-3 - 5}{2} \right) = \left(\frac{6}{2}, \frac{-8}{2} \right) = (3, -4)$$

Therefore, M' is the midpoint of $\overline{A'D'}$.

The properties observed for this specific example will be true for any figure and its image under a line reflection.

MODEL PROBLEM

Given $\triangle ABC$ whose vertices are $A(1, 3)$, $B(-2, 0)$, and $C(4, -3)$.

a. On one set of axes, draw $\triangle ABC$ and its image $\triangle A'B'C'$ under a reflection in the y-axis.

b. Find the coordinates of all points on the sides of $\triangle ABC$ that remain fixed under the given line reflection.

Solution:

a. In step 1, draw and label $\triangle ABC$. In step 2, find the images of the vertices of $\triangle ABC$ by using the rule $P(x,\ y) \rightarrow P'(-x,\ y)$. Draw and label $\triangle A'B'C'$.

$A(1, 3) \rightarrow A'(-1, 3)$
$B(-2, 0) \rightarrow B'(2, 0)$
$C(4, -3) \rightarrow C'(-4, -3)$

Step 1

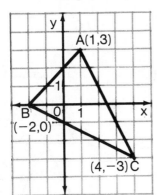

Step 2

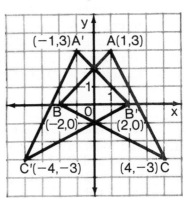

b. Under a reflection in the y-axis, points on the y-axis remain fixed. The sides of $\triangle ABC$ intersect the y-axis at $(0, 2)$ and $(0, -1)$. As seen in the graph, only these two points are common to both triangles.

Answer: a. See the graph labeled *Step 2.*   b. $(0, 2)$ and $(0, -1)$

EXERCISES

1. Under a reflection in the x-axis, the image of (x, y) is _____.
2. Under a reflection in the y-axis, the image of (x, y) is _____.
3. Under a reflection in the line $y = x$, the image of (x, y) is _____.

In 4–7, find the image of the point under a reflection in the x-axis.

4. $(5, 7)$ 5. $(6, -2)$ 6. $(-1, -4)$ 7. $(3, 0)$

In 8–11, find the image of the point under a reflection in the y-axis.

8. $(5, 7)$ 9. $(-4, 10)$ 10. $(0, 6)$ 11. $(-1, -6)$

In 12–15, find the image of the point under a reflection in the line $y = x$.

12. $(5, 7)$ 13. $(-3, 8)$ 14. $(0, -2)$ 15. $(6, 6)$

16. a. Using the rule $(x, y) \rightarrow (x, -y)$, find the images of $C(1, 4)$, $A(5, 1)$, and $T(4, 5)$, namely, C', A', and T'.
 b. On one set of axes, draw $\triangle CAT$ and $\triangle C'A'T'$.
 c. Find the lengths of $\overline{CA}$ and $\overline{C'A'}$.
 d. Is distance preserved under the given transformation?

17. a. Using the rule $(x, y) \rightarrow (-x, y)$, find the images of $D(2, 3)$, $O(0, 0)$, and $G(3, -2)$, namely, D', O', and G'.
 b. On one set of axes, draw $\triangle DOG$ and $\triangle D'O'G'$.
 c. Find the measures of $\angle DOG$ and $\angle D'O'G'$.
 (*Hint:* Look at the slopes.)
 d. Is angle measure preserved under the given transformation?

In 18–20, the vertices of $\triangle ABC$ are $A(1, 4)$, $B(3, 0)$, and $C(-3, -4)$.

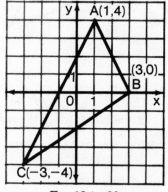

18. $\triangle ABC$ is reflected in the x-axis.
 a. Find the coordinates of the images A', B', and C' of the vertices.
 b. On one set of axes, draw $\triangle ABC$ and $\triangle A'B'C'$.
 c. Find the coordinates of all points on the sides of $\triangle ABC$ that remain fixed under the given reflection.

19. $\triangle ABC$ is reflected in the y-axis. Answer again parts **a, b,** and **c** of exercise 18 for this line reflection.

20. $\triangle ABC$ is reflected in the line $y = x$. Answer again parts **a, b,** and **c** of exercise 18 for this line reflection.

Ex. 18 to 20

In 21–24, the image of $\triangle ABC$ under a line reflection is $\triangle A'B'C'$.

a. Using the given coordinates, draw $\triangle ABC$ and $\triangle A'B'C'$ on one set of axes.
b. Find the equation of the line of reflection.

21. $\triangle ABC$: $A(2, 4)$, $B(2, 1)$, and $C(-1, 1)$.
 $\triangle A'B'C'$: $A'(4, 4)$, $B'(4, 1)$, and $C'(7, 1)$.

22. $\triangle ABC$: $A(1, 3)$, $B(2, 5)$, and $C(5, 3)$.
 $\triangle A'B'C'$: $A'(1, 1)$, $B'(2, -1)$, and $C'(5, 1)$.

23. $\triangle ABC$: $A(1, 4)$, $B(2, 1)$, and $C(4, 2)$.
 $\triangle A'B'C'$: $A'(-5, 4)$, $B'(-6, 1)$, and $C'(-8, 2)$.

24. $\triangle ABC$: $A(4, 2)$, $B(6, 2)$, and $C(2, -1)$.
 $\triangle A'B'C'$: $A'(2, 4)$, $B'(2, 6)$, and $C'(-1, 2)$.

25. The vertices of $\triangle RST$ are $R(0, 0)$, $S(1, 2)$, and $T(4, 1)$. If $\triangle RST$ is reflected in the line whose equation is $y = 3$, find the coordinates of the images R', S', and T'.

5-4 POINT REFLECTIONS AND POINT SYMMETRY

Point Reflections

In the diagram that follows, $\triangle ABC$ is reflected through point P, and its image $\triangle A'B'C'$ is formed. To find this image under a reflection through point P, the following steps are taken:

Step 1: From each vertex of $\triangle ABC$, a segment is drawn through point P to its image such that the distance from the vertex to point P is equal to the distance from point P to the image. Here, $\overline{AA'}$, $\overline{BB'}$, and $\overline{CC'}$ pass through point P so that $AP = PA'$, $BP = PB'$, and $CP = PC'$.

Step 2: The images A', B', and C' are connected to form $\triangle A'B'C'$, which is the reflection of $\triangle ABC$ through point P.

Given	*Step 1*	*Step 2*

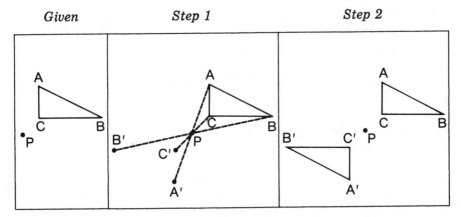

A reflection in a point P is indicated in symbols as R_P. To name the specific images under a reflection in point P, we write:

In Symbols   *In Words*

$R_P(A) = A'$.   Under a reflection in point P, the image of A is A'.
$R_P(B) = B'$.   Under a reflection in point P, the image of B is B'.
$R_P(C) = C'$.   Under a reflection in point P, the image of C is C'.

Under the point reflection just described, point P is the midpoint of each of the segments $\overline{AA'}$, $\overline{BB'}$, and $\overline{CC'}$, leading to our definition.

Notice, however, that one point remains fixed in the plane, namely, point P itself.

■ **DEFINITION.** A *reflection in a point P* is a transformation of the plane such that:

1. The image of the fixed point P is P.
2. For all other points, the image of K is K' where P is the midpoint of $\overline{KK'}$.

Point Symmetry

Imagine that every point in a figure moves to an image in the figure through a point of reflection located in the "center" of the figure.

This *point of reflection* is also called a *point of symmetry*, and each figure has *point symmetry*.

As shown here, a pinwheel and a playing card have point symmetry. Certain letters and words also have a point of symmetry.

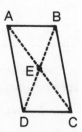

We can think of point symmetry as "turning the picture around" or "moving the picture through a half-turn." Try it. Turn the book upside down. Do the pictures look the same? They should, if they have point symmetry.

■ **DEFINITION.** *Point symmetry* occurs in a figure when the figure is its own image under a reflection in a point.

In the diagram, $ABCD$ is a parallelogram whose diagonals $\overline{AC}$ and $\overline{BD}$ intersect at point E. This point E is a point of symmetry for $\square ABCD$. Thus, $A \to C$, $B \to D$, $C \to A$, $D \to B$, interior points of $\overline{AB}$ have their images in $\overline{DC}$, and so forth. While $\square ABCD$ has point symmetry, it does *not* have line symmetry.

| **MODEL PROBLEM** |

Draw a geometric figure that has *both* line symmetry and point symmetry.

Answer: The square *EFGH* contains four lines of reflection, each of which demonstrates line symmetry.

    The four lines of reflection intersect at *P*, the point of symmetry.

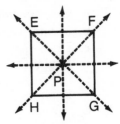

Note: other solutions are possible: rectangles, circles, regular polygons with an even number of sides.

EXERCISES

1. Tell which of the following words have point symmetry and, if such symmetry exists, locate the reflection point in the word.

    a. SIS   **b.** WOW   **c.** NOON   **d.** ZOO

    e. OX   **f.** SWIMS   **g.** un   **h.** pod

2. Which capital letters of the alphabet, when printed, have *both* point symmetry and line symmetry?

In 3–9, which refer to the accompanying diagram, the reflection of $\triangle ABC$ through point *B* is $\triangle DBE$.

3. What is the image of *A* under the point reflection?
4. $R_B(C) = ?$ **5.** $R_B(D) = ?$ **6.** $R_B(B) = ?$
7. What is the preimage of *C* under the point reflection?
8. $R_B(\angle CAB) = ?$ **9.** $R_B(\overline{CA}) = ?$

In 10–17: **a.** On your paper, copy the given figure, or sketch the geometric figure named. **b.** Tell whether or not the figure has a point of symmetry and, if it does, locate the point on your drawing.

10.

regular hexagon

11.

star

12. regular octagon
13. line segment
14. equilateral triangle
15. rhombus
16. regular pentagon
17. ellipse

18. **a.** On your paper, draw any triangle and label it *ABC*.
   b. Locate point *M*, the midpoint of $\overline{AC}$.
   c. Draw the reflection of $\triangle ABC$ through point *M*.
   d. Explain why the image of *A* is *C*, and the image of *C* is *A*.
   e. If the image of *B* is *B'*, what type of quadrilateral is *ABCB'*? Explain.

19. Crossword puzzles in the daily paper are symmetric. Cut out a crossword puzzle, study the pattern created by the black and white squares in the puzzle, and tell if the puzzle has point symmetry, line symmetry, or both.

5-5 POINT REFLECTIONS IN COORDINATE GEOMETRY

A common point reflection in coordinate geometry is a reflection in the origin.

□ EXAMPLE 1: In the diagram, the vertices of $\triangle ABC$ are *A*(1, 2), *B*(5, 5), and *C*(5, 2). These vertices are reflected in the origin (point *O*). Their images are connected to form $\triangle A'B'C'$.

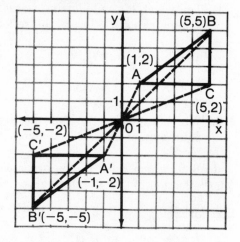

$$A(1, 2) \rightarrow A'(-1, -2)$$
$$B(5, 5) \rightarrow B'(-5, -5)$$
$$C(5, 2) \rightarrow C'(-5, -2)$$

From these examples we form a general rule.

■ **Under a reflection in point *O*, the origin:**

$$P(x, y) \rightarrow P'(-x, -y) \quad \text{OR} \quad R_o(x, y) = (-x, -y)$$

The *properties preserved under a point reflection* include all five properties listed previously for a line reflection, namely, distance, angle measure, parallelism, collinearity, and midpoints.

Compositions of Transformations

When two transformations occur, one following another, we have a *composition of transformations*. The first transformation produces an image, and the second transformation is performed on that image.

□ EXAMPLE 2: At the right, we again start with △ABC whose vertices are A(1, 2), B(5, 5), and C(5, 2). For the composition of transformations, we will consider two line reflections. First, by reflecting △ABC in the y-axis, we form △A'B'C', or △I. Then, by reflecting △A'B'C' in the x-axis, we form △A"B"C", or △II. Observe how the vertices behave:

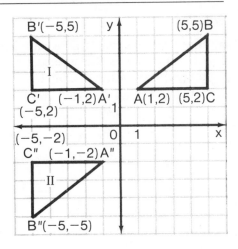

$A(1, 2) \rightarrow A'(-1, 2) \rightarrow A''(-1, -2)$
$B(5, 5) \rightarrow B'(-5, 5) \rightarrow B''(-5, -5)$
$C(5, 2) \rightarrow C'(-5, 2) \rightarrow C''(-5, -2)$

Now compare the original triangle, △ABC, with its final image, △A"B"C". How are these two triangles related? (As a hint, look at the diagram shown above in Example 1.) We observe:

■ **The composition of a line reflection in the y-axis, followed by a line reflection in the x-axis, is equivalent to a single transformation, namely, a reflection through point O, the origin.**

If we had reflected the triangle first in the x-axis and then in the y-axis, would this composition be equivalent to a reflection through point O, the origin? The answer is yes. However, not all compositions will act in the same way. In general, compositions of transformations are *not* commutative.

MODEL PROBLEM

The vertices of parallelogram ABCD are   A(1, 1),   B(3, 5), C(9, 5), and D(7, 1).

a. Find the coordinates of the point of symmetry for □ABCD.

b. Find the image of $\overline{AB}$ under this point reflection.

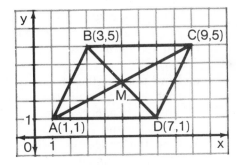

Solution:

a. Since the diagonals of a parallelogram bisect each other, the point of symmetry is the intersection point of the two diagonals, or the midpoint of either diagonal.

Using $\overline{AC}$, we find that the coordinates of midpoint M are:

$$\left(\frac{1+9}{2}, \frac{1+5}{2}\right) = \left(\frac{10}{2}, \frac{6}{2}\right) = (5, 3)$$

Using $\overline{BD}$, we find that the coordinates of midpoint M are:

$$\left(\frac{3+7}{2}, \frac{5+1}{2}\right) = \left(\frac{10}{2}, \frac{6}{2}\right) = (5, 3)$$

b. The image of $\overline{AB}$ under a reflection in point M is $\overline{CD}$, written $R_M(\overline{AB}) = \overline{CD}$.

Answer: **a.** (5, 3)   **b.** $\overline{CD}$

EXERCISES

1. Under a reflection in the origin, the image of (x, y) is _____.

In 2–5, find the image of the point under a reflection in the origin.

2. $(6, -3)$ **3.** $(-7, 1)$ **4.** $(0, 0)$ **5.** $(-3, 0)$

6. a. Using the rule $(x, y) \rightarrow (-x, -y)$, find the images of $A(2, 1)$, $B(4, 5)$, and $C(-1, 3)$, namely, A', B', and C'.
    b. On one set of axes, draw $\triangle ABC$ and $\triangle A'B'C'$.
    c. Find the coordinates of M, the midpoint of $\overline{AB}$.
    d. Using $\overline{AB}$ and M, show that a midpoint is preserved under this transformation.

7. a. Using the rule $(x, y) \rightarrow (-x, -y)$, find the images of $A(-5, -2)$, $B(-1, -2)$, $C(-1, 4)$, and $D(-3, 3)$, namely, A', B', C', and D'.
    b. On one set of axes, draw quadrilateral $ABCD$ and quadrilateral $A'B'C'D'$.
    c. Find the lengths of each side of $ABCD$ and $A'B'C'D'$ to show that distance is preserved under the given transformation.

8. a. On graph paper, draw $\triangle RST$ whose vertices are $R(-2, -1)$, $S(-2, 2)$, and $T(4, 2)$.
    b. Find the images of each of the vertices of $\triangle RST$, namely, R', S', and T', under a reflection through the origin.
    c. On the same graph used in part **a**, draw $\triangle R'S'T'$.
    d. Find the coordinates of all points on the sides of $\triangle RST$ that remain fixed under the point reflection.

9. **a.** On graph paper, draw the line whose equation is $y = \frac{1}{2}x + 2$.
   b. Name the coordinates of three points on this line, and call these points A, B, and C.
   c. Under a point reflection in the origin, name the coordinates of A', B', and C', the images of the three points found in **b**.
   d. On graph paper, draw the line containing A', B', and C'.
   e. What is the equation of the line drawn in **d**?

In 10–17, a line whose equation is given is reflected through the origin. What is the equation of the line that is its image? (*Hint:* See procedures in exercise 9.)

10. $x = 3$ 11. $y = -8$ 12. $y = x + 5$ 13. $y = -x + 1$
14. $y = x$ 15. $y = 2x - 3$ 16. $y = 3x$ 17. $y = -x$

18. The vertices of rhombus $ABCD$ are $A(0, 1)$, $B(3, 5)$, $C(8, 5)$, and $D(5, 1)$.

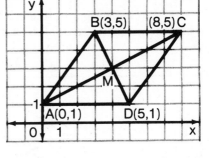

   a. Find the coordinates of M, the point of symmetry for rhombus $ABCD$.
   b. $R_M(A) = ?$ **c.** $R_M(\overline{BC}) = ?$
   d. $R_M(M) = ?$
   e. $R_M(\angle BCD) = ?$
   f. $R_M(\angle MCB) = ?$
   g. Find the length of $\overline{AB}$ and the length of its image. Is distance preserved in the given transformation?

In 19–22, the image of $\triangle ABC$ under a point reflection is $\triangle A'B'C'$.

a. Using the given coordinates, draw $\triangle ABC$ and $\triangle A'B'C'$ on one set of axes.
b. Find the coordinates of the point of reflection.

19. $\triangle ABC$: $A(2, 2)$, $B(5, 2)$, $C(5, 4)$
    $\triangle A'B'C'$: $A'(0, 2)$, $B'(-3, 2)$, $C'(-3, 0)$

20. $\triangle ABC$: $A(2, 5)$, $B(5, 6)$, $C(2, 3)$
    $\triangle A'B'C'$: $A'(2, 1)$, $B'(-1, 0)$, $C'(2, 3)$

21. $\triangle ABC$: $A(3, 1)$, $B(5, 5)$, $C(6, 3)$
    $\triangle A'B'C'$: $A'(3, 7)$, $B'(1, 3)$, $C'(0, 5)$

22. $\triangle ABC$: $A(-3, 6)$, $B(1, 7)$, $C(2, 4)$
    $\triangle A'B'C'$: $A'(-1, 0)$, $B'(-5, -1)$, $C'(-6, 2)$

23. The vertices of $\triangle DEF$ are $D(4, 3)$, $E(8, 1)$, and $F(8, 3)$. If $\triangle DEF$ is reflected through the point $(4, 1)$, find the coordinates of the images, D', E', and F'.

24. The vertices of $\triangle ABC$ are $A(1, 3)$, $B(1, 1)$, and $C(5, 1)$.

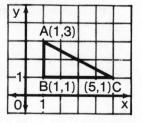

a. On graph paper, draw $\triangle ABC$ and its image $\triangle A'B'C'$ under a point reflection through the origin.

b. Using the same graph, reflect $\triangle A'B'C'$ in the x-axis to form its image $\triangle A''B''C''$.

c. What single transformation is equivalent to the composition of a point reflection through the origin followed by a reflection in the x-axis?

5-6 TRANSLATIONS

If a line reflection is like a "flip," and a point reflection is like a "half-turn," then a *translation* is like a "slide" or a "shift."

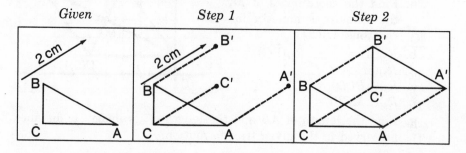

Given   *Step 1*   *Step 2*

In the preceding diagram, $\triangle ABC$ is moved 2 centimeters in the direction indicated by the arrow, and its image $\triangle A'B'C'$ is formed.

Step 1: From each vertex of $\triangle ABC$, a segment 2 centimeters long is drawn parallel to the arrow that indicates the direction of this "shift." Thus, $AA' = BB' = CC' = 2$ centimeters, and also $\overline{AA'} \parallel \overline{BB'} \parallel \overline{CC'}$.

Step 2: The images A', B', and C' are connected to form $\triangle A'B'C'$.

Any transformation of the plane that slides a figure as shown here is called a translation, symbolized by "T." Under a translation, if one point moves, then all points move and *no* point remains fixed.

■ **DEFINITION.** A *translation* is a transformation of the plane that shifts every point in the plane the same distance in the same direction to its image.

A rule for a translation is easily stated in coordinate geometry. At the right, the segment $\overline{AB}$ is translated to its image $\overline{A'B'}$ by moving each point 3 units to the right and 2 units down. Thus, by counting, we form the rule for the translation:

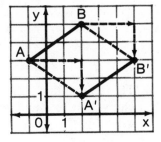

$$P(x, y) \rightarrow P'(x + 3, y - 2)$$

OR

$$T_{3,-2}(x, y) = (x + 3, y - 2)$$

■ **Under a translation of *a* units horizontally and *b* units vertically:**

$$P(x, y) \rightarrow P'(x + a, y + b) \quad \text{OR} \quad T_{a,b}(x, y) = (x + a, y + b)$$

The *properties preserved under a translation* include all five properties listed previously for a line reflection, namely, distance, angle measure, parallelism, collinearity, and midpoints.

| MODEL PROBLEM |

The vertices of $\triangle ABC$ are $A(-2, 1)$, $B(1, 4)$, and $C(2, 2)$.
a. On graph paper, draw and label $\triangle ABC$.
b. Using the same axes, graph $\triangle A'B'C'$, the image of $\triangle ABC$ under the translation $T_{5,-1}$.
c. Using the same axes, graph $\triangle A''B''C''$, the image of $\triangle A'B'C'$ under the translation $T_{-3,5}$.
d. Name a single transformation that is equivalent to $T_{5,-1}$ followed by $T_{-3,5}$.

Solution:

a. See graph.

b. By the first translation, $T_{5,-1}(x, y) = (x + 5, y - 1)$. Thus:

$$A(-2, 1) \rightarrow A'(3, 0)$$
$$B(1, 4) \rightarrow B'(6, 3)$$
$$C(2, 2) \rightarrow C'(7, 1)$$

$\triangle A'B'C'$ is shown.

c. By the second translation, $T_{-3,5}(x, y) = (x - 3, y + 5)$.

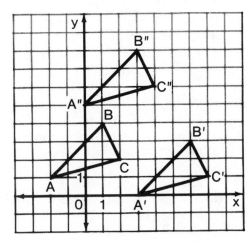

Thus: $A'(3, 0) \rightarrow A''(0, 5)$
         $B'(6, 3) \rightarrow B''(3, 8)$
         $C'(7, 1) \rightarrow C''(4, 6)$

$\triangle A''B''C''$ is shown.

d. The translation $(T_{5,-1})$ followed by a second translation $(T_{-3,5})$ is equivalent to a single translation $(T_{5-3,-1+5})$, or simply $(T_{2,4})$. *Ans.*

EXERCISES

In 1–4, use $(x, y) \rightarrow (x - 3, y + 12)$ to find the image of the point.

1. $(4, 4)$ **2.** $(2, 0)$ **3.** $(-3, -15)$ **4.** $(-5, -5)$

In 5–8, find the image of the point $(2, 7)$ under the given translation.

5. $T_{1,2}$ **6.** $T_{3,-6}$ **7.** $T_{-4,0}$ **8.** $T_{-2,-8}$

In 9–14, find the rule for the translation so that the image of A is A'.

9. $A(3, 8) \rightarrow A'(4, 6)$ **10.** $A(1, 0) \rightarrow A'(0, 1)$
11. $A(2, 5) \rightarrow A'(-1, 1)$ **12.** $A(-1, 2) \rightarrow A'(-2, -3)$
13. $A(0, -3) \rightarrow A'(-7, -3)$ **14.** $A(4, -7) \rightarrow A'(4, -2)$

15. A translation maps $B(0, 2)$ onto $B'(5, 0)$. Find the coordinates of C', the image of $C(-3, 1)$, under the same translation.

16. A translation maps the origin to the point $(-1, 7)$. What is the image of $(3, -7)$ under the same translation?

17. The vertices of nine congruent rectangles are marked in the accompanying diagram.

Under a given translation, if the image of D is M, find the image of:
a. C **b.** F **c.** $\overline{DH}$
d. $\overline{FG}$ **e.** $\overline{BD}$ **f.** $\angle BCG$

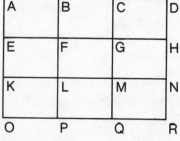

18. a. On graph paper, draw $\triangle HOT$ whose vertices are $H(-2, 0)$, $O(0, 0)$, and $T(0, 4)$.
b. Using the same axes, graph $\triangle H'O'T'$, the image of $\triangle HOT$ under the translation $T_{2,-3}$.
c. Using the same axes, graph $\triangle H''O''T''$, the image of $\triangle H'O'T'$ under the translation $T_{1,4}$.
d. Name a single transformation that is equivalent to $T_{2,-3}$ followed by $T_{1,4}$.

19. The vertices of $\triangle ICE$ are $I(-3, 1)$, $C(-1, 0)$, and $E(-1, 4)$.
    a. On graph paper, draw $\triangle ICE$ and its image $\triangle I'C'E'$ under a line reflection in the y-axis.
    b. Using the same graph, reflect $\triangle I'C'E'$ in a line whose equation is $x = 3$ to form its image $\triangle I''C''E''$.
    c. What single transformation is equivalent to a reflection in the y-axis followed by a reflection in the line $x = 3$?

5-7 ROTATIONS

Think of what happens to all the points on the steering wheel of a car as the wheel is turned. Except for the fixed point in the center of the wheel, every point moves through an arc so that the position of each point is changed by the same number of degrees. This transformation, which is like a "turn," is called a ***rotation***.

At the right, $\triangle ABC$ is rotated 70° counterclockwise about the fixed point P to form its image $\triangle A'B'C'$. By drawing rays $\overrightarrow{PA}$ and $\overrightarrow{PA'}$, we see that $m\angle APA' = 70°$. Notice that the distance from P to A is equal to the distance from P to A', or $PA = PA'$. Also, $m\angle BPB' = 70°$ and $PB = PB'$. In the same way, if the rays $\overrightarrow{PC}$ and $\overrightarrow{PC'}$ were drawn, we would see that $m\angle CPC' = 70°$ and $PC = PC'$.

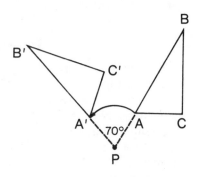

This rotation of 70° counterclockwise about point P is written in symbols as "$R_{P,70°}$." Thus, $R_{P,70°}(A) = A'$ indicates that the image of A is A' under the rotation. It is understood in this symbolism that a counterclockwise direction is being taken.

In the diagrams at the right, we observe the following:

1. The measure of the angle of rotation is positive when the rotation is counterclockwise.

2. The measure of the angle of rotation is negative when the rotation is clockwise.

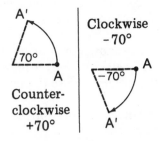

■ **DEFINITION.** A ***rotation*** is a transformation of the plane about a point P, called the *center of rotation*, and through an angle of ϕ degrees such that:

1. The image of the fixed point P is P.

2. For all other points, the image of K is K' where $m\angle KPK' = \phi$ and $PK = PK'$.

Note: The symbol for rotation is abbreviated in two cases:

1. Since a rotation of $180°$ about a point P is equivalent to a reflection in point P, we abbreviate $R_{P,180°}$ as R_P, the symbol for a point reflection.

2. If no point is mentioned, it is assumed that the rotation is taken about the origin. Thus, $R_{50°}$ means a $50°$ rotation about O, the origin.

The *properties preserved under a rotation* include all five properties listed previously for a line reflection, namely, distance, angle measure, parallelism, collinearity, and midpoints.

Rotational Symmetry

The geometric shapes at the right have rotational symmetry. If the equilateral triangle is rotated $120°$, it is its own image. Notice that $3(120°) = 360°$. The equilateral triangle is also its own image under a rotation of $240°$, which is a multiple of $120°$.

If the regular pentagon is rotated $72°$, it is its own image. Notice that $5(72°) = 360°$. The regular pentagon is also its own image under rotations of $144°$, $216°$, and $288°$, each of which is a multiple of $72°$.

■ **DEFINITION.** *Rotational symmetry* occurs in a figure if the figure is its own image under a rotation of ϕ degrees, and only the center point remains fixed.

The most common rotation, other than that of $180°$, is a rotation of $90°$, called a *quarter-turn*. We will state a rule for a quarter-turn about the origin and use this rule in the model problem to follow.

■ **Under a rotation of $90°$ counterclockwise through point O, the origin:**

$$P(x, y) \rightarrow P'(-y, x) \qquad \text{OR} \qquad R_{90°}(x, y) = (-y, x)$$

| MODEL PROBLEM |

The vertices of $\triangle ABC$ are $A(1, 3)$, $B(5, 1)$, and $C(1, 1)$. On one set of axes, draw $\triangle ABC$ and its image $\triangle A'B'C'$ under a rotation of 90° counterclockwise about the origin.

Solution:

1. $\triangle ABC$ is graphed at the right.

2. By the rule for a quarter-turn, $P(x, y) \rightarrow P'(-y, x)$. Thus:

   $A(1, 3) \rightarrow A'(-3, 1)$
   $B(5, 1) \rightarrow B'(-1, 5)$
   $C(1, 1) \rightarrow C'(-1, 1)$

3. Connect the images A', B', and C' to form $\triangle A'B'C'$, graphed at the right.

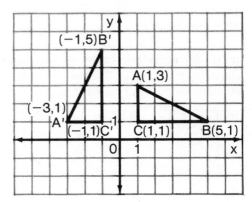

| EXERCISES |

1. Under a rotation of 90° counterclockwise about the origin, the image of (x, y) is ___.

2. Under a rotation of 180° about the origin, the image of (x, y) is ___.

In 3–9, which refer to the accompanying diagram, $\triangle ABC \rightarrow \triangle DEF$ by a quarter-turn about P.

3. What is the image of A under the quarter-turn?

4. $R_{P, 90°}(B) = ?$ 5. $R_{P, 90°}(P) = ?$

6. $R_{P, 90°}(\overline{CA}) = ?$ 7. $m\angle APD = ?$

8. What is the preimage of $\overline{FE}$ under the quarter-turn?

9. Does $m\angle BPE = m\angle CPF$? Why?

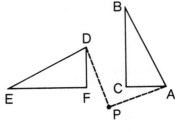

Ex. 3 to 9

In 10–15, $\triangle ABC$ is rotated 90° counterclockwise to its image $\triangle A'B'C'$.

a. Using the rule $(x, y) \rightarrow (-y, x)$ and the given coordinates, find the images of A, B, and C.

b. On one set of axes, draw $\triangle ABC$ and $\triangle A'B'C'$.

10. $A(1, 1)$, $B(1, 5)$, and $C(4, 5)$ **11.** $A(1, 2)$, $B(1, 6)$, and $C(3, 2)$
12. $A(4, 3)$, $B(4, -2)$, and $C(1, -1)$ **13.** $A(1, 3)$, $B(3, 5)$, and $C(3, 0)$
14. $A(-3, 3)$, $B(-1, 3)$, and $C(-1, -1)$
15. $A(2, 1)$, $B(3, -1)$, and $C(-2, -1)$

In 16–23: **a.** Tell whether or not the figure drawn or named has rotational symmetry. **b.** Where possible, find the degree measure of the smallest angle that will rotate the figure to be its own image.

16.   **17.**

regular octagon

18. rectangle
19. equilateral triangle
20. parallelogram
21. regular pentagon
22. regular hexagon
23. circle

24. Find the image of the point $(-3, 0)$ under the rotation $R_{180°}$.
25. What is the image of the point $(-1, -4)$ under a clockwise rotation of 90° about the origin, that is, under $R_{-90°}$?

5-8 DILATIONS

When a photograph is enlarged or reduced, a change, or transformation, takes place in its size. There is a constant ratio of the distances between points in the original photograph compared to the distances between their images in the enlargement or reduction. This type of transformation is called a **dilation**.

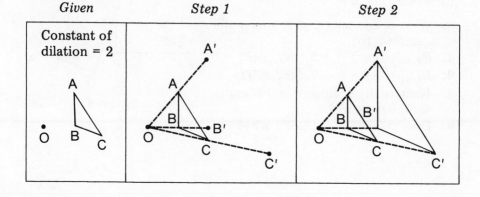

| *Given* | *Step 1* | *Step 2* |

In the preceding diagram, $\triangle ABC$ is to be dilated so that the *center of dilation* is point O and the *constant of dilation* is 2.

Step 1: From O, the center of dilation, rays are drawn to pass through each of the vertices of $\triangle ABC$. Using 2, the constant of dilation, the images (A', B', and C') are located on these rays so that $OA' = 2 \cdot OA$, $OB' = 2 \cdot OB$, and $OC' = 2 \cdot OC$.

Step 2: The images, A', B', and C', are connected to form $\triangle A'B'C'$.

This dilation, with a constant factor of 2, is written in symbols as D_2. Thus, $D_2(A) = A'$ indicates that the image of A under the dilation is A'.

Although any point may be chosen as the center of dilation, we will limit dilations in this chapter to those where point O, the origin, is the center of dilation.

■ **DEFINITION.** A *dilation* of k, where k is a positive number called the *constant of dilation*, is a transformation of the plane such that:

1. The image of point O, the center of dilation, is O.

2. For all other points, the image of P is P' where $\overrightarrow{OP}$ and $\overrightarrow{OP'}$ name the same ray and $OP' = k \cdot OP$.

A rule for a dilation is stated in coordinate geometry as follows:

■ **Under a dilation of k (a positive number) whose center of dilation is the origin:**

$$P(x,y) \rightarrow P'(kx, ky) \quad \text{OR} \quad D_k(x,y) = (kx, ky)$$

Compositions Involving Dilations

At the right, $\triangle ABC \rightarrow \triangle A'B'C'$ by a point reflection in the origin. Then, $\triangle A'B'C' \rightarrow \triangle A''B''C''$ by a dilation of 2. This composition of a point reflection followed by a dilation of 2 is equivalent to multiplying the x and y values in each of the coordinates by -2.

The vertices of $\triangle ABC$ are $A(1, 3)$, $B(3, 1)$, and $C(2, 0)$. Under the point reflection, $R_o(x, y) = (-x, -y)$. Thus:

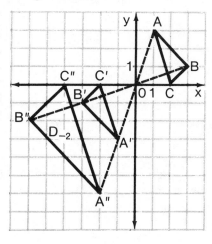

$A(1, 3) \rightarrow A'(-1, -3)$
$B(3, 1) \rightarrow B'(-3, -1)$
$C(2, 0) \rightarrow C'(-2, 0)$

Under a dilation of 2, $D_2(x, y) = (2x, 2y)$. Thus:

$$A'(-1, -3) \rightarrow A''(-2, -6)$$
$$B'(-3, -1) \rightarrow B''(-6, -2)$$
$$C'(-2, 0) \rightarrow C''(-4, 0)$$

Compare each original vertex to its final image. Here:

$$A(1, 3) \rightarrow A''(-2, -6)$$
$$B(3, 1) \rightarrow B''(-6, -2)$$
$$C(2, 0) \rightarrow C''(-4, 0)$$

Since these coordinates follow the rule $(x, y) \rightarrow (-2x, -2y)$, we will symbolize this composition of transformations as D_{-2}. This example illustrates the truth of the following general statement.

■ A composition of transformations consisting of a point reflection about the origin and a dilation of k, where k is a positive number, is equivalent to the single transformation:

$$D_{-k}(x, y) = (-kx, -ky)$$

Under a dilation there is generally only one fixed point, namely, the center of dilation. If the constant of dilation is 1, however, all points are fixed.

The *properties preserved under a dilation* include only four of the properties listed for the other transformations studied earlier in this chapter, namely, angle measure, parallelism, collinearity, and midpoints.

■ Distance is not preserved under a dilation.

| EXERCISES |

In 1-4, use the rule $(x, y) \rightarrow (4x, 4y)$ to find the image of the given point.

1. $(3, 5)$ 2. $(-3, 2)$ 3. $(7, 0)$ 4. $(-4, 9)$

In 5-8, find the image of the given point under a dilation of 5.

5. $(2, 1)$ 6. $(12, 20)$ 7. $(-9, -7)$ 8. $(0, -8)$

In 9-12, $D_{-3}(x, y) = (-3x, -3y)$ is a composition of a half-turn about the origin and a dilation of 3. Using D_{-3}, find the image of the given point.

9. $(6, -1)$ 10. $(-4, 0)$ 11. $(-3, -8)$ 12. $(10, -1)$

In 13-22, write a single rule for a dilation, or a composition involving a dilation, by which the image of A is A'.

13. $A(2, 5) \rightarrow A'(4, 10)$

14. $A(3, -1) \rightarrow A'(21, -7)$

15. $A(10, 4) \rightarrow A'(5, 2)$

16. $A(-20, 8) \rightarrow A'(-5, 2)$

17. $A(4, 6) \rightarrow A'(6, 9)$

18. $A(4, -3) \rightarrow A'(-8, 6)$

19. $A(-2, 5) \rightarrow A'(8, -20)$

20. $A(-12, 9) \rightarrow A'(-8, 6)$

21. $A(0, 9) \rightarrow A'(0, -3)$

22. $A(4, 0) \rightarrow A'(-5, 0)$

In 23-27, which refer to the accompanying diagram, O is the center of dilation and $D_k(\triangle OQR) = \triangle OPS$.

23. What is the image of R under the dilation?

24. $D_k(Q) = ?$ **25.** $D_k(\overline{OR}) = ?$

26. If $OP = PQ$, what is the constant of dilation k?

27. Using the value of k from exercise 26, find the value of $RQ : SP$.

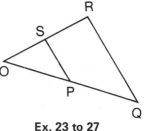

Ex. 23 to 27

28. a. On graph paper, draw $\triangle HOG$ whose vertices are $H(1, 2)$, $O(0, 0)$, and $G(1, 3)$.

b. Using the same set of axes, graph $\triangle H'OG'$ such that $D_3(\triangle HOG) = \triangle H'OG'$.

c. Using GH and $G'H'$, show why distance is *not* preserved under the dilation.

d. What is the ratio $GH : G'H'$?

29. The vertices of $\triangle HEN$ are $H(0, 4)$, $E(6, 4)$, and $N(4, -2)$. Under a dilation of $\frac{1}{2}$, the image of $\triangle HEN$ is $\triangle H'E'N'$.

a. On one set of axes, graph $\triangle HEN$ and $\triangle H'E'N'$.

b. Find the lengths of $\overline{HE}$ and $\overline{H'E'}$.

c. Find the ratio $HE : H'E'$.

d. Show that $\triangle HEN \sim \triangle H'E'N'$.

5-9 REVIEW EXERCISES

In 1–7, for each figure drawn or named:

a. Does the figure have *line symmetry*? If yes, how many lines of symmetry does the figure have?

b. Does the figure have *point symmetry*?

c. Does the figure have *rotational symmetry*? If yes, find the degree measure of the smallest angle that will rotate the figure to be its own image.

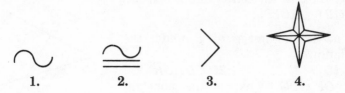

1.   **2.**   **3.**   **4.**

5. parallelogram   **6.** regular pentagon   **7.** equilateral triangle

In 8–20, which refer to the accompanying diagram, $ABCD$ is a square. The midpoints of sides $\overline{AB}$, $\overline{BC}$, $\overline{CD}$, and $\overline{DA}$ are E, F, G, and H, respectively. Lines $\overleftrightarrow{AC}$, $\overleftrightarrow{BD}$, $\overleftrightarrow{EG}$, and $\overleftrightarrow{FH}$ intersect at O.

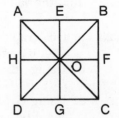

8. Under a reflection in the line $\overleftrightarrow{EG}$, which is the image of B?

9. $r_{\overleftrightarrow{HF}}(B) = ?$ **10.** $r_{\overleftrightarrow{AC}}(B) = ?$

11. $r_{\overleftrightarrow{DB}}(B) = ?$

12. Under a reflection through point O, what is the image of B?

13. Under a rotation of $90°$ about O, what is the image of B?

14. $R_{O,90°}(E) = ?$ **15.** $R_{O,90°}(\overline{CF}) = ?$ **16.** $R_{O,90°}(\angle AOH) = ?$

17. $R_{O,180°}(\overline{OH}) = ?$ **18.** $R_{O,270°}(C) = ?$ **19.** $R_{O,-90°}(\overline{GD}) = ?$

20. Under a translation, $E \rightarrow D$. What is the image of B under this translation?

In 21–32, find the image of $(5, 2)$ under the given transformation.

21. Reflection in the x-axis.   **22.** Reflection in the y-axis.

23. Reflection in the line $x = 4$.   **24.** Reflection in the line $y = 2$.

25. Reflection in the line $y = x$.   **26.** Reflection in the origin.

27. Quarter-turn about the origin.   **28.** The translation: $T_{3, -2}$.

29. Dilation of $1\frac{1}{2}$, center at origin.   **30.** D_{-3} **31.** $T_{-6,0}$

32. $R_{-90°}$, that is, a clockwise rotation of $90°$ about the origin.

In 33–38, name the single transformation that will assign a point to its image by the given rule.

33. $(x, y) \rightarrow (x, -y)$ **34.** $(x, y) \rightarrow (-x, -y)$ **35.** $(x, y) \rightarrow (y, x)$

36. $(x, y) \rightarrow (-x, y)$ **37.** $(x, y) \rightarrow (x - 1, y)$ **38.** $(x, y) \rightarrow (-y, x)$

39. The vertices of $\triangle AYE$ are $A(2, -3)$, $Y(5, 1)$, and $E(1, -1)$. Under the transformation whose rule is $(x, y) \rightarrow (-y, x)$, $\triangle AYE \rightarrow \triangle A'Y'E'$.
   a. On the same set of axes, graph $\triangle AYE$ and $\triangle A'Y'E'$.
   b. Name the transformation shown in part **a.**
   c. Using AY and $A'Y'$, show that distance is preserved here.

40. **a.** On graph paper, draw $\triangle AOK$ whose vertices are $A(1, 3)$, $O(0, 0)$, and $K(4, 2)$.
   b. Using the same axes, graph $\triangle A'O'K'$, the image of $\triangle AOK$ under $T_{2,4}$.
   c. Using the same axes, graph $\triangle A''O''K''$, the image of $\triangle A'O'K'$ under $T_{1, -3}$.
   d. What single translation is equivalent to $T_{2,4}$ followed by $T_{1, -3}$?

41. $\triangle YUP \rightarrow \triangle Y'U'P'$ by a reflection in the y-axis. Then, by a reflection in the line $x = 1$, $\triangle Y'U'P' \rightarrow \triangle Y''U''P''$. The vertices of $\triangle YUP$ are $Y(-4, 2)$, $U(-2, 3)$, and $P(-3, -1)$.
   a. Using one set of axes, graph $\triangle YUP$, $\triangle Y'U'P'$, and $\triangle Y''U''P''$.
   b. Name the single transformation by which $\triangle YUP \rightarrow \triangle Y''U''P''$.

42. **a.** On graph paper, draw $\triangle END$ whose vertices are $E(0, 2)$, $N(1, 4)$, and $D(4, 0)$.
   b. Using the same axes, graph $\triangle E'N'D'$, the image of $\triangle END$ under a point reflection in the origin.
   c. Using the same axes, graph $\triangle E''N''D''$, the image of $\triangle E'N'D'$ under a dilation of 2.
   d. Complete the statement: $\triangle END \rightarrow \triangle E''N''D''$ by the rule $(x, y) \rightarrow$ _____ .

Chapter **6**

Relations and Functions

6-1 RELATIONS

In our daily lives, we often "relate" one set of information to another. For example, the table at the right lists the heights of the first-string players on a school's basketball team. This relation of heights and players can also be stated as a set of ordered pairs:

Height (in centimeters)	Player
183	Florio
185	Laube
186	Richko
186	Andrews
188	Jones

{(183, Florio), (185, Laube), (186, Richko), (186, Andrews), (188, Jones)}

■ DEFINITION. A *relation* is a set of ordered pairs.

The *domain* of a relation is the set consisting of all first elements of the ordered pairs. In the given example, the domain is the set of heights, namely, {183, 185, 186, 188}.

The *range* of a relation is the set consisting of all second elements of the ordered pairs. In the given example, the range is the set of players, namely, {Florio, Laube, Richko, Andrews, Jones}.

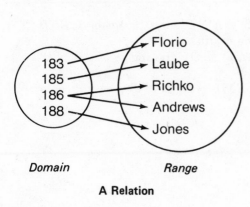

Domain   *Range*

A Relation

As shown at the left, the relation can also be displayed by means of an *arrow diagram*. Arrows are drawn from the elements in the domain to their corresponding elements in the range.

In a relation, an element in the domain may at times correspond to more than one element in the range. Here, for example, 186 → Richko and 186 → Andrews.

Relations and Finite Sets

When the number of ordered pairs in a relation is finite, there are many ways to display the relation. For example, let us consider the relation "is less than," using the set of numbers {1, 2, 3}.

1. *Set of Ordered Pairs*

Let x and y be elements of the set {1, 2, 3}. List all ordered pairs (x, y) where x is less than y. Since $1 < 2, 1 < 3$, and $2 < 3$, the relation is the set of ordered pairs:

$$\{(1, 2), (1, 3), (2, 3)\}$$

2. *Table of Values*

The ordered pairs that define the relation may also be listed as a table of values. Whether using ordered pairs or a table, we notice that:

x	y
1	2
1	3
2	3

   the domain (or set of x-coordinates) = {1, 2}
   the range (or set of y-coordinates) = {2, 3}

3. *Graph*

In a coordinate graph, the domain is a subset of the numbers on the x-axis, and the range is a subset of the numbers on the y-axis. We have listed only the elements 1, 2, and 3 on each axis so that they will correspond to the finite set {1, 2, 3}.

There are two ways to display this relation as a coordinate graph.

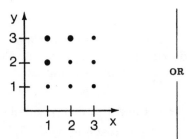   OR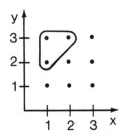

Here, heavy dots are placed on the points $(1, 2), (1, 3)$, and $(2, 3)$ to show the relation $x < y$.

Here, the points $(1, 2), (1, 3)$, and $(2, 3)$ are encircled to show the relation $x < y$.

4. *Arrow Diagram*

After indicating two sets, arrows are drawn from the elements in the domain to their corresponding elements in the range. Notice again that the domain is {1, 2} and the range is {2, 3}.

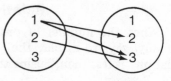

Domain = {1, 2}   Range = {2, 3}

5. *Number Lines*

Whether we use one number line (Fig. 1) or two number lines (Fig. 2), we once again draw arrows from the elements in the domain to their corresponding elements in the range to display the relation.

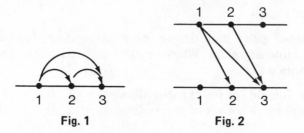

Fig. 1   **Fig. 2**

Note: A relation is sometimes identified by means of a single letter. For example, to name the relation we have just studied, we could write $R = \{(1, 2), (1, 3), (2, 3)\}$ or $r = \{(1, 2), (1, 3), (2, 3)\}$.

Relations and Infinite Sets

When the number of ordered pairs in a relation is infinite, we usually rely on one of two methods to indicate the relation.

1. *A Rule*

In some cases, a relation can be specified by a rule that makes it possible for us to determine the ordered pairs in the relation. Two such relations, one involving an equation and one involving an inequality, are stated here.

Relation $A = \{(x, y)|y = \frac{1}{2}x + 1$ and $x \in$ real numbers$\}$

Relation $B = \{(x, y)|x < y$ and $x \in$ real numbers$\}$

In each of the given relations, the domain is the set of real numbers. Let us make an agreement that allows us to abbreviate the set notation for each of these relations:

■ **If no set is specified, the domain will be the set of real numbers.**

We can now rewrite each of these relations as follows:

$$\text{Relation } A = \{(x, y)|y = \tfrac{1}{2}x + 1\}$$
$$\text{Relation } B = \{(x, y)|x < y\}$$

2. *Graph*

Since a coordinate graph is a picture of a set of ordered pairs, a relation can be shown as a graph. Elements of the domain are found on the x-axis, while elements of the range are read from the y-axis. The graphs of relations A and B stated above are presented here.

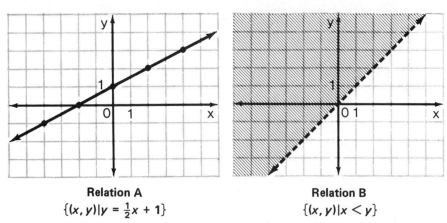

Relation A   **Relation B**
$\{(x, y)|y = \tfrac{1}{2}x + 1\}$ $\{(x, y)|x < y\}$

Every graph containing an infinite set of ordered pairs displays a relation, even if it is not possible to describe the relation by means of a rule, that is, an equation or an inequality.

┌─────── **KEEP IN MIND** ───────┐

A relation is a set of ordered pairs.

└──────────────────────────────┘

MODEL PROBLEMS

1. The accompanying arrow diagram shows the relation "is a divisor of" between the set of numbers $\{2, 4, 5\}$ and the set of numbers $\{1, 2, 3, 4, 5, 6\}$.
   a. Write the indicated relation as a set of ordered pairs.

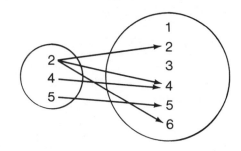

b. State the domain of the relation.
c. State the range of the relation.

Solution	*Answer*

a. Every arrow points from an *x*-coordinate to its corresponding *y*-coordinate(s).

a. {(2, 2), (2, 4), (2, 6), (4, 4), (5, 5)}

b. The domain is the set of *x*-coordinates in the relation.

b. Domain = {2, 4, 5}

c. The range is the set of *y*-coordinates in the relation. Notice that 1 and 3 are not in the relation.

c. Range = {2, 4, 5, 6}

2. For the relation shown in the accompanying diagram, state:
a. the domain   **b.** the range

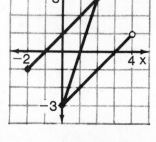

Solution

a. The domain is the set of *x*-coordinates. The point farthest to the left has an *x*-coordinate of -2. At the right, the graph approaches but *does not include* the point whose *x*-coordinate is 4. Since every real number between -2 and 4 is an *x*-coordinate for one or more points on the graph, the domain is $\{x | -2 \leq x < 4\}$.

b. The range is the set of *y*-coordinates. The highest point on the graph has a *y*-coordinate of 3, and the lowest point has a *y*-coordinate of -3. Since every real number between -3 and 3 inclusive is a *y*-coordinate for one or more points on the graph, the range is $\{y | -3 \leq y \leq 3\}$.

Answer: **a.** Domain = $\{x | -2 \leq x < 4\}$
       b. Range = $\{y | -3 \leq y \leq 3\}$

EXERCISES

In 1-4: **a.** State the domain of the relation. **b.** State the range of the relation.

1. {(3, 5), (4, 6), (5, 5), (6, 6)}   **2.** {(1, 2), (1, 1), (1, 0), (1, -1)}
3. {(2, 7), (3, 1), (2, 1), (3, 9)}   **4.** {(0, 2), (0, 4), (1, 2), (3, 4)}

In 5-17, a picture of a relation is given. a. Write the relation as a set of ordered pairs. b. State the domain of the relation. c. State the range of the relation.

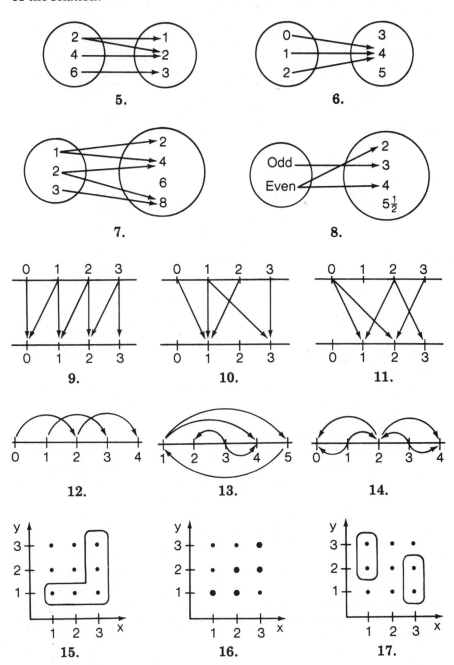

In 18–23, the x-coordinates and y-coordinates of the ordered pairs in the given relation are real numbers. **a.** State the domain of the relation. **b.** State the range of the relation.

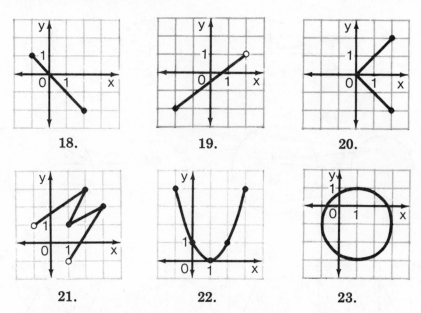

18.   19.   20.

21.   22.   23.

24. A relation R contains the following ordered pairs: $(-1, 3), (0, 4),$ $(1, 5), (2, 0), (2, 1), (2, 2), (2, 3), (2, 4), (2, 6), (3, 3), (3, 7),$ $(4, 3), (4, 6), (5, 4),$ and $(5, 5)$.
   a. Draw a coordinate graph of the relation R.
   b. Which elements of the domain correspond to more than one y-coordinate?
   c. What is the largest number in the range?
   d. State the range of the relation.
25. A relation $M = \{(x, y) | y < x - 5\}$ and its domain is the set of real numbers. **a.** Name two ordered pairs in the relation M, each of which has an x-coordinate of 6. **b.** Name two ordered pairs in the relation M, each of which has a y-coordinate of -5.

6-2 FUNCTIONS

Look carefully at the two relations that follow. While each relation matches fathers with their children, there is a very important difference between the two relations.

A father may have more than one child, but a child has one and only one father.

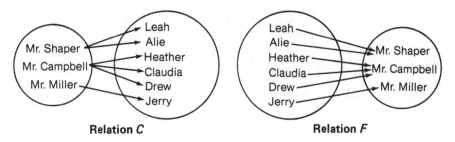

Relation *C*   **Relation *F***

In relation *C*, Mr. Shaper has 2 children, Mr. Campbell has 3 children, and Mr. Miller has 1 child. Thus, every father listed in the domain corresponds to 1, 2, or 3 children listed in the range.

In relation *F*, however, every child listed in the domain corresponds to *one and only one* father listed in the range. While both *C* and *F* are relations, only relation *F* is called a *function*.

■ **DEFINITION.** A *function* is a relation in which each element of the domain corresponds to one and only one element in the range.

□ EXAMPLE 1: At the right, every element in the domain is matched to an
element in the range by the rule $x \rightarrow x^2$.
Since every number has "one and only
one square," this relation is a function.
This function may be listed as a set of
ordered pairs:

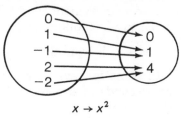

$x \rightarrow x^2$

A Function

$$\{(0, 0), (1, 1), (-1, 1), (2, 4), (-2, 4)\}$$

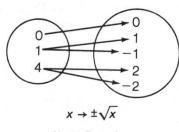

$x \rightarrow \pm\sqrt{x}$

Not a Function

□ EXAMPLE 2: At the left, every element in the domain is matched to its
square root(s) in the range by the rule
$x \rightarrow \pm\sqrt{x}$. Since every nonzero number
has two square roots, this is not a function. This relation may be listed as a set
of ordered pairs:

$$\{(0, 0), (1, 1), (1, -1), (4, 2), (4, -2)\}$$

In terms of ordered pairs, how do the two examples differ?

1. A Function: $\{(0, 0), (1, 1), (-1, 1), (2, 4), (-2, 4)\}$
2. Not a Function: $\{(0, 0), (1, 1), (1, -1), (4, 2), (4, -2)\}$

          Same first   Same first
          element   element

If two or more ordered pairs in a relation have the same first element, then some element in the domain corresponds to more than one element in the range. Thus, the relation is not a function, as seen in example 2. This leads to an alternate definition of function.

■ **DEFINITION.** A *function* is a relation in which no two ordered pairs have the same first element.

Functions and Graphs

We have learned that a relation whose domain and range are subsets of the real numbers can be displayed as a coordinate graph. We will now study a method that enables us to tell very quickly when a relation presented as a graph is a function.

The set of ordered pairs that names a function in example 1 on the preceding page is displayed in Fig. 1 at the right. Each ordered pair (x, y) is determined by the rule $x \rightarrow x^2$. Since the second element y of each ordered pair is equal to x^2, it follows that:

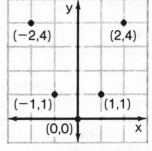

Fig. 1

$y = x^2$ is equivalent to the rule $x \rightarrow x^2$

In a function, every element x of the domain corresponds to one and only one element y in the range. Keep this in mind to understand the vertical-line test for a function.

■ *Vertical-Line Test for a Function:* If each vertical line drawn through the graph of a relation intersects the graph at one and only one point, then the relation is a function.

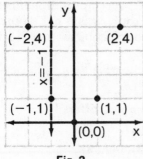

Fig. 2

□ Example 1: When the domain is {-2, -1, 0, 1, 2}, the five points whose coordinates fit the equation $y = x^2$ are graphed in Fig. 2. Is the indicated relation a function?

The vertical line whose equation is $x = -1$ intersects the graph of the relation at one and only one point, $(-1, 1)$. Thus, if $x = -1$, then $y = 1$. Similarly, each vertical line that passes through a point in the relation will intersect one and only one point of the relation.

Answer: The indicated relation is a function.

Note: If a domain had not been specified, the relation whose equation is $y = x^2$ would have as its domain the set of all real numbers, as graphed in Fig. 3. The vertical-line test demonstrates geometrically that every real number has one and only one square. Thus, we can claim that:

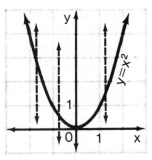

"$y = x^2$ is a function."
OR
"$y = x^2$ is a function of x."

Fig. 3

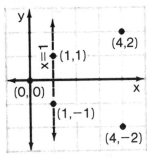

Fig. 4

□ EXAMPLE 2: The set of ordered pairs in Fig. 4 fits the rule $x \rightarrow \pm\sqrt{x}$. Is the indicated relation a function? While it was shown earlier that this relation is not a function, let us apply the vertical-line test.

The vertical line whose equation is $x = 1$ intersects the graph of the relation at two points: $(1, 1)$ and $(1, -1)$. Thus, if $x = 1$, then $y = 1$ or $y = -1$. This vertical-line test demonstrates that there is at least one element x of the domain that corresponds to more than one element y in the range.

Answer: The indicated relation is *not* a function.

Note: Since the rule $x \rightarrow \pm\sqrt{x}$ determines a set of ordered pairs (x, y), it follows that $y = \pm\sqrt{x}$. By squaring both members of $y = \pm\sqrt{x}$, we form the equation $y^2 = x$. The relation whose equation is $y^2 = x$ can be extended to include a domain of non-negative real numbers, as graphed in Fig. 5.

The vertical-line test demonstrates in Fig. 5 that every positive real number has *two* square roots. Thus, we state that "$y^2 = x$ is *not* a function of x." It is correct, however, to state

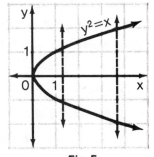

Fig. 5

that "$y^2 = x$ is a function of y." This can be demonstrated geometrically by rotating the graph 90° so that the y-axis is a horizontal line. Try it by turning the book sideways. In spite of this fact, we will make the following agreements to avoid confusion:

1. When testing for a function, unless otherwise stated, we will test to see if the relation is a function of x only.
2. If a relation is not a function of x, we will simply state that the relation is not a function.

Therefore, using the graph in Fig. 5, we state that "$y^2 = x$ is *not* a function."

Restricted Domains and Functions

If no set is specified as the domain of a relation, we have agreed that the domain will be the set of real numbers. A relation, however, might not be a function for the set of real numbers. For example, $y = \dfrac{1}{x}$ is not a function for the set of reals because no y-value corresponds to an x-value of 0. And yet, $y = \dfrac{1}{x}$ is a function for "the set of real numbers *less zero*," written in symbols as "Real numbers/$\{0\}$." Thus, the relation is a function for a *restricted domain*.

■ If no set is specified as the domain of a relation, it is often possible to state that the relation is a function by letting the domain be the largest possible subset of the real numbers for which a function exists, that is, for which every element of the domain corresponds to one and only one real number in the range.

For example, $y = \dfrac{1}{x-3}$ is a function whose domain = Real numbers/$\{3\}$, and $y = \sqrt{x}$ is a function whose domain = $\{x | x \geq 0\}$. We will study restricted domains in greater detail in section 4 of this chapter.

KEEP IN MIND

A function is a relation in which:

1. each element of the domain corresponds to one and only one element in the range

OR

2. no two ordered pairs have the same first element.

| MODEL PROBLEMS |

In 1–4, let the domain = {1, 2, 3}. State whether or not the indicated relation is a function.

1.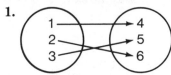

Solution: Every element in the domain corresponds to one and only one element in the range.

Answer: This relation *is* a function with range {4, 5, 6}.

2.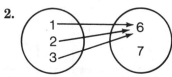

Solution: Every element in the domain corresponds to one and only one element in the range.

Answer: This relation *is* a function with range {6}.

3.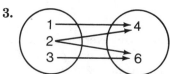

Solution: The element 2 in the domain corresponds to *more than one* element in the range.

Answer: This relation *is not* a function.

4.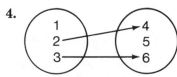

Solution: The element 1 in the domain corresponds to *no* elements in the range.

Answer: For the domain {1, 2, 3}, this relation *is not* a function.

5. Find the largest possible *restricted domain* that allows the relation in problem 4 to be a function.

Solution: Exclude the element 1 from the domain because this element corresponds to no element in the range. Thus, we restrict the domain to include only the elements 2 and 3, each of 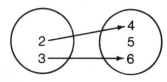 which corresponds to one and only one element in the range, {4, 6}.

Answer: A function exists for the restricted domain {2, 3}.

6. Which relation, if any, is a function?

$$A = \{(0, 3), (1, 8), (1, 5)\} \qquad B = \{(0, 2), (1, 2), (3, 2)\}$$

Solution: Relation A contains $(1, 8)$ and $(1, 5)$. Since 1 corresponds to both 8 and 5, A is not a function. In relation B, however, no two ordered pairs have the same first element.

Answer: B is a function.

7. Which graph represents a relation that is a function?

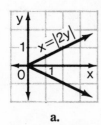
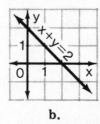
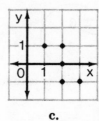
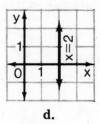

  a.   **b.**   **c.**   **d.**

Solution: By the vertical-line test, only in **b** will every vertical line intersect the graph of the relation at one and only one point. For example, consider the vertical line whose equation is $x = 2$:

In a, when $x = 2$, then $y = 1$ or $y = -1$. (*Not* a function.)

In b, when $x = 2$, then $y = 0$.

In c, when $x = 2$, then $y = 1$, $y = 0$, or $y = -1$. (*Not* a function.)

In d, when $x = 2$, then y may be any real number. (*Not* a function.)

Answer: **b**

┌ EXERCISES ┐

In 1-8, let the domain = {0, 1, 2}. a. State whether or not the indicated relation is a function. b. If the relation is not a function, explain why.

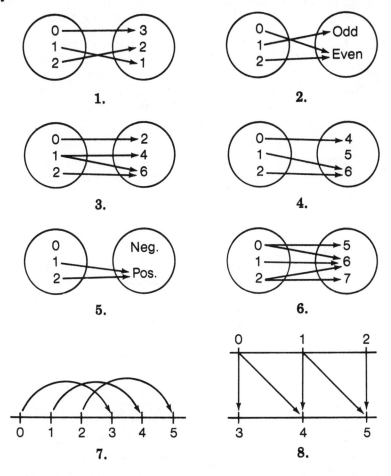

1.

2.

3.

4.

5.

6.

7.

8.

9. For those relations in exercises 1–8 that are *not* functions, state the exercise number and the largest possible *restricted domain* that allows the relation to be a function.

In 10–13: a. State the domain of the relation. b. State the range of the relation. c. Tell whether or not the relation is a function.

10. {(2, 3), (3, 5), (4, 7), (5, 9)}   11. {(3, 1), (2, 3), (1, 2), (3, 2)}
12. {(6, 6), (6, 3), (6, 2), (6, 1)}   13. {(5, 2), (6, 1), (7, 2), (8, 1)}

In 14–19, let the domain = {-1, 0, 1}.
a. Copy the encircled sets at the right, and draw an arrow diagram to indicate the ordered pairs in the relation whose rule is given. b. State the range of the relation. c. For the domain {-1, 0, 1}, is the relation a function?

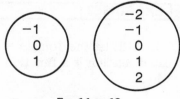

Ex. 14 to 19

14. $x \to x$ 15. $x \to x^2$ 16. $x \to x - 1$
17. $x \to 2x + 1$ 18. $x \to 1 - x$ 19. $x \to y$, where $y < x$

20. Let $A = \{(10, 3), (7, 2), (-2, 5), (x, 1)\}$. Name all possible values that x may *not* represent so that relation A can be a function.

21. The domain of a function F is $\{-2, \sqrt{2}, \sqrt{3}, 2\}$, and the rule of the function is $x \to x^2 + 3$. List all ordered pairs in this function.

22. The domain of a function B is $\{1, \sqrt{2}, \sqrt{3}, 2\}$, and the rule of the function is $x \to 3x - 2$. What is the range of function B?

In 23–34, determine whether or not the graph represents a function. (*Hint:* Use the vertical-line test.)

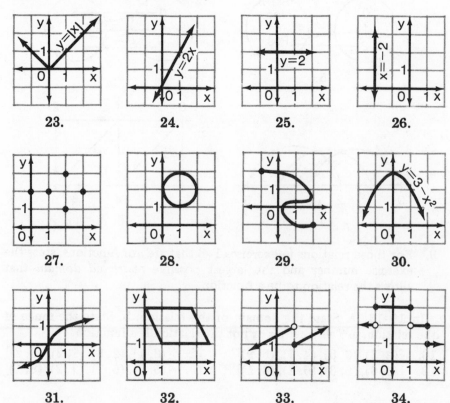

23.   24.   25.   26.

27.   28.   29.   30.

31.   32.   33.   34.

35. **a.** What is the domain of the relation whose graph is given in exercise 23? **b.** What is its range?
36. **a.** What is the domain of the relation whose graph is given in exercise 28? **b.** What is its range?
37. **a.** What is the domain of the relation whose graph is given in exercise 34? **b.** What is its range?

In 38–50, let $D \rightarrow R$ represent a domain D corresponding to a range R.
a. Is the relation a function? **b.** If not, explain why.

38. Children → mothers.
39. Sisters → brothers.
40. People → ages.
41. Heights → students.
42. Students → teachers.
43. States → capitals.
44. Homes in the U.S.A. → zip codes.
45. Cities in the U.S.A. → zip codes.
46. Birthdays → people.
47. School lockers → students.
48. Letters sent first class → cost of postage.
49. Cars being driven → drivers at the wheel.
50. Drivers' license registration numbers → licensed drivers.

51. Describe three functions that exist in the world about us, similar to the functions found in exercises 38 to 50.

6-3 FUNCTION NOTATION

If a function is specified by a rule or an equation, the rule may be expressed in a variety of ways. For example, let us consider the function f in which "every element in the range is 3 more than its corresponding element in the domain." This rule is indicated by any of the following expressions:

In Symbols	In Words
1. f: $x \rightarrow x + 3$	1. "Under function f, x maps to $x + 3$." OR "Under f, x is assigned to $x + 3$." OR "The image of x under f is $x + 3$."
2. $x \xrightarrow{\text{f}} x + 3$	2. Same as 1.
3. $f(x) = x + 3$	3. "f of x equals $x + 3$."
4. $f = \{(x, y) \mid y = x + 3\}$	4. "Function f is the set of ordered pairs (x, y) such that y equals $x + 3$."
5. $y = x + 3$	5. "y equals $x + 3$."

Notice that the last expression, $y = x + 3$, is simply an abbreviated form of the function notation stated in expression 4.

When we compare expressions 3 and 5, namely $f(x) = x + 3$ and $y = x + 3$, it becomes clear by substitution that:

$$f(x) = y$$

In other words, $f(x)$, read "f of x," is equal to the y-coordinate in an ordered pair. The symbol $f(x)$ does *not* indicate multiplication; rather, $f(x)$ indicates the substitution of an x-value to find its corresponding y-value, called $f(x)$. Compare the two expressions that follow:

$y = x + 3$	$f(x) = x + 3$
If $x = 2$, then $y = 2 + 3$	If $x = 2$, then $f(2) = 2 + 3$
$y = 5$	$f(2) = 5$
The ordered pair $(x, y) = (2, 5)$ is a member of the function $y = x + 3$.	The ordered pair $(x, f(x)) = (2, 5)$ is a member of the function $f(x) = x + 3$.

| MODEL PROBLEMS |

1. If a function g is defined by $g(x) = 2x + 15$, find the value of $g(-3)$.

How to Proceed	*Solution*
1. Write the rule of the function.	$g(x) = 2x + 15$
2. Substitute -3 for x, and simplify.	$g(-3) = 2(-3) + 15$
	$g(-3) = -6 + 15$
	$g(-3) = 9$ *Ans.*

2. The graph of function f is shown at the right.

   a. Find $f(-1)$.   b. Find $f(0)$.
   c. Find $f(1)$.   d. Find $f(3)$.

Solution

a. Since $f(x) = y$, the ordered pair $(-1, 2)$ indicates that $f(-1) = 2$.
b. Using the ordered pair $(0, 3)$, we see that $f(0) = 3$.
c. Using the ordered pair $(1, 3)$, we see that $f(1) = 3$.
d. Using the ordered pair $(3, 4)$, we see that $f(3) = 4$.

Answer: a. 2   b. 3   c. 3   d. 4

Note: While the domain of function f is $\{x|-1 \leq x \leq 3\}$, the range may be expressed in either of two ways:

$$\{y|1 < y \leq 4\} \text{ or } \{f(x)|1 < f(x) \leq 4\}$$

| EXERCISES |

In 1-6, the rule that defines function f is given.
a. Find f(1). **b.** Find f(2).

1. $f(x) = 4x$ **2.** $f(x) = x - 5$ **3.** $f(x) = x^2 - 1$

4. $f(x) = \dfrac{1}{x}$ **5.** $f(x) = 3x - 4$ **6.** $f(x) = \dfrac{x - 1}{x + 1}$

In 7-12, the rule that defines function g is given.
a. Find g(3). **b.** Find g(-1).

7. $g(x) = \dfrac{2x}{3}$ **8.** $g(x) = 8 - x$ **9.** $g(x) = \dfrac{x + 7}{x + 2}$

10. $g(x) = x^2 + x$ **11.** $g(x) = -x^2$ **12.** $g(x) = x^2 - 2x - 3$

13. If a function h is defined by $h(x) = \dfrac{x^2 - x}{4}$, find the value of:

     a. h(4)   **b.** h(-4)   **c.** h(2)   **d.** h(1)   **e.** h(1.2)

14. If a function k is defined by $k(x) = \dfrac{6 - x}{x - 3}$, find the value of:

     a. k(2)   **b.** k(9)   **c.** k(0)   **d.** k(6)   **e.** k(-3)

In 15-20, use the following functions m, p, and r:

$$x \xrightarrow{\text{m}} 2x \qquad p: x \longrightarrow x^2 \qquad x \xrightarrow{\text{r}} x + 2$$

15. Find m(3).   **16.** Find p(3).   **17.** Find r(3).
18. Under which function (m, p, or r) will 5 map to 7?
19. Under which function (m, p, or r) will the image of $\frac{1}{2}$ be 1?
20. For m, p, and r, which function(s) assign 2 in the domain to 4 in the range?

21. The graph of function f consists of the union of four line segments, as shown at the right.
     a. Find f(-1).   **b.** Find f(0).
     c. Find f(1).   **d.** Find f(2).
     e. Find $f(\frac{1}{2})$.   **f.** Find $f(2\frac{2}{3})$.
     g. State the domain of function f.
     h. State the range of function f.

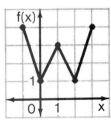
Ex. 21

22. The graph of function g is shown at the right.
    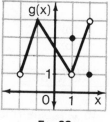
    a. Find g(-1).   **b.** Find g(1).
    c. Find g(2).   **d.** Find g(0).
    e. Find g($1\frac{1}{2}$).   **f.** Find g($-1\frac{1}{2}$).
    g. For what value(s) of x will g(x) = $2\frac{1}{2}$?
    h. State the domain of function g.
    i. State the range of function g.

    Ex. 22

In 23–26, select the numeral preceding the expression that best answers the question.

23. Which of the following functions contains the ordered pair (3, 1)?
    (1) $y = 3x$ (2) $y = x + 2$ (3) $y = x - 2$ (4) $y = 2x + 1$
24. In which of the following functions is 7 from the domain mapped to 10 in the range?
    (1) $\{(x, y) | y = x - 3\}$ (2) $\{(x, y) | y = x + 3\}$
    (3) $\{(x, y) | y = 7\}$ (4) $\{(x, y) | y = x + 7\}$
25. Which of the following is *not* a function?
    (1) $y = 2x$ (2) $x = 2y$ (3) $y = 2$ (4) $x = 2$
26. The domain of function F is $\{1, 0, -1\}$ and F: $x \to x^3 - x$. What is the range of F?
    (1) $\{0\}$ (2) $\{0, 2\}$ (3) $\{0, -2\}$ (4) $\{1, 0, -2\}$

27. A relation R = $\{(x, y) | y < \frac{1}{3}x - 2\}$. **a.** Name two ordered pairs in R, each having an x-coordinate of 3. **b.** Name two ordered pairs in R, each having a y-coordinate of -2. **c.** Is relation R a function? Explain why.
28. A car travels on a highway at an average rate of 55 miles per hour. For this particular car, the formula "distance = rate × time," or $d = rt$, can be stated as $d = 55t$.
    a. Does the rule $d = 55t$ describe a function? Explain why.
    b. *True* or *False:* Since $d = 55t$ and f(t) = $55t$ each describe the same rule, it can be said that $d = $ f(t), or "distance is a function of time."
    c. Find the distance traveled by the car: (1) in 1 hour   (2) in 2 hours   (3) in 3 hours 12 minutes.

6-4 TYPES OF FUNCTIONS

In this section, we will list some types of functions. Many of these functions should be familiar to you from earlier studies in mathematics.

1. *Linear Functions*

$$y = mx + b \qquad \text{OR} \qquad f(x) = mx + b$$

Any equation of the form $y = mx + b$, or $f(x) = mx + b$, is a function whose graph is a straight line. Recall that m is the slope of the line, and b is the y-intercept.

For example, the equation $y = -4x + 3$ is equivalent to $f(x) = -4x + 3$. Thus, if $x = 2$, then $y = -5$. Also, if $x = 2$, then $f(x) = f(2) = -5$. This relation is a function because, for every x-value, there is one and only one corresponding y-value, or one and only one corresponding $f(x)$.

2. *Constant Functions*

$$y = b \qquad \text{OR} \qquad f(x) = b$$

If the slope of a linear function is zero, or $m = 0$, then the linear function becomes $y = b$ or $f(x) = b$. This relation is called a constant function because every x corresponds to the same constant value, b.

For example, the graph of the constant function $y = 3$, or $f(x) = 3$, is shown at the right. Here, $f(1) = 3$, $f(2) = 3$, $f(-1) = 3$, and so forth. While the domain consists of all real numbers, the range consists of a single element, $\{3\}$.

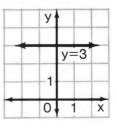

Note: The equation $x = b$ is *not* a constant function because its graph is a vertical line, indicating that x corresponds to more than one value of y.

3. *Quadratic Functions*

$$y = ax^2 + bx + c \text{ where } a \neq 0 \qquad \text{OR} \qquad f(x) = ax^2 + bx + c \text{ where } a \neq 0$$

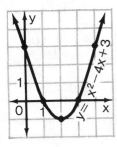

The quadratic equation whose graph appears at the left is $y = x^2 - 4x + 3$, or $f(x) = x^2 - 4x + 3$. Here, $f(0) = 3$, $f(1) = 0$, $f(2) = -1$, $f(3) = 0$, $f(4) = 3$, and so forth. Since every x-value corresponds to one and only one y-value, this relation is a function.

The vertical-line test can be used to demonstrate that quadratics of the form $y = ax^2 + bx + c$ where $a \neq 0$, or $f(x) = ax^2 + bx + c$ where $a \neq 0$, are functions.

Note: Quadratics of the form $x = ay^2 + by + c$ where $a \neq 0$ are *not* functions. For example, the quadratic $x = y^2$ contains two points whose coordinates are (4, 2) and (4, -2), indicating that an x-value corresponds to more than one y-value.

4. *Polynomial Functions*

$$y = a_n x^n + a_{n-1} x^{n-1} + a_{n-2} x^{n-2} + \ldots + a_0$$

OR

$$f(x) = a_n x^n + a_{n-1} x^{n-1} + a_{n-2} x^{n-2} + \ldots + a_0$$

For example, consider $f(x) = x^4 - 2x^3 + x^2 - 4x + 3$. If $x = 2$, then $f(x) = f(2) = 16 - 16 + 4 - 8 + 3 = -1$. In the same way, for each chosen value of x, there is one and only one corresponding $f(x)$. Thus, the given polynomial is a function.

The polynomial function can be a quadratic function, such as $f(x) = x^2 - 4x + 3$ when $n = 2$; or a linear function, such as $f(x) = -4x + 3$ when $n = 1$; or a constant function, such as $f(x) = 3$ when $n = 0$. We see, therefore, that the quadratic, linear, and constant functions are merely special cases of the more general polynomial function.

5. *Absolute-Value Functions*

$$y = |ax + b| \qquad \text{OR} \qquad f(x) = |ax + b|$$

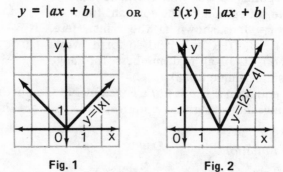

Fig. 1   Fig. 2

The graph of the absolute-value equation $y = |x|$, or $f(x) = |x|$, is shown in Fig. 1. Here, if $x = -2$, then $f(x) = f(-2) = |-2| = 2$.

The graph of the absolute-value equation $y = |2x - 4|$, or $f(x) = |2x - 4|$, is shown in Fig. 2. Here, if $x = 1$, $f(x) = f(1) = |2 \cdot 1 - 4| = |2 - 4| = |-2| = 2$.

In each equation, every value of x corresponds to one and only one $f(x)$, that is, one and only one y-value. Therefore, $y = |x|$ and $y = |2x - 4|$ are both functions. Remember that these functions may be written as $f(x) = |x|$ and $f(x) = |2x - 4|$, respectively.

Note: Absolute-value equations of the form $x = |ay + b|$ are *not* functions. For example, the equation $x = |y|$ contains two points whose coordinates are (3, 3) and (3, -3), indicating that an x-value corresponds to more than one y-value.

6. *Step Functions*

There are many examples of step functions, each of which has a graph that resembles a series of steps. Let us consider two examples.

☐ EXAMPLE 1: *The Greatest-Integer Function*

The symbol $[x]$ means "the greatest integer equal to or less than x." Therefore, the greatest-integer function is indicated by the rule $y = [x]$, or $f(x) = [x]$. A partial graph of this step function is shown at the right.

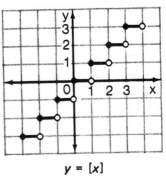

$y = [x]$

If $x = 2$, then $f(x) = f(2) = [2] = 2$.

If $x = 2\frac{2}{3}$, then $f(x) = f(2\frac{2}{3}) = [2\frac{2}{3}] = 2$.

If $x = 2.98$, $f(x) = f(2.98) = [2.98] = 2$.

However:

If $x = 3$, $f(x) = f(3) = [3] = 3$.

Therefore, if x is equal to or greater than 2 but less than 3, the greatest integer in x is 2. In symbols, we write: If $2 \leq x < 3$, then $[x] = 2$. This set of values is graphed on the line $y = 2$ as "●—○," a segment with one missing endpoint.

Similar reasoning is used to complete the graph of this step function. It should be clear that $[3.6] = 3$, $[1\frac{1}{3}] = 1$, and $[0.7] = 0$. Notice, if x is not an integer, then $[x]$ is less than x. Using this reasoning and thinking of a thermometer, we can see that $[-\frac{1}{2}] = -1$, $[-1.13] = -2$, and $[-2\frac{1}{4}] = -3$.

For the greatest-integer function, $y = [x]$, we observe that the domain is the set of real numbers and the range is the set of integers.

☐ EXAMPLE 2: *The Post-Office Function*

The cost of sending a letter by first-class mail is determined by the weight of the letter. In 1982, the U.S. Post Office charged $.20 for sending a first-class letter weighing 1 ounce or less, and $.17 for each additional ounce or part of an ounce. The table and graph on the next page show the beginning portion of this step function.

Weight of Letter (in ounces)	Cost of First-Class Mail (in dollars)
$0 < x \le 1$	.25 = .25
$1 < x \le 2$	.25 + .20 = .45
$2 < x \le 3$	.25 + 2(.20) = .65
$3 < x \le 4$	.25 + 3(.20) = .85
$4 < x \le 5$	.25 + 4(.20) = 1.05

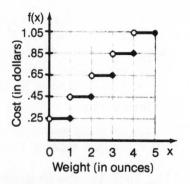

Weight (in ounces)

Since a letter cannot have a weight of 0 ounce or a negative weight, the domain of this function is $\{x|x \in \text{positive real numbers}\}$. The range consists of multiples of \$.20 added to the initial cost of \$.25.

Note: The post-office function is easily graphed from the given table. While it is not necessary to know the equation of the post-office function, we will mention the equation here because it is an interesting rule that involves a negation, the greatest integer, and absolute value. Can you see how the rule works? If x is the weight of a letter in ounces and $f(x)$ is the cost in dollars for first-class postage, then:

$$f(x) = .25 + .20(|[-x]| - 1)$$

7. *Other Functions*

A great number of other functions exist in mathematics. Many transformations are functions, as we will see in the next chapter. In later chapters of this book, we will study trigonometric functions, exponential functions, and logarithmic functions. Also, as we will see in the exercises that follow, there are many functions in the world about us.

Remember: A relation is sometimes a function under a restricted domain. If no domain is specified, we have agreed to let the domain be the largest possible subset of the real numbers for which a function exists.

For example, $f(x) = \dfrac{1}{x^2 - 9}$ is a function whose domain is the set Real numbers/$\{3, -3\}$, and $g(x) = \sqrt{x - 4}$ is a function whose domain is $\{x|x \ge 4\}$. Other examples can be found in model problems 3 and 4, which follow.

MODEL PROBLEMS

1. In the function $\{(x, y)|y = 2x - 5\}$, the domain is $\{x|0 \le x \le 4\}$. Find the range of the function.

Solution

Since $y = 2x - 5$ is a linear function, consider the extreme values of x in the domain.

When $x = 0$: $y = 2x - 5 = 2(0) - 5 = 0 - 5 = -5$.

When $x = 4$: $y = 2x - 5 = 2(4) - 5 = 8 - 5 = 3$.

In a linear function, the range consists of all real numbers between the extreme values of y. Here, the range is $-5 \leq y \leq 3$. (Although it is not necessary to graph the function, the graph is

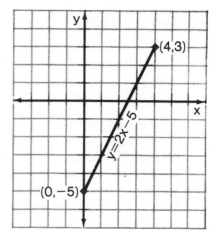

included here as an aid to understanding the solution.)

Answer: The range is $\{y \mid -5 \leq y \leq 3\}$.

2. The domain of $f(x) = 10 - x$ is $7 \leq x \leq 12$. What is the greatest value in the range of f?

Solution

Since $f(x) = 10 - x$ is a linear function, test the extreme values of x in the domain to find the extreme values of $f(x)$ in the range.

$$f(x) = 10 - x$$
When $x = 7$, then $f(7) = 10 - 7 = 3$.
When $x = 12$, then $f(12) = 10 - 12 = -2$.

The greatest value in the range of this linear function is 3.

Answer: 3

3. What is the domain of the function $k(x) = \dfrac{x - 3}{x - 6}$?

Solution

Every real number x corresponds to one and only one value $k(x)$, except when the denominator $x - 6$ is equal to zero.

If $x - 6 = 0$, then $x = 6$.

Thus, the domain of the function is the set of real numbers less 6.

Answer: Real numbers/$\{6\}$.

4. What is the domain of the function $f(x) = \dfrac{1}{\sqrt{x-2}}$?

Solution

If $x = 2$, then $f(x) = \dfrac{1}{\sqrt{x-2}} = \dfrac{1}{\sqrt{2-2}} = \dfrac{1}{\sqrt{0}} = \dfrac{1}{0}$ (undefined).

Also, if x is less than 2, then f(x) contains the square root of a negative number, that is, f(x) is not a real number. Since the range as well as the domain is to contain only real numbers, the domain of the function is restricted to those real numbers greater than 2.

Answer: $\{x \mid x > 2\}$.

| EXERCISES |

In 1-9: **a.** Identify the function as being a linear, constant, quadratic, absolute-value, or step function. **b.** Find f(3). **c.** Find $f(\frac{1}{2})$.

1. $f(x) = 2$ 2. $f(x) = x$ 3. $f(x) = [x]$
4. $f(x) = 2x^2$ 5. $f(x) = 2 - x$ 6. $f(x) = |2 - x|$
7. $f(x) = [2 - x]$ 8. $f(x) = x^2 - x$ 9. $f(x) = \pi$

In 10-12, for the given function and its domain, find the range.

10. $\{(x, y) \mid y = 3x - 2\}$; domain $= \{x \mid -2 \le x \le 2\}$
11. $f(x) = 5 - \frac{1}{2}x$; domain $= \{x \mid 0 \le x \le 16\}$
12. $g: x \to 6 - x$; domain $= \{x \mid x \ge 0\}$

13. If the domain of $y = 4x - 3$ is $\{x \mid 2 \le x \le 5\}$, what is the greatest value in its range?
14. The domain of $f(x) = 9 - 2x$ is $4 \le x \le 10$. What is the greatest value in the range of f?
15. Let the domain of the quadratic function $y = 1 + 4x - x^2$ be $0 \le x \le 4$. **a.** Graph the function, including all points whose x-coordinates are 0, 1, 2, 3, 4. **b.** What is the greatest value in the range of this function?

In 16-24, state the largest possible domain such that the given relation is a function.

16. $f(x) = \dfrac{2}{x - 2}$ 17. $g(x) = \dfrac{x - 3}{x - 9}$ 18. $h(x) = \dfrac{x}{x + 7}$

19. $m(x) = \dfrac{6}{x^2 - 16}$ **20.** $k : x \rightarrow \dfrac{1}{x^2 - 5x}$ **21.** $r : x \rightarrow 3x - 6$

22. $x \xrightarrow{f} \sqrt{x - 1}$ **23.** $x \xrightarrow{g} \dfrac{1}{\sqrt{x - 1}}$ **24.** $y = \dfrac{1}{x^2 + 1}$

In 25–30, for the given function: **a.** State the domain. **b.** State the range.

25. $y = 2x$ **26.** $y = x^2$ **27.** $y = \sqrt{x}$

28. $h(x) = |x - 5|$ **29.** $f(x) = 10$ **30.** $x \xrightarrow{g} x + 10$

31. Given the function $f(x) = \dfrac{2x - 6}{x - 3}$:

a. State the domain of the function.
b. Find $f(5)$. c. Find $f(38)$.
d. Find $f(0)$. e. Find $f(-2)$.
f. *True* or *False:* For every x in the domain stated in part a, $f(x) = 2$. Explain why.
g. State the range of the function.

32. Evaluate each expression, finding the greatest integer.
   a. $[17]$ b. $[27\frac{1}{2}]$ c. $[1.23]$ d. $[0.8]$
   e. $[-5\frac{1}{2}]$ f. $[-0.1]$ g. $[\frac{15}{4}]$ h. $[-\frac{15}{4}]$

In 33–35, select the numeral preceding the expression that best answers the question.

33. If $k(x) = \dfrac{x - 2}{x - 1}$, for what value of x will $k(x) = 0$?

   (1) 1 (2) 2 (3) both 1 and 2 (4) neither 1 nor 2

34. Which of the following ordered pairs is *not* an element of the greatest-integer function, $y = [x]$?
   (1) $(8, 8)$ (2) $(2.76, 2)$ (3) $(-3.6, -3)$ (4) $(-4.6, -5)$

35. Which of the following is *not* a function?
   (1) the line $y = 5x - 4$ (2) the parabola $y = x^2 - 3x$
   (3) the line $y = 2$. (4) the circle $x^2 + y^2 = 16$

In 36–41: **a.** Graph the given function for the domain $-3 \leq x \leq 3$. **b.** Using this domain, what is the range of the function?

36. $y = |x|$ **37.** $f(x) = |3x|$ **38.** $y = |x| + 2$

39. $f(x) = |x + 2|$ **40.** $y = 3 - |x|$ **41.** $f(x) = x + |x|$

In 42–47: **a.** Graph the given function for the domain $0 \leq x < 6$. **b.** Using this domain, what is the range of the function?

42. $f(x) = [x]$ **43.** $y = [x - 2]$ **44.** $g(x) = [3 - x]$

45. $y = [\frac{1}{2}x]$ **46.** $h(x) = [\frac{1}{3}x]$ **47.** $y = x - [x]$

Applications With Functions

48. A newspaper deliverer earns $.07 for each paper delivered. Thus, the earnings E is a function of the number n of newspapers delivered, or $E = f(n)$. The formula to determine the deliverer's earnings is $E = .07n$, or $f(n) = .07n$.

a. How much is earned when 20 papers are delivered? That is, find $f(20)$.

b. Find $f(15)$.

c. Find $f(32)$.

d. Find $f(57)$.

e. If Jennifer earns $2.66 for a daily delivery, how many newspapers does she deliver each day?

49. The accompanying chart shows an 8% sales tax to be collected on amounts from $.01 to $1.06. On sales over $1.06, the tax is computed by multiplying the amount of sale by .08, and rounding to the nearest whole cent.

The sales tax t is a function of the amount A of the sale, that is, $t = f(A)$.

Amount of Sale	8% Sales Tax
$.01 to $.10	NONE
.11 to .17	$.01
.18 to .29	.02
.30 to .42	.03
.43 to .54	.04
.55 to .67	.05
.68 to .79	.06
.80 to .92	.07
.93 to 1.06	.08

a. Find the sales tax on an item costing $.52, that is, find $f(\$.52)$.

b. Find $f(\$.89)$.   c. Find $f(\$.39)$.

d. Find $f(\$.08)$.   e. Find $f(\$9.95)$.

f. *True* or *False:* Sales tax is an example of a step function.

g. *True* or *False:* For amounts less than $1.00, the tax in the chart is equal to 8% of the amount, rounded to the nearest whole cent.

50. The cost of sending a telegram is \$7.45 for 15 words or less, and \$.25 for each additional word. Thus, the cost c is a function of the number of words w, or $c = f(w)$.

   a. Find the cost to send a telegram of 17 words, that is, find $f(17)$.

   b. Find $f(21)$.   c. Find $f(13)$.   d. Find $f(38)$.

   e. If a telegram costs \$10.45, how many words does it contain?

51. A mailgram costs \$3.90 for 50 words or less, and \$1.25 for each additional 50 words or part thereof. The cost c is a function of the number of words w, or $c = f(w)$.

   a. Find the cost of a mailgram containing 38 words, that is, find $f(38)$.

   b. Find $f(65)$.   c. Find $f(170)$.   d. Find $f(114)$.

   e. What is the maximum number of words in a mailgram costing \$8.90?

   f. What is the minimum number of words in a mailgram costing \$8.90?

52. If an object is dropped from a height, the distance d that it travels is a function of the time t for its fall, or $d = f(t)$. In Earth's gravity, this distance is found by the formula $d = 16t^2$, or $f(t) = 16t^2$, where d is distance in feet and t is time in seconds. Assume that an object is dropped from the top of a tall building.

   a. How many feet will the object travel in 1 second? Or, find $f(1)$.

   b. Find $f(2)$.   c. Find $f(3)$.   d. Find $f(4)$.

   e. If the object hits the ground in 7 seconds, how tall is the building?

53. A car-rental agency charges \$50 for a rental of 3 days or less and \$20 for each additional day or part thereof. A weekly rental costs \$100. The graph at the right represents the charges for rentals of one week or less.

   a. *True* or *False:* The rental fee r is a function of the number of days d of rental, or $r = f(d)$.

   b. Find the rental fee for 4 days, that is, find $f(4)$.

   c. Explain why $f(6) \neq \$110$.

   d. State the range of the function.

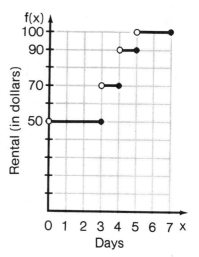

Hours	Rate
First 2 hours or less	$3.00
Each additional 1 hour or part	$1.00
Maximum for 24 hours	$7.00

54. The rates charged to park a car in a city lot are shown in the accompanying sign. The rate r is a function of the number of hours h that the car was parked, or $r = f(h)$.
   a. Graph the function for a 24-hour period.
   b. Find the rate to park a car for $2\frac{1}{2}$ hours, that is, find $f(2\frac{1}{2})$.
   c. Find $f(4\frac{1}{4})$.
   d. Find $f(7)$.
   e. Find $f(10\frac{1}{2})$.

55. A local telephone call from a home phone, under timed service, costs $.08 for the first 5 minutes or less and $.01 for each additional minute or any part thereof.
   a. Graph this step function for costs of telephone calls lasting 12 minutes or less.
   b. *True* or *False:* The cost is a function of the time of the call.
   c. Using these rates, find the cost of a call lasting:
      (1) $6\frac{1}{2}$ minutes
      (2) 10 minutes 45 seconds
      (3) 1 hour

56. The table at the right shows the rates for a taxicab ride in a large city. The rate r is a function of the mileage m, or $r = f(m)$.
   a. Graph the function for all rates on rides of 1 mile or less.
   b. Find $f(\frac{1}{2}$ mile$)$.   c. Find $f(0.7$ mile$)$.
   d. Find $f(\frac{1}{4}$ mile$)$.   e. Find $f(\frac{7}{8}$ mile$)$.
   f. What is the rate for a cab ride of 6 miles?

Mileage	Rate
First $\frac{1}{10}$ mile or less	$1.20
Each additional $\frac{1}{10}$ mile or part	$ .10

6-5 SPECIAL RELATIONS AND FUNCTIONS

In the last section, we classified some types of functions by defining them by their equations, such as $y = ax^2 + bx + c$ (where $a \neq 0$), a function whose graph is a parabola.

A second-degree equation in which both variables are squared can be written in the form $ax^2 + by^2 = c$. This equation is a relation that is not a function. There are three different graphs that have this equation; a circle, an ellipse, and a hyperbola. The values of a, b, and c determine the graph.

> *Circle*: $a = b$ and a, b, and c have the same sign.
> *Ellipse*: $a \neq b$ and a, b, and c have the same sign.
> *Hyperbola*: a and b have opposite signs.

Each of these second-degree relations can be defined as a locus of points.

Circle

We can restate, in terms of locus, the definition of a circle given in Chapter 3.

A *circle* is the locus of points equidistant from a fixed point.

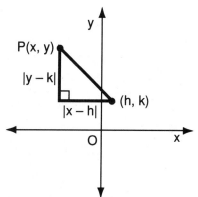

To write the equation of a circle whose center is at (h, k) and the length of whose radius is r, let $P(x, y)$ represent any point on the circle and use the distance formula:

$$\sqrt{(x_1 - x_2)^2 + (y_1 - y_2)^2} = d$$

$$\sqrt{(x - h)^2 + (y - k)^2} = r$$

or $\quad (x - h)^2 + (y - k)^2 = r^2$

| MODEL PROBLEM |

Write the equation of the locus of points 3 units from $(2, -1)$.

Solution

Use the formula: $(x - h)^2 + (y - k)^2 = r^2$
Substitute the given values: $h = 2$, $k = -1$, and $r = 3$

$$(x - 2)^2 + (y - (-1))^2 = 3^2$$

Answer: $(x - 2)^2 + (y + 1)^2 = 9$

EXERCISES

In 1–6, write an equation of the circle with the given length of the radius and the given point at center.

1. $r = 2$, $(0, 0)$ **2.** $r = 5$, $(1, 3)$ **3.** $r = 1$, $(-1, 5)$
4. $r = 3$, $(0, -3)$ **5.** $r = 12$, $(-4, 0)$ **6.** $r = \sqrt{6}$, $(-2, -2)$

In 7–10, find the coordinates of the center and the length of the radius for the circle whose equation is given.

7. $x^2 + y^2 = 16$ **8.** $(x - 3)^2 + (y - 1)^2 = 49$
9. $(x - 2)^2 + (y + 1)^2 = 4$ **10.** $(x + 6)^2 + (y + 5)^2 = 8$

11. Write an equation of the circle with center at $(2, -3)$ that is tangent to the x-axis.

12. A circle whose center is in the second quadrant is tangent to both axes. Write an equation of the circle if the length of its radius is 6.

13. Find the equation of a circle that contains the points $(1, -2)$, $(1, 4)$, and $(5, 6)$.

Parabola

A *parabola* is the locus of points equidistant from a fixed point and a fixed line. The fixed point is called the *focus* and the fixed line is called the *directrix.*

The simplest equation for a parabola can be derived by placing the focus on the y-axis and the directrix perpendicular to the y-axis so that the origin is equidistant from the focus and directrix.

Let the focus be the point $F(0, d)$ and the directrix be the line $y = -d$. Let $P(x, y)$ be any point on the locus. The distance from any point on the locus to the focus is PF. The distance from any point on the locus to the directrix is the length of the perpendicular segment from P to the directrix. If M is the projection of P on the directrix, the distance from P to the directrix is PM. Therefore:

$$PF = PM$$

$$\sqrt{(x - 0)^2 + (y - d)^2} = |y - (-d)|$$

$$(x - 0)^2 + (y - d)^2 = (y + d)^2$$

$$x^2 + y^2 - 2dy + d^2 = y^2 + 2dy + d^2$$

$$x^2 = 4dy$$

$$\frac{1}{4d} x^2 = y$$

The axis of symmetry of this parabola is the y-axis, the line through the focus perpendicular to the directrix. The turning point is the origin, the point on the axis of symmetry that is equidistant from the focus and directrix. When we compare the equation of this parabola with the general quadratic function $y = ax^2 + bx + c$, we see that $a = \dfrac{1}{4d}$, $b = 0$, and $c = 0$. The value of a is determined by the distance between the focus and directrix. The values of b and c depend on the coordinates of the focus. The following is an example of a parabola whose focus is not on the y-axis.

☐ EXAMPLE: Find an equation of the parabola whose focus is (4, 1) and whose directrix is $y = -3$.

Solution

Let $P(x, y)$ be any point of the locus. The coordinates of F are (4, 1) and the coordinates of M, the projection of P on $y = -3$, are $(x, -3)$. Therefore:

$$PF = PM$$
$$\sqrt{(x - 4)^2 + (y - 1)^2} = |y - (-3)|$$
$$(x - 4)^2 + (y - 1)^2 = (y + 3)^2$$
$$x^2 - 8x + 16 + y^2 - 2y + 1 = y^2 + 6y + 9$$
$$x^2 - 8x + 8 = 8y$$
$$\frac{1}{8}x^2 - x + 1 = y$$

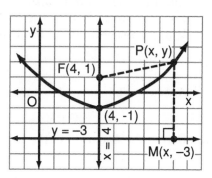

Parabola $y = \dfrac{1}{8}x^2 - x + 1$

The axis of symmetry of this parabola is the line $x = 4$, the line through the focus perpendicular to the directrix. The turning point is (4, −1), the point on the axis of symmetry that is equidistant from the focus and directrix.

The equation of a parabola is a function if the directrix is perpendicular to the y-axis. If the directrix is parallel to the y-axis or is an oblique line, the equation of the parabola is a relation that is not a function. (See model problem 2.)

MODEL PROBLEMS

1. Find the axis of symmetry and the turning point of the parabola whose focus is (4, −2) and whose directrix is $y = 6$.

   Solution

The axis of symmetry is perpendicular to the directrix. Therefore the axis of symmetry is a vertical line whose equation is $x = c$, where c is a constant. Since the focus is on the axis of symmetry, and the x-coordinate of the focus is 4, the equation of the axis of symmetry is $x = 4$.

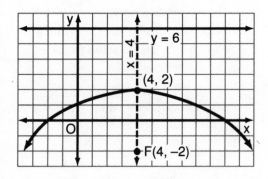

The turning point is also on the axis of symmetry. Therefore, its x-coordinate is 4. Since it is equidistant from the focus and directrix, its y-coordinate is the average of the y-coordinates of the focus and directrix, $\dfrac{-2 + 6}{2} = 2$.

Answer: The axis of symmetry is $x = 4$ and the turning point is (4, 2).

2. Derive the equation of the locus of points equidistant from the point (1, 0) and the line $x = -1$.

   Solution

The locus is a parabola with focus $F(1, 0)$ and directrix $x = -1$. Let $P(x, y)$ be any point of the locus.

$$PF = PM$$
$$\sqrt{(x - 1)^2 + (y - 0)^2} = |x - (-1)|$$
$$(x - 1)^2 + y^2 = (x + 1)^2$$
$$x^2 - 2x + 1 + y^2 = x^2 + 2x + 1$$
$$y^2 = 4x$$
$$\frac{1}{4} y^2 = x$$

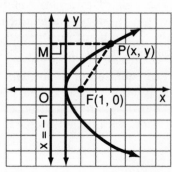

| EXERCISES |

In 1–6, derive an equation of the parabola with the given point as focus and the given line as directrix.

1. $(0, -3), y = 3$ **2.** $(0, 2), y = -2$ **3.** $(1, 1), y = 5$
4. $(1, 1), y = -3$ **5.** $(1, 0), x = -1$ **6.** $(1, 1), x = 5$

7. Write an equation of the axis of symmetry of a parabola whose focus is $(3, 4)$ and whose directrix is $y = -1$.
8. Write an equation of the axis of symmetry of a parabola whose focus is $(3, 4)$ and whose directrix is $x = -1$.
9. Write the coordinates of the turning point of a parabola whose focus is $(-2, 1)$ and whose directrix is $y = -3$.
10. Write the coordinates of the focus of a parabola if the turning point is $(2, -1)$ and the directrix is $y = 5$.

For a parabola, $PF = PM$, thus the ratio $PF:PM = 1$. If $PF \neq PM$, the ratio $PF:PM$ may be greater than 1 or it may be a positive number less than 1. Each of these conditions defines a different curve.

Ellipse

An *ellipse* is the locus of points such that the ratio $PF:PM < 1$ where P is any point of the locus, PF is the distance from P to a fixed point and PM is the distance from P to a fixed line. The value of the ratio is called the *eccentricity* of the ellipse.

In the following example, the fixed point and the fixed line have been chosen so that the graph will be symmetric with respect to the axes.

☐ EXAMPLE: Find an equation of the ellipse with focus $F(3, 0)$, directrix $x = 12$, and eccentricity $\frac{1}{2}$.

Solution

$$\frac{PF}{PM} = \frac{1}{2}$$

$$PF = \frac{1}{2} PM$$

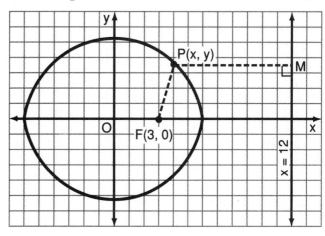

$$\sqrt{(x - 3)^2 + (y - 0)^2} = \frac{1}{2} |12 - x|$$

$$(x - 3)^2 + (y - 0)^2 = \frac{1}{4} (12 - x)^2$$

$$x^2 - 6x + 9 + y^2 = \frac{1}{4} (144 - 24x + x^2)$$

$$4x^2 - 24x + 36 + 4y^2 = 144 - 24x + x^2$$

$$3x^2 + 4y^2 = 108$$

If both sides of the equation are divided by the constant term, the x and y intercepts of the graph can easily be determined.

$$\frac{3x^2}{108} + \frac{4y^2}{108} = \frac{108}{108}$$

$$\frac{x^2}{36} + \frac{y^2}{27} = 1$$

If $x = 0$:
$$\frac{0^2}{36} + \frac{y^2}{27} = 1$$

$$y^2 = 27$$

$$y = \pm \sqrt{27}$$

The y-intercepts are $+\sqrt{27}$ and $-\sqrt{27}$.

If $y = 0$:
$$\frac{x^2}{36} + \frac{0^2}{27} = 1$$

$$x^2 = 36$$

$$x = \pm \sqrt{36}$$

$$x = \pm 6$$

The x-intercepts are $+6$ and -6.

To graph the ellipse, solve the equation for y in terms of x. Then choose values of x between $+6$ and -6 and find the corresponding values of y. Plot the points and connect them with a smooth curve.

$$3x^2 + 4y^2 = 108$$

$$4y^2 = 108 - 3x^2$$

$$y^2 = \frac{108 - 3x^2}{4}$$

$$y = \pm\sqrt{\frac{108 - 3x^2}{4}}$$

x	$\pm\sqrt{\dfrac{108 - 3x^2}{4}}$		=	y
0	$\pm\sqrt{\dfrac{108 - 3(0)^2}{4}}$	$= \pm\sqrt{\dfrac{108}{4}}$	=	$\pm\sqrt{27}$
± 2	$\pm\sqrt{\dfrac{108 - 3(\pm 2)^2}{4}}$	$= \pm\sqrt{\dfrac{96}{4}}$	=	$\pm\sqrt{24}$
± 4	$\pm\sqrt{\dfrac{108 - 3(\pm 4)^2}{4}}$	$= \pm\sqrt{\dfrac{60}{4}}$	=	$\pm\sqrt{15}$
± 6	$\pm\sqrt{\dfrac{108 - 3(\pm 6)^2}{4}}$	$= \pm\sqrt{\dfrac{0}{4}}$	=	0

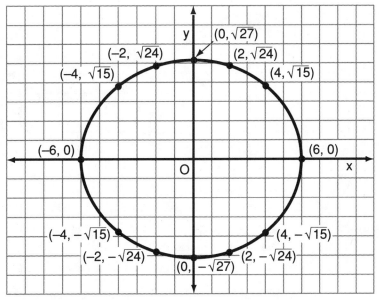

Ellipse $3x^2 + 4y^2 = 108$ or $\dfrac{x^2}{36} + \dfrac{y^2}{27} = 1$

Note that when x has an absolute value larger than 6, the value of y is the square root of a negative number, not a real number. For each value of x in the interval $-6 < x < 6$, there are two values of y, a positive value and a negative value. Thus the equation $3x^2 + 4y^2 = 108$ is a relation that is <u>not</u> a function. The domain is $\{x \mid -6 < x < 6\}$ and the range is $\{y \mid -\sqrt{27} < y < \sqrt{27}\}$.

There are two lines of symmetry for the ellipse, the x-axis and the y-axis; and one point of symmetry, the origin. The segments of the lines of symmetry in the interior of the ellipse are called the *major axis* and *minor axis* of the ellipse. The longer segment, whose endpoints are $(6, 0)$ and $(-6, 0)$, is the major axis. Its length is $|6 - (-6)|$ or 12. The shorter segment, whose endpoints are $(0, \sqrt{27})$ and $(0, -\sqrt{27})$, is the minor axis. Its length, $|\sqrt{27} - (-\sqrt{27})|$, is $2\sqrt{27}$ or $6\sqrt{3}$.

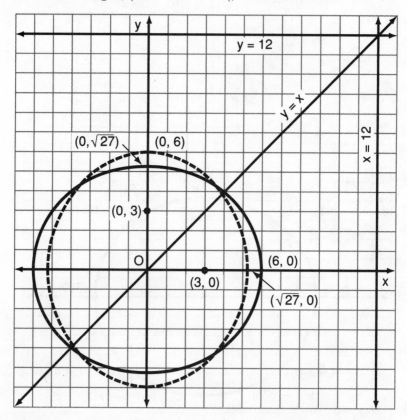

If we reflect the curve with its focus $(3, 0)$ and directrix $x = 12$ over the line $y = x$, the image of the focus is $(0, 3)$, the image of the directrix is $y = 12$, and the image of any point (x, y) of the locus is the point (y, x). Therefore the image of the ellipse $\dfrac{x^2}{36} + \dfrac{y^2}{27} = 1$ is the ellipse $\dfrac{y^2}{36} + \dfrac{x^2}{27} = 1$. The major axis is a segment of the y-axis and the minor axis is a segment of the x-axis.

| MODEL PROBLEM |

Sketch the graph of $3x^2 + 12y^2 = 12$.

Solution

In order to use the points of intersection with the x and y axes, write the equation in intercept from by dividing each side by the constant term.

$$\frac{3x^2}{12} + \frac{12y^2}{12} = \frac{12}{12}$$

$$\frac{x^2}{4} + \frac{y^2}{1} = 1$$

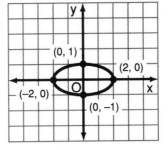

If $y = 0$: $x^2 = 4$
The x-intercepts are ± 2.

If $x = 0$: $y^2 = 1$
The y-intercepts are ± 1.

Using the four points at which the graph intersects the x and y axes, sketch the ellipse.

| EXERCISES |

In 1–6, use the x and y intercepts to sketch the ellipse.

1. $x^2 + 4y^2 = 4$ **2.** $25x^2 + 4y^2 = 100$ **3.** $x^2 + 9y^2 = 9$
4. $x^2 + 9y^2 = 36$ **5.** $16x^2 + 4y^2 = 64$ **6.** $4x^2 + 3y^2 = 12$

7. Write an equation of the ellipse whose major axis is a segment of the x-axis of length 12 and whose minor axis is a segment of the y-axis of length 8.
8. Find the eccentricity of an ellipse if the focus is at $(2, 0)$, the directrix is $x = 18$, and $(6, 0)$ is a point of the ellipse.

Hyperbola

A *hyperbola* is the locus of points such that the ratio $PF:PM > 1$, where P is any point of the locus, PF is the distance from P to a fixed point, and PM is the distance from P to a fixed line. The value of the ratio is the eccentricity of the hyperbola.

In the following example, the fixed point and the fixed line have been chosen so that the graph will be symmetric with respect to the axes.

☐ EXAMPLE: Find an equation of the hyperbola with focus $F(4, 0)$, directrix $x = 1$, and eccentricity 2.

Solution

$$\frac{PF}{PM} = 2$$

$$PF = 2\,PM$$

$$\sqrt{(x - 4)^2 + (y - 0)^2} = 2|x - 1|$$

$$(x - 4)^2 + (y - 0)^2 = 4(x - 1)^2$$

$$x^2 - 8x + 16 + y^2 = 4(x^2 - 2x + 1)$$

$$x^2 - 8x + 16 + y^2 = 4x^2 - 8x + 4$$

$$-3x^2 + y^2 = -12$$

$$3x^2 - y^2 = 12$$

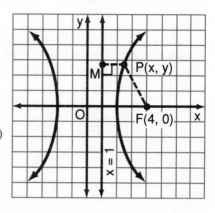

If both sides of the equation are divided by the constant term, the intercepts of the graph can easily be determined.

$$\frac{3x^2}{12} - \frac{y^2}{12} = \frac{12}{12}$$

$$\frac{x^2}{4} - \frac{y^2}{12} = 1$$

If $x = 0$: $\dfrac{0^2}{4} - \dfrac{y^2}{12} = 1$

$$y^2 = -12$$

Since there is no real number whose square is negative, the graph has no y-intercepts.

If $y = 0$: $\dfrac{x^2}{4} - \dfrac{0^2}{12} = 1$

$$x^2 = 4$$

$$x = \pm\sqrt{4}$$

$$x = \pm 2$$

The x-intercepts are $+2$ and -2.

To draw the graph of the hyperbola, solve the equation for y in terms of x. Then choose values of x that are greater than 2 or less than -2 and find the corresponding values of y. Plot the points, and connect the points with positive values of x and the points with negative values of x. The hyperbola will have two branches.

$$3x^2 - y^2 = 12$$
$$-y^2 = -3x^2 + 12$$
$$y^2 = 3x^2 - 12$$
$$y = \pm\sqrt{3x^2 - 12}$$

x	$\pm\sqrt{3x^2 - 12}$	=	y
± 2	$\pm\sqrt{3(\pm 2)^2 - 12} = \pm\sqrt{0}$	=	0
± 3	$\pm\sqrt{3(\pm 3)^2 - 12} = \pm\sqrt{15}$	=	$\pm\sqrt{15}$
± 4	$\pm\sqrt{3(\pm 4)^2 - 12} = \pm\sqrt{36}$	=	± 6

Note that for values of x between -2 and 2, the value of y is the square root of a negative number, not a real number. For each value of x that has an absolute value greater than 2, there are two values of y, a positive value and a negative value. Thus the equation $3x^2 - y^2 = 12$ is a relation that is not a function. The domain is $\{x \mid |x| \geq 2\}$ and the range is the set of real numbers.

There are two lines of symmetry for this hyperbola, the x-axis and the y-axis; and one point of symmetry, the origin. The segment of the line of symmetry that intersects the curve, whose endpoints are points of the curve, is the transverse axis. The endpoints of the transverse axis are $(2, 0)$ and $(-2, 0)$ and the length of the transverse axis is $|2 - (-2)|$ or 4.

Hyperbola $3x^2 - y^2 = 12$ or $\dfrac{x^2}{4} - \dfrac{y^2}{12} = 1$

If we reflect the curve with its focus $(4, 0)$ and directrix $x = 1$ over the line $y = x$, the image of the focus is $(0, 4)$, the image of the directrix is $y = 1$, and the image of any point (x, y) the locus is the point (y, x). Therefore the image of the hyperbola $\dfrac{x^2}{4} - \dfrac{y^2}{12} = 1$ is the hyperbola $\dfrac{y^2}{4} - \dfrac{x^2}{12} = 1$. The transverse axis is a segment of the y-axis.

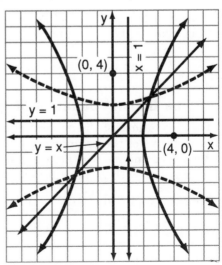

| MODEL PROBLEM |

Find the length of the transverse axis of the hyperbola $2x^2 - 8y^2 = -16$.

Solution

Write the equation in intercept form. Since the coefficient of x^2 is negative, there are no points of intersection with the x-axis. If $x = 0$, $y = \pm\sqrt{2}$. The transverse axis is a segment of the y-axis from $(0, \sqrt{2})$ to $(0, -\sqrt{2})$. The length of the transverse axis is $|\sqrt{2} - (-\sqrt{2})| = 2\sqrt{2}$.

Answer: $2\sqrt{2}$

$$\frac{2x^2}{-16} - \frac{8y^2}{-16} = \frac{-16}{-16}$$

$$-\frac{x^2}{8} + \frac{y^2}{2} = 1$$

$$0 + \frac{y^2}{2} = 1$$

$$y^2 = 2$$

$$y = \pm\sqrt{2}$$

| EXERCISES |

In 1–6: **a.** The transverse axis is a segment of which axis?
b. What is the length of the transverse axis?
c. What are the domain and range of the relation?

1. $4x^2 - y^2 = 16$ **2.** $3x^2 - 12y^2 = 24$ **3.** $y^2 - x^2 = 9$

4. $x^2 - 3y^2 = -9$ **5.** $x^2 - y^2 = 1$ **6.** $-8x^2 + y^2 = 64$

7. a. Sketch the graph of $4x^2 - 9y^2 = 36$.
  b. Sketch the reflection in the line $x = y$ of the graph drawn in part **a**.
  c. Write an equation of the reflection drawn in part **b**.

The General Equation of the Ellipse and Hyperbola

The derivations of the general equations of an ellipse and of a hyperbola in terms of the eccentricity are identical. Let e be the eccentricity of the curve and let a be an arbitrary positive constant. The graph will be symmetric with respect to the axes if we place the focus $F(ae, 0)$ on the x-axis and the directrix $x = \dfrac{a}{e}$ perpendicular to the x-axis.

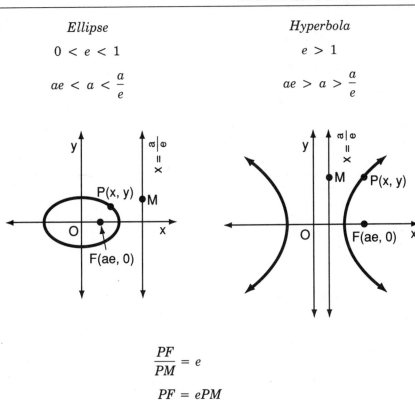

Ellipse	*Hyperbola*
$0 < e < 1$	$e > 1$
$ae < a < \dfrac{a}{e}$	$ae > a > \dfrac{a}{e}$

$$\frac{PF}{PM} = e$$

$$PF = ePM$$

$$\sqrt{(x - ae)^2 + (y - 0)^2} = e\left|x - \frac{a}{e}\right|$$

$$(x - ae)^2 + (y - 0)^2 = e^2\left(x - \frac{a}{e}\right)^2$$

$$x^2 - 2aex + a^2e^2 + y^2 = e^2\left(x^2 - 2\frac{a}{e}x + \frac{a^2}{e^2}\right)$$

$$x^2 - 2aex + a^2e^2 + y^2 = e^2x^2 - 2aex + a^2$$

$$x^2 - e^2x^2 + y^2 = a^2 - a^2e^2 \quad \text{Collect variable terms}$$
on the left side.

$$x^2(1 - e^2) + y^2 = a^2(1 - e^2)$$

This last equation applies to both the ellipse and the hyperbola. The difference between the two curves depends on whether the constant $(1 - e^2)$ is positive or negative, as seen in the following:

Ellipse	*Hyperbola*

Ellipse

$0 < e < 1$
$0 < e^2 < 1$
$0 > -e^2 > -1$
$1 > 1 - e^2 > 0$

The coefficients of x^2 and y^2 and the constant term are all positive.

Hyperbola

$e > 1$
$e^2 > 1$
$-e^2 < -1$
$1 - e^2 < 0$

The coefficients of x^2 and y^2 have opposite signs.

The intercept form of the equation of the curve is found by dividing each term by the constant term, $a^2 (1 - e^2)$.

$$\frac{x^2 (1 - e^2)}{a^2 (1 - e^2)} + \frac{y^2}{a^2 (1 - e^2)} = \frac{a^2 (1 - e^2)}{a^2 (1 - e^2)}$$

$$\frac{x^2}{a^2} + \frac{y^2}{a^2 (1 - e^2)} = 1$$

For the ellipse

$1 - e^2 > 0$
Let $a^2 (1 - e^2) = b^2$

$$\frac{x^2}{a^2} + \frac{y^2}{b^2} = 1, \; 0 < b < a$$

Focus: $(ae, 0)$

Directrix: $x = \dfrac{a}{e}$

The x-intercepts are a and $-a$.
The y-intercepts are b and $-b$.
Length of the major axis: $2a$
Length of the minor axis: $2b$

For the hyperbola

$1 - e^2 < 0$
Let $a^2 (1 - e^2) = -b^2$

$$\frac{x^2}{a^2} - \frac{y^2}{b^2} = 1$$

Focus: $(ae, 0)$

Directrix: $x = \dfrac{a}{e}$

The x-intercepts are a and $-a$.
There are no y-intercepts.
Length of the transverse axis: $2a$

If the focus is on the y-axis and the directrix is perpendicular to the y-axis, similar equations result.

For the ellipse

$$\frac{x^2}{b^2} + \frac{y^2}{a^2} = 1$$

For the hyperbola

$$-\frac{x^2}{b^2} + \frac{y^2}{a^2} = 1$$

If we reflect the ellipse or the hyperbola, with its focus and directrix, over the y-axis, the image of $F(ae, 0)$ is $F''(-ae, 0)$ and the image of the directrix $x = \dfrac{a}{e}$ is $x = -\dfrac{a}{e}$. F' and $x = -\dfrac{a}{e}$ are a new focus and directrix for the curve.

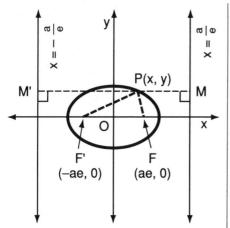 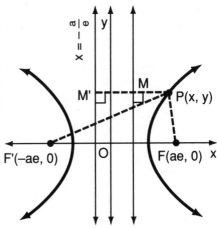

For the ellipse:	*For the hyperbola:*
$PF = ePM$	$PF = ePM$
$PF' = ePM'$	$PF' = ePM'$
$PF + PF' = ePM + ePM'$	$PF - PF' = |ePM - ePM'|$
$PF + PF' = e(PM + PM')$	$PF - PF' = e|PM - PM'|$

As seen in the diagram for the ellipse, $PM + PM'$ is the distance between the directrixes, that is, $\dfrac{a}{e} - \left(-\dfrac{a}{e}\right) = 2\dfrac{a}{e}$. Similarly, for the hyperbola, the distance $|PM - PM'| = 2\dfrac{a}{e}$. Thus:

$$PF + PF' = e\left(2\dfrac{a}{e}\right) = 2a \qquad \qquad PF - PF' = e\left(2\dfrac{a}{e}\right) = 2a$$

This result allows us to give a new definition of the ellipse and the hyperbola.

An *ellipse* is the locus of points such that the sum of the distances from any point on the locus to two fixed points is a constant.

This definition of an ellipse makes it possible to define the circle as a special case of an ellipse in which the two focal points are the same point.

A *circle* is the locus of points such that the distance from any point on the locus to a fixed point is a constant.

A *hyperbola* is the locus of points such that the absolute value of the difference of the distances from any point on the locus to two fixed points is a constant.

As defined earlier on page 262, the *parabola* is the locus of points equidistant from a fixed point and a fixed line.

The parabola, the ellipse, the circle, and the hyperbola are called **conic sections.** Each of these curves is the intersection of a right circular cone with a plane as shown in the diagrams.

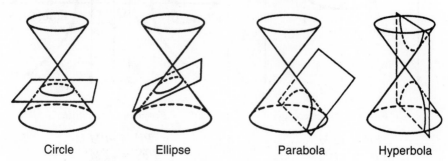

Circle   Ellipse   Parabola   Hyperbola

| **MODEL PROBLEMS** |

1. Name the curve that is the graph of the equation.
   a. $x^2 - 8y^2 = 16$ **b.** $x^2 = 1 - y^2$ **c.** $y = x^2 - 5x$

Solution

a. Since both x and y are squared, compare the equation to $ax^2 + by^2 = c$.
$a = 1$, $b = -8$, a and b have opposite signs.

Answer: hyperbola

b. Since both x and y are squared, compare the equation to $ax^2 + by^2 = c$.
Write the equation as $x^2 + y^2 = 1$.
$a = 1$, $b = 1$, $a = b$.

Answer: circle

c. Only one variable is squared.

Answer: parabola

2. Find the coordinates of the focus and the equation of the directrix of the hyperbola whose equation is $x^2 - y^2 = 2$.

How to Proceed	*Solution*
Write the equation in intercept form by dividing by the constant term.	$x^2 - y^2 = 2$ $$\frac{x^2}{2} - \frac{y^2}{2} = 1$$
Compare with the equation $\dfrac{x^2}{a^2} - \dfrac{y^2}{b^2} = 1$.	$a^2 = 2, b^2 = 2$
For the hyperbola: $a^2(1 - e^2) = -b^2$	$2(1 - e^2) = -2$
Solve for e, $e > 1$.	$2 - 2e^2 = -2$ $-2e^2 = -4$ $e^2 = 2$ $e = \sqrt{2}$
The focus is the point $(ae, 0)$.	$a = \sqrt{2},\ e = \sqrt{2}$ Focus: $(2, 0)$ *Ans.*
The directrix is the line $x = \dfrac{a}{e}$.	$\dfrac{a}{e} = \dfrac{\sqrt{2}}{\sqrt{2}} = 1$ Directrix: $x = 1$ *Ans.*

Note: This hyperbola is its own image under a reflection in the y-axis. The image of the focus $(2, 0)$ is another focus, $(-2, 0)$, and the image of the directrix $x = 1$ is another directrix, $x = -1$.

EXERCISES

In 1–9: **a.** Name the curve that is the graph of the equation.
    b. Find the x and y intercepts, if they exist.
    c. Sketch the graph.
    d. Is the relation a function?

1. $x^2 + y^2 = 1$ 2. $3x^2 - 2y^2 = 6$ 3. $y = -x^2 - 4$

4. $x^2 = 9 - 3y^2$ 5. $x = y^2 - 3y + 1$ 6. $x^2 = y^2 + 4$

7. $8y = x^2 + 2x$ 8. $3x^2 + 3y^2 = 27$ 9. $x^2 + 5y^2 = 20$

10. Write an equation of an ellipse that intersects the x-axis at $(+5, 0)$ and the y-axis at $(0, +3)$.

11. An ellipse is symmetric with respect to the x and y axes. The major axis is a segment of the x-axis with length 12 and the length of the minor axis is 4. Write an equation of the ellipse.

12. **a.** Derive the equation of the parabola whose focus is (1, 1) and whose directrix is the x-axis.
    b. Point $A(4, d)$ is on the parabola described in part **a**. Find d.
    c. Find the distance from A to the focus.
    d. Find the distance from A to the directrix.
    e. Is A equidistant from the focus and the directrix?

13. The equation of an ellipse is $9x^2 + 25y^2 = 225$.
    a. Find the x-intercepts.
    b. Find the y-intercepts.
    c. What is the length of the major axis?
    d. What is the length of the minor axis?
    e. What are the coordinates of the focal point?
    f. What is an equation of the directrix?

14. **a.** Derive the equation of the ellipse with focus at (4, 0), directrix at $x = 9$, and eccentricity $\frac{2}{3}$.
    b. Find the x-intercepts.　　**c.** Find the y-intercepts.
    d. What is the length of the major axis?
    e. What is the length of the minor axis?

15. **a.** Derive the equation of the ellipse with focus at (0, 4), directrix at $y = 9$, and eccentricity $\frac{2}{3}$.
    b. Find the x-intercepts.　　**c.** Find the y-intercepts.
    d. What is the length of the major axis?
    e. What is the length of the minor axis?

16. The ellipse whose equation is $5x^2 + 9y^2 = 180$ is reflected over the line $y = x$. Find the equation of the ellipse formed by this reflection.

17. **a.** Derive the equation of the hyperbola with focus at (9, 0), directrix at $x = 1$, and eccentricity 3.
    b. Find the x-intercepts, if they exist.
    c. Find the y-intercepts, if they exist.
    d. What is the length of the transverse axis?

18. **a.** Derive the equation of the hyperbola with focus at (0, 9), directrix at $y = 1$, and eccentricity 3.
    b. Find the x-intercepts, if they exist.
    c. Find the y-intercepts, if they exist.
    d. What is the length of the transverse axis?

19. The equation of a hyperbola is $3x^2 - y^2 = 12$.
   a. Find the x-intercepts, if they exist.
   b. Find the y-intercepts, if they exist.
   c. Find the length of the transverse axis.
   d. Find the coordinates of the focal point.
   e. Find the equation of the directrix.

20. Which of the following is a function?
   (1) $x^2 - y^2 = 10$ (2) $x^2 + 4y^2 = 10$
   (3) $x = y^2 - 10y$ (4) $y = x^2 - 10x$

21. Which of the following does not intersect the y-axis at $(0, 2)$?
   (1) $x^2 + y^2 = 4$ (2) $4x^2 + 6y^2 = 24$
   (3) $4x^2 - 6y^2 = 24$ (4) $y = x^2 + 2$

6-6 INVERSE VARIATION AND THE HYPERBOLA

The distance from the Hamlin Park entrance to the farthest picnic area is 8 miles. To walk that distance at 4 miles per hour takes 2 hours. Traveling that distance by bicycle at a rate of 8 miles per hour takes 1 hour. To drive 8 miles at a rate of 20 miles per hour takes $\frac{2}{5}$ of an hour.

rate	time
4	2
8	1
20	$\frac{2}{5}$

As the rate at which we travel increases, the time required to travel a constant distance decreases. We say that, for a constant distance, rate and time *vary inversely*.

Let x represent rate in miles per hour and y represent time in hours. Using the formula (rate)(time) = distance, we can write the equation $xy = 8$ to express the relationship between the rate and the time needed to travel 8 miles. This illustrates the following principle:

■ **If x and y *vary inversely*, then xy = a nonzero constant. The value xy is the *constant of variation*.**

□ EXAMPLE: The number of days (x) needed to complete a job varies inversely as the number of workers (y) assigned to the job. If the job can be completed by 2 workers in 10 days, then the constant of variation is the product 2(10) or 20. To find the number of workers needed to complete the job in 5 days, let $x = 5$ and solve the equation $xy = 20$ for y.

$$xy = 20$$

$$5y = 20$$

$$y = 4$$

Therefore 4 workers can complete the job in 5 days. Note that when the number of workers is increased by the factor 2, the number of days needed to complete the job was decreased by the reciprocal factor $\frac{1}{2}$.

x	y
10	2
5	4

Graphing Inverse Variation

To draw the graph of $xy = 20$, we find ordered pairs that are elements of the solution set. For the positive product 20, values of x and y can be either both positive or both negative.

x	y
1	20
2	10
4	5
5	4
10	2
20	1

x	y
-1	-20
-2	-10
-4	-5
-5	-4
-10	-2
-20	-1

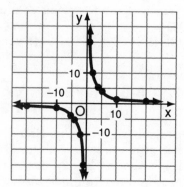

Hyperbola $xy = 20$

The graph, a hyperbola, consists of two parts. One part is in quadrant I where both x and y are positive and the other part is in quadrant III where both x and y are negative. Notice that there is no value of x for which y is 0 and no value of y for which x is 0. Therefore the graph has no x-intercept and no y-intercept. Since for every nonzero value of x there is exactly one value of y, the equation $xy = 20$ defines a function whose domain and range are the set of nonzero real numbers.

| **MODEL PROBLEMS** |

1. The cost of hiring a bus for a trip to Niagara Falls is $400. The cost per person (x) varies inversely as the number of persons (y) who will go on the trip.
   a. Find the cost per person if 25 persons go on the trip.
   b. Find the number of persons who are going if the cost per person is $12.50.

Solution

Since x and y vary inversely, we can write the equation $xy = 400$.

a.

$$xy = 400$$
$$x(25) = 400$$
$$x = \frac{400}{25}$$
$$x = 16$$

b.

$$xy = 400$$
$$12.50y = 400$$
$$y = \frac{400}{12.50}$$
$$y = 32$$

Answer: **a.** $16 per person. **b.** 32 persons.

2. Draw the graph of $xy = -12$.

Solution

Since the product of x and y is negative, when x is positive, y will be negative and when x is negative, y will be positive.

x	y
1	-12
2	-6
3	-4
4	-3
6	-2
8	-1.5
12	-1

x	y
-1	12
-2	6
-3	4
-4	3
-6	2
-8	1.5
-12	1

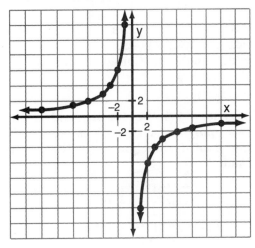

$$xy = -12$$

Note that the graph has both line symmetry and point symmetry. The lines $y = x$ and $y = -x$ are lines of symmetry. The origin $(0, 0)$ is the point of symmetry.

EXERCISES

In 1–8, draw the graph of the equation.

1. $xy = 6$ **2.** $xy = 15$ **3.** $xy = -4$ **4.** $xy = -2$

5. $y = \dfrac{10}{x}$ **6.** $y = -\dfrac{8}{x}$ **7.** $y = \dfrac{1}{x}$ **8.** $y = -\dfrac{1}{x}$

9. If 4 typists can complete the typing of a manuscript in 9 days, how long would it take 12 typists to complete the manuscript?
10. If 4 typists can complete the typing of a manuscript in 9 days, how many typists are needed to complete the typing in 6 days?
11. If a man can drive from his home to Albany in 6 hours at 45 mph, how long would it take him if he drove at 60 mph?
12. If a man can drive from his home to Albany in 6 hours at 45 mph, how fast did he drive if he made the trip in 5 hours?
13. Let S be the set of all rectangles that have an area of 600 cm^2. The length varies inversely as the width.
    a. What is the length of a rectangle from the set S whose width is 20 cm?
    b. What is the width of a rectangle from the set S whose length is 100 cm?

6-7 COMPOSITION OF FUNCTIONS

We are familiar with many operations in mathematics, such as addition and multiplication. In fact, we have performed these operations with functions. Consider the polynomial functions $f(x) = 2x$ and $g(x) = x + 4$.

When we add the polynomials $2x$ and $x + 4$, we actually add two polynomial functions to find a single function called the "sum."

$$f(x) = 2x$$
$$g(x) = x + 4$$
$$\overline{(f + g)(x) = 3x + 4}$$

When we multiply the polynomials $2x$ and $x + 4$, we actually multiply two polynomial functions to find a single function called the "product."

$$f(x) \cdot g(x) = 2x(x + 4)$$
$$(f \cdot g)(x) = 2x^2 + 8x$$

We will now learn a new operation to be performed on functions. Under this binary operation, called *composition*, one function "follows" another. That is, the second function is applied to the result of the first function.

Composition of Functions: Method 1

The composition of functions f and g is sometimes written in the form $f(g(x))$, read as "f of g of x." In this form, we treat the *innermost function first*.

Let $f(x) = 2x$, and $g(x) = x + 4$. Before we find the rule of the single function called the composition $f(g(x))$, let us study some numerical examples.

	☐ EXAMPLE 1:	☐ EXAMPLE 2:
How to Proceed	Find $f(g(3))$.	Find $f(g(\frac{1}{2}))$.
1. Evaluate $g(x)$. Here, $g(x) = x + 4$.	$= f(7)$	$= f(4\frac{1}{2})$
2. Evaluate $f(x)$. Since $f(x) = 2x$, let f(result) = *twice* the result.	$= 14$	$= 9$

In general, for any number x under the given functions:

$f(g(x))$

1. Function g adds 4 to each value in its domain. $= f(x + 4)$

2. Function f doubles each value in its domain. $= 2(x + 4)$

$= 2x + 8$

Thus, the function $f(g(x)) = 2x + 8$ states the rule for the composition where g is applied first, or where "f *follows* g."

Composition of Functions: Method 2

The expression $(f \circ g)(x)$ is read as "f composition g of x," or as "f *following* g of x." The symbol "$\circ$" indicating composition is an open circle, placed in a raised position between f and g. As in method 1, remember that g must be applied first. Let us examine the same functions, expressed in a different format.

$$x \xrightarrow{\text{g}} x + 4 \quad \text{AND} \quad x \xrightarrow{\text{f}} 2x$$

	☐ EXAMPLE 1:	☐ EXAMPLE 2:
How to Proceed	Find $(f \circ g)(3)$.	Find $(f \circ g)(\frac{1}{2})$.
1. Using g first, assign an element from the domain to its corresponding element in the range.	$3 \xrightarrow{\text{g}} 7 \xrightarrow{\text{f}} 14$ $f \circ g$	$\frac{1}{2} \xrightarrow{\text{g}} 4\frac{1}{2} \xrightarrow{\text{f}} 9$ $f \circ g$
2. Treat the element in the range of g as an element in the domain of f; apply the rule for f.		
3. Write the answer, using $(f \circ g)$.	*Answer:* $(f \circ g)(3) = 14$	*Answer:* $(f \circ g)(\frac{1}{2}) = 9$

This same procedure can be applied to the general case to find the rule for the single function that represents $(f \circ g)(x)$.

$$x \xrightarrow{\text{ g }} (x + 4) \xrightarrow{\text{ f }} 2(x + 4) = 2x + 8$$
$$f \circ g$$

Thus, $(f \circ g)(x) = 2x + 8$ states the rule for the composition where g is applied first, or where "f *follows* g."

These examples illustrate the fact that there are two equivalent forms of the composition "f *following* g," namely:

$$f(g(x)) = (f \circ g)(x)$$

Note: For the composition "f *following* g" to be meaningful, each element in the range of g must be an element in the domain of f.

Group Properties Under Composition

With certain restrictions (to be explained in the next section), it is true that a set of functions, under the operation of composition, is a *group*. Thus, four properties are satisfied:

1. *Closure.* The composition of two functions is a single function.
2. *Associativity.* For all functions f, g, and h:

$$(f \circ (g \circ h))(x) = ((f \circ g) \circ h)(x)$$

3. *Identity.* There is an identity function i under composition such that, for every function f:

$$(f \circ i)(x) = f(x) \quad \text{AND} \quad (i \circ f)(x) = f(x)$$

□ EXAMPLE:

Let $f(x) = 3x + 7$
Let $i(x) = x$
$\left.\begin{array}{}\\\\\\\end{array}\right\}$
$x \xrightarrow{\text{ i }} x \xrightarrow{\text{ f }} 3x + 7$ $\quad$ $x \xrightarrow{\text{ f }} 3x + 7 \xrightarrow{\text{ i }} 3x + 7$
$(f \circ i)(x) = f(x)$ $\qquad\quad$ $(i \circ f)(x) = f(x)$

■ The identity function for the operation of composition may be written in one of three forms:

$$x \xrightarrow{\text{ i }} x \qquad i(x) = x \qquad y = x$$

4. *Inverses.* For every function f (in a restricted set to be explained), for which there is an inverse function f^{-1} under composition:

$$(f \circ f^{-1})(x) = i(x) \qquad \text{AND} \qquad (f^{-1} \circ f)(x) = i(x)$$

We will study the restriction, as well as ways to find the inverse f^{-1}, in the next section.

┌─────────────────────── KEEP IN MIND ───────────────────────┐

In the composition "f *following* g," written:

$$f(g(x)) \qquad \text{OR} \qquad (f \circ g)(x)$$

function g is applied before function f.

└──┘

MODEL PROBLEMS

1. Using the functions $f(x) = 3x$ and $g(x) = x - 4$, demonstrate that composition of functions is *not* commutative.

Solution: Method 1

1. Find: $f(g(x))$
   $= f(x - 4)$
   $= 3(x - 4)$
   $= 3x - 12$

2. Find: $g(f(x))$
   $= g(3x)$
   $= 3x - 4$

3. Since $f(g(x)) = 3x - 12$ and $g(f(x)) = 3x - 4$, then $f(g(x)) \neq g(f(x))$. Thus, composition of functions is *not* commutative.

Solution: Method 2

1. Find $(f \circ g)(x)$.

$$x \xrightarrow{g} x - 4 \xrightarrow{f} 3(x - 4) = 3x - 12$$

$(f \circ g)(x) = 3x - 12$

2. Find $(g \circ f)(x)$.

$$x \xrightarrow{f} 3x \xrightarrow{g} 3x - 4$$

$(g \circ f)(x) = 3x - 4$

3. Since $(f \circ g)(x) \neq (g \circ f)(x)$, composition of functions is *not* commutative.

Note: The commutative property is *not* required in a group.

2. Let $h(x) = x^2$ and $r(x) = x + 3$. **a.** Evaluate $(h \circ r)(5)$. **b.** Find the rule of the function $(h \circ r)(x)$.

Solution

a. To evaluate $(h \circ r)(5)$, apply r first. Under r, $5 \to 8$. Under h, 8 is squared.

$$5 \xrightarrow{\ r\ } 8 \xrightarrow{\ h\ } 64$$

$(h \circ r)(5) = 64$ *Ans.*

b. Use the same process with $(h \circ r)(x)$. Under r, $x \to x + 3$. Under h, $x + 3$ is squared.

$$x \xrightarrow{\ r\ } x + 3 \xrightarrow{\ h\ } (x + 3)^2 = x^2 + 6x + 9$$

$(h \circ r)(x) = x^2 + 6x + 9$ *Ans.*

Note: By substituting $x = 5$ in the rule $(h \circ r)(x) = x^2 + 6x + 9$, we can show again that $(h \circ r)(5) = (5)^2 + 6(5) + 9 = 25 + 30 + 9 = 64$.

| **EXERCISES** |

In 1–8, using $f(x) = x + 5$ and $g(x) = 4x$, evaluate the composition.

1. $f(g(2))$ 2. $g(f(2))$ 3. $f(g(-1))$ 4. $g(f(-1))$
5. $f(g(0))$ 6. $g(f(0))$ 7. $g(f(\frac{1}{2}))$ 8. $f(g(\frac{1}{2}))$

In 9–16, using $f(x) = 3x$ and $g(x) = x - 2$, evaluate the composition.

9. $(f \circ g)(4)$ 10. $(g \circ f)(4)$ 11. $(f \circ g)(-2)$ 12. $(g \circ f)(-2)$
13. $(g \circ f)(0)$ 14. $(f \circ g)(0)$ 15. $(g \circ f)(\frac{2}{3})$ 16. $(f \circ g)(\frac{2}{3})$

In 17–24, using $h(x) = x^2$ and $p(x) = 2x - 3$, evaluate the composition.

17. $(h \circ p)(2)$ 18. $(p \circ h)(2)$ 19. $(p \circ h)(1)$ 20. $(h \circ p)(1)$
21. $(p \circ h)(-3)$ 22. $(h \circ p)(-3)$ 23. $(h \circ p)(1.5)$ 24. $(p \circ h)(1.5)$

25. Let $f(x) = x + 6$ and $g(x) = 3x$. **a.** Find the rule of the function $(f \circ g)(x)$. **b.** Find the rule of the function $(g \circ f)(x)$. **c.** Does $(f \circ g)(x) = (g \circ f)(x)$?
26. Let $r(x) = x - 8$ and $t(x) = x^2$. **a.** Find the rule of the function $(r \circ t)(x)$. **b.** Find the rule of the function $(t \circ r)(x)$. **c.** Does $(r \circ t)(x) = (t \circ r)(x)$?
27. Let $d(x) = 2x + 3$ and $c(x) = x - 3$. **a.** Evaluate $(d \circ c)(2)$. **b.** Find the rule of the function $(d \circ c)(x)$. **c.** Use the rule from part **b** to find the value of $(d \circ c)(2)$. **d.** Do the answers from parts **a** and **c** agree?

In 28-37, for the given functions $f(x)$ and $g(x)$, find the rule of the composition $(f \circ g)(x)$.

28. $f(x) = 6x$; $g(x) = x - 2$
29. $f(x) = x - 10$; $g(x) = 4x$
30. $f(x) = x$; $g(x) = 2x + 5$
31. $f(x) = x - 3$; $g(x) = x - 5$
32. $f(x) = 2x$; $g(x) = 5x$
33. $f(x) = 3x + 2$; $g(x) = x - 3$
34. $f(x) = \frac{1}{2}x - 3$; $g(x) = 4x + 6$
35. $f(x) = 5 - x$; $g(x) = x + 2$
36. $f(x) = x^2$; $g(x) = x - 5$
37. $f(x) = 4 - x^2$; $g(x) = x - 2$

38. If $f(x) = x + 8$, then the rule of the composition $(f \circ f)(x)$ is:
    (1) $x + 8$ (2) $x + 16$ (3) $2x + 8$ (4) $2x + 16$

39. If $g(x) = 2x + 5$, what is the rule of the composition $(g \circ g)(x)$?

40. Let $h(x) = x^2 + 2x$, and $g(x) = x - 3$. In a to e, evaluate the composition.
    a. $(h \circ g)(4)$ b. $(h \circ g)(3)$ c. $(h \circ g)(2)$ d. $(h \circ g)(1)$
    e. $(h \circ g)(-2)$ f. Find the rule of the function $(h \circ g)(x)$.

41. Let $f(x) = x + 5$, $g(x) = 2x$, and $h(x) = x - 2$.
    a. If $k(x) = (f \circ g)(x)$, find the rule of the function $k(x)$.
    b. Find the rule of $((f \circ g) \circ h)(x)$, that is, the rule of $((k) \circ h)(x)$.
    c. If $r(x) = (g \circ h)(x)$, find the rule of the function $r(x)$.
    d. Find the rule of $(f \circ (g \circ h))(x)$, that is, the rule of $(f \circ (r))(x)$.
    e. Using parts b and d, does $((f \circ g) \circ h)(x) = (f \circ (g \circ h))(x)$? If yes, what group property is demonstrated? If no, explain why.

42. Let $b(x) = |x|$, $d(x) = [x]$, $f(x) = \dfrac{1}{x}$, $g(x) = x - 3$, and $h(x) = 2x$.

    Find the rule of the given composition:
    a. $(f \circ b)(x)$ b. $(d \circ g)(x)$ c. $(b \circ g)(x)$
    d. $(g \circ d)(x)$ e. $(g \circ f)(x)$ f. $(h \circ g)(x)$
    g. $(f \circ (g \circ h))(x)$ h. $(d \circ (h \circ f))(x)$
    i. $(g \circ (h \circ b))(x)$ j. $(b \circ (f \circ h))(x)$

43. To find a 6% sales tax on an item, Ms. Reres reasoned as follows: Using function s, I multiply the price of the item by .06; then, using function r, I round the amount found to the nearest whole cent.
    a. *True or False:* If x is the price of an item, then the 6% sales tax on the item is found by the composition $(r \circ s)(x)$.
    b. Evaluate $(r \circ s)(\$1.39)$.   c. Evaluate $(r \circ s)(\$16.79)$.
    d. *True or False:* If the 6% sales tax is added to the price of the item by function t, the customer's bill is found by $(t \circ (r \circ s))(x)$.
    e. Evaluate $(t \circ (r \circ s))(\$3.15)$.
    f. Evaluate $(t \circ (r \circ s))(\$19.95)$.

6-8 INVERSE FUNCTIONS UNDER COMPOSITION

The relation f, shown as a set of ordered pairs and illustrated in an arrow diagram, is a function because every x corresponds to one and only one y.

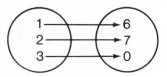

Function f

Function f = {(1, 6), (2, 7), (3, 0)}

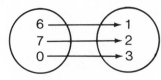

Inverse Function f^{-1}

Under function f, $x \to y$. There is another relation called f^{-1} in which $y \to x$. In this example, the relation f^{-1} is a function because every element of the domain corresponds to one and only one element in the range. The function f^{-1} is called the "inverse of f under composition."

Inverse function f^{-1} = {(6, 1), (7, 2), (0, 3)}

We have stated that the identity function under composition is $x \to x$ or $i(x) = x$ or $y = x$. Notice here that $(f^{-1} \circ f)(x) = i(x)$, and $(f \circ f^{-1})(x) = i(x)$.

$$1 \xrightarrow{f} 6 \xrightarrow{f^{-1}} 1$$
$$2 \longrightarrow 7 \longrightarrow 2$$
$$3 \longrightarrow 0 \longrightarrow 3$$

$$(f^{-1} \circ f)(x) = i(x)$$

$$6 \xrightarrow{f^{-1}} 1 \xrightarrow{f} 6$$
$$7 \longrightarrow 2 \longrightarrow 7$$
$$0 \longrightarrow 3 \longrightarrow 0$$

$$(f \circ f^{-1})(x) = i(x)$$

The relation g, shown as a set of ordered pairs and illustrated in an arrow diagram, is a function because every x corresponds to one and only one y.

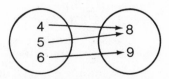

Function g

Function g = {(4, 8), (5, 8), (6, 9)}

The relation g^{-1}, however, in which the x and y elements are interchanged, is *not* a function.

Relation g^{-1} = {(8, 4), (8, 5), (9, 6)}

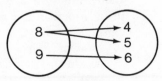

Relation g^{-1} is *not* a function

In g^{-1}, $8 \to 4$ and $8 \to 5$. Since one element 8 corresponds to more than one element in the range, g^{-1} is called a "one-to-many relation."

Conversely, in function g, $4 \to 8$ and $5 \to 8$. Since many elements correspond to the same one element, g is called a "many-to-one function."

As illustrated by this example, a many-to-one function does *not* have an inverse function under composition.

Consider function f for which an inverse function exists. For every x, there is one and only one y. Also, for every second element y, there is one and only one first element x. Function f is called a "one-to-one function."

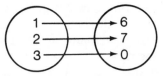

Function f is *one-to-one*

By definition, a function is a *one-to-one function* if and only if each second element corresponds to one and only one first element.

As illustrated in the first example containing f and f^{-1}, a one-to-one function *does* have an inverse function under composition.

■ For every one-to-one function f: The set of ordered pairs obtained by interchanging the first and second elements of each pair in f is f^{-1}, the inverse function under composition.

Recall that the set of real numbers less zero is a group under the operation of multiplication. Since zero has no inverse under multiplication, it is necessary to restrict the set of real numbers to those elements having inverses so that a group exists. In the same way, by restricting the set of functions to those having inverses, namely one-to-one functions, we can now say:

■ The set of one-to-one functions, under the operation of composition, is a group.

Ways to Find the Inverse Function

In the examples that follow, we will study the function whose rule is $f(x) = \frac{1}{2}x + 2$, or $y = \frac{1}{2}x + 2$. Whether stated as a rule or drawn on a coordinate graph, the function contains an infinite number of ordered pairs. Let us first examine a function with a finite number of ordered pairs by selecting some pairs of the function f: $(0, 2)$, $(4, 4)$, and $(6, 5)$.

1. *Ordered Pairs*

The inverse under composition of a one-to-one function is formed by interchanging the x-coordinate and y-coordinate of each pair in the function.

For example, if function f = {(0, 2), (4, 4), (6, 5)},
then inverse function f^{-1} = {(2, 0), (4, 4), (5, 6)}.

2. *Coordinate Graph*

In transformation geometry, we learned that $(x, y) \rightarrow (y, x)$ by a reflection in the line $y = x$. The inverse under composition of a one-to-one function f is graphed by reflecting the function f in the line $y = x$, that is, the line whose equation is the identity function i.

In the figure at the right, the identity $i(x) = x$, or $y = x$, is graphed as a broken line along the diagonal. The function $f(x) = \frac{1}{2}x + 2$, or $y = \frac{1}{2}x + 2$, is graphed.

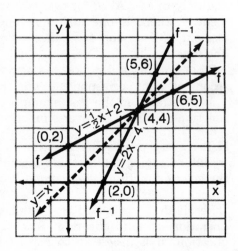

Notice that f contains the points whose coordinates are (0, 2), (4, 4), and (6, 5).

By reflecting f in the line $y = x$, we graph the inverse function f^{-1}. Notice that f^{-1} contains the points whose coordinates are (2, 0), (4, 4), and (5, 6). The equation of this inverse function is $f^{-1}(x) = 2x - 4$, or $y = 2x - 4$.

3. *Rule of the Inverse*

The rule of the inverse function f^{-1} can be found by interchanging x and y in the rule of the given function f. For example:

(1) Express $f(x) = \frac{1}{2}x + 2$ in terms of x and y.

$$f: \quad y = \tfrac{1}{2}x + 2$$

(2) Interchange x and y to form the inverse under composition.

$$f^{-1}: x = \tfrac{1}{2}y + 2$$

(3) Solve for y.

$$x - 2 = \tfrac{1}{2}y$$
$$2(x - 2) = 2(\tfrac{1}{2}y)$$
$$2x - 4 = y$$

Note: The rule of the inverse function f^{-1} may be written in one of two forms: $y = 2x - 4$ or $f^{-1}(x) = 2x - 4$.

To show that $y = 2x - 4$ is the inverse function of $y = \frac{1}{2}x + 2$, demonstrate that the composition of these functions is the identity function.

$$y = \tfrac{1}{2}x + 2, \text{ or } f(x) = \tfrac{1}{2}x + 2$$
$$y = 2x - 4, \text{ or } f^{-1}(x) = 2x - 4$$

$$x \xrightarrow{\;f\;} \tfrac{1}{2}x + 2 \xrightarrow{\;f^{-1}\;} 2(\tfrac{1}{2}x + 2) - 4 \qquad\Bigg|\qquad x \xrightarrow{\;f^{-1}\;} 2x - 4 \xrightarrow{\;f\;} \tfrac{1}{2}(2x - 4) + 2$$
$$= x + 4 - 4 \qquad\qquad\qquad = x - 2 + 2$$
$$\longrightarrow = x \qquad\qquad\qquad\qquad \longrightarrow = x$$
$$(f^{-1} \circ f)(x) = x \qquad\qquad\qquad (f \circ f^{-1})(x) = x$$

Restricted Domains and Inverse Functions

The domain of the function $g(x) = x^2$, or $y = x^2$, is the set of real numbers. If x and y are interchanged, the rule for this new relation is $x = y^2$, or $y = \pm\sqrt{x}$. Since function y is not one-to-one, the relation $x = y^2$ is not a function. (See Fig. 1.)

If the domain of g is restricted, however, so that g is a one-to-one function, then the inverse g^{-1} under composition is a function. By restricting the domain to the set of positive real numbers and zero, the inverse of the function $y = x^2$ is the function $x = y^2$. Since the range is restricted to the set of positive real numbers and zero, the function y is one-to-one, and can be written as $y = \sqrt{x}$. (See Fig. 2.)

Domain of g = {Real numbers}

| | Restricted Domain of g = {Positive reals and zero} |

Function g: $y = x^2$
Relation formed
by interchanging
x and y: $y = \pm\sqrt{x}$
(*not* a function)

Function g: $y = x^2$
Inverse Function g^{-1}: $y = \sqrt{x}$
Range of g^{-1} = {Positive reals and zero}

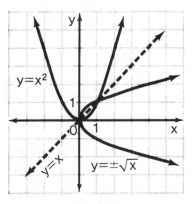

Fig. 1

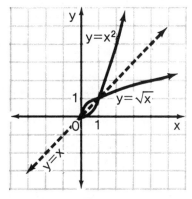

Fig. 2

┌─────────────────KEEP IN MIND─────────────────┐

1. Only a function f that is one-to-one has an inverse
   under composition. The domain of the inverse is
   the range of the original function.

2. This inverse f^{-1} is formed by interchanging x and
   y in the given function.

└───┘

MODEL PROBLEMS

1. What is the inverse of the function $y = 3x + 5$?

   Solution: If the inverse of a function is being sought and no operation is named, the inverse is assumed to be the *inverse under composition*.

   1. Write the given function f. $y = 3x + 5$

   2. To form f^{-1}, the inverse under composi- $x = 3y + 5$
      tion, interchange x and y.

   3. Solve for y, so that the answer can be writ- $x - 5 = 3y$
      ten in the form $y = mx + b$. $\dfrac{x - 5}{3} = y$

   Answer: $y = \dfrac{x - 5}{3}$ OR $y = \dfrac{1}{3}x - \dfrac{5}{3}$

2. What is the inverse of the function $\{(3, 7), (2, 1), (5, -4)\}$?

   Solution: Interchange the x-coordinate and y-coordinate of each
   pair in the given function to find its inverse under composition.

   Answer: $\{(7, 3), (1, 2), (-4, 5)\}$

EXERCISES

In 1–4, write the inverse of the given function.

1. $\{(1, 5), (2, 7), (3, -2), (4, -3)\}$ 2. $\{(0, 6), (4, 2), (-1, 7), (-2, 8)\}$
3. $\{(1, 1), (2, 2), (3, 3), (4, 4)\}$ 4. $\{(1, k), (2, k + 1), (3, k + 2)\}$

5. Let $f = \{(3, -2), (4, -2), (5, -1)\}$. **a.** Write the relation formed by
   interchanging the x-coordinate and y-coordinate of f. **b.** Is this
   new relation a function? If not, explain why.

6. *True* or *False:* The relation formed by interchanging x and y in each pair of a function is also a function.

7. *True* or *False:* The relation formed by interchanging x and y in each pair of a one-to-one function is also a one-to-one function.

In 8–16, write the equation of the inverse of the given function, solved for y.

8. $y = 3x$ 　　　　　　9. $y = x - 6$ 　　　　　10. $y = \frac{1}{4}x$

11. $y = \frac{1}{3}x + 1$ 　　　12. $y = 4x - 8$ 　　　13. $y = 5 - x$

14. $y = \sqrt{x}$ 　　　　　15. $y - 12 = 3x$ 　　　16. $y + 7 = x^3$

In 17–19 and 23, select the numeral preceding the expression that best completes the sentence or answers the question.

17. If $f(x) = 6x - 2$, then the inverse function $f^{-1}(x)$ is:

  (1) $\dfrac{x + 2}{6}$ 　　(2) $\dfrac{x}{6} + 2$ 　　(3) $\dfrac{x + 1}{3}$ 　　(4) $\dfrac{x}{3} + 1$

18. If $m(x) = 2x - 1$, then $m^{-1}(x)$ is:

  (1) $\frac{1}{2}x + 1$ 　　(2) $\frac{1}{2}x + \frac{1}{2}$ 　　(3) $2x + 1$ 　　(4) $-2x + 1$

19. The inverse of the function $y = -2x$ is:

  (1) $y = 2x$ 　　(2) $x = 2y$ 　　(3) $x = -2y$ 　　(4) $y = x - 2$

20. If $f(x) = 3x - 7$, evaluate: a. $f(2)$ 　　b. $f^{-1}(-1)$

21. If $g(x) = \frac{2}{3}x + 4$, evaluate: a. $g(-3)$ 　　b. $g^{-1}(2)$

22. If $h(x) = 5x - 2$, find the value of $(h^{-1} \circ h)(123)$.

23. What is the inverse of the function $\{(3, 1), (4, -1), (-2, 6)\}$?

  (1) $\{(3, -1), (4, 1), (-2, -6)\}$ 　　(2) $\{(-1, 3), (1, 4), (-6, -2)\}$

  (3) $\{(-3, -1), (-4, 1), (2, -6)\}$ 　　(4) $\{(1, 3), (-1, 4), (6, -2)\}$

In 24–26: a. On graph paper, copy the function f. b. Using the same axes, sketch the graph of f^{-1}, the inverse under composition. c. State the domain and range of f. d. State the domain and range of f^{-1}.

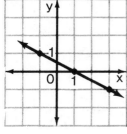
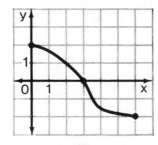
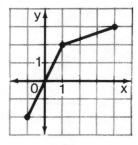

24. 　　　　　　　　25. 　　　　　　　　26.

27. Demonstrate that the identity function $i(x) = x$ is its own inverse under composition.

In 28–30: **a.** On graph paper, draw the graph of the function, including points whose x-coordinates are -2, -1, 0, 1, and 2. **b.** On the same axes, draw the graph of the relation that is the reflection of the given function in the line $y = x$. **c.** Is the relation in part **b** a function? **d.** State the equation of the relation drawn in part **b**.

28. $y = x^3$ **29.** $y = |x|$ **30.** $y = 3$

31. Under composition, the identity function $y = x$ is its own inverse. Name another function in the form $y = mx + b$ that is its own inverse.

6-9   REVIEW EXERCISES

In 1–9: **a.** State the domain of the relation. **b.** State the range of the relation. **c.** Is the relation a function? If not, explain why.

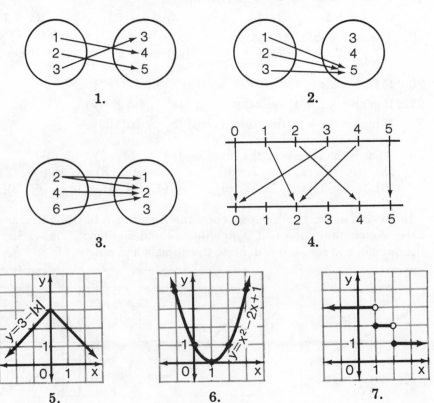

1.

2.

3.

4.

5.

6.

7.

8. $\{(2, 2), (3, 1), (1, 3)\}$ **9.** $\{(2, 4), (3, 1), (2, 3), (3, 2)\}$

10. If the function $f(x) = x^2 - 4x$, find the value of $f(-3)$.

11. If $g(x) = \sqrt[3]{x}$, find $g(-8)$.

In 12–14, select the numberal preceding the expression that best completes the sentence or answers the question.

12. Which of the following is *not* a function?
(1) $y = |5x|$ (2) $y = 5x^2$ (3) $y = 5$ (4) $x = 5$

13. The domain for $h(x) = 2x - 7$ is $-2 \le x \le 2$. The range is:
(1) $-3 \le y \le 3$ (2) $-11 \le y \le -3$
(3) $-11 \le y \le 3$ (4) $-11 \le y \le 11$

14. The domain for $p(x) = 6 - 2x$ is $\{x | -5 \le x \le 1\}$. The greatest value in the range is:
(1) 16   (2) 8   (3) -4 (4) 4

In 15–17, state the largest possible domain such that the given relation is a function.

15. $f(x) = \dfrac{x - 2}{x - 8}$ **16.** $x \xrightarrow{\text{g}} \dfrac{4}{x^2 + 4x}$ **17.** $y = \dfrac{1}{\sqrt{x - 8}}$

In 18–20, for the given function: **a.** State the domain. **b.** State the range.

18. $f(x) = 4x - 12$ **19.** $x \xrightarrow{\text{g}} [x]$ **20.** $y = \sqrt{x - 8}$

21. Let $m(x) = 5x$ and $d(x) = x - 4$.
a. Find $(m \circ d)\,(6)$.   **b.** Find $(d \circ m)\,(6)$.
c. Write the rule for $(m \circ d)\,(x)$.   **d.** Write the rule for $(d \circ m)\,(x)$.

In 22–25: **a.** Identify the graph of each of the following as a circle, an ellipse, a hyperbola, or a parabola. **b.** Sketch the graph.

22. $4x^2 + y^2 = 4$ **23.** $y = x^2 - 4$
24. $y^2 = x^2 - 4$ **25.** $y^2 = 4 - x^2$

26. An ellipse for which the x- and y-axes are lines of symmetry intersects the x-axis at $(6, 0)$ and the y-axis at $(0, 2)$. Write an equation of the ellipse.

27. Which of the following is the equation of the graph shown at the right?
(1) $4x^2 + y^2 = 16$
(2) $x^2 + 4y^2 = 16$
(3) $x^2 + 2y^2 = 4$
(4) $2x^2 + y^2 = 4$

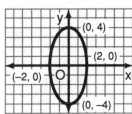

28. Which of the following is both a function and a hyperbola?

(1) $y^2 = 9 - x^2$ (2) $y^2 = x^2 - 9$ (3) $y = 9 - x^2$ (4) $y = \dfrac{9}{x}$

In 29–32, given functions f and g, find the rule of the composition $(f \circ g)(x)$.

29. $f(x) = x + 2$; $g(x) = x - 9$ **30.** $f(x) = 2x + 1$; $g(x) = x - 3$
31. $f(x) = x^2$; $g(x) = 2x + 1$ **32.** $f(x) = 3x + 6$; $g(x) = \frac{1}{3}x - 2$

33. Write the inverse of the function $\{(3, -8), (4, 1), (0, -5), (-2, 6)\}$.

In 34–36, write, in the form $y = mx + b$, the equation of the inverse of the given function.

34. $y = 6x$ **35.** $y = 2x - 14$ **36.** $y = 3 - \frac{1}{3}x$

37. What is the inverse of the function $y = \dfrac{3x - 2}{5}$?

(1) $y = \dfrac{3x + 2}{5}$ (2) $y = \dfrac{5x - 2}{3}$

(3) $y = \dfrac{5x + 2}{3}$ (4) $y = \dfrac{3x - 2}{5}$

38. Let $f(x) = 5x - 4$.
a. Write the rule for f^{-1}, the inverse under composition.
b. Find $f(2)$.   **c.** Find $f^{-1}(6)$.   **d.** Find $f^{-1}(0)$.
e. Find $(f^{-1} \circ f)(17)$.   **f.** Explain why f is a one-to-one function.

39. The identity function under composition is:

(1) $y = 0$ (2) $y = x$ (3) $y = 1$ (4) $y = \dfrac{1}{x}$

40. The charges for a telephone call from New York City to Philadelphia, dialed directly on a home phone, are $.52 for the first minute or less and $.35 for each additional minute or portion thereof.
a. Graph this step function for charges on calls lasting 6 minutes or less. Let x represent the time, and $f(x)$ the charge.
b. For the domain $0 < x \le 6$, what is the range?
c. What is the charge for a 1-hour phone call from New York City to Philadelphia?
d. If the charge for a New York City-Philadelphia call is $6.12, what was the maximum time in minutes for this call?
e. Is the relation formed by interchanging x and $f(x)$ from part **a** also a function? If not, explain why.

Chapter **7**

Functions and Transformation Geometry

7-1 TRANSFORMATIONS OF THE PLANE AS FUNCTIONS

In Chapter 5, we studied transformations of the plane and, in Chapter 6, we studied functions. Since a transformation of the plane is a one-to-one correspondence between the points in the plane, it is true that:

■ Every transformation of the plane is a one-to-one function.

For example, let us compare a familiar transformation of the plane, such as a line reflection, with a one-to-one algebraic function.

f: *A one-to-one algebraic function*

$$f(x) = 2x - 3$$

r_m : *A reflection in line m*

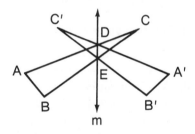

1. For every x in the domain, there is one and only one corresponding $f(x)$ in the range.
   For example:

$$f(10) = 17$$
$$f(6) = 9$$
$$f(3) = 3$$

Therefore, f is a *function*.

1. For every point in the plane, there is one and only one corresponding point (or image) in the plane.
   For example:

$$r_m (A) = A'$$
$$r_m (C) = C'$$
$$r_m (E) = E$$

Therefore, r_m is a *function*.

297

2. Since every f(x) corresponds to a unique value of x, that is, one and only one x, the function f is *one-to-one*.

   For example, if f(x) = 5, then x = 4.

3. For the function f:

   domain = {real numbers}
   range = {real numbers}

   Thus, function f is a one-to-one correspondence between elements of the set of real numbers.

2. Since every image has a unique preimage, that is, one and only one preimage, the function r_m is *one-to-one*.

   For example, given the image B', its preimage is B.

3. For the function r_m :

   domain = {points in plane}
   range = {points in plane}

   Thus, transformation r_m or function r_m is a one-to-one correspondence between the points in the plane.

Just as r_m , the reflection in line m, is a function, it is also true that every transformation of the plane is a function. For example:

R_P: A reflection in point P is a function.

$R_{M,\phi}$: A rotation about point M, through an angle of measure ϕ, is a function.

Coordinate Rules of Transformations

In Chapter 5, we learned the coordinate rules for many standard transformations. As we proceed into this chapter, it is helpful to know the function rules of the most common transformations.

Line Reflections:
1. A reflection in the x-axis: $r_{x\text{-axis}}(x,y) = (x,-y)$
2. A reflection in the y-axis: $r_{y\text{-axis}}(x,y) = (-x,y)$
3. A reflection in the line $y = x$: $r_{y=x}(x,y) = (y,x)$
4. A reflection in the line $y = -x$: $r_{y=-x}(x,y) = (-y,-x)$

Point Reflection in the Origin:
5. A reflection in the origin, or a half-turn, or a rotation of 180°: $R_O(x,y) = (-x,-y)$

Rotations About the Origin:
6. Counterclockwise rotation of 90° about O: $R_{90°}(x,y) = (-y,x)$
7. Counterclockwise rotation of 180° about O: $R_{180°}(x,y) = (-x,-y)$
8. Counterclockwise rotation of 270° about O: $R_{270°}(x,y) = (y,-x)$

Translation:
9. A translation of "*a*" units hori-
    zontally and "*b*" units vertically: $T_{a,b}(x, y) = (x + a, y + b)$

Dilation:
10. A dilation of k, where $k > 0$, and
    the origin is the center of dilation: $D_k(x, y) = (kx, ky)$

EXERCISES

1. Under function f: $(x, y) \rightarrow (x, -y)$.
   a. Function f is a reflection:
      (1) in the x-axis   (2) in the y-axis
      (3) in the line $x = y$ (4) in the line $x = -y$
   b. State the domain of f.   c. State the range of f.
2. a. If function g is a reflection in the line $x = y$, the rule of g is:
      (1) $(x, y) \rightarrow (x, -y)$ (2) $(x, y) \rightarrow (-x, y)$
      (3) $(x, y) \rightarrow (y, x)$ (4) $(x, y) \rightarrow (-y, -x)$
   b. State the domain of g.   c. State the range of g.
3. Match the name of the transformation of the plane in *Column A* with its function rule in *Column B*.

   Column A   *Column B*
   1. A reflection in the y-axis.   *a.* $(x, y) \rightarrow (x + 3, y)$
   2. A point reflection in the origin.   *b.* $(x, y) \rightarrow (3x, 3y)$
   3. A reflection in the line $y = -x$.   *c.* $(x, y) \rightarrow (-x, y)$
   4. A dilation of 3.   *d.* $(x, y) \rightarrow (-y, x)$
   5. A translation of 3 to the right.   *e.* $(x, y) \rightarrow (-x, -y)$
   6. A counterclockwise rotation of 90°   *f.* $(x, y) \rightarrow (-y, -x)$
      about the origin.

4. Let $f(x, y) = (x, 0)$, where the domain is the set of points in the plane.
   a. Find $f(5, 2)$.   b. Find $f(3, 7)$.   c. Find $f(3, 4)$.
   d. The range of f is:
      (1) the x-axis   (2) the y-axis
      (3) the set of points in the plane   (4) the line $y = x$
   e. Is f a function?   f. Is f a one-to-one function?
   g. Explain why f is *not* a transformation of the plane.

5. Given the rule $g(x, y) = (2, y)$.
   a. Find $g(5, 6)$.   b. Find $g(9, 7)$.   c. Find $g(-8, 7)$.
   d. State the domain of g.   e. State the range of g.
   f. Explain why g is a function.
   g. Explain why g is *not* a transformation of the plane.

7-2 COMPOSITIONS AND SYMMETRIES

We have learned that composition is an operation performed on the set of functions. Since every transformation of the plane is a one-to-one function, composition can be performed on these transformations.

Earlier, on page 217 of Chapter 5, we learned that a reflection in the y-axis followed by a reflection in the x-axis is a composition of transformations equivalent to a reflection through point O, the origin. Using the symbols learned in Chapter 6, we can now indicate this composition as:

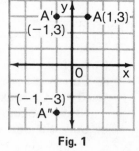

Fig. 1

$$r_{x\text{-axis}} \circ r_{y\text{-axis}} = R_O$$

Remember that the composition is read as "a reflection in the x-axis *following* a reflection in the y-axis." That is, for the given transformations, we reflect in the y-axis *first*.

To perform this composition on a point such as A, as shown in Fig. 1, we follow the rule outlined in the box at the right.

$$A \xrightarrow{\;r_{y\text{-axis}}\;} A' \xrightarrow{\;r_{x\text{-axis}}\;} A''$$
$$r_{x\text{-axis}} \circ r_{y\text{-axis}} (A) = A''$$

While we shall study many other compositions throughout this chapter, let us first consider compositions of transformations using symmetric figures.

Symmetries

Recall that *symmetry* occurs in a figure when the figure is its own image under a given transformation. In Fig. 2, the diagonals of square $ABCD$ meet at point M. We have learned three types of symmetry, each of which occurs in square $ABCD$:

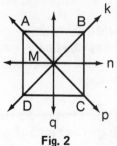

Fig. 2

1. *Line symmetry*, using r_k, r_n, r_p, or r_q.
2. *Point symmetry*, using R_M.
3. *Rotational symmetry*, using $R_{M,90°}$, $R_{M,180°}$, or $R_{M,270°}$.

In the model problems that follow, we shall use square $ABCD$ and its symmetries to study compositions involving transformations.

| MODEL PROBLEMS |

1. In the accompanying figure, k and n are symmetry lines for square $ABCD$. Find $r_n \circ r_k (A)$.

How to Proceed	*Solution*
1. First, reflect point A in line k to find its image C.	$A \xrightarrow{r_k} C \xrightarrow{r_n} B$
2. Then, reflect C in line n to find the image B.	$r_n \circ r_k (A) = B$
Answer: B	

2. In the accompanying diagram, p and q are lines of symmetry, and M is the midpoint of diagonal $\overline{AC}$. Find $R_M \circ r_p \circ r_q (D)$.

How to Proceed	*Solution*
1. First, reflect point D in line q to find its image C.	$D \xrightarrow{r_q} C \xrightarrow{r_p} C \xrightarrow{R_M} A$
2. Since C is a point on line p, the image of C under a reflection in line p is still C.	$R_M \circ r_p \circ r_q (D) = A$
3. Finally, reflect C through point M to find the image A.	
Answer: A	

Note: As seen in this example, we omit parentheses when writing the composition of three or more transformations to reduce the symbolism involved.

3. In square $ABCD$, p and n are symmetry lines, and M is the midpoint of $\overline{AC}$. What is $r_n \circ R_{M,90°} \circ r_p (\overline{AB})$?
    (1) $\overline{AB}$ (2) $\overline{BC}$ (3) $\overline{CD}$ (4) $\overline{DA}$

How to Proceed	*Solution*

1. Reflect $\overline{AB}$ in line p to find its image $\overline{AD}$.

2. Rotate $\overline{AD}$ counterclockwise through a 90° angle about M to find its image $\overline{DC}$.

3. Reflect $\overline{DC}$ in line n to find the image $\overline{AB}$.

Answer: (1) $\overline{AB}$

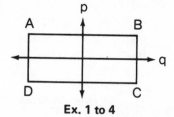

$$AB \xrightarrow{\;r_p\;} AD \xrightarrow{\;R_{M,90°}\;} DC \xrightarrow{\;r_n\;} AB$$

$$r_n \circ R_{M,90°} \circ r_p\,(\overline{AB}) = \overline{AB}$$

The answer is choice (1) $\overline{AB}$.

| EXERCISES |

In 1–4, p and q are symmetry lines for rectangle $ABCD$.

1. Find $r_p \circ r_q\,(A)$.
2. Find $r_q \circ r_p\,(D)$.
3. Find $r_q \circ r_p \circ r_q\,(C)$.
4. Find $r_p \circ r_q \circ r_p\,(B)$.

Ex. 1 to 4

In 5–10, regular hexagon $ABCDEF$ is inscribed in circle O, and O is the center of rotation.

5. Find $R_{60°} \circ R_{180°}\,(A)$.
6. Find $R_{240°} \circ R_{120°}\,(E)$.
7. Find $R_{300°} \circ R_{120°}\,(B)$.
8. Find $R_{120°} \circ R_{-240°}\,(F)$.
9. Find $R_{-120°} \circ R_{180°} \circ R_{-300°} \circ R_{240°}\,(C)$.
10. The composition $R_{-240°} \circ R_{180°} \circ R_{-60°}$ is equivalent to the transformation:
    (1) $R_{120°}$ (2) $R_{180°}$ (3) $R_{240°}$ (4) $R_{300°}$

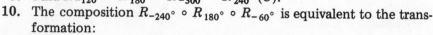

Ex. 5 to 10

In 11–18, k and m are symmetry lines for equilateral triangle ABC, and points A, B, and C are equidistant from point P.

11. Find $r_k \circ r_m\,(C)$.
12. Find $r_m \circ r_k\,(C)$.
13. Find $r_m \circ r_k \circ r_m\,(A)$.
14. Find $r_k \circ r_m \circ r_k\,(B)$.
15. Find $R_{P,120°} \circ R_{P,120°}\,(A)$.
16. Find $r_m \circ R_{P,240°}\,(A)$.

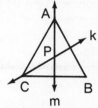

Ex. 11 to 18

17. Which of the vertices, A, B, or C, is its own image under the composition $r_k \circ R_{P,120°}$?

18. a. Find the image of each vertex, A, B, and C, under the composition $r_k \circ R_{P,120°} \circ r_m$.
    b. The composition $r_k \circ R_{P,120°} \circ r_m$ is equivalent to the single transformation:
    (1) r_k (2) r_m (3) $R_{P,120°}$ (4) $R_{P,240°}$

In 19-27, k and g are symmetry lines for square $ABCD$, and M is the midpoint of diagonal $\overline{AC}$.

19. a. Find $r_g \circ r_k (B)$.
    b. Find $r_k \circ r_g (B)$.
    c. The answers in parts **a** and **b** demonstrate that composition of transformations:

Ex. 19 to 27

    (1) is commutative   (2) is *not* commutative
    (3) is associative   (4) is *not* associative

20. Find $r_k \circ r_g \circ r_k (D)$.
21. Find $r_g \circ r_k \circ r_g (A)$.
22. Find $R_M \circ r_k \circ r_g (D)$.
23. Find $r_g \circ R_M \circ r_k (A)$.
24. Find $r_k \circ r_g (\overline{DA})$.
25. Find $r_g \circ r_k (\overline{BC})$.
26. Find $r_g \circ R_M \circ r_g (\overline{BC})$.
27. Find $r_k \circ r_g \circ R_M (\overline{AB})$.

In 28-36, m and p are lines of symmetry for regular hexagon $ABCDEF$.

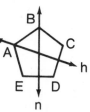

28. Find $r_p \circ r_m (D)$.
29. Find $r_m \circ r_p (E)$.
30. Find $r_m \circ r_p \circ r_m (B)$.
31. Find $r_p \circ r_m \circ r_p (C)$.
32. Find $r_m \circ r_p (\overline{AB})$.

Ex. 28 to 36

33. Find $r_p \circ r_m (\overline{AB})$.
34. Find $r_p \circ r_m \circ r_p (\overline{ED})$.
35. Find $r_m \circ r_p \circ r_m (\overline{CB})$.
36. *True* or *False*: The composition $r_m \circ r_p$ is equivalent to a counterclockwise rotation of $120°$ about the center of regular hexagon $ABCDEF$. (*Hint:* Find the image of each vertex by using $r_m \circ r_p$.)

In 37-43, h and n are two of the five possible lines of symmetry for regular pentagon $ABCDE$.

37. Find $r_h \circ r_n (E)$.   38. Find $r_n \circ r_h (E)$.
39. Find $r_h \circ r_n (C)$.   40. Find $r_n \circ r_h \circ r_n (D)$.
41. Find $r_n \circ r_h \circ r_n (\overline{CB})$.
42. Find $r_h \circ r_n \circ r_h (\overline{BA})$.
43. Every point on the pentagon is its own image under the composition:

Ex. 37 to 43

    (1) $r_h \circ r_n$ (2) $r_n \circ r_h$ (3) $r_n \circ r_n$ (4) $r_h \circ r_h \circ r_h$

In 44–50, p, q, and m are symmetry lines for the figure shown at the right.

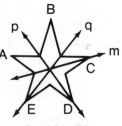

44. Find $r_p \circ r_q(B)$.   45. Find $r_q \circ r_m(A)$.
46. Find $r_m \circ r_p(C)$.   47. Find $r_m \circ r_q(D)$.
48. Find $r_q \circ r_p \circ r_m(D)$.
49. Find $r_p \circ r_q \circ r_m(B)$.
50. Under the composition $r_m \circ r_q \circ r_p$, which point (A, B, C, D, or E) is its own image?

Ex. 44 to 50

In 51–58, $ABCD$ is a square. The midpoints of sides $\overline{AB}, \overline{BC}, \overline{CD}$, and $\overline{DA}$ are E, F, G, and H, respectively. Lines $\overleftrightarrow{AC}, \overleftrightarrow{BD}, \overleftrightarrow{EG}$, and $\overleftrightarrow{HF}$ are drawn.

51. Find $r_{\overleftrightarrow{EG}} \circ r_{\overleftrightarrow{AC}}(B)$.   52. Find $r_{\overleftrightarrow{BD}} \circ r_{\overleftrightarrow{HF}}(D)$.
53. Find $r_{\overleftrightarrow{AC}} \circ r_{\overleftrightarrow{HF}} \circ r_{\overleftrightarrow{BD}}(A)$.   54. Find $r_{\overleftrightarrow{EG}} \circ r_{\overleftrightarrow{BD}} \circ r_{\overleftrightarrow{AC}}(H)$.
55. Find $r_{\overleftrightarrow{AC}} \circ r_{\overleftrightarrow{HF}} \circ r_{\overleftrightarrow{BD}}(F)$.   56. Find $r_{\overleftrightarrow{AC}} \circ r_{\overleftrightarrow{EG}} \circ r_{\overleftrightarrow{HF}}(\overline{AE})$.
57. Find $r_{\overleftrightarrow{EG}} \circ r_{\overleftrightarrow{BD}} \circ r_{\overleftrightarrow{AC}}(\overline{BF})$.   58. Find $r_{\overleftrightarrow{BD}} \circ r_{\overleftrightarrow{EG}} \circ r_{\overleftrightarrow{HF}}(\overline{AB})$.

7-3 MORE COMPOSITIONS WITH TRANSFORMATIONS

The composition of two or more functions, each of which is one-to-one, is a one-to-one function. Since every transformation of the plane is a one-to-one function, the composition of two or more such transformations is a single transformation of the plane.

In Chapter 5, we studied some examples of compositions of transformations. For example, the composition of two translations is a single translation. Now that we have learned to work with functions, we can determine rules for the compositions of transformations, as seen in the following examples.

☐ EXAMPLE 1:

Find the rule of the composition consisting of a dilation of 2 *following* a point reflection in the origin.

Solution: For any point on the plane, we first reflect the point in the origin. Then, we dilate its image by a factor of 2.

$$(x, y) \xrightarrow{\;R_O\;} (-x, -y) \xrightarrow{\;D_2\;} (-2x, -2y)$$
$$D_2 \circ R_O = D_{-2}$$

Answer: $D_{-2}(x, y) = (-2x, -2y)$

Note: The rule of this composition may be studied by considering a specific triangle. In the accompanying diagram, the coordinates of the vertices of $\triangle ABC$ are $A(1, 0), B(1, 2)$, and $C(0, 2)$. Thus:

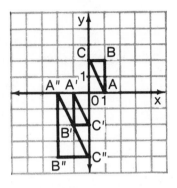

$$A(1,0) \xrightarrow{R_O} A'(-1,0) \xrightarrow{D_2} A''(-2,0)$$
$$B(1,2) \longrightarrow B'(-1,-2) \longrightarrow B''(-2,-4)$$
$$C(0,2) \longrightarrow C'(0,-2) \longrightarrow C''(0,-4)$$

$$D_{-2}(x,y) = (-2x,-2y)$$

☐ EXAMPLE 2:

Find the rule of the composition of transformations, $R_O \circ D_2$.

Solution: For any point on the plane, we first dilate the point by a factor of 2. Then, we reflect its image in the origin.

$$(x,y) \xrightarrow{D_2} (2x,2y) \xrightarrow{R_O} (-2x,-2y)$$

$$R_O \circ D_2 = D_{-2}$$

Answer: $D_{-2}(x,y) = (-2x,-2y)$

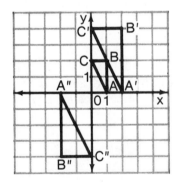

Note: The composition of transformations, $R_O \circ D_2$, is applied to $\triangle ABC$ in the diagram at the left. By using the coordinates $A(1, 0), B(1, 2)$, and $C(0, 2)$, we see that $\triangle ABC \rightarrow \triangle A''B''C''$ by the rule:

$$D_{-2}(x,y) = (-2x,-2y)$$

Examples 1 and 2 demonstrate that $D_2 \circ R_O = R_O \circ D_2$; that is, the composition of these specific transformations is commutative. Will this be true for the composition of any transformation?

☐ EXAMPLE 3:

Let $D_4(x, y) = (4x, 4y)$. Let $T_{3,0}(x, y) = (x + 3, y)$.
a. Find the rule of the composition of transformations, $T_{3,0} \circ D_4$.
b. Find the rule of the composition of transformations, $D_4 \circ T_{3,0}$.

Solution

a. In $T_{3,0} \circ D_4$, we *first* dilate the point by a factor of 4. Then, we translate its image 3 units to the right.

$$(x, y) \xrightarrow{\;D_4\;} (4x, 4y) \xrightarrow{\;T_{3,0}\;} (4x + 3, 4y)$$

$$T_{3,0} \circ D_4$$

Answer: $T_{3,0} \circ D_4 (x, y) = (4x + 3, 4y)$

b. In $D_4 \circ T_{3,0}$, we *first* translate the point 3 units to the right. Then, we dilate its image by a factor of 4.

$$(x, y) \xrightarrow{\;T_{3,0}\;} (x + 3, y) \xrightarrow{\;D_4\;} (4x + 12, 4y)$$

$$D_4 \circ T_{3,0}$$

Answer: $D_4 \circ T_{3,0} (x, y) = (4x + 12, 4y)$

The rules of the compositions found in parts a and b of example 3 just shown demonstrate that $T_{3,0} \circ D_4 \neq D_4 \circ T_{3,0}$. In general, it is true that:

■ **Composition of transformations is not commutative.**

Glide Reflections

A line reflection is like a "flip," and a translation is like a "shift" or a "glide." The pattern of footsteps seen in the following diagram represents a composition of a line reflection and a translation, called a *glide reflection*.

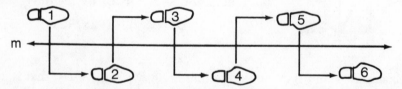

From footstep 1 to 2, we reflect the step in line m, then translate this image to the right. From footstep 2 to 3, we again reflect the step in line m, then translate this image to the right. Notice that the translation is parallel to the line of reflection.

If we had first translated the footstep to the right and then reflected its image in line m, the same pattern would occur. Thus, the composition of these specific transformations is commutative.

■ **DEFINITION.** A *glide reflection* is a transformation of the plane that represents the composition, in either order, of a line reflection and a translation that is parallel to the line of reflection.

KEEP IN MIND

The composition of any two transformations of the plane is itself a transformation of the plane.

MODEL PROBLEMS

1. The coordinates of the vertices of $\triangle ABC$ are $A(2, -6)$, $B(2, -2)$, and $C(4, -5)$.
   a. On graph paper, draw and label $\triangle ABC$.
   b. Find the coordinates of the vertices of $\triangle A'B'C'$, the image of $\triangle ABC$ after a translation of $T_{0,7}$.
   c. Find the coordinates of the vertices of $\triangle A''B''C''$, the image of $\triangle A'B'C'$ after a reflection in the y-axis.
   d. The single transformation that maps $\triangle ABC$ onto $\triangle A''B''C''$ is a:
      (1) line reflection   (2) glide reflection   (3) point reflection
      (4) rotation

Solution

a. $\triangle ABC$ is graphed at the right.

b. Under the given translation:

$$T_{0,7}(x, y) = (x, y + 7)$$

Thus: $A(2, -6) \rightarrow A'(2, 1)$
$B(2, -2) \rightarrow B'(2, 5)$
$C(4, -5) \rightarrow C'(4, 2)$
$\triangle A'B'C'$ is graphed at the right.

c. Under a reflection in the y-axis:

$$r_{y\text{-axis}}(x, y) = (-x, y)$$

Thus: $A'(2, 1) \rightarrow A''(-2, 1)$
$B'(2, 5) \rightarrow B''(-2, 5)$
$C'(4, 2) \rightarrow C''(-4, 2)$
$\triangle A''B''C''$ is graphed at the right.

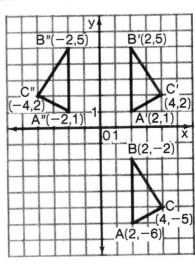

d. The translation $T_{0,7}$ is parallel to the y-axis, which is the line of reflection. The composition of a line reflection and a translation parallel to the line of reflection is a *glide reflection* (choice 2).

Answer: **a.** See graph. **b.** $A'(2, 1), B'(2, 5)$, and $C'(4, 2)$
        c. $A''(-2, 1), B''(-2, 5)$, and $C''(-4, 2)$
        d. (2) glide reflection

2. Triangle ABC has the vertices $A(0, 3)$, $B(2, 3)$, and $C(4, 5)$.
  a. Graph $\triangle ABC$ and its image, $\triangle A'B'C'$, after a reflection in the line $y = x$.
  b. Graph $\triangle A''B''C''$, the image of $\triangle A'B'C'$ after the translation $T_{-3,3}$.
  c. Describe, by name, the single transformation that maps $\triangle ABC$ onto $\triangle A''B''C''$.
  d. Find the rule of the transformation equivalent to $T_{-3,3} \circ r_{y=x}$.

Solution

a. Under a reflection in the line $y = x$:

$$r_{y=x}(x, y) = (y, x)$$

Thus: $A(0, 3) \rightarrow A'(3, 0)$
     $B(2, 3) \rightarrow B'(3, 2)$
     $C(4, 5) \rightarrow C'(5, 4)$

$\triangle ABC$ and $\triangle A'B'C'$ are graphed at the right.

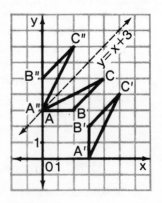

b. Under the translation $T_{-3,3}$:

$$T_{-3,3}(x, y) = (x - 3, y + 3)$$

Thus: $A'(3, 0) \rightarrow A''(0, 3)$
     $B'(3, 2) \rightarrow B''(0, 5)$
     $C'(5, 4) \rightarrow C''(2, 7)$

$\triangle A''B''C''$ is graphed at the right.

c. $\triangle ABC$ maps onto $\triangle A''B''C''$ by a reflection in the line $y = x + 3$, as sketched in the coordinate graph.

  Note: The composition of a reflection in the line $y = x$, followed by the translation $T_{-3,3}$, is *not* a glide reflection because the translation is *not parallel* to the line of reflection. In this specific case, the composition is a line reflection. In other cases, other transformations may occur.

d. For any point:

$$(x, y) \xrightarrow{\ r_{y=x}\ } (y, x) \xrightarrow{\ T_{-3,3}\ } (y - 3, x + 3)$$

$$T_{-3,3} \circ r_{y=x}(x, y) = (y - 3, x + 3)$$

Answer: **a** and **b.** See graph. **c.** A reflection in the line $y = x + 3$.

d. $T_{-3,3} \circ r_{y=x}(x, y) = (y - 3, x + 3)$

OR

$$(x, y) \rightarrow (y - 3, x + 3)$$

3. In the accompanying diagram, find the coordinates of $r_m \circ r_k(A)$.

Solution

1. First, reflect point A in line k (the line $y = -x$). Recall that $r_{y=-x}(x, y) = (-y, -x)$.

2. Then, reflect this image in line m (the line $y = 2$). Graph paper is helpful in plotting these points. Therefore:

$$(-3, -2) \xrightarrow{\ r_k\ } (2, 3) \xrightarrow{\ r_m\ } (2, 1)$$

$$r_m \circ r_k(-3, -2) = (2, 1)$$

Answer: $(2, 1)$

4. Name the single transformation that is equivalent to the composition $r_{x\text{-axis}} \circ r_{y=x}$.

How to Proceed

1. First, reflect the point (x, y) in the line $y = x$.

2. Then, reflect the image found in the x-axis.

3. The transformation that maps (x, y) to $(y, -x)$ is a rotation of $270°$ about the origin.

Solution

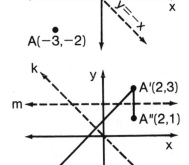

$$(x, y) \xrightarrow{\ r_{y=x}\ } (y, x) \xrightarrow{\ r_{x\text{-axis}}\ } (y, -x)$$

$$r_{x\text{-axis}} \circ r_{y=x} = R_{270°}$$

Answer: $R_{270°}$, or a rotation of $270°$ about the origin.

EXERCISES

1. The coordinates of the vertices of $\triangle ABC$ are $A(1, 3)$, $B(1, 6)$, and $C(3, 3)$.
   a. On graph paper, draw and label $\triangle ABC$.
   b. Find the coordinates of the vertices of $\triangle A'B'C'$, the image of $\triangle ABC$ under a point reflection in the origin.
   c. Find the coordinates of $\triangle A''B''C''$, the image of $\triangle A'B'C'$ after a reflection in the y-axis.
   d. Name the single transformation that maps $\triangle ABC$ onto $\triangle A''B''C''$.

2. The vertices of $\triangle ABC$ are $A(-6, -2)$, $B(-4, 1)$, and $C(-1, 4)$.
   a. Graph $\triangle ABC$ and its image, $\triangle A'B'C'$, under the translation $T_{6,0}$.
   b. Graph $\triangle A''B''C''$, the image of $\triangle A'B'C'$ after a reflection in the x-axis.
   c. The single transformation that maps $\triangle ABC$ onto $\triangle A''B''C''$ is a:
      (1) line reflection   (2) point reflection   (3) glide reflection
      (4) rotation

3. The coordinates of the vertices of $\triangle CDE$ are $C(0, 1)$, $D(4, 1)$, and $E(4, 3)$.
   a. Graph $\triangle CDE$ and its image, $\triangle C'D'E'$, under a reflection in the x-axis.
   b. Graph $\triangle C''D''E''$, the image of $\triangle C'D'E'$ under a reflection in the line $y = x$.
   c. The composition $r_{y=x} \circ r_{x\text{-axis}}$, which maps $\triangle CDE$ onto $\triangle C''D''E''$, is equivalent to the transformation:
      (1) $r_{y\text{-axis}}$ (2) $r_{y=-x}$ (3) $R_{90°}$ (4) $R_{270°}$

4. The vertices of $\triangle ACE$ are $A(0, 2)$, $C(5, 4)$, and $E(6, 2)$.
   a. Graph $\triangle ACE$ and $\triangle A'C'E'$ where $R_{90°}(\triangle ABC) = \triangle A'B'C'$. Recall that the rule for a rotation of $90°$ about the origin is $(x, y) \rightarrow (-y, x)$.
   b. Graph $\triangle A''B''C''$ such that $R_{180°}(\triangle A'B'C') = \triangle A''B''C''$.
   c. Graph $\triangle A'''B'''C'''$ such that $r_{x\text{-axis}}(\triangle A''B''C'') = \triangle A'''B'''C'''$.
   d. The composition $r_{x\text{-axis}} \circ R_{180°} \circ R_{90°}$ is equivalent to the single transformation:
      (1) $r_{y=x}$ (2) $r_{y=-x}$ (3) $r_{y\text{-axis}}$ (4) $R_{270°}$

5. The coordinates of the vertices of $\triangle BUG$ are $B(1, 1)$, $U(1, 4)$, and $G(7, 1)$. Let $r_{y\text{-axis}}(\triangle BUG) = \triangle B'U'G'$, and $r_{y=x}(\triangle B'U'G') = \triangle B''U''G''$.
   a. On graph paper, draw and label $\triangle BUG$, $\triangle B'U'G'$, and $\triangle B''U''G''$.
   b. State the coordinates of the vertices of $\triangle B'U'G'$.

   c. State the coordinates of the vertices of $\triangle B''U''G''$.

   d. Name the single transformation that maps $\triangle BUG$ onto $\triangle B''U''G''$.

6. Triangle BCD has the vertices $B(-1, 2)$, $C(-1, 4)$, and $D(3, 2)$.

   a. Graph $\triangle BCD$ and its image, $\triangle B'C'D'$, after a reflection in the line $y = -x$.

   b. Graph $\triangle B''C''D''$, the image of $\triangle B'C'D'$ after the translation $T_{5,5}$.

   c. The single transformation that maps $\triangle BCD$ onto $\triangle B''C''D''$ is a:

     (1) line reflection   (2) glide reflection

     (3) point reflection   (4) rotation

7. The vertices of $\triangle JAM$ have the coordinates $J(1, 0)$, $A(1, 4)$, and $M(3, 1)$.

   a. Let $\triangle JAM \rightarrow \triangle J'A'M'$ by a dilation of 2 with the origin as the center of dilation. Let $\triangle J'A'M' \rightarrow \triangle J''A''M''$ by a reflection in the y-axis. On graph paper, draw and label $\triangle JAM$, $\triangle J'A'M'$, and $J''A''M''$.

   b. Let $\triangle JAM \rightarrow \triangle J'''A'''M'''$ by a reflection in the y-axis. Let $\triangle J'''A'''M''' \rightarrow \triangle J''''A''''M''''$ by a dilation of 2 with the origin as the center of dilation. Graph $\triangle J'''A'''M'''$ and $\triangle J''''A''''M''''$.

   c. *True* or *False*: $r_{y\text{-axis}} \circ D_2 = D_2 \circ r_{y\text{-axis}}$.

In 8–10, the coordinates of point P are $(1, -2)$, and the equations of lines m and q are stated in the accompanying diagrams. Find the coordinates of: a. $r_m \circ r_q(P)$.   b. $r_q \circ r_m(P)$

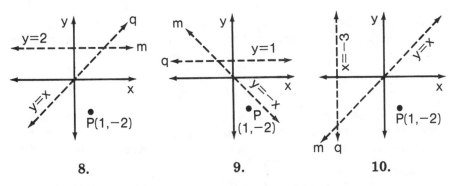

        8.   **9.**   **10.**

In 11–16, write the rule of the single transformation that is equivalent to the stated composition.

11. $r_{x\text{-axis}} \circ D_3$ 12. $r_{y=x} \circ T_{2,5}$

13. $T_{5,-4} \circ r_{y\text{-axis}}$ 14. $T_{1,-6} \circ D_4$

15. $D_2 \circ T_{-1,2} \circ R_{180°}$ 16. $R_{180°} \circ T_{-1,2} \circ D_2$

In 17–22: **a.** Write the rule of the single transformation that is equivalent to the given composition. **b.** Name the transformation whose rule is found in part **a.**

17. $r_{x\text{-axis}} \circ r_{y\text{-axis}}$

18. $R_{180°} \circ R_{90°}$

19. $R_{180°} \circ r_{x\text{-axis}}$

20. $r_{y=x} \circ r_{y=-x}$

21. $r_{y=x} \circ R_{90°}$

22. $R_{180°} \circ r_{y=-x}$

7-4 ISOMETRIES, ORIENTATION, AND OTHER PROPERTIES

In Chapter 5, we learned that certain properties are preserved under given transformations. Under a *line reflection*, a *point reflection*, a *rotation*, and a *translation*, the following five properties are preserved.

1. Distance is preserved.
2. Angle measure is preserved.
3. Parallelism is preserved.
4. Collinearity is preserved.
5. A midpoint is preserved.

It can be shown that other properties are preserved under these transformations, such as betweenness of points, and area.

Under a *dilation*, distance is *not* preserved, and area is *not* preserved, while the other properties mentioned are preserved.

■ **DEFINITION.** A transformation that preserves distance is called an *isometry*.

Thus, every line reflection, point reflection, rotation, and translation is an isometry. Since a dilation does not preserve distance, a dilation is not an isometry.

Orientation (or Order)

In each of the figures that follow, the vertices of $\triangle ABC$ appear in a *clockwise order*, or *orientation*. In other words, as we go from vertex A to vertex B to vertex C, we follow a clockwise direction, indicated in the figure by the symbol " $\curvearrowright$."

Under a translation (Fig. 1), this clockwise order, or orientation, is preserved because the images A', B', and C' appear in a clockwise order. Under a rotation (Fig. 2), orientation is again preserved because the images A', B', and C' appear in a clockwise order. However, under a line reflection (Fig. 3), orientation is *not* preserved because the images A', B', and C' appear in a *counterclockwise order*, indicated in the figure by the symbol " $\curvearrowleft$."

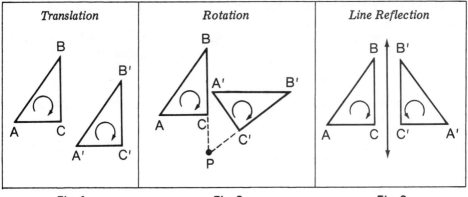

| Fig. 1 | Fig. 2 | Fig. 3 |

■ **DEFINITION.** A *direct isometry* is one that preserves order (or orientation).

Every translation, rotation, and point reflection is a direct isometry. In other words, each of these transformations preserves both order (orientation) and distance.

While a dilation preserves order (orientation), remember that a dilation is not an isometry because it fails to preserve distance.

■ **DEFINITION.** An *opposite isometry* is one that changes the order, or orientation, from clockwise to counterclockwise, or vice-versa.

Every line reflection is an opposite isometry. That is, every line reflection preserves distance but fails to preserve order.

Properties Under a Composition

Recall that a glide reflection is a *composition* of a line reflection and a translation parallel to the line of reflection. In the figure at the right, $\triangle ART \rightarrow \triangle A''R''T''$ by a glide reflection. What properties are preserved under this composition?

It can be shown that a glide reflection preserves: distance, angle measure, parallelism, collinearity, and midpoints. Since these properties are preserved by both a line reflection and a translation, it should be clear that the composition of such transformations will also preserve these same properties.

Notice, however, that orientation is *not* preserved under a glide reflection. In the figure at the right, the vertices A, R, and T are in a counterclockwise order, while their images A'', R'', and T''

are in a clockwise order. Therefore, *a glide reflection is an opposite isometry*.

Since a translation is a direct isometry and both a line reflection and a glide reflection are opposite isometries, the example just studied illustrates the fact that:

■ **The composition of a direct isometry and an opposite isometry is an opposite isometry.**

In the next section, we will examine many other compositions and study properties preserved under different given conditions.

| MODEL PROBLEMS |

1. The coordinates of the vertices of △ABC are A(0, 3), B(1, 1), and C(4, 1). Given the transformations:

$$K(x, y) = (x, 3y)$$
$$L(x, y) = (-x - 2, y - 3)$$
$$M(x, y) = (-y - 2, x + 3)$$

   a. Sketch △ABC and its image, △A'B'C', after the transformation K.
   b. Sketch △A"B"C", the image of △ABC after the transformation L.
   c. Sketch △A'''B'''C''', the image of △ABC after the transformation M.
   d. Which transformation given above is *not* an isometry?
   e. Which transformation given above does *not* preserve orientation?

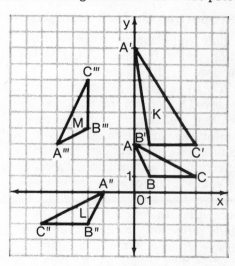

Solution

No composition is involved in this problem. In each part, we transform the original triangle, that is, $\triangle ABC$.

a. Since $K(x,y) = (x, 3y)$:

   $A(0, 3) \rightarrow A'(0, 9)$
   $B(1, 1) \rightarrow B'(1, 3)$
   $C(4, 1) \rightarrow C'(4, 3)$
   Graph $\triangle ABC$ and $\triangle A'B'C'$.

b. Since $L(x,y) = (-x - 2, y - 3)$:

   $A(0, 3) \rightarrow A''(-2, 0)$
   $B(1, 1) \rightarrow B''(-3, -2)$
   $C(4, 1) \rightarrow C''(-6, -2)$
   Graph $\triangle A''B''C''$.

c. Since $M(x,y) = (-y - 2, x + 3)$:

   $A(0, 3) \rightarrow A'''(-5, 3)$
   $B(1, 1) \rightarrow B'''(-3, 4)$
   $C(4, 1) \rightarrow C'''(-3, 7)$
   Graph $\triangle A'''B'''C'''$.

d. Distance is not preserved under transformation K. For example, $AB = \sqrt{5}$ and its image $A'B' = \sqrt{37}$. Therefore, K is not an isometry.

e. The vertices A, B, and C appear in a counterclockwise order. Under transformation L, their images A'', B'', and C'' appear in a clockwise order. Therefore, L does not preserve orientation.

Answer: **a, b,** and **c.** See graph.   **d.** K **e.** L

2. Which transformation is an example of a direct isometry? (1) line reflection   (2) glide reflection   (3) point reflection   (4) dilation

Solution

Since an isometry preserves distance, eliminate choice (4) dilation. A direct isometry preserves orientation (or order) as well as distance.

As illustrated in the point reflection at the right, the vertices A, B, C, and their images, A', B', and C', appear in the same clockwise order. Thus, a point reflection is a direct isometry.

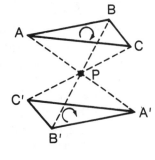

Answer: (3) point reflection

| EXERCISES |

1. The vertices of $\triangle ABC$ are $A(1, 2)$, $B(2, 5)$, and $C(5, 2)$. Under transformation P, $(x, y) \rightarrow (2 - x, 3 - y)$.
   a. Sketch $\triangle ABC$ and its image, $\triangle A'B'C'$, after transformation P.

    b. Transformation P preserves:
      (1) distance only   (2) order only   (3) both distance and order   (4) neither distance nor order

2. The vertices of $\triangle RST$ are $R(-1, 2)$, $S(0, 5)$, and $T(3, 2)$. Under transformation K, $(x, y) \to (y + 2, x - 4)$.
    a. Sketch $\triangle RST$ and its image, $\triangle R'S'T'$, after transformation K.
    b. Is transformation K a direct isometry, an opposite isometry, or not an isometry at all? Explain why.

3. Triangle ABC has the vertices $A(1, 1)$, $B(-1, 2)$, and $C(1, 6)$. Given the transformations: $F(x, y) = (2x, 2y)$
$$G(x, y) = (2y - 1, x - 1)$$
    a. Graph $\triangle ABC$ and its image, $\triangle A'B'C'$, after transformation F.
    b. Graph $\triangle A''B''C''$, the image of $\triangle ABC$ after transformation G.
    c. Which of the transformations given above are isometries?
      (1) F only   (2) G only
      (3) both F and G (4) neither F nor G

4. The coordinates of the vertices of $\triangle ABC$ are $A(-3, -4)$, $B(-2, -1)$, and $C(0, -5)$. Under transformation X, $\triangle ABC \to \triangle A'B'C'$ by a reflection in the x-axis. Under transformation Y, $\triangle ABC \to \triangle A''B''C''$ by a reflection in the origin. Under transformation Z, $\triangle ABC \to \triangle A'''B'''C'''$ by a reflection in point C.
    a. Graph and label $\triangle ABC$, $\triangle A'B'C'$, $\triangle A''B''C''$, and $\triangle A'''B'''C'''$.
    b. Which of the given transformations is *not* a direct isometry?

5. The coordinates of the vertices of $\triangle ACE$ are $A(1, 3)$, $C(2, 0)$, and $E(3, 6)$.
    a. Graph and label $\triangle ACE$.
    b. Find the coordinates of the vertices of $\triangle A'C'E'$, the image of $\triangle ACE$ under a reflection in the line $y = x$. Graph $\triangle A'C'E'$.
    c. Find the coordinates of the vertices of $\triangle A''C''E''$, the image of $\triangle ACE$ under the composition $r_{x\text{-axis}} \circ r_{y\text{-axis}}$. Graph $\triangle A''C''E''$.
    d. Find the coordinates of the vertices of $\triangle A'''C'''E'''$, the image of $\triangle ACE$ after a translation that maps $P(0, 0)$ to $P'''(2, -5)$. Graph $\triangle A'''C'''E'''$.
    e. Which of the given transformations is an opposite isometry?

6. The vertices of $\triangle ABC$ are $A(2, 1)$, $B(4, 1)$, and $C(0, -3)$. Given the transformations: $P(x, y) = (1 - x, 5 - y)$
$$Q(x, y) = (4 - y, x + 1)$$
$$R(x, y) = (-2x, y - 2)$$
    a. Graph $\triangle ABC$ and its image, $\triangle A'B'C'$, after transformation P.
    b. Graph $\triangle A''B''C''$, the image of $\triangle ABC$ after transformation Q.
    c. Graph $\triangle A'''B'''C'''$, the image of $\triangle ABC$ after transformation R.
    d. Which transformation given above is *not* an isometry?
    e. Which transformation given above does *not* preserve order?

7. Given: Y is the transformation $(x, y) \rightarrow (-y, -x)$.
    E is the transformation $(x, y) \rightarrow (x - 1, -y)$.
    S is the transformation $(x, y) \rightarrow (2x, 2y)$.
    The coordinates of the vertices of $\triangle ABC$ are $A(1, 2)$, $B(5, 1)$, and $C(3, -1)$.
    a. Sketch $\triangle ABC$ and its image, $\triangle A'B'C'$, after the transformation Y.
    b. Sketch $\triangle A''B''C''$, the image of $\triangle A'B'C'$ after the transformation E.
    c. Sketch $\triangle A'''B'''C'''$, the image of $\triangle A''B''C''$ after the transformation S.
    d. Which transformation, Y, E, or S, is *not* an isometry?
    e. Which transformation, Y, E, or S, preserves orientation?

In 8–14, select the numeral preceding the expression that best answers the question.

8. Which transformation is *not* an example of an isometry?
    (1) line reflection   (2) rotation   (3) translation   (4) dilation
9. Which transformation does *not* preserve orientation?
    (1) line reflection   (2) rotation   (3) translation   (4) dilation
10. Which property is *not* preserved under a glide reflection?
    (1) collinearity   (2) distance
    (3) orientation   (4) betweenness
11. Which property is *not* preserved under a dilation?
    (1) angle measure   (2) parallelism   (3) area   (4) midpoint
12. Which properties are preserved under a rotation?
    (1) distance only
    (2) angle measure only
    (3) both distance and angle measure
    (4) neither distance nor angle measure
13. Which transformation is an example of a direct isometry?
    (1) $r_{x\text{-axis}}$ (2) $r_{y = x}$ (3) $R_{90°}$ (4) D_5
14. Which rule indicates a transformation that is a direct isometry?
    (1) $(x, y) \rightarrow (x, -y)$ (2) $(x, y) \rightarrow (-x, -y)$
    (3) $(x, y) \rightarrow (-x, y)$ (4) $(x, y) \rightarrow (y, x)$

15. The coordinates of the vertices of rectangle $SEAN$ are $S(1, 1)$, $E(1, 3)$, $A(5, 3)$, and $N(5, 1)$. The transformation B maps (x, y) to $(x + y, -y)$.
    a. Graph rectangle $SEAN$.
    b. Graph quadrilateral $S'E'A'N'$, the image of $\square SEAN$ after the transformation B, and state the coordinates of S', E', A', and N'.

c. Which of the following seven properties are preserved by transformation B: distance, angle measure, collinearity, parallelism, midpoint, orientation, area?

d. For each property from part c that is *not* preserved by transformation B, cite an example to show that the property does not hold.

16. Given $P(3, 2)$ and $Q(7, 4)$ and the transformations A, B, C, D whose rules are stated below.

$$A: (x, y) \rightarrow (x + 2, y - 4) \qquad C: (x, y) \rightarrow (-x, y)$$
$$B: (x, y) \rightarrow (4 - y, x) \qquad D: (x, y) \rightarrow (y, x + 1)$$

a. Graph $\overline{PQ}$ and its image $\overline{P'Q'}$ after the transformation A.

b. Graph $\overline{P''Q''}$, the image of $\overline{PQ}$ after the transformation B.

c. Graph $\overline{P'''Q'''}$, the image of $\overline{PQ}$ after the transformation C.

d. Graph $\overline{P''''Q''''}$, the image of $\overline{PQ}$ after the transformation D.

e. Compare the slopes of the pairs of segments listed below, and tell whether these slopes are *equal*, *reciprocals*, *additive inverses*, or *negative reciprocals*.

(i) $\overline{PQ}$ and $\overline{P'Q'}$ (ii) $\overline{PQ}$ and $\overline{P''Q''}$

(iii) $\overline{PQ}$ and $\overline{P'''Q'''}$ (iv) $\overline{PQ}$ and $\overline{P''''Q''''}$

7-5 COMPOSITIONS WITH LINE REFLECTIONS

In this section, we will limit all compositions of transformations to those involving line reflections only. Why? The reason may be surprising but, in the next few pages, we will demonstrate that:

■ **Every isometry (or distance-preserving transformation) can be expressed as the composition of either two or three line reflections.**

Case I. Translations

In Fig. 1, line k is *parallel* to line m, $r_k(\triangle ABC) = \triangle A'B'C'$, and $r_m(\triangle A'B'C') = \triangle A''B''C''$. Therefore:

$$r_m \circ r_k(\triangle ABC) = \triangle A''B''C''$$

By studying the diagram, we can observe certain facts:

1. Since $\triangle ABC \rightarrow \triangle A''B''C''$ so that every point in the plane

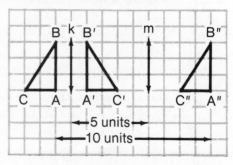

Fig. 1

moves the same distance in the same direction to its image, the composition $r_m \circ r_k$ where $k \| m$ is a *translation*.

2. The distance between lines k and m is 5 units. The distance of the translation, measured from A to A'', or from B to B'', or from C to C'', is 10 units. Thus, the distance of the translation is twice the distance between the parallel lines of reflection.

3. The direction of the translation is perpendicular to lines k and m, and this direction is the same as the direction from the first line of reflection, k, to the second, m.

These three facts will be true for the composition of any two line reflections in which the lines are parallel to each other. Suppose we had reflected first in line m and then in line k. This situation is displayed in Fig. 2, which follows.

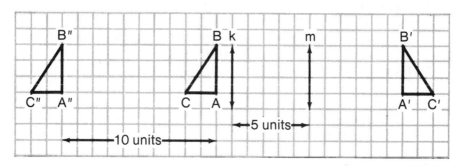

Fig. 2

In Fig. 2, where $k \| m$, $r_m(\triangle ABC) = \triangle A'B'C'$, and $r_k(\triangle A'B'C') = \triangle A''B''C''$. Therefore, $r_k \circ r_m(\triangle ABC) = \triangle A''B''C''$. Again, we observe:

1. The composition $r_k \circ r_m$, where $k \| m$, is a *translation*.
2. The distance of the translation is twice the distance between the parallel lines of reflection.
3. The direction of the translation is perpendicular to lines k and m, but the direction of the translation is now the same as the direction from the first line of reflection, m, to the second, k.

While Fig. 1 and Fig. 2 indicate that $r_m \circ r_k \neq r_k \circ r_m$, each composition is a translation whose direction and distance are determined by the slope of the lines and the distance between them.

The General Case for Translations

In Fig. 3, we observe a general situation where $k \| m$ and where $r_m \circ r_k(\triangle ABC) = \triangle A''B''C''$. To understand why this composition is a translation, let us study its direction and distance.

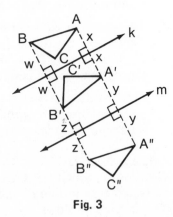

Fig. 3

Direction: In Fig. 3, by the definition of a line reflection, $\overline{AA'} \perp k$, $\overline{BB'} \perp k$, and $\overline{CC'} \perp k$. Also, $\overline{A'A''} \perp m$, $\overline{B'B''} \perp m$, and $\overline{C'C''} \perp m$. Since $k \| m$, certain points must be collinear: $\overline{AA'A''}$, $\overline{BB'B''}$, and $\overline{CC'C''}$. Each of these segments is perpendicular to lines k and m, and each indicates the direction of the translation (as in A to A'').

Distance: In Fig. 3, if x represents the distance from A to line k, then x is also the distance from A' to k. If y represents the distance from A' to line m, then y is also the distance from A'' to m.

Therefore, the distance between lines k and m is $(x + y)$, and the distance of the translation from A to A'' is $x + x + y + y = 2x + 2y = 2(x + y)$.

In a similar way, if w is the distance from B to line k, then w is also the distance from B' to k. If z is the distance from B' to line m, then z is also the distance from B'' to m. Using $\overline{BB'B''}$, we find that the distance between lines k and m is $(w + z)$, and the distance of the translation from B to B'' is $w + w + z + z = 2w + 2z = 2(w + z)$. Notice, however, that $(w + z) = (x + y)$. Therefore, the distance of the translation from B to B'' is $2(w + z) = 2(x + y)$. By a similar argument, the distance of the translation from C to C'' is also $2(x + y)$.

These examples demonstrate that every composition of two line reflections in parallel lines is a translation. Conversely, it can be shown that:

■ **Every translation is equivalent to the composition of two line reflections, $r_m \circ r_k$, where $k \| m$.**

For example, in Fig. 4, $\triangle ABC \to \triangle A''B''C''$ by a translation. By choosing two lines k and m such that $k \perp \overleftrightarrow{AA''}$, $m \perp \overline{AA''}$, and the distance between lines k and m is $\frac{1}{2}(AA'')$, we illustrate that the translation is equivalent to the composition $r_m \circ r_k$, where $k \| m$. Notice that lines k and m are *not unique*. That is, there are many lines k and m that allow us to map $\triangle ABC$ onto $\triangle A''B''C''$ by a composition of two line reflections.

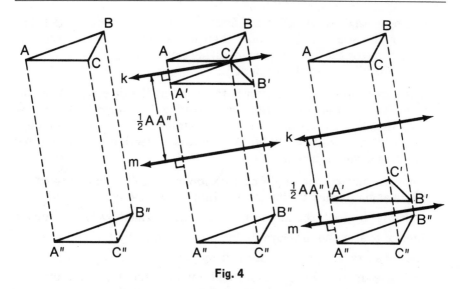

Fig. 4

The examples we have just studied illustrate another truth concerning order, or orientation. Since a line reflection is an opposite isometry and a translation is a direct isometry, we observe that:

■ **The composition of two opposite isometries is a direct isometry.**

Case II. Rotations

In Fig. 5, lines k and m inter-sect at point P, $r_k(\triangle ABC) = \triangle A'B'C'$, and $r_m(\triangle A'B'C') = \triangle A''B''C''$. Therefore:

$$r_m \circ r_k(\triangle ABC) = \triangle A''B''C''$$

Consider the case where k and m intersect to form an acute angle. By studying the diagram, we can observe certain facts:

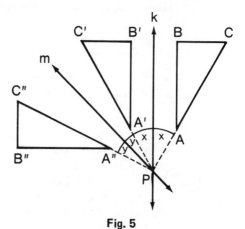

Fig. 5

1. The composition $r_m \circ r_k$, where lines k and m intersect at point P, is a *rotation* about P.
   Hint: To prove that $\triangle ABC \to \triangle A''B''C''$ by a rotation about P:
   a. Show that $PA = PA''$, $PB = PB''$, and $PC = PC''$. For example, since P is a point on each line of reflection, $PA = PA'$ and $PA' = PA''$. Thus, $PA = PA''$.

 b. Show that m∠*APA"* = m∠*BPB"* = m∠*CPC"*. See step 3 below for the start of this proof; then use a plan similar to the one shown for the general case involving translations.

(The proof is left to the student.)

2. The direction of the rotation is the same as the direction from the first line of reflection to the second. In Fig. 5, using $r_m \circ r_k$, we see that the direction from *k* to *m* is counterclockwise, and all angle measures are positive. *Note:* A clockwise direction of rotation has negative angle measures.

3. The measure of the angle of rotation is twice the measure of the acute angle formed by the intersecting lines *k* and *m*.

 In Fig. 5, if *x* is the measure of the angle formed by $\overrightarrow{PA}$ and line *k*, then *x* is also the measure of the angle formed by $\overrightarrow{PA'}$ and *k*. If *y* is the measure of the angle formed by $\overrightarrow{PA'}$ and line *m*, then *y* is also the measure of the angle formed by $\overrightarrow{PA''}$ and *m*. Therefore, the measure of the acute angle formed by lines *k* and *m* is (*x* + *y*), and the measure of the angle of rotation, ∠*APA"*, is *x* + *x* + *y* + *y* = 2*x* + 2*y* = 2(*x* + *y*).

 Conversely, it can be shown that:

■ **Every rotation about a point *P* is equivalent to the composition of two line reflections, $r_m \circ r_k$, where lines *k* and *m* intersect at *P*.**

Therefore, given a rotation of *x*° about point *P*, we can demonstrate that $R_{P,x°}$ is equivalent to $r_m \circ r_k$ by choosing lines *k* and *m* so that *k* and *m* intersect at *P* to form an angle whose measure is $\frac{1}{2}x°$. Note that lines *k* and *m* are not unique.

Note: In Fig. 6, lines *x* and *y* are perpendicular; that is, the lines intersect at an angle of 90°. Thus, $r_x \circ r_y$ is a rotation of 180° about point *O*, the point where lines *x* and *y* intersect. From this example, which is a special case involving rotations, we conclude that:

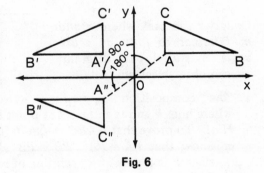

Fig. 6

■ Every reflection through a point P is equivalent to the composition of two line reflections, $r_m \circ r_k$, where $k \perp m$ at point P.

Case III. Glide Reflections

In Fig. 7, $\triangle ABC \rightarrow \triangle A'''B'''C'''$ by a glide reflection. Recall that a glide reflection is the composition of a line reflection and a translation parallel to the line of reflection. Since a translation is the composition of two line reflections (for example, $\triangle A'B'C' \rightarrow \triangle A'''B'''C'''$ by $r_{k_3} \circ r_{k_2}$), we conclude that:

$$r_{k_3} \circ r_{k_2} \circ r_{k_1} (\triangle ABC) = \triangle A'''B'''C'''$$

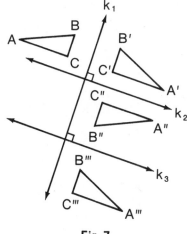

Fig. 7

■ Every glide reflection is equivalent to the composition of three line reflections.

As seen in Fig. 7, two of these lines of reflection are parallel ($k_2 \parallel k_3$), and the remaining line is perpendicular to the first two ($k_1 \perp k_2$, and $k_1 \perp k_3$). The lines k_1, k_2, and k_3 are not unique.

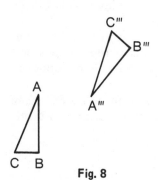

Fig. 8

Case IV. Any Isometry (Distance-Preserving Transformation)

Every isometry (or distance-preserving transformation) can be displayed by showing the effect of the transformation on three non-collinear points. In Fig. 8, for example, there is one and only one isometry by which $A \rightarrow A'''$, $B \rightarrow B'''$, and $C \rightarrow C'''$. The following sequence of steps and accompanying diagrams demonstrates that $\triangle ABC \rightarrow \triangle A'''B'''C'''$ by a composition of no more than three line reflections. This procedure is called:

■ *The Three-Line Reflection Theorem.* Every isometry is equivalent to the composition of no more than three line reflections.

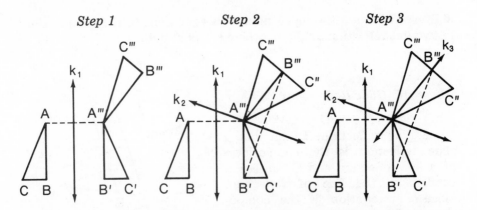

Step 1: Let $A \to A'''$ by a reflection in line k_1. Construct k_1 as the perpendicular bisector of $\overline{AA'''}$. Thus, $r_{k_1}(\triangle ABC) = \triangle A'''B'C'$.

Step 2: Let $B' \to B'''$ by a reflection in line k_2. Construct k_2 as the perpendicular bisector of $\overline{B'B'''}$. Since $A'''B' = A'''B'''$, then A''' is a point on line k_2. Thus, $r_{k_2}(\triangle A'''B'C') = \triangle A'''B'''C''$.

Step 3: Let $C'' \to C'''$ by a reflection in line k_3. Construct k_3 as the perpendicular bisector of $\overline{C''C'''}$. Since $A'''C'' = A'''C'''$ and $B'''C'' = B'''C'''$, then both A''' and B''' are points on line k_3. Thus, $r_{k_3}(\triangle A'''B'''C'') = \triangle A'''B'''C'''$.

We conclude that:

$$r_{k_3} \circ r_{k_2} \circ r_{k_1}(\triangle ABC) = \triangle A'''B'''C'''$$

The lines of reflection, k_1, k_2, and k_3, are not unique. This can be demonstrated by mapping B to B''' as the first step.

Notice that $\triangle ABC$ has clockwise orientation, while $\triangle A'''B'''C'''$ has counterclockwise orientation. This demonstrates that the composition of three line reflections, or three opposite isometries, is an opposite isometry.

If $\triangle A'''B'''C'''$ had the same order or orientation as $\triangle ABC$, then $\triangle ABC$ could map onto $\triangle A'''B'''C'''$ by a composition of only two line reflections.

KEEP IN MIND

Every transformation that preserves distance can be expressed as the composition of no more than three line reflections.

| EXERCISES |

In 1–5, select the numeral preceding the expression that best completes the sentence.

1. If line $g \parallel$ line h, then the composition $r_g \circ r_h$ is equivalent to a:
   (1) translation   (2) point reflection   (3) glide reflection
   (4) line reflection
2. If line $g \perp$ line h, then the composition $r_g \circ r_h$ is equivalent to a:
   (1) translation   (2) point reflection   (3) glide reflection
   (4) line reflection
3. If line g intersects line h, then the composition $r_g \circ r_h$ is equivalent to a:
   (1) translation   (2) glide reflection   (3) rotation
   (4) line reflection
4. The composition of two line reflections is *always*:
   (1) a translation   (2) a rotation   (3) a direct isometry
   (4) an opposite isometry
5. If the composition $r_p \circ r_q$ is equivalent to a rotation of 80°, then lines p and q intersect to form an angle whose measure is:
   (1) 40°   (2) 80°   (3) 90°   (4) 160°

In 6–14, copy $\triangle ABC$ and the appropriate lines onto graph paper. The coordinates of the vertices of $\triangle ABC$ are $A(0, 4)$, $B(1, 2)$, and $C(3, 2)$. The equation of line k is $x = 3$, of line m is $x = 6$, and of line p is $y = 4$. Let r_x represent a reflection in the x-axis, and let r_y represent a reflection in the y-axis.

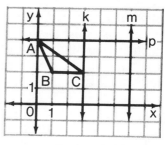

Ex. 6 to 14

6. a. Find the coordinates of the vertices of $\triangle A''B''C''$, the image of $\triangle ABC$ under the composition $r_m \circ r_k$.
   b. The composition $r_m \circ r_k$ is equivalent to the isometry:
      (1) $T_{0,6}$ (2) $T_{6,0}$ (3) $T_{0,-6}$ (4) $T_{-6,0}$
7. a. Find the coordinates of the vertices of $\triangle A''B''C''$, the image of $\triangle ABC$ under the composition $r_k \circ r_m$.
   b. The composition $r_k \circ r_m$ is equivalent to the translation:
      (1) $T_{0,6}$ (2) $T_{6,0}$ (3) $T_{0,-6}$ (4) $T_{-6,0}$
8. a. Find the coordinates of the vertices of $\triangle A''B''C''$, the image of $\triangle ABC$ under the composition $r_x \circ r_p$.
   b. The composition $r_x \circ r_p$ is equivalent to the translation:
      (1) $T_{0,8}$ (2) $T_{8,0}$ (3) $T_{0,-8}$ (4) $T_{-8,0}$

9. a. If $r_p \circ r_x(\triangle ABC) = \triangle A''B''C''$, find the coordinates of A'', B'', and C''.

    b. Name the single transformation equivalent to $r_p \circ r_x$.

10. a. If $r_x \circ r_y(\triangle ABC) = \triangle A''B''C''$, find the coordinates of A'',B'', and C''.

    b. Name the single transformation equivalent to $r_x \circ r_y$.

11. a. If $r_p \circ r_k(\triangle ABC) = \triangle A''B''C''$, find the coordinates of A'',B'', and C''.

    b. The composition $r_p \circ r_k$ is equivalent to a reflection through the point:

    (1) $(3, 4)$ (2) $(4, 3)$ (3) $(6, 4)$ (4) $(0, 0)$

12. A point reflection through A is equivalent to the composition:

    (1) $r_x \circ r_y$ (2) $r_y \circ r_p$ (3) $r_x \circ r_p$ (4) $r_p \circ r_x$

13. Which composition is *not* a translation to the right?

    (1) $r_m \circ r_k$ (2) $r_m \circ r_y$ (3) $r_k \circ r_y$ (4) $r_y \circ r_k$

14. a. If $r_p \circ r_m \circ r_k(\triangle ABC) = \triangle A'''B'''C'''$, find the coordinates of A''',B''', and C'''.

    b. The isometry equivalent to $r_p \circ r_m \circ r_k$ is a:

    (1) translation   (2) point reflection   (3) glide reflection
    (4) line reflection

In 15–17, copy the two given congruent triangles onto your paper in their given positions. Demonstrate that one triangle maps onto the other by a composition of no more than three line reflections by locating reflection lines.

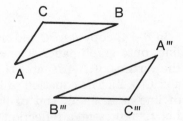

15. Demonstrate how $\triangle ABC \rightarrow \triangle A'''B'''C'''$.

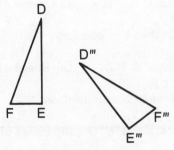

16. Demonstrate how $\triangle DEF \rightarrow \triangle D'''E'''F'''$.

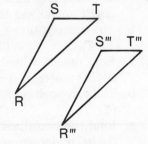

17. Demonstrate how $\triangle RST \rightarrow \triangle R'''S'''T'''$.

18. a. On graph paper, draw and label the congruent triangles, $\triangle ABC$ and $\triangle A'''B'''C'''$, whose vertices have the coordinates $A(3, 6)$, $B(1, 6)$, $C(0, 3)$, $A'''(7, 2)$, $B'''(7, 4)$, and $C'''(10, 5)$.
    b. Demonstrate that $\triangle ABC \rightarrow \triangle A'''B'''C'''$ by a composition of three line reflections by locating reflection lines.
    c. *Challenge:* If $A \rightarrow A'''$ by the first line of reflection and $B' \rightarrow B'''$ by the second, find the equations of the three lines of reflection.

19. *True* or *False*: If $\triangle RST \rightarrow \triangle R'S'T'$ by a direct isometry, then $\triangle RST$ can map onto $\triangle R'S'T'$ by a composition of *exactly two* line reflections. Explain your reasoning.

7-6 GROUPS OF TRANSFORMATIONS

In Chapter 6, we learned that the set of one-to-one functions, under the operation of composition, is a group. Earlier in this chapter, we demonstrated that a transformation of the plane is a one-to-one function. Therefore:

■ **The set of transformations of the plane, under the operation of composition, is a group.**

In symbols, we write (Transformations of the plane, ∘) is a group. This statement indicates that four properties are true:

1. *Closure.* The composition of two transformations is a single transformation.
2. *Associativity.* For all transformations f, g, and h, and for any point P:
$$f \circ (g \circ h)(P) = (f \circ g) \circ h(P)$$
3. *Identity.* There is an identity transformation I that leaves every point on the plane fixed. Thus, $I(P) = P$, or $I(x, y) = (x, y)$. This identity transformation I may be expressed in various forms:

As a translation:	As a dilation whose center is the origin:	As a rotation about a point P:
$I = T_{0,0}$	$I = D_1$	$I = R_{P,0°}$
where	where	where
$T_{0,0}(x, y) = (x + 0, y + 0)$	$D_1(x, y) = (1x, 1y)$	$R_{P,0°}(x, y) = (x, y)$

4. *Inverses.* For every transformation f, there is an inverse transformation f^{-1} such that:
$$f^{-1} \circ f(P) = P \quad \text{AND} \quad f \circ f^{-1}(P) = P$$

Other Groups of Transformations

Selected sets of transformations also form groups under the operation of composition. Among them:

1. The set of all isometries, or distance-preserving transformations, is a group under composition. In symbols, (Isometries, ∘) is a group.
2. (Direct isometries, ∘) is a group.
3. (Translations, ∘) is a group.
4. (Rotations about a point P, ∘) is a group.

■ **DEFINITION.** If S is a *subset* of G and both $(S, *)$ and $(G, *)$ are *groups*, then $(S, *)$ is called a *subgroup* of $(G, *)$.

For example, (Translations, ∘) is a *subgroup* of (Transformations of the plane, ∘) because both systems are groups, and the set of translations is a subset of the set of transformations of the plane.

| MODEL PROBLEMS |

1. Demonstrate that the set of translations under the operation of composition is a group.

Solution

To demonstrate that (Translations, ∘) is a group, we must verify four properties in general terms:

1. *Closure.*

$$(x, y) \xrightarrow{T_{a,b}} (x + a, y + b) \xrightarrow{T_{c,d}} (x + a + c, y + b + d)$$

$$T_{c,d} \circ T_{a,b} = T_{a+c,b+d}$$

Therefore, the composition of two translations is a translation.

2. *Associativity.* Since the set of all one-to-one functions is associative under composition and translations are one-to-one functions, the set of translations is associative under composition.

3. *Identity.* The identity translation is $T_{0,0}(x, y) = (x + 0, y + 0)$, or simply $T_{0,0}(x, y) = (x, y)$, as demonstrated below.

$$(x, y) \xrightarrow{T_{a,b}} (x + a, y + b) \xrightarrow{T_{0,0}} (x + a, y + b)$$

$$T_{0,0} \circ T_{a,b} = T_{a,b}$$

AND

$$(x, y) \xrightarrow{T_{0,0}} (x, y) \xrightarrow{T_{a,b}} (x + a, y + b)$$

$$T_{a,b} \circ T_{0,0} = T_{a,b}$$

4. *Inverses.* The inverse translation of $T_{a,b}(x, y) = (x + a, y + b)$ is $T_{-a,-b}(x, y) = (x - a, y - b)$, as demonstrated below.

$$(x, y) \xrightarrow{T_{a,b}} (x + a, y + b) \xrightarrow{T_{-a,-b}} (x, y)$$

$$T_{-a,-b} \circ T_{a,b} = T_{0,0}$$

AND

$$(x, y) \xrightarrow{T_{-a,-b}} (x - a, y - b) \xrightarrow{T_{a,b}} (x, y)$$

$$T_{a,b} \circ T_{-a,-b} = T_{0,0}$$

2. Give as many reasons as possible to indicate why the set of half-turns about a point P is *not* a group under composition.

Solution

(Half-turns about point P, $\circ$) is not a group because:

1. It is *not closed.* The composition of two half-turns about point P is equivalent to $(x, y) \rightarrow (x, y)$, which is *not* a half-turn.
2. There is *no identity* in the set of half-turns. That is, there is no half-turn whose rule is $(x, y) \rightarrow (x, y)$, leaving every point on the plane fixed.
3. If a system has no identity, then it also has *no inverses.*

EXERCISES

1. Let $(R, \circ)$ represent the set of all rotations about the origin under the operation of composition.
   a. Find $R_{40°} \circ R_{80°}$ b. Find $R_{-75°} \circ R_{184°}$
   c. Name the group property that is demonstrated by the statement $R_\theta \circ R_\gamma = R_{\theta+\gamma}$.
   d. Explain why $(R, \circ)$ is associative.
   e. Name the identity rotation in $(R, \circ)$.
   f. Find $R_\theta \circ R_{-\theta}$ g. Explain why $(R, \circ)$ is a group.
2. Let $D_k(x, y) = (kx, ky)$, where $k > 0$. Then, $(D_k, \circ)$ represents the set of all dilations under the operation of composition.
   a. Find $D_5 \circ D_3$ b. Find $D_8 \circ D_{\frac{1}{4}}$.   c. Find $D_2 \circ D_{\frac{5}{2}}$.

   d. *True* or *False*: $D_a \circ D_b = D_{ab}$, where $a > 0$ and $b > 0$, demonstrates that the set of dilations is closed under composition.
   e. What is the identity element in $(D_k, \circ)$?
   f. What is the inverse for D_a in this system?
   g. Is $(D_k, \circ)$ a group? Explain why.

In 3–6: **a.** Is the given set of transformations closed under the operation of composition? **b.** If the answer to part a is "no," explain why.

3. Isometries   **4.** Direct isometries
5. Line reflections   **6.** Glide reflections

7. Let R_{60k}, where k is an integer, represent the set of rotations about the origin through angles whose measures are $60k$ degrees. Included in this set are $R_{-60°}$, $R_{0°}$, $R_{60°}$, $R_{120°}$, and so forth. Demonstrate that this set of rotations under the operation of composition is a group.

8. Let R_{30k}, where k is an integer, represent the set of rotations about the origin through angles whose measures are $30k$ degrees.
   a. Is $(R_{30k}, \circ)$ a group? Explain why.
   b. Is $(R_{60k}, \circ)$ a subgroup of $(R_{30k}, \circ)$? Explain why.

9. Give as many reasons as possible to indicate why the set of line reflections is *not* a group under composition.

10. Give as many reasons as possible to indicate why the set of opposite isometries is *not* a group under composition.

11. **a.** Let $D_a(x, y) = (ax, ay)$, where a is any real number. Find the reason why $(D_a, \circ)$ is *not* a group.
   b. Let $D_b(x, y) = (bx, by)$, where $b \neq 0$. Is $(D_b, \circ)$ a group? Explain why.

12. *True* or *False*: (Translations, $\circ$) is a subgroup of (Isometries, $\circ$).

13. *True* or *False*: (Rotations, $\circ$) is a subgroup of (Direct isometries, $\circ$).

14. *True* or *False*: (Opposite isometries, $\circ$) is a subgroup of (Isometries, $\circ$).

7-7 REVIEW EXERCISES

1. Let k represent any transformation of the plane.
   a. State the domain of k.   **b.** State the range of k.
   c. *True* or *False*: The transformation k is a one-to-one function.

2. Given the transformations $F(x, y) = (2x, y)$
$$G(x, y) = (x + 3, -y)$$
   a. Write the coordinate rule of the composition, $G \circ F$.
   b. Write the coordinate rule of the composition, $F \circ G$.
   c. Write the coordinate rule of the composition, $G \circ G$.

In 3–13, k and n are symmetry lines for square $ABCD$, and M is the midpoint of diagonal $\overline{BD}$.

Ex. 3 to 13

3. Find $r_k \circ r_n (B)$.
4. Find $r_n \circ r_k (B)$.
5. Find $r_n \circ r_k \circ r_n (D)$.
6. Find $r_k \circ r_n \circ r_k (D)$.
7. Find $R_M \circ r_k (C)$.
8. Find $r_n \circ R_M (A)$.
9. Find $r_k \circ r_n (\overline{BC})$.
10. Find $r_n \circ r_k (\overline{BC})$.
11. Find $R_{M,270°} \circ r_k (\overline{DA})$.
12. Find $r_n \circ R_{M,90°} (\overline{BA})$.
13. By finding the image for each vertex of the square under the given composition, we can show that $R_M \circ r_k \circ r_n$ is equivalent to the single transformation:
    (1) $R_{M,90°}$ (2) $R_{M,270°}$ (3) r_k (4) r_n

14. The coordinates of the vertices of $\triangle ABC$ are $A(1, -4), B(3, 1)$, and $C(3, -3)$.
    a. On graph paper, draw and label $\triangle ABC$.
    b. Find the coordinates of the vertices of $\triangle A'B'C'$, the image of $\triangle ABC$ reflected over the y-axis. Graph $\triangle A'B'C'$.
    c. Find the coordinates of the vertices of $\triangle A''B''C''$, the image of $\triangle A'B'C'$ reflected over the line $y = -x$. Graph $\triangle A''B''C''$.
    d. Find the coordinates of the vertices of $\triangle A'''B'''C'''$, the image of $\triangle A''B''C''$ under the translation $T_{7,-2}$. Graph $\triangle A'''B'''C'''$.

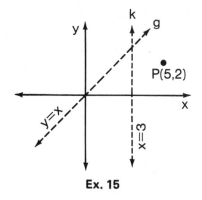

Ex. 15

15. The coordinates of point P are $(5, 2)$, the equation of line k is $x = 3$, and the equation of line g is $y = x$. Find the coordinates of:
    a. $r_k \circ r_g (P)$ b. $r_k \circ r_x (P)$
    c. $r_g \circ r_y (P)$ d. $r_g \circ r_k (P)$

In 16–19, write the rule of the single transformation that is equivalent to the stated composition.

16. $r_{y\text{-axis}} \circ T_{2,3}$
17. $D_2 \circ r_{y=x}$
18. $r_{y=-x} \circ R_{180°}$
19. $r_{x\text{-axis}} \circ T_{-1,-2} \circ D_3$

20. The single transformation that is equivalent to the composition $r_{y=x} \circ r_{x\text{-axis}}$ is:
    (1) $R_{90°}$ (2) $R_{180°}$ (3) $R_{270°}$ (4) $r_{y=-x}$

21. The vertices of △ABC are A(1, 2), B(4, 4), and C(4, 2). Given the transformations:

$$F(x, y) = (4 - x, -y)$$
$$G(x, y) = (-x, y - 1)$$
$$H(x, y) = (2x, y + 3)$$

   a. Graph △ABC and its image, △A'B'C', after transformation F.
   b. Graph △A"B"C", the image of △ABC after transformation G.
   c. Graph △A'''B'''C''', the image of △ABC after transformation H.
   d. Which transformation given above does *not* preserve order?
   e. Which transformation given above is *not* an isometry?

In 22-25, select the numeral preceding the expression that best completes the sentence or answers the question.

22. Which transformation is an example of a direct isometry?
   (1) $r_{x\text{-axis}}$ (2) $r_{x=5}$ (3) D_2 (4) $R_{40°}$
23. A property *not* preserved under a line reflection is:
   (1) collinearity   (2) angle measure   (3) orientation   (4) distance
24. If lines a and b are parallel, then the composition $r_a \circ r_b$ is equivalent to a:
   (1) rotation   (2) translation   (3) glide reflection   (4) dilation
25. If line c intersects line d, the composition $r_c \circ r_d$ is equivalent to a:
   (1) rotation   (2) translation   (3) glide reflection   (4) dilation

26. Copy the two given congruent triangles onto your paper in their given positions. Demonstrate that △ABC → △A'''B'''C''' by a composition of no more than three line reflections by locating reflection lines.

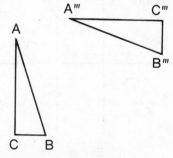

Ex. 26

27. *True or False*: Since a line reflection is an opposite isometry, the composition of two line reflections is also an opposite isometry.

In 28-31: a. Tell whether the given set of transformations is a group under the operation of composition. b. If the system is *not* a group, explain why.
28. (Isometries, ∘)   29. (Line reflections, ∘)
30. (Translations, ∘)   31. (Rotations about point P, ∘)

Chapter 8

Trigonometric Functions

8-1 THE RIGHT TRIANGLE

The word *trigonometry*, which is Greek in origin, means "measurement of triangles." Although the study of trigonometry includes much more than the measurement of triangles, we will begin by recalling some relationships in the right triangle.

In a right triangle, the **hypotenuse**, which is the longest side, is opposite the right angle. The other two sides, which are contained in the rays of the right angle, are called **legs**. These legs are identified as the **opposite leg** and the **adjacent leg** to describe their positions relative to the acute angles of the triangle. (See $\triangle ABC$ in Fig. 1 and $\triangle ABC$ in Fig. 2. Notice that they are the same triangle.)

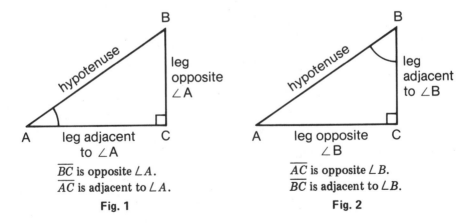

$\overline{BC}$ is opposite $\angle A$.
$\overline{AC}$ is adjacent to $\angle A$.

Fig. 1

$\overline{AC}$ is opposite $\angle B$.
$\overline{BC}$ is adjacent to $\angle B$.

Fig. 2

Similar Triangles

In Fig. 3, $\triangle ACB$, $\triangle AED$, and $\triangle AGF$ are right triangles. Since each triangle contains $\angle A$ and a right angle, the triangles are similar by a.a. $\cong$ a.a.

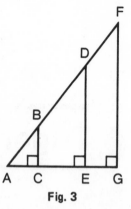

$$\triangle ACB \sim \triangle AED \sim \triangle AGF$$

Corresponding sides of similar triangles are in proportion. Thus:

$$\frac{CB}{AC} = \frac{ED}{AE} = \frac{GF}{AG}$$

Each of these ratios compares the length of the opposite leg to the length of the adjacent leg and has a constant value for every right triangle with a specific acute angle. This ratio is called the *tangent* of the angle.

Fig. 3

■   tangent of $\angle A = \dfrac{\text{length of side opposite } \angle A}{\text{length of side adjacent to } \angle A}$

OR

$$\tan A = \frac{\text{opp}}{\text{adj}}$$

Using the same similar triangles, we can write other ratios.

$$\frac{CB}{AB} = \frac{ED}{AD} = \frac{GF}{AF}$$

Each of these ratios compares the length of the opposite leg to the length of the hypotenuse. This ratio is called the *sine* of the angle.

■   sine of $\angle A = \dfrac{\text{length of side opposite } \angle A}{\text{length of hypotenuse}}$

OR

$$\sin A = \frac{\text{opp}}{\text{hyp}}$$

Using these same similar triangles, we can write a third set of ratios.

$$\frac{AC}{AB} = \frac{AE}{AD} = \frac{AG}{AF}$$

Each of these ratios compares the length of the adjacent leg to the length of the hypotenuse. This ratio is called the *cosine* of the angle.

■   cosine of $\angle A$ = $\dfrac{\text{length of side adjacent to } \angle A}{\text{length of hypotenuse}}$

OR

$$\cos A = \frac{\text{adj}}{\text{hyp}}$$

| MODEL PROBLEM |

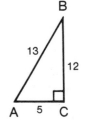

In the diagram, $\triangle ABC$ is a right triangle with m$\angle C$ = 90, AC = 5, BC = 12, and AB = 13. Find the value of:
a. tan A b. sin A c. cos A
d. tan B e. sin B f. cos B

Solution

a. tan $A = \dfrac{\text{opp}}{\text{adj}} = \dfrac{BC}{AC} = \dfrac{12}{5}$ *Ans.*   d. tan $B = \dfrac{\text{opp}}{\text{adj}} = \dfrac{AC}{BC} = \dfrac{5}{12}$ *Ans.*

b. sin $A = \dfrac{\text{opp}}{\text{hyp}} = \dfrac{BC}{AB} = \dfrac{12}{13}$ *Ans.*   e. sin $B = \dfrac{\text{opp}}{\text{hyp}} = \dfrac{AC}{AB} = \dfrac{5}{13}$ *Ans.*

c. cos $A = \dfrac{\text{adj}}{\text{hyp}} = \dfrac{AC}{AB} = \dfrac{5}{13}$ *Ans.*   f. cos $B = \dfrac{\text{adj}}{\text{hyp}} = \dfrac{BC}{AB} = \dfrac{12}{13}$ *Ans.*

| EXERCISES |

In 1-6, $\triangle RST$ is a right triangle with m$\angle S$ = 90, RS = 15, ST = 8, and RT = 17. Give the value of the ratio as a fraction.

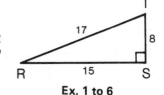

1. sin R 2. cos R 3. tan R
4. sin T 5. cos T 6. tan T

Ex. 1 to 6

In 7 and 8, $\triangle ABC$ is an equilateral triangle with $AB = 10$ and $\overline{BD} \perp \overline{AC}$.

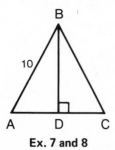

Ex. 7 and 8

7. a. Find AD. **b.** Find BD.
   c. Find $\sin A$. **d.** Find $\cos A$.
   e. Find $\tan A$. **f.** Find $\cos 60°$.

8. a. Find $m\angle CBD$. **b.** Find $\sin 30°$.
   c. Find $\cos 30°$. **d.** Find $\tan 30°$.

8-2 ANGLES AS ROTATIONS

In our study of geometry, we limited the measures of angles to values greater than $0°$ and less than or equal to $180°$. There are, however, many situations in which the measure of an angle can be less than $0°$ or greater than $180°$.

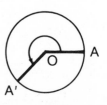

Fig. 1

When a wheel turns, the change in position of any one spoke generates an angle of any size. For example, if a wheel makes $\frac{5}{8}$ of a complete rotation in the counterclockwise direction, a spoke that was initially in the position $\overrightarrow{OA}$ (see Fig. 1) is rotated to the position $\overrightarrow{OA'}$. The rotation, $\angle AOA'$, measures $225°$ ($\frac{5}{8}$ of $360°$, a complete rotation).

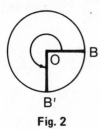

Fig. 2

If a wheel makes $\frac{3}{4}$ of a complete rotation in the counterclockwise direction, a spoke that was initially in the position $\overrightarrow{OB}$ (see Fig. 2) is rotated to the position $\overrightarrow{OB'}$. The rotation, $\angle BOB'$, measures $270°$ ($\frac{3}{4}$ of $360°$).

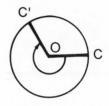

Fig. 3

If a wheel makes $\frac{2}{3}$ of a complete rotation in the clockwise direction, a spoke that was initially in the position $\overrightarrow{OC}$ is rotated to the position $\overrightarrow{OC'}$ (see Fig. 3). This is a rotation through $240°$, but in the direction opposite to that of $\angle AOA'$ and $\angle BOB'$. On the number line, we use positive and negative values to indicate opposite directions. We will do the same with angles. Since we let a rotation in the counterclockwise direction be positive, we will call a rotation in the clockwise direction negative. Thus, the rotation, $\angle COC'$, measures $-240°$.

■ An angle formed by a counterclockwise rotation has a positive measure. An angle formed by a clockwise rotation has a negative measure.

Classifying Angles by Quadrant

The ray from which a rotation begins is called the *initial side* of the angle.

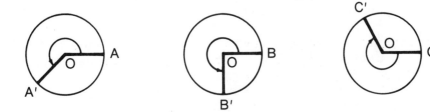

In the diagrams, $\overrightarrow{OA}$ is the initial side of $\angle AOA'$, $\overrightarrow{OB}$ is the initial side of $\angle BOB'$, and $\overrightarrow{OC}$ is the initial side of $\angle COC'$.

The ray at which the rotation ends is called the *terminal side* of the angle.

In the diagrams, $\overrightarrow{OA'}$ is the terminal side of $\angle AOA'$, $\overrightarrow{OB'}$ is the terminal side of $\angle BOB'$, and $\overrightarrow{OC'}$ is the terminal side of $\angle COC'$.

In studying angles as rotations, we will find it convenient to associate angles with the coordinate plane by placing the vertex of an angle at the origin. We say that an angle is in *standard position* when its vertex is at the origin and its initial side coincides with the non-negative ray of the x-axis.

The x-axis and y-axis divide the plane into four *quadrants*, as shown in the diagram. An angle in standard position is classified by the quadrant in which the terminal side lies.

Examples of four angles in standard position are given at the top of the next page.

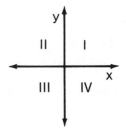

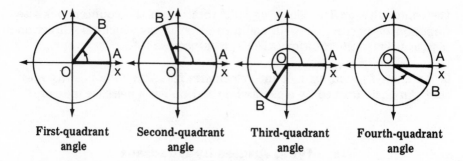

First-quadrant angle   Second-quadrant angle   Third-quadrant angle   Fourth-quadrant angle

If $0 < \text{m}\angle AOB < 90$, $\angle AOB$ is in the *first* quadrant.
If $90 < \text{m}\angle AOB < 180$, $\angle AOB$ is in the *second* quadrant.
If $180 < \text{m}\angle AOB < 270$, $\angle AOB$ is in the *third* quadrant.
If $270 < \text{m}\angle AOB < 360$, $\angle AOB$ is in the *fourth* quadrant.

An angle in standard position whose terminal side lies on one of the axes is a *quadrantal angle*. The measure of a quadrantal angle is a multiple of 90 degrees.

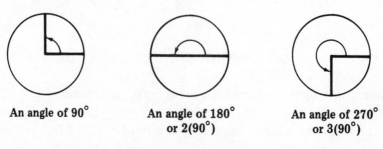

An angle of 90°   An angle of 180°   An angle of 270°
                         or 2(90°)   or 3(90°)

Coterminal Angles

When ray $\overrightarrow{OD}$ is rotated 300° in the counterclockwise direction (+300°), the terminal side of the angle, $\overrightarrow{OD'}$, lies in the fourth quadrant, as shown in the diagram. When $\overrightarrow{OD}$ is rotated 60° in the clockwise direction (-60°), the terminal side of the angle is also $\overrightarrow{OD'}$. In standard position, an angle of 300° and an angle of -60° have the same terminal side. Angles in standard position having the same terminal side are *coterminal angles*. If two

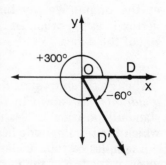

angles are coterminal, the difference of their measures is 360° or a multiple of 360°. Notice that 300° - (-60°) = 360°.

When a wheel continues to turn, it makes more than one rotation, thus generating angles of more than 360°.

   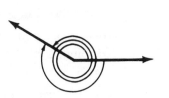   

An angle of 430° is coterminal with an angle of 70° because:

430° - 70° = 360°

An angle of 870° is coterminal with an angle of -210° because:

870° - (-210°) = 1080° = 3(360°)

An angle of 320° is coterminal with an angle of -400° because:

320° - (-400°) = 720° = 2(360°)

If x and y are the measures of coterminal angles, then $x - y = 360k$ for some integer, k. Therefore, $x - 360k = y$. For example, to find angles coterminal with an angle of 910°, add or subtract multiples of 360°.

910° - 360° = 550°
550° - 360° = 190°, that is, 910° - 360°(2) = 190°
190° - 360° = -170°, that is, 910° - 360°(3) = -170°

OR

910° + 360° = 1270°, that is, 910° - 360°(-1) = 1270°
1270° + 360° = 1630°, that is, 910° - 360°(-2) = 1630°

Thus, angles of 1630°, 1270°, 910°, 550°, 190°, and -170° are all coterminal angles.

| MODEL PROBLEMS |

1. In which quadrant does an angle of the given measure lie?
   a. 200°   b. -40°

   Solution

   a. Since $180 < 200 < 270$, the angle is in the third quadrant. *Ans.*

   b. An angle coterminal with an angle of -40° has a measure of
      -40 + 360 = 320, or 320°. Since $270 < 320 < 360$, the angle is
      in the fourth quadrant. *Ans.*

2. Find the angle of smallest positive measure that is coterminal with
   an angle of the given measure.   a. 790°   b. -100°

   Solution

   a. Subtract 360° or a multiple of   790° - 360° = 430°
      360° from the given measure   430° - 360° = 70°   *Ans.*
      until a measure between 0° and
      360° is obtained.

   b. Add 360°, that is, -360°(-1)   -100° + 360° = 260°   *Ans.*
      to the given measure.

| EXERCISES |

In 1-15, determine the quadrant in which an angle of the given measure lies.

1. 140°	2. 210°	3. 97°	4. 315°	5. 80°
6. 240°	7. -168°	8. -200°	9. -260°	10. 175°
11. -340°	12. -380°	13. 500°	14. -475°	15. 420°

In 16-30, find the angle of smallest positive measure coterminal with
an angle of the given measure.

16. 400°	17. 520°	18. 710°	19. 375°	20. 580°
21. 450°	22. 790°	23. 840°	24. -30°	25. -110°
26. -370°	27. -540°	28. -400°	29. -75°	30. -800°

In 31-40, draw an angle of the given measure, indicating the direction
of the rotation by an arrow.

31. 100°	32. 210°	33. 270°	34. 315°	35. 360°
36. 540°	37. 700°	38. -50°	39. -140°	40. -200°

8-3 SINE AND COSINE AS COORDINATES

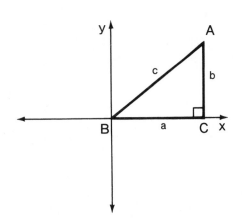

Let $\triangle ABC$ in the diagram be a right triangle with B at the origin and C on the x-axis of the coordinate plane. If $AB = c$, $AC = b$, $BC = a$, and $m\angle BCA = 90$, then

$$\sin B = \frac{b}{c} \text{ and } \cos B = \frac{a}{c}.$$

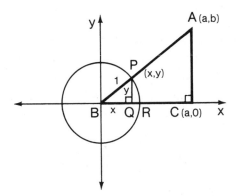

In the diagram, we see a circle with its center at the origin and its radius of length 1. Such a circle is called the **unit circle**. Since $\overrightarrow{BA}$ intersects the circle at P, and $\overline{BP}$ is a radius of the unit circle, $BP = 1$. Under a dilation, $D_{\frac{1}{c}}$:

$$A(a, b) \rightarrow P(x, y)$$
$$C(a, 0) \rightarrow Q(x, 0)$$
$$B(0, 0) \rightarrow B(0, 0)$$

Since right $\triangle PBQ$ is similar to right $\triangle ABC$, we can express the sine and cosine of angle B, using $\triangle PBQ$. Therefore:

$$\sin B = \frac{PQ}{BP} = \frac{y}{1} = y \qquad \cos B = \frac{BQ}{BP} = \frac{x}{1} = x$$

The sine and cosine of an angle can be expressed in terms of the coordinates of the point at which the terminal side of the angle in standard position intersects the unit circle. This relationship between the coordinates of a point on the terminal side of the angle and the sine and cosine of the angle will make it possible to define sine and cosine for angles that are not the angles of a right triangle.

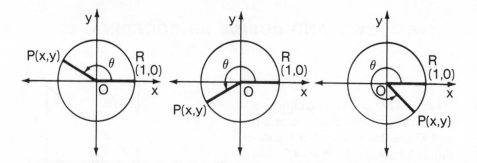

■ In a unit circle, if m∠*ROP* = *θ*, then sin *θ* = *y*, and cos *θ* = *x*.

Note: Greek letters are often used as the measures of angles in trigonometry. For example: *θ*, theta; *φ*, phi; and *ρ*, rho.

To find the sine and cosine of an angle, we must think of the length of a line segment as a directed distance.

In geometry, we always considered the length of a line segment to be positive. When we are defining the trigonometric functions, it is often convenient to associate the length of a line segment with the coordinates of a point. Thus, a vertical line segment will be considered positive if it is above the *x*-axis and negative if it is below the *x*-axis. A horizontal line segment is positive if it is to the right of the *y*-axis and negative if it is to the left of the *y*-axis.

8-4   THE SINE AND COSINE FUNCTIONS

The Sine Function

A circle with a radius of measure 1 (a unit circle), has its center, *O*, at the origin of the coordinate plane. Let *R* be the point (1, 0) at which the circle intersects the *x*-axis, and let *P* be any point on the circle. Angle *ROP* is an angle in standard position. Draw $\overline{PQ}$ ⊥ to the *x*-axis such that *Q* is the foot of the perpendicular. If the coordinates of *P* are (*a*, *b*), then *PQ* = *b*.

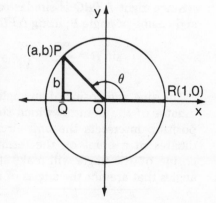

■ The sine function assigns to every θ that is the measure of an angle in standard position a unique value b that is the y-coordinate of the point where the terminal side of the angle intersects a unit circle with its center at the origin.

$$\theta \xrightarrow{\text{sine}} b$$

The value b is called "the sine of the angle whose measure is θ," written:

$$b = \sin \theta$$

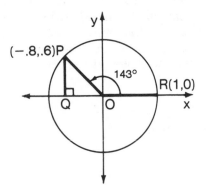

In the diagram, $\angle ROP$ is an angle of $143°$, and the coordinates of point P on circle O are $(-.8, .6)$. Therefore:

$$\sin 143° = .6$$

Thus, $PQ = .6$, or $PQ = \sin 143°$.

The diagrams that follow show an angle in each quadrant, where $m\angle ROP = \theta$ and $\sin \theta = PQ$.

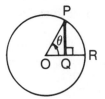 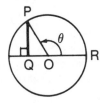 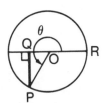 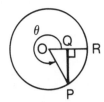

First quadrant	Second quadrant	Third quadrant	Fourth quadrant
PQ is *positive.*	*PQ* is *positive.*	*PQ* is *negative.*	*PQ* is *negative.*
Sin θ is *positive.*	Sin θ is *positive.*	Sin θ is *negative.*	Sin θ is *negative.*

The Cosine Function

Look again at the unit circle with its center, O, at the origin and points $R(1, 0)$ and $P(a, b)$ on the circle. Draw $\overline{PQ}$ perpendicular to the x-axis so that Q is the foot of the perpendicular. The x-coordinate of P is $a = OQ$, a directed distance.

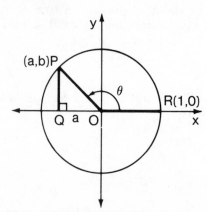

■ The cosine function assigns to every θ that is the measure of an angle in standard position a unique value a that is the x-coordinate of the point where the terminal side of the angle intersects a unit circle with its center at the origin.

$$\theta \xrightarrow{\text{cosine}} a$$

The value a is called "the cosine of the angle whose measure is θ," written:

$$a = \cos \theta$$

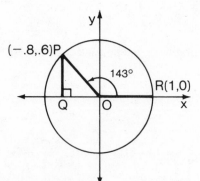

In the diagram, $\angle ROP$ is an angle of $143°$, and the coordinates of P on circle O are $(-.8, .6)$. Therefore:

$$\cos 143° = -.8$$

Notice that $OQ = -.8$, a negative value because it is measured to the left of the origin.

The diagrams that follow show an angle in each quadrant, where $m\angle ROP = \theta$ and $\cos \theta = OQ$.

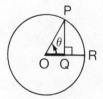

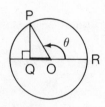

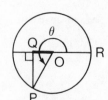

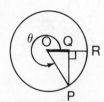

First quadrant	Second quadrant	Third quadrant	Fourth quadrant
OQ is *positive.*	*OQ* is *negative.*	*OQ* is *negative.*	*OQ* is *positive.*
Cos θ is *positive.*	Cos θ is *negative.*	Cos θ is *negative.*	Cos θ is *positive.*

| MODEL PROBLEM |

Find: **a.** $\sin 180°$ **b.** $\cos 180°$

Solution

In the diagram, $\angle ROP$ is an angle of 180°, and the coordinates of point P on the unit circle are $(x, y) = (-1, 0)$. Therefore:

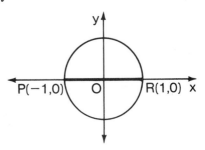

a. $\sin 180° = y$ **b.** $\cos 180° = x$
     $\sin 180° = 0$ $\cos 180° = -1$
        Ans.   *Ans.*

| EXERCISES |

In 1-4, $m\angle ROP = \theta$ and $OR = 1$. Given the coordinates of point P on circle O, find: **a.** $\sin \theta$ **b.** $\cos \theta$

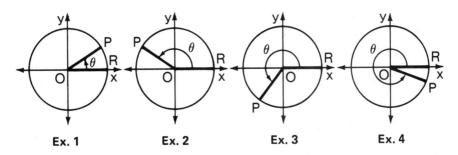

Ex. 1   Ex. 2   Ex. 3   Ex. 4

1. $P(.96, .28)$ **2.** $P\left(-\dfrac{3}{4}, \dfrac{\sqrt{7}}{4}\right)$ **3.** $P(-.6, -.8)$ **4.** $P\left(\dfrac{2\sqrt{6}}{5}, -\dfrac{1}{5}\right)$

5. Name the quadrants in which an angle of measure θ could lie when:
   a. $\cos \theta > 0$ **b.** $\cos \theta < 0$

6. Name the quadrants in which an angle of measure θ could lie when:
   a. $\sin \theta > 0$ **b.** $\sin \theta < 0$

7. Name the quadrant in which an angle of measure θ could lie when:
   a. $\sin \theta > 0$ and $\cos \theta > 0$ **b.** $\sin \theta < 0$ and $\cos \theta > 0$
   c. $\sin \theta < 0$ and $\cos \theta < 0$ **d.** $\sin \theta > 0$ and $\cos \theta < 0$

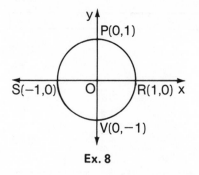

Ex. 8

8. Circle O is a unit circle that intersects the x-axis at $R(1, 0)$ and $S(-1, 0)$ and the y-axis at $P(0, 1)$ and $V(0, -1)$. Find:

a. $m\angle ROR$ b. $\cos 0°$

c. $\sin 0°$ d. $m\angle ROP$

e. $\cos 90°$ f. $\sin 90°$

g. $m\angle ROS$ h. $\cos 180°$

i. $\sin 180°$ j. $m\angle ROV$

k. $\cos 270°$ l. $\sin 270°$

8-5 THE TANGENT FUNCTION

Let us look again at right triangle ABC with B at the origin and C on the positive ray of the x-axis in the coordinate plane. If $AB = c$, $AC = b$, $BC = a$, and $m\angle BCA = 90$, then:

$$\tan B = \frac{b}{a}$$

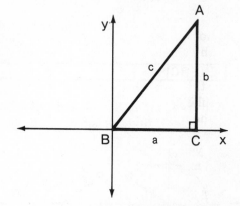

In the diagram, we see a circle with its center at the origin and its radius of length 1. A line is drawn tangent to the circle at point R. Thus, $\overrightarrow{BA}$ intersects the unit circle at P and the tangent line at T.

Under a dilation, $D_{\frac{1}{c}}$:

$$A(a, b) \to P(x, y)$$
$$C(a, 0) \to Q(x, 0)$$
$$B(0, 0) \to B(0, 0)$$

Under a dilation, $D_{\frac{1}{a}}$:

$$A(a, b) \to T\left(1, \frac{b}{a}\right)$$
$$C(a, 0) \to R(1, 0)$$
$$B(0, 0) \to B(0, 0)$$

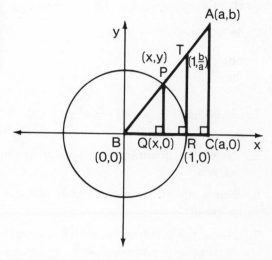

Since right $\triangle ABC \sim$ right $\triangle TBR \sim$ right $\triangle PBQ$, we can express the tangent of angle B by using any of these triangles.

In $\triangle ABC$, $\tan B = \dfrac{AC}{BC} = \dfrac{b}{a}$.

In $\triangle TBR$, $\tan B = \dfrac{TR}{BR} = \dfrac{\left(\dfrac{b}{a}\right)}{1} = \dfrac{b}{a}$.

In $\triangle PBQ$, $\tan B = \dfrac{PQ}{BQ} = \dfrac{y}{x}$. Thus, $\dfrac{y}{x} = \dfrac{b}{a}$.

We can use the ratio of the coordinates of the point where the terminal side of an angle intersects the unit circle to define tangent.

Look again at the unit circle with its center, O, at the origin and points $R(1, 0)$ and $P(a, b)$ on the circle. Draw $\overline{PQ}$ perpendicular to the x-axis so that Q is the foot of the perpendicular. The x-coordinate of P is $a = OQ$, the y-coordinate of P is $b = PQ$, and $\mathrm{m}\angle ROP = \theta$.

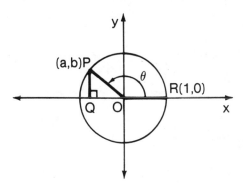

■ The tangent function assigns to every θ that is the measure of an angle in standard position a unique value $\dfrac{b}{a}$ that is the ratio of the y-coordinate to the x-coordinate of the point where the terminal side of the angle intersects a unit circle with its center at the origin.

$$\theta \xrightarrow{\text{tangent}} \dfrac{b}{a}$$

The value $\dfrac{b}{a}$ is called "the tangent of the angle whose measure is θ," written:

$$\dfrac{b}{a} = \tan \theta$$

The tangent function is defined for all values of θ where $a \neq 0$.

In the diagram, $\angle ROP$ is an angle of $143°$, and the coordinates of point P on circle O are $(-.8, .6)$. Therefore:

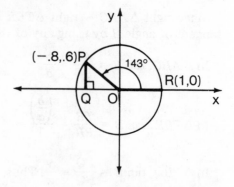

$$\tan 143° = \frac{.6}{-.8}, \text{ or}$$

$$\tan 143° = \frac{-3}{4} = -.75$$

The diagrams that follow show $\angle ROP$ in each of the four quadrants. A tangent is drawn to the unit circle O at $R(1, 0)$. Line $\overleftrightarrow{OP}$ intersects the tangent line at T.

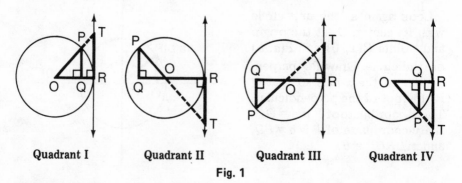

| Quadrant I | Quadrant II | Quadrant III | Quadrant IV |

Fig. 1

In each of the four diagrams in Fig. 1, right $\triangle POQ \sim$ right $\triangle TOR$ by a.a. $\cong$ a.a. Thus, $\dfrac{QP}{OQ} = \dfrac{RT}{OR}$. We will use this proportion to develop two statements regarding the tangent of an angle.

In each diagram, $\angle ROP$ is an angle in standard position in the unit circle. Let m$\angle ROP = \theta$. A *tangent segment* $\overline{RT}$ is a segment of the tangent line from R, the point of tangency on the initial side of the angle, to T, the point where the line of the terminal side of the angle intersects the tangent. Notice where $\overline{RT}$ lies in each of the four diagrams in Fig. 1.

Observation 1: The length of the tangent segment $\overline{RT}$, when considered as a directed distance, is equal to $\tan \theta$.

1. Since $\triangle POQ \sim \triangle TOR$, then corresponding sides of similar triangles are in proportion.	1. $\dfrac{QP}{OQ} = \dfrac{RT}{OR}$
2. If $m\angle ROP = \theta$, then $\dfrac{QP}{OQ} = \tan\theta$ by definition.	2. $\tan\theta = \dfrac{RT}{1}$
Also, in the unit circle, $OR = 1$. Substitute these values in the given proportion.	
3. Simplify.	3. $\tan\theta = RT$

Observation 2: For an angle of measure θ, $\dfrac{\sin\theta}{\cos\theta} = \tan\theta$.

1. Corresponding sides of similar triangles are in proportion.	1. $\dfrac{QP}{OQ} = \dfrac{RT}{OR}$
2. In the unit circle, $OR = 1$.	2. $\dfrac{QP}{OQ} = \dfrac{RT}{1}$
3. If $m\angle ROP = \theta$, then, by definition, $\sin\theta = QP$, $\cos\theta = OQ$, and $\tan\theta = RT$. Substitute these values in the proportion, and simplify.	3. $\dfrac{\sin\theta}{\cos\theta} = \tan\theta$

(*Note:* If $\cos\theta = 0$, then $\tan\theta$ is undefined.)

Study the chart that follows, and note the signs of the trigonometric functions for angles in each quadrant. Compare these values, and the direction of RT, with the four diagrams in Fig. 1 seen earlier.

Quadrant	Sin θ	Cos θ	$\dfrac{\text{Sin }\theta}{\text{Cos }\theta}$	Tan θ	Direction of $\overline{RT}$
I	+	+	$\dfrac{+}{+}$	+	above the x-axis
II	+	–	$\dfrac{+}{-}$	–	below the x-axis
III	–	–	$\dfrac{-}{-}$	+	above the x-axis
IV	–	+	$\dfrac{-}{+}$	–	below the x-axis

Quadrantal Angles

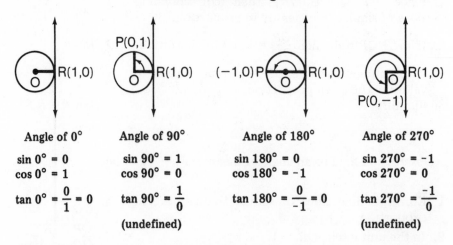

Angle of 0°	Angle of 90°	Angle of 180°	Angle of 270°
sin 0° = 0	sin 90° = 1	sin 180° = 0	sin 270° = -1
cos 0° = 1	cos 90° = 0	cos 180° = -1	cos 270° = 0
$\tan 0° = \dfrac{0}{1} = 0$	$\tan 90° = \dfrac{1}{0}$	$\tan 180° = \dfrac{0}{-1} = 0$	$\tan 270° = \dfrac{-1}{0}$
	(undefined)		(undefined)

For an angle of 0° (such as $\angle ROR$ in the first diagram of the figure above), and an angle of 180°, the line of the terminal side of the angle is the x-axis. Since this line intersects the tangent to circle O at R, tan 0° = 0 and tan 180° = 0.

For angles of 90° and 270°, the line of the terminal side of the angle is the y-axis. Since the y-axis is parallel to the tangent line drawn to circle O at R, the y-axis does not intersect the tangent line, and tan 90° and tan 270° are undefined.

Function Values of Coterminal Angles

Since we have defined the sine, cosine, and tangent of an angle in terms of the coordinates of the point at which the terminal side intersects the unit circle, two angles that are coterminal have the same function values.

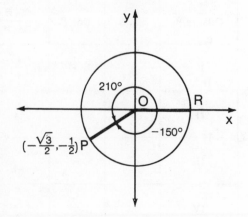

In the diagram, $\overrightarrow{OP}$ is the terminal side of both an angle of 210° and an angle of -150°.

Therefore:

$$\sin 210° = \sin(-150°) = -\frac{1}{2}$$

$$\cos 210° = \cos(-150°) = -\frac{\sqrt{3}}{2}$$

$$\tan 210° = \tan(-150°) = \frac{-\dfrac{1}{2}}{-\dfrac{\sqrt{3}}{2}} = +\frac{\dfrac{1}{2} \cdot 2\sqrt{3}}{\dfrac{\sqrt{3}}{2} \cdot 2\sqrt{3}} = \frac{\sqrt{3}}{3}$$

| MODEL PROBLEMS |

1. A circle whose center is at the origin intersects the x-axis at $A(1, 0)$ and $\overrightarrow{OB}$ at $B(\frac{5}{13}, -\frac{12}{13})$. If $m\angle AOB = \theta$, find:
   a. $\sin\theta$ b. $\cos\theta$ c. $\tan\theta$

Solution

Sketch the circle.

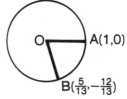

a. Since $\sin\theta$ is the y-coordinate of point B, $\sin\theta = -\frac{12}{13}$. *Ans.*

b. Since $\cos\theta$ is the x-coordinate of point B, $\cos\theta = \frac{5}{13}$. *Ans.*

c. Then $\tan\theta = \dfrac{\sin\theta}{\cos\theta} = \dfrac{-\frac{12}{13}}{\frac{5}{13}} \cdot \dfrac{13}{13} = -\dfrac{12}{5}$. *Ans.*

2. In what quadrant is an angle of measure θ if $\sin\theta > 0$ and $\cos\theta < 0$.

Solution

If $\sin\theta > 0$, the angle could be in quadrant I or II.
If $\cos\theta < 0$, the angle could be in quadrant II or III.
Therefore, only in quadrant II are both conditions satisfied.

Answer: quadrant II

| EXERCISES |

1. In the figure, a unit circle is drawn with the length of the radius, $OA = 1$, and $m\angle COA = \theta$. Name the line segment whose length (or directed distance) is equal to: **a.** $\sin \theta$ **b.** $\cos \theta$ **c.** $\tan \theta$

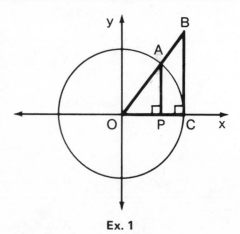

Ex. 1

2. Copy the four circles and central angles shown below. If the measure of the radius of each circle is 1, draw and name the line segments that represent: **a.** $\sin \theta$ **b.** $\cos \theta$ **c.** $\tan \theta$

Ex. 2

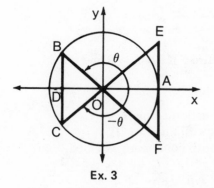

Ex. 3

3. In the diagram shown, $OA = 1$, $m\angle AOB = \theta$, and $m\angle AOC = -\theta$. Name the line segment whose directed measure is the value of:
   a. $\sin \theta$ **b.** $\cos \theta$
   c. $\tan \theta$ **d.** $\sin (-\theta)$
   e. $\cos (-\theta)$ **f.** $\tan (-\theta)$

In 4–9, name two quadrants in which $\angle A$ may lie.

4. Sin A is positive. 5. Cos A is positive. 6. Tan A is positive.
7. Sin A is negative. 8. Cos A is negative. 9. Tan A is negative.

In 10–15, name the quadrant in which $\angle A$ lies.

10. Tan $A > 0$, cos $A < 0$.
11. Sin $A < 0$, cos $A > 0$.
12. Cos $A < 0$, tan $A < 0$.
13. Sin $A > 0$, tan $A > 0$.
14. Tan $A < 0$, sin $A < 0$.
15. Cos $A < 0$, sin $A < 0$.

16. Copy and complete the following table, giving the sign of the function value in each quadrant.

Quadrant	Sin A	Cos A	Tan A
I			
II			
III			
IV			

17. Copy and complete the following table, giving the function value of each quadrantal angle. If a function is not defined for an angle measure, write "undefined."

A	$0°$	$90°$	$180°$	$270°$	$360°$
Sin A					
Cos A					
Tan A					

18. If $0° \leq \theta \leq 360°$, for what values of θ is tan θ not defined?
19. If $(\sin \theta)(\cos \theta) > 0$, name all quadrants in which an angle of measure θ can lie.
20. Points $A(1, 0)$ and $B(.6, -.8)$ are points on a unit circle O. If $m\angle AOB = \theta$, find: **a.** sin θ **b.** cos θ **c.** tan θ
21. A circle with center O intersects the x-axis at $C(1, 0)$ and $D(-1, 0)$. If $m\angle COD = \phi$, find: **a.** sin ϕ **b.** cos ϕ **c.** tan ϕ
22. Points $R(1, 0)$ and $P(-\frac{4}{5}, -\frac{3}{5})$ are points on a unit circle O. If $m\angle ROP = \theta$, find: **a.** sin θ **b.** cos θ **c.** tan θ

In 23–26, select the numeral preceding the expression that best completes the sentence or answers the question.

23. If tan $\phi < 0$ and sin $\phi = .7$, then the angle of measure ϕ is in quadrant:
    (1) I   (2) II   (3) III   (4) IV
24. Points A and B are on unit circle O. The coordinates of A are $(1, 0)$ and of B are $\left(-\frac{1}{2}, \frac{\sqrt{3}}{2}\right)$. If $m\angle AOB = 120°$, then tan $120°$ equals:

    (1) $-\sqrt{3}$ (2) $-\frac{\sqrt{3}}{2}$ (3) $-\frac{\sqrt{3}}{3}$ (4) $\frac{\sqrt{3}}{2}$

25. If $\cos A = \dfrac{\sqrt{2}}{2}$ and $\tan A = 1$, then the value of $\sin A$ is:

(1) $\dfrac{1}{2}$ (2) 2 (3) $\dfrac{\sqrt{2}}{2}$ (4) $\sqrt{2}$

26. Which of the following is *false*?
   (1) $\sin 300° = \sin(-60°)$ (2) $\cos 210° = \cos(-150°)$
   (3) $\cos 90° = \cos(-90°)$ (4) $\sin 90° = \sin(-90°)$

8-6 FUNCTION VALUES OF SPECIAL ANGLES

Some angles, such as those with measures that are multiples of 30° and 45°, occur frequently in the applications of trigonometry. We can use some of the relationships that we learned in the study of geometry to find exact function values for these angles.

Angles of 30° and 60°

In the diagram, equilateral $\triangle ABC$ is separated into two congruent triangles, $\triangle ACD$ and $\triangle BCD$, by $\overline{CD} \perp \overline{AB}$. Therefore, $m\angle A = 60$, $m\angle ACD = 30$, and $m\angle CDA = 90$. If $AC = 1$, then $AD = \frac{1}{2}$.

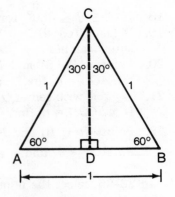

Using the Pythagorean Theorem:

$$(CD)^2 + (AD)^2 = (AC)^2$$
$$(CD)^2 + \left(\frac{1}{2}\right)^2 = (1)^2$$
$$(CD)^2 + \frac{1}{4} = 1$$
$$(CD)^2 = \frac{3}{4}$$
$$CD = \pm\frac{\sqrt{3}}{2}$$

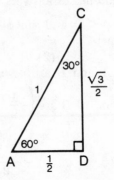

Previously, we always chose the positive value to be the measure of a line segment. When a length represents a trigonometric function value, we will use directed line segments, which may be positive or negative.

Let $\angle ROP$ be a central angle of the unit circle with its center at the origin. If $m\angle ROP = 30$, and $OP = 1$, then $PQ = \frac{1}{2}$ and $OQ = \frac{\sqrt{3}}{2}$. Therefore:

$$\sin 30° = \frac{1}{2}$$

$$\cos 30° = \frac{\sqrt{3}}{2}$$

$$\tan 30° = \frac{\frac{1}{2}}{\frac{\sqrt{3}}{2}} \cdot \frac{2}{2} = \frac{1}{\sqrt{3}} \cdot \frac{\sqrt{3}}{\sqrt{3}} = \frac{\sqrt{3}}{3}$$

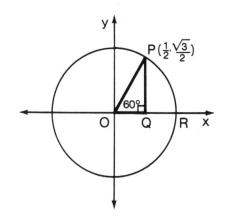

Again, let $\angle ROP$ be a central angle of the unit circle with its center at the origin. If $m\angle ROP = 60$, and $OP = 1$, then $PQ = \frac{\sqrt{3}}{2}$ and $OQ = \frac{1}{2}$. Therefore:

$$\sin 60° = \frac{\sqrt{3}}{2}$$

$$\cos 60° = \frac{1}{2}$$

$$\tan 60° = \frac{\frac{\sqrt{3}}{2}}{\frac{1}{2}} \cdot \frac{2}{2} = \frac{\sqrt{3}}{1} = \sqrt{3}$$

Angles of 45°

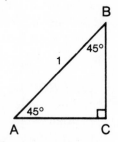

In the diagram, $\triangle ABC$ is an isosceles right triangle with a right angle at C. Therefore, $m\angle A = m\angle B = 45$, and $AC = BC$. If $AB = 1$, let $AC = x$ and $BC = x$. Using the Pythagorean Theorem:

$$(AC)^2 + (BC)^2 = (AB)^2$$
$$x^2 + x^2 = 1^2$$
$$2x^2 = 1$$
$$x^2 = \frac{1}{2}$$
$$x = \pm\sqrt{\frac{1}{2}} = \pm\frac{\sqrt{1}}{\sqrt{2}}$$
$$x = \pm\frac{1}{\sqrt{2}} \cdot \frac{\sqrt{2}}{\sqrt{2}} = \pm\frac{\sqrt{2}}{2}$$

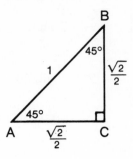

Remember, in trigonometry, directed line segments may have positive or negative lengths.

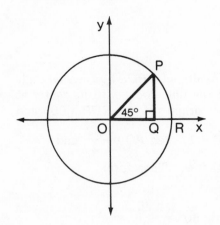

Let $\angle ROP$ be a central angle of the unit circle with its center at the origin. If $m\angle ROP = 45$ and $OP = 1$, then $PQ = \frac{\sqrt{2}}{2}$ and $OQ = \frac{\sqrt{2}}{2}$.

Therefore:

$$\sin 45° = \frac{\sqrt{2}}{2}$$

$$\cos 45° = \frac{\sqrt{2}}{2}$$

$$\tan 45° = \frac{\dfrac{\sqrt{2}}{2}}{\dfrac{\sqrt{2}}{2}} = 1$$

Summary

θ	0°	30°	45°	60°	90°
sin θ	0	$\dfrac{1}{2}$	$\dfrac{\sqrt{2}}{2}$	$\dfrac{\sqrt{3}}{2}$	1
cos θ	1	$\dfrac{\sqrt{3}}{2}$	$\dfrac{\sqrt{2}}{2}$	$\dfrac{1}{2}$	0
tan θ	0	$\dfrac{\sqrt{3}}{3}$	1	$\sqrt{3}$	undefined

| MODEL PROBLEM |

Find the exact numerical value of (sin 60°)(cos 30°) + tan 60°.

How to Proceed	*Solution*
1. Write the expression.	(sin 60°)(cos 30°) + tan 60°
2. Substitute values.	$\left(\dfrac{\sqrt{3}}{2}\right)\left(\dfrac{\sqrt{3}}{2}\right) + \sqrt{3}$
3. Multiply.	$\dfrac{3}{4} + \sqrt{3}$
4. To add, express the terms with a common denominator.	$\dfrac{3}{4} + \dfrac{4\sqrt{3}}{4}$
	$\dfrac{3 + 4\sqrt{3}}{4}$ *Ans.*

| EXERCISES |

In 1–26, find the exact numerical value of the expression.

1. sin 30°   2. cos 45°   3. tan 60°   4. cos 30°
5. tan 45°   6. sin 60°   7. cos 60°   8. sin 45°
9. sin 30° + cos 60°   10. tan 45° + sin 30°

11. $\sin 45° + \cos 45°$
12. $(\tan 60°)(\tan 30°)$
13. $(\sin 60°)(\cos 60°)$
14. $\tan 45° + 2 \cos 60°$
15. $(\cos 45°)^2$
16. $(\tan 30°)^2$
17. $(\cos 30°)^2 + (\sin 30°)^2$
18. $(\sin 30° + \tan 60°)^2$
19. $(\sin 30°)(\cos 60°) + (\cos 30°)(\sin 60°)$
20. $(\tan 45° + \tan 30°)^2$
21. $\sin 90° + \cos 0°$
22. $\cos 180° + \sin 270°$
23. $(\tan 60°)^2 + \sin 180°$
24. $(\sin 0°)(\tan 30°) + \cos 0°$
25. $(\cos 180° + 2 \tan 45°)^2$
26. $\cos 60°(\cos 0° - \cos 180°)$

In 27–30, select the numeral preceding the expression that best completes the sentence or answers the question.

27. If θ is the measure of an acute angle and $\sin \theta = \dfrac{\sqrt{3}}{2}$, then:

    (1) $\cos \theta = \dfrac{\sqrt{3}}{2}$ (2) $\tan \theta = 1$ (3) $\cos \theta = \dfrac{1}{2}$ (4) $\tan \theta = \dfrac{\sqrt{3}}{3}$

28. If $\phi = 30°$, which expression has the largest numerical value?
    (1) $\sin \phi$ (2) $\cos \phi$ (3) $\tan \phi$ (4) $(\cos \phi)(\tan \phi)$
29. If $\sin \theta = \sqrt{3} \cos \theta$, then θ can equal:
    (1) $0°$ (2) $30°$ (3) $45°$ (4) $60°$
30. The value of $2(\sin 30°)(\cos 30°)$ is equal to the value of:
    (1) $\sin 60°$ (2) $\cos 60°$ (3) $\sin 90°$ (4) $\tan 30°$

8-7 USING A TABLE OF VALUES OF TRIGONOMETRIC FUNCTIONS

Finding a Trigonometric Function Value

We have found function values for angles of $30°$, $45°$, and $60°$, as well as for the quadrantal angles. To find function values for other angles requires methods that are studied in advanced mathematics courses. The results of these methods are recorded in a table on pages 750–754 of this book.

In the table "Values of Trigonometric Functions," the sine, cosine, and tangent function values are given for every measure from $0°$ to $90°$ at intervals of 10 minutes (written $10'$). A minute is $\frac{1}{60}$ of a degree. The measures from $0°00'$ to $45°00'$ are listed in the left-hand column. To find these values, we read the table from top to bottom, using the names of the functions listed at the top of the column.

For example, to find sin 35°40′, we find 35° in the left-hand column and read down to 40 for the minutes. Now we find the column that reads "Sin" at the top, and locate the entry in this column that is to the right of 35°40′, as shown in the table below.

$$\sin 35°40′ = .5831$$

Values of Trigonometric Functions

Angle	Sin	Cos	Tan	Cot	
33° 00′	.5446	.8387	.6494	1.5399	57° 00′
10	.5471	.8371	.6536	1.5301	50
20	.5495	.8355	.6577	1.5204	40
30	.5519	.8339	.6619	1.5108	30
40	.5544	.8323	.6661	1.5013	20
50	.5568	.8307	.6703	1.4919	10
34° 00′	.5592	.8290	.6745	1.4826	56° 00′
10	.5616	.8274	.6787	1.4733	50
20	.5640	.8258	.6830	1.4641	40
30	.5664	.8241	.6873	1.4550	30
40	.5688	.8225	.6916	1.4460	20
50	.5712	.8208	.6959	1.4370	10
35° 00′	.5736	.8192	.7002	1.4281	55° 00′
10	.5760	.8175	.7046	1.4193	50
20	.5783	.8158	.7089	1.4106	40
30	.5807	.8141	.7133	1.4019	30
40	.5831	.8124	.7177	1.3934	20
50	.5854	.8107	.7221	1.3848	10
36° 00′	.5878	.8090	.7265	1.3764	54° 00′
	Cos	Sin	Cot	Tan	Angle

The measures from 45°00′ to 90°00′ are listed in the right-hand column. To find these values, we read the table from bottom to top, using the names of the functions listed at the bottom of the column.

For example, to find tan 56°50′, we find 56° in the right-hand column and read up to 50 for the minutes. Now we find the column that reads "Tan" at the bottom, and locate the entry in this column to the left of 56°50'.

$$\tan 56°50' = 1.5301$$

The values that we read from this table have been rounded to the nearest ten thousandth.

Finding Angle Measure From the Table

If we know the function value of an acute angle and want to find the angle measure, we use the reverse of the procedure just described.

To find θ, the measure of an acute angle when $\cos \theta = .9502$, we must examine two columns in the table (the column with "Cos" at the top and the column with "Cos" at the bottom). Since .9502 is located in the column that has "Cos" at the top, the measure of the angle appears at the left. Reading the left-hand column from the top down, as shown in the table below, we see:

If $\cos \theta = .9502$, then $\theta = 18°10'$.

Values of Trigonometric Functions

Angle	Sin	Cos	Tan	Cot	
18° 00'	.3090	.9511	.3249	3.0777	72° 00'
10←	.3118	.9502	.3281	3.0475	50
20	.3145	.9492	.3314	3.0178	40
30	.3173	.9483	.3346	2.9887	30
40	.3201	.9474	.3378	2.9600	20
50	.3228	.9465	.3411	2.9319	10
19° 00'	.3256	.9455	.3443	2.9042	71° 00'
10	.3283	.9446	.3476	2.8770	50
20	.3311	.9436	.3508	2.8502	40
30	.3338	.9426	.3541	2.8239	30
40	.3365	.9417	.3574	2.7980	20
50	.3393	.9407	.3607	2.7725	10
20° 00'	.3420	.9397	.3640	2.7475	70° 00'
10	.3448	.9387	.3673	2.7228	50
20	.3475	.9377	.3706	2.6985	40
30	.3502	.9367	.3739	2.6746	30
40	.3529	.9356	.3772	2.6511	20
50	.3557	.9346	.3805	2.6279	10
	Cos	Sin	Cot	Tan	Angle

To find ϕ, the measure of an acute angle when tan ϕ = 2.7980, we must examine the two "Tan" columns in the table. Since 2.7980 is located in the column that has "Tan" at the bottom, the measure of the angle appears at the right. Reading the right-hand column from the bottom up, as shown in the preceding table, we see:

If tan ϕ = 2.7980, then ϕ = 70°20'.

MODEL PROBLEMS

1. Find cos 52°20'.

   Solution: Locate 52° in the table on pages 750–754. Since 52° is in the column to the right, we will read the table from bottom to top. Move up to find 20' above 52°, and read the value in the column that has "Cos" at the bottom.

   Answer: .6111

2. If tan A = 0.4592, find the measure of acute $\angle A$.

   Solution: Locate 0.4592 in the column that reads "Tan" at the top of the table on pages 750–754. The angle measure is found in the left-hand column, which is read from the top down.

   Answer: 24°40'

EXERCISES

To answer these exercises, use the table "Values of Trigonometric Functions" found on pages 750–754.

In 1–12, express the function value as a four-place decimal.

1. sin 37°20'	2. tan 22°40'	3. cos 12°10'	4. sin 42°50'
5. tan 76°30'	6. sin 88°50'	7. cos 49°10'	8. tan 8°30'
9. cos 54°40'	10. sin 60°20'	11. tan 50°00'	12. cos 85°50'

In 13–24, find the measure of the acute angle, expressed in degrees and minutes.

13. cos A = .9881	14. tan A = .3281	15. sin A = .0987
16. tan A = 1.3111	17. sin A = .9520	18. cos A = .4120
19. sin A = .8572	20. tan A = 4.2747	21. cos A = .6905
22. sin A = .5200	23. tan A = 1.2349	24. cos A = .2447

25. As the measure of $\angle A$ increases from $0°$ to $90°$, tell whether:
   a. $\sin A$ increases or decreases
   b. $\cos A$ increases or decreases
   c. $\tan A$ increases or decreases

In 26–31: **a.** Express the value of the trigonometric function as a four-place decimal. **b.** Find the measure of the acute angle, expressed in degrees and minutes.

26. $\sin B = \frac{51}{100}$ **27.** $\cos B = \frac{27}{100}$ **28.** $\tan B = \frac{37}{50}$
29. $\cos A = \frac{21}{40}$ **30.** $\tan A = \frac{159}{100}$ **31.** $\sin A = \frac{363}{400}$

In 32 and 33, select the numeral preceding the expression that best completes the sentence or answers the question.

32. Which statement correctly shows the relationship between degrees and minutes?
   (1) $1° = 60'$ (2) $1° = 100'$ (3) $1' = 60°$ (4) $1' = 100°$
33. An angle of $26\frac{1}{2}°$ is equal in measure to an angle of:
   (1) $26°5'$ (2) $26°30'$ (3) $26°50'$ (4) $26°60'$

In 34–37: **a.** Tell whether the given statement is *true* or *false*. **b.** Explain your answer.

34. If $\angle A$ is an acute angle, then $\sin A < \cos A$.
35. If θ is the measure of an acute angle and θ is greater than $45°$, then $\tan \theta > 1$.
36. If $\angle B$ is an acute angle, then $\sin B < \tan B$.
37. If θ and ϕ are the measures of acute angles and $\theta < \phi$, then $\cos \theta < \cos \phi$.

8-8 TRIGONOMETRIC FUNCTIONS INVOLVING ANGLE MEASURES TO THE NEAREST MINUTE

Using Interpolation to Find Values of Trigonometric Functions

To find the value of a trigonometric function of an acute angle whose measure is not given in the table, we use a proportion to approximate the value. This procedure, called *interpolation*, is used only within small intervals, such as an interval of 10 minutes.

☐ EXAMPLE 1: Find the value of sin 34°27′ to four decimal places.

Step 1: In the table on pages 750–754, find two angle measures closest to 34°27′, namely, 34°20′ and 34°30′. Then find the sines of these measures, and arrange the information in a table as shown below.

Step 2: The arrows to the left of the table show differences in angle measures, taken from a starting point of 34°20′. The ratio of these differences is $\frac{7'}{10'}$, written in simplified form as $\frac{7}{10}$. The arrows to the right of the table show differences in sine values, taken from a starting point of .5640. The ratio of these differences is $\frac{.0001k}{.0024}$, written in simplified form as $\frac{k}{24}$.

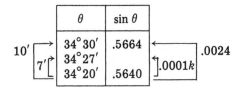

Step 3: Since the ratio of the differences of the angle measures is approximately equal to the ratio of the corresponding differences of their sine values, we write and solve a proportion involving these ratios. (From this point on, we will write only the *simplified proportion*, such as $\frac{7}{10} = \frac{k}{24}$, when we interpolate.)

$$\frac{7'}{10'} = \frac{.0001k}{.0024}$$

$$\frac{7}{10} = \frac{k}{24}$$

$$10k = 168$$

$$k = 16.8$$

$$k = 17$$

Just as 5640 + 17 = 5657, so does sin 34°27′ = .5640 + .0017 = .5657

Step 4: When rounded to the nearest integer, the value of k is 17. Therefore, the difference between the sine value we are finding and .5640 is .0001 (17) or .0017. As θ increases from 34°20′ to 34°27′, sin θ increases. Thus, we add .0017 to .5640 to find the value of 34°27′.

Answer: sin 34°27′ = .5657

Note: In the table just seen, the larger sine value (that is, the larger decimal) is placed at the top of the table to make subtraction easier to perform. In this interval, we observe that *as θ increases, sin θ increases.*

☐ EXAMPLE 2: Find the value of cos 57°22' to four decimal places.

Step 1: In the table on pages 750–754, the angle measures closest to 57°22' are 57°20' and 57°30'. These angle measures and the corresponding cosines of these measures are arranged in a table so that the larger cosine value (larger decimal) is placed at the top. In this interval, *as θ increases, cos θ decreases,* and *as θ decreases, cos θ increases.*

Step 2: From a starting point of 57°30', the ratio of the differences in angle measures is $\frac{8}{10}$ in simplified form.

From the corresponding starting point of cos 57°30', or .5373, the ratio of the differences in cosine values is $\frac{k}{25}$ in simplified form.

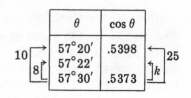

Step 3: We write and solve the simplified proportion, using the ratios of the differences found in step 2.

$$\frac{8}{10} = \frac{k}{25}$$

$$10k = 200$$

$$k = 20$$

Step 4: Here, *k* is 20, an integer. Since cos θ is *increasing* from the starting point of .5373 and moving toward .5398, we *add* .0020 to .5373 to find the value of cos 57°22'.

Just as 5373 + 20 = 5393, so does

cos 57°22' = .5373 + .0020

= .5393

Answer: cos 57°22' = .5393

Using Interpolation To Find Angle Measures

To find the measure of an acute angle when the value of its trigonometric function is not given in the table, we again use *interpolation* within a small interval.

☐ EXAMPLE 3: If cos A = .8247, find the measure of acute angle *A*, correct to the nearest minute.

Step 1: The cosine values closest to .8247 are .8258 and .8241. These cosine values and their corresponding angle measures are arranged in a table. The larger cosine value is placed at the top.

Step 2: From a starting point of 34°30′, the ratio of the differences in angle measures is $\frac{k}{10}$ in simplified form. From a corresponding starting point of .8241, the ratio of the differences in cosine values is $\frac{6}{17}$ in simplified form.

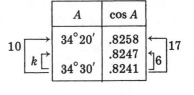

Step 3: We write and solve the simplified proportion, using the ratios of the differences found in step 2.

$$\frac{k}{10} = \frac{6}{17}$$

$$17k = 60$$

Step 4: The value of k is 4, rounded to the nearest integer. Since m∠A is *decreasing* from the starting point of 34°30′ and moving toward 34°20′, we *subtract* 4′ from 34°30′ to find the angle measure.

$$k = \frac{60}{17} = 3\frac{9}{17}$$

$$k = 4$$

$$m\angle A = 34°30′ - 4′$$
$$= 34°26′$$

Answer: m∠A = 34°26′

MODEL PROBLEM

If tan A = 1.3450, refer to the table on pages 750–754 and find the measure of acute angle A: **a.** to the nearest ten minutes **b.** to the nearest minute.

Solution

a. Interpolation is not needed when finding angle measures to the nearest ten minutes. In the table on pages 750–754, since 1.3450 is closer to 1.3432 than to 1.3514, the measure of ∠A is closer to 53°20′ than to 53°30′. Thus, the measure of ∠A to the nearest ten minutes is 53°20′.

A	tan A	
53°30′	1.3514	.0064
	1.3450	.0018
53°20′	1.3432	(closer)

b. **1.** From the table on pages 750–754, the tangent values closest to 1.3450 and their corresponding angle measures are arranged in a table, as shown on top of the next page.

2. In simplified form, the ratio of differences in the angle measures is $\dfrac{k}{10}$ and the ratio of differences in the tangent values is $\dfrac{18}{82}$.

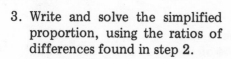

	A	tan A
10 →	53°30′	1.3514
k		1.3450
	53°20′	1.3432

82

18

3. Write and solve the simplified proportion, using the ratios of differences found in step 2.

$$\frac{k}{10} = \frac{18}{82}$$

$$82k = 180$$

4. The value of k is 2, rounded to the nearest integer. From the starting point of 53°20′ and moving toward 53°30′, m∠A increases. Thus, we add 2′ to 53°20′ to find the angle measure.

$$k = \frac{180}{82} = 2\frac{16}{82}$$

$$k = 2$$

$$m\angle A = 53°20′ + 2′ = 53°22′$$

Answer: a. 53°20′   b. 53°22′

EXERCISES

In 1–50, refer to the table on pages 750–754.

In 1–18, use interpolation to find the trigonometric function value to four decimal places.

1. sin 12°42′   2. sin 82°58′   3. tan 38°13′
4. cos 67°36′   5. sin 42°28′   6. tan 71°46′
7. cos 27°27′   8. cos 50°44′   9. sin 9°21′
10. tan 56°19′   11. tan 46°8′   12. cos 8°7′
13. sin 17°42′   14. cos 23°16′   15. sin 31°52′
16. tan 15°33′   17. cos 53°46′   18. tan 3°47′

In 19–24, find the measure of acute ∠A to the nearest ten minutes.

19. cos A = .9583   20. sin A = .5060   21. tan A = .7465
22. sin A = .7203   23. tan A = 1.2345   24. cos A = .5772

In 25–36, use interpolation to find the measure of acute ∠A to the nearest minute.

25. tan A = .1175   26. cos A = .8215   27. sin A = .3213
28. sin A = .7486   29. tan A = .4460   30. cos A = .5915

31. tan A = 1.3902	32. cos A = .1170	33. sin A = .2685
34. cos A = .6816	35. sin A = .8836	36. tan A = 3.1217

In 37–42, find the measure of θ (a) to the nearest ten minutes (b) to the nearest minute.

37. cos θ = .7440	38. sin θ = .2988	39. cos θ = .9883
40. sin θ = .9712	41. tan θ = 2.1994	42. tan θ = 5.1056

43. Find the value of tan $37°17'$ to four decimal places.
44. If tan B = 0.4750, find the measure of acute $\angle B$ to the nearest minute.

In 45–50: a. Express the value of the trigonometric function as a four-place decimal. b. Find the measure of the acute angle to the nearest minute.

45. sin $C = \frac{3}{4}$	46. cos $A = \frac{1}{5}$	47. tan $B = \frac{7}{40}$
48. cos $B = \frac{5}{8}$	49. tan $\theta = \frac{12}{5}$	50. sin $A = \frac{1}{3}$

8-9 FINDING REFERENCE ANGLES

Since the standard table of values of trigonometric functions gives values only for angles whose measures are between $0°$ and $90°$, we will now learn how to relate function values of angles in quadrants II, III, and IV to values of trigonometric functions of angles in quadrant I.

Second-Quadrant Angles

In Fig. 1, let $\angle ROP$ be any second-quadrant angle in standard position in the unit circle. If m$\angle ROP = \theta$ and P is a point on the unit circle, then the coordinates of P are $(-a, b)$ or $(\cos \theta, \sin \theta)$.

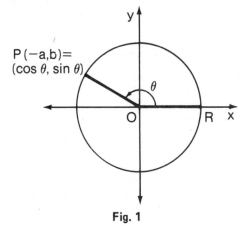

Fig. 1

In Fig. 2, we draw $\overline{PQ} \perp \overleftrightarrow{RO}$, and we reflect $\triangle POQ$ in the y-axis. The image of $\triangle POQ$ is $\triangle P'OQ'$ with P' a point on the unit circle. Notice that $\angle ROP'$, called the *reference angle*, is formed in the first quadrant.

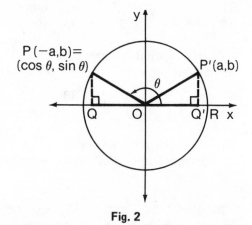

Fig. 2

To find the measure of the reference angle, study Fig. 3. Since $\angle ROP$ and $\angle QOP$ are supplementary, m$\angle QOP = 180° -$ m$\angle ROP = 180° - \theta$. Because angle measure is preserved under a line reflection, m$\angle ROP' =$ m$\angle QOP = 180° - \theta$. Therefore:

■ **For every angle in the second quadrant whose degree measure is θ, there is a reference angle in the first quadrant whose measure is $180° - \theta$.**

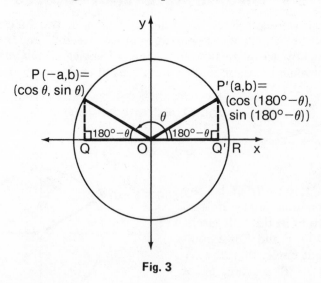

Fig. 3

Under a reflection in the y-axis, $P(-a, b) \rightarrow P'(a, b)$. The coordinates of $P'(a, b) = (\cos (180° - \theta), \sin (180° - \theta))$, as shown in Fig. 3. Using the definitions of sine, cosine, and tangent, we find that the following relationships hold:

Since $\sin \theta = b$ and $\sin (180° - \theta) = b$: $\sin \theta = \sin (180° - \theta)$

Since $\cos \theta = -a$ and $\cos (180° - \theta) = a$: $\cos \theta = -\cos (180° - \theta)$

Since $\tan \theta = \dfrac{b}{-a}$ and $\tan (180° - \theta) = \dfrac{b}{a}$: $\tan \theta = -\tan (180° - \theta)$

☐ EXAMPLE 1: Find $\cos 130°$.

1. For an angle of measure θ in the second quadrant, the cosine is negative and the measure of the reference angle is $180° - \theta$. Therefore, $\cos \theta = -\cos (180° - \theta)$.
2. By substitution, $\cos 130° = -\cos (180° - 130°) = -\cos 50°$.
3. From the table on pages 750–754, $\cos 50° = .6428$. Therefore, $-\cos 50° = -.6428$.

Answer: $\cos 130° = -.6428$

☐ EXAMPLE 2: Find $\sin 173°$.

1. For an angle of measure θ in the second quadrant, $\sin \theta = \sin (180° - \theta)$.

2. $\sin 173° = \sin (180° - 173°)$
          $= \sin 7°$

3. $\sin 7° = .1219$

Answer: $\sin 173° = .1219$

EXAMPLE 3: Find $\tan 118°$.

1. For an angle of measure θ in the second quadrant, $\tan \theta = -\tan (180° - \theta)$.

2. $\tan 118° = -\tan (180° - 118°)$
          $= -\tan 62°$

3. $-\tan 62° = -1.8807$

Answer: $\tan 118° = -1.8807$

Third-Quadrant Angles

In Fig. 1, let $\angle ROP$ be any third-quadrant angle in standard position in the unit circle. If $m\angle ROP = \theta$ and P is a point on the unit circle, the coordinates of P are $(-a, -b)$ or $(\cos \theta, \sin \theta)$.

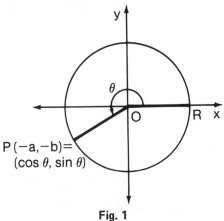

Fig. 1

In Fig. 2, we draw $\overline{PQ} \perp \overleftrightarrow{RO}$, and we reflect $\triangle POQ$ in the origin, O. The image of $\triangle POQ$ is $\triangle P'OQ'$ with P' a point on the unit circle. Notice that the *reference angle*, $\angle ROP'$, is formed in the first quadrant.

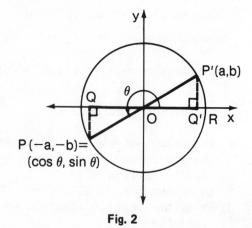

Fig. 2

To find the measure of the reference angle, study Fig. 3. Since $m\angle ROP = 180° + m\angle QOP$, then $m\angle QOP = \theta - 180°$. Because angle measure is preserved under a point reflection, $m\angle ROP' = m\angle QOP = \theta - 180°$. Therefore:

■ **For every angle in the third quadrant whose degree measure is θ, there is a reference angle in the first quadrant whose measure is $\theta - 180°$.**

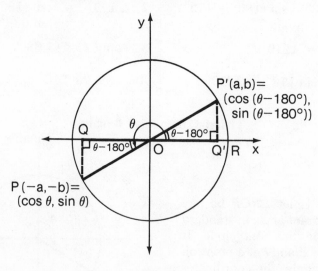

Fig. 3

Under a reflection in the origin, $P(-a, -b) \rightarrow P'(a, b)$. The coordinates of $P'(a, b) = (\cos(\theta - 180°), \sin(\theta - 180°))$, as shown in Fig. 3. Using the definitions of sine, cosine, and tangent, we find that the following relationships hold:

Since $\sin \theta = -b$ and $\sin (\theta - 180°) = b$: $\sin \theta = -\sin (\theta - 180°)$

Since $\cos \theta = -a$ and $\cos (\theta - 180°) = a$: $\cos \theta = -\cos (\theta - 180°)$

Since $\tan \theta = \dfrac{-b}{-a} = \dfrac{b}{a}$ and $\tan (\theta - 180°) = \dfrac{b}{a}$: $\tan \theta = \tan (\theta - 180°)$

☐ EXAMPLE 1: Find sin 200°.

1. For an angle of measure θ in the third quadrant, the sine is negative and the measure of the reference angle is $\theta - 180°$. Therefore, $\sin \theta = -\sin (\theta - 180°)$.
2. By substitution, $\sin 200° = -\sin (200° - 180°) = -\sin 20°$.
3. From the table on pages 750–754, $\sin 20° = .3420$. Thus, $-\sin 20° = -.3420$.

Answer: $\sin 200° = -.3420$

☐ EXAMPLE 2: Find cos 266°.

1. For an angle of measure θ in the third quadrant, $\cos \theta = -\cos (\theta - 180°)$.
2. $\cos 266° = -\cos (266° - 180°)$
   $= -\cos 86°$
3. $-\cos 86° = -.0698$

Answer: $\cos 266° = -.0698$

☐ EXAMPLE 3: Find tan 253°.

1. For an angle of measure θ in the third quadrant, $\tan \theta = \tan (\theta - 180°)$.
2. $\tan 253° = \tan (253° - 180°)$
   $= \tan 73°$
3. $\tan 73° = 3.2709$

Answer: $\tan 253° = 3.2709$

Fourth-Quadrant Angles

In Fig. 1, let $\angle ROP$ be any fourth-quadrant angle in standard position in the unit circle. If $m\angle ROP = \theta$ and P is a point on the unit circle, then the coordinates of P are $(a, -b) = (\cos \theta, \sin \theta)$.

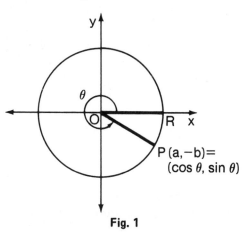

Fig. 1

In Fig. 2, we draw $\overline{PQ} \perp \overleftrightarrow{RO}$, and we reflect $\triangle POQ$ in the x-axis. The image of $\triangle POQ$ is $\triangle P'OQ$ with P' a point on the unit circle. Notice that the *reference angle*, $\angle ROP'$, is formed in the first quadrant.

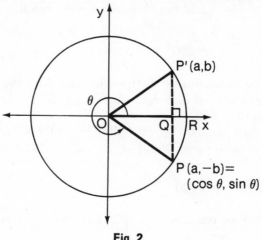

Fig. 2

To find the measure of the reference angle, study Fig. 3. Here the counterclockwise rotation $\angle ROP$ and $\angle POQ$ makes a complete rotation. Thus, $m\angle ROP + m\angle POQ = 360°$, or $m\angle POQ = 360° - m\angle ROP = 360° - \theta$. Since angle measure is preserved under a line reflection, $m\angle ROP' = m\angle POQ = 360° - \theta$.

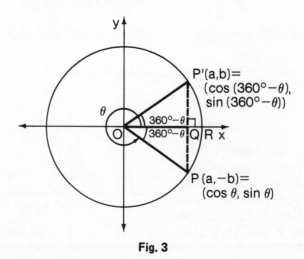

Fig. 3

■ For every angle in the fourth quadrant whose degree measure is θ, there is a reference angle in the first quadrant whose measure is $360° - \theta$.

Under a reflection in the x-axis, $P(a, -b) \rightarrow P'(a, b)$. The coordinates of $P'(a, b) = (\cos (360° - \theta), \sin (360° - \theta))$, as shown in Fig. 3. Using the definitions of sine, cosine, and tangent, we find that the following relationships hold:

Since $\sin \theta = -b$ and $\sin (360° - \theta) = b$: $\sin \theta = -\sin (360° - \theta)$

Since $\cos \theta = a$ and $\cos (360° - \theta) = a$: $\cos \theta = \cos (360° - \theta)$

Since $\tan \theta = \dfrac{-b}{a}$ and $\tan (360° - \theta) = \dfrac{b}{a}$: $\tan \theta = -\tan (360° - \theta)$

☐ EXAMPLE 1: Find tan 320°.

1. For an angle of measure θ in the fourth quadrant, the tangent is negative and the measure of the reference angle is $360° - \theta$. Therefore, $\tan \theta = -\tan (360° - \theta)$.
2. By substitution, $\tan 320° = -\tan (360° - 320°) = -\tan 40°$.
3. From the table on pages 750–754, $\tan 40° = .8391$. Therefore, $-\tan 40° = -.8391$.

Answer: $\tan 320° = -.8391$

☐ EXAMPLE 2: Find sin 348°.

1. For an angle of measure θ in the fourth quadrant, $\sin \theta = -\sin (360° - \theta)$.
2. $\sin 348° = -\sin (360° - 348°)$
      $= -\sin 12°$
3. $-\sin 12° = -.2079$

Answer: $\sin 348° = -.2079$

☐ EXAMPLE 3: Find cos 285°.

1. For an angle of measure θ in the fourth quadrant, $\cos \theta = \cos (360° - \theta)$.
2. $\cos 285° = \cos (360° - 285°)$
      $= \cos 75°$
3. $\cos 75° = .2588$

Answer: $\cos 285° = .2588$

Summary

If θ is the measure of an angle greater than $90°$ but less than $360°$:

$90° < \theta < 180°$	$180° < \theta < 270°$	$270° < \theta < 360°$
quadrant II	quadrant III	quadrant IV
$\sin \theta = \sin (180° - \theta)$ $\cos \theta = -\cos (180° - \theta)$ $\tan \theta = -\tan (180° - \theta)$	$\sin \theta = -\sin (\theta - 180°)$ $\cos \theta = -\cos (\theta - 180°)$ $\tan \theta = \tan (\theta - 180°)$	$\sin \theta = -\sin (360° - \theta)$ $\cos \theta = \cos (360° - \theta)$ $\tan \theta = -\tan (360° - \theta)$

If θ is the measure of an angle greater than $360°$ or less than $0°$, we first find a coterminal angle whose measure is between $0°$ and $360°$. In this way, we can find the value of a trigonometric function of an angle of any degree measure.

| MODEL PROBLEMS |

1. Express each of the following as a function of a positive acute angle:
   a. $\cos 260°$ b. $\sin 715°$ c. $\tan (-110°)$

Solution

a. Determine the quadrant: Since $180° < 260° < 270°$, the angle lies in the third quadrant.

Determine the reference angle: $\theta - 180° = 260° - 180° = 80°$.

Determine the sign: The cosine of a third-quadrant angle is negative.

$$\cos 260° = -\cos 80°$$

b. Determine the angle of smallest positive measure that is coterminal with an angle of $715°$: $715° - 360° = 355°$

Determine the quadrant: Since $270° < 355° < 360°$, the angle lies in the fourth quadrant.

Determine the reference angle: Here $360° - \theta = 360° - 355° = 5°$.

Determine the sign: The sine of a fourth-quadrant angle is negative.

$$\sin 715° = -\sin 5°$$

c. Determine the angle of smallest positive measure that is coterminal with an angle of $-110°$: $-110° + 360° = 250°$

Determine the quadrant: Since $180° < 250° < 270°$, the angle lies in the third quadrant.

Determine the reference angle: $\theta - 180° = 250° - 180° = 70°$.

Determine the sign: The tangent of a third-quadrant angle is positive.

$$\tan (-110°) = \tan 70°$$

Answer: a. $-\cos 80°$ b. $-\sin 5°$ c. $\tan 70°$

2. Find cos 168°20′.

Solution

Determine the quadrant: Since 90° < 168°20′ < 180°, the angle lies in the second quadrant.

Determine the reference angle: Here 180° − θ = 180° − 168°20′
$$= 179°60′ − 168°20′$$
$$= 11°40′$$

Determine the sign: The cosine of a second-quadrant angle is negative.

$$\cos 168°20′ = -\cos 11°40′$$

Find the function value in the table on pages 750–754: −cos 11°40′ = −.9793

Answer: cos 168°20′ = −.9793

EXERCISES

In 1–24, express the given function as a function of a positive acute angle.

1. sin 100°	2. cos 150°	3. sin 340°
4. tan 300°	5. cos 190°	6. tan 215°
7. cos 290°	8. tan 145°	9. sin 248°
10. cos 305°	11. sin 200°	12. tan 237°
13. sin 98°	14. tan 345°	15. sin 500°
16. cos 690°	17. tan 620°	18. sin 650°
19. sin (−20°)	20. cos (−200°)	21. sin (−340°)
22. cos (−250°)	23. tan (−80°)	24. sin (−158°)

In 25–30: a. Express the given function as a function of a positive acute angle. b. Find the exact function value.

25. sin 150°	26. cos 300°	27. tan 225°
28. cos 240°	29. tan 120°	30. cos 405°

In 31–42, find the exact function value.

31. cos 315°	32. sin 135°	33. tan 330°
34. sin 240°	35. sin 390°	36. tan 600°
37. cos 570°	38. sin (−45°)	39. cos (−30°)
40. tan (−120°)	41. cos (−300°)	42. sin (−135°)

In 43–48, find the exact value of the given expression.

43. sin 210° + cos 120°   44. tan 135° + sin 330°
45. cos 135° + cos 225°   46. sin 300° + sin (−240°)
47. tan (−315°) + tan 135°   48. (sin 60°)(cos 150°) − tan (−45°)

In 49–66: **a.** Express the function as a function of a positive acute angle. **b.** Using the table on pages 750–754, find the value of the given function to four decimal places.

49. sin 172°   50. cos 252°   51. tan 292°
52. cos 347°   53. tan 103°   54. sin 196°
55. tan 221°   56. cos 400°   57. tan 737°
58. sin (−80°)   59. sin (−299°)   60. cos (−239°)
61. cos 216°30′   62. tan 208°20′   63. sin 152°40′
64. tan (−53°20′)   65. sin (−126°10′)   66. cos (−244°50′)

In 67–71, select the numeral preceding the expression that best completes the sentence or answers the question.

67. The value of tan 150° is equal to the value of:
    (1) tan 30°   (2) tan 60°   (3) −tan 30°   (4) −tan 60°
68. The value of tan 135° is equal to the value of:
    (1) cos 90°   (2) sin 90°   (3) cos 270°   (4) sin 270°
69. The value of cos 390° is equal to the value of:
    (1) sin 30°   (2) sin 60°   (3) −sin 30°   (4) −sin 60°
70. Which of the following has the largest numerical value?
    (1) sin 150°   (2) cos 225°   (3) cos 270°   (4) tan 315°
71. Which of the following has the smallest numerical value?
    (1) cos 120°   (2) tan 225°   (3) sin 240°   (4) cos 315°

8-10 RADIAN MEASURE

Just as we can measure a line segment by using different units of length, such as inches or centimeters, so we can measure an angle by using a unit of measure other than a degree.

■ **DEFINITION.** A *radian* is the measure of an angle that, when drawn as a central angle of a circle, would intercept an arc whose length is equal to the length of a radius of the circle.

In the figure, the measure of a radius of circle O is r. The distance from P to Q along the circle is r. The measure of $\angle POQ$ is *1 radian*.

In a unit circle, an angle of 1 radian intercepts an arc whose length is 1.

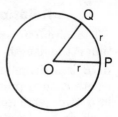

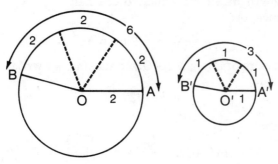

Fig. 1   Fig. 2

In Fig. 1, radius $\overline{OA}$ has a length of 2, $\angle AOB$ is a central angle, and the intercepted arc $\overset{\frown}{AB}$ has a length of 6. Since $\overset{\frown}{AB}$ can be divided into 3 parts, the length of each being equal to the length of the radius, 2, then $m\angle AOB = 3$ radians.

In Fig. 2, radius $O'A'$ has a length of 1, $\angle A'O'B'$ is a central angle, and the intercepted arc $\overset{\frown}{A'B'}$ has a length of 3. Since $\overset{\frown}{A'B'}$ can be divided into 3 parts, the length of each being equal to the length of the radius, 1, then $m\angle A'O'B' = 3$ radians. These examples illustrate the following relationship:

$$\text{measure of an angle in radians} = \frac{\text{length of the intercepted arc}}{\text{length of the radius}}$$

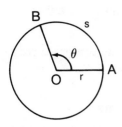

In general, if θ is the measure of a central angle in radians, s is the length of the intercepted arc, and r is the length of a radius, then:

$$\theta = \frac{s}{r}$$

If both members of this equation are multiplied by r, the rule is stated as $s = r\theta$.

A Relationship Between Degrees and Radians

The circumference of a circle is equal to 2π times the length of its radius, or $C = 2\pi r$. If $\overline{AOB}$ is a diameter of circle O with a radius of length r, then points A and B separate the circle into two semicircles. The length of each semicircle is the length of an arc equal to one-half the circumference, or:

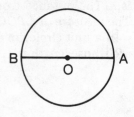

$$s = \tfrac{1}{2}C = \tfrac{1}{2}(2\pi r) = \pi r$$

To find the measure of straight $\angle AOB$ in radians, we write:

$$\theta = \frac{s}{r}$$

Then, substitute:

$$m\angle AOB = \frac{\pi r}{r} = \pi \text{ radians}$$

Since the measure of a straight angle, such as $\angle AOB$, is π radians and since the measure of a straight angle is also expressed as $180°$, the following relationship is true:

$$\pi \text{ radians} = 180°$$

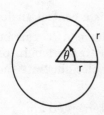

$\theta = 1 \text{ radian} = \dfrac{180°}{\pi}$

$1 \text{ radian} \approx 57°18'$

To find the degree measure of an angle of 1 radian, we can divide both sides of the equation just stated by π. Thus:

$$1 \text{ radian} = \frac{180°}{\pi} \text{ (an irrational number)}$$

By substituting an approximate rational value for π, such as 3.1416, we find an approximate degree measure for an angle of 1 radian.

$$1 \text{ radian} = \frac{180°}{\pi} \approx \frac{180°}{3.1416} \approx 57.3° \text{ or } 57°18'$$

Thus, an angle of 1 radian has a degree measure slightly larger than $57°$, or 1 radian = $57°$ (to the nearest degree).

Changing From Degrees to Radians

We have seen that π radians = $180°$. Just as the measure of a straight angle can be expressed in degrees or in radians, the measure of any angle (such as $\angle A$) can be expressed in degrees or in radians. Thus, we form the proportion:

$$\frac{m\angle A \text{ in degrees}}{m\angle A \text{ in radians}} = \frac{\text{measure of a straight angle in degrees}}{\text{measure of a straight angle in radians}}$$

Since 180° and π radians are the measures of a straight angle, the proportion can be rewritten as:

$$\frac{m\angle A \text{ in degrees}}{m\angle A \text{ in radians}} = \frac{180°}{\pi \text{ radians}}$$

OR

$$\frac{\textbf{measure in degrees}}{\textbf{measure in radians}} = \frac{\textbf{180}}{\pi}$$

☐ EXAMPLE: Express in radian measure an angle of 75°.

How to Proceed	*Solution*
1. Identify the variable.	Let x = the measure in radians of an angle of 75°.
2. Write the proportion.	$\dfrac{\text{measure in degrees}}{\text{measure in radians}} = \dfrac{180}{\pi}$
3. Substitute 75° and x radians for the angle in question.	$\dfrac{75}{x} = \dfrac{180}{\pi}$
4. Solve for x.	$180x = 75\pi$ $x = \dfrac{75}{180}\pi$
5. Express in simplest form.	$x = \dfrac{5}{12}\pi$ *Ans.*

Note: Although the measure of the angle is $\frac{5}{12}\pi$ radians, mathematicians generally agree that the word "radian" or any symbol for the word need not be written when stating a radian measure. Thus, the radian measure of the angle is simply written as $\frac{5}{12}\pi$.

This agreement can cause some confusion. For example, if $m\angle A = 2$, do we mean 2° or 2 radians? To ease this confusion, we observe:

1. If an angle measure is found by the rule $\theta = \dfrac{s}{r}$, then a radian measure is being found.

2. If the situation is unclear, we will identify the type of angle measure being used, either by words or by symbols, as in $m\angle A = 2°$ and $m\angle B = 2$ radians.

Changing From Radians to Degrees

To find the degree measure of an angle whose radian measure is known, we use the same proportion developed earlier.

□ EXAMPLE: Find the degree measure of an angle of $\frac{\pi}{4}$ radians.

How to Proceed	*Solution*
1. Identify the variable.	Let x = the degree measure of an angle of $\frac{\pi}{4}$ radians.
2. Write the proportion.	$\dfrac{\text{measure in degrees}}{\text{measure in radians}} = \dfrac{180}{\pi}$
3. Substitute $x°$ and $\frac{\pi}{4}$ radians for the angle in question.	$\dfrac{x}{\frac{\pi}{4}} = \dfrac{180}{\pi}$
4. Solve for x.	$\pi x = \dfrac{\pi}{4}(180)$
	$\pi x = 45\pi$
	$x = 45$

Answer: The measure of the angle is $45°$.

MODEL PROBLEMS

1. In a circle, the length of a radius is 4 cm. Find the length of an arc intercepted by a central angle whose measure is 1.5 radians.

How to Proceed	*Solution*
1. Write the rule that shows that the radian measure θ of a central angle is equal to the length s of the intercepted arc divided by the length r of a radius.	$\theta = \dfrac{s}{r}$
2. Substitute the given values.	$1.5 = \dfrac{s}{4}$
3. Solve for s.	$s = 4(1.5)$
	$s = 6$

Answer: 6 cm

2. Express in radian measure an angle of 135°.

How to Proceed	*Solution*
1. Identify the variable.	Let x = the radian measure of an angle of 135°.
2. Write the proportion.	$$\frac{\text{measure in degrees}}{\text{measure in radians}} = \frac{180}{\pi}$$
3. Substitute 135° and x radians for the angle in question.	$$\frac{135}{x} = \frac{180}{\pi}$$
4. Solve for x.	$$180x = 135\pi$$ $$x = \frac{135}{180}\pi$$
5. Express in simplest form.	$$x = \frac{3}{4}\pi$$

Answer: $\frac{3}{4}\pi$, or $\frac{3\pi}{4}$

3. Express $\frac{7\pi}{3}$ radians in degrees.

Method 1	*Method 2*
Let y = the degree measure of an angle of $\frac{7\pi}{3}$ radians.	Substitute 180° for π radians, and simplify.
Use the proportion, and solve for y.	$$\frac{7\pi}{3} \text{ radians} = \frac{7}{3}(180°)$$
$$\frac{\text{measure in degrees}}{\text{measure in radians}} = \frac{180}{\pi}$$	$$= \frac{7}{\cancel{3}}(\cancel{180°}^{\,60°}) = 420°$$
$$\frac{y}{\frac{7\pi}{3}} = \frac{180}{\pi}$$	
$$\pi y = \frac{7\pi}{3}(180)$$	
$$\pi y = 420\pi$$	
$$y = 420$$	
Answer: 420°	

EXERCISES

In 1-15, find the radian measure of an angle of the given degree measure.

1. 30°	2. 90°	3. 45°	4. 120°	5. 160°
6. 180°	7. 210°	8. 225°	9. 270°	10. 300°
11. 315°	12. 100°	13. 198°	14. 99°	15. 396°

In 16-30, find the degree measure of an angle of the given radian measure.

16. $\dfrac{\pi}{3}$ 17. $\dfrac{\pi}{9}$ 18. $\dfrac{\pi}{10}$ 19. $\dfrac{2\pi}{5}$ 20. $\dfrac{\pi}{2}$

21. $\dfrac{5\pi}{6}$ 22. π 23. $\dfrac{4\pi}{3}$ 24. $\dfrac{10\pi}{9}$ 25. $\dfrac{3\pi}{4}$

26. $\dfrac{11\pi}{6}$ 27. 3π 28. $\dfrac{7\pi}{2}$ 29. $\dfrac{5\pi}{18}$ 30. $\dfrac{2\pi}{27}$

In 31-38, θ is the measure of a central angle that intercepts an arc of length s in a circle with a radius of length r.

31. If $s = 12$ and $r = 4$, find θ.
32. If $s = 6$ and $r = 1$, find θ.
33. If $s = 10$ and $r = 2.5$, find θ.
34. If $s = 12$ and $\theta = 6$, find r.
35. If $s = 12$ and $\theta = .5$, find r.
36. If $\theta = 2.5$ and $r = 4$, find s.
37. If $\theta = \frac{1}{3}$ and $s = 3$, find r.
38. If $\theta = 4$ and $r = 1.25$, find s.

39. In a circle, a central angle of $\frac{1}{3}$ radian intercepts an arc of 3 centimeters. Find the length, in centimeters, of a radius of the circle.
40. A circle has a radius of 1.7 inches. Find the length of an arc intercepted by a central angle whose measure is 2 radians.
41. In a circle whose radius measures 5 cm, a central angle intercepts an arc of length 12 cm. Find the radian measure of the central angle.
42. In a circle, a central angle of 4.2 radians intercepts an arc whose length is 6.3 meters. Find the length of a radius in meters.
43. A central angle intercepts an arc on a circle equal in length to a diameter of the circle. Find the measure in radians of the central angle.

44. On a clock, the length of the pendulum is 30 centimeters. A swing of the pendulum determines an angle of 0.8 radian. Find, in centimeters, the distance traveled by the tip of the pendulum during this swing.

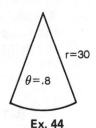

Ex. 44

45. As the pendulum of a clock swings through an angle of 30°, the tip of the pendulum travels along an arc whose length is 4π inches.
    a. Express the angle of 30° in radian measure.
    b. Find the length of the pendulum in inches.

46. Copy and complete the table.

Amount of Rotation	$\frac{1}{2}$	$\frac{1}{12}$	$\frac{1}{8}$	$\frac{1}{6}$	$\frac{1}{4}$	$\frac{3}{4}$
Degree Measure	180°					
Radian Measure	π					

47. Express $\dfrac{10\pi}{3}$ radians in degrees.

48. Find the radian measure of an angle of 306°.

In 49–53, select the numeral preceding the expression that best completes the sentence.

49. One radian is approximately equal to:
    (1) 45°   (2) 50°   (3) 57°   (4) 60°
50. The number of radians in a complete rotation is approximately:
    (1) 6.28   (2) 3.14   (3) 10   (4) 36
51. Two-thirds of a rotation determines an angle whose radian measure is:
    (1) $\dfrac{2\pi}{3}$ (2) $\dfrac{4\pi}{3}$ (3) $\dfrac{8\pi}{3}$ (4) $\dfrac{3\pi}{2}$

52. Three-eighths of a rotation determines an angle whose measure is:
    (1) $\dfrac{3\pi}{8}$ (2) 140°   (3) $\dfrac{3\pi}{4}$ (4) 150°

53. A wheel whose radius measures 10 inches is rotated. If a point on the circumference of the wheel moves a distance of 5 feet, then the point travels through an angle whose radian measure is:

(1) $\frac{1}{2}$ (2) 2   (3) $\frac{1}{6}$ (4) 6

8-11 TRIGONOMETRIC FUNCTIONS INVOLVING RADIAN MEASURE

Since angle measure can be expressed in radians as well as in degrees, we can find values of trigonometric functions of angles expressed in radian measure. To do this, we convert the radian measure to a degree measure and follow the procedures learned earlier.

| MODEL PROBLEMS |

1. Find the value of $\sin \frac{\pi}{3}$.

Solution

1. Find the degree measure of an angle of $\frac{\pi}{3}$ radians by using the proportion method or by using substitution as shown here.

$$\frac{\pi}{3} \text{ radians} = \frac{1}{3}(\pi \text{ radians}) = \frac{1}{3}(180°) = 60°$$

2. Find the value of the function.

$$\sin \frac{\pi}{3} = \sin 60° = \frac{\sqrt{3}}{2}$$

Answer: $\dfrac{\sqrt{3}}{2}$

2. If a function f is defined as $f(x) = \cos 2x + \sin x$, find the numerical value of $f\left(\frac{\pi}{2}\right)$.

How to Proceed	*Solution*
1. Write the function.	$f(x) = \cos 2x + \sin x$
2. Let $x = \frac{\pi}{2}$, and simplify the terms in the expression.	$f\left(\frac{\pi}{2}\right) = \cos\left(2 \cdot \frac{\pi}{2}\right) + \sin\left(\frac{\pi}{2}\right)$
	$= \cos \pi + \sin \frac{\pi}{2}$

3. Change radian measures to degrees in the expression. $= \cos 180° + \sin 90°$

4. Evaluate and simplify. $= -1 + 1 = 0$

Answer: $f\left(\dfrac{\pi}{2}\right) = 0$

EXERCISES

In 1–16, find the exact value of the trigonometric function.

1. $\cos \dfrac{\pi}{3}$ 2. $\tan \dfrac{\pi}{4}$ 3. $\sin \dfrac{2\pi}{3}$ 4. $\cos \dfrac{4\pi}{3}$

5. $\sin \dfrac{5\pi}{4}$ 6. $\sin \dfrac{7\pi}{6}$ 7. $\tan \dfrac{4\pi}{3}$ 8. $\cos \dfrac{11\pi}{6}$

9. $\tan \dfrac{5\pi}{6}$ 10. $\cos \dfrac{3\pi}{4}$ 11. $\sin \dfrac{5\pi}{3}$ 12. $\tan 3\pi$

13. $\sin \dfrac{15\pi}{4}$ 14. $\tan(-\pi)$ 15. $\cos\left(\dfrac{-\pi}{6}\right)$ 16. $\sin\left(\dfrac{-5\pi}{6}\right)$

17. If a function f is defined as $f(x) = \cos 3x$, find the numerical value

   of: a. $f\left(\dfrac{\pi}{2}\right)$ b. $f\left(\dfrac{\pi}{6}\right)$ c. $f(\pi)$

18. If a function f is defined as $f(x) = \sin\left(\dfrac{x}{2}\right)$, find the numerical value

   of: a. $f(3\pi)$ b. $f\left(\dfrac{\pi}{2}\right)$ c. $f\left(\dfrac{5\pi}{3}\right)$

In 19–24, find the numerical value of $f(\pi)$ for the given function f.

19. $f(x) = \sin 2x$ 20. $f(x) = \cos \frac{1}{4}x$

21. $f(x) = \tan\left(\dfrac{x}{3}\right)$ 22. $f(x) = \sin x + \cos 2x$

23. $f(x) = \tan 2x - \sin\left(\dfrac{x}{2}\right)$ 24. $f(x) = \sin x \cos 2x$

In 25–30, find the numerical value of $f\left(\dfrac{\pi}{3}\right)$ for the given function f.

25. $f(x) = \sin 2x$ 26. $f(x) = \tan 5x$

27. $f(x) = \cos \frac{1}{2}x$ 28. $f(x) = \sin\left(\dfrac{7x}{2}\right)$

29. $f(x) = \sin \frac{1}{2}x + \cos 5x$ 30. $f(x) = \tan x \tan 2x$

In 31–34, select the numeral preceding the expression that best completes the sentence or answers the question.

31. If $f(x) = \tan 5x + \cos 2x$, then $f\left(\dfrac{\pi}{4}\right)$ equals:

   (1) 1   (2) 2   (3) 0   (4) $\dfrac{\sqrt{2}}{2}$

32. If $f(x) = \cos x + \tan \dfrac{x}{3}$, then $f(\pi)$ is:

   (1) $\dfrac{\sqrt{3} + 3}{3}$ (2) $\dfrac{\sqrt{3} - 3}{3}$ (3) $\sqrt{3} + 1$ (4) $\sqrt{3} - 1$

33. Which of the following functions has the largest numerical value when $x = 2\pi$?

   (1) $f(x) = \sin \dfrac{x}{2}$ (2) $g(x) = \cos \dfrac{x}{2}$

   (3) $h(x) = \sin \dfrac{x}{4}$ (4) $k(x) = \cos \dfrac{x}{4}$

34. What is the value of $\sin \dfrac{4\pi}{3} \sin \dfrac{\pi}{3}$?

   (1) $-\dfrac{3}{4}$ (2) $\dfrac{3}{4}$ (3) $-\dfrac{3}{2}$ (4) $\dfrac{3}{2}$

8-12 THE RECIPROCAL TRIGONOMETRIC FUNCTIONS

We have defined three trigonometric functions, namely, the sine, cosine, and tangent functions. Three other trigonometric functions can be defined in terms of the sine, cosine, and tangent.

■ The *secant function* assigns to every angle measure θ, for which $\cos \theta \neq 0$, a unique value that is the reciprocal of $\cos \theta$.

$$\theta \xrightarrow{\text{secant}} \frac{1}{\cos \theta}$$

Using "sec" for secant, we write the reciprocal identity as:

$$\sec \theta = \frac{1}{\cos \theta}$$

Since $\sec \theta$ is the reciprocal of $\cos \theta$, $\sec \theta$ is undefined when $\cos \theta = 0$. For example:

$$\text{Cos } 90° = 0. \text{ Thus, } \sec 90° = \frac{1}{\cos 90°} = \frac{1}{0} \text{ (undefined).}$$

The secant function is undefined when sec $\theta = \frac{1}{0}$, that is, when $\theta = 90°$, 270°, etc.

■ The *cosecant function* assigns to every θ for which sin $\theta \neq 0$ a unique value that is the reciprocal of sin θ.

$$\theta \xrightarrow{\text{cosecant}} \frac{1}{\sin \theta}$$

The abbreviation for cosecant is "csc," not the first three letters of the word as in the other functions. This reciprocal identity is:

$$\csc \theta = \frac{1}{\sin \theta}$$

Since csc θ is the reciprocal of sin θ, csc θ is undefined when sin $\theta = 0$. For example:

$$\text{Sin } 180° = 0. \text{ Thus, csc } 180° = \frac{1}{\sin 180°} = \frac{1}{0} \text{ (undefined).}$$

The cosecant function is undefined when csc $\theta = \frac{1}{0}$, that is, when $\theta = 0°$, 180°, 360°, etc.

■ The *cotangent function* assigns to every θ for which tan $\theta \neq 0$ a unique value that is the reciprocal of tan θ.

$$\theta \xrightarrow{\text{cotangent}} \frac{1}{\tan \theta}$$

Using "cot" for cotangent, we write the reciprocal identity as:

$$\cot \theta = \frac{1}{\tan \theta}$$

Using the quotient identity tan $\theta = \dfrac{\sin \theta}{\cos \theta}$, we can express cot θ in terms of sin θ and cos θ, forming another quotient identity.

$$\cot \theta = \frac{1}{\tan \theta} = \frac{1}{\dfrac{\sin \theta}{\cos \theta}} = \frac{\cos \theta}{\sin \theta}$$

Using either $\dfrac{1}{\tan \theta}$ or $\dfrac{\cos \theta}{\sin \theta}$ as an expression for cot θ, we find that cot θ is undefined when tan $\theta = 0$. For example:

$$\text{Tan } 180° = 0 \text{ and sin } 180° = 0.$$

$$\text{Thus, cot } 180° = \frac{1}{\tan 180°} = \frac{\cos 180°}{\sin 180°} = \frac{1}{0} \text{ (undefined).}$$

The cotangent function is undefined when $\cot \theta = \frac{1}{0}$, that is, when $\theta = 0°$, $180°$, $360°$, etc. Notice, however, that the cotangent function is defined when $\theta = 90°$ and $270°$ because:

$$\text{Cot } 90° = \frac{\cos 90°}{\sin 90°} = \frac{0}{1} = 0, \text{ and } \cot 270° = \frac{\cos 270°}{\sin 270°} = \frac{0}{-1} = 0.$$

Function Values as Lengths of Line Segments

The following diagrams represent each trigonometric function value as the length of a line segment. Angle ROP is an angle in standard position in a unit circle whose center is at the origin. The tangent to the circle at $R(1, 0)$ intersects the line of the terminal ray at T. Point C is the point of intersection of the circle with the non-negative ray of the y-axis. The tangent to the circle at C intersects the line of the terminal ray at S.

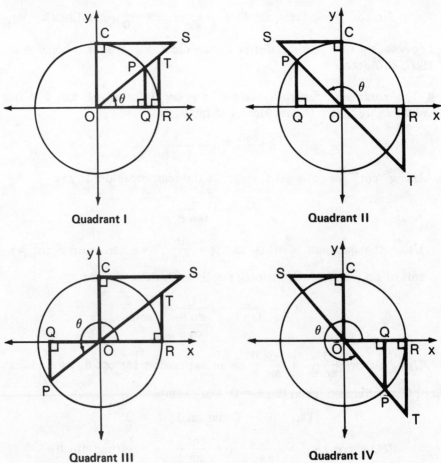

Quadrant I

Quadrant II

Quadrant III

Quadrant IV

In each diagram $\overline{CS} \perp y$-axis, and $\overline{RT}$ and $\overline{PQ} \perp x$-axis. Since $\angle OSC \cong \angle POQ \cong \angle TOR$ and $\angle OCS \cong \angle PQO \cong \angle TRO$, then $\triangle OSC \sim \triangle POQ \sim \triangle TOR$ by a.a. $\cong$ a.a. If $m\angle ROP = \theta$,

$$\sin \theta = PQ$$
$$\cos \theta = OQ$$
$$\tan \theta = TR$$

$$\sec \theta = OT \text{ since } \frac{OT}{OR} = \frac{OP}{OQ} \text{ or } \frac{\sec \theta}{1} = \frac{1}{\cos \theta}$$

$$\csc \theta = OS \text{ since } \frac{OS}{OC} = \frac{PO}{PQ} \text{ or } \frac{\csc \theta}{1} = \frac{1}{\sin \theta}$$

$$\cot \theta = CS \text{ since } \frac{CS}{OC} = \frac{RO}{TR} \text{ or } \frac{\cot \theta}{1} = \frac{1}{\tan \theta}$$

A number and its reciprocal have the same sign. Recall that a vertical line segment is positive if it is above the x-axis and negative if it is below the x-axis. A horizontal line segment is positive if it is to the right of the y-axis and negative if it is to the left of the y-axis. A line segment that is part of the line of the terminal side of a central angle is positive if it is part of the terminal ray and negative if it is part of the opposite ray of the terminal side.

Quadrant	I	II	III	IV
$\sin \theta = PQ$	+	+	-	-
$\csc \theta = OS$	+ part of terminal ray, $\overrightarrow{OP}$	+ part of terminal ray, $\overrightarrow{OP}$	- part of ray opposite $\overrightarrow{OP}$	- part of ray opposite $\overrightarrow{OP}$
$\cos \theta = OQ$	+	-	-	+
$\sec \theta = OT$	+ part of terminal ray, $\overrightarrow{OP}$	- part of ray opposite $\overrightarrow{OP}$	- part of ray opposite $\overrightarrow{OP}$	+ part of terminal ray, $\overrightarrow{OP}$
$\tan \theta = TR$	+	-	+	-
$\cot \theta = CS$	+ right of y-axis	- left of y-axis	+ right of y-axis	- left of y-axis

| MODEL PROBLEMS |

1. Write each expression in terms of $\sin \theta$, $\cos \theta$, or both. Simplify wherever possible.   a. $\sec \theta \cdot \cot \theta$ b. $\dfrac{\tan \theta}{\csc \theta}$

Solution

a. Since $\sec \theta = \dfrac{1}{\cos \theta}$ and $\cot \theta = \dfrac{\cos \theta}{\sin \theta}$, substitute these values in the expression.

$$\sec \theta \cdot \cot \theta = \frac{1}{\cos \theta} \cdot \frac{\cos \theta}{\sin \theta} = \frac{1}{\cancel{\cos \theta}} \cdot \frac{\overset{1}{\cancel{\cos \theta}}}{\sin \theta} = \frac{1}{\sin \theta} \quad Ans.$$

b. Since $\tan \theta = \dfrac{\sin \theta}{\cos \theta}$ and $\csc \theta = \dfrac{1}{\sin \theta}$, substitute these values in the expression. Then, simplify the complex fraction.

$$\frac{\tan \theta}{\csc \theta} = \frac{\left(\dfrac{\sin \theta}{\cos \theta}\right)}{\left(\dfrac{1}{\sin \theta}\right)} = \frac{\left(\dfrac{\sin \theta}{\cos \theta}\right) \cdot \sin \theta}{\left(\dfrac{1}{\cancel{\sin \theta}}\right) \cdot \cancel{\sin \theta}} = \frac{\dfrac{\sin^2 \theta}{\cos \theta}}{1} = \frac{\sin^2 \theta}{\cos \theta} \quad Ans.$$

Note: The product $\sin \theta \cdot \sin \theta$ is equal to $(\sin \theta)^2$, which is usually written as $\sin^2 \theta$ to indicate that the sine function value is being squared.

2. Find the exact numerical value of $\cot 45° \cdot \csc 45°$.

Solution

Since $\tan 45° = 1$, $\cot 45° = 1$ and since $\sin 45° = \dfrac{\sqrt{2}}{2}$,

$\csc 45° = \dfrac{2}{\sqrt{2}}$.

Therefore:

$$\cot 45° \cdot \csc 45° = 1 \cdot \frac{2}{\sqrt{2}} = \frac{2}{\sqrt{2}}$$

$$= \frac{2}{\sqrt{2}} \cdot \frac{\sqrt{2}}{\sqrt{2}} = \frac{2\sqrt{2}}{2} = \sqrt{2} \quad Ans.$$

| EXERCISES |

In 1–12, write each expression in terms of $\sin \theta$, $\cos \theta$, or both. Simplify wherever possible.

1. $\tan \theta$ 2. $\cot \theta$ 3. $\sec \theta$ 4. $\csc \theta$

5. $\tan \theta \cdot \csc \theta$ 6. $\dfrac{\tan \theta}{\sec \theta}$ 7. $\dfrac{\cot \theta}{\csc \theta}$ 8. $\dfrac{\cos \theta}{\sec \theta}$

9. $\dfrac{\sin \theta}{\csc \theta}$ 10. $\dfrac{\tan \theta}{\cot \theta}$ 11. $\dfrac{\sec \theta}{\cot \theta}$ 12. $\dfrac{\sec \theta}{\csc \theta}$

13. Copy and complete the following table. If a function is undefined for an angle measure, write "undefined."

θ	0°	30°	45°	60°	90°
$\sec \theta$					
$\csc \theta$					
$\cot \theta$					

In 14–31, find the exact numerical value of the expression.

14. $\csc 150°$ 15. $\sec 240°$ 16. $\cot 315°$ 17. $\csc 120°$

18. $\sec 2\pi$ 19. $\cot \dfrac{\pi}{6}$ 20. $\csc \dfrac{3\pi}{2}$ 21. $\sec \dfrac{5\pi}{4}$

22. $\sin 30° \cdot \csc 30°$ 23. $\tan 45° \cdot \sec 30°$

24. $\sec^2 60° + \csc^2 60°$ 25. $\tan^2 60° + \cot 45°$

26. $\sin \dfrac{\pi}{3} \cdot \tan \dfrac{\pi}{6}$ 27. $\cot \dfrac{\pi}{4} + \sec^2 \dfrac{\pi}{6}$

28. $\cos^2 \dfrac{\pi}{4} + \sec^2 \dfrac{\pi}{4}$ 29. $\csc \dfrac{\pi}{2} + \sec \pi$

30. $\cot \dfrac{\pi}{2} + \sec 0$ 31. $\cot^2 \dfrac{\pi}{3} + \csc \dfrac{\pi}{2}$

In 32–37, name the quadrants in which $\angle A$ may lie.

32. $\csc A > 0$ 33. $\cot A < 0$ 34. $\sec A < 0$

35. $\cot A > 0$ 36. $\csc A < 0$ 37. $\sec A > 0$

In 38–43, name the quadrant in which $\angle B$ must lie.

38. $\cot B < 0$ and $\sin B > 0$ 39. $\sec B < 0$ and $\tan B > 0$

40. $\sec B < 0$ and $\csc B > 0$ **41.** $\cot B < 0$ and $\sec B > 0$

42. $\csc B < 0$ and $\tan B > 0$ **43.** $\csc B < 0$ and $\sec B > 0$

44. If $\sin A = \frac{3}{5}$ and $\cos A = -\frac{4}{5}$, find: **a.** the quadrant in which $\angle A$ lies   **b.** $\tan A$ **c.** $\sec A$ **d.** $\csc A$ **e.** $\cot A$

45. If $\sin \theta = -\frac{2}{3}$ and $\cos \theta = -\frac{\sqrt{5}}{3}$, find: **a.** the quadrant in which the angle whose measure is θ lies   **b.** $\tan \theta$ **c.** $\sec \theta$ **d.** $\csc \theta$ **e.** $\cot \theta$

46. If $\sin \phi = -\frac{1}{\sqrt{5}}$ and $\cos \phi = \frac{2}{\sqrt{5}}$, find: **a.** the quadrant in which the angle whose measure is ϕ lies   **b.** $\tan \phi$ **c.** $\sec \phi$ **d.** $\csc \phi$ **e.** $\cot \phi$

47. If $\sec \theta = \sqrt{3}$ and $\csc \theta = \frac{\sqrt{6}}{2}$, find: **a.** the quadrant in which the angle whose measure is θ lies   **b.** $\cos \theta$ **c.** $\sin \theta$ **d.** $\tan \theta$ **e.** $\cot \theta$

In 48–55, select the numeral preceding the expression that best completes the sentence or answers the question.

48. If $\cos A > 0$, then which must always be true?
(1) $\sin A > 0$ (2) $\tan A > 0$ (3) $\sec A > 0$ (4) $\csc A > 0$

49. If $\csc B < 0$, then which must always be true?
(1) $\sin B < 0$ (2) $\cos B < 0$ (3) $\tan B < 0$ (4) $\cot B < 0$

50. In which quadrant are cotangent and cosecant both negative?
(1) I   (2) II   (3) III   (4) IV

51. If $\cot x > 0$ and $\sec x < 0$, which must be true?
(1) $\tan x < 0$ (2) $\sin x > 0$ (3) $\cos x > 0$ (4) $\sin x < 0$

52. If $\sin y \sec y > 0$ and $\sin y < 0$, which is true?
(1) $\cos y < 0$ (2) $\tan y < 0$ (3) $\cot y < 0$ (4) $\sec y > 0$

53. If $\sin A \cot A > 0$ and $\sin A < 0$, which must be true?
(1) $\cos A > 0$ (2) $\tan A > 0$ (3) $\sec A < 0$ (4) $\csc A > 0$

54. The value of $\sec \frac{\pi}{6} \div \cot \frac{\pi}{6}$ is:

(1) $\frac{1}{3}$ (2) $\frac{2}{3}$ (3) $\sqrt{3}$ (4) $\frac{\sqrt{3}}{3}$

55. The value of $\csc \frac{\pi}{3} + \sec \frac{\pi}{3}$ is:

(1) $\frac{2\sqrt{3} + 2}{3}$ (2) $\frac{2\sqrt{3} + 6}{3}$ (3) $\frac{2 + \sqrt{6}}{3}$ (4) $\frac{4\sqrt{3}}{3}$

8-13 THE PYTHAGOREAN IDENTITIES

If circle O is a unit circle whose center is at the origin, the equation of the circle is:

$$x^2 + y^2 = 1$$

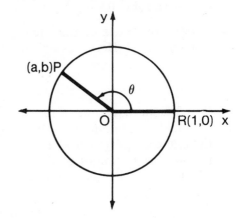

In the diagram, P is any point on circle O. If the coordinates of point P are (a, b), then:

$$a^2 + b^2 = 1$$

Since $m\angle ROP = \theta$, we know that $a = \cos\ \theta$ and $b = \sin\ \theta$. Therefore, the equation becomes:

$$(\cos\theta)^2 + (\sin\theta)^2 = 1 \quad \text{OR} \quad \cos^2\theta + \sin^2\theta = 1$$

Recall that the square of the sine of an angle $(\sin\theta)^2$ is written without parentheses as $\sin^2\ \theta$ to emphasize the fact that it is the function value, not the measure of the angle, which is being squared. Also $(\cos\theta)^2 = \cos^2\ \theta$, $(\tan\theta)^2 = \tan^2\ \theta$, $(\sec\theta)^2 = \sec^2\ \theta$, etc.

■ **DEFINITION.** An *identity* is an equation that is true for all values of the variable.

To demonstrate that $\cos^2\ \theta + \sin^2\ \theta = 1$ is an identity, we may select any values for θ and show that the equation is true. For example:

If $\theta = 60°$, $\cos\theta = \dfrac{1}{2}$ and $\sin\theta = \dfrac{\sqrt{3}}{2}$.

$$\text{Then, } \cos^2\theta + \sin^2\theta = \left(\frac{1}{2}\right)^2 + \left(\frac{\sqrt{3}}{2}\right)^2 = \frac{1}{4} + \frac{3}{4} = 1.$$

If $\theta = 135°$, $\cos\theta = -\dfrac{\sqrt{2}}{2}$ and $\sin\theta = \dfrac{\sqrt{2}}{2}$.

$$\text{Then, } \cos^2\theta + \sin^2\theta = \left(-\frac{\sqrt{2}}{2}\right)^2 + \left(\frac{\sqrt{2}}{2}\right)^2 = \frac{2}{4} + \frac{2}{4} = 1.$$

If $\theta = \pi$, $\cos\theta = -1$ and $\sin\theta = 0$.

$$\text{Then, } \cos^2\ \theta + \sin^2\ \theta = (-1)^2 + (0)^2 = 1 + 0 = 1.$$

From this basic identity, $\cos^2\ \theta + \sin^2\ \theta = 1$, we can derive other identities.

Derivation 1

1. Write the identity involving $\cos^2 \theta$ and $\sin^2 \theta$.

$$\cos^2 \theta + \sin^2 \theta = 1$$

2. Divide both sides of the equation by $\cos^2 \theta$. Here, $\cos \theta \neq 0$.

$$\frac{\cos^2 \theta + \sin^2 \theta}{\cos^2 \theta} = \frac{1}{\cos^2 \theta}$$

$$\frac{\cos^2 \theta}{\cos^2 \theta} + \frac{\sin^2 \theta}{\cos^2 \theta} = \frac{1}{\cos^2 \theta}$$

3. Rewrite the equation in terms of other trigonometric functions, and simplify. The identity is true for all values of θ for which $\tan \theta$ and $\sec \theta$ are defined.

$$\left(\frac{\cos \theta}{\cos \theta}\right)^2 + \left(\frac{\sin \theta}{\cos \theta}\right)^2 = \left(\frac{1}{\cos \theta}\right)^2$$

$$(1)^2 + (\tan \theta)^2 = (\sec \theta)^2$$

$$1 + \tan^2 \theta = \sec^2 \theta$$

Derivation 2

1. Write the identity involving $\cos^2 \theta$ and $\sin^2 \theta$.

$$\cos^2 \theta + \sin^2 \theta = 1$$

2. Divide both sides of the equation by $\sin^2 \theta$. Here, $\sin \theta \neq 0$.

$$\frac{\cos^2 \theta + \sin^2 \theta}{\sin^2 \theta} = \frac{1}{\sin^2 \theta}$$

$$\frac{\cos^2 \theta}{\sin^2 \theta} + \frac{\sin^2 \theta}{\sin^2 \theta} = \frac{1}{\sin^2 \theta}$$

3. Rewrite the equation in terms of other trigonometric functions, and simplify. The identity is true for all values of θ for which $\cot \theta$ and $\csc \theta$ are defined.

$$\left(\frac{\cos \theta}{\sin \theta}\right)^2 + \left(\frac{\sin \theta}{\sin \theta}\right)^2 = \left(\frac{1}{\sin \theta}\right)^2$$

$$(\cot \theta)^2 + (1)^2 = (\csc \theta)^2$$

$$\cot^2 \theta + 1 = \csc^2 \theta$$

If we look at these function values as lengths of line segments, we can see how these identities can be found by using the Pythagorean Theorem.

The diagram shows the unit circle with its center at the origin, where $\overleftrightarrow{RT}$ is tangent to the circle at R, $\overleftrightarrow{CS}$ is tangent to the circle at C, and $\overleftrightarrow{PQ} \perp x$-axis at Q. If $m\angle ROP = \theta$, then:

$PQ = \sin \theta \quad\quad OQ = \cos \theta \quad\quad RT = \tan \theta$

$OS = \csc \theta \quad\quad OT = \sec \theta \quad\quad CS = \cot \theta$

In right $\triangle OPQ$, $(OQ)^2 + (PQ)^2 = (OP)^2$
$$\cos^2 \theta + \sin^2 \theta = 1$$

In right $\triangle OTR$, $(OR)^2 + (RT)^2 = (OT)^2$
$$1 + \tan^2 \theta = \sec^2 \theta$$

In right $\triangle OSC$, $(CS)^2 + (OC)^2 = (OS)^2$
$$\cot^2 \theta + 1 = \csc^2 \theta$$

Using the diagrams for the angles in other quadrants given on page 388 of this chapter, we can verify that these identities are true for an angle in any quadrant.

Summary

In this chapter, we have defined eight basic trigonometric identities.

Reciprocal Identities	Quotient Identities	Pythagorean Identities
$\csc \theta = \dfrac{1}{\sin \theta}$	$\tan \theta = \dfrac{\sin \theta}{\cos \theta}$	$\cos^2 \theta + \sin^2 \theta = 1$
$\sec \theta = \dfrac{1}{\cos \theta}$	$\cot \theta = \dfrac{\cos \theta}{\sin \theta}$	$1 + \tan^2 \theta = \sec^2 \theta$
$\cot \theta = \dfrac{1}{\tan \theta}$		$\cot^2 \theta + 1 = \csc^2 \theta$

MODEL PROBLEMS

1. If $\sec A = -3$ and $\angle A$ is in quadrant II, find $\tan A$.

Solution

1. Use the Pythagorean identity: $1 + \tan^2 A = \sec^2 A$

2. Substitute the given value. $1 + \tan^2 A = (-3)^2$

3. Solve for $\tan A$.

$$1 + \tan^2 A = 9$$
$$\tan^2 A = 8$$
$$\tan A = \pm\sqrt{8} = \pm 2\sqrt{2}$$

4. In quadrant II, the tangent is negative. $\tan A = -\sqrt{8}$, or $-2\sqrt{2}$

Answer: $-\sqrt{8}$, or $-2\sqrt{2}$

2. Write the expression $1 + \cot^2 \theta$ in terms of $\sin \theta$, $\cos \theta$, or both. Then, express the result in simplest form.

How to Proceed	*Solution*
1. Write an equivalent expression for $\cot^2 \theta$.	$1 + \cot^2 \theta = 1 + \dfrac{\cos^2 \theta}{\sin^2 \theta}$
2. To add fractions, obtain like denominators. Then, add the numerators and maintain the common denominator.	$= 1\left(\dfrac{\sin^2 \theta}{\sin^2 \theta}\right) + \dfrac{\cos^2 \theta}{\sin^2 \theta}$ $= \dfrac{\sin^2 \theta}{\sin^2 \theta} + \dfrac{\cos^2 \theta}{\sin^2 \theta}$ $= \dfrac{\sin^2 \theta + \cos^2 \theta}{\sin^2 \theta}$
3. Since $\sin^2 \theta + \cos^2 \theta = 1$, the numerator can be replaced by 1.	$= \dfrac{1}{\sin^2 \theta}$ *Ans.*

EXERCISES

In 1–12, name the trigonometric function of the angle A that, when written in the blank, will make the equation an identity.

1. $\sin^2 A + (\underline{\qquad})^2 = 1$
2. $1 + (\underline{\qquad})^2 = \sec^2 A$
3. $(\underline{\qquad})^2 + 1 = \csc^2 A$
4. $(\underline{\qquad})^2 = 1 - \cos^2 A$
5. $\sin A \cot A = \underline{\qquad}$
6. $\sec A \cos^2 A = \underline{\qquad}$
7. $\pm\sqrt{1 + \tan^2 A} = \underline{\qquad}$
8. $\underline{\qquad} = \pm\sqrt{\cot^2 A + 1}$
9. $\csc A \tan A = \underline{\qquad}$
10. $\sin A \sec A = \underline{\qquad}$
11. $\cos A \div \cot A = \underline{\qquad}$
12. $\cot A \sec A = \underline{\qquad}$

In 13–20, use a Pythagorean identity to find the required function value.

13. If $\sin A = .6$ and $\angle A$ is in quadrant II, find $\cos A$.
14. If $\tan B = -\frac{3}{4}$ and $\angle B$ is in quadrant IV, find $\sec B$.
15. If $\csc C = \frac{13}{5}$ and $\angle C$ is in quadrant I, find $\cot C$.
16. If $\cos A = -\frac{1}{3}$ and $\angle A$ is in quadrant III, find $\sin A$.
17. If $\sec B = -\sqrt{5}$ and $\angle B$ is in quadrant II, find $\tan B$.
18. If $\cot C = -\sqrt{15}$ and $\angle C$ is in quadrant IV, find $\csc C$.
19. If $\tan A = 3$ and $\angle A$ is in quadrant III, find $\sec A$.
20. If $\cos B = \dfrac{\sqrt{5}}{3}$ and $\angle B$ is in quadrant I, find $\sin B$.

21. a. If $\theta = \dfrac{\pi}{6}$ radians, find the values of $\sin \theta$ and $\cos \theta$.

    b. Demonstrate that $\sin^2 \theta + \cos^2 \theta = 1$ when $\theta = \dfrac{\pi}{6}$ radians.

22. a. If $\theta = 225°$, find the values of $\tan \theta$ and $\sec \theta$.
    b. Demonstrate that $\tan^2 \theta + 1 = \sec^2 \theta$ when $\theta = 225°$.
23. a. If $\theta = 300°$, find the values of $\cot \theta$ and $\csc \theta$.
    b. Demonstrate that $\cot^2 \theta + 1 = \csc^2 \theta$ when $\theta = 300°$.
24. What is the value of $\sin^2 63° + \cos^2 63°$?

In 25–36, write the given expression in terms of $\sin A$, $\cos A$, or both. Then express the result in simplest form.

25. $\sec A \cot A$ 26. $\csc A \tan A$ 27. $\sec A \cos^2 A$
28. $\cot^2 A \tan A$ 29. $1 + \tan^2 A$ 30. $\tan A + \cot A$
31. $\sec^2 A + \csc^2 A$ 32. $\sec^2 A - 1$ 33. $\csc^2 A - 1$

34. $\dfrac{\tan A}{\sec A}$ 35. $\dfrac{\cot A}{\csc A}$ 36. $\dfrac{\sec A \cos A}{\tan A \cot A}$

In 37–40, select the numeral preceding the expression that best completes the statement.

37. The expression $\dfrac{1}{\sec \theta}(\tan \theta + \sec \theta)$ equals:

    (1) $\sin \theta$ (2) $\sin \theta + 1$ (3) $\cos \theta$ (4) $\cos \theta + 1$
38. The product $(1 + \csc \theta)(1 - \csc \theta)$ equals:
    (1) $\tan^2 \theta$ (2) $-\tan^2 \theta$ (3) $\cot^2 \theta$ (4) $-\cot^2 \theta$

39. The expression $\dfrac{\sec \theta - \csc \theta}{\sec \theta}$ is equal to:

    (1) $\dfrac{\sin \theta - \cos \theta}{\sin \theta}$ (2) $\dfrac{\sin \theta - \cos \theta}{\cos \theta}$

    (3) $\dfrac{\cos \theta - \sin \theta}{\sin \theta}$ (4) $\dfrac{\cos \theta - \sin \theta}{\cos \theta}$

40. The product $(1 - \sec B)(1 + \cos B)$ equals:

    (1) 0 (2) 2 (3) $\dfrac{\cos^2 B - 1}{\cos B}$ (4) $\dfrac{\cos B - 1}{\cos B}$

8-14 FINDING THE REMAINING TRIGONOMETRIC FUNCTION VALUES OF ANY ANGLE WHEN ONE FUNCTION VALUE IS KNOWN

If we know one trigonometric function value and the quadrant in which the angle lies, it is possible to find the remaining five trigonometric function values of the angle.

Method 1: Using Identities

☐ EXAMPLE: If $\sin \theta = \frac{5}{13}$ and θ is the measure of an angle in the second quadrant, find the values of: a. $\cos \theta$ b. $\tan \theta$ c. $\csc \theta$ d. $\sec \theta$ e. $\cot \theta$

Solution

a. 1. Use the Pythagorean identity: $\qquad\qquad \cos^2 \theta + \sin^2 \theta = 1$

2. Substitute the given value. $\qquad\qquad \cos^2 \theta + (\frac{5}{13})^2 = 1$

3. Solve for $\cos \theta$. $\qquad\qquad\qquad\quad \cos^2 \theta + \frac{25}{169} = 1$

4. Since an angle in the second quadrant has a negative cosine, then $\cos \theta$ is the negative square root of $\frac{144}{169}$.

$$\cos^2 \theta + \frac{25}{169} = \frac{169}{169}$$
$$\cos^2 \theta = \frac{144}{169}$$
$$\cos \theta = -\frac{12}{13}$$

b. 1. Write the quotient identity (shown below) for $\tan \theta$.
2. Substitute the known values, and simplify.

$$\tan \theta = \frac{\sin \theta}{\cos \theta} = \frac{\frac{5}{13}}{-\frac{12}{13}} = \frac{\frac{5}{13}}{-\frac{12}{13}} \cdot \frac{13}{13} = \frac{5}{-12} = -\frac{5}{12}$$

In c, d, and e: 1. Write a reciprocal identity (shown below).

2. Substitute the known value, and simplify.

c. $\csc \theta = \dfrac{1}{\sin \theta} = \dfrac{1}{\frac{5}{13}} = \dfrac{1}{\frac{5}{13}} \cdot \dfrac{13}{13} = \dfrac{13}{5}$

d. $\sec \theta = \dfrac{1}{\cos \theta} = \dfrac{1}{-\frac{12}{13}} = \dfrac{1}{-\frac{12}{13}} \cdot \dfrac{13}{13} = \dfrac{13}{-12} = -\dfrac{13}{12}$

e. $\cot \theta = \dfrac{1}{\tan \theta} = \dfrac{1}{-\dfrac{5}{12}} = \dfrac{1}{-\dfrac{5}{\cancel{12}}} \cdot \dfrac{\cancel{12}}{\cancel{12}} = \dfrac{12}{-5} = -\dfrac{12}{5}$

Answer: In the second quadrant:

When $\sin \theta = \dfrac{5}{13}$: a. $\cos \theta = -\dfrac{12}{13}$ b. $\tan \theta = -\dfrac{5}{12}$

c. $\csc \theta = \dfrac{13}{5}$ d. $\sec \theta = -\dfrac{13}{12}$ e. $\cot \theta = -\dfrac{12}{5}$

Method 2: Using Right Triangles and Directed Distances

In section 3 of this chapter, we saw that a dilation could be used to transform a given right triangle into a similar right triangle in the unit circle. The hypotenuse of the right triangle in the unit circle has a length of 1 because the radius of the circle has a length of 1.

We can also use a dilation to transform a right triangle in the unit circle into a similar right triangle whose hypotenuse does not have a length of 1. Just as the lengths of sides of a right triangle in the unit circle are treated as directed distances, so too are the lengths of sides in the newly formed similar triangle treated as *directed distances* in a coordinate plane.

For example, if θ is the measure of an angle in the fourth quadrant where $\cos \theta = \frac{3}{5}$ and $\sin \theta = -\frac{4}{5}$, we can construct right $\triangle OQP$ in the unit circle (see Fig. 1). Here, $OQ = \frac{3}{5}$, $QP = -\frac{4}{5}$, and hypotenuse $\overline{OP}$ has a length of 1.

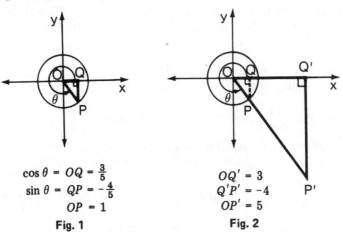

$\cos \theta = OQ = \frac{3}{5}$
$\sin \theta = QP = -\frac{4}{5}$
$OP = 1$
Fig. 1

$OQ' = 3$
$Q'P' = -4$
$OP' = 5$
Fig. 2

Under a dilation of 5 with the origin as the fixed point, right $\triangle OQP$ will have as its image right $\triangle OQ'P'$ (see Fig. 2). The length of each side of $\triangle OQ'P'$ will be *five times* the length of its corresponding side in $\triangle OQP$.

Thus, under a dilation of 5, or under D_5:

$$\overline{OQ} \to \overline{OQ'}, \text{ where } OQ = \tfrac{3}{5} \text{ and } OQ' = 5(\tfrac{3}{5}) = 3.$$

$$\overline{QP} \to \overline{Q'P'}, \text{ where } QP = -\tfrac{4}{5} \text{ and } Q'P' = 5(-\tfrac{4}{5}) = -4.$$

$$\overline{OP} \to \overline{OP'}, \text{ where } OP = 1 \text{ and } OP' = 5(1) = 5.$$

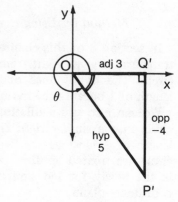

In right $\triangle OQ'P'$, the acute angle whose vertex is at the origin, that is, $\angle Q'OP'$, is directly related to the fourth-quadrant angle whose measure is θ (see Fig. 3). Using acute $\angle Q'OP'$, we can identify the sides of right $\triangle OQ'P'$ as follows:

$\overline{Q'P'}$ is *opposite* the acute $\angle Q'OP'$.

$\overline{OQ'}$ is *adjacent* to the acute $\angle Q'OP'$.

$\overline{OP'}$ is the *hypotenuse* of the triangle.

Recall that the sine, cosine, and tangent functions were originally defined as ratios involving the lengths of sides of a

Fig. 3

right triangle. We now extend these definitions to include directed distances for any right triangle, either in the unit circle or similar to that right triangle. We name the sides of the right triangle as they relate to the acute angle whose vertex is at the origin and which is determined by the angle of any given measure θ. Thus, using right $\triangle OQ'P'$, we now state:

$$\cos \theta = \frac{\text{adj}}{\text{hyp}} = \frac{3}{5} \qquad \sin \theta = \frac{\text{opp}}{\text{hyp}} = -\frac{4}{5} \qquad \tan \theta = \frac{\text{opp}}{\text{adj}} = -\frac{4}{3}$$

Since $\sec \theta = \dfrac{1}{\cos \theta}$, $\csc \theta = \dfrac{1}{\sin \theta}$, and $\cot \theta = \dfrac{1}{\tan \theta}$, these reciprocal identities allow us to redefine these functions as ratios involving the lengths of sides of a right triangle. Thus, for right $\triangle OQ'P'$:

$$\sec \theta = \frac{\text{hyp}}{\text{adj}} = \frac{5}{3} \qquad \csc \theta = \frac{\text{hyp}}{\text{opp}} = -\frac{5}{4} \qquad \cot \theta = \frac{\text{adj}}{\text{opp}} = -\frac{3}{4}$$

Let us now apply this method to solving the problem stated earlier in this section.

□ EXAMPLE. If $\sin \theta = \frac{5}{13}$ and θ is the measure of an angle in the second quadrant, find the values of: **a.** $\cos \theta$ **b.** $\tan \theta$ **c.** $\csc \theta$ **d.** $\sec \theta$ **e.** $\cot \theta$

Solution

1. Draw right $\triangle OQ'P'$ in quadrant II. Since $\sin \theta = \dfrac{5}{13} = \dfrac{\text{opp}}{\text{hyp}}$, let $Q'P' = 5$ and $OP' = 13$. Also, let $OQ' = x$. (Note that right $\triangle OQ'P'$ is the image under a dilation of 13 of $\triangle OQP$, a right triangle in quadrant II of the unit circle where $\sin \theta = \frac{5}{13}$.)

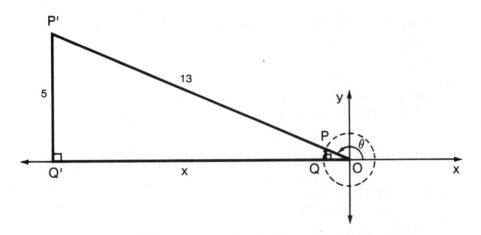

2. Solve for x by using the Pythagorean Theorem:
$$x^2 + 5^2 = 13^2$$
$$x^2 + 25 = 169$$
$$x^2 = 144$$
$$x = -12$$

   In quadrant II, x is a negative value.

3. To find the remaining trigonometric functions, use the definitions involving ratios of lengths of sides in a right triangle. Remember to use directed distances.

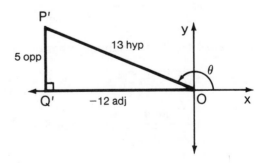

Answer: **a.** $\cos \theta = \dfrac{\text{adj}}{\text{hyp}} = \dfrac{-12}{13}$ **b.** $\tan \theta = \dfrac{\text{opp}}{\text{adj}} = \dfrac{5}{-12} = -\dfrac{5}{12}$

   c. $\csc \theta = \dfrac{\text{hyp}}{\text{opp}} = \dfrac{13}{5}$ **d.** $\sec \theta = \dfrac{\text{hyp}}{\text{adj}} = \dfrac{13}{-12} = -\dfrac{13}{12}$

   e. $\cot \theta = \dfrac{\text{adj}}{\text{opp}} = \dfrac{-12}{5}$

MODEL PROBLEMS

1. Express each of the five remaining trigonometric functions in terms of $\sin \theta$.

Solution

To express one function in terms of another, we make use of the *identities* for trigonometric functions.

1. Use the Pythagorean identity that includes $\sin \theta$, and solve for $\cos \theta$. Since no quadrant is specified, $\cos \theta$ may be positive or negative.

$$\cos^2 \theta + \sin^2 \theta = 1$$
$$\cos^2 \theta = 1 - \sin^2 \theta$$
$$\cos \theta = \pm\sqrt{1 - \sin^2 \theta}$$

2. Use reciprocal identities to find $\csc \theta$ and $\sec \theta$.

$$\csc \theta = \frac{1}{\sin \theta} \qquad \sec \theta = \frac{1}{\cos \theta} = \frac{1}{\pm\sqrt{1 - \sin^2 \theta}}$$

3. Use quotient identities to find $\tan \theta$ and $\cot \theta$.

$$\tan \theta = \frac{\sin \theta}{\cos \theta} = \frac{\sin \theta}{\pm\sqrt{1 - \sin^2 \theta}} \qquad \cot \theta = \frac{\cos \theta}{\sin \theta} = \frac{\pm\sqrt{1 - \sin^2 \theta}}{\sin \theta}$$

Answer: The remaining functions expressed in terms of $\sin \theta$ are:

$$\cos \theta = \pm\sqrt{1 - \sin^2 \theta} \qquad \csc \theta = \frac{1}{\sin \theta} \qquad \sec \theta = \pm\frac{1}{\sqrt{1 - \sin^2 \theta}}$$

$$\tan \theta = \pm\frac{\sin \theta}{\sqrt{1 - \sin^2 \theta}} \qquad \cot \theta = \pm\frac{\sqrt{1 - \sin^2 \theta}}{\sin \theta}$$

2. If $\tan A = \dfrac{\sqrt{7}}{3}$ and $\sin A < 0$, find $\cos A$.

Solution

1. Since $\tan A$ is positive and $\sin A$ is negative, the terminal side of $\angle A$ must lie in quadrant III.

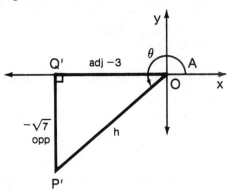

2. Draw right $\triangle OQ'P'$ in quadrant III. Here, x and y are both negative values. Since $\tan A = \dfrac{\sqrt{7}}{3}$ and x and y are negative, let

    $\tan A = \dfrac{-\sqrt{7}}{-3} = \dfrac{\text{opp}}{\text{adj}}$.

    Thus, $Q'P' = -\sqrt{7}$ (opposite acute $\angle Q'OP'$).
    $OQ' = -3$ (adjacent to acute $\angle Q'OP'$).

3. Let h = the length of hypotenuse $\overline{OP'}$. Use the Pythagorean Theorem to find the value of h.

    $$h^2 = (-\sqrt{7})^2 + (-3)^2$$
    $$h^2 = 7 + 9$$
    $$h^2 = 16$$
    $$h = 4$$

4. Therefore, $\cos A = \dfrac{\text{adj}}{\text{hyp}} = \dfrac{-3}{4}$ *Ans.*

Alternate Solution

1. Use the Pythagorean iden- $\tan^2 A + 1 = \sec^2 A$
    tity that includes $\tan A$.

2. Substitute the value of $\tan A$. $\left(\dfrac{\sqrt{7}}{3}\right)^2 + 1 = \sec^2 A$

3. Simplify the left member of the equation. $\dfrac{7}{9} + \dfrac{9}{9} = \sec^2 A$

4. Since $\tan A$ is positive and $\sin A$ is negative, $\angle A$ lies in quadrant III. Thus, $\sec A$ is negative. Solve for the negative value of $\sec A$.

   $\dfrac{16}{9} = \sec^2 A$

   $-\dfrac{4}{3} = \sec A$

5. Use the reciprocal identity to find $\cos A$. $\cos A = \dfrac{1}{\sec A} = \dfrac{1}{-\frac{4}{3}} = -\dfrac{3}{4}$

Answer: $\cos A = -\dfrac{3}{4}$

EXERCISES

In 1–4, a right triangle is drawn in one of the quadrants relating to an angle whose measure is θ. Using the lengths of the sides of the triangle, indicated as directed distances, find: **a.** $\sin \theta$ **b.** $\cos \theta$ **c.** $\tan \theta$ **d.** $\csc \theta$ **e.** $\sec \theta$ **f.** $\cot \theta$

1.

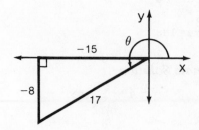

2.

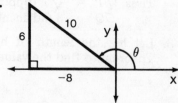

3.

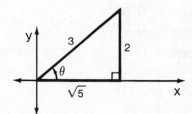

4.

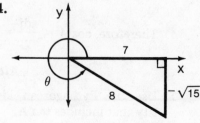

5. If $\cos \theta = \frac{3}{5}$ and $\sin \theta < 0$, find: a. $\sin \theta$ b. $\sec \theta$ c. $\csc \theta$ d. $\tan \theta$ e. $\cot \theta$

6. If $\tan A = -\frac{3}{4}$ and $\cos A < 0$, find: a. $\sec A$ b. $\cos A$ c. $\sin A$ d. $\csc A$ e. $\cot A$

7. If $\tan \theta = \dfrac{\sqrt{11}}{5}$ and $\sec \theta > 0$, find: a. $\sec \theta$ b. $\cos \theta$ c. $\sin \theta$ d. $\csc \theta$ e. $\cot \theta$

8. If $\sin B = -\frac{5}{13}$ and $\tan B > 0$, find: a. $\cos B$ b. $\tan B$ c. $\cot B$ d. $\csc B$ e. $\sec B$

9. If $\cot \theta = \frac{3}{2}$ and $\csc \theta < 0$, find: a. $\csc \theta$ b. $\tan \theta$ c. $\sin \theta$ d. $\cos \theta$ e. $\sec \theta$

10. If $\csc A = -\dfrac{\sqrt{10}}{3}$ and $\cot A < 0$, find: a. $\cot A$ b. $\sin A$ c. $\tan A$ d. $\sec A$ e. $\cos A$

In 11–15, express each of the five remaining trigonometric functions in terms of the given function.

11. $\cos \theta$ 12. $\tan \theta$ 13. $\cot \theta$ 14. $\sec \theta$ 15. $\csc \theta$

16. If $\sin A = \dfrac{\sqrt{3}}{2}$ and $\tan A$ is negative, find $\cos A$.

17. If $\cos A = -\dfrac{\sqrt{2}}{2}$ and $\cot A > 0$, find $\sin A$.

18. If $\tan B = -\frac{6}{8}$ and $\sin B < 0$, find $\cos B$.

19. If $\sin A = \dfrac{\sqrt{24}}{7}$ and $\sec A > 0$, find $\tan A$.

In 20 and 21, select the numeral preceding the expression that best completes the sentence.

20. If $\tan \theta = \frac{2}{5}$ and $\sin \theta < 0$, then $\cos \theta$ is equal to:

   (1) $\dfrac{5}{\sqrt{29}}$ (2) $-\dfrac{\sqrt{29}}{5}$ (3) $-\dfrac{5}{\sqrt{29}}$ (4) $-\dfrac{5}{29}$

21. If $\tan A = -\frac{1}{3}$ and $\cos A < 0$, then $\sin A$ equals:

   (1) $\dfrac{\sqrt{10}}{10}$ (2) $-\dfrac{3}{\sqrt{10}}$ (3) $-\dfrac{\sqrt{10}}{10}$ (4) $\dfrac{3\sqrt{10}}{10}$

8-15 COFUNCTIONS

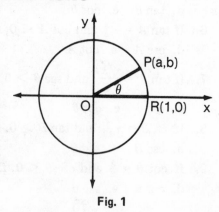

Fig. 1

In Fig. 1, $\angle ROP$ is a central angle of a unit circle whose center is at the origin. Points $R(1, 0)$ and $P(a, b)$ are on the circle. If the measure of acute angle ROP is θ, then $\cos \theta = a$ and $\sin \theta = b$.

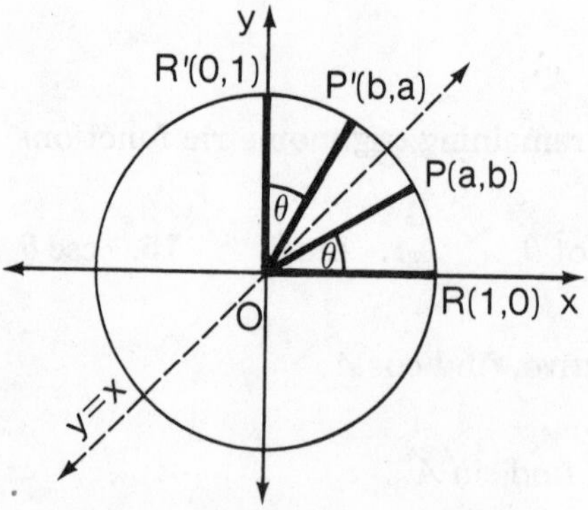

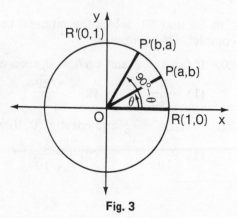

Fig. 2

In Fig. 2, $\angle ROP$ is reflected in the line $y = x$:

$$r_{y=x}(x, y) = (y, x)$$
$$P(a, b) \to P'(b, a)$$
$$R(1, 0) \to R'(0, 1)$$
$$O(0, 0) \to O(0, 0)$$

Since a line reflection preserves distance, $OP' = OP = 1$ and P' is a point on the unit circle. Since a line reflection preserves angle measure, $m\angle R'OP' = m\angle ROP = \theta$. Thus:

$$m\angle ROP' = m\angle ROR' - m\angle ROP$$
$$= 90° - \theta$$

In Fig. 3, we see that $\angle ROP'$ is an angle in standard position, and the coordinates of P' are (b, a). Therefore, $\cos (90° - \theta) = b$ and $\sin (90° - \theta) = a$.

Fig. 3

Since sin θ = b and cos $(90° - \theta) = b$: sin θ = cos $(90° - \theta)$
Since cos θ = a and sin $(90° - \theta) = a$: cos θ = sin $(90° - \theta)$

Thus, the sine of an acute angle is equal to the cosine of its complement, and the cosine of an acute angle is equal to the sine of its complement. The sine and cosine functions are called *cofunctions*.

Other cofunctions exist in trigonometry. For example:

Since tan θ = $\dfrac{b}{a}$ and cot $(90° - \theta) = \dfrac{b}{a}$: tan θ = cot $(90° - \theta)$

Since cot θ = $\dfrac{a}{b}$ and tan $(90° - \theta) = \dfrac{a}{b}$: cot θ = tan $(90° - \theta)$

Thus, the tangent of an acute angle is equal to the cotangent of its complement, and the cotangent of an acute angle is equal to the tangent of its complement. The tangent and cotangent functions are cofunctions.

Since sec θ = $\dfrac{1}{a}$ and csc $(90° - \theta) = \dfrac{1}{a}$: sec θ = csc $(90° - \theta)$

Since csc θ = $\dfrac{1}{b}$ and sec $(90° - \theta) = \dfrac{1}{b}$: csc θ = sec $(90° - \theta)$

Thus, the secant of an acute angle is equal to the cosecant of its complement, and the cosecant of an acute angle is equal to the secant of its complement.

These observations allow us to make the following general statement:

■ Any trigonometric function of an acute angle is equal to the cofunction of its complement.

Notice how the prefix *co-* allows us to identify easily the pairs of functions that are cofunctions:
Sine and *co*sine are *co*functions.
Tangent and *co*tangent are *co*functions.
Secant and *co*secant are *co*functions.
Note that the prefix *co-* matches the first two letters of the word "complement," a concept that is basic in the definition of cofunctions.
The cofunction relationship is used in the construction of the table of trigonometric values. Each angle whose measure is listed in the column at the left is the complement of the angle whose measure is listed on the same line in the column at the right. Also, each column that names a function at the top names the cofunction at the bottom.

| MODEL PROBLEMS |

1. If x and $(x + 20°)$ are the measures of two acute angles and $\sin x = \cos (x + 20°)$, find x.

Solution

If the function of one acute angle is equal to the cofunction of another acute angle, then the angles are complementary. Thus:

$$x + (x + 20) = 90$$
$$2x + 20 = 90$$
$$2x = 70$$
$$x = 35$$

Check by substitution: Does $\sin 35° = \cos (35° + 20°)$ or $\sin 35° = \cos 55°$? (True.)

Answer: 35°

2. Express $\sin 285°$ as the function of an angle whose measure is less than 45°.

Solution

1. Determine the quadrant: An angle of 285° is in the fourth quadrant.

2. Determine the reference angle: Here, $360° - \theta = 360° - 285° = 75°$.

3. Determine the sign: The sine of a fourth-quadrant angle is negative.

4. Use cofunctions: $\sin 285° = -\sin 75° = -\cos 15°$.

Answer: $-\cos 15°$

| EXERCISES |

In 1–12, write the expression as a function of an acute angle whose measure is less than 45°.

1. $\sin 80°$	2. $\tan 72°$	3. $\sin 50°$
4. $\cos 67°$	5. $\sec 83°$	6. $\cot 65°$
7. $\csc 58°$	8. $\cos 75°$	9. $\sin 88°$
10. $\tan 56°30'$	11. $\cot 87°20'$	12. $\cos 63°50'$

In 13-23, the equation contains the measures of two acute angles. Find a value of θ for which the statement is true.

13. $\sin 10° = \cos \theta$ 14. $\tan 48° = \cot \theta$ 15. $\sec 70° = \csc \theta$
16. $\sin \theta = \cos \theta$ 17. $\sin \theta = \cos 2\theta$ 18. $\tan \theta = \cot 5\theta$
19. $\sec \theta = \csc (\theta + 60°)$ 20. $\sin 2\theta = \cos (\theta + 15°)$
21. $\cos \theta = \sin (2\theta + 15°)$ 22. $\tan (\theta + 5°) = \cot (2\theta - 20°)$
23. $\sec (\theta + 8°) = \csc (90° - 2\theta)$

In 24-38, write the expression as the function of an acute angle whose measure is less than $45°$.

24. $\sin 280°$ 25. $\cos 110°$ 26. $\tan 265°$.
27. $\sec 125°$ 28. $\cot 95°$ 29. $\tan 310°$
30. $\cos 258°$ 31. $\sin 420°$ 32. $\sec 490°$
33. $\cos 635°$ 34. $\tan 600°$ 35. $\cos (-50°)$
36. $\sin (-80°)$ 37. $\sin (-100°)$ 38. $\cot (-277°)$

In 39-43, select the numeral preceding the expression that best completes the sentence.

39. If x and y are the measures of two acute angles and $\tan x = \cot y$, then:
    (1) $x = y + 90°$ (2) $x = y - 90°$
    (3) $x = 90° - y$ (4) $y = x - 90°$
40. If θ is the measure of an acute angle and $\cos \theta = \sin 60°$, then $\cos \theta$ equals:

    (1) $30°$ (2) $60°$ (3) $\dfrac{\sqrt{3}}{2}$ (4) $\dfrac{1}{2}$

41. If x is the measure of an acute angle and $\sin (x + 15°) = \cos 45°$, then $\sin x$ equals:

    (1) $\dfrac{1}{2}$ (2) $\dfrac{\sqrt{2}}{2}$ (3) $\dfrac{\sqrt{3}}{2}$ (4) $30°$

42. If r and t are the measures of two acute angles so that $r + t = 90°$, then $\cos r$ equals:
    (1) $\sin (90° - t)$ (2) $\sin (t + 90°)$ (3) $\sin (t - 90°)$ (4) $\sin t$
43. If b is the measure of an acute angle and $\cos b = .75$, then:
    (1) $\sin (b - 90°) = .75$ (2) $\sin (90° - b) = .75$
    (3) $\sin b = .75$ (4) $\sin b = .25$

44. If k is the measure of an acute angle and $\cos k = \sin (2k + 30°)$, demonstrate that k may equal $20°$ or $60°$.

8-16 REVIEW EXERCISES

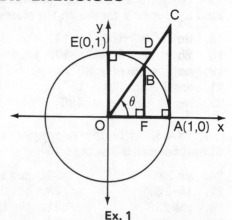

Ex. 1

1. In the diagram, $\overleftrightarrow{CA}$ is tangent to circle O at $A(1, 0)$ and $\overleftrightarrow{ED}$ is tangent to circle O at $E(0, 1)$. Point B is on circle O, points D and C are on $\overleftrightarrow{OB}$, and $\overline{BF} \perp \overleftrightarrow{OA}$. If $m\angle AOB = \theta$, name the line segment whose measure is each of the following:
   a. $\sin \theta$ b. $\cos \theta$
   c. $\tan \theta$ d. $\sec \theta$
   e. $\csc \theta$ f. $\cot \theta$

2. Express each of the following degree measures in radian measure:
   a. $90°$ b. $120°$ c. $30°$ d. $100°$ e. $135°$
   f. $-180°$ g. $240°$ h. $315°$ i. $450°$ j. $-200°$

3. Express each of the following radian measures in degree measure:
   a. $\dfrac{\pi}{2}$ b. $\dfrac{2\pi}{5}$ c. $\dfrac{5\pi}{6}$ d. $\dfrac{3\pi}{2}$ e. $\dfrac{11\pi}{12}$
   f. $\dfrac{7\pi}{3}$ g. $-\dfrac{3\pi}{4}$ h. $-\dfrac{9\pi}{10}$ i. $\dfrac{17\pi}{6}$ j. $\dfrac{8\pi}{15}$

4. In a circle whose radius has length r, a central angle whose radian measure is θ intercepts an arc of length s.
   a. If $s = 10$ and $r = 2$, find θ.   b. If $s = 5\pi$ and $r = 10$, find θ.
   c. If $\theta = 3$ and $s = 9$, find r.   d. If $\theta = \pi$ and $s = 2\pi$, find r.
   e. If $\theta = \frac{1}{2}$ and $r = 3$, find s.   f. If $\theta = \dfrac{\pi}{3}$ and $r = 9$, find s.

5. Find the exact value of each expression.
   a. $\sin 120°$ b. $\cos 300°$ c. $\tan 405°$ d. $\cos 135°$
   e. $\tan (-120°)$ f. $\sin \dfrac{5\pi}{3}$ g. $\sec (-\pi)$ h. $\cot \dfrac{3\pi}{2}$

6. Express each of the following as a function of an acute angle:
   a. $\cos 190°$ b. $\tan 305°$ c. $\sin 138°$ d. $\sec 92°$
   e. $\csc 350°$ f. $\sin (-165°)$ g. $\cos (-284°)$ h. $\tan (-142°)$

7. Find each function value to four decimal places.
   a. $\sin 43°20'$ b. $\cos 77°50'$ c. $\sin 61°18'$
   d. $\tan 39°46'$ e. $\cos 12°8'$ f. $\tan 82°33'$

8. Find the measure of θ to the nearest minute.
   a. $\sin \theta = .9390$ b. $\tan \theta = .2732$ c. $\cos \theta = .4472$
   d. $\tan \theta = 1.1868$ e. $\cos \theta = .1113$ f. $\sin \theta = .5555$

9. If $\tan \theta = -\frac{3}{4}$ and $\sin \theta > 0$, find:
   a. $\sec \theta$ b. $\cos \theta$ c. $\sin \theta$ d. $\csc \theta$ e. $\cot \theta$

10. If $\cos \theta = -\frac{2}{3}$ and $\sin \theta < 0$, find $\tan \theta$.

11. Express each of the five remaining trigonometric functions in terms of $\cos \theta$.

12. Write each of the given expressions in terms of $\sin \theta$, $\cos \theta$, or both. Express the result in simplest form.

    a. $\csc \theta \sin^2 \theta$ b. $\sec \theta \cot \theta$ c. $\cot^2 \theta + 1$

    d. $\dfrac{\sec \theta}{\tan \theta}$ e. $\dfrac{\tan \theta}{\cot \theta}$ f. $\sec^2 \theta + \csc^2 \theta$

13. If $f(x) = \sin x$, find $f\left(\dfrac{\pi}{3}\right)$.   14. If $g(x) = \cos 2x$, find $g\left(\dfrac{\pi}{4}\right)$.

In 15-20, select the numeral preceding the expression that best completes the sentence or answers the question.

15. If $\sin x \cos x < 0$, x must be the measure of an angle in quadrants:
    (1) I or III   (2) II or IV   (3) I or IV   (4) II or III

16. An angle whose measure is $\dfrac{\pi}{3}$ has the same terminal side as an angle whose measure is:
    (1) $\dfrac{2\pi}{3}$ (2) $\dfrac{5\pi}{3}$ (3) $-\dfrac{\pi}{3}$ (4) $-\dfrac{5\pi}{3}$

17. The degree measure of an angle of 1 radian is:
    (1) $57°$ (2) between $57°$ and $58°$ (3) $\dfrac{1}{\pi}°$ (4) $\dfrac{\pi}{180}°$

18. If $\sin x < 0$, which must also be true?
    (1) $\cos x < 0$ (2) $\tan x < 0$ (3) $\sec x < 0$ (4) $\csc x < 0$

19. The value of $\tan \dfrac{3\pi}{4}$ is equal to the value of:
    (1) $\sin \dfrac{\pi}{2}$ (2) $\cos \dfrac{\pi}{2}$ (3) $\sin \dfrac{3\pi}{2}$ (4) $\cos \dfrac{3\pi}{2}$

20. The expression $\dfrac{\csc \theta - \sin \theta}{\cot \theta}$ is equivalent to:
    (1) $\cos \theta$ (2) $\sin \theta$ (3) $1 - \sin^2 \theta$ (4) $\cos^2 \theta$

21. Given that each equation contains the measures of two acute angles, find a value of x for which the statement is true.
    a. $\sin x = \cos (2x + 45)$ b. $\sin (x + 20) = \cos x$
    c. $\cot (x + 10) = \tan 3x$ d. $\sec (x + 12) = \csc (x + 8)$

22. Write each expression as a function of an acute angle whose measure is less than $45°$.
    a. $\sin 125°$ b. $\cos 108°$ c. $\tan 297°$ d. $\sin (-105°)$

Chapter 9

Trigonometric Graphs

9-1 THE WRAPPING FUNCTION

Let us think of a vertical number line that is tangent to a unit circle so that the point representing 0 on the number line coincides with the point $(1, 0)$ on the unit circle. Imagine that we could wrap the number line about the circle so that each point on the line coincides with a point on the circle. The diagram shows some corresponding points as the positive ray of the number line makes a single rotation about the circle.

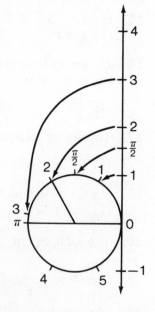

Each point on the circle determines the terminal ray of a central angle of the circle. The radian measure of an angle in standard position is equal to the coordinate of the corresponding point on the number line.

For example, a point whose coordinate is 2 on the number line corresponds to a point on the unit circle that determines an angle whose radian measure is 2. Similarly, a point whose coordinate is $\frac{\pi}{2}$ (approximately 1.5708) on the number line corresponds to a point on the unit circle that determines an angle whose radian measure is $\frac{\pi}{2}$, that is, a right angle.

We can continue to wrap the positive ray of the number line about the circle in a counterclockwise direction. In the same way, we can wrap the negative ray of the number line about the circle in a clockwise direction. Under this *wrapping function*, every point on the real-number line corresponds to one and only one point on the unit circle.

412

There is an infinite number of points on the real-number line, however, that correspond to the same point on the unit circle. For example, the points that represent the real numbers $0, 2\pi, -2\pi, 4\pi, -4\pi, 6\pi, -6\pi$, etc. all correspond to the point $(1, 0)$ on the circle, that is, the point that determines a central angle of 0 radian. This set of real numbers is indicated by the expression $2\pi k$, where k represents any integer.

In the same way, think of the point on the unit circle that determines the terminal ray of an angle in standard position of 2 radians. The real numbers $2, 2 + 2\pi, 2 + 4\pi, 2 - 2\pi$, and so forth have points on the number line that correspond to this point on the circle. Thus, the set of points whose real numbers are of the form $2 + 2\pi k$ for all integral values of k corresponds to a point on the unit circle that determines a central angle of 2 radians.

We will use this correspondence between points on the real-number line and the radian measures of angles of a unit circle to graph trigonometric functions.

| EXERCISES |

In 1 and 2, select the numeral preceding the expression that best answers the question.

1. Which number does *not* have a point on the real-number line corresponding to the point on the unit circle that determines an angle measure of 0 radian?
   (1) 6π (2) 2π (3) 3π (4) 4π
2. A point P on the unit circle determines an angle of 3 radians. Which real number does *not* have a point on the number line that corresponds to point P on the circle?
   (1) $3 + 2\pi$ (2) $3 - 2\pi$ (3) $3 + \pi$ (4) $3 + 10\pi$

3. Name four real numbers whose points on the number line correspond to the point on the unit circle whose coordinates are $(-1, 0)$.

In 4–8, name three real numbers whose points on the number line correspond to the point on the unit circle that determines an angle whose radian measure is given.

4. $\dfrac{\pi}{2}$ 5. 5   6. $\dfrac{\pi}{3}$ 7. 1.23   8. $\dfrac{7\pi}{6}$

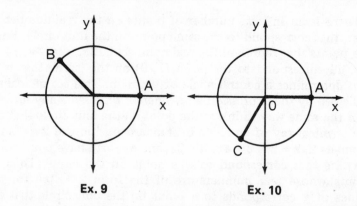

Ex. 9   Ex. 10

9. In a unit circle, $\angle AOB$ indicates an angle associated with $\frac{3}{8}$ of a rotation, A and B are points on the circle, and A lies on the positive x-axis. Name four real numbers whose points on the number line correspond to point B on the circle.

In 10, select the numeral preceding the expression that best answers the question.

10. In a unit circle, $\angle AOC$ indicates an angle associated with $\frac{2}{3}$ of a rotation, and point C is on the circle, as shown in the accompanying diagram. If k represents any integer, which of the following expressions names the infinite set of real numbers that correspond to point C on the circle?

(1) $\frac{2}{3} + 2\pi k$ (2) $\frac{2\pi}{3} + 2\pi k$ (3) $\frac{4\pi}{3} + 2\pi k$ (4) $\frac{4}{3} + 2\pi k$

In 11–18, three of the four given numbers have points on the real-number line that correspond to the same point on the unit circle. Which real number has a point that does *not* belong to this set?

11. $8\pi, 10\pi, 11\pi, 12\pi$

12. $5\pi, 10\pi, 15\pi, -5\pi$

13. $\frac{3\pi}{2}, \frac{5\pi}{2}, \frac{7\pi}{2}, \frac{15\pi}{2}$

14. $\frac{-\pi}{6}, \frac{-13\pi}{6}, \frac{5\pi}{6}, \frac{11\pi}{6}$

15. $\frac{\pi}{4}, \frac{9\pi}{4}, \frac{-7\pi}{4}, \frac{-3\pi}{4}$

16. $\frac{5\pi}{3}, \frac{-5\pi}{3}, \frac{-\pi}{3}, \frac{11\pi}{3}$

17. $4 - 2\pi, 4 + \pi, 4 + 2\pi, 4 - 4\pi$

18. $\frac{1}{3}, \frac{2\pi + 1}{3}, \frac{6\pi + 1}{3}, \frac{1 - 12\pi}{3}$

9-2   GRAPH OF $y = \sin x$

A function is a set of ordered pairs and can be represented as a set of points in the coordinate plane when the domain and range of the function are subsets of the real numbers. This set of points is called the graph of the function. In *Courses I and II*, we learned to graph algebraic functions such as $y = 2x - 1$ and $y = x^2 - 3x + 7$. Now we will learn to graph $y = \sin x$.

In order to graph $y = \sin x$, we will use as the x-axis the number line that we have associated with points on the unit circle. We can select from the domain that is the set of all real numbers any convenient set of values for x and find the corresponding value of $\sin x$ or y. One possible set with which to begin is the multiples of $\dfrac{\pi}{6}$ from 0 to 2π, which includes the measure of each of the quadrantal angles and two angles in each quadrant.

x	0	$\dfrac{\pi}{6}$	$\dfrac{\pi}{3}$	$\dfrac{\pi}{2}$	$\dfrac{2\pi}{3}$	$\dfrac{5\pi}{6}$	π	$\dfrac{7\pi}{6}$	$\dfrac{4\pi}{3}$	$\dfrac{3\pi}{2}$	$\dfrac{5\pi}{3}$	$\dfrac{11\pi}{6}$	2π
$\sin x$	0	.5	.87	1	.87	.5	0	$-.5$	$-.87$	-1	$-.87$	$-.5$	0

The value of $\sin \dfrac{\pi}{6} = \dfrac{1}{2}$ or .5, and the value of $\sin \dfrac{\pi}{3} = \dfrac{\sqrt{3}}{2} = .87$, to the nearest hundredth.

After choosing a scale along the y-axis, we let 3 be an approximation for π on the x-axis as a convenient way to present the graph as closely as possible to its actual scale.

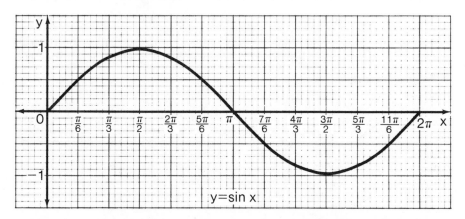

The curve shown above is the *basic sine curve*, or *sine wave*. If we divide the interval from 0 to 2π into four equal parts, the curve increases from 0 to 1 in the first quarter, decreases from 1 to 0 and from 0 to -1 in the second and third quarters, and then increases from -1

to 0 in the fourth quarter. In the unit circle, we see this same pattern for $\sin \theta$. As θ increases from 0 to $\frac{\pi}{2}$ radians, $\sin \theta$ increases from 0 to 1.

Then, in quadrant II, as θ increases from $\frac{\pi}{2}$ to π radians, $\sin \theta$ decreases from 1 to 0. The similarities continue for quadrants III and IV, as shown in the model problem at the end of this section.

This, of course, is only a part of the graph of $y = \sin x$. If we choose values of x from -2π to 0, we have the following set of values:

x	-2π	$-\frac{11\pi}{6}$	$-\frac{5\pi}{3}$	$-\frac{3\pi}{2}$	$-\frac{4\pi}{3}$	$-\frac{7\pi}{6}$	$-\pi$	$-\frac{5\pi}{6}$	$-\frac{2\pi}{3}$	$-\frac{\pi}{2}$	$-\frac{\pi}{3}$	$-\frac{\pi}{6}$	0
$\sin x$	0	.5	.87	1	.87	.5	0	-.5	-.87	-1	-.87	-.5	0

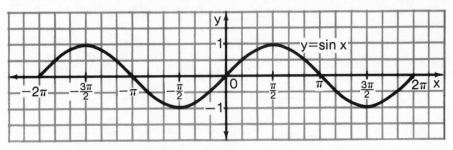

The graph of $y = \sin x$ is drawn over an interval from -2π to 2π. The set of values for $\sin x$ from $x = -2\pi$ to $x = 0$ duplicates the values for $\sin x$ from $x = 0$ to $x = 2\pi$. Thus, the graph of $y = \sin x$ in the interval $-2\pi \leq x \leq 0$ repeats the basic sine curve. We observe:

$$\sin x = \sin (x + 2\pi k) \text{ for any integer } k$$

The graph of $y = \sin x$ has translational symmetry under the translation $T_{2\pi,0}$; that is, the graph of $y = \sin x$ is its own image under the translation $T_{2\pi,0}$.

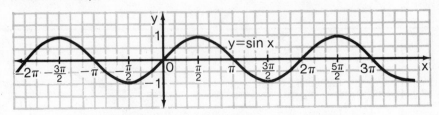

$$y = \sin x$$

Domain = $\{x \mid x \in \text{Real numbers}\}$
Range = $\{y \mid -1 \leq y \leq 1\}$

A function, f, is a *periodic function* if there exists a nonzero constant, p, such that for every x in the domain of f, $f(x + p) = f(x)$. The smallest positive value of p is called the **period** of the function. In other words, the period is the length of the interval between successive repetitions of the curve. For example, when $x = \dfrac{\pi}{2}$, sin $x = 1$. Then, when $x = \dfrac{5\pi}{2}$, sin $x = 1$. Therefore, the difference between $\dfrac{5\pi}{2}$ and $\dfrac{\pi}{2}$ shows an interval length of $\dfrac{5\pi}{2} - \dfrac{\pi}{2} = \dfrac{4\pi}{2} = 2\pi$, or 2π is the period of the function $y = \sin x$.

| MODEL PROBLEM |

From the graph of $y = \sin x$, determine whether sin x increases or decreases in each quadrant.

Solution

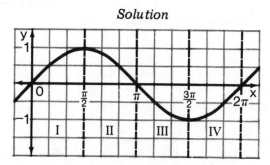

Sketch the graph and divide the interval from 0 to 2π into quadrants. Study the change in y-values in each quadrant.

Answer: In quadrant I, sin x increases from 0 to 1.
In quadrant II, sin x decreases from 1 to 0.
In quadrant III, sin x decreases from 0 to -1.
In quadrant IV, sin x increases from -1 to 0.

| EXERCISES |

1. Sketch the graph of $y = \sin x$ from $x = -2\pi$ to $x = 2\pi$.
2. What is the period of $y = \sin x$?
3. **a.** What is the largest value of sin x? **b.** What is the smallest value of sin x?
4. What is the range of $y = \sin x$?
5. For what values of x in the interval $-2\pi \le x \le 2\pi$ is sin $x = 1$?

6. For what values of x in the interval $-2\pi \le x \le 2\pi$ is $\sin x = -1$?
7. Between what values of x in the interval $-2\pi \le x \le 2\pi$ is $\sin x$
   a. increasing?   b. decreasing?
8. Name three real numbers that are *not* elements of the range of the function $y = \sin x$.

9-3 GRAPH OF $y = \cos x$

We can sketch the graph of $y = \cos x$ by making a table of values, as we did for $y = \sin x$, using multiples of $\frac{\pi}{6}$ from 0 to 2π.

x	0	$\frac{\pi}{6}$	$\frac{\pi}{3}$	$\frac{\pi}{2}$	$\frac{2\pi}{3}$	$\frac{5\pi}{6}$	π	$\frac{7\pi}{6}$	$\frac{4\pi}{3}$	$\frac{3\pi}{2}$	$\frac{5\pi}{3}$	$\frac{11\pi}{6}$	2π
$\cos x$	1	.87	.5	0	$-.5$	$-.87$	-1	$-.87$	$-.5$	0	.5	.87	1

Once again, we have chosen a scale along the y-axis and we have let 3 be an approximation for π on the x-axis as a convenient way to present the graph as closely as possible to its actual scale.

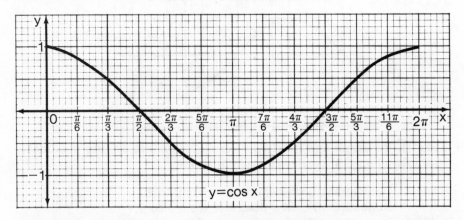

The curve shown above is the *basic cosine curve*. If we divide the interval from 0 to 2π into four equal parts, the curve decreases from 1 to 0 and then from 0 to -1 in the first and second quarters and then increases from -1 to 0 and from 0 to 1 in the third and fourth quarters. In the unit circle, we see this same pattern of values for $\cos \theta$. As θ increases from 0 to $\frac{\pi}{2}$ radians, $\cos \theta$ decreases from 1 to 0. Then, in quadrant II, as θ increases from $\frac{\pi}{2}$ to π radians, $\cos \theta$ decreases from 0 to -1. The similarities continue for quadrants III and IV, as shown in the model problem at the end of this section.

Like the sine function, the cosine function is a periodic function with a period of 2π.

$$\cos x = \cos (x + 2\pi k) \text{ for any integer } k$$

When $x = \pi$, $\cos x = -1$ and when $x = 3\pi$, $\cos x = -1$. The difference between 3π and π shows an interval length of $3\pi - \pi = 2\pi$, or 2π is the period of the function $y = \cos x$.

The graph of $y = \cos x$ has translational symmetry with respect to the translation $T_{2\pi,0}$, and its graph is an endless repetition of the basic cosine curve drawn on the preceding page.

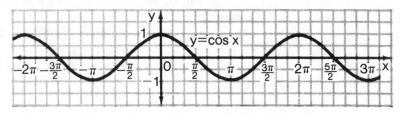

$y = \cos x$

Domain $= \{x \mid x \in \text{Real numbers}\}$

Range $= \{y \mid -1 \leq y \leq 1\}$

MODEL PROBLEM

From the graph of $y = \cos x$, determine whether $\cos x$ increases or decreases in each quadrant.

Solution

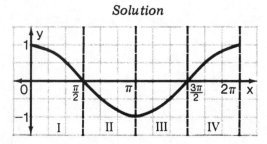

Sketch the graph and divide the interval from 0 to 2π into quadrants. Study the change in y-values in each quadrant.

Answer: In quadrant I, $\cos x$ decreases from 1 to 0.
In quadrant II, $\cos x$ decreases from 0 to -1.
In quadrant III, $\cos x$ increases from -1 to 0.
In quadrant IV, $\cos x$ increases from 0 to 1.

| **EXERCISES** |

1. Sketch the graph of $y = \cos x$ from $x = -2\pi$ to $x = 2\pi$.
2. What is the period of $y = \cos x$?
3. a. What is the largest value of $\cos x$? b. What is the smallest value of $\cos x$?
4. What is the range of $y = \cos x$?
5. For what values of x in the interval $-2\pi \leq x \leq 2\pi$ is $\cos x = 1$?
6. For what values of x in the interval $-2\pi \leq x \leq 2\pi$ is $\cos x = -1$?
7. Between what values of x in the interval $-2\pi \leq x \leq 2\pi$ is $\cos x$
   a. increasing?   b. decreasing?
8. a. Sketch the graph of $y = \cos x$ in the interval $-2\pi \leq x \leq 2\pi$.
   b. On the same set of axes, sketch the graph of $y = \sin x$.
   c. For what value of q is $y = \sin x$ the image of $y = \cos x$ under the translation $T_{q,0}$?

9-4 AMPLITUDE

In order to draw the graph of $y = a \sin x$ for some constant, a, we must first find the values of $\sin x$ and then multiply these values by a. For example, the table at the right shows us how to find four rational approximations for y when the function is $y = 2 \sin x$. To abbreviate our work, study the chart that follows. This compact chart gives the values we might use to graph $y = 2 \sin x$ and $y = \frac{1}{2} \sin x$ over the interval $0 \leq x \leq 2\pi$.

x	$2 \sin x$	$=$	y
0	$2 \sin 0 = 2(0)$	$=$	0
$\dfrac{\pi}{6}$	$2 \sin \dfrac{\pi}{6} = 2(.5)$	$=$	1.0
$\dfrac{\pi}{3}$	$2 \sin \dfrac{\pi}{3} = 2(.866)$	$=$	1.732
$\dfrac{\pi}{2}$	$2 \sin \dfrac{\pi}{2} = 2(1)$	$=$	2

x	0	$\dfrac{\pi}{6}$	$\dfrac{\pi}{3}$	$\dfrac{\pi}{2}$	$\dfrac{2\pi}{3}$	$\dfrac{5\pi}{6}$	π	$\dfrac{7\pi}{6}$	$\dfrac{4\pi}{3}$	$\dfrac{3\pi}{2}$	$\dfrac{5\pi}{3}$	$\dfrac{11\pi}{6}$	2π
$\sin x$	0	$.5$	$.866$	1	$.866$	$.5$	0	$-.5$	$-.866$	-1	$-.866$	$-.5$	0
$2 \sin x$	0	1.0	1.732	2	1.732	1.0	0	-1.0	-1.732	-2	-1.732	-1.0	0
$\frac{1}{2} \sin x$	0	$.25$	$.433$	$.5$	$.433$	$.25$	0	$-.25$	$-.433$	$-.5$	$-.433$	$-.25$	0

After finding the values of $\sin x$ in row 2, we multiply those values by 2 to find the values of $2 \sin x$ in row 3. To draw the graph of $y = 2 \sin x$, we use values of x from row 1 with the corresponding values of y from row 3. In the same way, we multiply the values of $\sin x$ in row 2 by $\frac{1}{2}$

to find the values of $\frac{1}{2}\sin x$ in row 4. To draw the graph of $y = \frac{1}{2}\sin x$, we use values of x from row 1 with the corresponding values of y from row 4. These graphs are shown below.

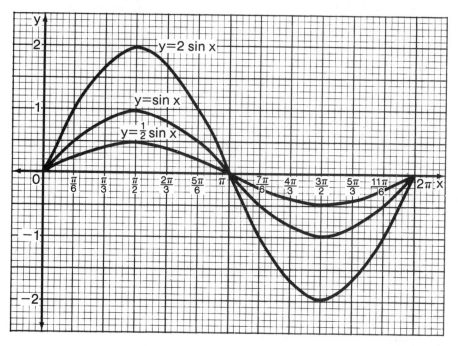

Notice that each graph follows the pattern of the basic sine curve. Each graph intersects the x-axis at 0, π, and 2π; reaches its highest point at $\frac{\pi}{2}$; and reaches its lowest point at $\frac{3\pi}{2}$. Like the graph of $y = \sin x$, the graph of $y = a \sin x$ has translational symmetry under the translation $T_{2\pi,0}$, and the complete graph repeats the basic pattern endlessly.

The graph of $y = -2 \sin x$ is the reflection in the x-axis of the graph of $y = 2 \sin x$, as shown in the figure below.

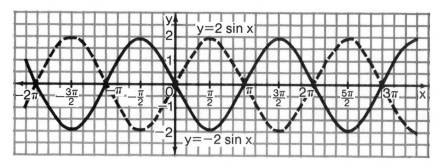

The maximum value of $y = \sin x$ is 1 and the minimum value is -1. The range of $y = \sin x$ is $-1 \leq y \leq 1$. The maximum value of $y = 2 \sin x$ is 2 and the minimum value is -2. The range of $y = 2 \sin x$ is $-2 \leq y \leq 2$. The maximum value of $y = \frac{1}{2} \sin x$ is $\frac{1}{2}$ and the minimum value is $-\frac{1}{2}$. The range of $y = \frac{1}{2} \sin x$ is $-\frac{1}{2} \leq y \leq \frac{1}{2}$.

Notice from the graph of $y = -2 \sin x$ (the reflection of $y = 2 \sin x$ in the x-axis) that the maximum value is 2 and the minimum value is -2. The range of $y = -2 \sin x$ is $-2 \leq y \leq 2$.

The *amplitude* of a periodic function is one-half the difference between the maximum value and the minimum value.

If $y = \dfrac{1}{2} \sin x$, amplitude $= \dfrac{\frac{1}{2} - (-\frac{1}{2})}{2} = \dfrac{1}{2}$.

If $y = -2 \sin x$, amplitude $= \dfrac{2 - (-2)}{2} = 2$.

If $y = a \sin x$, amplitude $= \dfrac{|a| - (-|a|)}{2} = |a|$.

The cosine function, like the sine function, has as its range the set of real numbers from -1 to 1. The same principles that were discussed for $y = a \sin x$ apply to the functions of the form $y = a \cos x$. For example, if $y = 4 \cos x$, the range of the function is $-4 \leq y \leq 4$ and the amplitude is 4.

■ In general, for the functions $y = a \sin x$ and $y = a \cos x$:

$$\text{the amplitude} = |a|$$

MODEL PROBLEM

Sketch the graph of $y = 2 \cos x$ in the interval $0 \leq x \leq 2\pi$.

Solution

Choose a convenient scale on the y-axis and, using this same scale, locate the points that are convenient approximations of $\dfrac{\pi}{2}$, π, $\dfrac{3\pi}{2}$, and 2π on the x-axis. These four values and 0 can be used to find the maximum, minimum, and 0 values of $2 \cos x$, as shown in the chart. Plot

x	$2 \cos x$	$=$	y
0	$2 \cos 0 = 2(1)$	$=$	2
$\dfrac{\pi}{2}$	$2 \cos \dfrac{\pi}{2} = 2(0)$	$=$	0
π	$2 \cos \pi = 2(-1)$	$=$	-2
$\dfrac{3\pi}{2}$	$2 \cos \dfrac{3\pi}{2} = 2(0)$	$=$	0
2π	$2 \cos 2\pi = 2(1)$	$=$	2

the points, using values from the chart. Using the shape of the basic cosine curve as a guide, sketch the curve.

Answer:

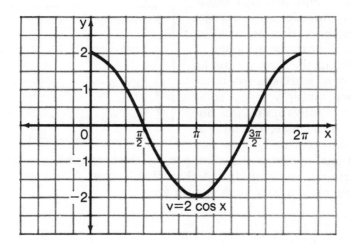

$v = 2 \cos x$

EXERCISES

In 1-6, state the amplitude of the given function.

1. $y = 3 \sin x$ 2. $y = \frac{1}{2} \cos x$ 3. $y = 2 \cos x$
4. $y = \frac{1}{4} \sin x$ 5. $y = -3 \sin x$ 6. $y = -.6 \cos x$

In 7-12, sketch the graph from $x = 0$ to $x = 2\pi$.

7. $y = \frac{1}{2} \cos x$ 8. $y = 2 \sin x$ 9. $y = 3 \cos x$
10. $y = .8 \sin x$ 11. $y = 4 \cos x$ 12. $y = -2 \sin x$

In 13-18, state the range of the given function.

13. $y = \sin x$ 14. $y = \cos x$ 15. $y = 2 \sin x$
16. $y = \frac{1}{2} \cos x$ 17. $y = 8 \cos x$ 18. $y = -3 \sin x$

19. a. State three values of x in the interval $0 \le x \le 2\pi$ for which $3 \sin x = 0$.
    b. For what value of x in the interval $0 \le x \le 2\pi$ is $3 \sin x$ a maximum value?
    c. For what value of x in the interval $0 \le x \le 2\pi$ is $3 \sin x$ a minimum value?

9-5 PERIOD

We saw that the graphs of $y = a \sin x$ have different amplitudes as the value of a changes but that the period remains 2π. Now we shall study the graph of a function of the form $y = \sin bx$ and observe what changes result when we multiply x by some constant before finding the sine value. For example, the table at the right shows us how to find

x	$\sin 2x$	$=$	y
0	$\sin 2 \cdot 0 = \sin 0$	$=$	0
$\dfrac{\pi}{6}$	$\sin 2 \cdot \dfrac{\pi}{6} = \sin \dfrac{\pi}{3}$	$=$	$.87$
$\dfrac{\pi}{4}$	$\sin 2 \cdot \dfrac{\pi}{4} = \sin \dfrac{\pi}{2}$	$=$	1
$\dfrac{\pi}{3}$	$\sin 2 \cdot \dfrac{\pi}{3} = \sin \dfrac{2\pi}{3}$	$=$	$.87$

four rational approximations for y when the function is $y = \sin 2x$. To abbreviate our work, study the chart that follows. This chart gives the values we might use to graph $y = \sin 2x$.

x	0	$\dfrac{\pi}{6}$	$\dfrac{\pi}{4}$	$\dfrac{\pi}{3}$	$\dfrac{\pi}{2}$	$\dfrac{2\pi}{3}$	$\dfrac{3\pi}{4}$	$\dfrac{5\pi}{6}$	π	$\dfrac{7\pi}{6}$	$\dfrac{5\pi}{4}$	$\dfrac{4\pi}{3}$	$\dfrac{3\pi}{2}$	$\dfrac{5\pi}{3}$	$\dfrac{7\pi}{4}$	$\dfrac{11\pi}{6}$	2π
$2x$	0	$\dfrac{\pi}{3}$	$\dfrac{\pi}{2}$	$\dfrac{2\pi}{3}$	π	$\dfrac{4\pi}{3}$	$\dfrac{3\pi}{2}$	$\dfrac{5\pi}{3}$	2π	$\dfrac{7\pi}{3}$	$\dfrac{5\pi}{2}$	$\dfrac{8\pi}{3}$	3π	$\dfrac{10\pi}{3}$	$\dfrac{7\pi}{2}$	$\dfrac{11\pi}{3}$	4π
$\sin 2x$	0	$.87$	1	$.87$	0	$-.87$	-1	$-.87$	0	$.87$	1	$.87$	0	$-.87$	-1	$-.87$	0

To draw the graph of $y = \sin 2x$, we use values of x from row 1 with the corresponding values of y from row 3.

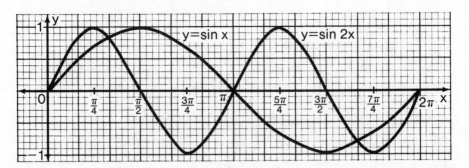

In $y = \sin x$, the basic sine curve appears once over an interval of 2π. In $y = \sin 2x$, the sine curve appears twice over an interval from 0 to 2π. In other words, given $y = \sin 2x$, the *frequency* of the curve is 2 and its period is π. In general, the *frequency* of a periodic function is the number of times the function repeats itself in a given interval. Let us agree that the interval to be considered is 2π for all trigonometric functions.

The period of a function is equal to the length of the interval, 2π, divided by the frequency. For example, in $y = \sin 2x$, the frequency is 2 and the period $= \dfrac{2\pi}{2} = \pi$.

This relationship can also be seen by studying translations. Since the graph of $y = \sin 2x$ completes a full cycle in a period of π, we can write:

$$\sin 2x = \sin (2x + 2\pi) = \sin 2(x + \pi)$$
$$T_{\pi,0}\text{: } \sin 2x \rightarrow \sin 2(x + \pi)$$

The curve $y = \sin 2x$ has translational symmetry under $T_{\pi,0}$ and has a period of π.

In general, if $y = \sin bx$ or $y = \cos bx$, where b is positive:

$$\sin bx = \sin (bx + 2\pi) = \sin b\left(x + \frac{2\pi}{b}\right)$$

$$T_{\frac{2\pi}{b},0}\text{: } \sin bx \rightarrow \sin b\left(x + \frac{2\pi}{b}\right)$$

$$\cos bx = \cos (bx + 2\pi) = \cos b\left(x + \frac{2\pi}{b}\right)$$

$$T_{\frac{2\pi}{b},0}\text{: } \cos bx \rightarrow \cos b\left(x + \frac{2\pi}{b}\right)$$

Therefore, $y = \sin bx$ and $y = \cos bx$ have translational symmetry under $T_{\frac{2\pi}{b},0}$. They are periodic functions with period $\dfrac{2\pi}{b}$.

■ In general, for the functions $y = \sin bx$ and $y = \cos bx$:

$$\text{the period} = \frac{2\pi}{|b|}$$

Note: The absolute-value symbol is used in writing $\dfrac{2\pi}{|b|}$ to ensure that the period is stated as a positive number.

| **MODEL PROBLEMS** |

1. **a.** Sketch the graph of $y = \cos \frac{1}{2}x$ from $x = 0$ to $x = 4\pi$.
   b. Find the value(s) of x in the interval $0 \le x \le 4\pi$ so that the value of $\cos \frac{1}{2}x$ is (i) a maximum (ii) a minimum (iii) 0.

Solution

a. 1. Compare $y = \cos \frac{1}{2}x$ to $y = \cos bx$ to find $b = \frac{1}{2}$.

Thus, the period $= \dfrac{2\pi}{|b|} = \dfrac{2\pi}{\frac{1}{2}} = \dfrac{2\pi}{\frac{1}{2}} \cdot \dfrac{2}{2} = 4\pi$.

2. Since a complete cycle occurs over a period of 4π, divide the period into quarters and find the values of y for the values of x shown in the table.

3. Plot these points and draw a smooth curve through them.

x	$\cos \dfrac{1}{2}x$	$=$	y
0	$\cos \dfrac{1}{2} \cdot 0 = \cos 0$	$=$	1
π	$\cos \dfrac{1}{2} \cdot \pi = \cos \dfrac{\pi}{2}$	$=$	0
2π	$\cos \dfrac{1}{2} \cdot 2\pi = \cos \pi$	$=$	-1
3π	$\cos \dfrac{1}{2} \cdot 3\pi = \cos \dfrac{3\pi}{2}$	$=$	0
4π	$\cos \dfrac{1}{2} \cdot 4\pi = \cos 2\pi$	$=$	1

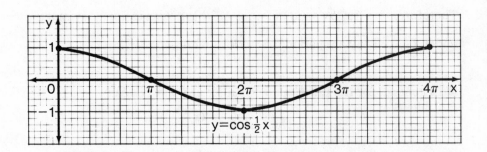

b. Maximum, minimum, and zero values can be read from the graph.

(i) The maximum value of $\cos \frac{1}{2}x$ occurs when $x = 0$ and when $x = 4\pi$.

(ii) The minimum value of $\cos \frac{1}{2}x$ occurs when $x = 2\pi$.

(iii) The expression $\cos \frac{1}{2}x$ equals 0 when $x = \pi$ and when $x = 3\pi$.

Answer: **a.** See the graph.

　　　　b. (i) 0 and 4π

　　　　　　(ii) 2π

　　　　　　(iii) π and 3π

2. Which is the equation of the function sketched in the accompanying diagram?

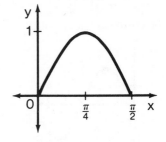

(1) $y = \sin 2x$ (2) $y = \cos 2x$

(3) $y = \sin \frac{1}{2}x$ (4) $y = \cos \frac{1}{2}x$

Solution

1. Since the sketch above is the start of the *sine* curve, reject choices (2) and (4).

2. Since the first half of the sine curve occurs over an interval from 0 to $\frac{\pi}{2}$, the period of the full curve is $2 \cdot \frac{\pi}{2} = \pi$. In choice (1), $y = \sin 2x$, the period $= \frac{2\pi}{|b|} = \frac{2\pi}{2} = \pi$.

Answer: (1) $y = \sin 2x$

| EXERCISES |

In 1–8, give the period of each function.

1. $y = \cos 2x$ 2. $y = \sin \frac{1}{2}x$ 3. $y = \cos 3x$ 4. $y = \cos x$
5. $y = \sin 4x$ 6. $y = \sin x$ 7. $y = 3 \sin 2x$ 8. $y = \frac{1}{2}\cos 6x$

In 9–11, sketch the graph from $x = 0$ to $x = 2\pi$.

9. $y = \cos 2x$ 10. $y = \sin \frac{1}{2}x$ 11. $y = \sin 3x$

In 12–14, sketch the graph in the interval $-2\pi \le x \le 2\pi$.

12. $y = \sin 2x$ 13. $y = \cos \frac{1}{2}x$ 14. $y = \cos 3x$

15. Find the values of x between 0 and 2π inclusive for which $y = \sin 2x$ is: **a.** a maximum **b.** a minimum **c.** 0

16. Find the values of x between 0 and 2π inclusive for which $y = \cos 2x$ is: **a.** a maximum **b.** a minimum **c.** 0

17. The graph of $y = \sin 4x$ is drawn for values of x in the interval $0 \le x \le 2\pi$. **a.** How many times will the basic sine curve appear? **b.** What is the frequency of the graph of $y = \sin 4x$?

18. The graph of $y = \cos \frac{1}{3}x$ is drawn for values of x in the interval $0 \le x \le 2\pi$. **a.** What fractional part of the basic cosine curve will appear? **b.** What is the frequency of the graph of $y = \cos \frac{1}{3}x$?

In 19-24, the graph shows $y = \sin bx$ or $y = \cos bx$. **a.** What is the frequency of the curve? **b.** What is the equation of the function?

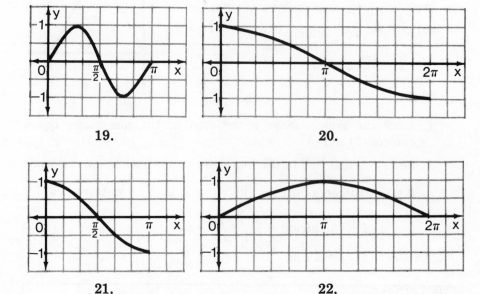

19.   20.

21.   22.

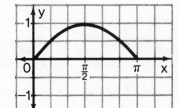

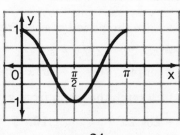

23.   24.

9-6 SKETCHING SINE AND COSINE CURVES

To summarize what we have learned in the last few sections, we can say:

■ For every function $y = a \sin bx$ and $y = a \cos bx$:

the amplitude $= |a|$

the frequency $= |b|$

the period $= \dfrac{2\pi}{|b|}$

| MODEL PROBLEMS |

1. Sketch the graph of $y = 3 \sin 2x$ in the interval $0 \leq x \leq 2\pi$.

Solution

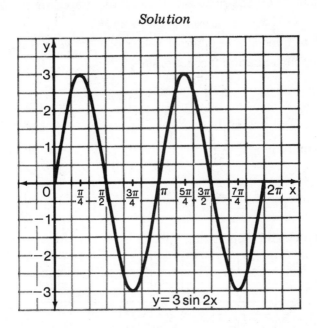

1. Determine the amplitude and the period of the curve $y = 3 \sin 2x$. By a comparison to $y = a \sin bx$, we see that $a = 3$ and $b = 2$. Therefore:

$$\text{Amplitude} = |a| = |3| = 3. \qquad \text{Period} = \frac{2\pi}{|b|} = \frac{2\pi}{|2|} = \frac{2\pi}{2} = \pi.$$

2. Since the amplitude is 3, the maximum value of the curve is 3 and the minimum value is -3.

3. Since the period is π, divide the interval on the x-axis from 0 to π into four quarters at the points $0, \frac{\pi}{4}, \frac{\pi}{2}, \frac{3\pi}{4}$, and π. In a sine curve, the zeros occur at the first, middle, and last of these points, that is, at $0, \frac{\pi}{2}$, and π for $y = 3 \sin 2x$. The maximum value, 3, occurs when $x = \frac{\pi}{4}$; and the minimum value, -3, occurs when $x = \frac{3\pi}{4}$ for the given equation.

    4. Repeat the process in step 3 for the interval from π to 2π.

    5. Draw a smooth curve through the points that have been plotted.

2.   **a.** Sketch, on the same set of axes, the graphs of $y = 2\cos x$ and $y = \sin \frac{1}{2}x$ as x varies from 0 to 2π.

    b. Determine the number of values of x between 0 and 2π that satisfy the equation $2\cos x = \sin \frac{1}{2}x$.

Solution

$y = 2\cos x$	$y = \sin \frac{1}{2}x$
amplitude = 2	amplitude = 1
frequency = 1	frequency = $\frac{1}{2}$
period $= \dfrac{2\pi}{1} = 2\pi$	period $= \dfrac{2\pi}{\frac{1}{2}} = 4\pi$

Sketch and label each graph.

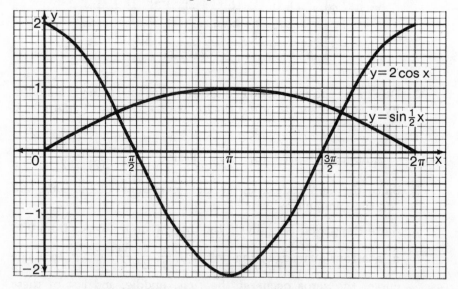

    b. The values of x that satisfy the equation $2\cos x = \sin \frac{1}{2}x$ are the x-coordinates of the points of intersection of $y = 2\cos x$ and $y = \sin \frac{1}{2}x$. Therefore, there are two values of x between 0 and 2π that satisfy the equation.

Answer: **a.** See the graph.   **b.** two

| EXERCISES |

In 1–6: **a.** State the amplitude of the graph. **b.** State the period of the graph.

1. $y = 3 \sin 2x$ **2.** $y = \frac{1}{2} \sin x$ **3.** $y = 2 \cdot \cos \frac{1}{2} x$

4. $y = 2 \cos 2x$ **5.** $y = \sin \frac{1}{2} x$ **6.** $y = -2 \cos x$

7. If $f(x) = \sin 2x$, find: **a.** $f(\pi)$ **b.** $f\left(\dfrac{\pi}{2}\right)$ **c.** $f\left(\dfrac{\pi}{6}\right)$

8. If $f(x) = \cos \dfrac{1}{2} x$, find: **a.** $f(\pi)$ **b.** $f\left(\dfrac{\pi}{2}\right)$ **c.** $f(3\pi)$

9. **a.** Sketch the graph of $y = \sin 2x$ from $x = 0$ to $x = 2\pi$.
     b. On the same set of axes, sketch the graph of $y = \frac{1}{2} \cos x$.
     c. For how many values of x in the interval $0 \leq x \leq 2\pi$ does $\sin 2x = \frac{1}{2} \cos x$?

10. **a.** Sketch the graph of $y = 2 \cos \frac{1}{2} x$ for values of x in the interval $-2\pi \leq x \leq 2\pi$.
     b. On the same set of axes, sketch the graph of $y = 2 \sin x$.
     c. For how many values of x in the interval $-2\pi \leq x \leq 2\pi$ does $2 \cos \frac{1}{2} x = 2 \sin x$?

11. **a.** On the same set of axes, sketch the graphs of $y = \cos 2x$ and $y = \sin x$ for values of x in the interval $0 \leq x \leq 2\pi$.
     b. For what values of x in the interval $0 \leq x \leq 2\pi$ does $\cos 2x = \sin x$?

12. **a.** On the same set of axes, sketch the graphs of $y = \cos x$ and $y = 2 \sin 2x$ for values of x in the interval $-\pi \leq x \leq \pi$.
     b. For how many values of x in the interval $-\pi \leq x \leq \pi$ does $\cos x = 2 \sin 2x$?

In 13–18: **a.** Sketch the graphs of the two functions on one set of axes, using the interval $0 \leq x \leq 2\pi$. **b.** Solve the given equation for x in the interval $0 \leq x \leq 2\pi$.

13. **a.** Graph $y = \sin x$
           $y = \cos x$
     b. Solve $\sin x = \cos x$

14. **a.** Graph $y = \cos \frac{1}{2} x$
           $y = -\sin x$
     b. Solve $\cos \frac{1}{2} x = -\sin x$

15. **a.** Graph $y = 3 \cos x$
           $y = \sin 2x$
     b. Solve $3 \cos x = \sin 2x$

16. **a.** Graph $y = -\cos x$
           $y = \sin x$
     b. Solve $-\cos x = \sin x$

17. **a.** Graph $y = \dfrac{1}{2} \sin x$

   $y = -2 \sin x$

   b. Solve $\dfrac{1}{2} \sin x = -2 \sin x$

18. **a.** Graph $y = \sin \dfrac{x}{2}$

   $y = -\cos x$

   b. Solve $\sin \dfrac{x}{2} = -\cos x$

19. **a.** On the same set of axes, sketch the graph of $y = \sin \frac{1}{2}x$ and $y = 2 \cos x$ for values of x in the interval $-2\pi \le x \le 0$.

   b. For how many values of x in the interval $-2\pi \le x \le 0$, does $\sin \frac{1}{2}x = 2 \cos x$?

20. **a.** Sketch the graph of $y = 2 \cos 2x$ in the interval $0 \le x \le 2\pi$.

   b. On the same set of axes, draw the graph of $y = 1$.

   c. What are the coordinates of the points of intersection of the graphs drawn in parts a and b?

21. An oscilloscope is a device that presents pictures of sound waves. The function $y = .002 \sin (200\pi x)$ describes the sound produced by a tuning fork, where x represents the time in seconds.

   a. What is the amplitude of the given function?

   b. What is its period?

   c. Complete the table, finding values of y to four decimal places:

x	0	$\dfrac{1}{800}$	$\dfrac{2}{800}$	$\dfrac{3}{800}$	$\dfrac{4}{800}$	$\dfrac{5}{800}$	$\dfrac{6}{800}$	$\dfrac{7}{800}$	$\dfrac{8}{800}$
y									

   d. Graph the picture of the sine wave produced by the tuning fork over an interval of time from 0 to $\frac{8}{800}$ (or $\frac{1}{100}$) second.

   e. How many complete "sine waves" will be produced by the tuning fork in an interval of one second?

22. Research the terms Amplitude Modulation and Frequency Modulation, commonly abbreviated as A.M. and F.M., and explain the terms in relation to practical applications of trigonometric graphs.

9-7 GRAPH OF $y = \tan x$

We can draw the graph of $y = \tan x$, as we did the graphs of $y = \sin x$ and $y = \cos x$, by making a table of values. Recall from our work with the unit circle that the tangent function is undefined at $\pm\dfrac{\pi}{2}$, $\pm\dfrac{3\pi}{2}$, or any odd multiple of $\dfrac{\pi}{2}$. This is indicated in the table by a dash (—).

x	-2π	$\dfrac{-11\pi}{6}$	$\dfrac{-7\pi}{4}$	$\dfrac{-5\pi}{3}$	$\dfrac{-3\pi}{2}$	$\dfrac{-4\pi}{3}$	$\dfrac{-5\pi}{4}$	$\dfrac{-7\pi}{6}$	$-\pi$	$\dfrac{-5\pi}{6}$	$\dfrac{-3\pi}{4}$	$\dfrac{-2\pi}{3}$	$\dfrac{-\pi}{2}$	$\dfrac{-\pi}{3}$	$\dfrac{-\pi}{4}$	$\dfrac{-\pi}{6}$	0
$\tan x$	0	.58	1	1.7	—	-1.7	-1	-.58	0	.58	1	1.7	—	-1.7	-1	-.58	0

x	0	$\dfrac{\pi}{6}$	$\dfrac{\pi}{4}$	$\dfrac{\pi}{3}$	$\dfrac{\pi}{2}$	$\dfrac{2\pi}{3}$	$\dfrac{3\pi}{4}$	$\dfrac{5\pi}{6}$	π	$\dfrac{7\pi}{6}$	$\dfrac{5\pi}{4}$	$\dfrac{4\pi}{3}$	$\dfrac{3\pi}{2}$	$\dfrac{5\pi}{3}$	$\dfrac{7\pi}{4}$	$\dfrac{11\pi}{6}$	2π
$\tan x$	0	.58	1	1.7	—	-1.7	-1	-.58	0	.58	1	1.7	—	-1.7	-1	-.58	0

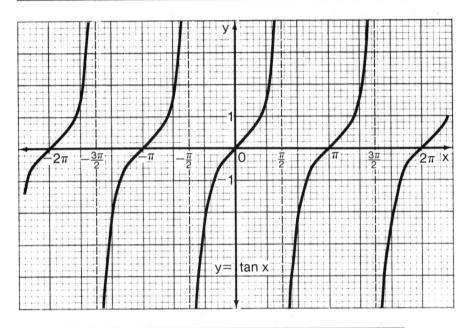

$$y = \tan x$$

Domain $= \left\{ x \,\middle|\, x \in \text{Real numbers and} \right.$

$\left. x \neq \dfrac{(2k + 1)\pi}{2} \text{ for integral values of } k \right\}$

Range $= \{ y \,|\, y \in \text{Real numbers}\}$

Notice that, unlike the sine and cosine curves, which are continuous, the graph of $y = \tan x$ is discontinuous at $x = \dfrac{\pi}{2}$, $x = \dfrac{3\pi}{2}$, $x = \dfrac{-\pi}{2}$, or at any value of $x = \dfrac{(2k + 1)\pi}{2}$ for integral values of k. By choosing points with x-coordinates sufficiently close to these odd multiples of $\frac{\pi}{2}$, we can make the value of $\tan x$ as large or as small as we choose. Therefore, $y = \tan x$ has no maximum or minimum value. The tangent function does not have an amplitude.

Like the sine and cosine functions, the tangent function is periodic. Notice that the shape of the portion of the curve from $x = \frac{-3\pi}{2}$ to $x = \frac{-\pi}{2}$ is repeated in the interval from $x = \frac{-\pi}{2}$ to $x = \frac{\pi}{2}$ and again from $\frac{\pi}{2}$ to $\frac{3\pi}{2}$. Since $\frac{3\pi}{2} - \frac{\pi}{2} = \frac{2\pi}{2} = \pi$, the period of the tangent function is π.

$$\tan x = \tan (x + \pi) \text{ for all } x$$

The graph of $y = \tan x$ has translational symmetry with respect to the translation $T_{\pi,0}$.

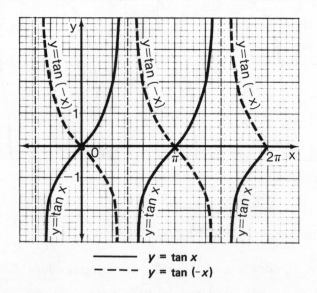

———— $y = \tan x$
– – – – $y = \tan (\text{-}x)$

The graph shows the reflection of the curve $y = \tan x$ in the x-axis. If this image is reflected in the y-axis, the image is the curve $y = \tan x$. Since a reflection in the x-axis followed by a reflection in the y-axis is a reflection in the origin, the graph of $y = \tan x$ has point symmetry with respect to a reflection in the origin.

| EXERCISES |

1. Sketch the graph of $y = \tan x$ from $x = -2\pi$ to $x = 2\pi$.
2. What is the period of $y = \tan x$?
3. What is the domain of $y = \tan x$?
4. What is the range of $y = \tan x$?

5. Between what values of x in the interval $-\dfrac{\pi}{2} < x < \dfrac{\pi}{2}$ is tan x

   a. increasing?   b. decreasing?

In 6, select the numeral preceding the expression that best answers the question.

6. Which is *not* an element of the domain of $y = \tan x$?

   (1) π (2) 2π (3) $\dfrac{\pi}{2}$ (4) $-\pi$

7. a. On the same set of axes, sketch the graph of $y = 2 \sin x$ and $y = \tan x$ for values of x in the interval $-\pi \leq x \leq \pi$.
   b. How many values of x in the interval $-\pi \leq x \leq \pi$ are solutions of the equation $\tan x = 2 \sin x$?

9-8 GRAPHS OF THE RECIPROCAL FUNCTIONS

The graphs of $y = \csc x$, $y = \sec x$, and $y = \cot x$ can be drawn by making a table of values as we did for $y = \sin x$, $y = \cos x$, and $y = \tan x$. Each of the diagrams that follow shows a pair of reciprocal functions.

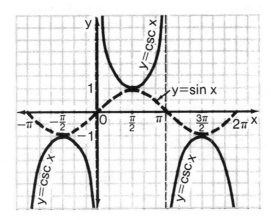

Fig. 1

In Fig. 1, $y = \sin x$ is shown as a dotted line and $y = \csc x$ is shown as a solid line. Notice that:

As $\sin x$ increases, $\csc x$ decreases and, as $\sin x$ decreases, $\csc x$ increases.

The values of $\sin x$ and $\csc x$ always have the same sign.

When $\sin x = 1$, $\csc x = 1$ and when $\sin x = -1$, $\csc x = -1$.

When $\sin x = 0$, $\csc x$ is undefined.

$$y = \csc x$$

Domain = $\{x \mid x \in \text{Real numbers and } x \neq k\pi \text{ for integral values of } k\}$
Range = $\{y \mid |y| \geq 1\}$

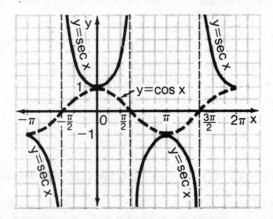

Fig. 2

In Fig. 2, $y = \cos x$ is shown as a dotted line and $y = \sec x$ is shown as a solid line. We can make observations about these functions that are similar to those we made for the graphs of the sine and cosecant functions.

$$y = \sec x$$

Domain = $\left\{ x \mid x \in \text{Real numbers and} \right.$

$\left. x \neq \dfrac{(2k + 1)\pi}{2} \text{ for integral values of } k \right\}$

Range = $\{y \mid |y| \geq 1\}$

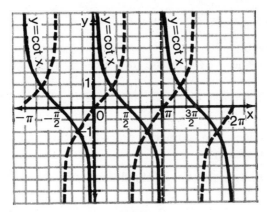

Fig. 3

In Fig. 3, $y = \tan x$ is shown as a dotted line and $y = \cot x$ is shown as a solid line. In addition to the observations made before, notice that $\cot x$ is undefined when $\tan x$ is 0 and that $\cot x$ is 0 when $\tan x$ is undefined.

$$y = \cot x$$

Domain = $\{x \mid x \in \text{Real numbers and } x \neq k\pi \text{ for integral values of } k\}$
Range = $\{y \mid y \in \text{Real numbers}\}$

| **EXERCISES** |

1. If $f(x) = \sec x$, find: **a.** $f(0)$ **b.** $f\left(\dfrac{\pi}{3}\right)$ **c.** $f(\pi)$

2. If $f(x) = \csc x$, find: **a.** $f\left(\dfrac{\pi}{2}\right)$ **b.** $f\left(\dfrac{7\pi}{6}\right)$ **c.** $f(-\pi)$

3. If $f(x) = \cot x$, find: **a.** $f\left(\dfrac{\pi}{4}\right)$ **b.** $f\left(\dfrac{\pi}{2}\right)$ **c.** $f\left(\dfrac{-\pi}{2}\right)$

4. State two values of x in the interval $\dfrac{-\pi}{2} \leq x \leq \dfrac{3\pi}{2}$ for which $\csc x$ is undefined.

5. For what value of x in the interval $0 \leq x \leq \pi$ is the value of $\csc x$ a minimum?

6. State two values of x in the interval $0 \leq x \leq 2\pi$ for which $\sec x$ is undefined.

7. State two values of x in the interval $\dfrac{-\pi}{2} \le x \le \dfrac{3\pi}{2}$ for which $\cos x = \sec x$.

8. State two values of x in the interval $0 \le x \le \pi$ for which $\tan x = \cot x$.

9. State two values of x in the interval $-\pi \le x \le \pi$ for which $\cot x = 0$.

In 10–12, select the numeral preceding the expression that best answers the question.

10. Which is *not* an element of the domain of $y = \cot x$?

    (1) 0   (2) $\dfrac{\pi}{2}$ (3) $\dfrac{3\pi}{2}$ (4) $\dfrac{-\pi}{2}$

11. Which is *not* an element of the range of $y = \sec x$?

    (1) 1   (2) 2   (3) –1   (4) $\frac{1}{2}$

12. Which is *not* an element of the range of $y = \csc x$?

    (1) $\sqrt{2}$ (2) –2   (3) $\dfrac{\sqrt{2}}{2}$ (4) 4

9-9 REFLECTION OVER THE LINE $y = x$

In Chapter 5, we discussed the reflection of a set of points in the line $y = x$:

$$P(x, y) \rightarrow P'(y, x) \qquad \text{OR} \qquad r_{y=x}(x, y) = (y, x)$$

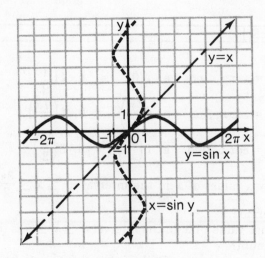

When the set of points that is the graph of $y = \sin x$ is reflected in the line $y = x$, the set of image points is the graph of $x = \sin y$. We can see from the graph that each x does not have a unique y. The set of ordered pairs defined by $x = \sin y$ is not a function of x since many ordered pairs have the same first element.

When we express a set of ordered pairs in terms of its equation, it is usually convenient to express the second element, y, in terms of the first element, x. Therefore, we want to rewrite $x = \sin y$ so that it is solved for y. We will begin by expressing the relationship in words.

$$x = \sin y$$

y is the angle whose sine is x

Since the measure of a central angle of a unit circle is equal to the measure of the arc, we could also say:

y is the arc whose sine is x

To abbreviate this statement and write it in symbolic form, we will pick out essential words.

y _is_ the _arc_ whose _sine_ is x

Using just these essential words, we can write the sentence as $y = \text{arc} \sin x$.

$$y = \text{arc} \sin x \leftrightarrow x = \sin y$$

☐ EXAMPLE 1: $\dfrac{\pi}{2} = \text{arc} \sin 1$ is equivalent to $1 = \sin \dfrac{\pi}{2}$.

Also, $90° = \text{arc} \sin 1$ is equivalent to $1 = \sin 90°$.

☐ EXAMPLE 2: $\dfrac{\pi}{6} = \text{arc} \sin \dfrac{1}{2}$ is equivalent to $\dfrac{1}{2} = \sin \dfrac{\pi}{6}$.

Also, $30° = \text{arc} \sin \dfrac{1}{2}$ is equivalent to $\dfrac{1}{2} = \sin 30°$.

Since $\dfrac{\pi}{2}$ radians and $90°$ are measures of the same angle, the two statements in example 1 reflect the same equality. Since $\dfrac{\pi}{6}$ radians and $30°$ are measures of the same angle, the two statements in example 2 reflect the same equality.

If we consider the values of θ for which $\theta = \sin \frac{1}{2}$, there are infinitely many solutions.

$$\theta = \text{arc} \sin \tfrac{1}{2} \leftrightarrow \sin \theta = \tfrac{1}{2}$$

Since θ is the measure of an angle whose sine is positive, θ can be the measure of an angle in quadrant I or quadrant II. In quadrant I, θ can be 30°, or any value that differs from 30° by a multiple of 360°. For example, 30° + 360° = 390°, 30° + 2(360°) = 750°, and 30° - 360° = -330°. In quadrant II, θ = 180° - 30° = 150°. Again, θ can be any value that differs from 150° by a multiple of 360°, such as 150° + 360° = 510°, 150° + 2(360°) = 870°, and 150° - 360° = -210°.

In terms of *degree measure*, the solution of the equation θ = arc sin $\frac{1}{2}$ is:

$$\theta = 30° + 360°k \qquad \text{OR} \qquad \theta = 150° + 360°k, \text{ where } k \text{ is any integer}$$

In terms of *radian measure*, the solution of the equation θ = arc sin $\frac{1}{2}$ is:

$$\theta = \frac{\pi}{6} + 2\pi k \qquad \text{OR} \qquad \theta = \frac{5\pi}{6} + 2\pi k, \text{ where } k \text{ is any integer}$$

We can use a similar notation for the reflection of the other trigonometric functions over the line $y = x$.

When the set of points whose equation is $y = \cos x$ is reflected over the line $y = x$, the image is the set of points whose equation is $x = \cos y$.

$$y = \text{arc } \cos x \leftrightarrow x = \cos y$$

When the set of points whose equation is $y = \tan x$ is reflected over the line $y = x$, the image is the set of points whose equation is $x = \tan y$.

$$y = \text{arc } \tan x \leftrightarrow x = \tan y$$

| **MODEL PROBLEMS** |

1. If θ = arc tan (-1) and $0° \leq \theta \leq 360°$, find θ.

Solution

1. Rewrite θ = arc tan (-1) as $\tan \theta = -1$.

2. Since $\tan \theta$ is negative, θ is in the second or fourth quadrant. Since $\tan \theta = -\tan 45°$, the measure of the reference angle is 45°. Therefore, in quadrant II, θ = 180° - 45° = 135°; and, in quadrant IV, θ = 360° - 45° = 315°.

Answer: 135° and 315°

2. Find cos (arc sin .6)

Solution

Here we are asked to find the cosine of an angle measure whose sine is .6.

1. Let θ = arc sin .6; that is, sin θ = .6.
2. Use the Pythagorean identity.　　$\cos^2 \theta = 1 - \sin^2 \theta$
3. Substitute.　　$\cos^2 \theta = 1 - (.6)^2$
4. Simplify.　　$\cos^2 \theta = 1 - .36$
　　　　　　　　$\cos^2 \theta = .64$
5. Solve for cos θ.　　$\cos \theta = \pm.8$

Answer: $\pm.8$

3. Find the value of cot (arc tan $\frac{8}{5}$).

Solution

Here we are asked to find the cotangent of an angle measure whose tangent is $\frac{8}{5}$.

1. Let ϕ = arc tan $\frac{8}{5}$; that is, tan ϕ = $\frac{8}{5}$.

2. Since $\cot \phi = \dfrac{1}{\tan \phi}$

$$\cot \phi = \frac{1}{\frac{8}{5}} = \frac{1}{\frac{8}{5}} \cdot \frac{5}{5} = \frac{5}{8}$$

Answer: $\dfrac{5}{8}$

EXERCISES

In 1–6, rewrite the statement as an equivalent statement, using arc sin, arc cos, or arc tan.

1. $\sin \theta = \dfrac{1}{2}$　　　　**2.** $\cos \theta = -\dfrac{\sqrt{3}}{2}$　　　　**3.** $\tan \theta = 2$

4. $\sin \theta = .1$　　　　**5.** $\tan \theta = -\frac{4}{5}$　　　　**6.** $\cos \theta = 1$

In 7–12, rewrite the statement as an equivalent statement, using sin θ, cos θ, or tan θ.

7. θ = arc cos $\dfrac{1}{2}$　　**8.** θ = arc tan $(-\sqrt{3})$　　**9.** θ = arc sin $\dfrac{\sqrt{2}}{2}$

10. θ = arc tan 1　　**11.** θ = arc sin (-1)　　**12.** θ = arc cos 0

In 13–21: **a.** Rewrite the statement as an equivalent statement, using sin θ, cos θ, or tan θ. **b.** Find all values of θ between 0° and 360° inclusive.

13. $\theta = \text{arc sin } 1$

14. $\theta = \text{arc cos}\left(-\dfrac{\sqrt{3}}{2}\right)$

15. $\theta = \text{arc tan } 0$

16. $\theta = \text{arc cos } \dfrac{\sqrt{2}}{2}$

17. $\theta = \text{arc sin } (.9903)$

18. $\theta = \text{arc cos } (.7314)$

19. $\theta = \text{arc tan } (-3.7321)$

20. $\theta = \text{arc cos } (-.8988)$

21. $\theta = \text{arc sin } (-.3907)$

In 22–27, find all values of θ in radians for $0 \leq \theta \leq 2\pi$.

22. $\theta = \text{arc sin } (-\frac{1}{2})$

23. $\theta = \text{arc tan } (-1)$

24. $\theta = \text{arc cos } \dfrac{\sqrt{3}}{2}$

25. $\theta = \text{arc sin } \dfrac{\sqrt{3}}{2}$

26. $\theta = \text{arc tan } \sqrt{3}$

27. $\theta = \text{arc cos}\left(-\dfrac{\sqrt{2}}{2}\right)$

28. Find sin (arc cos $\frac{1}{2}$).

29. Find tan (arc sin 0).

30. Find cos (arc sin (-1)).

31. Find sec (arc cos .3).

32. Find sin (arc csc 4).

33. Find tan (arc cot $\frac{3}{2}$).

34. Find the positive value of cos (arc sin $\frac{4}{5}$).

35. Find the positive value of sin (arc cos $(-\frac{5}{13})$).

36. Find the positive value of cos (arc sin .28).

37. Express cos (arc sin a) in terms of a.

38. Express sin (arc sin b) in terms of b.

In 39–42, select the numeral preceding the expression that best completes the sentence or answers the question.

39. If $x = \text{arc sin } (-\frac{1}{2})$, then x can be equal to:
(1) 30°   (2) 60°   (3) 300°   (4) 330°

40. If $y = \text{arc cos } (-\frac{1}{2})$, then y can be equal to:
(1) $\dfrac{\pi}{6}$ (2) $\dfrac{\pi}{3}$ (3) $\dfrac{2\pi}{3}$ (4) $\dfrac{5\pi}{6}$

41. If $\theta = \text{arc sin } \dfrac{\sqrt{2}}{2}$, then θ can *not* equal:
(1) 45°   (2) 135°   (3) 405°   (4) 585°

42. If $\theta = \text{arc cos } 1$, then θ can *not* equal which radian measure?
(1) 0   (2) π (3) 2π (4) -4π

9-10 INVERSE TRIGONOMETRIC FUNCTIONS

When the sets of points $y = \sin x$, $y = \cos x$, and $y = \tan x$ are reflected in the line $y = x$, the images are the sets of points $y = \text{arc sin } x$, $y = \text{arc cos } x$, and $y = \text{arc tan } x$. These sets of image points are not functions of x since each value of x is paired with many values of y. However, by limiting the range, that is, the allowable values of y, we can define functions that are inverse functions of the trigonometric functions.

Inverse Sine Function

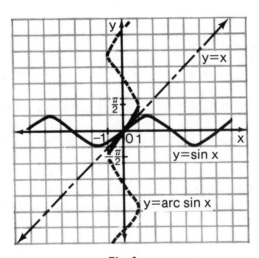

Fig. 1

In Fig. 1, two graphs are drawn: $y = \sin x$ and $y = \text{arc sin } x$ (the image of $y = \sin x$ under a reflection in the line $y = x$). We observe:

$y = \sin x$ is a function whose domain = $\{x | x \in \text{Real numbers}\}$ and whose range = $\{y | -1 \leq y \leq 1\}$.

However, $y = \text{arc sin } x$ is *not* a function of x. In $y = \text{arc sin } x$, the domain = $\{x | -1 \leq x \leq 1\}$ and the range = $\{y | y \in \text{Real numbers}\}$.

If we select values of y so that each value of x is assigned to a single value of y by the equation $y = \text{arc sin } x$, the resulting set of points will be a function. To do this, we must limit the range to the values of y from $-\dfrac{\pi}{2}$ to $\dfrac{\pi}{2}$ so that the values of x will vary from -1 to 1 just once.

This set of points (seen as the solid portion of the graph of $y = $ arc sin x in Fig. 1 and reproduced in Fig. 2) is a function.

To distinguish the relation $y = $ arc sin x from the function $y = $ Arc sin x just formed, we use the capital letter A to write the word "arc" in the function. Just as the range in $y = $ Arc sin x is restricted, notice that we must work with only a portion of the domain of the sine function (that is, a restricted domain of $y = $ sin x) to define the inverse sine function.

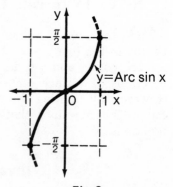

Fig. 2

A Subset of the Sine Function	Inverse Sine Function
$y = $ sin x	$y = $ Arc sin x
Restricted Domain $= \left\{ x \middle\| -\dfrac{\pi}{2} \le x \le \dfrac{\pi}{2} \right\}$	Domain $= \{ x \| -1 \le x \le 1 \}$
Range $= \{ y \| -1 \le y \le 1 \}$	Range $= \left\{ y \middle\| -\dfrac{\pi}{2} \le y \le \dfrac{\pi}{2} \right\}$

Another notation for $y = $ Arc sin x is $y = $ Sin^{-1} x, but we must be careful not to confuse this notation with the exponent -1 that indicates a reciprocal.

☐ EXAMPLE: Find Arc sin $\left(-\dfrac{\sqrt{2}}{2} \right)$.

If $\theta = $ Arc sin $\left(-\dfrac{\sqrt{2}}{2} \right)$, then sin $\theta = -\dfrac{\sqrt{2}}{2}$ and $-\dfrac{\pi}{2} \le \theta \le \dfrac{\pi}{2}$. Since sin $\dfrac{\pi}{4} = \dfrac{\sqrt{2}}{2}$, the reference angle is an angle of measure $\dfrac{\pi}{4}$. Since sin θ is negative, sin $\left(-\dfrac{\pi}{4} \right) = -\dfrac{\sqrt{2}}{2}$ and $\theta = -\dfrac{\pi}{4}$.

Answer: Arc sin $\left(-\dfrac{\sqrt{2}}{2} \right) = -\dfrac{\pi}{4}$

Inverse Cosine Function

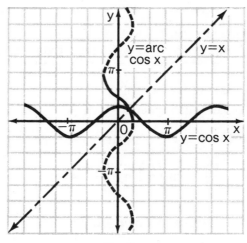

Fig. 3

In Fig. 3, two graphs are drawn: $y = \cos x$ and $y = \text{arc} \cos x$ (the image of $y = \cos x$ under a reflection in the line $y = x$). We observe:

$y = \cos x$ is a function whose domain $= \{x | x \in \text{Real numbers}\}$ and whose range $= \{y | -1 \leq y \leq 1\}$.

However, $y = \text{arc} \cos x$ is *not* a function of x. In $y = \text{arc} \cos x$, the domain $= \{x | -1 \leq x \leq 1\}$ and the range $= \{y | y \in \text{Real numbers}\}$.

As we did with the relation $y = \text{arc} \sin x$, we can select values of y so that each value of x is assigned to a single value of y by the equation $y = \text{arc} \cos x$. The resulting set of points will be a function. To do this, we must limit the range to the values of y from 0 to π so that the values of x will vary from -1 to 1 just once. This set of points (seen as the solid portion of the graph of $y = \text{arc} \cos x$ in Fig. 3 and reproduced in Fig. 4) is a function.

Just as with the inverse sine function, we use the capital A in writing the word "arc" to distinguish the relation $y = \text{arc} \cos x$ from the function $y = \text{Arc} \cos x$. The function $y = \text{Arc} \cos x$ is the inverse of a subset of the cosine function, $y = \cos x$, formed by restricting the domain.

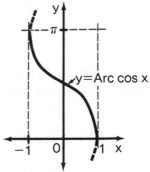

Fig. 4

A Subset of the Cosine Function	Inverse Cosine Function
$y = \cos x$	$y = \text{Arc } \cos x$
Restricted Domain = $\{x \mid 0 \le x \le \pi\}$ Range = $\{y \mid -1 \le y \le 1\}$	Domain = $\{x \mid -1 \le x \le 1\}$ Range = $\{y \mid 0 \le y \le \pi\}$

☐ EXAMPLE: Find $\sin (\text{Arc } \cos \frac{1}{2})$.

We are asked to find $\sin \theta$ when $\theta = \text{Arc } \cos \frac{1}{2}$, that is, when $\cos \theta = \frac{1}{2}$ and $0 \le \theta \le \pi$. Since $\cos \dfrac{\pi}{3} = \dfrac{1}{2}$, then $\theta = \dfrac{\pi}{3}$ and $\sin \dfrac{\pi}{3} = \dfrac{\sqrt{3}}{2}$. Therefore, $\sin \left(\text{Arc } \cos \dfrac{1}{2} \right) = \sin \left(\dfrac{\pi}{3} \right) = \dfrac{\sqrt{3}}{2}$.   *Ans.*

Inverse Tangent Function

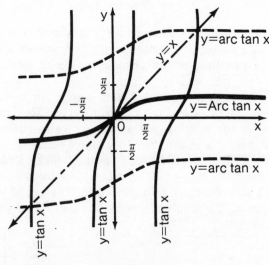

Fig. 5

In Fig. 5, two graphs are drawn: $y = \tan x$ and $y = \text{arc } \tan x$ (the image of $y = \tan x$ under a reflection in the line $y = x$). Recall that $\tan x$ is not defined for values of x that are odd multiples of $\dfrac{\pi}{2}$. Therefore, $y = \tan x$ is a function whose domain =

$$\left\{ x \,\middle|\, x \in \text{Real numbers and } x \ne \frac{(2k + 1)\pi}{2} \text{ for integral values of } k \right\}$$

and whose range = $\{y|y \in \text{Real numbers}\}$. However, $y = \text{arc tan } x$ is *not* a function of x.

For $y = \text{arc tan } x$, the domain = $\{x|x \in \text{Real numbers}\}$ and the range =

$$\left\{y\middle|y \in \text{Real numbers and } y \neq \frac{(2k + 1)\pi}{2} \text{ for integral values of } k\right\}.$$

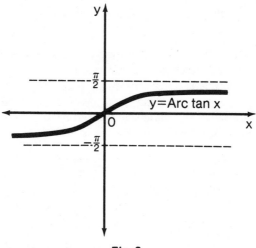

Fig. 6

If we select values of y so that each value of x is assigned to a single value of y by the equation $y = \text{arc tan } x$, the resulting set of points will be a function. To do this, we must limit the range to the values of y from $-\frac{\pi}{2}$ to $\frac{\pi}{2}$, but not including those values, so that the values of x will assume every real number as a value just once. This set of points (seen as a heavy line in Fig. 5 and reproduced in Fig. 6) is a function. The function $y = \text{Arc tan } x$ is the inverse of a subset of the tangent function, $y = \tan x$, formed by restricting the domain.

A Subset of the Tangent Function	Inverse Tangent Function
$y = \tan x$	$y = \text{Arc tan } x$
Restricted Domain = $\left\{x\middle|-\frac{\pi}{2} < x < \frac{\pi}{2}\right\}$	Domain = $\{x|x \in \text{Real numbers}\}$
Range = $\{y|y \in \text{Real numbers}\}$	Range = $\left\{y\middle|-\frac{\pi}{2} < y < \frac{\pi}{2}\right\}$

☐ EXAMPLE: Find cos (Arc tan $(-\sqrt{3})$).

We are asked to find cos θ when θ = Arc tan $(-\sqrt{3})$, that is, when tan $\theta = -\sqrt{3}$ and $-\frac{\pi}{2} < \theta < \frac{\pi}{2}$. Since tan $\frac{\pi}{3} = \sqrt{3}$, the angle whose measure is $\frac{\pi}{3}$ is the reference angle. Since tan θ is negative, tan $\left(-\frac{\pi}{3}\right) = -\sqrt{3}$ and $\theta = -\frac{\pi}{3}$. Thus, cos (Arc tan $(-\sqrt{3})$) = cos $\left(-\frac{\pi}{3}\right) = \frac{1}{2}$. *Ans.*

The functions y = Arc sec x, y = Arc csc x, and y = Arc cot x can be defined by a similar restriction of the range of the corresponding relation. In turn, we must restrict the domain of the functions y = sec x, y = csc x, and y = cot x to define these inverse trigonometric functions. The restricted domains and ranges are listed here for students who wish to investigate these functions.

A Subset of the Secant Function	Inverse Secant Function
y = sec x	y = Arc sec x
Restricted Domain = $\left\{x \middle\vert 0 \le x \le \pi, \text{and} \right.$ $\left. x \ne \frac{\pi}{2}\right\}$	Domain = $\{x\vert\ \vert x\vert \ge 1\}$ Range = $\left\{y \middle\vert 0 \le y \le \pi, \text{and} \right.$ $\left. y \ne \frac{\pi}{2}\right\}$
Range = $\{y\vert\ \vert y\vert \ge 1\}$	
A Subset of the Cosecant Function	Inverse Cosecant Function
y = csc x	y = Arc csc x
Restricted Domain = $\left\{x \middle\vert 0 < \vert x\vert \le \frac{\pi}{2}\right\}$	Domain = $\{x\vert\ \vert x\vert \ge 1\}$ Range = $\left\{y \middle\vert 0 < \vert y\vert \le \frac{\pi}{2}\right\}$
Range = $\{y\vert\ \vert y\vert \ge 1\}$	
A Subset of the Cotangent Function	Inverse Cotangent Function
y = cot x	y = Arc cot x
Restricted Domain = $\left\{x \middle\vert 0 < \vert x\vert \le \frac{\pi}{2}, \text{and} \right.$ $\left. x \ne -\frac{\pi}{2}\right\}$	Domain = $\{x\vert x \in \text{Real numbers}\}$ Range = $\left\{y \middle\vert 0 < \vert y\vert \le \frac{\pi}{2}, \text{and} \right.$ $\left. y \ne -\frac{\pi}{2}\right\}$
Range = $\{y\vert y \in \text{Real numbers}\}$	

Degree Measure of the Inverse Trigonometric Functions

When the expression Arc sin a, Arc cos a, or Arc tan a is evaluated, the value may be expressed in degrees as well as in radians.

$$\theta = \text{Arc sin } a \leftrightarrow \sin \theta = a \text{ and } -\frac{\pi}{2} \leq \theta \leq \frac{\pi}{2} \text{ in radians}$$

$$\sin \theta = a \text{ and } -90° \leq \theta \leq 90° \text{ in degrees}$$

$$\theta = \text{Arc cos } a \leftrightarrow \cos \theta = a \text{ and } 0 \leq \theta \leq \pi \text{ in radians}$$

$$\cos \theta = a \text{ and } 0° \leq \theta \leq 180° \text{ in degrees}$$

$$\theta = \text{Arc tan } a \leftrightarrow \tan \theta = a \text{ and } -\frac{\pi}{2} < \theta < \frac{\pi}{2} \text{ in radians}$$

$$\tan \theta = a \text{ and } -90° < \theta < 90° \text{ in degrees}$$

□ EXAMPLE: Evaluate Arc tan (-1).

Let Arc tan $(-1) = \theta$. Then $\tan \theta = -1$ and $-\frac{\pi}{2} < \theta < \frac{\pi}{2}$ in radians or $-90° < \theta < 90°$ in degrees.

Since $\tan \frac{\pi}{4} = 1$ and $\tan 45° = 1$, the reference angle is an angle of $\frac{\pi}{4}$ radians or $45°$. Therefore, $\tan \left(-\frac{\pi}{4}\right) = -1$ and $\tan (-45°) = -1$. The value of Arc tan (-1) can be given as $-\frac{\pi}{4}$ radians or as $-45°$.

Answer: Arc tan $(-1) = -\frac{\pi}{4}$, or Arc tan $(-1) = -45°$

| MODEL PROBLEMS |

1. If $\theta = \text{Arc cos } 0$, what is the value of θ in radians?

Solution

We are asked to find the value of θ when $\cos \theta = 0$ and $0 \leq \theta \leq \pi$.

Since $\cos \frac{\pi}{2} = 0$, $\theta = \frac{\pi}{2}$.

Answer: $\frac{\pi}{2}$

2. If θ = Arc sin $(-\frac{1}{2})$, find the value of θ in degrees.

Solution

We are asked to find θ when sin $\theta = -\frac{1}{2}$ and $-90° \le \theta \le 90°$. Since sin $30° = \frac{1}{2}$, the degree measure of the reference angle is $30°$. Therefore, sin $(-30°) = -\frac{1}{2}$ and $\theta = -30°$.

Answer: $-30°$

3. Find the value of Arc tan $\dfrac{\sqrt{3}}{3}$.

Solution

We are asked to find the measure of an angle whose tangent is $\dfrac{\sqrt{3}}{3}$ when that measure is between $-\dfrac{\pi}{2}$ and $\dfrac{\pi}{2}$. Since tan $\dfrac{\pi}{6} = \dfrac{\sqrt{3}}{3}$, Arc tan $\dfrac{\sqrt{3}}{3} = \dfrac{\pi}{6}$.

However, the measure of the angle could also have been given in degrees. Since tan $30° = \dfrac{\sqrt{3}}{3}$ and $-90° < 30° < 90°$, it is also true that Arc tan $\dfrac{\sqrt{3}}{3} = 30°$.

Answer: $\dfrac{\pi}{6}$ radians or $30°$.

| EXERCISES |

In 1–12, find the value of θ in degrees.

1. θ = Arc cos $\frac{1}{2}$ 2. θ = Arc sin 0 3. θ = Arc tan 1

4. θ = Arc sin $\frac{1}{2}$ 5. θ = Arc cos (-1) 6. θ = Arc cos $\left(-\dfrac{\sqrt{3}}{2}\right)$

7. θ = Arc tan $(-\sqrt{3})$ 8. θ = Arc sin (-1)

9. θ = Arc tan 0 10. θ = Arc sin $\left(-\dfrac{\sqrt{2}}{2}\right)$

11. θ = Arc cos 0 12. θ = Arc cos 1

In 13-24, find the value of θ in radians.

13. $\theta = \text{Arc sin } \dfrac{\sqrt{3}}{2}$

14. $\theta = \text{Arc tan } \sqrt{3}$

15. $\theta = \text{Arc cos } \left(-\dfrac{\sqrt{2}}{2}\right)$

16. $\theta = \text{Arc sin } 1$

17. $\theta = \text{Arc sin } (-\frac{1}{2})$

18. $\theta = \text{Arc tan } (-1)$

19. $\theta = \text{Arc cos } \dfrac{\sqrt{2}}{2}$

20. $\theta = \text{Arc tan } \left(-\dfrac{\sqrt{3}}{3}\right)$

21. $\theta = \text{Arc cos } (-\frac{1}{2})$

22. $\theta = \text{Arc sec } 2$

23. $\theta = \text{Arc cot } (-\sqrt{3})$

24. $\theta = \text{Arc csc } \sqrt{2}$

In 25-33, find the value of the given expression.

25. $\sin (\text{Arc cos } (-1))$

26. $\tan (\text{Arc sin } 0)$

27. $\tan (\text{Arc cos } \frac{4}{5})$

28. $\cos (\text{Arc tan } (-\frac{5}{12}))$

29. $\sec (\text{Arc cos } \frac{1}{6})$

30. $\sin (\text{Arc csc } (-\sqrt{2}))$

31. $\cot (\text{Arc tan } 1)$

32. $\cos (\text{Arc sin } \frac{1}{3})$

33. $\sec (\text{Arc tan } \sqrt{3})$

In 34-41, select the numeral preceding the expression that best completes the sentence.

34. If $\theta = \text{Arc sec } \sqrt{2}$, then θ equals:
(1) 30°   (2) 45°   (3) 60°   (4) 90°

35. The value of $\frac{1}{2} \cot (\text{Arc tan } \frac{8}{5})$ is:
(1) $\frac{5}{4}$ (2) $\frac{5}{16}$ (3) $\frac{4}{5}$ (4) $\frac{8}{5}$

36. The value of Arc sin $(-\frac{1}{2})$ is:
(1) 30°   (2) -60°   (3) -30°   (4) 210°

37. If $\theta = \text{Arc cos } \left(-\dfrac{\sqrt{3}}{2}\right)$, then θ equals:

(1) $-\dfrac{\pi}{6}$ (2) $\dfrac{\pi}{6}$ (3) $-\dfrac{5\pi}{6}$ (4) $\dfrac{5\pi}{6}$

38. A value of y that is *not* in the range of the function $y = \text{Arc sin } x$ is:

(1) 0   (2) $\dfrac{\pi}{2}$ (3) π (4) $-\dfrac{\pi}{2}$

39. The value of $2 \cos (\text{Arc sin } (-.6))$ is:
(1) 1.6   (2) -1.6   (3) .8   (4) -.8

40. If $f(x) = \text{Arc sin } x$, then $f(-\frac{1}{2})$ is:

(1) $\frac{\pi}{3}$ (2) $-\frac{\pi}{3}$ (3) $\frac{\pi}{6}$ (4) $-\frac{\pi}{6}$

41. A number that is *not* an element of the range of $y = \text{Arc tan } x$ is:

(1) $\frac{\pi}{3}$ (2) $\frac{\pi}{2}$ (3) 0   (4) $-\frac{\pi}{6}$

9-11 REVIEW EXERCISES

In 1-8, for the given function, state: **a.** its amplitude, if possible **b.** its period

1. $y = \sin x$ **2.** $y = 3 \cos 2x$ **3.** $y = \tan x$

4. $y = \frac{1}{2} \sin x$ **5.** $y = 2 \cos \frac{1}{2}x$ **6.** $y = 4 \sin 3x$

7. $y = -2 \cos \frac{x}{3}$ **8.** $y = -\frac{1}{2} \sin 2x$

In 9-16, for the given function, state: **a.** the domain   **b.** the range

9. $y = \sin x$ **10.** $y = 2 \cos x$ **11.** $y = \tan x$

12. $y = 3 \sin 2x$ **13.** $y = -2 \cos \frac{1}{2}x$ **14.** $y = \text{Arc sin } x$

15. $y = \text{Arc cos } x$ **16.** $y = \text{Arc tan } x$

17. **a.** Sketch the graph of $y = 2 \sin x$ for values of x in the interval $0 \leq x \leq 2\pi$.
     b. On the same set of axes, sketch the graph of $y = \cos 2x$.
     c. How many values of x in the interval $0 \leq x \leq 2\pi$ are solutions of the equation $2 \sin x = \cos 2x$?

18. **a.** Sketch the graph of $y = \cos \frac{1}{2}x$ for values of x in the interval $-2\pi \leq x \leq 2\pi$.
     b. On the same set of axes, sketch the graph of $y = \tan x$.
     c. For how many values of x in the interval $-2\pi \leq x \leq 2\pi$ does $\cos \frac{1}{2}x = \tan x$?

19. **a.** On the same set of axes, sketch the graphs of $y = \sin 2x$ and $y = 2 \cos x$ for values of x in the interval $-\pi \leq x \leq \pi$.
     b. How many values of x in the interval $-\pi \leq x \leq \pi$ are solutions of $\sin 2x = 2 \cos x$?

In 20-28, find all values of θ in the interval $0° \leq \theta \leq 360°$ that make the statement true.

20. $\theta = \text{arc sin } .5$ **21.** $\theta = \text{arc cos } (-1)$ **22.** $\theta = \text{arc tan } 1$

23. $\theta = \text{arc cos } 0$ **24.** $\theta = \text{arc sin}\left(-\dfrac{\sqrt{3}}{2}\right)$ **25.** $\theta = \text{arc tan }(-\sqrt{3})$

26. $\theta = \text{arc sec }(-2)$ **27.** $\theta = \text{arc csc }(-\sqrt{2})$ **28.** $\theta = \text{arc cot } 0$

In 29–34, write the value of the expression in radian measure.

29. Arc cos $\frac{1}{2}$ **30.** Arc sin (-1) **31.** Arc tan (-1)

32. Arc cos $\left(-\dfrac{\sqrt{3}}{2}\right)$ **33.** Arc cot 0 **34.** Arc csc 1

In 35–40, find the value of the given expression.

35. sin (Arc cos (-1))   **36.** cos (Arc sin .6)
37. tan (Arc cos .8)   **38.** csc (Arc sin $\frac{1}{3}$)
39. sec (Arc cos $(-\frac{2}{3})$)   **40.** tan (Arc cot (-4))

In 41–51, select the numeral preceding the expression that best completes the sentence or answers the question.

41. When x increases from $x = 0$ to $x = \pi$, the value of cos x:
(1) decreases   (2) decreases, then increases
(3) increases   (4) increases, then decreases

42. When x increases from $x = 0$ to $x = \pi$, the value of sin x:
(1) decreases   (2) decreases, then increases
(3) increases   (4) increases, then decreases

43. When x increases from $x = -\dfrac{\pi}{3}$ to $x = \dfrac{\pi}{3}$, the value of tan x:

(1) decreases   (2) decreases, then increases
(3) increases   (4) increases, then decreases

44. If $x = \text{arc cos}\left(-\dfrac{\sqrt{2}}{2}\right)$, then x can be equal to:

(1) $-45°$ (2) $45°$ (3) $135°$ (4) $315°$

45. If $f(x) = 2 \sin 2x$, then the value of $f\left(\dfrac{\pi}{4}\right)$ is:

(1) 1   (2) 2   (3) 0   (4) $2\sqrt{2}$

46. Which is *not* an element of the domain of $y = \tan x$?

(1) π (2) 0   (3) $-\pi$ (4) $-\dfrac{\pi}{2}$

47. The period of the curve whose equation is $y = \frac{1}{3} \cos 2x$ is:
(1) $\frac{1}{3}$ (2) 2   (3) π (4) 4π

48. Under which translation is the graph of $y = \sin x$ the image of the graph of $y = \cos x$?

(1) $T_{\frac{\pi}{2},0}$ (2) $T_{\pi,0}$ (3) $T_{2\pi,0}$ (4) $T_{-\pi,0}$

49. What is the maximum value of the function $y = 3 \sin 2x$?

(1) 1   (2) 2   (3) 3   (4) 6

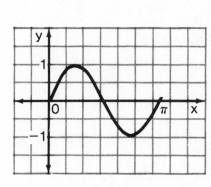

Ex. 50

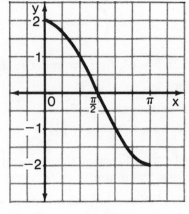

Ex. 51

50. The diagram shows the graph of:

(1) $y = \sin x$ (2) $y = \sin 2x$ (3) $y = \cos x$ (4) $y = \cos 2x$

51. The diagram shows the graph of:

(1) $y = 2 \sin x$ (2) $y = \cos 2x$ (3) $y = \sin 2x$ (4) $y = 2 \cos x$

Chapter 10

Exponential Functions

10-1 LAWS OF EXPONENTS

In Chapter 1, we reviewed some laws of exponents learned in earlier courses. The following rules summarize operations on powers with like bases:

■ If a, b, and c are positive integers:

1. *Multiplication Law.* $x^a \cdot x^b = x^{a+b}$
2. *Division Law.* $x^a \div x^b = x^{a-b}$ ($x \neq 0$ and $a > b$)
3. *Power Law.* $(x^a)^c = x^{ac}$

For example:

1. $x^3 \cdot x^2 = x^5$ AND $4^3 \cdot 4^2 = 4^5$
2. $y^7 \div y^4 = y^3$ AND $3^7 \div 3^4 = 3^3$
3. $(\sin^2 \theta)^3 = \sin^6 \theta$ AND $(10^2)^3 = 10^6$

In each of these examples, we were working with powers of like bases. Let us consider some examples of powers that have unlike bases, such as xy^3 and $(xy)^3$. Recall that:

$$xy^3 = x \cdot y \cdot y \cdot y, \text{ while}$$
$$(xy)^3 = (xy)(xy)(xy) = x \cdot x \cdot x \cdot y \cdot y \cdot y = x^3 y^3$$

For example, if $x = 2$ and $y = 3$, then:

1. $xy^3 = 2 \cdot 3^3 = 2 \cdot 3 \cdot 3 \cdot 3 = 54$
2. $(xy)^3 = (2 \cdot 3)^3 = 6^3 = 6 \cdot 6 \cdot 6 = 216$, or
   $(xy)^3 = x^3 y^3 = 2^3 \cdot 3^3 = 8 \cdot 27 = 216$

Let us consider the expressions $-x^4$ and $(-x)^4$. Since $-x = -1 \cdot x$, it follows that $-x^4 = -1 \cdot x^4$. However, $(-x)^4 = (-x)(-x)(-x)(-x) = x^4$.

For example, if $x = 2$, then:

1. $-x^4 = -1 \cdot x^4 = -1 \cdot 2^4 = -1 \cdot 2 \cdot 2 \cdot 2 \cdot 2 = -16$
2. $(-x)^4 = (-1 \cdot x)^4 = (-1 \cdot 2)^4 = (-2)^4 = (-2)(-2)(-2)(-2) = 16$, or
   $(-x)^4 = (-1 \cdot x)^4 = (-1)^4 \cdot (x)^4 = (-1)^4 \cdot (2)^4 = 1 \cdot 16 = 16$

When dividing powers with unlike bases, we observe a similar pattern. Recall that:

$$\frac{a^3}{b} = \frac{a \cdot a \cdot a}{b}, \text{ while}$$

$$\left(\frac{a}{b}\right)^3 = \left(\frac{a}{b}\right)\left(\frac{a}{b}\right)\left(\frac{a}{b}\right) = \frac{a^3}{b^3}$$

For example, if $a = 6$ and $b = 4$, then:

1. $\dfrac{a^3}{b} = \dfrac{6^3}{4} = \dfrac{6 \cdot 6 \cdot 6}{4} = \dfrac{216}{4} = 54$

2. $\left(\dfrac{a}{b}\right)^3 = \dfrac{a^3}{b^3} = \dfrac{6^3}{4^3} = \dfrac{6 \cdot 6 \cdot 6}{4 \cdot 4 \cdot 4} = \dfrac{216}{64} = \dfrac{27}{8}$

From the examples we have just seen, we now formulate two more laws of exponents.

■ In general, if a is a positive integer:

4. *Power-of-a-Product Law.* $(xy)^a = x^a y^a$

5. *Power-of-a-Quotient Law.* $\left(\dfrac{x}{y}\right)^a = \dfrac{x^a}{y^a}$ $(y \neq 0)$

| MODEL PROBLEMS |

1. Simplify: $\dfrac{2^8 \cdot 2^4}{(2^5)^2}$

How to Proceed   *Solution*

1. Use the rule for multiplying powers with like $\dfrac{2^8 \cdot 2^4}{(2^5)^2} = \dfrac{2^{12}}{2^{10}}$
   bases to simplify the numerator. Use the rule
   for finding the power of a power to simplify
   the denominator.

2. Use the rule for dividing powers with like bases and simplify. $= 2^2$
                                                                      $= 4$

Answer: 4

2. Find the value of $\dfrac{(xy)^4}{xy^3}$ when $x = 3$ and $y = 5$.

How to Proceed	*Solution*
1. Write the numerator without parentheses.	$\dfrac{(xy)^4}{xy^3} = \dfrac{x^4 y^4}{xy^3}$
2. Divide powers with like bases.	$= x^3 y$
3. Substitute.	$= 3^3 \cdot 5$
4. Simplify.	$= 27 \cdot 5$
	$= 135$

Answer: 135

EXERCISES

In 1–40, simplify. The literal exponents represent positive integers; $x \neq 0$ and $y \neq 0$.

1. $x^7 \cdot x^2$ 2. $x \cdot x^5$ 3. $x^9 \div x^4$ 4. $a^5 \div a$

5. $(a^3)^2$ 6. $(y^4)^3$ 7. $2^5 \cdot 2^2$ 8. $3^4 \div 3^3$

9. $\dfrac{x^8 \cdot x^2}{x^7}$ 10. $\dfrac{10^5 \cdot 10^4}{10^6}$ 11. $\dfrac{(x^5)^3}{x^{10}}$ 12. $\dfrac{(7^3)^4}{7^{10}}$

13. $\dfrac{y^4 \cdot y^3}{(y^2)^3}$ 14. $\dfrac{5^4 \cdot 5^7}{(5^3)^3}$ 15. $\dfrac{-x^5}{(-x)^4}$ 16. $\dfrac{-9^{10}}{(-9^2)^3}$

17. $\dfrac{(xy)^5}{xy^3}$ 18. $\dfrac{(2x)^5}{2x^3}$ 19. $\dfrac{(3a)^5}{3^3}$ 20. $\dfrac{-4y^7}{(2y)^2}$

21. $5^a \cdot 5^b$ 22. $3^4 \cdot 3^a$ 23. $x^b \cdot x$ 24. $y^{n+2} \cdot y^{n+1}$

25. $\dfrac{6^c}{6^b}$ 26. $\dfrac{10^3}{10^a}$ 27. $\dfrac{y^{4c}}{y^c}$ 28. $\dfrac{x^{k+2}}{x^k}$

29. $(\tfrac{1}{2} x^2)^3$ 30. $(4y^b)^2$ 31. $(-x^c)^2$ 32. $(x^2 y)^k$

33. $\dfrac{(2 \cdot 5)^4}{5^4}$ 34. $\dfrac{2 \cdot 5^4}{5^4}$ 35. $\dfrac{(3 \cdot 7)^a}{7^a}$ 36. $\dfrac{3 \cdot 7^a}{7^a}$

37. $3^4 \cdot 2^4$ 38. $2^c \cdot 5^c$ 39. $\dfrac{12^4}{6^4}$ 40. $\dfrac{6^a}{2^a}$

In 41–48, evaluate the expression when $a = 5$ and $b = 2$.

41. ab^2 42. $(ab)^2$ 43. $-ab^2$ 44. $(-ab)^2$
45. $-(ab)^2$ 46. $ab(ab)^3$ 47. $-a^4$ 48. $-a^3 \div (-a)^2$

49. Express each number as a power of 10.
    a. 100   b. 1000   c. 1,000,000   d. $10^5 \cdot 100^2$ e. $100^3 \div 10^4$
50. Express each number as a power of 4.
    a. 64   b. 256   c. $4^5 \cdot 16$ d. $4^{10} \div 16$ e. 16^3

In 51–54, select the numeral preceding the expression that best completes the sentence.

51. The expression 2^6 is equal to:
    (1) 4^5 (2) 4^2 (3) 4^3 (4) 4^4
52. The expression $3 \cdot 3^7$ is equal to:
    (1) 3^8 (2) 3^7 (3) 9^8 (4) 9^7
53. If $3^a = b$, then 3^{a+1} equals:
    (1) $b + 1$ (2) $3b$ (3) $b + 3$ (4) b^2
54. If $10^x = c$, then $100c$ equals:
    (1) 10^{x+2} (2) 10^{2x} (3) 1000^x (4) 100^{2x}

10-2 EXPONENTS THAT ARE NOT POSITIVE

Zero Exponents

We have defined a power as the product of equal factors. This definition requires that the exponent be a positive integer. Now we will define powers having exponents that are not positive in a way that is consistent with the rules for powers already established.

Recall the rule for dividing powers with like bases:

$$x^a \div x^b = x^{a-b} \qquad (x \neq 0)$$

If we do not require $a > b$, then a may be equal to b. When $a = b$:

$$x^a \div x^b = x^a \div x^a = x^{a-a} = x^0$$
$$\text{But } x^a \div x^a = 1$$

Therefore, in order for x^0 to be meaningful, we must make the following definition:

$$x^0 = 1 \qquad (x \neq 0)$$

Since the definition $x^0 = 1$ is based upon division, and division by 0 is not possible, we have stated that $x \neq 0$. Actually, the expression 0^0 (zero to the zero power) is one of several *indeterminate* expressions in mathematics. It is not possible to assign a value to an indeterminate expression.

Before we accept x^0 as a power, we must show that it satisfies the rules for powers. For example:

1. $x^a \cdot x^0 = x^{a+0} = x^a$ AND $x^a \cdot x^0 = x^a \cdot 1 = x^a$
2. $x^a \div x^0 = x^{a-0} = x^a$ AND $x^a \div x^0 = x^a \div 1 = x^a$
3. $(x^0 \cdot x^a)^b = (x^{a+0})^b = x^{ab}$ AND $(x^0 \cdot x^a)^b = (1 \cdot x^a)^b = x^{ab}$

Since x can be any base except 0, any nonzero base to the zero power is 1.

$$4^0 = 1 \qquad 12^0 = 1 \qquad (2ab)^0 = 1 \text{ if } a \neq 0 \text{ and } b \neq 0$$

Care must be taken when using the zero exponent, as seen in the following examples:

1. $(-4)^0 = 1$ 2. $-4^0 = -1 \cdot 4^0 = -1 \cdot 1 = -1$
3. $(3x)^0 = 1$ 4. $3x^0 = 3 \cdot x^0 = 3 \cdot 1 = 3$

Negative Exponents

Recall again the rule for dividing powers with like bases:

$$x^a \div x^b = x^{a-b} \qquad (x \neq 0)$$

If we do not require $a > b$, then a may be less than b. When $a < b$ and $x \neq 0$, we may use the division law as follows:

$$\frac{x^3}{x^5} = x^{3-5} = x^{-2}$$

It is also true that $\dfrac{x^3}{x^5} = \dfrac{\overset{1}{\cancel{x}} \cdot \overset{1}{\cancel{x}} \cdot \overset{1}{\cancel{x}}}{\underset{1}{\cancel{x}} \cdot \underset{1}{\cancel{x}} \cdot \underset{1}{\cancel{x}} \cdot x \cdot x} = \dfrac{1}{x^2}$

Therefore, $x^{-2} = \dfrac{1}{x^2}$.

We can use the division law for powers with like bases and the definition of the zero exponent to show that this relationship holds for any exponent b and any base $x \neq 0$.

$$\frac{1}{x^b} = \frac{x^0}{x^b} = x^{0-b} = x^{-b}$$

We can therefore make the following definition:

$$x^{-b} = \frac{1}{x^b} \qquad (x \neq 0)$$

In order to accept this as a valid definition, we must show that powers with negative exponents satisfy the rules for exponents. For example:

1. $x^5 \cdot x^{-3} = x^{5+(-3)} = x^2$ AND $x^5 \cdot x^{-3} = x^5 \cdot \frac{1}{x^3} = x^2$

2. $x^4 \div x^{-2} = x^{4-(-2)} = x^6$ AND $x^4 \div x^{-2} = x^4 \div \frac{1}{x^2} = x^4 \cdot x^2 = x^6$

3. $(x^{-3})^2 = x^{-3(2)} = x^{-6}$ AND $(x^{-3})^2 = \left(\frac{1}{x^3}\right)^2 = \frac{1}{x^6} = x^{-6}$

4. Since x^{-b} is the reciprocal of x^b, the base $x \neq 0$. Just as $x^{-b} = \frac{1}{x^b}$, it can be shown that $\frac{1}{x^{-n}} = x^n$. Consider this example:

$$\frac{1}{x^{-4}} = \frac{x^0}{x^{-4}} = x^{0-(-4)} = x^4$$

OR

$$\frac{1}{x^{-4}} = 1 \div x^{-4} = 1 \div \frac{1}{x^4} = 1 \cdot x^4 = x^4$$

5. $\left(\frac{2}{3}\right)^{-n} = \frac{2^{-n}}{3^{-n}} = 2^{-n} \div 3^{-n} = \frac{1}{2^n} \div \frac{1}{3^n} = \frac{1}{2^n} \cdot 3^n = \frac{3^n}{2^n} = \left(\frac{3}{2}\right)^n$

These last examples illustrate the truth of the following generalization:

$$\left(\frac{a}{b}\right)^{-n} = \left(\frac{b}{a}\right)^n \qquad (a \neq 0, b \neq 0)$$

| MODEL PROBLEMS |

1. Find the value of $3a^0 + a^{-2}$ if $a = 4$.

How to Proceed	*Solution*
1. Use the definitions $a^0 = 1$ and $a^{-2} = \dfrac{1}{a^2}$.	$3a^0 + a^{-2} = 3(1) + \dfrac{1}{a^2}$
2. Substitute the value of a.	$= 3(1) + \dfrac{1}{4^2}$
3. Simplify.	$= 3 + \dfrac{1}{16}$
	$= 3\dfrac{1}{16}$　*Ans.*

2. Write the expression $\dfrac{2c^2}{a^3 c^{-2}}$ without a denominator.

How to Proceed	*Solution*
1. Rewrite the expression, grouping powers with like bases.	$\dfrac{2c^2}{a^3 c^{-2}} = 2 \cdot \dfrac{1}{a^3} \cdot \dfrac{c^2}{c^{-2}}$
2. Use the definition of a negative exponent and the rule for dividing powers with like bases.	$= 2a^{-3} c^{2-(-2)}$
3. Simplify.	$= 2a^{-3} c^4$　*Ans.*

3. Using only positive exponents, write an expression equivalent to $\dfrac{5x^{-4}}{y^{-2}}$.

How to Proceed	*Solution*
1. Rewrite the expression, using a separate factor for each base.	$\dfrac{5x^{-4}}{y^{-2}} = 5 \cdot x^{-4} \cdot \dfrac{1}{y^{-2}}$
2. Rewrite each power that has a negative exponent, using the rules $x^{-b} = \dfrac{1}{x^b}$ and $\dfrac{1}{x^{-n}} = x^n$.	$= 5 \cdot \dfrac{1}{x^4} \cdot y^2$
	$= \dfrac{5y^2}{x^4}$　*Ans.*

EXERCISES

In 1–16, find the value of the given expression if $a \neq 0, y \neq 0$.

1. a^0
2. $2a^0$
3. $(2a)^0$
4. 4^0
5. -4^0
6. $(-4)^0$
7. $(5y)^0$
8. $5y^0$
9. $-5y^0$
10. $(-5y)^0$
11. 10^0
12. -10^0
13. $(-10)^0$
14. $(\frac{1}{4})^0$
15. $.2^0$
16. $(\frac{2}{3})^0$

In 17–44, write an equivalent expression, using only positive exponents. The bases are not equal to 0.

17. x^{-2}
18. a^{-5}
19. c^{-1}
20. $4b^{-2}$
21. $3k^{-4}$
22. $7y^{-1}$
23. $8x^{-3}$
24. $(2b)^{-1}$
25. $(3a)^{-3}$
26. $(9r)^{-2}$
27. $(2x)^{-4}$
28. $(.1b)^{-5}$

29. $\dfrac{1}{a^{-9}}$
30. $\dfrac{1}{x^{-4}}$
31. $\dfrac{1}{c^{-7}}$
32. $\dfrac{1}{d^{-3}}$

33. $\dfrac{3}{x^{-1}}$
34. $\dfrac{5}{y^{-10}}$
35. $\dfrac{x}{y^{-2}}$
36. $\dfrac{a}{c^{-6}}$

37. $\dfrac{b^2}{a^{-8}}$
38. $\dfrac{a^{-3}}{b^{-5}}$
39. $\dfrac{b^{-7}}{b^{-2}}$
40. $\dfrac{x^{-1}}{x^{-3}}$

41. $\dfrac{2y^{-4}}{y^4}$
42. $\dfrac{c^{-3}}{3c}$
43. $\dfrac{3d^{-5}}{6d}$
44. $\dfrac{x^{10}}{2x^{-1}}$

In 45–56, simplify and write the result without a denominator. The bases are not equal to 0.

45. $\dfrac{ab^2}{a^5 b^4}$
46. $\dfrac{x^3}{x^3 y^5}$
47. $\dfrac{(x^3 y)^3}{xy^3}$
48. $\dfrac{12x^2}{3x^{-7}}$

49. $\dfrac{4b^{-2}}{4b^{-3}}$
50. $\dfrac{9x^{-5}}{3x^{-8}}$
51. $\dfrac{4d^{-3}}{cd^{-1}}$
52. $\dfrac{(a^{-1})^{-2}}{a^4 b^2}$

53. $\dfrac{(x^{-2})^{-3}}{x^3 y^6}$
54. $\dfrac{12(y^2)^{-1}}{(12y)^{-1}}$
55. $\dfrac{(12y^2)^{-1}}{(12y)^{-1}}$
56. $\dfrac{3^{-2}}{(3a^2)^{-2}}$

In 57–84, find the value of the given expression.

57. 7^{-1}
58. 2^{-3}
59. 3^{-2}
60. 6^{-3}
61. $(\frac{1}{4})^2$
62. $(\frac{1}{4})^{-2}$
63. $(\frac{1}{10})^{-3}$
64. $(-\frac{1}{10})^3$
65. $5^0(2)^{-1}$
66. $12^0(4)^{-3}$
67. $8^{-1}(7)^0$
68. $10^{-2}(5)^2$
69. $(-8)^{-2}$
70. -8^{-2}
71. $(-9)^{-1}$
72. -9^{-1}
73. $(-5)^{-3}$
74. -5^{-3}
75. $(-10)^{-4}$
76. -10^{-4}

77. $\dfrac{8^0}{8^{-2}}$ **78.** $\dfrac{3^0}{3^{-4}}$ **79.** $\dfrac{5^0}{5^{-3}}$ **80.** $\dfrac{4^{-4}}{4^{-5}}$

81. $\dfrac{(2 \cdot 3)^2}{2 \cdot 3^{-1}}$ **82.** $\dfrac{(5 \cdot 2)^3}{5 \cdot 2^{-2}}$ **83.** $\dfrac{(7 \cdot 3)^{-1}}{7 \cdot 3^{-4}}$ **84.** $\dfrac{3 \cdot 8^{-2}}{(3 \cdot 8)^{-3}}$

85. Find the value of $7a^0 + (5a)^0$ if $a \neq 0$.
86. Find the value of $3b^{-2} + 3b^0$ when $b = 6$.
87. Find the value of $(\frac{3}{4})^{-2} \cdot (4)^{-3}$.
88. If $f(x) = 2x^{-1}$, find $f(\frac{2}{3})$.
89. If $f(x) = 2x^{-1} + (2x)^{-1}$, find $f(8)$.
90. If $f(x) = 3x^{-2}$, find $f(6)$.
91. If $f(x) = x^{-1} + (2x)^{-1}$, find $f(3)$.
92. If the function $g(x) = 2x^{-2} + x^{-1} + 3x^0$, find $g(2)$.

In 93–97, select the numeral preceding the expression that best completes the sentence.

93. The expression $2a^{-2}b^3$ is equivalent to:

(1) $\dfrac{b^3}{2a^2}$ (2) $\dfrac{a^{-2}}{2b^{-3}}$ (3) $\dfrac{1}{2a^2b^{-3}}$ (4) $\dfrac{2b^3}{a^2}$

94. The expression $\dfrac{12x^{-2}}{(3x)^{-1}}$ is equivalent to:

(1) $4x^{-3}$ (2) $36x^{-3}$ (3) $36x^{-1}$ (4) $4x^{-1}$

95. The expression $(9a^{-1})(3a)^{-2}$ is equivalent to:

(1) a^{-3} (2) $81a^{-3}$ (3) $27a^{-3}$ (4) $27a$

96. The expression $4a^0 + 4^{-1}$ is equal to:

(1) 1 (2) $1\frac{1}{4}$ (3) $4\frac{1}{4}$ (4) $\frac{1}{4}$

97. The expression $\dfrac{(2a)^0}{2a^{-2}}$, when $a = 3$, is equal to:

(1) 9 (2) $\frac{9}{2}$ (3) $\frac{1}{18}$ (4) 18

10-3   SCIENTIFIC NOTATION

In order to write and compute with very large or very small numbers, we find it convenient to use *scientific* (or *standard*) *notation*.

■ A number is in scientific notation if it is written in the form $a \times 10^n$ where $1 \leq a < 10$ and n is an integer.

The number $93,000,000 = 9.3 \times 10,000,000$ and is written in scientific notation as 9.3×10^7. The number $0.000562 = 5.62 \times .0001$ and is written in scientific notation as 5.62×10^{-4}.

Some Integral Powers of Ten		
$10^8 = 100,000,000$	$10^3 = 1000$	$10^{-2} = \frac{1}{100} = .01$
$10^7 = 10,000,000$	$10^2 = 100$	$10^{-3} = \frac{1}{1000} = .001$
$10^6 = 1,000,000$	$10^1 = 10$	$10^{-4} = \frac{1}{10000} = .0001$
$10^5 = 100,000$	$10^0 = 1$	$10^{-5} = \frac{1}{100000} = .00001$
$10^4 = 10,000$	$10^{-1} = \frac{1}{10} = 0.1$	$10^{-6} = \frac{1}{1000000} = .000001$

To understand how a number is changed from one notation to the other, recall that each time a number is multiplied by 10, the decimal point in the number is moved one place to the right. Each time a number is divided by 10 (that is, multiplied by 10^{-1} or by .1), the decimal point in the number is moved one place to the left. Therefore:

$$3.46 \times 10^4 = 3.4600. = 34,600$$

4 places

AND

$$1.89 \times 10^{-2} = \frac{1.89}{10^2} = .01.89 = .0189$$

2 places

To change a number from ordinary decimal notation to scientific notation, we must first divide it by a power of 10 to obtain a, the factor that is greater than or equal to 1, but less than 10. To do this, we use the place value of the first nonzero digit.

☐ EXAMPLE 1: Write 8790 in scientific notation.

1. To change 8790 to scientific notation, notice that the first digit, 8, is in the 1000 or 10^3 place.

$$\frac{8790}{10^3} = 8.790$$

2. Therefore: $8790 = \frac{8790}{10^3} \times 10^3 = 8.790 \times 10^3$ *Ans.*

☐ EXAMPLE 2: Write 0.0546 in scientific notation.

1. The first nonzero digit, 5, is in the .01 or 10^{-2} place.

$$\frac{0.0546}{10^{-2}} = 0.0546 \times 10^2 = 5.46$$

2. Therefore: $0.0546 = \dfrac{0.0546}{10^{-2}} \times 10^{-2} = 5.46 \times 10^{-2}$ *Ans.*

Notice that the number of decimal places that the decimal point is moved is the absolute value of the exponent of 10 in the scientific notation.

$$8{,}790 = 8.79 \times 10^3$$
3 places

$$0.05\,46 = 5.46 \times 10^{-2}$$
2 places

We will use this concept to change numbers from ordinary decimal notation to scientific notation, and vice versa, as seen in the examples that follow.

☐ EXAMPLE 3: Change 0.0000092 to scientific notation.

How to Proceed	*Solution*

1. Place a caret (∧) after the first nonzero digit. Count the number of places from the caret to the decimal point. This number (6) is the absolute value of the exponent of 10 when the number is written in scientific notation.

$0.000009\,2$
6 places to the left

2. Since we counted to the left, the exponent is negative.

$0.0000092 = 9.2 \times 10^{-6}$ *Ans.*

☐ EXAMPLE 4: Change 5.41×10^5 to ordinary decimal notation.

How to Proceed	*Solution*

In 10^5, the exponent, 5, is positive. Thus, to multiply by 10^5, move the decimal point in the factor 5.41 exactly 5 places to the right, annexing as many zeros to the right of the number as are needed.

5.41×10^5
$= 5\,41000.$
5 places to the right
$= 541{,}000$ *Ans.*

Many calculators can compute by using scientific notation, or they will display the result of a computation in scientific notation when the number of digits exceeds the display capacity of the calculator. For example, the product 120,000 × 3,000,000 might be displayed on some calculators in either of the following ways:

$$\boxed{\mathsf{3.6 \qquad \qquad 11}} \quad \text{OR} \quad \boxed{\mathsf{3.6 \qquad \qquad E\ 11}}$$

This means that the product is 3.6 × 10^{11}.

When performing a computation using scientific notation, we use the rules for multiplying and dividing powers with like bases.

☐ EXAMPLE 5: Express $\dfrac{(7.5 \times 10^5) \times (2.8 \times 10^{-8})}{1.5 \times 10^{-5}}$ as a single number

in scientific notation.

How to Proceed	*Solution*
1. Use the commutative and associative properties of multiplication.	$\dfrac{(7.5 \times 2.8) \times (10^5 \times 10^{-8})}{1.5 \times 10^{-5}}$
2. Use the rules for powers with like bases to multiply and divide powers of 10.	$= \dfrac{21.00 \times 10^{-3}}{1.5 \times 10^{-5}}$
	$= \quad 14.0 \times 10^2$
3. Change 14.0 to scientific notation and multiply the powers of 10.	$= 1.4 \times 10^1 \times 10^2$
	$= \quad 1.4 \times 10^3$ *Ans.*

MODEL PROBLEMS

1. Write 42,700 in scientific notation.

How to Proceed	*Solution*
1. Place a caret after the first nonzero digit. Count the number of places from the caret to the decimal point.	4⌃2700.⟶ 4 places to the right
2. Replace the caret with a decimal point so that 4.27 is the first factor. Since we counted 4 places to the right, the exponent of 10 is positive 4, that is, the second factor is 10^4.	$= 4.27 \times 10^4$

Answer: 4.27 × 10^4

2. Write 8.63×10^{-2} in ordinary decimal notation.

How to Proceed	*Solution*
In 10^{-2}, the exponent, -2, is negative. To multiply by 10^{-2}, move the decimal point in the factor 8.63 two places to the left, annexing as many zeros to the left of the number as necessary.	8.63×10^{-2} $= .08\overset{\frown}{63}$ 2 places to the left $= .0863$ *Ans.*

EXERCISES

In 1-6, the number can be written in scientific notation as 6.93×10^n. Find the value of n.

1. 693
2. 69.3
3. 0.0693
4. 0.000693
5. 693000
6. 6.93

In 7-12, write the number in scientific notation.

7. 40.7
8. 0.0053
9. 0.81
10. 3920
11. 9.05
12. 0.00000007

In 13-21, write the number in ordinary decimal notation.

13. 1.87×10^4
14. 5.2×10^1
15. 2.91×10^{-6}
16. 3.55×10^{-1}
17. 7.6×10^2
18. 6.82×10^{-3}
19. 8.76×10^0
20. 1.25×10^{-8}
21. 3.6×10^7

In 22-29, calculate and express the results (a) in scientific notation (b) in ordinary decimal notation.

22. $(1.5 \times 10^3)(3 \times 10^2)$

23. $(1.2 \times 10^5)(1.5 \times 10^{-3})$

24. $\dfrac{7.2 \times 10^4}{6 \times 10^2}$

25. $\dfrac{1.25 \times 10^{-3}}{2.5 \times 10^{-7}}$

26. $\dfrac{(5.4 \times 10^3)(3 \times 10^2)}{1.8 \times 10^4}$

27. $\dfrac{(2.4 \times 10^{-7})(7.5 \times 10^{-2})}{2 \times 10^{-3}}$

28. $\dfrac{(9 \times 10^{-5})^2}{3 \times 10^{-8}}$

29. $\dfrac{(1.2 \times 10^{-3})^2}{4 \times 10^3}$

In 30-34, write the number in ordinary decimal notation.

30. Light travels 3×10^8 meters per second.
31. The rest mass of a proton is 1.67×10^{-24} gram.
32. There are 6.02×10^{23} molecules in a mole.

33. The mass of the earth is 6×10^{27} grams.
34. There are 3.6×10^3 seconds in an hour.

In 35-41, write the number in scientific notation.

35. In a year, light travels approximately 9,500,000,000,000,000 meters.
36. The approximate distance from the earth to the sun is 93,000,000 miles.
37. One gram is about .035 ounce.
38. The wave length of violet light is .000016 inch.
39. The Pacific Ocean covers 70,000,000 sq. mi. of the earth's surface.
40. One micron equals 0.00003937 inch.
41. A *googol* is a large number defined in mathematics as one followed by one hundred zeros.

In 42-45, the given number shows a display from a calculator. Write the number (a) in scientific notation (b) in ordinary decimal notation.

42. $\boxed{2.157 \quad E8}$ 43. $\boxed{9.64 \quad E\text{-}7}$

44. $\boxed{3.011 \quad E\text{-}9}$ 45. $\boxed{5.4966E12}$

10-4 FRACTIONAL EXPONENTS

In Chapter 4, we defined the nth root of a number as one of the n equal factors of the number so that if $x > 0$ and $x^n = k$, then $\sqrt[n]{k} = x$. Since this definition implies a connection between roots and powers, we will look for a way to express a root as a power.

Case I.

1. For $x \geq 0$, we have learned that: $\quad\quad\quad \sqrt{x} \cdot \sqrt{x} = x$
2. If there exists a number, p, such that $x^p = \sqrt{x}$
   then it must be true that: $\quad\quad\quad\quad\quad x^p \cdot x^p = x$
3. In order for this statement to be consistent with the multiplication law for powers with like bases:

$$x^{p+p} = x^1$$
$$x^{2p} = x^1$$
$$2p = 1$$
$$p = \tfrac{1}{2}$$

4. Therefore, we conclude for $x \geq 0$: $\quad\quad\quad x^{\frac{1}{2}} = \sqrt{x}$

For example, $3^{\frac{1}{2}} = \sqrt{3}$ because:

$$3^{\frac{1}{2}} \cdot 3^{\frac{1}{2}} = 3^{\frac{1}{2}+\frac{1}{2}} = 3^{\frac{2}{2}} = 3$$

AND

$$\sqrt{3} \cdot \sqrt{3} = 3$$

Case II.

1. For any real number x:
2. If there exists a number, q, such that $x^q = \sqrt[3]{x}$, then:
3. In order for this statement to be consistent with the multiplication law for powers with like bases:

$$\sqrt[3]{x} \cdot \sqrt[3]{x} \cdot \sqrt[3]{x} = \sqrt[3]{x^3} = x$$

$$x^q \cdot x^q \cdot x^q = x$$

$$x^{q+q+q} = x^1$$
$$x^{3q} = x^1$$
$$3q = 1$$
$$q = \tfrac{1}{3}$$

4. Therefore, we conclude:

$$x^{\frac{1}{3}} = \sqrt[3]{x}$$

For example, $(-2)^{\frac{1}{3}} = \sqrt[3]{-2}$ because:

$$(-2)^{\frac{1}{3}} \cdot (-2)^{\frac{1}{3}} \cdot (-2)^{\frac{1}{3}} = (-2)^{\frac{1}{3}+\frac{1}{3}+\frac{1}{3}} = (-2)^1 = -2$$

AND

$$\sqrt[3]{-2} \cdot \sqrt[3]{-2} \cdot \sqrt[3]{-2} = \sqrt[3]{-8} = -2$$

These two cases lead to a definition of a power with the exponent $\dfrac{1}{n}$.

■ In general, if n is a counting number, then:

$$x^{\frac{1}{n}} = \sqrt[n]{x}$$

Note: There is no real number that is an even root of a negative number, as in $\sqrt{-9} = \sqrt[2]{-9}$ and in $\sqrt[4]{-1}$. Therefore, *if n is even, the base, x, must be non-negative*, that is, $x \geq 0$. If n is an odd number, however, the base, x, can be negative. Just as $\sqrt[3]{-8} = -2$, we can now state that $(-8)^{\frac{1}{3}} = -2$.

Study the following examples and compare the use of radicals with the use of fractional exponents, sometimes called *rational* exponents.

☐ EXAMPLE 1: Simplify $\dfrac{3}{\sqrt{3}}$.

Radicals	*Fractional Exponents*
$\dfrac{3}{\sqrt{3}} = \dfrac{3}{\sqrt{3}} \cdot \dfrac{\sqrt{3}}{\sqrt{3}} = \dfrac{3\sqrt{3}}{3} = \sqrt{3}$	$\dfrac{3}{3^{\frac{1}{2}}} = 3^{1-\frac{1}{2}} = 3^{\frac{1}{2}}$

☐ EXAMPLE 2: Simplify $\sqrt{2} \cdot \sqrt{3}$.

Radicals	*Fractional Exponents*
$\sqrt{2} \cdot \sqrt{3} = \sqrt{2 \cdot 3} = \sqrt{6}$	$2^{\frac{1}{2}} \cdot 3^{\frac{1}{2}} = (2 \cdot 3)^{\frac{1}{2}} = 6^{\frac{1}{2}}$

☐ EXAMPLE 3: Simplify $\dfrac{2}{\sqrt[3]{2}}$.

Radicals	*Fractional Exponents*
$\dfrac{2}{\sqrt[3]{2}} = \dfrac{2}{\sqrt[3]{2}} \cdot \dfrac{\sqrt[3]{4}}{\sqrt[3]{4}} = \dfrac{2\sqrt[3]{4}}{\sqrt[3]{8}}$ $= \dfrac{2\sqrt[3]{4}}{2}$ $= \sqrt[3]{4}$	$\dfrac{2}{2^{\frac{1}{3}}} = 2^{1-\frac{1}{3}} = 2^{\frac{2}{3}}$

In example 3, the expressions $\sqrt[3]{4}$ and $2^{\frac{2}{3}}$ represent the same real number. We can show that this is true by rewriting $2^{\frac{2}{3}}$, using the rule for the power of a power.

$$2^{\frac{2}{3}} = 2^{2 \cdot \frac{1}{3}} = (2^2)^{\frac{1}{3}} = \sqrt[3]{2^2} = \sqrt[3]{4}$$

This example suggests the following general definition, where n is a counting number:

$$x^{\frac{a}{n}} = \sqrt[n]{x^a} \qquad \text{AND} \qquad x^{\frac{a}{n}} = (\sqrt[n]{x})^a$$

Remember, if n is even, then $x \geq 0$.

☐ EXAMPLE 4: Evaluate $125^{\frac{2}{3}}$.

Method 1:
$$125^{\frac{2}{3}} = \sqrt[3]{125^2}$$
$$= \sqrt[3]{15625}$$
$$= 25$$

Method 2:
$$125^{\frac{2}{3}} = (\sqrt[3]{125})^2$$
$$= (5)^2$$
$$= 25$$

In most cases, it is easier to perform a computation involving roots and powers by finding the root before raising to a power, since the resulting numbers are smaller.

| MODEL PROBLEMS |

1. Evaluate $a^0 + a^{\frac{1}{3}} + a^{-2}$ when $a = 8$.

How to Proceed	*Solution*
1. Use the definitions of zero, negative, and fractional exponents.	$a^0 + a^{\frac{1}{3}} + a^{-2}$
	$= 1 + \sqrt[3]{a} + \dfrac{1}{a^2}$
2. Substitute the value of a.	$= 1 + \sqrt[3]{8} + \dfrac{1}{8^2}$
3. Simplify.	$= 1 + 2 + \dfrac{1}{64}$
	$= 3\dfrac{1}{64}$

Answer: $3\dfrac{1}{64}$

Note: It is possible to interchange steps 1 and 2 in this process. Thus,
$$a^0 + a^{\frac{1}{3}} + a^{-2} = 8^0 + 8^{\frac{1}{3}} + 8^{-2} = 1 + \sqrt[3]{8} + \frac{1}{8^2} = 1 + 2 + \frac{1}{64} = 3\frac{1}{64}.$$

2. If $f(x) = x^{-\frac{3}{2}}$, find $f(16)$.

Solution

$$f(16) = 16^{-\frac{3}{2}} = \frac{1}{16^{\frac{3}{2}}} = \frac{1}{(16^{\frac{1}{2}})^3} = \frac{1}{(\sqrt{16})^3} = \frac{1}{4^3} = \frac{1}{64} \quad Ans.$$

3. If $m = 8$, find the value of $(8m^0)^{\frac{2}{3}}$.

Solution

$$(8m^0)^{\frac{2}{3}} = (8 \cdot 8^0)^{\frac{2}{3}} = (8 \cdot 1)^{\frac{2}{3}} = 8^{\frac{2}{3}} = (\sqrt[3]{8})^2 = 2^2 = 4 \quad Ans.$$

4. Write $x^{-\frac{3}{4}}$, using radicals and positive integral exponents. $(x > 0)$

How to Proceed	*Solution*
1. Use the definition of a negative exponent.	$x^{-\frac{3}{4}} = \dfrac{1}{x^{\frac{3}{4}}}$
2. Use the definition of a fractional exponent.	$= \dfrac{1}{\sqrt[4]{x^3}} \quad Ans.$

EXERCISES

In 1–8, write the given expression, using a radical sign. Let the variables represent positive numbers.

1. $x^{\frac{1}{2}}$ 2. $a^{\frac{1}{3}}$ 3. $b^{\frac{1}{5}}$ 4. $y^{\frac{2}{3}}$

5. $3a^{\frac{1}{2}}$ 6. $(3a)^{\frac{1}{2}}$ 7. $ab^{\frac{1}{4}}$ 8. $(ab)^{\frac{1}{4}}$

In 9–16, use exponents to write the radical expression. Let the variables represent positive numbers.

9. $\sqrt{2}$ 10. $\sqrt{3a}$ 11. $\sqrt[3]{5x}$ 12. $2(\sqrt[5]{b})$

13. $7\sqrt{5}$ 14. $x\sqrt{y^3}$ 15. $-\sqrt{8}$ 16. $-\sqrt[3]{2}$

In 17–48, evaluate the given expression.

17. $25^{\frac{1}{2}}$ 18. $27^{\frac{1}{3}}$ 19. $16^{\frac{1}{4}}$ 20. $16^{-\frac{1}{2}}$

21. $125^{-\frac{1}{3}}$ 22. $(-125)^{\frac{1}{3}}$ 23. $100^{-\frac{1}{2}}$ 24. $81^{-\frac{1}{4}}$

25. $9^{\frac{3}{2}}$ **26.** $1000^{\frac{2}{3}}$ **27.** $32^{\frac{3}{5}}$ **28.** $(-8)^{\frac{2}{3}}$

29. $64^{\frac{1}{3}}$ **30.** $64^{\frac{3}{2}}$ **31.** $4^{-\frac{3}{2}}$ **32.** $27^{-\frac{2}{3}}$

33. $-4^{\frac{5}{2}}$ **34.** $81^{\frac{3}{4}}$ **35.** $(36^{-1})^{\frac{1}{2}}$ **36.** $(8^0)^{\frac{1}{3}}$

37. $(\frac{4}{25})^{\frac{1}{2}}$ **38.** $(\frac{9}{16})^{-\frac{1}{2}}$ **39.** $(-\frac{1}{27})^{\frac{1}{3}}$ **40.** $(\frac{1}{27})^{-\frac{1}{3}}$

41. $2^{\frac{1}{2}} \cdot 2^{\frac{3}{2}}$ **42.** $4^2 \cdot 4^{\frac{1}{2}}$ **43.** $9^{\frac{3}{2}} \cdot 9^{-\frac{1}{2}}$ **44.** $8^{\frac{1}{3}} \cdot 8^{-\frac{2}{3}}$

45. $5^0 + 5^{-1}$ **46.** $-4^0 + 4^{-\frac{1}{2}}$ **47.** $(8^0 + 8)^{\frac{1}{2}}$ **48.** $(3^0 + 3)^{-\frac{1}{2}}$

49. Find the value of $2a^{\frac{1}{3}}$ when $a = \frac{27}{8}$.

50. Evaluate $a^0 + a^{-\frac{1}{2}}$ when $a = 9$.

51. If $k = 4$, find the value of $(9k^0)^{\frac{3}{2}}$.

52. Find the value of $(m^{-1})^{\frac{3}{4}}$ when $m = 16$.

53. If $f(x) = x^{\frac{2}{3}}$, find the value of $f(-216)$.

54. If $f(x) = x^{-\frac{3}{2}}$, find the value of $f(100)$.

55. If the function $g(x) = 4x^{-\frac{1}{2}}$, find the value of $g(25)$.

In 56–60, select the numeral preceding the expression that best completes the sentence.

56. The expression $\dfrac{2x^{-\frac{1}{2}}}{x^{-1}}$ is equivalent to:

(1) $2\sqrt{x}$ (2) $\sqrt{2x}$ (3) $2\sqrt{x^3}$ (4) $\sqrt{2x^3}$

57. The expression $3\sqrt[4]{3}$ can be written as:

(1) $3^{\frac{3}{4}}$ (2) $3^{\frac{4}{3}}$ (3) $3^{\frac{5}{4}}$ (4) $3^{\frac{4}{5}}$

58. If $b = -64$, then $b^{-\frac{1}{3}}$ is equal to:

(1) 4 (2) -4 (3) $\frac{1}{4}$ (4) $-\frac{1}{4}$

59. If $a = -9$, then $a^{-\frac{1}{2}}$ is:

(1) 3 (2) $\frac{1}{3}$ (3) $-\frac{1}{3}$ (4) not a real number

60. If $x = \frac{8}{9}$, then $x^{-\frac{1}{3}}$ is equal to:

(1) $\dfrac{2}{3}$ (2) $\dfrac{3}{2}$ (3) $\dfrac{\sqrt[3]{9}}{2}$ (4) $\dfrac{2}{\sqrt[3]{9}}$

10-5 EXPONENTIAL FUNCTIONS

Linear Growth

A piece of paper is one layer thick. Place another paper on top of it, and the total is two layers thick. Place another paper on top again, and the total is now three layers thick.

As this process continues, we can describe the number of layers (y) in terms of the number of sheets added (x) by the *linear equation* $y = 1 + x$. For example, by adding 5 sheets, we have a total of 6 sheets of paper. This is an example of *linear growth* because the growth is described by means of a linear function.

Exponential Growth

Let us consider another experiment. A piece of paper is one layer thick. If the paper is folded in half, it is now two layers thick. If it is folded in half again, the paper is now four layers thick. If it is folded in half again, the paper is now eight layers thick.

| **0 folds** | **1 fold** | **2 folds** | **3 folds** |
| **1 layer** | **2 layers** | **4 layers** | **8 layers** |

Although we will reach a point where it is impossible to fold the paper, imagine that this process can continue. We can describe the number of layers of paper (y) in terms of the number of folds (x) by means of an *exponential function*, namely, $y = 2^x$. The table at the right shows some values for the exponential function $y = 2^x$.

For example, after 5 folds, the paper is 2^5 or 32 layers thick. After 9 folds, the paper is 2^9 or 512

x	2^x	$= y$
0	$2^0 =$	1
1	$2^1 =$	2
2	$2^2 =$	4
3	$2^3 =$	8
4	$2^4 =$	16
5	$2^5 =$	32

layers thick; in other words, it would be thicker than this book! This is an example of *exponential growth* because the growth is described by means of an exponential function.

There are many examples of exponential growth in the world around us. Over an interval of time, certain populations grow exponentially, whether they are a population of bacteria or rabbits or the people of a nation. Compound interest is based on exponential growth. In all these cases, growth is defined by an equation that involves a power b^x where b is a positive number and the exponent x is a variable.

In the paper-folding experiment described by the exponential function $y = 2^x$, we let the domain of x values be whole numbers. To study the function $y = 2^x$ on a wider scale, we can expand the domain to include all *rational values* of x, as defined earlier in this chapter. Here, we expand the table of values for $y = 2^x$ by considering some convenient negative and fractional values of x.

x	2^x		$= y$
-3	$2^{-3} = \dfrac{1}{2^3}$	$=$	$\dfrac{1}{8}$
-2	$2^{-2} = \dfrac{1}{2^2}$	$=$	$\dfrac{1}{4}$
-1	$2^{-1} = \dfrac{1}{2^1}$	$=$	$\dfrac{1}{2}$
$\dfrac{1}{2}$	$2^{\frac{1}{2}} = \sqrt{2}$	$=$	$\sqrt{2} \approx 1.4$
$\dfrac{3}{2}$	$2^{\frac{3}{2}} = \sqrt{2^3}$	$=$	$\sqrt{8} \approx 2.8$
$\dfrac{5}{2}$	$2^{\frac{5}{2}} = \sqrt{2^5}$	$=$	$\sqrt{32} \approx 5.7$

In Fig. 1, the pairs of numbers $(x, 2^x)$ or (x, y) are represented as points in the coordinate plane. They lie in a pattern that suggests a smooth curve.

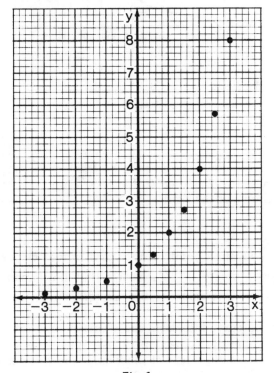

Fig. 1

If we assume that the curve drawn through these points in Fig. 2 is the graph of $y = 2^x$, then 2^x is defined for all real values of x.

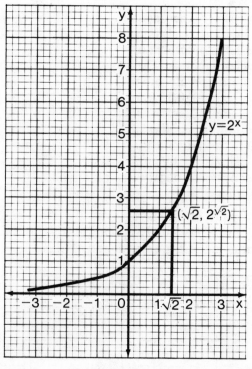

Fig. 2

The length of the diagonal of a square each of whose sides measures 1 is used to locate $\sqrt{2}$ on the x-axis, as shown at the right.

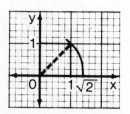

In Fig. 2, we locate a point on the curve whose x-coordinate is $\sqrt{2}$ and whose y-coordinate is $2^{\sqrt{2}}$. Notice that the exponent $\sqrt{2}$ is an *irrational number*. The value of $2^{\sqrt{2}}$, as seen on the graph, is approximately 2.66.

Notice that the curve intersects the y-axis at $(0, 1)$. As the value of x decreases, the value of 2^x becomes closer and closer to 0. The curve approaches but does not intersect the x-axis. The x-axis is an *asymptote* of the curve whose equation is $y = 2^x$.

Let us consider a different exponential function, $y = 5^x$. The table below shows some pairs that were used to draw the graph of $y = 5^x$.

x	5^x		$= y$
-1	$5^{-1} = \frac{1}{5}$	$=$	$.2$
0	5^0	$=$	1
$\frac{1}{2}$	$5^{\frac{1}{2}} = \sqrt{5}$	$\approx$	2.24
1	5^1	$=$	5

Notice, in Fig. 3, how the graph of $y = 5^x$ has the same basic shape as the graph of $y = 2^x$.

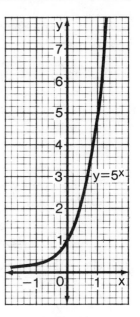

Fig. 3

Let us consider an exponential function in which the value of base b is positive but less than 1. The graph of $y = (\frac{1}{2})^x$ can be drawn, using the pairs of values shown in the table below.

x	$(\frac{1}{2})^x$		$= y$
-3	$(\frac{1}{2})^{-3} = 2^3$	$=$	8
-2	$(\frac{1}{2})^{-2} = 2^2$	$=$	4
-1	$(\frac{1}{2})^{-1} = 2^1$	$=$	2
0	$(\frac{1}{2})^0$	$=$	1
1	$(\frac{1}{2})^1$	$=$	$\frac{1}{2}$
2	$(\frac{1}{2})^2$	$=$	$\frac{1}{4}$
3	$(\frac{1}{2})^3$	$=$	$\frac{1}{8}$

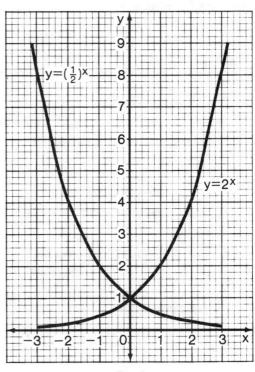

Fig. 4

In Fig. 4, we see both the graph of $y = (\frac{1}{2})^x$ and the reflection of this graph over the y-axis. Since $r_{y\text{-axis}}\,(x, y) = (-x, y)$, the equation of the reflection of $y = \left(\dfrac{1}{2}\right)^x$ over the y-axis is $y = \left(\dfrac{1}{2}\right)^{-x}$. Since $\left(\dfrac{1}{2}\right)^{-x} = \dfrac{1^{-x}}{2^{-x}} = \dfrac{1}{2^{-x}} = 2^x$, the reflection over the y-axis of $y = \left(\dfrac{1}{2}\right)^x$ is $y = 2^x$.

■ **Summary.**

1. The equation $y = b^x$, where $b > 0$ and $b \neq 1$, defines an exponential function.

$$\text{Domain} = \{x \mid x \in \text{Real numbers}\}$$
$$\text{Range} \;\; = \{y \mid y \in \text{positive Real numbers}\}$$

2. The reflection in the y-axis of the graph of $y = b^x$ is the graph of $y = \left(\dfrac{1}{b}\right)^x$.

Note: Although b^x is defined for $b = 1$, the function $y = 1^x$ is the constant function $y = 1$, which is *not* an exponential function.

MODEL PROBLEM

a. Sketch the graph of $y = (\frac{3}{2})^x$ in the interval $-3 \leq x \leq 3$.

b. Sketch the graph of $y = (\frac{2}{3})^x$ in the interval $-3 \leq x \leq 3$.

c. Use the graph to find an approximate rational value of $(\frac{3}{2})^{1.4}$ to the nearest tenth.

How to Proceed   *Solution*

a. 1. Make a table of values for integral values of x from $x = -3$ to $x = 3$.

   2. Plot the points corresponding to the pairs of values in the table.

   3. Draw a smooth curve through these points.

x	$(\frac{3}{2})^x$		y
-3	$(\frac{3}{2})^{-3} = (\frac{2}{3})^3$	$=$	$\frac{8}{27}$
-2	$(\frac{3}{2})^{-2} = (\frac{2}{3})^2$	$=$	$\frac{4}{9}$
-1	$(\frac{3}{2})^{-1} = (\frac{2}{3})^1$	$=$	$\frac{2}{3}$
0	$(\frac{3}{2})^0$	$=$	1
1	$(\frac{3}{2})^1$	$=$	$\frac{3}{2}$
2	$(\frac{3}{2})^2$	$=$	$\frac{9}{4}$
3	$(\frac{3}{2})^3$	$=$	$\frac{27}{8}$

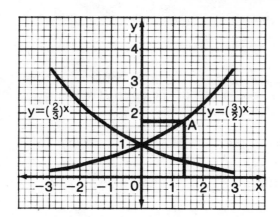

b. 1. The graph of $y = (\frac{2}{3})^x$ is the reflection in the y-axis of the graph of $y = (\frac{3}{2})^x$. For each point (x, y) on the graph of $y = (\frac{3}{2})^x$, locate a point $(-x, y)$ on the graph of $y = (\frac{2}{3})^x$.

2. Draw a smooth curve through these points.

c. 1. Draw a vertical line at $x = 1.4$ to locate the point $(1.4, (\frac{3}{2})^{1.4})$ on the graph (point A).

2. Draw a horizontal line from point A to the y-axis. Approximate the value to the nearest tenth.

Answer: a and b on graph   c. 1.8

EXERCISES

1. a. Sketch the graph of $y = 3^x$ in the interval $-3 \leq x \leq 3$.
   b. From the graph, approximate the value of $3^{2.2}$ to the nearest tenth.
2. a. Sketch the graph of $y = 4^x$ in the interval $-2 \leq x \leq 2$.
   b. From the graph, approximate the value of $4^{\sqrt{2}}$ to the nearest tenth.
3. a. Sketch the graph of $y = (\frac{1}{4})^x$ in the interval $-2 \leq x \leq 2$.
   b. For the domain of real numbers, what is the range of the function $y = (\frac{1}{4})^x$?
   c. In what quadrants does the graph of $y = (\frac{1}{4})^x$ lie?
4. a. Sketch the graph of $y = (\frac{5}{2})^x$ in the interval $-3 \leq x \leq 3$.
   b. Sketch the reflection in the y-axis of the graph of $y = (\frac{5}{2})^x$.
   c. What is an equation of the graph drawn in answer to part b?

In 5–8, find the function value when $f(x) = 6^x$.

5. $f(-1)$ **6.** $f(0)$ **7.** $f(\frac{1}{2})$ **8.** $f(\frac{3}{2})$

9. For what values of b does $y = b^x$ define an exponential function?
10. Name the coordinates of the point at which the graphs of $y = 5^x$ and $y = 8^x$ intersect.

In 11–13, select the numeral preceding the expression that best completes the sentence.

11. The equation $y = a^x$ defines an exponential function when a equals:
   (1) 1   (2) 2   (3) 0   (4) −1
12. The graph of $y = b^x$, where $b > 0$ and $b \neq 1$, lies in quadrants:
   (1) I and II   (2) I and IV   (3) II and III   (4) II and IV
13. If $b > 0$, then the graph of $y = b^x$ must contain the point:
   (1) $(-1, 0)$ (2) $(1, 0)$ (3) $(0, 1)$ (4) $(0, 0)$

14. If $f(x) = -2^x$, find a. $f(0)$ b. $f(2)$ c. $f(-2)$.
15. a. Sketch the graph of $y = 2^x$ in the interval $-3 \leq x \leq 3$.
   b. On the same set of axes, sketch the graph of $y = -2^x$.
   c. What is the equation of the line of reflection such that the graph of $y = -2^x$ is the image of the graph of $y = 2^x$?
16. In 1950, the population of a city was 10,000 people. If the population has doubled every 10 years since that time, find the population of this city   a. in 1960   b. in 1970   c. in 1980.
   d. If this exponential growth continues, what will be the population of the city in the year 2000?
17. If a bank compounds interest annually (once each year), then the amount of money A in a bank account is determined by the formula $A = P(1 + r)^x$. Here, P = the principal, or the amount invested, r = the rate of interest, and x = the number of years involved. If \$100 is invested at 6% interest compounded annually, then the amount A in the account is found by $A = \$100(1.06)^x$. Find the amount of money in this account at the end of a. 1 year
   b. 2 years   c. 3 years   d. 4 years.
18. The thickness of a sheet of paper is 0.004 inch. If x represents the number of times that this sheet of paper is folded in half over itself, then $y = 2^x$ determines the number of layers of paper, and $y = .004(2)^x$ determines the thickness of all the layers of paper. What is the thickness of all the layers if such a sheet of paper is folded in half over itself exactly 15 times?

10-6 EQUATIONS WITH FRACTIONAL OR NEGATIVE EXPONENTS

In raising a power to a power, we multiply exponents.

$$(x^a)^c = x^{ac}$$

If a and c are reciprocals, their product is 1 and the resulting power is x^1 or x.

$$(x^2)^{\frac{1}{2}} = x^1 = x$$
$$(x^{\frac{3}{4}})^{\frac{4}{3}} = x^1 = x$$
$$(x^{-\frac{1}{2}})^{-2} = x^1 = x$$

This observation will simplify the solution of equations containing powers with fractional or negative exponents.

| MODEL PROBLEM |

Solve for x and check: **a.** $2x^{-\frac{1}{3}} = 6$ **b.** $x^{\frac{3}{2}} + 1 = 9$

How to Proceed   *Solution*

1. Change the equation into an equivalent equation that has only the variable term with coefficient 1 in the left member.

a. $\quad 2x^{-\frac{1}{3}} = 6$
$\quad\quad x^{-\frac{1}{3}} = 3$

b. $\quad x^{\frac{3}{2}} + 1 = 9$
$\quad\quad x^{\frac{3}{2}} = 8$

2. Raise both members of the equation to a power, using the reciprocal of the given exponent.

$(x^{-\frac{1}{3}})^{-3} = 3^{-3}$

$(x^{\frac{3}{2}})^{\frac{2}{3}} = 8^{\frac{2}{3}}$

3. Simplify both members of the equation.

$x^1 = \dfrac{1}{3^3}$

$x = \dfrac{1}{27}$

$x^1 = \sqrt[3]{8^2}$

$x = (2)^2$

$x = 4$

4. Check the solution in the original equation.

$2x^{-\frac{1}{3}} = 6$

$2(\frac{1}{27})^{-\frac{1}{3}} \stackrel{?}{=} 6$

$2\sqrt[3]{27} \stackrel{?}{=} 6$

$2(3) \stackrel{?}{=} 6$

$6 = 6$

(True)

$x^{\frac{3}{2}} + 1 = 9$

$4^{\frac{3}{2}} + 1 \stackrel{?}{=} 9$

$(\sqrt{4})^3 + 1 \stackrel{?}{=} 9$

$8 + 1 \stackrel{?}{=} 9$

$9 = 9$

(True)

Answer: **a.** $\frac{1}{27}$ **b.** 4

| EXERCISES |

In 1-4, find the value of a for which the expression is equal to x.

1. $(x^{\frac{1}{2}})^a$ 2. $(x^{\frac{3}{4}})^a$ 3. $(x^{-6})^a$ 4. $(x^{-\frac{3}{5}})^a$

In 5-16, solve and check. All variables represent positive numbers.

5. $x^{\frac{1}{2}} = 7$ 6. $x^{\frac{1}{3}} = 5$ 7. $x^{\frac{2}{3}} = 4$

8. $y^{-2} = 9$ 9. $a^{-\frac{1}{4}} = 2$ 10. $b^{-\frac{1}{2}} = \frac{1}{3}$

11. $x^{\frac{4}{3}} - 1 = 15$ 12. $y^{-\frac{3}{2}} + 2 = 10$ 13. $2x^{\frac{4}{3}} = 162$

14. $2x^{-\frac{1}{2}} = 3$ 15. $2x^{-\frac{1}{4}} + 3 = 4$ 16. $4x^{\frac{2}{3}} - 5 = 20$

17. Find the root of the equation $x \cdot x^{\frac{1}{2}} = 8$. (Simplify the left member by using the rule for multiplying powers with like bases.)

18. Solve for y and check: $2y \cdot y^{\frac{1}{3}} = .0002$

In 19 and 20, select the numeral preceding the expression that best completes the sentence.

19. A root of $(x + 2)^{\frac{2}{3}} = 4$ is:
    (1) 6   (2) 2   (3) 8   (4) 4

20. A root of $3y^{\frac{3}{2}} = 6$ is:
    (1) $\sqrt{8}$ (2) $\sqrt[3]{2}$ (3) $\sqrt[3]{12}$ (4) $\sqrt[3]{4}$

In 21-26, solve the equation in which the variable represents a positive number. (*Hint:* Before solving, change the radical expression to a power having a fractional exponent.)

21. $\sqrt[3]{x^2} = 16$ 22. $\sqrt{x^3} = 64$ 23. $\sqrt[3]{y^4} = 625$

24. $\sqrt{y^5} = 32$ 25. $\sqrt{m} = 15$ 26. $\sqrt[4]{b} = 9$

10-7 EXPONENTIAL EQUATIONS

An equation in which the variable appears in an exponent is called an *exponential equation*. Simple exponential equations involving powers of like bases can be solved by using the following observation:

$$\text{For } b \neq 0 \text{ and } b \neq 1, b^x = b^y \leftrightarrow x = y.$$

In the model problems that follow, we will solve the equations by equating exponents of like bases. If the bases in the given equation are not equal (see model problems 2, 3, and 4), we will change one or both bases as a first step.

| MODEL PROBLEMS |

1. Solve and check: $5^{x+1} = 5^4$

How to Proceed	Solution	Check
1. Write the equation.	$5^{x+1} = 5^4$	$5^{x+1} = 5^4$
2. Since the bases are alike, equate the exponents.	$x + 1 = 4$	$5^{3+1} \stackrel{?}{=} 5^4$ $5^4 = 5^4$
3. Solve the resulting equation.	$x = 3$	(True)

Answer: $x = 3$

2. Solve and check: $2^{x-1} = 8^2$

How to Proceed	Solution	Check
1. Write the equation.	$2^{x-1} = 8^2$	$2^{x-1} = 8^2$
2. Change the right member to base 2, using $8 = 2^3$.	$2^{x-1} = (2^3)^2$	$2^{7-1} \stackrel{?}{=} 8^2$
3. Simplify the right member.	$2^{x-1} = 2^6$	$2^6 \stackrel{?}{=} 8^2$
4. Equate the exponents of like bases.	$x - 1 = 6$	$64 = 64$ (True)
5. Solve the resulting equation.	$x = 7$	

Answer: $x = 7$

3. Solve and check: $9^{x+1} = 27^x$

How to Proceed	Solution	Check
1. Write the equation.	$9^{x+1} = 27^x$	$9^{x+1} = 27^x$
2. Change each member to base 3, using $9 = 3^2$ and $27 = 3^3$.	$(3^2)^{x+1} = (3^3)^x$	$9^{2+1} \stackrel{?}{=} 27^2$ $9^3 \stackrel{?}{=} 27^2$
3. Simplify each member.	$3^{2x+2} = 3^{3x}$	$729 = 729$ (True)
4. Equate exponents of like bases.	$2x + 2 = 3x$	
5. Solve the resulting equation.	$2 = x$	

Answer: $x = 2$

4. Solve and check: $(\frac{1}{4})^x = 8^{1-x}$

How to Proceed	*Solution*	*Check*
1. Write the equation.	$(\frac{1}{4})^x = 8^{1-x}$	$(\frac{1}{4})^x = 8^{1-x}$
2. Change each member to base 2, using $\frac{1}{4} = \frac{1}{2^2} = 2^{-2}$ and $8 = 2^3$.	$(2^{-2})^x = (2^3)^{1-x}$	$(\frac{1}{4})^3 \stackrel{?}{=} 8^{1-3}$
3. Simplify each member.	$2^{-2x} = 2^{3-3x}$	$(\frac{1}{4})^3 \stackrel{?}{=} 8^{-2}$
4. Equate exponents of like bases.	$-2x = 3 - 3x$	$\frac{1}{64} = \frac{1}{64}$
5. Solve the resulting equation.	$x = 3$	(True)

Answer: x = 3

Note: It is not always possible to express each member as an integral power of the same base. We will learn how to solve equations such as $2^x = 3$ in Chapter 11.

EXERCISES

In 1-15, express the number as a power with an integral exponent and the smallest possible positive integral base.

1. 36   2. 25   3. 16   4. 27   5. 64
6. 81   7. 125   8. 32   9. 1000   10. .1
11. $\frac{1}{3}$ 12. $\frac{1}{2}$ 13. $\frac{1}{8}$ 14. $\frac{1}{25}$ 15. $\frac{1}{27}$

In 16-36, solve and check.

16. $4^{x+2} = 4^3$ 17. $3^{2x-1} = 3^{x+2}$ 18. $7^{x-4} = 7$
19. $12^{2x-10} = 12^{x-5}$ 20. $4^{x-2} = 4^{3x}$ 21. $2^x = 4$
22. $5^{x-1} = 125$ 23. $49^x = 7^{x+1}$ 24. $36^x = 6^{x-1}$
25. $64^x = 4^{x+2}$ 26. $9^{2x} = 3^{3x+1}$ 27. $8^{2x} = 2^{2x+2}$
28. $5^{3x} = 25^{x+1}$ 29. $16^{x-3} = 4^{x-3}$ 30. $9^{x+1} = 3^x$
31. $8^x = 4^{x-1}$ 32. $27^x = 9^{x+2}$ 33. $(\frac{1}{2})^x = 8^{2-x}$
34. $(\frac{1}{3})^{1-x} = 9^x$ 35. $125^x = 25$ 36. $32^x = 4$

In 37 and 38, select the numeral preceding the expression that best completes the sentence.

37. The solution set of $2^{x^2+2} = 2^{3x}$ is:
   (1) $\{1\}$ (2) $\{2\}$ (3) $\{1, 2\}$ (4) $\{8, 64\}$

38. The solution set of $3^{x^2-3} = 3^{2x}$ is:
   (1) $\{-1\}$ (2) $\{3, -1\}$ (3) $\{3\}$ (4) $\{729, \frac{1}{9}\}$

10-8 REVIEW EXERCISES

In 1-6, simplify and write the expression without a denominator. All variables represent positive numbers.

1. $\dfrac{a^2b^{-1}}{ab^{-3}}$

2. $\dfrac{20x^{-2}y^5}{4x^{-2}y^4}$

3. $(2c^{\frac{1}{2}}d)(c^{\frac{3}{2}}d^{-1})$

4. $(4x^{-1}y^{\frac{2}{3}})^{\frac{3}{2}}$

5. $\dfrac{(2a^2b^4)^2}{2a^3b^{-5}}$

6. $\left(\dfrac{1}{x^2y}\right)^{-3}$

In 7-18, find the value of the expression.

7. 25^0

8. 15^{-1}

9. $100^{\frac{1}{2}}$

10. -8^{-2}

11. $(-8)^{-2}$

12. $64^{\frac{2}{3}}$

13. $.008^{\frac{1}{3}}$

14. $-64^{\frac{1}{2}}$

15. $(\frac{3}{4})^{-2}$

16. $125^{-\frac{2}{3}}$

17. $4^{-\frac{3}{2}}$

18. $(12^{\frac{1}{5}})^0$

19. Find the value of $3a^0 + a^{\frac{1}{2}}$ when $a = 9$.

20. Find the value of $(5b)^0 + (2b)^{\frac{3}{2}}$ when $b = 8$.

21. Find the value of $2c^{-\frac{1}{3}} + c^0$ when $c = 27$.

22. Find the value of $(x + 1)^{-1} + (x + 2)^{\frac{3}{4}} + (x + 3)^0$ when $x = 14$.

23. If $f(x) = x^{-\frac{2}{3}}$, find $f(27)$.

24. If $f(x) = 9x^{-2}$, find $f(\frac{3}{5})$.

25. a. Sketch the graph of $y = 3^x$ in the interval $-2 \le x \le 2$.
    b. Use the graph of $y = 3^x$ to determine the value of $3^{1.6}$ to the nearest tenth.

26. For what value of b is the graph of $y = b^x$ the reflection in the y-axis of the graph of $y = 6^x$?

In 27 and 28, select the numeral preceding the expression that best completes the sentence.

27. The equation that defines the same function as $y = 5^x$ is:
   (1) $y = 5^{-x}$ (2) $y = -5^{-x}$ (3) $y = (\frac{1}{5})^{-x}$ (4) $y = (\frac{1}{5})^x$

28. The graphs of $y = 3^x$ and $y = 3^{-x}$ intersect at:
    (1) $(0, 1)$ only   (2) $(1, 0)$ only
    (3) $(0, 1)$ and $(1, 3)$ (4) no point

In 29–34, solve and check. All variables represent positive numbers.

29. $x^{\frac{5}{3}} = 32$ 30. $a^{-2} + 2 = 27$ 31. $y^{\frac{3}{4}} = .008$

32. $4x^{\frac{1}{3}} = 12$ 33. $3b^{-\frac{1}{2}} = 10$ 34. $2c^{\frac{3}{4}} + 1 = 55$

In 35–40, solve and check. All variables represent positive numbers.

35. $6^{x-1} = 6^{2x-4}$ 36. $5^{x+1} = 25^x$ 37. $32^x = 8$

38. $8^x = 4^{x+1}$ 39. $1000^x = 100^{x-1}$ 40. $(\frac{1}{3})^{x-1} = 9$

In 41–45, select the numeral preceding the expression that best completes the sentence or answers the question.

41. The expression $\dfrac{2\sqrt{x}}{x}$ is equivalent to:

    (1) $2x^{\frac{1}{2}}$ (2) $(2x)^{\frac{1}{2}}$ (3) $2x^{-\frac{1}{2}}$ (4) $(2x)^{-\frac{1}{2}}$

42. The expression $(4 - x)^{-\frac{1}{2}}$ represents a real number:
    (1) for all x (2) for $x > 4$ only
    (3) for $x \leq 4$ only   (4) for $x < 4$ only

43. Which of the following is *not* a real number?

    (1) $-64^{\frac{1}{2}}$ (2) $(-64)^{\frac{1}{2}}$ (3) -64^0 (4) $-64^{\frac{1}{3}}$

44. The product $8 \cdot 8^5$ is *not* equal to:
    (1) 2^{18} (2) 4^9 (3) 8^6 (4) 64^5

45. The product $2^{\frac{1}{2}} \cdot 8^{\frac{1}{3}}$ is *not* equal to:

    (1) $16^{\frac{1}{6}}$ (2) $8^{\frac{1}{2}}$ (3) $2^{\frac{3}{2}}$ (4) $2\sqrt{2}$

46. If $x^n = 3x^{n-1}$ for all n, what is the value of x?

47. If $n^x = (n + 1)^x$ for all $n \neq 0$, what is the value of x?

Chapter **11**

Logarithmic Functions

11-1 EXPONENTIAL FUNCTIONS AND THEIR INVERSES

In Chapter 10, we studied the graphs of exponential functions of the form $y = b^x$. In this section, we will study the inverses of these exponential functions.

The figure below shows the graph of $y = 2^x$ and its reflection in the line $y = x$. Since $r_{y=x}(x, y) = (y, x)$, the equation of the reflection of $y = 2^x$ is $x = 2^y$. This equation defines a function that is the inverse of $y = 2^x$ under composition of functions.

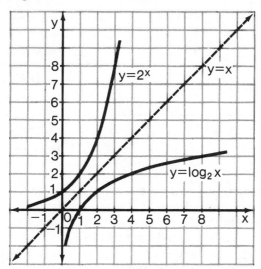

We prefer to write the equation that defines a function by expressing y in terms of x. Therefore, we want to rewrite $x = 2^y$ so that it is solved for y. We begin by describing y in the equation $x = 2^y$ in words:

y is the exponent to the base 2 needed to obtain x.

Since the word "logarithm" means exponent, we can write this sentence as:

y is the logarithm to the base 2 of x.

487

To abbreviate this sentence and write it in symbolic form, we will pick out essential parts, using "log" for logarithm.

<u>y</u> is the <u>logarithm to the base 2 of x</u>.
$$y = \log_2 x$$

Notice that the base, 2, is written a half-line below the word "log."
Therefore:

$$x = 2^y \qquad \leftrightarrow \qquad y = \log_2 x$$

Thus, the equation $x = 2^y$ is *equivalent* to the equation $y = \log_2 x$; these equations name the same set of points, some of which are included in the chart that follows.

Point (x, y)	Exponential Form $x = 2^y$	Logarithmic Form $y = \log_2 x$	Logarithmic Form Is Read As
$(8, 3)$	$8 = 2^3$	$3 = \log_2 8$	3 is the log to the base 2 of 8.
$(4, 2)$	$4 = 2^2$	$2 = \log_2 4$	2 is the log to the base 2 of 4.
$(\frac{1}{2}, -1)$	$\frac{1}{2} = 2^{-1}$	$-1 = \log_2 \frac{1}{2}$	-1 is the log to the base 2 of $\frac{1}{2}$.
$(\frac{1}{4}, -2)$	$\frac{1}{4} = 2^{-2}$	$-2 = \log_2 \frac{1}{4}$	-2 is the log to the base 2 of $\frac{1}{4}$.

Recall that the exponential equation $y = 2^x$ is a function whose graph lies entirely in quadrants I and II. Under a reflection in the line $y = x$, we see that the graph of its *inverse* (that is, the graph of $x = 2^y$ or $y = \log_2 x$) lies entirely in quadrants I and IV.

We can see from the graph of $y = \log_2 x$ that, since no vertical line can intersect the curve in more than one point, no two pairs have the same first element. The set of ordered pairs defined by $y = \log_2 x$ is a function whose domain is the set of positive real numbers and whose range is the set of all real numbers.

What we have observed using the base 2 can be applied to any base, b, for which the exponential function $y = b^x$ is defined.

For $b > 0$ and $b \neq 1$: $x = b^y \leftrightarrow y = \log_b x$.

The equation $y = \log_b x$ defines a logarithmic function that is the inverse under composition of the exponential function $y = b^x$.

A function f(x) is an *exponential* *function*:	Its inverse $f^{-1}(x)$ is a *logarithmic* *function*:
$$y = b^x$$	$$y = \log_b x$$
Domain = $\{x \mid x \in \text{Real numbers}\}$ Range = $\{y \mid y > 0\}$	Domain = $\{x \mid x > 0\}$ Range = $\{y \mid y \in \text{Real numbers}\}$

MODEL PROBLEM

a. Sketch the graph of $y = 5^x$ to include the points whose ordered pairs are found in the table at the right.

x	-2	-1	0	1	2
$y = 5^x$	$\frac{1}{25}$	$\frac{1}{5}$	1	5	25

b. Name the ordered pairs that are the images of those pairs in the table of part a, under a reflection in the line $y = x$.

c. Sketch the graph of the inverse of $y = 5^x$, using the pairs from part b.

d. State the equation of the curve graphed in part c, using some form of the word "logarithm."

Solution

a. Plot the points whose ordered pairs are given in the table, and draw a smooth curve through them.

b. Under a reflection in the line $y = x$, $(x, y) \to (y, x)$. The equation formed is $x = 5^y$.

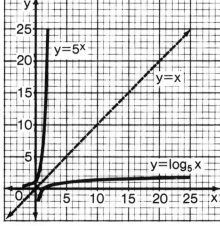

x	$\frac{1}{25}$	$\frac{1}{5}$	1	5	25
y	-2	-1	0	1	2

Ans.

c. Plot the points whose ordered pairs are given in the table in part b. Draw a smooth curve through these points, as shown on the graph above.

d. Since $x = b^y \leftrightarrow y = \log_b x$ and here $b = 5$, then $x = 5^y \leftrightarrow y = \log_5 x$. *Answer:* The equation of the graph in part c is $y = \log_5 x$.

EXERCISES

1. **a.** Copy and complete the table at the right by finding, for each of the given values of x, the corresponding value of $y = 3^x$.

x	-2	-1	0	1	2
$y = 3^x$					

   b. Sketch the graph of $y = 3^x$ to include the points whose ordered pairs were found in part a.

   c. Name the ordered pairs that are the images of those pairs in the table of part a, under a reflection in the line $y = x$.

   d. Sketch the graph of the inverse of $y = 3^x$ by using the pairs from part c.

   e. State the equation of the curve graphed in part d, using some form of the word "logarithm."

2. **a.** Copy and complete the table at the right by finding, for each of the given values of x, the corresponding value of $y = 4^x$.

x	$-\frac{3}{2}$	-1	$-\frac{1}{2}$	0	$\frac{1}{2}$	1	$\frac{3}{2}$
$y = 4^x$							

   b. Sketch the graph of $y = 4^x$ to include the points whose ordered pairs were found in part a.

   c. Name seven ordered pairs found in the graph of the inverse of $y = 4^x$.

   d. Sketch the graph of $y = \log_4 x$ to include the points named in part c.

3. The table at the right lists three selected pairs of the function $y = 5^x$. Note that values of y are stated as rational approximations to the nearest tenth.

x	5^x		$= y$
$-\dfrac{1}{2}$	$5^{-\frac{1}{2}} = \dfrac{1}{\sqrt{5}}$	$= \dfrac{\sqrt{5}}{5}$	$\approx .5$
$\dfrac{1}{2}$	$5^{\frac{1}{2}} = \sqrt{5}$		≈ 2.2
$\dfrac{3}{2}$	$5^{\frac{3}{2}} = (\sqrt{5})^3 = 5\sqrt{5}$		≈ 11.2

   a. Sketch the graph of $y = 5^x$ in the interval $-1 \leq x \leq \frac{3}{2}$, including points where $x = -1$, $-\frac{1}{2}, 0, \frac{1}{2}, 1$, and $\frac{3}{2}$.

   b. Sketch the graph of $y = \log_5 x$ (that is, the reflection of $y = 5^x$ in the line whose equation is $y = x$).

4. **a.** Name five ordered pairs of the function $y = 6^x$.
   b. Name five pairs in the function $y = \log_6 x$. (*Hint:* The functions named in parts a and b are inverse functions.)
   c. Sketch the graph of $y = \log_6 x$.

5. **a.** Sketch the graph of $y = 8^x$, including points for which $x = -1$, $-\frac{2}{3}, -\frac{1}{3}, 0, \frac{1}{3}, \frac{2}{3}, 1$.
   b. Sketch the graph of $y = \log_8 x$, including the images of the ordered pairs in part a under a reflection in the line $y = x$.

6. **a.** Sketch the graph of $y = (\frac{5}{2})^x$ in the interval $-1 \leq x \leq 2$.
   b. Sketch the graph of $y = \log_{\frac{5}{2}} x$ in the interval $\frac{2}{5} \leq x \leq \frac{25}{4}$.
   c. What is the y-intercept of $y = (\frac{5}{2})^x$?
   d. What is the x-intercept of $y = \log_{\frac{5}{2}} x$?

7. **a.** Sketch the graph of $y = \log_{10} x$ in the interval $\frac{1}{10} \leq x \leq 10$. (It will be helpful to choose $\sqrt{10} \approx 3.2$ as a value of x.)
   b. On the same set of axes, draw the graph of $y = 1$.
   c. What are the coordinates of the point of intersection of the graphs of $y = \log_{10} x$ and $y = 1$?
   d. In what quadrants does the graph of $y = \log_{10} x$ lie?

8. The graph of $y = 7^x$ is reflected in the line $y = x$. What is the equation of the set of image points?

9. When drawn on the same set of axes, at what point, if any, do the graphs of $y = \log_5 x$ and $y = \log_3 x$ intersect?

In 10–15, write the equation of $f^{-1}(x)$, the inverse of $f(x)$.

10. $f(x) = 2^x$ 11. $f(x) = 6^x$ 12. $f(x) = 12^x$
13. $f(x) = \log_3 x$ 14. $f(x) = \log_7 x$ 15. $f(x) = \log_9 x$

In 16–23, select the numeral preceding the expression that best completes the sentence or answers the question.

16. The graph of $y = \log_8 x$ lies entirely in quadrants:
    (1) I and II   (2) II and III   (3) I and III   (4) I and IV

17. The function that is the inverse of $y = 5^x$ under composition is:
    (1) $y = -5^x$ (2) $y = 5^{-x}$ (3) $y = \log_5 x$ (4) $x = \log_5 y$

18. At what point does the graph of $y = \log_5 x$ intersect the x-axis?
    (1) $(1, 0)$ (2) $(0, 1)$ (3) $(5, 0)$ (4) There is no point of intersection.

19. A number that is *not* in the domain of the function $y = \log_{10} x$ is:
    (1) 1   (2) 0   (3) $\frac{1}{2}$ (4) 10

20. If $y = \log_{10} x$ and $x > 1$, then y is:
    (1) positive   (2) zero   (3) negative   (4) not a real number
21. If $y = \log_{10} x$ and $x = 1$, then y is:
    (1) positive   (2) zero   (3) negative   (4) not a real number
22. If $y = \log_{10} x$ and $0 < x < 1$, then y is:
    (1) positive   (2) zero   (3) negative   (4) not a real number
23. If $y = \log_{10} x$ and $x \leq 0$, then y is:
    (1) positive   (2) zero   (3) negative   (4) not a real number

11-2 LOGARITHMIC FORM OF AN EQUATION

In the last section, we saw that an exponential equation and a logarithmic equation are two ways of expressing a statement about a power. The basic statement made by the equation $\log_5 125 = 3$ is that the log (that is, the exponent) is 3.

$$\log_5 125 = 3$$
$$\downarrow \quad\quad \downarrow\downarrow$$

exponent is 3

The base is the number 5, written to the right of and a half-line below the word "log."

$$\log_5 125 = 3 \quad\quad \leftrightarrow \quad\quad 5^3 = 125$$

■ In general: $\log_b c = a \quad\quad \leftrightarrow \quad\quad b^a = c \; (b > 0 \text{ and } b \neq 1)$

Logarithmic Form	Exponential Form
$\log_7 49 = 2$ $\downarrow \quad \downarrow\downarrow$ exponent is 2	$7^2 = 49$
$\log_{10} .1 = -1$	$10^{-1} = .1$
$\log_4 2 = \frac{1}{2}$	$4^{\frac{1}{2}} = 2$
$\log_9 27 = \frac{3}{2}$	$9^{\frac{3}{2}} = 27$

| **MODEL PROBLEMS** |

1. Write $8^2 = 64$ in logarithmic form.

Solution

1. The basic statement is that the loga- $\qquad$ log = 2
rithm (or exponent) is 2.

2. The base, 8, is written to the right of $\qquad$ $\log_8 = 2$
and a half-line below the word "log."

3. The power, 64, follows the word "log." $\qquad$ $\log_8 64 = 2$ *Ans.*

2. Write $\log_3 81 = 4$ in exponential form.

Solution

Since $\log_b c = a$ is equivalent to $b^a = c$, we know that $\log_3 81 = 4$ is equivalent to $3^4 = 81$.

Answer: $3^4 = 81$

3. Solve for x: $\log_x 8 = 3$

Solution

1. Rewrite the logarithmic equation in expo- $\qquad$ $\log_x 8 = 3$
nential form. The base is x and the exponent $\qquad$ $x^3 = 8$
is 3.

2. Use the reciprocal of the exponent to solve $\qquad$ $(x^3)^{\frac{1}{3}} = 8^{\frac{1}{3}}$
for x.

3. Simplify. $\qquad$ $x = \sqrt[3]{8}$
$x = 2$

Answer: $x = 2$

4. Find the value of $\log_{25} 125$.

Solution

Here we are being asked to find a when $\log_{25} 125 = a$.

1. Write the equation in exponential form. $\qquad$ $\log_{25} 125 = a$
The base is 25 and the exponent is a. $\qquad$ $25^a = 125$

2. Change each power to base 5: $25 = 5^2$ $\qquad$ $(5^2)^a = 5^3$
and $125 = 5^3$.

3. Equate exponents of like bases. $\qquad$ $2a = 3$

4. Solve for a. $\qquad$ $a = \frac{3}{2}$

Answer: $\log_{25} 125 = \frac{3}{2}$

| EXERCISES |

In 1–12, write the exponential equation in logarithmic form.

1. $2^4 = 16$ 2. $3^2 = 9$ 3. $10^{-2} = .01$ 4. $12^0 = 1$

5. $8^{\frac{1}{3}} = 2$ 6. $5^{\frac{1}{2}} = \sqrt{5}$ 7. $6^{-1} = \frac{1}{6}$ 8. $9^{-\frac{1}{2}} = \frac{1}{3}$

9. $4^a = b$ 10. $x^3 = y$ 11. $5^c = d$ 12. $b^a = 8$

In 13–24, write the logarithmic equation in exponential form.

13. $\log_2 64 = 6$ 14. $\log_7 49 = 2$ 15. $\log_{10} 1000 = 3$

16. $\log_{27} 9 = \frac{2}{3}$ 17. $\log_4 32 = \frac{5}{2}$ 18. $\log_{15} 1 = 0$

19. $\log_{\frac{1}{9}} 3 = -\frac{1}{2}$ 20. $\log_{\frac{1}{8}} 2 = -\frac{1}{3}$ 21. $\log_{10} .001 = -3$

22. $\log_a \sqrt[3]{a} = \frac{1}{3}$ 23. $\log_a \frac{1}{a} = -1$ 24. $\log_b c = \frac{1}{5}$

In 25–33, solve for x.

25. $\log_2 x = 4$ 26. $\log_5 x = 2$ 27. $\log_3 x = -2$

28. $\log_7 x = 0$ 29. $\log_{81} x = \frac{1}{2}$ 30. $\log_9 x = -\frac{1}{2}$

31. $\log_6 x = -2$ 32. $\log_{10} x = 7$ 33. $\log_{64} x = \frac{2}{3}$

In 34–42, solve for n.

34. $\log_3 27 = n$ 35. $\log_2 16 = n$ 36. $\log_{10} 10{,}000 = n$

37. $\log_5 \frac{1}{25} = n$ 38. $\log_8 32 = n$ 39. $\log_{100} .1 = n$

40. $\log_{36} 216 = n$ 41. $\log_{49} \frac{1}{7} = n$ 42. $\log_{27} \frac{1}{9} = n$

In 43–51, solve for b.

43. $\log_b 27 = 3$ 44. $\log_b 49 = 2$ 45. $\log_b \frac{1}{2} = -1$

46. $\log_b 125 = \frac{3}{2}$ 47. $\log_b 4 = \frac{1}{2}$ 48. $\log_b 6 = \frac{1}{3}$

49. $\log_b .1 = -\frac{1}{2}$ 50. $\log_b 8 = -3$ 51. $\log_b \sqrt{5} = \frac{1}{4}$

52. If $f(x) = \log_4 x$, find $f(64)$.

53. If $f(x) = \log_3 x$, find $f(1)$.

54. If $f(x) = \log_2 x$, find $f(\sqrt{2})$.

55. If $\log_{10} x = -3$, find x.

In 56–58, select the numeral preceding the expression that best completes the sentence.

56. The value of $\log_{25} 5$ is: (1) $\frac{1}{2}$ (2) 2   (3) $\frac{1}{5}$ (4) 5

57. If $\log_8 x = \frac{2}{3}$, then x equals:

   (1) $\frac{16}{3}$ (2) $\frac{64}{3}$ (3) $16\sqrt{2}$ (4) 4

58. If $\log_2 a = \log_3 a$, then a equals:

   (1) 1   (2) 2   (3) 3   (4) 0

11-3 COMPUTATIONS WITH LOGARITHMS

Over the years, we have learned procedures that allow us to perform operations in arithmetic, such as multiplication and division. In the last chapter on exponents, we learned rules for working with powers of like bases, such as the powers of 2 in the accompanying table.

Powers of 2	Logarithms to the Base 2
$2^0 = 1$	$\log_2 1 = 0$
$2^1 = 2$	$\log_2 2 = 1$
$2^2 = 4$	$\log_2 4 = 2$
$2^3 = 8$	$\log_2 8 = 3$
$2^4 = 16$	$\log_2 16 = 4$
$2^5 = 32$	$\log_2 32 = 5$
$2^6 = 64$	$\log_2 64 = 6$
$2^7 = 128$	$\log_2 128 = 7$
$2^8 = 256$	$\log_2 256 = 8$
$2^9 = 512$	$\log_2 512 = 9$
$2^{10} = 1024$	$\log_2 1024 = 10$
$2^{11} = 2048$	$\log_2 2048 = 11$
$2^{12} = 4096$	$\log_2 4096 = 12$

To the right of each power of 2 in the table is an equivalent form that involves a logarithm to the base 2.

In this section, we will see that problems can be solved by various methods. These methods will include the use of exponents, logarithms, and standard arithmetic.

In order to use logarithms in calculations, we will derive rules for logarithms based on the rules for working with powers of like bases.

Product Rule

If $b > 0$ and $b \neq 1$:

$$b^x = A \leftrightarrow \log_b A = x$$
$$b^y = B \leftrightarrow \log_b B = y$$
$$b^{x+y} = AB \leftrightarrow \log_b AB = x + y$$

By substitution: $\log_b AB = \log_b A + \log_b B$

For example, $\log_2 (4 \cdot 8) = \log_2 32 = 5$.

By the product rule, $\log_2 (4 \cdot 8) = \log_2 4 + \log_2 8 = 2 + 3 = 5$.

| MODEL PROBLEM |

If $N = 8 \cdot 32$, find N.

Solution

Method 1: Exponential Form

1. Write the expression to be evaluated.

$N = 8 \cdot 32$

2. Express each number as a power of 2.

$N = 2^3 \cdot 2^5$

3. Add exponents when multiplying powers with like bases.

$N = 2^{3+5}$
$N = 2^8$

4. Evaluate, or find the value from the table.

$N = 256$ *Ans.*

Method 2: Logarithmic Form

1. Write the expression to be evaluated.

$N = 8 \cdot 32$

2. Write the logarithm to the base 2 of each side of the equation.

$\log_2 N = \log_2 (8 \cdot 32)$

3. Use the product rule.

$\log_2 N = \log_2 8 + \log_2 32$

4. Evaluate the right-hand member (see the table), and simplify.

$\log_2 N = 3 + 5$
$\log_2 N = 8$

5. Find N by referring to the table. (Since $\log_2 256 = 8, N = 256$.)

$N = 256$ *Ans.*

Method 3: Arithmetic

This method should serve only as a *check* for the procedures just followed. By multiplication, $N = 8 \cdot 32 = 256$.   *Ans.*

Notice that since exponential form and logarithmic form are simply two different ways of writing the same numerical relationship, methods 1 and 2 are different only in the way in which the computation is written.

Quotient Rule

If $b > 0$ and $b \neq 1$:

$$b^x = A \leftrightarrow \log_b A = x$$
$$b^y = B \leftrightarrow \log_b B = y$$

$$b^{x-y} = \frac{A}{B} \leftrightarrow \log_b \frac{A}{B} = x - y$$

By substitution: $$\log_b \frac{A}{B} = \log_b A - \log_b B$$

For example, $\log_2 (128 \div 8) = \log_2 16 = 4$.
By the quotient rule, $\log_2 (128 \div 8) = \log_2 128 - \log_2 8 = 7 - 3 = 4$.

MODEL PROBLEMS

1. Evaluate $\dfrac{128^2}{256}$

Solution

Method 1: Exponential Form

Let $A = \dfrac{128^2}{256} = \dfrac{128 \cdot 128}{256}$

$A = \dfrac{2^7 \cdot 2^7}{2^8}$

$A = \dfrac{2^{14}}{2^8}$

$A = 2^{14-8}$
$A = 2^6$
$A = 64$ *Ans.*

Method 2: Logarithmic Form

$A = \dfrac{128^2}{256} = \dfrac{128 \cdot 128}{256}$

$\log_2 A = \log_2 \left(\dfrac{128 \cdot 128}{256} \right)$

$\log_2 A = \log_2 (128 \cdot 128) - \log_2 256$
$\log_2 A = \log_2 128 + \log_2 128 - \log_2 256$
$\log_2 A = 7 + 7 - 8$
$\log_2 A = 14 - 8$
$\log_2 A = 6$
$A = 64$ *Ans.*

Method 3: Arithmetic

$$\frac{128^2}{256} = \frac{\overset{1}{\cancel{128}} \cdot 128}{\underset{2}{\cancel{256}}} = \frac{128}{2} = 64 \quad Ans.$$

2. Express $\log_2 \dfrac{ab}{c}$ in terms of $\log_2 a$, $\log_2 b$, and $\log_2 c$.

Solution

1. Use the quotient rule. $\log_2 \dfrac{ab}{c} = \log_2 ab - \log_2 c$

2. Use the product rule. $\log_2 \dfrac{ab}{c} = \log_2 a + \log_2 b - \log_2 c$ *Ans.*

Power Rule

If $b > 0$ and $b \neq 1$:
$$b^x = A \ \leftrightarrow \ \log_b A = x$$
$$b^{xc} = A^c \ \leftrightarrow \ \log_b A^c = xc = cx$$
By substitution: $\log_b A^c = c \log_b A$

We can find the values of some logarithmic expressions by two methods. For example, the product rule tells us that:

$$\log_2 (4^3) = \log_2 (4 \cdot 4 \cdot 4) = \log_2 4 + \log_2 4 + \log_2 4 = 2 + 2 + 2 = 6$$

By the power rule, we know that:

$$\log_2 (4^3) = 3 \log_2 4 = 3 \cdot 2 = 6$$

Notice that $\log_2 4 + \log_2 4 + \log_2 4 = 3 \log_2 4$.

Not all logarithmic expressions involving powers can be evaluated by the product rule. For example, $\log_2 \sqrt[3]{2}$ can be simplified only by using the power rule. Here, $\log_2 \sqrt[3]{2} = \log_2 (2^{\frac{1}{3}}) = \frac{1}{3} \log_2 2 = \frac{1}{3} \cdot 1 = \frac{1}{3}$.

MODEL PROBLEM

Find $\sqrt[3]{512}$.

Solution

Method 1: Exponential Form

$$\text{Let } x = \sqrt[3]{512}$$
$$x = (512)^{\frac{1}{3}}$$
$$x = (2^9)^{\frac{1}{3}}$$
$$x = 2^{9 \cdot \frac{1}{3}}$$
$$x = 2^3$$
$$x = 8 \quad Ans.$$

Method 2: Logarithmic Form

$$\text{Let } x = \sqrt[3]{512}$$
$$\log_2 x = \log_2 \sqrt[3]{512}$$
$$\log_2 x = \log_2 (512)^{\frac{1}{3}}$$
$$\log_2 x = \frac{1}{3} \log_2 512$$
$$\log_2 x = \frac{1}{3} \cdot 9$$
$$\log_2 x = 3$$
$$x = 8 \quad Ans.$$

Method 3: Arithmetic

We have not learned an algorithm to find the cube root of a number. Since $8 \cdot 8 \cdot 8 = 512$, however, we can state that $\sqrt[3]{512} = 8$. *Ans.*

The problems that have been used to illustrate the rules derived in this section have been limited to values in the table given at the beginning of this section. The rules, however, apply to logarithms having any base.

Later in this chapter, we will see that operations involving logarithms can be extended to all numbers. When this happens, we will see how very difficult problems in arithmetic can be handled easily by the use of logarithms to the base 10. Logarithms to the base 10 are called *common logarithms*. When no base is written in the logarithmic form of a number, the base is understood to be 10.

Thus, $\log 100 = 2$ means $\log_{10} 100 = 2$.

MODEL PROBLEMS

1. Find the value of $\log_6 12 + \log_6 3$.

How to Proceed	*Solution*
1. Write the given expression.	$\log_6 12 + \log_6 3$
2. Use the product rule.	$= \log_6 (12 \cdot 3)$
3. Simplify.	$= \log_6 36$
4. Evaluate the result.	$= 2$ *Ans.*

2. If $2 \log_3 9 + \log_3 x = \log_3 27$, find x.

    How to Proceed

    1. Write the given equation.

    2. Simplify the left-hand member: Use the power rule, and use the product rule.

    3. Equate the powers and solve for x.

    Solution

    $2 \log_3 9 + \log_3 x = \log_3 27$

    $\log_3 9^2 + \log_3 x = \log_3 27$

    $\log_3 (9^2 \cdot x) = \log_3 27$

    $\log_3 (81x) = \log_3 27$

    $81x = 27$

    $x = \frac{27}{81} = \frac{1}{3}$ *Ans.*

Alternate Solution

1. Write the given equation. $2 \log_3 9 + \log_3 x = \log_3 27$

2. Evaluate each term if possible. $2 \cdot 2 + \log_3 x = 3$
   $4 + \log_3 x = 3$

3. Subtract 4 from both members of the equation. $\log_3 x = -1$

4. Rewrite in exponential form and simplify. $3^{-1} = x$
   $\frac{1}{3} = x$ *Ans.*

3. If $x = \dfrac{a}{\sqrt{b}}$, express $\log x$ in terms of $\log a$ and $\log b$.

Solution

1. Express $\sqrt{b}$ as a power. $x = \dfrac{a}{\sqrt{b}}$

   $x = \dfrac{a}{b^{\frac{1}{2}}}$

2. Write the common logarithm of each member of the equation. $\log x = \log \left(\dfrac{a}{b^{\frac{1}{2}}} \right)$

3. Use the quotient rule. $\log x = \log a - \log b^{\frac{1}{2}}$

4. Use the power rule. $\log x = \log a - \frac{1}{2} \log b$ *Ans.*

4. The expression $\log_3 a^5 b$ is equivalent to:
   (1) $5 \log_3 ab$ (2) $5 \log_3 a + \log_3 b$ (3) $\log_3 5ab$
   (4) $\log_3 5a + \log_3 b$

Solution

1. Use the product rule. $\log_3 a^5 b = \log_3 a^5 + \log_3 b$

2. Use the power rule. $\log_3 a^5 b = 5 \log_3 a + \log_3 b$

Answer: (2) $5 \log_3 a + \log_3 b$

| EXERCISES |

In 1–18, evaluate the given expression by using: **a.** powers of 2 in exponential form **b.** logarithms to the base 2 **c.** standard arithmetic. (*Note:* The table on page 495 may be used for these exercises.)

1. $64 \cdot 16$ **2.** $1024 \div 256$ **3.** 16^3

4. $\sqrt[5]{32}$ **5.** $\sqrt{1024}$ **6.** $\sqrt[4]{256}$

7. $4^3 \cdot 8$ **8.** $8^3 \div 2$ **9.** $16\sqrt{64}$

10. $\dfrac{\sqrt{4096}}{16}$ **11.** $\sqrt{\dfrac{4096}{16}}$ **12.** $\dfrac{8^4}{16^2}$

13. $\dfrac{128 \cdot 4^2}{512}$ **14.** $\dfrac{64\sqrt[3]{64}}{256}$ **15.** $\dfrac{\sqrt[3]{4096}}{16}$

16. $\sqrt{\dfrac{2048}{32}}$ **17.** $\dfrac{32\sqrt[3]{512}}{64}$ **18.** $32\sqrt[3]{\dfrac{512}{64}}$

In 19–24, write the given expression in terms of $\log a$ and $\log b$.

19. $\log ab$ **20.** $\log (a \div b)$ **21.** $\log (a^2 b)$

22. $\log \dfrac{\sqrt{a}}{b}$ **23.** $\log a\sqrt{b}$ **24.** $\log (ab)^2$

In 25–30, solve for n.

25. $\log_3 9 + \log_3 3 = \log_3 n$ **26.** $\log_4 64 - \log_4 16 = \log_4 n$

27. $3 \log_2 4 = \log_2 n$ **28.** $\log_6 216 - \frac{1}{2} \log_6 36 = \log_6 n$

29. $\log 1000 - 2 \log 100 = \log n$ **30.** $\log_3 n - \log_3 \frac{1}{3} = \log_3 9$

In 31–34, select the numeral preceding the expression that best completes the sentence.

31. The expression $\log_5 \dfrac{a}{c^2}$ is equivalent to:

    (1) $\log_5 a - \frac{1}{2}\log_5 c$ (2) $\dfrac{\log_5 a}{2 \log_5 c}$ (3) $2(\log_5 a - \log_5 c)$

    (4) $\log_5 a - 2 \log_5 c$

32. If $x = (ab)^2$, then $\log x$ equals:

    (1) $2 \log a + \log b$ (2) $2 \log a + 2 \log b$ (3) $\log a + 2 \log b$

    (4) $\log 2a + \log 2b$

33. The expression $\log \sqrt{\dfrac{x}{y}}$ is equivalent to:

    (1) $\dfrac{1}{2} \log x - \log y$ (2) $\dfrac{1}{2}\left(\dfrac{\log x}{\log y}\right)$ (3) $\log \dfrac{1}{2}x - \log \dfrac{1}{2}y$

    (4) $\dfrac{1}{2}(\log x - \log y)$

34. The value of $\log_4 8 + \log_4 2$ is:
    (1) 1 (2) 2 (3) 16 (4) 4

11-4 COMMON LOGARITHMS

Aids in Computation

From the beginnings of the use of numbers, people have devised aids in computation. One of the earliest such tools is the abacus, still in use today in some parts of the world. In the 17th century, the development of logarithms greatly aided the calculations necessary in the study of astronomy and helped advance that science. The slide rule, a mechanical device that uses a scale based on logarithms, was for many years the familiar tool of scientists and engineers. Although calculators and computers today have replaced many older calculating devices, the logarithmic functions remain important in mathematics, science, engineering, and business. Knowing how to compute using logarithms will help you to understand and work with functions and equations involving logarithms and arising from physical situations.

Common Logarithms and Scientific Notation

In section 3, we saw how a table of powers of 2 and a table of logarithms to the base 2 allowed us to perform operations by different methods. We were limited, however, to just those numbers in the table. Is it possible to make a table that can be used to express every number as a power of some base and use that table to simplify all computation?

Since there are infinitely many positive real numbers, we obviously cannot make a table that lists all of them. Every positive number, however, can be written in scientific notation as $a \times 10^n$ where $1 \leq a < 10$ and n is an integer. We can easily express integral powers of 10 by using *common logarithms*, that is, logarithms to the base 10. We must therefore learn how to use a *table of common logarithms* for numbers whose values are $1 \leq a < 10$. This table and the integral powers of 10 will enable us to find the common logarithm of any number.

Integral Powers of Ten

To obtain the common logarithm of an integral power of 10, we need only note the exponent. Recall that when the base is omitted in writing the logarithmic form, the base is understood to be 10.

Exponential Form (Powers of 10)	Logarithmic Form (Common Logarithms)
10^4 = 10 000	log 10 000 = 4
10^3 = 1 000	log 1 000 = 3
10^2 = 100	log 100 = 2
10^1 = 10	log 10 = 1
10^0 = 1	log 1 = 0
10^{-1} = 0.1	log 0.1 = -1
10^{-2} = 0.01	log 0.01 = -2
10^{-3} = 0.001	log 0.001 = -3

Numbers Between 1 and 10

In the preceding table, we observe that as the numbers increase from 0.001 to 10 000, the logarithms of those numbers increase from -3 to 4. Between every two consecutive integral powers of 10, there exists an infinite number of powers of 10. Between 10^0 and 10^1, some powers include $10^{\frac{1}{2}}$, $10^{.6}$, $10^{.7011}$, and so forth.

Let $10^M = a$ where $0 \leq M < 1$.
Then: $10^0 \leq 10^M < 10^1$,
or $1 \leq a < 10$.
Thus: $\log 1 \leq \log a < \log 10$,
or $0 \leq \log a < 1$.
In other words, if a is a number greater than or equal to 1 but

Exponential Form	Logarithmic Form
10^1 = 10	log 10 = 1
$10^M = a$	log a = M
10^0 = 1	log 1 = 0

less than 10, then log a is greater than or equal to 0 but less than 1.

The common logarithm of a number between 1 and 10 can be approximated to any required degree of accuracy by methods of advanced mathematics. The values of these logarithms, written as four-place decimals, are given in the table of Common Logarithms of Numbers on pages 748 and 749 of this book. The table lists an approximate value of the logarithm of any number having three significant digits from 1.00 to 9.99. A section of the table of common logarithms is shown on the next page.

Common Logarithms of Numbers

N	0	1	2	3	4	5	6	7	8	9
10	0000	0043	0086	0128	0170	0212	0253	0294	0334	0374
11	0414	0453	0492	0531	0569	0607	0645	0682	0719	0755
12	0792	0828	0864	0899	0934	0969	1004	1038	1072	1106
13	1139	1173	1206	1239	1271	1303	1335	1367	1399	1430
14	1461	1492	1523	1553	1584	1614	1644	1673	1703	1732
15	1761	1790	1818	1847	1875	1903	1931	1959	1987	2014
16	2041	2068	2095	2122	2148	2175	2201	2227	2253	2279
→17	2304	2330	2355	2380	2405	(2430)	2455	2480	2504	2529
18	2553	2577	2601	2625	2648	2672	2695	2718	2742	2765
19	2788	2810	2833	2856	2878	2900	2923	2945	2967	2989

To find the common logarithm of 1.75, look for the first two digits, 17, in the first column under N and look for the third digit, 5, in the row next to N at the top of the table. Read the entry in the corresponding row and column, as shown in the figure. The entry 2430 represents .2430, a number between 0 and 1.

$$\log 1.75 \approx .2430 \text{ or } 10^{.2430} \approx 1.75$$
$$\log 1.75 = .2430 \text{ to the nearest ten-thousandth}$$

We can also read from the portion of the table shown above that:

$$\log 1.08 \approx .0334 \text{ or } 10^{.0334} \approx 1.08$$
$$\log 1.40 \approx .1461 \text{ or } 10^{.1461} \approx 1.40$$

Using the complete table on pages 748 and 749, we see that:

$$\log 5.38 \approx .7308 \text{ or } 10^{.7308} \approx 5.38$$
$$\log 9.27 \approx .9671 \text{ or } 10^{.9671} \approx 9.27$$

The approximate value of the common logarithm of every number between 1 and 10 can be expressed as a four-place decimal between 0 and 1. The table of Common Logarithms of Numbers given in this book lists, as four-place decimals, the logarithms of numbers having three or fewer significant digits. We will write log 6.92 = .8401 to mean that log 6.92 = .8401 to the nearest ten-thousandth.

The Logarithm of Any Number

The common logarithm of any integral power of 10 can be found by inspection. For example, $\log 100 = \log 10^2 = 2$ and $\log 0.001 = \log 10^{-3} = -3$. The table of common logarithms shows the logarithm of any number between 1 and 10 containing three significant digits. Now scientific notation enables us to find the logarithm of any number having three significant digits by using two parts of the logarithm.

We know that $5720 = 5.72 \times 10^3$.
From the table: log 5.72 = .7574 or $10^{.7574} = 5.72$
By inspection: $\log 10^3 = 3$
Therefore:

1. $5720 = 5.72 \times 10^3$ $\longleftrightarrow$ $\log 5720 = \log (5.72 \times 10^3)$
$= 10^{.7574} \times 10^3$ $= \log 5.72 + \log 10^3$
$= 10^{.7574+3}$ $= .7574 + 3$
$= 10^{3.7574}$ $= 3.7574$

Notice that $10^3 < 5720 < 10^4$ and $3 < \log 5720 < 4$.

Let us repeat these steps with some other numbers.

2. $57.2 = 5.72 \times 10^1$ $\longleftrightarrow$ $\log 57.2 = \log (5.72 \times 10^1)$
$= 10^{.7574} \times 10^1$ $= \log 5.72 + \log 10^1$
$= 10^{.7574+1}$ $= .7574 + 1$
$= 10^{1.7574}$ $= 1.7574$

Notice that $10^1 < 57.2 < 10^2$ and $1 < \log 57.2 < 2$.

3. $0.0572 = 5.72 \times 10^{-2}$ $\longleftrightarrow$ $\log 0.0572 = \log (5.72 \times 10^{-2})$
$= 10^{.7574} \times 10^{-2}$ $= \log 5.72 + \log 10^{-2}$
$= 10^{.7574-2}$ $= .7574 - 2$

The log of 0.0572 is the sum of a positive decimal value and a negative integer. We will not combine these two parts into a single negative number since it is more convenient to work with them in the form .7574 - 2.
Notice that $10^{-2} < .0572 < 10^{-1}$ and $-2 < \log .0572 < -1$.
We see that in each case the logarithm can be thought of as the sum of two parts.

1. The integral part of a common logarithm is called the "characteristic." The *characteristic* of log a is the greatest integer less than or equal to log a.

2. The positive decimal part of a common logarithm is called the "mantissa." The *mantissa* of log a is the difference between log a and the characteristic of log a.

The common logarithm of a number is the sum of the characteristic and the mantissa.
For example, consider the three problems we have just studied.

Logarithm	Characteristic	Mantissa
$\log 5720 = \log (5.72 \times 10^3)$ $= 3.7574$	3	.7574
$\log 57.2 = \log (5.72 \times 10^1)$ $= 1.7574$	1	.7574
$\log 0.0572 = \log (5.72 \times 10^{-2}) = .7574 - 2$	-2	.7574

We saw above that log 0.0572 = .7574 - 2, that is, the sum of the mantissa .7574 and the characteristic - 2. It is often convenient to write a negative characteristic as the sum of a positive part and a negative part. For example, we can write - 2 as 8 - 10.

Thus, log 0.0572 = .7574 - 2 = .7574 + 8 - 10 = 8.7574 - 10.

It is also possible to write the characteristic - 2 as other sums.

log 0.0572 = .7574 - 2	log 0.0572 = 4.7574 - 6
log 0.0572 = 8.7574 - 10	log 0.0572 = 28.7574 - 30

In all of these cases, notice that the characteristic is - 2.

To determine the characteristic of a common logarithm, we can use one of the following methods:

Place Value	Scientific Notation
⑧46 $100 = 10^2$ characteristic is 2	$846 = 8.46 \times 10^2$ characteristic is 2

■ To find the common logarithm of a number:

1. Find the characteristic, using place value or scientific notation.

2. Locate the mantissa in the table of Common Logarithms of Numbers.

| MODEL PROBLEMS |

1. Find log 27.9.

Solution

1. Find the characteristic.

Method 1: Place Value	*Method 2: Scientific Notation*
②7.9 $10 = 10^1$	$27.9 = 2.79 \times 10^1$

The characteristic is 1.

2. Find the mantissa. In the table of Common Logarithms of Numbers, locate the entry in row 27 under the column headed 9. The mantissa is .4456.

3. Add the characteristic and mantissa.

$$\log 27.9 = 1.4456 \quad Ans.$$

2. Find log 0.0096.

Solution

1. Find the characteristic.

$$\log 0.0096 = \qquad -3$$
$$= 7. \qquad -10$$

2. Find the mantissa. Locate the entry in row 96 under the column headed 0.

$$\log 0.0096 = .9823 - 3$$
$$7.9823 - 10$$

Answer: Any logarithm whose characteristic is -3 and whose mantissa is .9823. Two possible ways of writing the solution are .9823 $-$ 3, and 7.9823 $-$ 10.

3. Find the integral value of n such that $n < \log 0.582 < n + 1$.

Solution

Method 1:

1. Locate 0.582 between two consecutive powers of 10.

$$.1 < 0.582 < 1$$
$$10^{-1} < 0.582 < 10^0$$

2. Write the logarithm of each member of the inequality.

$$\log 10^{-1} < \log 0.582 < \log 10^0$$
$$-1 < \log 0.582 < 0$$

3. Therefore, by comparing this result with the given statement, we see that $n = -1$ *Ans.*

Method 2:

Since the characteristic of $\log a$ is the greatest integer less than or equal to $\log a$, then n is the characteristic of log 0.582. Therefore, $n = -1$.   *Ans.*

| **EXERCISES** |

In 1–12, find the characteristic of the common logarithm of the given number.

1. 279	2. 56.8	3. 9280	4. 7.65
5. 0.824	6. 0.00039	7. 0.021	8. 42,800
9. 7.8	10. 12.7	11. 0.005	12. 97

In 13-24, find the mantissa of the common logarithm of the given number.

13. 6.82   14. 43.9   15. 1780   16. 9.86
17. 0.597   18. 0.00203   19. 0.076   20. 15
21. 3000   22. 7   23. 10.8   24. 0.03

In 25-36, find the common logarithm of the given number.

25. 58.6   26. 767   27. 0.178   28. 3.59
29. 9540   30. 0.083   31. 0.009   32. 68.9
33. 4.71   34. 0.000013   35. 0.105   36. 5

37. If log 4.81 = .6821, find the value of each given logarithm without using the table of common logarithms.
    a. log 481   b. log 48100   c. log 0.000481
38. If log 85.3 = 1.9309, find the value of each given logarithm without using a table.
    a. log 853   b. log 0.853   c. log 8.53
39. If $10^{0.6021}$ = 4, find the value of each given power of 10.
    a. $10^{1.6021}$ b. $10^{3.6021}$ c. $10^{0.6021-2}$ d. $10^{9.6021-10}$
40. If 10^x = 47.5, find x to the nearest ten-thousandth.
41. If n is an integer and $n < \log 346 < n + 1$, find n.
42. Find the integral value of a such that $a < \log 0.02 < a + 1$.
43. If 10^y = 148, find y to four decimal places.

In 44-46, select the numeral preceding the expression that best completes the sentence or answers the question.

44. If $3 < \log a < 4$, then the characteristic of log a is:
    (1) 1   (2) 0   (3) 3   (4) 4
45. If $-2 < \log x < -1$, then the characteristic of log x is:
    (1) -1   (2) -2   (3) .1   (4) .01
46. If $0 < \log b < 1$, which of the following is true?
    (1) $0 < b < 1$ (2) $1 < b < 10$
    (3) $10^0 < \log b < 10^1$ (4) $10^1 < \log b < 10^{10}$

11-5 ANTILOGARITHMS

If log x = y, then x is the number whose common logarithm is y. The number x is called the *antilogarithm* of y.

For example, if log N = 1.4048, then the antilogarithm of 1.4048 is the number N. Since log 25.4 = 1.4048, the antilogarithm of 1.4048 is 25.4.

A logarithm is the sum of a mantissa and a characteristic. The antilogarithm is the product $a \times 10^n$ where $1 \leq a < 10$. The value of a is found by using the table of Common Logarithms of Numbers. The value of n is the characteristic of the log.

| MODEL PROBLEMS |

1. Find N if $\log N = 2.8537$.

Solution

1. Write the form for N in scientific notation.

             $\log N = 2.8537$
             $N = a \times 10^n$

2. Find a by looking for .8537 among the mantissas in the table of Common Logarithms of Numbers. Since .8537 is in row 71 under 4, $a = 7.14$.

             $N = 7.14 \times 10^n$

3. Since the characteristic of $\log N$ is 2, $n = 2$.

             $N = 7.14 \times 10^2$

4. Write N in ordinary notation.

             $N = 714$

Answer: $N = 714$

2. Find the antilogarithm of $9.6860 - 10$ to three significant digits.

Solution

1. Write the general form of a number in scientific notation to represent the antilogarithm.

             $a \times 10^n$

2. Find a by looking for .6860 among the mantissas in the table of common logarithms.
   $\log 4.86 = .6866$

           .6860 $\rceil$ closer
   $\log 4.85 = .6857 \leftarrow$

             4.85×10^n

Since the mantissa .6860 lies between .6857 and .6866 and it is closer to .6857, let $a = 4.85$ (the number whose logarithm is .6857).

3. Since the characteristic is $9 - 10$ or -1, $n = -1$.

             4.85×10^{-1}

4. Write the antilogarithm in ordinary decimal notation.

             0.485 *Ans.*

EXERCISES

In 1-12, find N to three significant digits.

1. $\log N = 0.6884$
2. $\log N = 3.9294$
3. $\log N = 1.7612$
4. $\log N = .2201 - 2$
5. $\log N = .6365 - 1$
6. $\log N = 2.0170$
7. $\log N = 9.9643 - 10$
8. $\log N = 1.4942$
9. $\log N = 7.3784 - 10$
10. $\log N = 2.7372$
11. $\log N = 8.9886 - 10$
12. $\log N = 4.3856$

In 13-18, find the antilogarithm of the given logarithm to three significant digits.

13. 2.1367
14. $9.4014 - 10$
15. 0.9042
16. $.6263 - 3$
17. 0.0086
18. $.9881 - 1$

19. If $\log x = 1.8650$, find x to the nearest tenth.
20. Find, to the nearest thousandth, the antilogarithm of $9.3183 - 10$.
21. If $10^{3.7924} = a$, find a to the nearest hundred.
22. If $10^{.3384-2} = b$, find b to the nearest ten-thousandth.
23. Find the antilogarithm of 2.7891 to the nearest integer.
24. Find the antilogarithm of $.5567 - 1$ to three decimal places.
25. Find the antilogarithm of $8.0343 - 10$ to three significant digits.

11-6 APPLYING THE PRODUCT RULE

In section 3 of this chapter, we used the rule for multiplying powers with like bases to derive the product rule for logarithms.

$$\log AB = \log A + \log B$$

This rule can be applied to any number of factors.

$$\log (AB)C = \log (AB) + \log C = \log A + \log B + \log C$$

MODEL PROBLEMS

1. Use logarithms to find, to the nearest tenth, the value of N when $N = 467 \times 0.109$.

How to Proceed	Solution

1. Use the product rule to write an equation involving logarithms.

$N = 467 \times 0.109$
$\log N = \log 467 + \log 0.109$

2. Find the logarithm of each number, and add to find log N in simplest form.

$$\begin{aligned} \log 467 &= 2.6693 \\ + \log 0.109 &= \underline{.0374 - 1} \\ \log N &= 2.7067 - 1 \\ \log N &= 1.7067 \end{aligned}$$

3. Find the antilogarithm N, expressed in scientific notation.

$$N = 5.09 \times 10^1$$

4. Write N in ordinary decimal notation.

$$N = 50.9$$

Answer: 50.9

Note: As a check, $467 \times 0.109 = 50.903 = 50.9$, to the nearest tenth.

2. If $\log x = a$, then $\log .01x$ equals (1) $2 + a$ (2) $-2 + a$ (3) $.01a$ (4) $.01 + a$

Solution

1. Use the product rule to express $\log .01x$ in terms of $\log .01$ and $\log x$.

$$\log .01x = \log .01 + \log x$$

2. Since $.01 = 10^{-2}$, $\log .01 = -2$. Substitute this value and the given value, $\log x = a$.

$$\log .01x = -2 + a$$

Answer: (2) $-2 + a$

3. If $\log 2 = a$ and $\log 7 = b$, write an expression for $\log 28$ in terms of a and b.

Solution

1. Write 28 in terms of its factors, 2 and 7.

$$\begin{aligned} 28 &= 4 \times 7 \\ &= 2 \times 2 \times 7 \end{aligned}$$

2. Write an equation using logarithms.

$$\log 28 = \log (2 \times 2 \times 7)$$

3. Use the product rule to simplify the right-hand member.

$$\log 28 = \log 2 + \log 2 + \log 7$$

4. Substitute the given values for log 2 and log 7, and combine like terms.

$$\begin{aligned} \log 28 &= a + a + b \\ \log 28 &= 2a + b \quad Ans. \end{aligned}$$

EXERCISES

In 1–10, find the product to three significant digits.

1. 8.49×72.1 2. 51.7×8.30 3. $278 \times .453$
4. $3.14 \times .065$ 5. 1.06×3500 6. 411×1.56
7. $587 \times 0.00639 \times 22.1$ 8. $8.24 \times 0.281 \times 0.0467$
9. $0.419(55.2)(0.00153)$ 10. $2(3.14)(0.00878)$

11. If $\log x = \log a + \log b$, express x in terms of a and b.

In 12–15, select the numeral preceding the expression that best completes the sentence or answers the question.

12. If $\log a = c$, then $\log 100a$ equals:
    (1) $100c$ (2) $2c$ (3) $2 + c$ (4) $2 + \log c$
13. If $\log 7.11 = b$, then $\log 7110$ equals:
    (1) $1000b$ (2) $3b$ (3) $3 + b$ (4) $1000 + b$
14. If $\log 2.38 = k$, then $\log .0238$ equals:

    (1) $k + 2$ (2) $k - 2$ (3) $\dfrac{k}{100}$ (4) $k - 100$

15. If $\log 6.38 = x$, which expression must equal $x + 1$?
    (1) $\log 7.38$ (2) $\log 5.38$ (3) $\log 63.8$ (4) $\log .638$

In 16–23, if $\log 2 = a$, $\log 3 = b$, and $\log 10 = 1$, write the given expression in terms of a and b.

16. $\log 6$ 17. $\log 9$ 18. $\log 4$ 19. $\log 12$
20. $\log 20$ 21. $\log 30$ 22. $\log 60$ 23. $\log 90$

In 24–27, solve each problem by using logarithms.

24. Find, to the nearest integer, the number of square kilometers in a rectangular field 1.6 km by 8.71 km.
25. Find, to the nearest hundredth, the circumference of a circular disk whose diameter measures .744 ft. (Let $\pi = 3.14$.)
26. If the radius of a circle measures 17.8 cm, find, to the nearest centimeter, the circumference of the circle. (Let $\pi = 3.14$.)
27. Find, to the nearest integer, the volume of a rectangular solid if $l = 3.7$, $w = 1.6$, and $h = 5.4$. ($V = lwh$)

11-7 APPLYING THE QUOTIENT RULE

In section 3 of this chapter, we used the rule for dividing powers with like bases to derive the quotient rule for logarithms.

$$\log \frac{A}{B} = \log A - \log B$$

When a smaller number is to be divided by a larger one, the result of subtracting a larger logarithm from a smaller one is a negative logarithm. A negative logarithm is written as the sum of a positive mantissa and a negative characteristic. How this is done is shown in the following example:

☐ EXAMPLE: Using logarithms, find $N = 4 \div 5$.

Solution: Using arithmetic, we know that $4 \div 5 = \frac{4}{5} = \frac{8}{10} = 0.8$.
Let us see how the same answer is obtained using logarithms.

1. Use the quotient rule to write an equation involving logarithms.

$$N = 4 \div 5$$
$$\log N = \log 4 - \log 5$$

2. Find the logarithm of each number. To obtain a *positive mantissa* in log N, add 10 and subtract 10 from log 4 *before* performing the subtraction.

$$
\begin{array}{rll}
\log 4 = & 0.6021 = & 10.6021 - 10 \\
-\log 5 = & -0.6990 = & -\ .6990 \\
\hline
\log N = & & 9.9031 - 10
\end{array}
$$

3. Find the value of the antilogarithm N.

$$N = 8.00 \times 10^{-1}$$
$$N = 0.800$$

Answer: $N = 0.8$

MODEL PROBLEMS

1. Use logarithms to find $5680 \div 272$ to the nearest tenth.

Solution

1. Use the quotient rule to write an equation involving logarithms.

$$N = 5680 \div 272$$
$$\log N = \log 5680 - \log 272$$

2. Find the logarithm of each number. Then subtract to find log N.

$$
\begin{array}{rl}
\log 5680 = & 3.7543 \\
-\log\ \ 272 = & -2.4346 \\
\hline
\log N = & 1.3197
\end{array}
$$

3. The mantissa .3197 is closest to .3201 in the table. Using this value, find the antilogarithm N.

$$N = 2.09 \times 10^1$$
$$N = 20.9 \quad Ans.$$

2. Find the value of $\dfrac{175}{.025}$ using logarithms.

Solution

1. Use the quotient rule to write an equation involving logarithms.

$N = \dfrac{175}{.025}$

$\log N = \log 175 - \log .025$

2. Find the logarithm of each number, and subtract to find $\log N$.

$$\begin{array}{rcl} \log 175 = & 2.2430 & = 2.2430 \\ -\log .025 = & -(.3979 - 2) & = -.3979 + 2 \\ \hline \log N = & & 1.8451 + 2 \end{array}$$

Thus, $\log N = 3.8451$

3. Find the value of the antilogarithm N.

$N = 7.00 \times 10^3$

$N = 7000$ *Ans.*

3. Use logarithms to find x to the nearest thousandth: $x = \dfrac{57.1}{456 \times 0.839}$

Solution

$$x = \dfrac{57.1}{456 \times 0.839}$$

$\log x = \log 57.1 - \log (456 \times 0.839)$
$\log x = \log 57.1 - (\log 456 + \log 0.839)$

$$\begin{array}{rll} \log 57.1 = & 1.7566 = & 11.7566 - 10 \\ -\log \text{denominator} = & -2.5828 = & -2.5828 \\ \hline \log x = & & 9.1738 - 10 \end{array}$$

$$\begin{array}{rl} \log 456 = & 2.6590 \\ +\log 0.839 = & 9.9238 - 10 \\ \hline \log \text{denominator} = & 12.5828 - 10 \\ \log \text{denominator} = & 2.5828 \end{array}$$

$x = 1.49 \times 10^{-1}$
$x = 0.149$ *Ans.*

Note: The problem can also be solved using an equation without parentheses, namely, $\log x = \log 57.1 - \log 456 - \log 0.839$.

| EXERCISES |

1. Evaluate $\dfrac{8 \times 16}{4}$: a. using arithmetic   b. using logarithms.

2. Evaluate $\dfrac{24 \times 15}{48}$: a. using arithmetic   b. using logarithms.

In 3–15, using logarithms, find the value of each expression to three significant digits.

3. $384 \div 156$ 4. $79.3 \div 2.89$ 5. $4.52 \div 0.107$
6. $72.9 \div 261$ 7. $1.97 \div 35.8$ 8. $0.192 \div 0.00852$

9. $\dfrac{0.427}{0.897}$ 10. $\dfrac{14.9}{0.00731}$ 11. $\dfrac{0.0628}{0.0989}$

12. $\dfrac{47.6(0.0172)}{35.6}$ 13. $\dfrac{5980(0.626)}{78.7}$

14. $\dfrac{859}{(19.7)(2.48)}$ 15. $\dfrac{3.72}{(4.81)(97.6)}$

16. If $\log x = \log a - \log b$, express x in terms of a and b.

In 17-19, select the numeral preceding the expression that best completes the sentence.

17. If $\log N = \log a + \log b - \log c$, then N equals:

     (1) $a + b - c$ (2) $\dfrac{ab}{c}$ (3) $ab - c$ (4) $\dfrac{a + b}{c}$

18. If $\log 3.59 = x$, then $\log .0359$ equals:

     (1) $x - 2$ (2) $x + 2$ (3) $\dfrac{x}{100}$ (4) $\dfrac{x}{.01}$

19. If $\log N = \log a - (\log b + \log c)$, then N equals:

     (1) $a - b + c$ (2) $a - (b + c)$ (3) $\dfrac{a}{bc}$ (4) $\dfrac{a}{b + c}$

In 20-27, if $\log 2 = a$, $\log 3 = b$, and $\log 10 = 1$, write the given expression in terms of a and b.

20. $\log \frac{2}{3}$ 21. $\log \frac{3}{2}$ 22. $\log \frac{6}{10}$ 23. $\log \frac{20}{3}$

24. $\log \frac{30}{2}$ 25. $\log \frac{10}{3}$ 26. $\log \frac{9}{20}$ 27. $\log \frac{10}{9}$

11-8 APPLYING THE POWER RULE

In section 3 of this chapter, we used the rule for raising a power to a power to derive the power rule for logarithms.

$$\log A^c = c \log A$$

The power rule applies to both integral and fractional values of c, as seen in the following examples:

1. $\log 1.96^4 = 4 \log 1.96 = 4(0.2923) = 1.1692$

2. $\log \sqrt{7.32} = \log 7.32^{\frac{1}{2}} = \frac{1}{2} \log 7.32$

                 $= \frac{1}{2} (0.8645)$

                 $= 0.43225$

                 $= 0.4323$ (The mantissa is rounded to four decimal places.)

3. $\log \sqrt{0.732} = \log 0.732^{\frac{1}{2}} = \frac{1}{2} \log 0.732$
    $$= \frac{1}{2} (0.8645 - 1)$$

Here, some other form of the negative characteristic must be used because the characteristic must be an integer after multiplication by $\frac{1}{2}$. The characteristic of $\log 0.732$ is -1, which may be written as $9 - 10$ or as $1 - 2$ or as any other combination that equals -1 and has an even negative part.

$$\log \sqrt{0.732} = \frac{1}{2} (9.8645 - 10) = 4.9323 - 5$$

OR

$$\log \sqrt{0.732} = \frac{1}{2} (1.8645 - 2) = 0.9323 - 1$$

The characteristic of $\log \sqrt{0.732}$ is also -1, written here as $4 - 5$ and as $0 - 1$.

| MODEL PROBLEMS |

1. Use logarithms to find $\sqrt[3]{0.932}$ to the nearest thousandth.

Solution

1. Write an equation involving logarithms, and use the power rule to rewrite it in terms of $\log 0.932$.

    $N = \sqrt[3]{0.932}$
    $\log N = \log \sqrt[3]{0.932} = \log 0.932^{\frac{1}{3}}$
    $\log N = \frac{1}{3} \log 0.932$

2. Find the value of $\log 0.932$, and write the characteristic, -1, so that it has a multiple of 3 as its negative part.

    $\log N = \frac{1}{3}(.9694 - 1)$
    or $\log N = \frac{1}{3}(2.9694 - 3) = 0.9898 - 1$
    $\log N = \frac{1}{3}(29.9694 - 30) = 9.9898 - 10$

3. Find the value of the antilogarithm, N.

    $N = 9.77 \times 10^{-1}$
    $N = 0.977$ *Ans.*

2. If $A = \pi r^2$, find A to the nearest tenth when $r = 2.50$ and $\pi = 3.14$.

Solution

1. Write an equation involving logarithms.

$$A = \pi r^2$$
$$\log A = \log \pi + 2 \log r$$

2. Substitute the given values.

$$\log A = \log 3.14 + 2 \log 2.50$$

3. Find the logarithms of the numbers, and perform the computations.

$$\log 2.50 = 0.3979$$
$$\times 2$$
$$2 \log 2.50 = \overline{0.7958}$$
$$+\log 3.14 = +0.4969$$
$$\log A = \overline{1.2927}$$

4. Find the value of the antilogarithm, A.

$$A = 1.96 \times 10^1$$
$$A = 19.6 \quad Ans.$$

3. Find $\log \sqrt{.331}$ to four decimal places.

Solution

1. Use the power rule to write an expression for $\log \sqrt{.331}$.

$$\log \sqrt{.331} = \log .331^{\frac{1}{2}}$$
$$\log \sqrt{.331} = \tfrac{1}{2} \log .331$$

2. Find the value of $\log .331$. Write the characteristic, -1, so that it has an even number as its negative part.

$$\log \sqrt{.331} = \tfrac{1}{2} (.5198 - 1)$$
$$\log \sqrt{.331} = \tfrac{1}{2} (9.5198 - 10)$$
$$\log \sqrt{.331} = 4.7599 - 5$$

(*Note:* We do not find the antilogarithm because the problem asks for the value of a logarithm.)

Answer: Any logarithm whose characteristic is -1 and whose mantissa is .7599. One possible solution is 4.7599 - 5. Two other common forms are .7599 - 1 or 9.7599 - 10.

4. Solve for x: $x^{.7} = 12$

Solution

1. Write the given equation.

$$x^{.7} = 12$$

2. Write an equation involving logarithms.

$$\log x^{.7} = \log 12$$

3. Use the power rule to simplify the equation.

$$.7 \log x = \log 12$$

4. Solve for $\log x$.

$$\log x = \frac{\log 12}{.7}$$

5. Find log 12, and simplify the right-hand member.

$$\log x = \frac{1.0792}{.7}$$

$$\log x = \frac{10.792}{7}$$

$$\log x = 1.5417$$

6. Find the antilogarithm, x.

$$x = 3.48 \times 10^1$$
$$x = 34.8 \quad Ans.$$

| EXERCISES |

In 1-14, using logarithms, find the value of each expression to three significant digits.

1. 45.7^2 2. 1.38^3 3. 0.07^5 4. 1.08^8

5. $\sqrt{346}$ 6. $\sqrt[3]{1.09}$ 7. $\sqrt[3]{0.850}$ 8. $\sqrt[5]{0.553}$

9. $93.0\sqrt{.761}$ 10. $18.0\sqrt[3]{.301}$ 11. $\dfrac{(6.92)^4}{155}$ 12. $\dfrac{3.14(4.81)^2}{0.671}$

13. $\sqrt{\dfrac{2 \times 1.67}{59.8}}$ 14. $\dfrac{27.6\sqrt{3}}{2}$

In 15-18, express x in terms of a, b, and c.

15. $\log x = \frac{1}{2}(\log a + \log b - \log c)$
16. $\log x = 2 \log a - \frac{1}{2}(\log b + \log c)$
17. $\log x = \frac{1}{2}(\log a - (\log b + \log c))$
18. $\log x = \frac{1}{2}\log a - (\log b + \frac{1}{2}\log c)$

In 19-24, express $\log N$ in terms of $\log x$, $\log y$, and $\log z$.

19. $N = x^2yz$ 20. $N = x\sqrt{yz}$ 21. $N = x^2y\sqrt{z}$

22. $N = \dfrac{xy}{z^3}$ 23. $N = \sqrt{\dfrac{xy}{z}}$ 24. $N = \dfrac{x\sqrt{y}}{z^2}$

25. Find $\sqrt{272}$ to the nearest tenth: a. using logarithms   b. using any other acceptable method in mathematics.

26. Find $\sqrt[3]{0.852}$ to the nearest hundredth.

27. If $V = \dfrac{\pi r^2 h}{3}$, find V to the nearest integer when $\pi = 3.14$, $r = 5.27$, and $h = 7.96$.

28. If $V = \dfrac{4\pi r^3}{3}$, find V to the nearest tenth when $\pi = 3.14$ and $r = 2.15$.

29. Find $\log \sqrt[3]{17}$ to four decimal places.

30. Find $\log 8.6^3$ to four decimal places.

31. Find $\log \sqrt{0.0072}$ to four decimal places.

In 32-40, select the numeral preceding the expression that best completes the sentence or answers the question.

32. The expression $\log \sqrt{\dfrac{a}{b}}$ is equivalent to:

    (1) $\frac{1}{2} \log a - \log b$ (2) $\frac{1}{2} (\log a - \log b)$

    (3) $\log \frac{1}{2} a - \log b$ (4) $\log \frac{1}{2} a - \log \frac{1}{2} b$

33. The expression $2 \log a + \frac{1}{3} \log b$ is equivalent to:

    (1) $\log \dfrac{a^2 b}{3}$ (2) $\log \dfrac{2ab}{3}$ (3) $\log \sqrt[3]{a^2 b}$ (4) $\log a^2 (\sqrt[3]{b})$

34. If $x = a\sqrt{b}$, then $\log x$ is equivalent to:

    (1) $\log a + \log \frac{1}{2} b$ (2) $\frac{1}{2} (\log a + \log b)$

    (3) $\log a + \frac{1}{2} \log b$ (4) $\log (a + \frac{1}{2} b)$

35. If $x = \dfrac{a}{2b}$, then $\log x$ is equivalent to:

    (1) $\log a - 2 \log b$ (2) $\log a - \log 2 + \log b$

    (3) $\frac{1}{2} (\log a - \log b)$ (4) $\log a - (\log 2 + \log b)$

36. If $K = \dfrac{\sqrt[3]{6}}{5^2}$, which of the following is equivalent to $\log K$?

    (1) $3 \log 6 - 2 \log 5$ (2) $\frac{1}{3} \log 6 - 2 \log 5$

    (3) $\dfrac{\frac{1}{3} \log 6}{2 \log 5}$ (4) $\dfrac{3 \log 6}{2 \log 5}$

37. If $N = \sqrt{\frac{5}{12}}$, then $\log N$ is equivalent to:

    (1) $\frac{1}{2} (\log 5 - \log 12)$ (2) $\frac{1}{2} \log 5 - \log 12$

    (3) $\dfrac{1}{2} \cdot \dfrac{\log 5}{\log 12}$ (4) $\sqrt{\dfrac{\log 5}{\log 12}}$

38. The expression $\log \sqrt{7^3}$ is equivalent to:

(1) $\frac{2}{3} \log 7$ (2) $\frac{3}{2} \log 7$ (3) $3 \log \frac{7}{2}$ (4) $\frac{1}{2} \log 3 \cdot 7$

39. The expression $\log 5^{n+1}$ is equivalent to:

(1) $n \log 5 + 5$ (2) $5 \log 5^n$
(3) $n \log 5 + 1$ (4) $n \log 5 + \log 5$

40. The expression $\log a^{n-2}$ is equivalent to:

(1) $n \log a - 2$ (2) $\dfrac{n \log a}{\log 2}$

(3) $n \log a - 2 \log a$ (4) $\dfrac{n \log a}{2}$

In 41–44, find x to the nearest integer.

41. $x^{1.2} = 6.9$ **42.** $x^{.5} = 2.65$ **43.** $x^{1.5} = 14.7$ **44.** $x^{2.2} = 237$

11-9 INTERPOLATION

Logarithms of Numbers Having Four Significant Digits

In Chapter 8, we learned to use interpolation to find a trigonometric function value between two values that are given in the table. We can apply that same method of approximation to find the logarithm of a number having four significant digits.

☐ EXAMPLE 1:   Find $\log 6786$.

1. Since the mantissas of numbers having three significant digits can be found in the table of Common Logarithms of Numbers (pages 748–749), determine the numbers with three significant digits that are closest to 6786, namely, 6780 and 6790. Find the mantissas for these numbers, and arrange them as shown in the chart in step 3 on the next page. Notice that each mantissa in the chart is written without its appropriate decimal point.

2. Find the differences between the smallest number and each of the other two numbers, as indicated in the chart. Form a ratio of these differences. Find the differences between the corresponding mantissas, and form a ratio.

3. Since these ratios are approximately equal within a small interval, write and solve a proportion using these ratios.

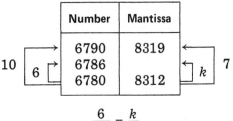

Number	Mantissa
6790	8319
6786	
6780	8312

$$\frac{6}{10} = \frac{k}{7}$$

4. Round the value of k to the nearest integer. Since k is the difference between the mantissa we are finding and the smaller mantissa in the chart, add 8312 + 4 to find the required mantissa.

$$10k = 42$$
$$k = 4.2$$
$$k \approx 4$$
$$8312 + 4 = 8316$$

5. The value of log 6786 is written by combining its characteristic with the mantissa just found.

$$\log 6786 = 3.8316$$

Note:

1. As the number increases, the logarithm of the number increases. By putting the smallest number at the bottom and the largest number at the top, we make subtraction easier.

2. Since the position of the decimal point in a number affects the characteristic but not the mantissa, we may omit the decimal point from the number when interpolating. For example, finding the mantissa for 678.6, 67.86, 6.786, or 0.6786 can be done as in the above example.

Using Interpolation to Find an Antilogarithm

To find a number to four significant digits when the mantissa of its logarithm is not given in the table, we use the same process of interpolation.

☐ EXAMPLE 2: If log N = 2.3547, find N to the nearest tenth.

1. To find the sequence of digits in the antilogarithm of 2.3547, find the two mantissas closest to .3547 in the table of Common Logarithms of Numbers (pages 748–749). Arrange these mantissas, 3541 and 3560, with their corresponding three-digit numbers, 226 and 227, in a chart as shown in step 3 on the next page. Place a 0 as the fourth digit after each number.

2. Find the differences between the smallest mantissa and the other two mantissas, as indicated in the chart. Form a ratio of these differences. Find the differences between the corresponding numbers and form a ratio.

3. Since these ratios are approximately equal within a small interval, write and solve a proportion using these ratios.

Number	Mantissa
2270	3560
	3547
2260	3541

$$\frac{k}{10} = \frac{6}{19}$$

$$19k = 60$$

$$k = 3\tfrac{3}{19}$$

$$k \approx 3$$

4. Round the value of k to the nearest integer. Since k is the difference between the number we are finding and the smaller number in the chart, add $2260 + 3$ to find the required number.

$$2260 + 3 = 2263$$

5. Use the characteristic of the logarithm to determine the place value of the number.

$$N = 2.263 \times 10^2$$
$$N = 226.3$$

When to Use Interpolation

The result of a computation can be no more accurate than the least accurate number on which it is based.

If $N = \dfrac{(37.2)^2}{3.80 \times .709}$, each number has three significant digits. Do not interpolate to find N.

If $N = \dfrac{(37.23)^2}{3.800 \times .7090}$, each number has four significant digits. When finding the antilogarithm, interpolate to find N to four significant digits.

Numbers that are not obtained by measurement or approximation are usually exact and do not change the accuracy of a computation. For example, when we are using $C = 2\pi r$ to find the circumference (C) of a circle, the number 2 is exact. The accuracy of C depends on the accuracy of r. The approximate value of π that is used in the computation should have at least as many significant digits as the value of r.

MODEL PROBLEM

Interpolate to find the value of N to four significant digits if $\log N = 9.7261 - 10$.

Solution

$\log N = 9.7261 - 10$
$N = 5.323 \times 10^{-1}$
$N = 0.5323$

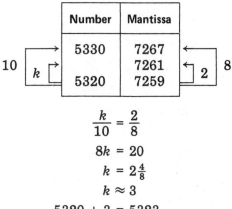

$$\frac{k}{10} = \frac{2}{8}$$
$$8k = 20$$
$$k = 2\tfrac{4}{8}$$
$$k \approx 3$$
$$5320 + 3 = 5323$$

Answer: 0.5323

EXERCISES

In 1–16, find the logarithm of the given number to four decimal places.

1. 3.875	**2.** 97.66	**3.** 0.4208	**4.** 1.077
5. 8254	**6.** 631.7	**7.** 0.05872	**8.** 574.3
9. 21.27	**10.** 60.05	**11.** 789.3	**12.** 0.8229
13. 0.05394	**14.** 110.6	**15.** 0.004671	**16.** 3.924

In 17–28, find, to four significant digits, the number whose logarithm is given.

17. 1.6083	**18.** 3.9273	**19.** 2.5830	**20.** 3.2441
21. 1.1799	**22.** 0.7843	**23.** 8.6623 - 10	**24.** 7.9928 - 10
25. 9.3027 - 10	**26.** .8778 - 1	**27.** .4718 - 3	**28.** .6993 - 2

29. Find $\sqrt[3]{0.7923}$ to the nearest ten-thousandth.
30. If $10^{2.9311} = x$, find x to the nearest tenth.
31. If $A = 2500(1.035)^{12}$, find A to the nearest integer.

32. The surface area of a sphere is given by the formula $A = 4\pi r^2$. Find A to the nearest integer when $r = 12.25$. (Use $\pi = 3.142$.)

33. The formula $r = \sqrt{\dfrac{3V}{\pi h}}$ gives the radius r of a right circular cone in terms of the volume V and the height h. Find r to the nearest hundredth if $V = 2528$, $h = 12.70$, and $\pi = 3.142$.

In 34–36, the amount of money, A, in a bank account is determined by the formula $A = P\left(1 + \dfrac{r}{n}\right)^{nt}$ where P is the principal (or the amount invested), r is the yearly rate of interest, n is the number of times each year that interest is compounded, and t is the number of years involved.

34. Find to the nearest dollar the amount of money in an account after 5 years if \$500 was deposited at 8% interest compounded quarterly (4 times a year). *Hint:* The equation $A = 500\left(1 + \dfrac{.08}{4}\right)^{4 \cdot 5}$ can be simplified to $A = 500(1.02)^{20}$ before solving by using logarithms.

35. When Lynn was born, her grandmother invested \$5000 for her at 7% compounded semiannually (twice a year). Find the value of the investment to the nearest hundred dollars when Lynn is 20 years old.

36. One hundred years ago, your great grandfather deposited the first \$25 he earned at 4% interest compounded annually. If ownership of the account was passed from generation to generation of your family, how much is it worth to you now? (Assume that no changes in rate of interest have taken place.) Find the value to the nearest dollar.

37. Depreciation (or decline in cash value) on a car can be determined by a formula $V = C(1 - r)^n$, where V is the value of a car after n years, C is the original cost, and r is the rate of depreciation. If a car originally costs \$8000 and the rate of depreciation is 30%, find the value of the car to the nearest dollar after 3 years.

11-10 LOGARITHMS OF TRIGONOMETRIC FUNCTION VALUES

If $a = 37.5 \sin 35°10'$, we could use logarithms to multiply. To do so, we need to find log sin 35°10'. It is not necessary first to find $\sin 35°10' = .5760$ and then to find log $.5760 = .7604 - 1$. The table of Logarithms of Trigonometric Functions on pages 755–759 will give the log sin 35°10′ directly. The table is read in the same way as the table of Values of Trigonometric Functions, as shown below.

Logarithms of Trigonometric Functions

↓

Angle	L Sin	L Cos	L Tan	L Cot	
34° 00′	9.7476	9.9186	9.8290	10.1710	56° 00′
10	9.7494	9.9177	9.8317	10.1683	50
20	9.7513	9.9169	9.8344	(10.1656)	←→ 40
30	9.7531	9.9160	9.8371	10.1629	30
40	9.7550	9.9151	9.8398	10.1602	20
50	9.7568	9.9142	9.8425	10.1575	10
→ 35° 00′	9.7586	9.9134	9.8452	10.1548	55° 00′ ←
↳ 10 →	(9.7604)	9.9125	9.8479	10.1521	50
20	9.7622	9.9116	9.8506	10.1494	40
30	9.7640	9.9107	9.8533	10.1467	30
40	9.7657	9.9098	9.8559	10.1441	20
50	9.7675	9.9089	9.8586	10.1414	10
36° 00′	9.7692	9.9080	9.8613	10.1387	54° 00′
	L Cos	L Sin	L Cot	L Tan	Angle

When reading angle measures between 0° and 45° at the left, we use the headings at the top of the column. When reading angle measures between 45° and 90° at the right, we use the labels at the bottom of the column. The characteristic of each logarithm has been increased by 10 for each entry in the table. Therefore:

■ When using the table of Logarithms of Trigonometric Functions, we must subtract 10 from each stated logarithmic value to give the correct characteristic of the logarithm.

Thus, we read from the table:

$$\log \sin 35°10' = 9.7604 - 10$$
$$\log \tan 55°40' = 10.1656 - 10$$

The problem stated above is completed in the following example.

☐ Example: If $a = 37.5 \sin 35°10'$, find the value of a to the nearest tenth.

$$\log a = \log 37.5 + \log \sin 35°10'$$
$$\log 37.5 = 1.5740$$
$$\underline{\log \sin 35°10' = 9.7604 - 10}$$
$$\log a = 11.3344 - 10 = 1.3344$$
$$a = 2.16 \times 10^1$$
$$a = 21.6 \quad Ans.$$

The following correspondence exists between the accuracy of degree measure and linear measure.

$$\text{nearest } 10' \rightarrow 3 \text{ significant digits}$$
$$\text{nearest } 1' \rightarrow 4 \text{ significant digits}$$

In the above example, since both 37.5 and 35°10′ are measures with the accuracy of 3 significant digits, the value of the product, a, is written with 3 significant digits.

Finding Angle Measures

When the logarithm of a trigonometric function value is given, the degree measure of the angle can be found by using the table of Logarithms of Trigonometric Functions.

If log cos A = 9.8394 - 10, we look in the table for the entries in a "L Cos" column closest to 9.8394, the log value increased by 10. We find that 9.8394 lies between 9.8391 and 9.8405, as shown below.

Logarithms of Trigonometric Functions

Angle	L Sin	L Cos	L Tan	L Cot	
43° 00′	9.8338	9.8641	9.9697	10.0303	47° 00′
10	9.8351	9.8629	9.9722	10.0278	50
20	9.8365	9.8618	9.9747	10.0253	40
30	9.8378	9.8606	9.9772	10.0228	30
40	9.8391	9.8594	9.9798	10.0202	←→ 20
50	9.8405	9.8582	9.9823	10.0177	←→ 10
44° 00′	9.8418	9.8569	9.9848	10.0152	46° 00′ ←
10	9.8431	9.8557	9.9874	10.0126	50
20	9.8444	9.8545	9.9899	10.0101	40
30	9.8457	9.8532	9.9924	10.0076	30
40	9.8469	9.8520	9.9949	10.0051	20
50	9.8482	9.8507	9.9975	10.0025	10
45° 00′	9.8495	9.8495	10.0000	10.0000	45° 00′
	L Cos	L Sin	L Cot	L Tan	Angle

If the measure of A is to be given to the nearest 10 minutes or to the nearest degree, we choose the closer entry and read the corresponding degree measure. Therefore, if log cos A = 9.8394 - 10:

$$m\angle A = 46°20' \text{ to the nearest 10 minutes}$$
$$m\angle A = 46° \text{ to the nearest degree}$$

If the measure of $\angle A$ is to be given to the nearest minute, we must interpolate. Recall that, as the measure of an angle increases from 0° to 90°, the value of the cosine and, therefore, of the log of the cosine, decreases. In arranging our chart for interpolation, we will place the smallest logarithm at the bottom to make subtraction easier.

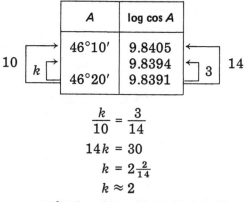

$$\frac{k}{10} = \frac{3}{14}$$
$$14k = 30$$
$$k = 2\frac{2}{14}$$
$$k \approx 2$$

46°20' - 2' = 46°18' (Since the angle measures are decreasing, subtract 2' from the larger angle measure.)
$$m\angle A = 46°18' \text{ to the nearest minute}$$

MODEL PROBLEM

If $\sin A = \dfrac{52.50 \sin 48°12'}{74.60}$, find the measure of acute angle A to the nearest minute.

Solution

Express the equation in logarithmic form.

$$\sin A = \frac{52.50 \sin 48°12'}{74.60}$$

$$\log \sin A = \log 52.50 + \log \sin 48°12' - \log 74.60$$

$$\log \sin A = \log 52.50 + \log \sin 48°12' - \log 74.60$$

$$\begin{aligned}
\log 52.50 &= 1.7202 \\
+\log \sin 48°12' &= +9.8724 - 10 \quad \text{(See interpolation below.)} \\
\hline
\log \text{numerator} &= 11.5926 - 10 \\
-\log 74.60 &= -1.8727 \\
\hline
\log \sin A &= 9.7199 - 10 \\
A &= 31°39' \quad \text{(See interpolation below.)}
\end{aligned}$$

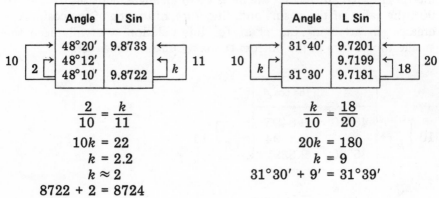

Angle	L Sin
48°20′	9.8733
48°12′	
48°10′	9.8722

$$\frac{2}{10} = \frac{k}{11}$$

$$10k = 22$$

$$k = 2.2$$

$$k \approx 2$$

$$8722 + 2 = 8724$$

Angle	L Sin
31°40′	9.7201
	9.7199
31°30′	9.7181

$$\frac{k}{10} = \frac{18}{20}$$

$$20k = 180$$

$$k = 9$$

$$31°30' + 9' = 31°39'$$

Answer: 31°39′

EXERCISES

In 1–15, find the logarithm of the given function value.

1. sin 27°	2. cos 32°	3. tan 15°
4. cos 44°20′	5. sin 68°40′	6. tan 71°50′
7. sin 82°10′	8. tan 51°30′	9. cos 48°50′
10. cos 21°26′	11. tan 38°42′	12. sin 8°37′
13. tan 66°52′	14. sin 56°06′	15. cos 77°15′

In 16–23, find θ to the nearest 10 minutes.

16. $\log \sin \theta = 9.8004 - 10$	17. $\log \tan \theta = 9.0453 - 10$
18. $\log \cos \theta = 9.9901 - 10$	19. $\log \sin \theta = 9.9404 - 10$
20. $\log \cos \theta = 9.8850 - 10$	21. $\log \tan \theta = 10.2100 - 10$
22. $\log \sin \theta = 9.9815 - 10$	23. $\log \cos \theta = 9.6650 - 10$

In 24–31, find θ to the nearest minute.

24. $\log \sin \theta = 9.4002 - 10$	25. $\log \cos \theta = 9.9605 - 10$
26. $\log \tan \theta = 9.5175 - 10$	27. $\log \sin \theta = 9.8918 - 10$

28. $\log \cos \theta = 9.7537 - 10$ 29. $\log \sin \theta = 9.6892 - 10$
30. $\log \tan \theta = 10.0415 - 10$ 31. $\log \cos \theta = 9.5498 - 10$

32. If $x = 14.3 \sin 12°30'$, find x to the nearest hundredth.
33. If $a = 1.78 \cos 57°20'$, find a to the nearest thousandth.

34. If $r = \dfrac{56.9}{\tan 32°10'}$, find r to the nearest tenth.

35. If $k = 1.75 \sin 78°40'$, find k to the nearest hundredth.

36. If $\sin A = \dfrac{12.4 \sin 56°40'}{15.6}$, find the measure of $\angle A$ to the nearest 10 minutes.

37. If $\cos C = \dfrac{17.85}{21.60}$, find the measure of acute $\angle C$ to the nearest minute.

38. When the length of the slant height of a cone is equal to the measure of the diameter of the base of the cone, the volume, V, is given by the equation $V = \frac{1}{3}\pi r^3 \tan 60°$. Find the volume of such a cone when the length of the radius (r) is 1.25 and $\pi = 3.14$. Express your answer to the nearest hundredth.

39. If $s = ab \sin C$, find s to the nearest integer when $a = 12.7$, $b = 18.6$, and $m\angle C = 27°10'$.

11-11 EXPONENTIAL EQUATIONS

In Chapter 10, we solved equations in which the exponent was a variable, such as $4^x = 8$, by writing each number as an integral power of the same base. At that time we were not able to solve $3^x = 5$ because 3 and 5 are not integral powers of the same base. Common logarithms make it possible to write any number as an approximate power to the base 10, however, thus enabling us to solve any exponential equation.

$$\text{If } \log 3 = x, \text{ then } 10^x = 3 \text{ or } 10^{\log 3} = 3.$$
$$\text{If } \log 5 = y, \text{ then } 10^y = 5 \text{ or } 10^{\log 5} = 5.$$

Compare the solutions of $4^x = 8$ and $3^x = 5$.

$4^x = 8$	$3^x = 5$
$(2^2)^x = 2^3$	$(10^{\log 3})^x = 10^{\log 5}$
$2^{2x} = 2^3$	$10^{x \log 3} = 10^{\log 5}$
$2x = 3$	$x \log 3 = \log 5$
$x = \dfrac{3}{2} = 1.5$	$x = \dfrac{\log 5}{\log 3} = \dfrac{0.6990}{0.4771} \approx 1.46$

The solution of $3^x = 5$ can be more briefly written as follows:

$$3^x = 5$$
$$\log 3^x = \log 5$$
$$x \log 3 = \log 5$$
$$x = \frac{\log 5}{\log 3} = \frac{0.6990}{0.4771} \approx 1.46$$

| MODEL PROBLEM |

If $9^x = 14$, find x to the nearest tenth.

Solution

1. Write the equation. $9^x = 14$

2. Write the log of each member of the $\log 9^x = \log 14$
   equation.

3. Use the power rule to simplify the left- $x \log 9 = \log 14$
   hand member.

4. Solve the equation for x. $x = \dfrac{\log 14}{\log 9}$

5. Simplify, rounding to the nearest tenth. $x = \dfrac{1.1461}{0.9542} = 1.2$

Answer: 1.2

| EXERCISES |

In 1–12, find x to the nearest tenth.

1. $2^x = 7$ 2. $3^x = 15.6$ 3. $7^x = 615$
4. $5^x = 47.6$ 5. $15^x = 295$ 6. $4.7^x = 10.2$
7. $1.06^x = 1.14$ 8. $4.5^x = 1.57$ 9. $3.8^x = 2.9$
10. $2.5^x = 9.88$ 11. $4.08^x = 2.02$ 12. $2.75^x = 3.72$

In 13–15, use the compound-interest formula $A = P\left(1 + \dfrac{r}{n}\right)^{nt}$. See

page 524 for an explanation of the formula.

13. How long must \$500 be left in an account that pays 7% interest
    compounded annually in order that the value of the account be

$750? *Hint:* The equation $750 = 500 \left(1 + \dfrac{.07}{1}\right)^{1t}$ may be simplified to $750 = 500 \, (1.07)^t$ before solving by using logarithms.

14. How long must a sum of money be left in an account at 6% interest compounded semiannually (twice a year) in order to double? (*Hint:* Let $P = 1$ and $A = 2$.)

15. How long must $100 be left at 8% interest compounded quarterly (four times a year) in order to acquire the value $1000?

16. The thickness of a sheet of paper is 0.004 inch. If x represents the number of times that this sheet of paper is folded in half over itself, then $y = 2^x$ determines the number of layers of paper, and $y = .004(2)^x$ determines the thickness of all the layers of paper. Calculate the number of folds that would produce a stack of paper closest to a mile high. (*Hint:* 1 mile = 63,360 inches.)

17. When Patty was in first grade, her mother gave her 10 cents a week to spend. In the second grade, Patty received 20 cents a week, double her first-grade allowance. In the third grade, Patty suggested to her mother that her allowance be doubled every year, but her mother was wise enough to refuse. If Patty's suggestion had been followed, what would her weekly allowance be in grade 11? (*Hint:* Use the formula $y = .10(2)^x$.)

18. If $\log_4 3 = x$: **a.** Rewrite the equation in exponential form. **b.** Solve the equation written in part a, using common logarithms. State the value of x to the nearest tenth.

In 19–21, solve for x. Use the method of exercise 18.

19. $\log_8 12 = x$ 　　　　 20. $\log_6 4 = x$ 　　　　 21. $\log_5 2 = x$

11-12   REVIEW EXERCISES

1. **a.** Sketch the graph of $y = \log_3 x$ in the interval $0 < x \leq 9$.
   b. On the same set of axes, sketch the graph that is the reflection in the line $y = x$ of $y = \log_3 x$.
   c. What is the equation of the graph drawn in part b?
2. **a.** State the domain of the function $y = \log_5 x$.
   b. State the range of the function $y = \log_5 x$.

In 3–8, write the expression in exponential form.

3. $3 = \log_2 8$ 　　　　 4. $0 = \log_7 1$ 　　　　 5. $\log .01 = -2$

6. $\log_5 .2 = -1$ 　　　　 7. $1.5 = \log_4 8$ 　　　　 8. $\log_3 \sqrt{3} = \tfrac{1}{2}$

In 9-14, write the expression in logarithmic form.

9. $7^2 = 49$ **10.** $5^3 = 125$ **11.** $27^{\frac{1}{3}} \doteq 3$

12. $0.001 = 10^{-3}$ **13.** $(\sqrt{5})^4 = 25$ **14.** $6^{-1} = \frac{1}{6}$

In 15-20, solve for x.

15. $x = \log 0.0001$ **16.** $\log_9 3 = x$ **17.** $\log_x \frac{1}{5} = -1$

18. $\log_2 x = 6$ **19.** $\log_x 9 = \frac{2}{3}$ **20.** $\log_6 x = 2$

21. If $f(x) = \log x$, find $f(100)$.

22. If $g(x) = \log_4 x$, find $g(8)$.

In 23-26, express x in terms of a and b.

23. $\log_2 x = 2 \log_2 a + \log_2 b$ **24.** $\log_5 x = 3 \log_5 a - \frac{1}{2} \log_5 b$

25. $\log x = \frac{1}{2} (\log a - \log b)$ **26.** $\log x = \log b + \frac{1}{3} \log a$

In 27-38, find the logarithm to four decimal places.

27. $\log 5.34$ **28.** $\log 106$ **29.** $\log .382$
30. $\log \cos 72°10'$ **31.** $\log \tan 12°40'$ **32.** $\log \sin 36°20'$
33. $\log 46.78$ **34.** $\log 0.01682$ **35.** $\log 8.786$
36. $\log \sin 62°43'$ **37.** $\log \tan 48°12'$ **38.** $\log \cos 28°37'$

In 39-41, find N to three significant digits.

39. $\log N = 2.5539$ **40.** $\log N = .9096 - 1$ **41.** $\log N = 7.7824 - 10$

In 42-43, find θ to the nearest 10 minutes.

42. $\log \cos \theta = 9.8653 - 10$ **43.** $\log \tan \theta = 9.7100 - 10$

In 44-46, find x to four significant digits.

44. $\log x = 1.8276$ **45.** $\log x = 8.4582 - 10$ **46.** $\log x = .2702 - 1$

In 47-48, find θ to the nearest minute.

47. $\log \tan \theta = 0.5255$ **48.** $\log \cos \theta = 9.8320 - 10$

49. Using logarithms, find $\sqrt{42}$ to the nearest hundredth.

50. Using logarithms, find $\sqrt[3]{0.873}$ to the nearest thousandth.

51. Using logarithms, find N to the nearest thousandth if $N = \sqrt{\dfrac{12.6}{54.3}}$.

52. Using logarithms, find $m\angle C$ to the nearest minute when:

$$\sin C = \frac{2(18.64)}{(6.120)(7.250)}$$

53. If $\log a = 1.7866$, find: **a.** $\log \sqrt{a}$ **b.** $\log 100a$ **c.** $\log a^2$
54. If $10^{1.6184} = x$, find x to the nearest hundredth.

In 55–60, find x to the nearest tenth.

55. $3^x = 4$ 56. $5^x = 3.6$ 57. $x^{1.5} = 15$
58. $2^x = 1.5$ 59. $4^x = 21$ 60. $x^{.4} = 1.6$

61. Find, to the nearest tenth, $\log_2 7$.
62. The First National Bank pays $8\frac{1}{2}\%$ interest compounded annually. City Bank pays 8% interest compounded quarterly (four times a year). In 5 years, what is the difference in the amounts of interest paid on \$500 by the two banks? Use $A = P\left(1 + \dfrac{r}{n}\right)^{nt}$. (See page 524 for an explanation of the formula.)

In 63–67, select the numeral preceding the expression that best completes the sentence.

63. If $A = \pi r^2$, then $\log A$ equals:
    (1) $2\pi \log r$ (2) $\log \pi + 2 \log r$ (3) $2\,(\log \pi + \log r)$
    (4) $\log 2 + \log \pi + \log r$
64. If $C = 2\pi r$, then $\log C$ equals:
    (1) $2\pi \log r$ (2) $\log \pi + 2 \log r$ (3) $2 \log \pi + \log r$
    (4) $\log 2 + \log \pi + \log r$
65. If $\log 3.87 = a$, then $\log 387$ equals:
    (1) $100a$ (2) $a + 2$ (3) $a + 100$ (4) $2a$
66. If $\log 3 = a$, then $\log 90$ equals:
    (1) $10 + 2a$ (2) $10a^2$ (3) $1 + 2a$ (4) $1 + a^2$
67. If $10^{2.5527} = 357$, then $10^{0.5527}$ equals:
    (1) 355 (2) $35{,}700$ (3) $\frac{357}{2}$ (4) 3.57

Chapter 12

Trigonometric Applications

In the study of geometry, we learned that the congruence of certain pairs of sides and angles of two given triangles can be used to prove that the triangles are congruent. This means that there is exactly one size and shape that those triangles can have.

The size and shape of a triangle are determined by any one of the following:

1. two sides and the included angle (s.a.s.).
2. three sides (s.s.s.).
3. two angles and the included side (a.s.a.).
4. two angles and a side opposite one of them (a.a.s.).

For example, if every student in a mathematics class were to construct a triangle with sides that measure 2 cm, 2.5 cm, and 3 cm, all of the triangles would have the same size and shape as the triangle at the right because the measures of three sides determine the size and shape of a triangle.

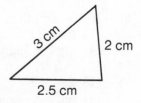

In this chapter, we will derive and use laws or formulas that enable us to find the measures of the other sides and angles of a triangle when any of the combinations listed above are known.

12-1 RECTANGULAR COORDINATES IN TRIGONOMETRIC FORM

Before we derive the formulas that enable us to determine the measures of the sides and angles of a triangle, we should recall the relationship between the trigonometric function values of the measure of an angle in standard position and the coordinates of a point on the terminal ray of that angle.

In Chapter 8, we learned that if P is a point of the unit circle on the terminal ray of an angle in standard position whose measure is θ, then the coordinates of P are $(\cos \theta, \sin \theta)$.

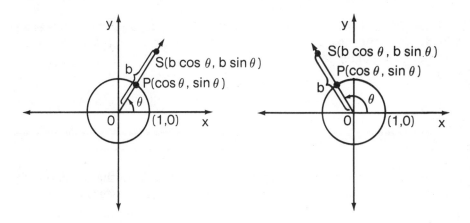

In the above diagrams, point P is mapped to point S by a dilation $D_b(x, y) = (bx, by)$. Therefore, $(b \cos \theta, b \sin \theta)$ are the coordinates of point S. That is, point S is at a distance b from the origin and on a ray that makes an angle of measure θ with the positive x-axis.

We can also derive this relationship by using similar triangles. Let P be a point of the unit circle with center at the origin, O. Let S be a point at a distance r from O on $\overrightarrow{OP}$. Draw $\overline{PQ}$ perpendicular to the x-axis at Q and $\overline{SR}$ perpendicular to the x-axis at R. In $\triangle QOP$, $OP = 1$, $PQ = \sin \theta$, $OQ = \cos \theta$, $m\angle QOP = \theta$, and $m\angle PQO = 90°$. In $\triangle ROS$, $OS = r$, $m\angle ROS = \theta$, and $m\angle SRO = 90°$. Therefore:

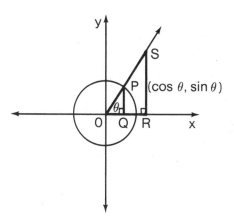

$$\triangle QOP \sim \triangle ROS \text{ by a.a.} \cong \text{a.a.}$$

Since the lengths of corresponding sides of similar triangles are in proportion, we can say:

$\dfrac{OQ}{OP} = \dfrac{OR}{OS}$	$\dfrac{PQ}{OP} = \dfrac{SR}{OS}$
$\dfrac{\cos \theta}{1} = \dfrac{OR}{r}$	$\dfrac{\sin \theta}{1} = \dfrac{SR}{r}$
$OR = r \cos \theta$	$SR = r \sin \theta$

Therefore, if S is a point r units from the origin, O, and $\overrightarrow{OS}$ makes an angle with the positive x-axis whose measure is θ, then S is the point whose coordinates are:

$$(r \cos \theta, r \sin \theta)$$

| MODEL PROBLEM | |

Find the coordinates of a point that is 4 units from the origin on the terminal ray of an angle in standard position whose measure is 150°.

Solution

1. The coordinates of point $A = (x, y) = (r \cos \theta, r \sin \theta)$. Here, $r = 4$ and $\theta = 150°$.

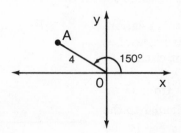

2. Recall: $\cos 150° = -\cos (180° - 150°) = -\cos 30° = \dfrac{-\sqrt{3}}{2}$

AND

$$\sin 150° = \sin (180° - 150°) = \sin 30° = \frac{1}{2}$$

3. Thus, $x = r \cos \theta$ $y = r \sin \theta$

     $x = 4 \cos 150°$ $y = 4 \sin 150°$

     $x = 4 \left(\dfrac{-\sqrt{3}}{2} \right)$ $y = 4 \left(\dfrac{1}{2} \right)$

     $x = -2\sqrt{3}$ $y = 2$

Answer: The coordinates of point A are $(-2\sqrt{3}, 2)$.

EXERCISES

In 1-9, find the rectangular coordinates (x, y) of a point that is r units from the origin and on the terminal ray of an angle in standard position whose measure is θ.

1. $r = 6, \theta = 30°$ 2. $r = 3, \theta = 90°$ 3. $r = 8, \theta = 135°$

4. $r = 9, \theta = 180°$ 5. $r = 2, \theta = 240°$

6. $r = \frac{1}{2}, \theta = 270°$ 7. $r = 1, \theta = 300°$

8. $r = 10, \theta = \text{Arc sin } \frac{3}{5}$ 9. $r = 5, \theta = \text{Arc cos } \left(\frac{-7}{25}\right)$

In 10-15: a. Write the coordinates of the point in simplest form. b. Locate the point on a coordinate graph.

10. $A(6 \cos 60°, 6 \sin 60°)$ 11. $B(2 \cos 90°, 2 \sin 90°)$

12. $C(\sqrt{2} \cos 135°, \sqrt{2} \sin 135°)$ 13. $D(4 \cos \pi, 4 \sin \pi)$

14. $E\left(8 \cos \frac{7\pi}{6}, 8 \sin \frac{7\pi}{6}\right)$ 15. $F\left(3 \cos \frac{3\pi}{2}, 3 \sin \frac{3\pi}{2}\right)$

In 16-19, triangle ABC is drawn in the coordinate plane with point A at the origin and point B on the positive ray of the x-axis. State the coordinates of point C.

16. $m\angle A = 45°, AC = 14$ 17. $m\angle A = 120°, AC = 10$

18. $A = \text{Arc sin } \frac{5}{13}, AC = 39$ 19. $A = \text{Arc cos } \left(\frac{-3}{5}\right), AC = 15$

20. The locus of points at a distance r from the origin is a circle whose center is the origin and whose radius is r. Demonstrate that the coordinates of all such points where $(x, y) = (r \cos \theta, r \sin \theta)$ satisfy the equation of the circle, $x^2 + y^2 = r^2$.

12-2 THE LAW OF COSINES

When the measures of two sides and the included angle of triangle ABC are known, the size and shape of the triangle are determined. We should be able to determine the measure of the third side. To do this, draw $\triangle ABC$ so that A is at the origin of the coordinate plane and B is on the positive ray of the x-axis, as shown on the next page. In Fig. 1, where $\angle A$ is acute, point C lies in the first quadrant. In Fig. 2, where $\angle A$ is obtuse, point C lies in the second quadrant.

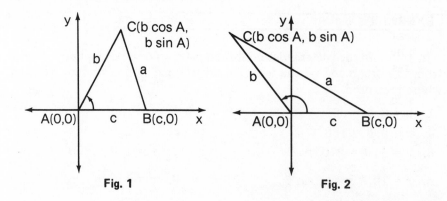

Fig. 1   Fig. 2

In each $\triangle ABC$, $AB = c$, $AC = b$, and $BC = a$. Therefore, the coordinates of point B on the positive ray of the x-axis are $(c, 0)$. The coordinates of point C at a distance b from the origin and on the terminal ray of $\angle A$ in standard position are $(b \cos A, b \sin A)$.

Thus, the coordinates of points B and C are given in terms of the measures of two sides and the included angle of the triangle. The distance formula can be used to express a, the distance from B to C in the coordinate plane, in terms of b, c, and $m\angle A$.

Proof: In $\triangle ABC$, $B(x_1, y_1) = (c, 0)$ and $C(x_2, y_2) = (b \cos A, b \sin A)$.

1. Write the distance formula.	$d^2 = (x_2 - x_1)^2 + (y_2 - y_1)^2$
2. Substitute the given coordinates of points B and C to find the distance a.	$a^2 = (b \cos A - c)^2 + (b \sin A - 0)^2$ $a^2 = (b \cos A - c)^2 + (b \sin A)^2$
3. Square each term in the right-hand member.	$a^2 = b^2 \cos^2 A - 2bc \cos A + c^2 + b^2 \sin^2 A$
4. Group the terms containing b^2 and factor.	$a^2 = b^2 \cos^2 A + b^2 \sin^2 A + c^2 - 2bc \cos A$ $a^2 = b^2 (\cos^2 A + \sin^2 A) + c^2 - 2bc \cos A$
5. Apply the identity $\cos^2 A + \sin^2 A = 1$ to simplify the equation.	$a^2 = b^2 (1) + c^2 - 2bc \cos A$ $a^2 = b^2 + c^2 - 2bc \cos A$

This result is called the *Law of Cosines*. It states that the square of the measure of one side of a triangle is equal to the sum of the squares of the measures of the other two sides minus twice the product of the measures of these two sides and the cosine of the angle between them.

This formula can be rewritten in terms of any two sides and their included angle.

Law of Cosines
$a^2 = b^2 + c^2 - 2bc \cos A$
$b^2 = a^2 + c^2 - 2ac \cos B$
$c^2 = a^2 + b^2 - 2ab \cos C$

Note: If $\triangle ABC$ is a right triangle with $m\angle C = 90°$, then:

$$c^2 = a^2 + b^2 - 2ab \cos 90°$$
$$c^2 = a^2 + b^2 - 2ab(0)$$
$$c^2 = a^2 + b^2 - 0$$
$$c^2 = a^2 + b^2$$

Thus, if $\triangle ABC$ is a right triangle, the Law of Cosines gives us the familiar relationship of the Pythagorean Theorem. This is not an unexpected result since the Pythagorean Theorem was used to prove both the distance formula and the basic identity $\cos^2 A + \sin^2 A = 1$, which were used to prove the Law of Cosines.

| MODEL PROBLEMS |

1. In $\triangle ABC$, if $a = 4$, $c = 6$, and $\cos B = \frac{1}{16}$, find b.

How to Proceed	*Solution*
1. Write the Law of Cosines, expressing b^2 in terms of a, c, and $\cos B$.	$b^2 = a^2 + c^2 - 2ac \cos B$
2. Substitute the given values.	$b^2 = 4^2 + 6^2 - 2(4)(6)(\frac{1}{16})$
3. Simplify.	$b^2 = 16 + 36 - 3$ $b^2 = 49$
4. Express b as a positive length.	$b = 7$ *Ans.*

2. In $\triangle RST$, $r = 11$, $s = 12$, and $m\angle T = 120°$. Find t to the nearest integer.

How to Proceed	*Solution*
1. Write the Law of Co-sines, expressing t^2 in terms of r, s, and $\cos T$.	$t^2 = r^2 + s^2 - 2rs \cos T$
2. Substitute the given values.	$t^2 = 11^2 + 12^2 - 2(11)(12)(\cos 120°)$
3. Since $\angle T$ is obtuse, $\cos 120° = -\cos 60°$ $= -\frac{1}{2}$	$t^2 = 11^2 + 12^2 - 2(11)(12)(-\frac{1}{2})$
4. Simplify.	$t^2 = 121 + 144 + 132$ $t^2 = 397$
5. Express t as a positive length.	$t = \sqrt{397}$
6. Find the square root to the nearest integer.	$t = 20$ *Ans.*

EXERCISES

1. In $\triangle PAT$, express p^2 in terms of a, t, and $\cos P$.
2. In $\triangle CAR$, express r^2 in terms of c, a, and $\cos R$.
3. In $\triangle SAD$, express a^2 in terms of s, d, and $\cos A$.
4. In $\triangle ABC$, if $a = 2$, $b = 3$, and $\cos C = \frac{1}{3}$, find c.
5. In $\triangle ABC$, if $b = 8$, $c = 5$, and $\cos A = \frac{1}{10}$, find a.
6. In $\triangle ABC$, if $a = 10$, $c = 7$, and $\cos B = \frac{1}{5}$, find b.
7. In $\triangle PQR$, if $p = 12$, $q = 8$, and $\cos R = \frac{1}{3}$, find r.
8. In $\triangle CAP$, if $c = 7$, $a = 6$, and $\cos P = (-\frac{3}{7})$, find p.
9. In $\triangle QRS$, if $s = 5$, $r = 7$, and $\cos Q = (-\frac{1}{10})$, find q.
10. In $\triangle ABC$, if $a = 5$, $b = 6$, and $\cos C = (-\frac{1}{3})$, find c.
11. In $\triangle BAD$, if $b = 3\sqrt{3}$, $a = 6$, and $m\angle D = 30°$, find d.

In 12, 13, and 15, select the numeral preceding the expression that best completes the sentence.

12. In $\triangle BCD$, if $b = 5, c = 4$, and m$\angle D = 60°$, then d is:

(1) $\frac{1}{2}$ (2) $\sqrt{21}$ (3) 3   (4) $\sqrt{41}$

13. In $\triangle BAR$, if $b = 1, a = \sqrt{3}$, and m$\angle R = 30°$, then r is:

(1) 1   (2) $\sqrt{2}$ (3) $\sqrt{3}$ (4) $\frac{1}{2}\sqrt{3}$

14. In $\triangle CDE$, if $c = 2, d = \sqrt{2}$, and m$\angle E = 45°$, find e.

15. In $\triangle ABC$, if $b = 3, c = 6$, and m$\angle A = 120°$, then a is:

(1) $\sqrt{63}$ (2) $\sqrt{53}$ (3) $\sqrt{37}$ (4) $\sqrt{27}$

16. In $\triangle PQR$, if $p = 3, q = 5$, and m$\angle R = 120°$, find r.
17. In $\triangle RST$, if $r = 3\sqrt{2}, s = 1$, and m$\angle T = 135°$, find t.
18. Using the Law of Cosines, find the length of a diagonal of a rectangle if the lengths of two adjacent sides are 5 and 12.
19. Find the length of the longer diagonal of a parallelogram if the lengths of two adjacent sides are 6 and 10 and the measure of an angle is $120°$.
20. Find to the nearest integer the measure of the base of an isosceles triangle if the measure of the vertex angle is $84°$ and the measure of each leg is 12.
21. The measures of two sides of a triangle are 20.0 and 12.0 and the measure of the included angle is $58°40'$. Find to the nearest tenth the measure of the third side of the triangle.

22. Find the length of one side of an equilateral triangle inscribed in a circle if the measure of a radius of the circle is $10\sqrt{3}$.

Ex. 22

23. The vertices of a triangle inscribed in a circle separate the circle into arcs whose measures are in the ratio $2:3:4$. If the measure of a radius of the circle is 10, find to the nearest integer the measure of the longest side of the triangle.
24. A kite is in the shape of a quadrilateral with two pairs of congruent adjacent sides. If the measures of two sides of the kite are 25 inches and 40 inches and the measure of the angle between the sides that are not congruent is $140°$, find to the nearest inch the length of the kite (i.e., the length of the longer diagonal).

12-3 USING THE LAW OF COSINES
TO FIND ANGLE MEASURE

If the measures of the three sides of a triangle are given, the size and shape of the triangle are determined. The Law of Cosines can be used to express the cosine of any angle of a triangle in terms of the measures of the sides of the triangle. Compare the steps that are used below to express $\cos C$ in terms of a, b, and c with those used to find the value of $\cos C$ when $a = 3$, $b = 4$, and $c = 2$.

The General Case

$$c^2 = a^2 + b^2 - 2ab \cos C$$

$$+ 2ab \cos C - c^2 = \qquad\qquad + 2ab \cos C - c^2$$

$$\overline{2ab \cos C = a^2 + b^2 \qquad\qquad - c^2}$$

$$\frac{2ab \cos C}{2ab} = \frac{a^2 + b^2 - c^2}{2ab}$$

$$\cos C = \frac{a^2 + b^2 - c^2}{2ab}$$

A Specific Example

$$c^2 = a^2 + b^2 - 2ab \cos C$$

$$2^2 = 3^2 + 4^2 - 2(3)(4) \cos C$$

$$2^2 = 3^2 + 4^2 - 24 \cos C$$

$$+ 24 \cos C - 2^2 = \qquad\qquad + 24 \cos C - 2^2$$

$$\overline{24 \cos C = 3^2 + 4^2 \qquad\qquad - 2^2}$$

$$\frac{24 \cos C}{24} = \frac{3^2 + 4^2 - 2^2}{24}$$

$$\cos C = \frac{9 + 16 - 4}{24} = \frac{21}{24} = \frac{7}{8}$$

The formula in the last line of the *General Case* expresses the value of $\cos C$ in terms of the measures of the sides of a triangle ABC. The equation in the last line of the *Specific Example* gives the cosine of the smallest angle of a triangle whose sides measure 2, 3, and 4. Since $\angle C$ is an angle of a triangle, $0° < m\angle C < 180°$. In this example, since the cosine of $\angle C$ is positive, $\angle C$ must be an acute angle. To find the measure of $\angle C$, we write $\cos C$ as a decimal and use the table of trigonometric function values.

$$\cos C = \tfrac{7}{8} = .8750$$
$$m\angle C = 29° \text{ (to the nearest degree)}$$

This relationship can be stated for the cosine of any angle in $\triangle ABC$, as illustrated in the following rules:

$$\cos A = \frac{b^2 + c^2 - a^2}{2bc}$$

$$\cos B = \frac{a^2 + c^2 - b^2}{2ac}$$

$$\cos C = \frac{a^2 + b^2 - c^2}{2ab}$$

In the following examples, notice that $\angle C$ may be acute, right, or obtuse, depending on the measures of the sides of $\triangle ABC$. In each case, find $m\angle C$ to the nearest degree. (Example 1 was studied earlier.)

☐ EXAMPLE 1:

In $\triangle ABC$, let $a = 3$, $b = 4$, and $c = 2$.

$\cos C = \dfrac{a^2 + b^2 - c^2}{2ab}$

$= \dfrac{3^2 + 4^2 - 2^2}{2(3)(4)}$

$= \dfrac{9 + 16 - 4}{24}$

$= \dfrac{21}{24} = \dfrac{7}{8}$

Here, $a^2 + b^2 > c^2$, cos C is positive, and $\angle C$ is acute.

Since cos $C = .8750$, $m\angle C \approx 29°$.

☐ EXAMPLE 2:

In $\triangle ABC$, let $a = 3$, $b = 4$, and $c = 5$.

$\cos C = \dfrac{a^2 + b^2 - c^2}{2ab}$

$= \dfrac{3^2 + 4^2 - 5^2}{2(3)(4)}$

$= \dfrac{9 + 16 - 25}{24}$

$= \dfrac{0}{24} = 0$

Here, $a^2 + b^2 = c^2$, cos C is zero, and $\angle C$ is right.

Since cos $C = 0$, $m\angle C = 90°$.

☐ EXAMPLE 3:

In $\triangle ABC$, let $a = 3$, $b = 4$, and $c = 6$.

$\cos C = \dfrac{a^2 + b^2 - c^2}{2ab}$

$= \dfrac{3^2 + 4^2 - 6^2}{2(3)(4)}$

$= \dfrac{9 + 16 - 36}{24}$

$= \dfrac{-11}{24}$

Here, $a^2 + b^2 < c^2$, cos C is negative, and $\angle C$ is obtuse.

Since cos $C \approx -.4583$ and cos $63° \approx .4583$, $m\angle C \approx (180° - 63°)$ $m\angle C \approx 117°$

If $\angle C$ is a right angle, then $\triangle ABC$ is a right triangle where:
1. c is the measure of the hypotenuse.
2. a is the measure of the leg adjacent to $\angle B$.
3. b is the measure of the leg adjacent to $\angle A$.

In the right $\triangle ABC$, $c^2 = a^2 + b^2$. Observe what happens when $a^2 + b^2$ is substituted for c^2 in the following situations:

$$\cos A = \frac{b^2 + c^2 - a^2}{2bc}$$

$$\cos A = \frac{b^2 + (a^2 + b^2) - a^2}{2bc}$$

$$\cos A = \frac{2b^2}{2bc}$$

$$\cos A = \frac{b}{c}$$

$$\cos A = \frac{\text{measure of leg adjacent to } \angle A}{\text{measure of hypotenuse}}$$

$$\cos B = \frac{a^2 + c^2 - b^2}{2ac}$$

$$\cos B = \frac{a^2 + (a^2 + b^2) - b^2}{2ac}$$

$$\cos B = \frac{2a^2}{2ac}$$

$$\cos B = \frac{a}{c}$$

$$\cos B = \frac{\text{measure of leg adjacent to } \angle B}{\text{measure of hypotenuse}}$$

| MODEL PROBLEMS |

1. In $\triangle ABC$, $a = 5$, $b = 7$, and $c = 10$. Find $\cos B$.

Solution

1. Write the Law of Cosines solved for $\cos B$.

$$\cos B = \frac{a^2 + c^2 - b^2}{2ac}$$

2. Substitute the given values.

$$\cos B = \frac{5^2 + 10^2 - 7^2}{2(5)(10)}$$

3. Express the value of $\cos B$ as a fraction in simplest form, or as a decimal.

$$\cos B = \frac{25 + 100 - 49}{100}$$

$$\cos B = \frac{76}{100} = \frac{19}{25}$$

Answer: $\cos B = \dfrac{19}{25}$, or $\cos B = .76$

2. In isosceles triangle RED, $RE = ED = 5$ and $RD = 8$. Find the measure of the vertex angle, $\angle E$, to the nearest degree.

Solution

1. Sketch the triangle. Let $RE = d = 5$, $ED = r = 5$, and $RD = e = 8$.

2. Write the formula for $\cos E$ in terms of d, r, and e.

$$\cos E = \frac{d^2 + r^2 - e^2}{2dr}$$

3. Substitute the given values, and simplify the value of cos E. Express cos E as a decimal value.

$$\cos E = \frac{5^2 + 5^2 - 8^2}{2(5)(5)}$$

$$\cos E = \frac{25 + 25 - 64}{50}$$

$$\cos E = \frac{-14}{50} = -.28$$

4. Since cos E is negative, $\angle E$ is an obtuse angle. Find the reference angle, the acute angle whose cosine is .28.

$$\cos 74° \approx .28$$
$$m\angle E \approx (180° - 74°) = 106°$$

Answer: $m\angle E = 106°$ to the nearest degree

| **EXERCISES** |

1. In $\triangle PQR$, express cos P in terms of p, q, and r.
2. In $\triangle SAD$, express cos D in terms of s, a, and d.
3. In $\triangle ABC$, $a = 5$, $b = 7$, and $c = 8$. Find cos C.
4. In $\triangle ABC$, $a = 5$, $b = 9$, and $c = 12$. Find cos A.
5. In $\triangle ABC$, $a = 5$, $b = 12$, and $c = 13$. Find cos B.
6. Find the cosine of the largest angle of a triangle if the measures of the sides of the triangle are 5, 6, and 7.
7. Find the cosine of the largest angle of a triangle if the measures of the sides of the triangle are 2, 3, and 4.
8. In $\triangle RST$, $r = 5$, $s = 7$, and $t = 8$. Find $m\angle S$.
9. In $\triangle CDE$, $c = 1$, $d = 2$, and $e = \sqrt{3}$. Find $m\angle E$.
10. In $\triangle ABC$, the measures of the sides are 3, 5, and 7. Find the measure of the largest angle in the triangle.

In 11–13, select the numeral preceding the expression that best completes the sentence or answers the question.

11. In $\triangle BDE$, $b = 3$, $e = 5$, and $d = \sqrt{7}$. What is the value of cos D?
    (1) $-\frac{1}{2}$ (2) $\frac{1}{30}$ (3) $\frac{9}{10}$ (4) $\frac{4}{5}$
12. In $\triangle RPM$, $r = 9$, $p = 12$, and $m = 15$. The cosine of the largest angle in the triangle is:
    (1) 1   (2) 0   (3) $\frac{4}{5}$ (4) 90
13. In $\triangle ABC$, $a = 6$, $b = 6$, and $c = 6\sqrt{2}$. The measure of $\angle B$ is:
    (1) 45°   (2) 60°   (3) 120°   (4) 135°

14. In $\triangle FTG$, $t = 10$, $g = 14$, and $f = 20$. Find to the nearest degree the measure of $\angle T$.

15. Find to the nearest degree the measure of the vertex angle of a triangle whose sides measure 3, 3, and 5.

16. Find to the nearest degree the measure of a base angle of an isosceles triangle whose equal sides each measure 3 and whose base measures 4.

17. Find to the nearest degree the measure of the largest angle of a triangle whose sides measure 7, 9, and 12.

18. The lengths of the sides of a triangle are 9, 40, and 41. Find the measure of the largest angle.

19. A cross-country ski trail is laid out in the shape of a triangle. The lengths of the three paths that make up the trail are 2000 m, 1200 m, and 1800 m. Find to the nearest degree the measure of the smallest angle of the trail.

20. The lengths of the sides of a triangle inscribed in a circle are 9, 5, and 10. Find to the nearest degree the measure of: **a.** the smallest angle in the triangle. **b.** the smallest arc of the circle whose endpoints are vertices of the triangle.

21. Use the formula $\cos A = \dfrac{b^2 + c^2 - a^2}{2bc}$ to show that the measure of an angle of an equilateral triangle is $60°$. (*Hint:* Let the measure of each side of the triangle = s.)

12-4 AREA OF A TRIANGLE

Since the measures of two sides and the included angle of a triangle determine the size and shape of the triangle, we are able to use these measures to find the area of a triangle.

Draw triangle ABC in three positions in the coordinate plane. In each case, place one vertex at the origin and another vertex on the positive x-axis.

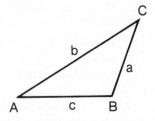

In Fig. 1, point C is a distance b from the origin on the terminal ray of $\angle A$. Therefore, the coordinates of C are $(b \cos A, b \sin A)$. The measure of the base, $\overline{AB}$, of the triangle is c. The measure of the altitude, h, is the y-coordinate of point C, namely, $b \sin A$. Therefore:

Area of $\triangle ABC = \frac{1}{2}$ (base)(height)

$$K = \frac{1}{2} c \cdot b \sin A$$

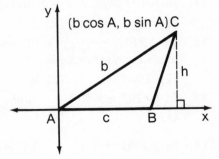

Fig. 1

In Fig. 2 and Fig. 3, $\triangle ABC$ is repositioned so that the sides whose measures are a and b, respectively, act as the base. In each case, we again find the area of the triangle.

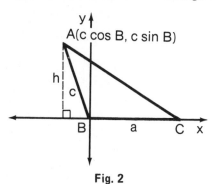

Fig. 2

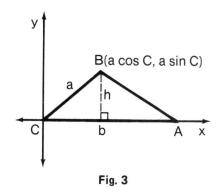

Fig. 3

| Area of $\triangle ABC = \frac{1}{2}$ (base)(height) | Area of $\triangle ABC = \frac{1}{2}$ (base)(height) |
| $K = \frac{1}{2}a \cdot c \sin B$ | $K = \frac{1}{2}b \cdot a \sin C$ |

The area of a triangle, K, is equal to one-half the product of the measures of two sides and the sine of the angle between them.

$$\text{Area of } \triangle ABC = \tfrac{1}{2}ab \sin C = \tfrac{1}{2}bc \sin A = \tfrac{1}{2}ca \sin B$$

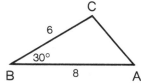

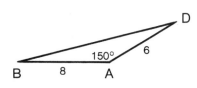

☐ EXAMPLE 1: Find the area of $\triangle ABC$ if $c = 8$, $a = 6$, and m$\angle B =$ 30°.

$$K = \tfrac{1}{2}ac \sin B$$
$$K = \tfrac{1}{2}(6)(8) \sin 30°$$
$$K = \tfrac{1}{2}(6)(8)(\tfrac{1}{2})$$
$$K = 12$$

☐ EXAMPLE 2: Find the area of $\triangle ABD$ if $AB = 8$, $AD = 6$, and m$\angle DAB = 150°$.

$$K = \tfrac{1}{2}(AB)(AD) \sin \angle DAB$$
$$K = \tfrac{1}{2}(8)(6) \sin 150°$$
$$K = \tfrac{1}{2}(8)(6)(\tfrac{1}{2})$$
$$K = 12$$

The diagonals of a parallelogram divide the parallelogram into two congruent triangles. For example, in parallelogram $ABCD$, adjacent sides measure 6 and 8 and consecutive angles measure 30° and

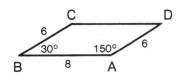

150°. Note on the next page how the triangles formed by diagonals $\overline{AC}$ and $\overline{BD}$, respectively, relate to the triangles in examples 1 and 2 above.

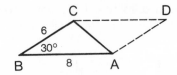

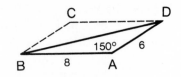

From example 1:

    Area of $\triangle ABC$

    $= \frac{1}{2}(AB)(BC) \sin B = 12$

Thus,

    area of $\square ABCD$

    $= 2 \cdot$ area of $\triangle ABC$

    $= (AB)(BC) \sin B = 24$

From example 2:

    Area of $\triangle ABD$

    $= \frac{1}{2}(AB)(AD) \sin A = 12$

Thus,

    area of $\square ABCD$

    $= 2 \cdot$ area of $\triangle ABD$

    $= (AB)(AD) \sin A = 24$

These examples illustrate that the area of a parallelogram is equal to the product of the measures of two adjacent sides and the sine of the angle between them.

MODEL PROBLEMS

1. Find the area of an equilateral triangle if the measure of one side is 4.

Solution

The measure of each side of the equilateral triangle is 4 and the measure of each angle is 60°.

$$K = \frac{1}{2}ab \sin C$$
$$K = \frac{1}{2}(4)(4) \sin 60°$$
$$K = \frac{1}{2}(4)(4)\left(\frac{\sqrt{3}}{2}\right)$$
$$K = 4\sqrt{3} \quad Ans.$$

2. Find to the nearest hundred the number of square feet in the area of a triangular lot at the intersection of two streets if the angle of intersection is $76°10'$ and the frontage along the streets is 220 ft. and 156 ft.

Solution

1. Draw and label a diagram.

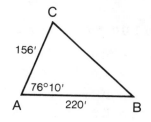

2. Write the formula in terms of the diagram.

$K = \frac{1}{2} bc \sin A$

3. Substitute the given values.

$K = \frac{1}{2}(156)(220) \sin 76°10'$

4. Use logarithms, a calculator, or ordinary arithmetic for the calculations, rounding the answer to the nearest hundred.

$K = 16{,}700$

Answer: 16,700

3. The area of a parallelogram is 20. Find the measures of the angles of the parallelogram if the measures of two adjacent sides are 8 and 5.

Solution

Let *ABCD* be the parallelogram with $AB = 8$ and $BC = AD = 5$. Here, $\angle A$ is acute and $\angle B$ is obtuse.

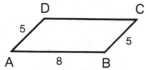

$K = (AB)(AD)(\sin A)$
$20 = (8)(5)(\sin A)$
$20 = 40 \sin A$
$\dfrac{20}{40} = \dfrac{40 \sin A}{40}$
$.5 = \sin A$

Since $\angle A$ is acute,
$m\angle A = 30°$.

$K = (AB)(BC)(\sin B)$
$20 = (8)(5)(\sin B)$
$20 = 40 \sin B$
$\dfrac{20}{40} = \dfrac{40 \sin B}{40}$
$.5 = \sin B$

Since $\angle B$ is obtuse,
$m\angle B = 180° - 30° = 150°$.

Answer: 30° and 150°

EXERCISES

In 1–10, find the area of triangle ABC.

1. $a = 6, b = 7, \sin C = \frac{1}{3}$
2. $b = 12, c = 14, \sin A = \frac{3}{4}$
3. $a = 9, b = 10, \sin C = \frac{2}{3}$
4. $a = 15, c = 12, \sin B = \frac{3}{5}$
5. $b = 13, c = 21, \sin A = \frac{2}{7}$
6. $b = 8, c = 20, m\angle A = 30°$
7. $a = 7, b = 5, m\angle C = 90°$
8. $a = \sqrt{3}, b = 8, m\angle C = 60°$
9. $a = 2\sqrt{3}, c = 3, m\angle B = 120°$
10. $a = 12, c = 9, m\angle B = 150°$

11. In $\triangle ABC$, $a = 3\sqrt{2}$, $b = 5$, and $m\angle C = 45°$. Find the area of $\triangle ABC$.
12. In $\triangle DEF$, $d = 8$, $f = 4\sqrt{2}$, and $m\angle E = 135°$. Find the area of $\triangle DEF$.
13. In isosceles triangle RST, $RS = ST = 6$. If the measure of the vertex angle is $150°$, what is the area of $\triangle RST$?
14. Find the area of an isosceles triangle if the measure of a base angle is $75°$ and the measure of each of the congruent sides is 10.
15. Find to the nearest integer the area of an isosceles triangle if the measure of the vertex angle is $42°$ and the measure of each of the congruent sides is 12.
16. Find to the nearest tenth the area of a triangle if the measures of two sides and the included angle are 2.6, 5.2, and $67°$.
17. Find to the nearest integer the number of square meters in the area of a triangle if the lengths of two adjacent sides are 71.9 and 14.3 and the measure of the angle between them is $38°40'$.
18. In right triangle ABC, $m\angle C = 90°$, $a = 5$, and $b = 12$. Find the area of the triangle, using: a. $K = \frac{1}{2}$ base · height   b. $K = \frac{1}{2} ab \sin C$

In 19, select the numeral preceding the expression that best completes the sentence.

19. In parallelogram $ABCD$, $AB = 4$, $AD = 5\sqrt{3}$, and $m\angle A = 60°$. The area of $\square ABCD$ is:

    (1) 15   (2) 30   (3) $5\sqrt{3}$ (4) $20\sqrt{3}$

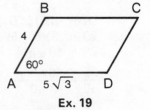

Ex. 19

20. In rhombus $PQRS$, the length of each side is 8 and $m\angle PQR = 30°$. Find the area of rhombus $PQRS$.
21. Find to the nearest integer the area of a parallelogram if the measures of a pair of adjacent sides are 20.0 and 24.0 and the measure of an angle is $32°20'$.
22. If the area of $\triangle ABC$ is 12, find $\sin C$ when $a = 5$ and $b = 6$.
23. If the area of $\triangle CAP$ is 75, find $\sin A$ when $c = 30$ and $p = 20$.

24. Find to the nearest degree the measure of the acute angle of a triangle between two sides of measures 10 and 15 if the area of the triangle is 50.

25. In triangle ROC, $\angle O$ is an obtuse angle, $OR = 32$, $OC = 36$, and the area of the triangle is 240. Find the measure of $\angle O$ to the nearest degree.

26. If the measure of the side of an equilateral triangle is represented by s, show that the area of the triangle is $\dfrac{s^2}{4}\sqrt{3}$.

In 27–30, the lengths of the sides of a triangle are given. a. Find the measure of the largest angle of the triangle to the nearest degree. b. Using the answer from part a, find the area of the triangle to the nearest square centimeter.

27. 5 cm, 6 cm, 7 cm   28. 5 cm, 6 cm, 4 cm
29. 5 cm, 6 cm, 2 cm   30. 5 cm, 6 cm, 8 cm

31. In a parallelogram, two consecutive sides measure 2 cm and 10 cm, and the length of the longer diagonal is 11 cm. a. Find the measure of the largest angle of the parallelogram to the nearest degree. b. Using the answer from part a, find the area of the parallelogram to the nearest square centimeter.

12-5   THE LAW OF SINES

In section 4, we derived three forms of the formula for the area of a triangle.

$$\text{Area of } \triangle ABC = \tfrac{1}{2}\, bc \sin A = \tfrac{1}{2}\, ac \sin B = \tfrac{1}{2}\, ab \sin C$$

If we divide each of the last three terms of the equality by $\tfrac{1}{2}abc$, the result is a triple equality.

$$\frac{\tfrac{1}{2}\, bc \sin A}{\tfrac{1}{2}\, abc} = \frac{\tfrac{1}{2}\, ac \sin B}{\tfrac{1}{2}\, abc} = \frac{\tfrac{1}{2}\, ab \sin C}{\tfrac{1}{2}\, abc}$$

$$\frac{\sin A}{a} = \frac{\sin B}{b} = \frac{\sin C}{c}$$

This equality, also written in terms of the reciprocals of these ratios, is called the *Law of Sines*.

$$\frac{a}{\sin A} = \frac{b}{\sin B} = \frac{c}{\sin C}$$

The Law of Sines can be used to find the measure of a side of a triangle when the measures of two angles and a side are known (a.a.s or a.s.a.).

□ Example: In $\triangle ABC$, $a = 10$, m$\angle A = 30°$, and m$\angle B = 50°$. Find b to the nearest integer.

1. Choose the ratios that use a and b.

$$\frac{a}{\sin A} = \frac{b}{\sin B}$$

2. Substitute and solve for b.

$$\frac{10}{\sin 30°} = \frac{b}{\sin 50°}$$

$$\frac{10}{.5} = \frac{b}{.766}$$

$$.5b = 10(.766)$$
$$.5b = 7.66$$

$$b = \frac{7.66}{.5} = \frac{76.6}{5} = 15.32$$

3. Round the value of b to the nearest integer.

$$b = 15$$

Answer: $b = 15$

If $\angle C$ is a right angle, then $\triangle ABC$ is a right triangle where:

1. c is the measure of the hypotenuse.
2. a is the measure of the leg opposite $\angle A$.
3. b is the measure of the leg opposite $\angle B$.

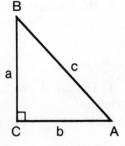

Observe what happens when $\sin C = \sin 90° = 1$.

$$\frac{c}{\sin C} = \frac{a}{\sin A}$$

$$c \sin A = a \sin C$$

$$\frac{c \sin A}{c} = \frac{a \sin C}{c}$$

$$\sin A = \frac{a \sin 90°}{c}$$

$$\sin A = \frac{a(1)}{c}$$

$$\sin A = \frac{a}{c}$$

$$\sin A = \frac{\text{measure of leg opposite } \angle A}{\text{measure of hypotenuse}}$$

$$\frac{c}{\sin C} = \frac{b}{\sin B}$$

$$c \sin B = b \sin C$$

$$\frac{c \sin B}{c} = \frac{b \sin C}{c}$$

$$\sin B = \frac{b \sin 90°}{c}$$

$$\sin B = \frac{b(1)}{c}$$

$$\sin B = \frac{b}{c}$$

$$\sin B = \frac{\text{measure of leg opposite } \angle B}{\text{measure of hypotenuse}}$$

In section 3 of this chapter, we showed that in right triangle ABC with $m\angle C = 90°$:

$$\cos A = \frac{b}{c} = \frac{\text{measure of leg adjacent to } \angle A}{\text{measure of hypotenuse}}$$

Therefore:

$$\tan A = \frac{\sin A}{\cos A} = \frac{\frac{a}{c}}{\frac{b}{c}} = \frac{a}{c} \cdot \frac{c}{b} = \frac{a}{b} = \frac{\text{measure of leg opposite } \angle A}{\text{measure of leg adjacent to } \angle A}$$

If we let "opp" represent "measure of the leg opposite the angle," "adj" represent "measure of the leg adjacent to the angle," and "hyp" represent "measure of the hypotenuse," then:

In right triangle ABC with $m\angle C = 90°$:

$$\sin A = \frac{a}{c} = \frac{\text{opp}}{\text{hyp}} \qquad \cos A = \frac{b}{c} = \frac{\text{adj}}{\text{hyp}} \qquad \tan A = \frac{a}{b} = \frac{\text{opp}}{\text{adj}}$$

MODEL PROBLEMS

1. In triangle ABC, $a = 12$, $\sin A = \frac{1}{3}$, and $\sin C = \frac{1}{4}$. Find c.

Solution

1. Choose the two ratios of the Law of Sines that use a, the side whose measure we know, and c, the side whose measure we want to know.

$$\frac{c}{\sin C} = \frac{a}{\sin A}$$

2. Substitute the given values.

$$\frac{c}{\frac{1}{4}} = \frac{12}{\frac{1}{3}}$$

3. Solve for c.

$$\frac{1}{3}c = \frac{1}{4}(12)$$
$$\frac{1}{3}c = 3$$
$$c = 9 \quad Ans.$$

2. In triangle DAT, $m\angle D = 27°$, $m\angle A = 105°$, and $t = 21$. Find d to the nearest integer.

Solution

1. Since we must use the ratios involving t and d, we must find $m\angle T$.

$$m\angle T = 180° - (m\angle A + m\angle D)$$
$$m\angle T = 180° - (105° + 27°)$$
$$m\angle T = 48°$$

2. Write a proportion, using the two ratios of the Law of Sines in terms of t and d.

$$\frac{d}{\sin D} = \frac{t}{\sin T}$$

3. Substitute the given values.

$$\frac{d}{\sin 27°} = \frac{21}{\sin 48°}$$

4. Solve for d.

$$d = \frac{21 \sin 27°}{\sin 48°}$$

5. Use logarithms, a calculator, or ordinary arithmetic for the calculations. Round the result to the nearest integer. (Arithmetic is shown here.)

$$d = \frac{21(.4540)}{.7431}$$

$$d = \frac{9.5340}{.7431}$$

$$d = 12.8$$
$$d = 13 \quad Ans.$$

3. In $\triangle ABC$, $m\angle A = 30°$ and $m\angle B = 45°$. Find the ratio $a:b$.

Solution

1. Write the two ratios of the Law of Sines that use a and b.

$$\frac{a}{\sin A} = \frac{b}{\sin B}$$

2. Substitute the given values.

$$\frac{a}{\sin 30°} = \frac{b}{\sin 45°}$$

3. Write the values for $\sin 30°$ and $\sin 45°$.

$$\frac{a}{\frac{1}{2}} = \frac{b}{\frac{\sqrt{2}}{2}}$$

4. Solve for $\frac{a}{b}$, and simplify the result.

$$\frac{\sqrt{2}}{2}a = \frac{1}{2}b$$

$$\frac{\frac{\sqrt{2}}{2}a}{\frac{\sqrt{2}}{2}b} = \frac{\frac{1}{2}b}{\frac{\sqrt{2}}{2}b}$$

$$\frac{a}{b} = \frac{\frac{1}{2}}{\frac{\sqrt{2}}{2}} = \frac{\frac{1}{2} \cdot \sqrt{2}}{\frac{\sqrt{2}}{2} \cdot \sqrt{2}}$$

$$= \frac{\frac{\sqrt{2}}{2}}{\frac{2}{2}} = \frac{\sqrt{2}}{2} \quad Ans.$$

4. In right $\triangle ABC$, $m\angle C = 90°$, $m\angle A = 56°$, and $BC = 8.7$. Find AB to the nearest tenth.

Solution

Method 1: Law of Sines

1. Since $BC = a$ and $AB = c$, write the Law of Sines in terms of a and c.

$$\frac{c}{\sin C} = \frac{a}{\sin A}$$

2. Substitute the given values.

$$\frac{c}{\sin 90°} = \frac{8.7}{\sin 56°}$$

3. Substitute the values of the sines of the angles, and perform the computation. Round the value of c to the nearest tenth.

$$\frac{c}{1} = \frac{8.7}{.8290}$$

$$c = \frac{8.700\,00}{.829}$$

$$c = 10.49$$
$$c = 10.5 \quad Ans.$$

Method 2: Sine ratio of a right triangle

1. Write the sine ratio for a right triangle.

$$\sin A = \frac{a}{c}$$

2. Substitute the given values.

$$\sin 56° = \frac{8.7}{c}$$

$$.8290 = \frac{8.7}{c}$$

3. Solve for c.

$$.8290c = 8.7$$

4. Perform the computation, and round the value of c to the nearest tenth.

$$c = \frac{8.7}{.8290}$$

$$c = 10.5 \quad Ans.$$

| EXERCISES |

1. In $\triangle ABC$, $a = 6$, $\sin A = .2$, and $\sin B = .3$. Find b.
2. In $\triangle ABC$, $b = 12$, $\sin B = .6$, and $\sin C = .9$. Find c.
3. In $\triangle ABC$, $c = 8$, $\sin C = \frac{1}{4}$, and $\sin B = \frac{3}{8}$. Find b.
4. In $\triangle ABC$, $b = 30$, $\sin B = .6$, and $\sin A = .8$. Find a.
5. In $\triangle DEN$, $d = 12$, $\sin D = .4$, and $\sin N = .3$. Find n.
6. In $\triangle CAP$, $c = \sqrt{6}$, $m\angle C = 45°$, and $m\angle P = 60°$. Find p.
7. In $\triangle REM$, $r = 3\sqrt{2}$, $m\angle R = 135°$, and $m\angle M = 30°$. Find m.

8. In $\triangle ABC$, $a = 6\sqrt{2}$, $m\angle A = 45°$, and $m\angle B = 105°$. Find c.
9. In $\triangle CAR$, $a = 24$, $m\angle A = 27°$, and $m\angle C = 83°$. Find c to the nearest integer.
10. In $\triangle PAR$, $p = 8.5$, $m\angle P = 72°$, and $m\angle A = 68°$. Find r to the nearest tenth.

In 11, 12, and 19, select the numeral preceding the expression that best completes the sentence.

11. In an isosceles triangle, the length of each of the congruent sides is 12 and the measure of a base angle is 30°. The length of the base of the triangle is:

    (1) $12\sqrt{3}$ (2) $12\sqrt{2}$ (3) 12   (4) $6\sqrt{3}$

12. In an isosceles triangle, the vertex angle measures 90° and the length of each congruent leg is $5\sqrt{2}$. The length of the base of the triangle is:

    (1) 5   (2) 10   (3) $10\sqrt{2}$ (4) $10\sqrt{3}$

13. Use the Law of Sines to find to the nearest integer the measure of the base of an isosceles triangle if the measure of the vertex angle is 70° and the measure of each of the congruent sides is 15.
14. Find to the nearest tenth the length of one of the congruent sides of an isosceles triangle if the measure of the base is 24.6 and the measure of each base angle is 72°10′.
15. Find to the nearest tenth the length of one of the congruent sides of an isosceles triangle if the measure of the base is 48.9 and the measure of the vertex angle is 57°40′.
16. A ladder that is 10 ft. long leans against a wall so that the top of the ladder just reaches the top of the wall. Find to the nearest tenth of a foot the height of the wall if the foot of the ladder makes an angle of 72° with the ground.
17. A wire that is 8.5 meters long runs in a straight line from the top of a telephone pole to a stake in the ground. If the wire makes an angle of 68° with the ground, find the height of the pole to the nearest tenth of a meter.
18. A straight road slopes upward at an angle of 15° from the horizontal. How long is the section of the road that rises a vertical distance of 250 feet? (Express the answer to the nearest hundred feet.)

19. In $\triangle ABC$, $m\angle A = 45°$ and $m\angle B = 30°$. The ratio of $a:b$ is:

    (1) $\sqrt{2}:1$ (2) 2:1   (3) $\sqrt{2}:2$ (4) 1:2

20. In $\triangle RST$, $\sin R = 0.4$ and $m\angle S = 30°$. Find the ratio $r:s$.
21. In $\triangle PQR$, $\sin P = 0.15$ and $\sin Q = 0.5$. Find the ratio $q:p$.

22. In $\triangle ABC$, m$\angle A$ = 24°50', m$\angle B$ = 65°30', and BC = 25.6. Find AC to the nearest tenth.
23. In $\triangle DEF$, m$\angle E$ = 31°20', m$\angle F$ = 18°40', and DF = 72.3. Find DE to the nearest tenth.

12-6   THE AMBIGUOUS CASE

It would seem that if we know the measures of two sides of a triangle and the angle opposite one of them, we could use the Law of Sines to solve for the measure of another angle. But we know that the measures of two sides and the angle opposite one of them do not suffice to determine the size and shape of a triangle (see the figures below).

In $\triangle ABC$ and $\triangle A'B'C'$, m$\angle A$ = m$\angle A'$ = 30°, AB = $A'B'$ = 12, and BC = $B'C'$ = 7. But $\triangle ABC$ is not congruent to $\triangle A'B'C'$.

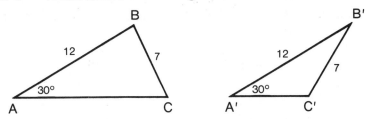

We will use the Law of Sines to study the number of possible triangles that can be constructed when the measures of two sides and the angle opposite one of them are known.

First, we will consider the cases, like the one given above, when the given angle is acute.

Case I. $c > a > c \sin A$

If a = 7, m$\angle A$ = 30°, and c = 12, find m$\angle C$.

$$\frac{a}{\sin A} = \frac{c}{\sin C}$$

$$\frac{7}{\sin 30°} = \frac{12}{\sin C}$$

$$\frac{7}{\frac{1}{2}} = \frac{12}{\sin C}$$

$$7 \sin C = \tfrac{1}{2}(12)$$

$$\sin C = \tfrac{6}{7} \approx .8571$$

$$\text{m}\angle C \approx 59°$$

$$\text{and m}\angle C' \approx (180° - 59°) = 121°$$

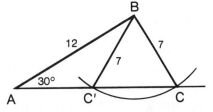

In $\triangle ABC$, there are two possible angles whose sine is .8571, namely, a first-quadrant angle and a second-quadrant angle.

Conclusion: In $\triangle ABC$, if $\angle A$ is acute and $c > a > c \sin A$, then two possible triangles can be drawn.

Case II. $a = c \sin A$

If $a = 6$, $m\angle A = 30°$, and $c = 12$, find $m\angle C$.

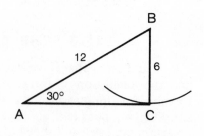

$$\frac{a}{\sin A} = \frac{c}{\sin C}$$

$$\frac{6}{\sin 30°} = \frac{12}{\sin C}$$

$$\frac{6}{\frac{1}{2}} = \frac{12}{\sin C}$$

$$6 \sin C = \tfrac{1}{2}(12)$$

$$\sin C = \tfrac{6}{6} = 1$$

$$m\angle C = 90°$$

Conclusion: In $\triangle ABC$, if $\angle A$ is acute and $a = c \sin A$, then one right triangle can be drawn.

Case III. $a > c > c \sin A$

If $a = 15$, $m\angle A = 30°$, and $c = 12$, find $m\angle C$.

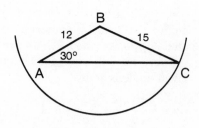

$$\frac{a}{\sin A} = \frac{c}{\sin C}$$

$$\frac{15}{\sin 30°} = \frac{12}{\sin C}$$

$$\frac{15}{\frac{1}{2}} = \frac{12}{\sin C}$$

$$15 \sin C = \tfrac{1}{2}(12)$$

$$\sin C = \tfrac{6}{15} = .4$$

$$m\angle C \approx 24°$$

and $m\angle C' \approx (180° - 24°) = 156°$

But $m\angle A + m\angle C' = 30° + 156° = 186°$. Therefore, $\angle C'$ is not a possible solution. If $a = c$, it can be shown in a similar way that only one triangle can be drawn, namely, an isosceles triangle.

Conclusion: In $\triangle ABC$, if $\angle A$ is acute and $a \geq c > c \sin A$, then only one triangle can be drawn.

Case IV. $a < c \sin A$

If $a = 4$, m$\angle A = 30°$, and $c = 12$, find m$\angle C$.

$$\frac{a}{\sin A} = \frac{c}{\sin C}$$

$$\frac{4}{\sin 30°} = \frac{12}{\sin C}$$

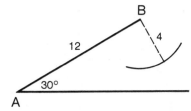

$$\frac{4}{\frac{1}{2}} = \frac{12}{\sin C}$$

$$4 \sin C = \tfrac{1}{2}(12)$$

$$\sin C = \tfrac{6}{4} = 1.5$$

(There is no angle whose sine is greater than 1.)

Conclusion: In $\triangle ABC$, if $\angle A$ is acute and $a < c \sin A$, then no triangle can be drawn.

If the given angle is obtuse, the measure of the side opposite the obtuse angle must be greater than the measure of any other side. In the next two cases, $\angle A$ is obtuse.

Case V. $a > c$

If $a = 15$, m$\angle A = 150°$, and $c = 12$, find m$\angle C$.

$$\frac{a}{\sin A} = \frac{c}{\sin C}$$

$$\frac{15}{\sin 150°} = \frac{12}{\sin C}$$

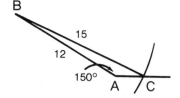

$$\frac{15}{\frac{1}{2}} = \frac{12}{\sin C}$$

$$15 \sin C = \tfrac{1}{2}(12)$$

$$\sin C = \tfrac{6}{15} = .4$$

$$m\angle C \approx 24°$$

(There can be no more than one obtuse angle in a triangle. Therefore, $\angle C$ must be acute.)

Conclusion: In $\triangle ABC$, if $\angle A$ is obtuse and $a > c$, then only one triangle can be drawn.

Case VI. $a < c$

If $a = 8$, $m\angle A = 150°$, and $c = 12$, find $m\angle C$.

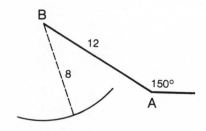

$$\frac{a}{\sin A} = \frac{c}{\sin C}$$

$$\frac{8}{\sin 150°} = \frac{12}{\sin C}$$

$$\frac{8}{\frac{1}{2}} = \frac{12}{\sin C}$$

$$8 \sin C = \tfrac{1}{2}(12)$$

$$\sin C = \tfrac{6}{8} = .75$$

$$m\angle C \approx 49°$$

or $m\angle C' \approx (180° - 49°) = 131°$

But $m\angle A + m\angle C = 150° + 49° = 199°$ and $m\angle A + m\angle C' = 150° + 131° = 281°$. Therefore, neither value is a solution. If $a = c$, it can also be shown that no triangle can be drawn.

Conclusion: In $\triangle ABC$, if $\angle A$ is obtuse and $a \leq c$, then no triangle can be drawn.

	Summary	
$\angle A$ is acute	$c > a > c \sin A$	2 triangles
	$a = c \sin A$	1 triangle
	$a \geq c > c \sin A$	1 triangle
	$a < c \sin A$	0 triangles
$\angle A$ is obtuse	$a > c$	1 triangle
	$a \leq c$	0 triangles

In any situation, after applying the Law of Sines to the given data, we can discover the number of possible triangles by:

1. A check to see that the sum of the measures of two angles of the triangle(s) is less than 180°.

OR

2. An accurate drawing, using the given data.

MODEL PROBLEMS

1. State the number of possible triangles that can be constructed if $a = 10$, $b = 12$, and $m\angle B = 20°$.

Solution

1. Write the Law of Sines in terms of a and b.

$$\frac{a}{\sin A} = \frac{b}{\sin B}$$

2. Substitute the given values.

$$\frac{10}{\sin A} = \frac{12}{\sin 20°}$$

$$\frac{10}{\sin A} = \frac{12}{.3420}$$

3. Solve for $\sin A$.

$$12 \sin A = 10(.3420)$$

$$\sin A = \frac{3.420}{12}$$

$$\sin A = .285$$

4. Find to the nearest degree the measures of an acute angle and an obtuse angle whose sine is approximately equal to .285.

$$m\angle A = 17°$$
$$m\angle A = (180° - 17°) = 163°$$

5. Combine the measures of $\angle A$ and $\angle B$ to determine if a third angle is possible.

$m\angle A = 17°$	$m\angle A = 163°$
$m\angle B = 20°$	$m\angle B = 20°$
$m\angle C = 143°$	
A triangle is possible using this value for $m\angle A$.	No triangle is possible using this value for $m\angle A$.

Answer: one triangle

2. In $\triangle PRQ$, $p = 12$, $\sin P = .6$, and $r = 8$. Find $\sin R$.

Solution

1. Write the Law of Sines in terms of p and r.

$$\frac{p}{\sin P} = \frac{r}{\sin R}$$

2. Substitute the given values.

$$\frac{12}{.6} = \frac{8}{\sin R}$$

3. Solve the equation for $\sin R$.

$$12 \sin R = .6(8)$$

Note: Since we are asked to find $\sin R$, it is not necessary to determine $m\angle R$ or to decide whether the angle is acute or obtuse.

$$\sin R = \frac{4.8}{12}$$

$$\sin R = .4 \quad Ans.$$

EXERCISES

In 1–8, how many non-congruent triangles can be constructed with the given measures?

1. $a = 4, b = 6$, and $m\angle A = 30°$
2. $a = \sqrt{2}, b = 3$, and $m\angle A = 45°$
3. $a = 4, b = 6$, and $m\angle A = 150°$
4. $a = 4, b = 6$, and $m\angle B = 150°$
5. $a = 15, b = 12$, and $m\angle B = 45°$
6. $a = 15, b = 12$, and $m\angle A = 135°$
7. $a = 5, b = 8$, and $m\angle A = 40°$
8. $a = 9, b = 12$, and $m\angle A = 35°$

9. How many distinct triangles can be constructed if the measures of two sides are to be 35 and 70 and the measure of the angle opposite the smaller of these sides is to be 30°?

In 10–12, select the numeral preceding the expression that best completes the sentence.

10. If $a = 6, b = 8$, and $m\angle A = 30°$, the number of distinct triangles that can be constructed is:
    (1) 1   (2) 2   (3) 3   (4) 0

11. In $\triangle ABC$, if $m\angle A = 30°, a = 5$, and $b = 10$, then $\triangle ABC$:
    (1) must be an acute triangle
    (2) must be a right triangle
    (3) must be an obtuse triangle
    (4) may be an acute or an obtuse triangle

12. In $\triangle PQR$, if $m\angle P = 30°, p = 5$, and $r = 8$, then $\triangle PQR$:
    (1) must be an acute triangle
    (2) must be a right triangle
    (3) must be an obtuse triangle
    (4) may be an acute or an obtuse triangle

13. In $\triangle ABC, a = 18, b = 12$, and $\sin A = .6$. Find $\sin B$.
14. In $\triangle ABC, a = 10, c = 15$, and $\sin C = .3$. Find $\sin A$.
15. In $\triangle ABC, b = 8, c = 18$, and $\sin C = \frac{1}{4}$. Find $\sin B$.
16. In $\triangle ABC, b = 24, c = 30$, and $m\angle C = 30°$. Find $\sin B$.
17. The measures of two sides of a triangle are 34 and 22 and the measure of the angle opposite the smaller of these sides is 30°. Find to the nearest degree two possible measures of the angle opposite the larger of these sides.

18. **a.** Use the Law of Cosines to find two possible measures of c in $\triangle ABC$ when $a = 7$, $b = 8$, and m$\angle A = 60°$. **b.** Use the Law of Cosines and the results obtained in part **a** to find to the nearest degree two possible measures of $\angle B$ in $\triangle ABC$.

12-7 FORCES

If two forces act to push or pull an object in the same direction, such as two children pulling a sled together, the effect is that of a single force equal to the sum of the applied forces and in the direction of the two forces.

If two forces act to push or pull an object in opposite directions, such as two children who want to go in opposite directions with their sled, the effect is that of a single force equal to the difference between the applied forces and in the direction of the larger force.

If the two children pull in directions that form an angle other than a straight angle with each other, the effect is that of a single force acting in a direction in the interior of the angle between the applied forces.

A force is an example of a vector quantity. A *vector quantity* is a quantity that has both magnitude (size) and direction. A force is represented by a directed line segment, or *vector*, in which the length of the line segment represents the magnitude and an arrowhead indicates the direction. When two forces act at a point, the single force that has the same effect as the combination of the applied forces is called the *resultant*.

In the accompanying figure, $\overrightarrow{PQ}$ and $\overrightarrow{PS}$ are vectors that represent two forces applied to a body at point P. The resultant of these two forces is represented by the vector $\overrightarrow{PR}$. The figure illustrates the following experimental result for two forces acting on a body at an angle other than a straight angle.

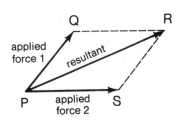

■ The vectors that represent the applied forces form two adjacent sides of a parallelogram, and the vector that represents the resultant force is the diagonal of this parallelogram.

Note: The resultant does not bisect the angle between two applied forces that are unequal in magnitude.

| MODEL PROBLEMS |

1. Two forces of 25 and 15 pounds act on a body so that the angle between them is an angle of 75°. Find the magnitude of the resultant to the nearest pound.

Solution

1. Draw and label the parallelogram *ABCD*. Let $\overrightarrow{AD}$ and $\overrightarrow{AB}$ represent the given forces, and $\overrightarrow{AC}$ represent the resultant. (*Note:* In $\square ABCD$, *AD* = *BC* = 15.)

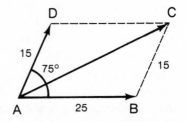

2. Consecutive angles of a parallelogram are supplementary. Find m∠*B*.

m∠*B* = 180° − m∠*DAB*
m∠*B* = 180° − 75°
m∠*B* = 105°

3. In △*ABC*, the measures of two sides and the included angle are known. Use the Law of Cosines to find the measure of the third side of the triangle (that is, the magnitude of the resultant $\overrightarrow{AC}$).

$b^2 = a^2 + c^2 - 2ac \cos B$

4. Substitute given values.

$b^2 = 15^2 + 25^2$
$\quad - 2(15)(25) \cos 105°$

5. Since ∠*B* is obtuse, cos 105° = −cos (180° − 105°) = −cos 75°.

$b^2 = 15^2 + 25^2$
$\quad - 2(15)(25)(-\cos 75°)$

6. Perform the computation.

$$b^2 = 15^2 + 25^2$$
$$- 2(15)(25)(-.2588)$$
$$b^2 = 225 + 625 + 194.1$$
$$b^2 = 1044.1$$

7. Write the positive value of b to the nearest integer.

$b = 32$ *Ans.*

2. A resultant force of 143 pounds is needed to move a heavy box. Two applied forces act at angles of $35°40'$ and $47°30'$ with the resultant. Find to the nearest pound the magnitude of the larger force.

Solution

1. Draw and label the parallelogram $ABCD$. If $\overrightarrow{AC}$ represents the resultant, then $AC = 143$, $m\angle BAC = 35°40'$, and $m\angle DAC = 47°30'$.

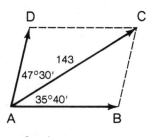

2. If parallel lines ($\overleftrightarrow{AD}$ and $\overleftrightarrow{BC}$) are cut by a transversal ($\overleftrightarrow{AC}$), the alternate interior angles ($\angle DAC$ and $\angle ACB$) are congruent. Therefore, $m\angle ACB = m\angle DAC = 47°30'$.

3. In $\triangle ABC$, the longer of sides $\overline{AB}$ and $\overline{BC}$ is opposite the larger angle. Therefore, $\overrightarrow{AB}$ represents the larger applied force.

4. To use the Law of Sines, we must know the measure of $\angle B$, the angle opposite side $\overline{AC}$, whose measure is known. Here:

$$m\angle B = 180° - (m\angle BAC + m\angle ACB)$$
$$= 180° - (35°40' + 47°30') = 96°50'$$

5. Write the Law of Sines, and substitute known values.

$$\frac{c}{\sin C} = \frac{b}{\sin B}$$

$$\frac{AB}{\sin 47°30'} = \frac{143}{\sin 96°50'}$$

6. Solve for AB.

$$AB = \frac{143 \sin 47°30'}{\sin 96°50'}$$

7. Use substitution: $\sin 96°50' = \sin (180° - 96°50') = \sin 83°10'$

$$AB = \frac{143 \sin 47°30'}{\sin 83°10'}$$

8. Compute and round the value of AB to the nearest integer.

$AB = 106$ *Ans.*

EXERCISES

1. Two forces of 12 pounds and 20 pounds act on a body with an angle of 60° between them. Find the magnitude of the resultant to the nearest pound.

2. Find to the nearest tenth the magnitude of the resultant force if two forces of 2.5 and 4.0 pounds act with an angle of 40° between them.

3. If two forces of 30 pounds and 40 pounds act on a body with an angle of 120° between them, find the magnitude of the resultant to the nearest pound.

4. Two forces act on a body so that the resultant is a force of 50 pounds. The measures of the angles between the resultant and the forces are 25° and 38°. Find the magnitude of the larger applied force to the nearest pound.

5. The measures of the angles between the resultant and two applied forces are 65° and 42°. If the magnitude of the resultant is 24 pounds, find to the nearest pound the magnitude of the smaller force.

6. Two forces act on a body. The measure of the angle between the 34-pound force and the 40-pound resultant is 60°. Find the magnitude of the other force to the nearest pound.

7. When forces of 12 pounds and 8 pounds act on a body, the magnitude of the resultant is 15 pounds. Find to the nearest degree the measure of the angle between the resultant and the larger force.

8. When forces of 20 pounds and 25 pounds act on a body, the magnitude of the resultant is 30 pounds. Find the measure of the angle between the resultant and the smaller force to the nearest degree.

9. Find to the nearest degree the measure of the angle between two applied forces of 9 and 11 pounds if the resultant is a force of 14 pounds.

10. Find to the nearest degree the measure of the angle between two applied forces of 8 and 10 pounds if the resultant is a force of 5 pounds.

11. In still air, Mr. Chafer's plane flies 400 miles per hour. When the wind is blowing to the north at 40 miles per hour, find to the nearest degree the angle at which he must set his course east of due north in order to keep to a course that is 30° east of due north. (*Hint:* First determine the measure of the angle between the course he must set and the course along which the plane actually flew.)

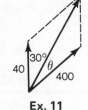

Ex. 11

12. By setting a course slightly west of due south, Mr. Chafer flew to an airport due south of where he took off. The wind was blowing at 30 miles per hour at an angle of 40° east of due south. If it took an hour to fly the 420 miles from takeoff to landing, find to the nearest 10 miles per hour the speed at which the plane would have been traveling in still air.

13. For Joe to cross the bay to a point directly east of his camp, he must set a course 6° south of due east so as to compensate for a current that flows to the northeast, 30° north of due east. If Joe's boat travels 20 mph in still water, what is the rate of the current to the nearest integer?

12-8 SOLVING TRIANGLES

If the measures of a side and any two other parts of a triangle are known, it is possible to find the measures of the other sides and angles.

Summary of Methods of Solution		
Known Measures	**Law to Be Used**	**Measure to Be Found**
Two sides and the included angle (s.a.s.).	1. First, the Law of Cosines. 2. Then, the Law of Sines (or the Law of Cosines). 3. Finally, repeat step 2 (or sum of the angle measures = 180°).	The third side. The second angle. The third angle.
Three sides (s.s.s.).	1. First, the Law of Cosines. 2. Then, the Law of Sines (or the Law of Cosines). 3. Finally, repeat step 2 (or sum of the angle measures = 180°).	The first angle. The second angle. The third angle.
Two angles and the included side (a.s.a.).	1. First, sum of the angle measures of a triangle = 180°. 2. Then, Law of Sines. 3. Finally, the Law of Sines (or the Law of Cosines).	The third angle. The second side. The third side.

Known Measures	Law to Be Used	Measure to Be Found
Two angles and the side opposite one of them (a.a.s.).	Do 1 and 2 in *either order.* 1. First, the Law of Sines. 2. Then, sum of the angle measures of a triangle = 180°. 3. Finally, the Law of Sines (or the Law of Cosines).	The second side. The third angle. The third side.
Two sides and an angle opposite one of them (s.s.a.).	1. First, the Law of Sines. If there is a solution, then: 2. Sum of the angle measures = 180°. 3. Finally, the Law of Sines (or the Law of Cosines).	The second angle (0, 1, or 2 solutions). The third angle. The third side.

The preceding summary can be used to solve any triangle, including the right triangle. However, we have seen that the right triangle can also be solved by using special ratios.

In right triangle ABC with $m\angle C = 90°$
$\sin A = \dfrac{a}{c} = \dfrac{\text{measure of leg opposite } \angle A}{\text{measure of hypotenuse}} = \dfrac{\text{opp}}{\text{hyp}}$
$\cos A = \dfrac{b}{c} = \dfrac{\text{measure of leg adjacent to } \angle A}{\text{measure of hypotenuse}} = \dfrac{\text{adj}}{\text{hyp}}$
$\tan A = \dfrac{a}{b} = \dfrac{\text{measure of leg opposite } \angle A}{\text{measure of leg adjacent to } \angle A} = \dfrac{\text{opp}}{\text{adj}}$

In problems involving the use of angle measure that can be solved using the laws of trigonometry, an angle is often described as an angle of elevation or an angle of depression.

An *angle of elevation* is an angle formed by rays that are parts of a horizontal line and a line of sight that is elevated, or raised upward, from the horizontal. If a person standing at point A looks up to point B at the top of a cliff, $\overline{BC}$, the angle of elevation of B from A is $\angle CAB$.

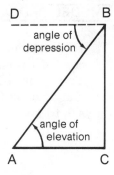

An *angle of depression* is an angle formed by rays that are parts of a horizontal line and a line of sight that is depressed, or lowered, from the horizontal. If a person standing at the top of a cliff at point B looks down to point A, the angle of depression of A from B is $\angle DBA$.

To solve a problem involving sides and angles of a triangle:
1. Read the problem carefully.
2. Draw a diagram and label it.
3. Note the relationship of the sides and angles whose measures are given to the side or angle whose measure is to be found.
4. Choose the appropriate law and write it in terms of the letters that were used to label the diagram.
5. Determine the required measure.
6. Check the reasonableness of the answer.

— KEEP IN MIND —

The laws that follow may be rewritten to include different angles of a triangle. Here, each law is written to include $\angle C$ of $\triangle ABC$.

Law of Cosines: $c^2 = a^2 + b^2 - 2ab \cos C$

Law of Cosines: $\cos C = \dfrac{a^2 + b^2 - c^2}{2ab}$

Law of Sines: $\dfrac{c}{\sin C} = \dfrac{b}{\sin B}$ or $\dfrac{c}{\sin C} = \dfrac{a}{\sin A}$

Area of a Triangle: $K = \frac{1}{2}ab \sin C$

| MODEL PROBLEMS |

1. To determine the distance between two points A and B on opposite sides of a swampy region, a surveyor chose a point C that was 350 m from point A and 400 m from point B. If the measure of $\angle ACB$ was found to be $105°40'$, find the distance across the swampy region, AB, to the nearest meter.

Solution

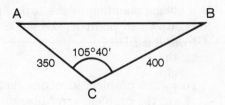

1. Draw and label a diagram.

2. Since the measures of two sides and the included angle are known, use the Law of Cosines: $c^2 = a^2 + b^2 - 2ab \cos C$.

3. Substitute the given values, and express $\cos 105°40'$ in terms of the cosine of an acute angle:

$$\cos 105°40' = -\cos(180° - 105°40') = -\cos 74°20'$$

4. Compute, and round c to the nearest positive integer.

$c^2 = a^2 + b^2 - 2ab \cos C$
$c^2 = 400^2 + 350^2 - 2(400)(300) \cos 105°40'$
$c^2 = 400^2 + 350^2 - 2(400)(300)(-\cos 74°20')$
$c^2 = 400^2 + 350^2 - 2(400)(300)(-.2700)$
$c^2 = 160,000 + 122,500 + 75,600$
$c^2 = 358,100$
$c = 598.4$
$c = 598$

Answer: 598 m

2. From a point A at the edge of a river, the measure of the angle of elevation of the top of a tree on the opposite bank is $37°$. From a point B that is 50 ft. from the edge of the river and in line with point A and the foot of the tree, the measure of the angle of elevation of the top of the tree is $22°$. Find to the nearest foot the width of the river.

Solution

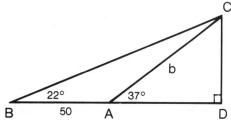

1. Draw and label a diagram. Note that *AD* = the width of the river.

2. Since ∠*BAC* and ∠*CAD* are supplementary, it is possible to determine the measures of the angles of △*ABC*. Use the Law of Sines to find the measure of $\overline{AC}$, a side common to △*ABC* and △*ACD*. (*Note:* The value of *b* is found to the nearest tenth. Values are rounded to the degree of accuracy required by the problem only in the last step of the solution.)

   m∠*BAC* = 180° - m∠*CAD*
   m∠*BAC* = 180° - 37°
   m∠*BAC* = 143°
   m∠*BCA* = 180° - (m∠*B* + m∠*BAC*)
   m∠*BCA* = 180° - (22° + 143°)
   m∠*BCA* = 180° - 165°
   m∠*BCA* = 15°

   $$\frac{b}{\sin B} = \frac{BA}{\sin \angle BCA}$$

   $$\frac{b}{\sin 22°} = \frac{50}{\sin 15°}$$

   $$b = \frac{50 \sin 22°}{\sin 15°}$$

   $$b = 72.4$$

3. Triangle *ACD* is a right triangle. Use the cosine ratio to find *AD*.

   $$\cos \angle CAD = \frac{\text{adj}}{\text{hyp}}$$

   $$\cos 37° = \frac{AD}{b}$$

   $$AD = b \cos 37°$$

4. Let *b* = 72.4, and compute the value of *AD*. (The computation may be performed by using the logarithmic equation: log *AD* = log 72.4 + log cos 37°.)

   $$AD = 72.4(.7986)$$
   $$AD = 57.8$$
   $$AD = 58$$

Answer: The width of the river is 58 ft.

| EXERCISES |

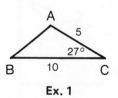

Ex. 1

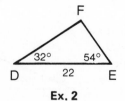

Ex. 2

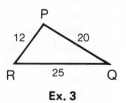

Ex. 3

1. In $\triangle ABC$, $a = 10$, $b = 5$, and m$\angle C = 27°$. a. Find c to the nearest integer. b. Find m$\angle B$ to the nearest degree. c. Find m$\angle A$ to the nearest degree.

2. In $\triangle DEF$, m$\angle D = 32°$, m$\angle E = 54°$, and $DE = 22$. a. Find m$\angle F$. b. Find EF to the nearest integer. c. Find DF to the nearest integer.

3. In $\triangle PQR$, $p = 25$, $q = 12$, and $r = 20$. Find to the nearest degree the measure of each angle of the triangle.

4. In $\triangle ABC$, $a = 31.6$, $b = 17.8$, and m$\angle A = 112°40'$.
   a. Find to the nearest 10 minutes the measure of $\angle B$.
   b. Find to the nearest 10 minutes the measure of $\angle C$.
   c. Find c to the nearest tenth.

5. In right triangle ABC, m$\angle C = 90°$, m$\angle A = 37°$, and $AC = 43$. Find AB to the nearest integer.

6. Find to the nearest foot the height of a tree that casts a 24-foot shadow when the angle of elevation of the sun is $52°$.

7. Find to the nearest meter the height that a kite has reached if 120 meters of string have been let out and the string makes an angle of $68°$ with the ground. (Assume that the string makes a straight line.)

8. Each morning the Beckebredes jog north along a straight path for .8 mi., turn at an angle of $60°$ and jog to the southwest for 1.2 mi., then turn again to the southeast and jog back to their starting point. What is the total distance that they jog each morning to the nearest tenth of a mile?

9. In order to know how much seed to buy for a triangular plot of land, Kevin needs to know the area of the plot. The lengths of the sides are 15 ft., 25 ft., and 30 ft.
   a. Find the measure of the smallest angle to the nearest degree.
   b. Using the measure of the angle found in part a, find the area of the plot of land to the nearest square foot.

10. The ladder to the top of the slide at the playground is 6 feet long and the distance down the slide is 16 feet. If the ladder and the slide make an angle of $95°$ with each other, what is the distance to the nearest foot from the bottom of the ladder to the end of the slide?

11. From points A and B that are 150 m apart, a third point C is sighted such that m$\angle CAB$ is 42°20′ and m$\angle CBA$ is 81°50′. Find the distance from A to C to the nearest meter.

12. Birdsong nature trail consists of three straight paths forming a triangle. The first two sections of the trail are .5 km and .8 km in length and make an angle of 100° with each other. Find to the nearest tenth of a kilometer the distance back to the starting point from the end of the second section.

13. A young tree is braced by two wires extending in straight lines from the same point on the trunk of the tree to points on the ground on opposite sides of the tree. The wires are fastened to stakes in the ground 32 inches apart and make angles of 35° and 40° with the ground. Find to the nearest inch the lengths of the wires.

14. The pasture gate has begun to sag so that it is now in the shape of a parallelogram with sides that measure 60 inches and 25 inches and an angle of 95°. Find to the nearest inch the length of a board needed to brace the gate along its larger diagonal.

15. a. The measures of two sides of a triangle are 12 cm and 16 cm, and the measure of the included angle is 54°. Find the measure of the third side to the nearest centimeter.
    b. Find the area of the triangle in part a to the nearest cm².

16. From points A and B that are 150 m apart, a third point C is sighted so that m$\angle CAB$ = 86°40′ and m$\angle CBA$ = 28°30′. Find to the nearest meter the distance from B to C.

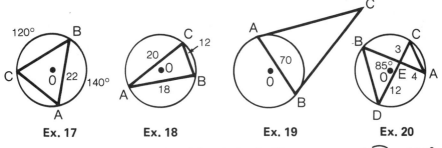

Ex. 17   Ex. 18   Ex. 19   Ex. 20

17. Triangle ABC is inscribed in circle O. The measure of $\overarc{BC}$ = 120°, the measure of $\overarc{AB}$ = 140°, and AB = 22. a. Find AC to the nearest integer. b. Find the area of $\triangle ABC$ to the nearest integer.

18. In $\triangle ABC$ inscribed in circle O, AB = 18, BC = 12, and AC = 20. Find to the nearest degree the measures of arcs $\overarc{AB}$, $\overarc{BC}$, and $\overarc{CA}$.

19. From point C, two lines are drawn tangent to circle O at points A and B. Chord $\overline{AB}$ is drawn. Points A and B separate the circle into two arcs whose measures are in the ratio 5:4, and AB = 70.
    a. Find the measure of $\angle ACB$. b. Find AC to the nearest integer.
    c. Find the area of $\triangle ABC$ to the nearest hundred.

20. Chords $\overline{AB}$ and $\overline{CD}$ intersect at E in circle O. The measure of $\angle DEB$ is $85°$, $DE = 12$, $EC = 3$, and $EA = 4$. a. Find EB. b. Find AC to the nearest integer. c. Find the area of $\triangle DEB$ to the nearest integer.

21. Quadrilateral $ABCD$ is an isosceles trapezoid with diagonal $\overline{BD}$ drawn. The measure of $\angle ADB$ is $85°$, $m\angle DBA = 20°$, $\overline{AD} \cong \overline{BC}$, and $DB = 18$. a. Find BC to the nearest integer. b. Find the area of the trapezoid $ABCD$. (*Hint:* Find the sum of the areas of $\triangle ABD$ and $\triangle DBC$.)

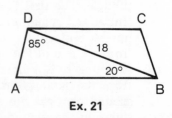

Ex. 21

22. From point A, the angle of elevation of the top of a building measures $32°10'$. From B, a point that is 125 ft. closer to the building, the angle of elevation at the top of the building measures $46°30'$. Find to the nearest foot the height of the building.

23. In rhombus $ABCD$, $m\angle DAB = 121°20'$ and the length of each side is 20.0.
    a. Find the length of diagonal $\overline{DB}$ to the nearest tenth.
    b. Find the area of $ABCD$ to the nearest integer.

24. In trapezoid $ABCD$, $\overline{AB}$ and $\overline{DC}$ are parallel bases and $\overline{BC} \perp \overline{DC}$. Find AB to the nearest integer when $m\angle CBD = 37°$, $m\angle CDA = 130°$, and $BC = 52$.

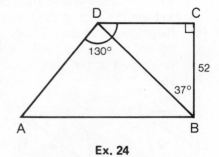

Ex. 24

12-9 REVIEW EXERCISES

1. In $\triangle ABC$, $b = 12$, $c = 9$, and $\sin B = .4$. Find $\sin C$.
2. In $\triangle ABC$, $\sin A = \frac{1}{5}$, $\sin B = \frac{3}{8}$, and $b = 15$. Find a.
3. In $\triangle ABC$, $m\angle A = 30°$, $m\angle B = 45°$, and $b = 10\sqrt{2}$. Find a.
4. In $\triangle ABC$, $a = 8$, $c = 16$, and $m\angle A = 30°$. Find $m\angle C$.
5. In $\triangle ABC$, $a = 5$, $b = 8$, and $m\angle C = 60°$. Find c.
6. In $\triangle ABC$, $a = 10$, $b = 4$, and $c = 8$. Find $\cos A$.
7. In $\triangle ABC$, $a = 6$, $c = 2$, and $m\angle B = 120°$. Find b.
8. In $\triangle ABC$, $a = 10$, $b = 10\sqrt{3}$, and $c = 10$. Find $m\angle C$.
9. In $\triangle ABC$, $b = 10$, $c = 12$, and $m\angle A = 30°$. Find the area of the triangle.

10. Find the area of $\triangle ABC$ if $a = 5$, $b = 3\sqrt{2}$, and $m\angle C = 45°$.

11. Find the area of $\triangle ABC$ if $a = 8$, $c = 12$, and $\sin B = \frac{1}{4}$.

12. In right triangle ABC, $m\angle C = 90°$, $b = 8$, and $c = 16$. Find the measure of $\angle A$.

13. In right triangle ABC, $m\angle C = 90°$, $b = 2\sqrt{2}$, and $c = 4$. Find the measure of $\angle B$.

14. In $\triangle ABC$, $a = 13.7$, $m\angle A = 15°40'$, and $m\angle B = 65°30'$. Find b to the nearest tenth.

15. Find to the nearest degree the obtuse angle of a parallelogram if the measures of two consecutive sides are 21 and 16 and the measure of the longer diagonal is 28.

16. Two men move a heavy box by pulling on two ropes attached to the box at the same point. The ropes make an angle of $24°30'$ with each other and the applied forces are 15.0 and 21.0 pounds. Find to the nearest tenth of a pound the magnitude of the resultant force.

17. Forces of 42 and 53 pounds act on a body so that the angle between the resultant and the larger force is $47°$. Find to the nearest degree two possible measures of the angle between the resultant and the smaller force.

18. From a point on level ground that is 25 feet from the foot of a flagpole, the angle of elevation of the top of the pole is $62°$. Find the height of the flagpole to the nearest foot.

19. A signal tower has two lights that are 30 feet apart, one directly above the other. From a boat, the angle of elevation of the lower light is measured to be $14°$ and the angle of elevation of the upper light is measured to be $32°$.

    a. Find to the nearest foot the distance from the boat to the lower light.

    b. The boat and the foot of the tower are in the same horizontal plane. Find to the nearest foot the distance from the boat to the foot of the tower.

20. A line $\overleftrightarrow{PC}$ is tangent to circle O at C, and a secant $\overrightarrow{PAB}$ intersects circle O at A and at B, as shown in the diagram. Chords $\overline{AC}$ and $\overline{BC}$ are drawn. The measure of the secant segment $\overline{PB}$ is 18, and the measure of its external segment $\overline{PA}$ is 8.

    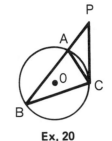

    Ex. 20

    a. Find the measure of the tangent segment $\overline{PC}$.

    b. If $m\overset{\frown}{BC}$ is $160°$, find $m\angle BAC$ and $m\angle CAP$.

    c. Find the measure of $\angle ACP$ to the nearest degree.

    d. Find the length of chord $\overline{AC}$ to the nearest integer.

21. The three sides of a triangular plot of land measure 40.0, 60.0, and 35.0 meters.
    a. Find to the nearest degree the measure of the smallest angle between two sides of the plot.
    b. Using the answer from part a, find the area of the plot to the nearest square meter.

22. Find to the nearest integer the radius of a circle if a chord of length 12 intercepts an arc of $42°$.

In 23-30, select the numeral preceding the expression that best completes the sentence or answers the question.

23. If the measures of two sides of a triangle are to be 16 and 18 and the measure of the angle opposite the shorter of these two sides is to be $30°$, then:
    (1) one acute triangle can be constructed   (2) two triangles can be constructed   (3) no triangle can be constructed   (4) one right triangle can be constructed

24. If two sides of a triangle are to be 6 and 10 and the measure of the angle opposite the side whose measure is 6 is to be $150°$, then:
    (1) one obtuse triangle can be constructed   (2) two triangles can be constructed   (3) no triangle can be constructed   (4) one acute triangle can be constructed

25. In $\triangle ABC$, $a = 6$, $c = 6\sqrt{3}$, and $m\angle A = 30°$. Then the measure of $\angle C$ is:
    (1) $30°$ only   (2) $60°$ only   (3) $120°$ only   (4) $60°$ or $120°$

26. In $\triangle ABC$, if $B = \text{Arc sin } \frac{2}{3}$, $a = 3$, and $b = 2$, then $\triangle ABC$:
    (1) must be a right triangle (2) must be an acute triangle (3) must be an obtuse triangle   (4) may be either an acute or an obtuse triangle

27. In $\triangle ABC$, $m\angle A = 40°$, $a = 10$, and $b = 5$. Triangle ABC:
    (1) must be a right triangle (2) must be an acute triangle (3) must be an obtuse triangle   (4) may be either an acute or an obtuse triangle

28. If in $\triangle ABC$, $a = 2\sqrt{2}$, $m\angle A = 30°$, and $m\angle B = 45°$, then b is:
    (1) 1   (2) 2   (3) $2\sqrt{2}$ (4) 4

29. In $\triangle ABC$, $a = 4$, $b = 6$, and $c = 8$. What is the value of $\cos C$?
    (1) $-\frac{1}{16}$ (2) $\frac{1}{16}$ (3) $-\frac{1}{4}$ (4) $\frac{1}{4}$

30. In right $\triangle ABC$, $m\angle C = 90°$, $a = 12$, and $c = 15$. What is the value of $\sin A$?
    (1) 1   (2) $\frac{4}{5}$ (3) $\frac{3}{5}$ (4) $\frac{5}{4}$

Chapter 13

Trigonometric Equations and Identities

13-1 TYPES OF EQUATIONS

In our study of mathematics, we have worked with equations in which the variable represents a real number. That real number may be a trigonometric function value. The set of replacements for the variable that makes the equation true is the solution set of the equation.

An *identity* is an equation whose solution set is the set of all possible replacements of the variable for which each member of the equation is defined.

Algebraic Identities	*Trigonometric Identities*
$3x + 4x = 7x$	$\sin^2 \theta + \cos^2 \theta = 1$
$\dfrac{5x^2}{x} = 5x \quad (x \neq 0)$	$\dfrac{\cos \theta}{\sin \theta} = \cot \theta \quad (\theta \neq k\pi$ for all integral values of k)

A *conditional equation* is an equation whose solution set does not contain all possible replacements of the variable.

Algebraic Conditional Equations	*Trigonometric Conditional Equations*
$3x + 4 = 7x$	$2 \sin \theta + 1 = 0$
$\dfrac{x}{3} = x^2$	$\dfrac{\sin \theta}{2} = \sin^2 \theta$

To prove that an equality is not an identity, we need to find only one value of the variable for which the statement is not true. For example:

$3x + 4 = 7x$ is not an identity because for $x = 2$, $3(2) + 4 \neq 7(2)$, or $10 \neq 14$

To prove that an equality is an identity, however, we cannot substitute all possible replacements of the variable to show that the statement is always true. In this chapter, we will learn how to use principles of real numbers, valid substitutions, and operations to prove that certain statements are identities.

| MODEL PROBLEM |

Which of the following is an identity?
a. $2x + 3 = 5x$ b. $\cos 2\theta = 2\cos\theta$ c. $\sin\theta + \cos\theta = 1$
d. $2x + 3x = 5x$

Solution

a. $$2x + 3 \overset{?}{=} 5x$$
If $x = 2$, then $2(2) + 3 \neq 5(2)$ Thus, $2x + 3 = 5x$ is not an identity.

b. $$\cos 2\theta \overset{?}{=} 2\cos\theta$$
If $\theta = 30°$, then $\cos 2(30°) \overset{?}{=} 2\cos 30°$
$$\cos 60° \overset{?}{=} 2\cos 30°$$

$$\frac{1}{2} \neq \overset{1}{\cancel{2}}\left(\frac{\sqrt{3}}{\underset{1}{\cancel{2}}}\right)$$ Thus, $\cos 2\theta = 2\cos\theta$ is not an identity.

c. $$\sin\theta + \cos\theta \overset{?}{=} 1$$
If $\theta = 30°$, then $\sin 30° + \cos 30° \overset{?}{=} 1$

$$\frac{1}{2} + \frac{\sqrt{3}}{2} \neq 1$$ Thus, $\sin\theta + \cos\theta = 1$ is not an identity.

d. $$2x + 3x \overset{?}{=} 5x$$
If $x = 2$, then $2(2) + 3(2) = 5(2)$ (True)

By the distributive property, $2x + 3x = (2 + 3)x = 5x$. Therefore, the equation is true, not just for $x = 2$, but for all replacements of x.

Answer: d. $2x + 3x = 5x$ is an identity.

| EXERCISES |

In 1-6, prove that the equation is not an identity by giving one value of x that makes the equation false.

1. $x^2 + 1 = (x + 1)^2$

2. $2(x + 1) = 2x + 1$

3. $\dfrac{3x + 2}{2} = 3x$

4. $\tan x + \cot x = 2$

5. $\sin 2x = 2 \sin x$

6. $\sin x = 1 - \cos x$

In 7-15, state whether the equation is an identity or a conditional equation.

7. $x + 1 = 1 + x$

8. $\sin^2 \theta - 1 = 0$

9. $2x + 1 = 3x$

10. $\tan^2 \theta + 1 = \sec^2 \theta$

11. $3(x + 2) = 3x + 6$

12. $(x + 2)^2 = x^2 + 4$

13. $\dfrac{\sin \theta}{\cos \theta} = \tan \theta$

14. $\dfrac{x^2 - 1}{x - 1} = x + 1$

15. $\sin\left(\dfrac{\pi}{2} - \theta\right) = 1 - \sin \theta$

16. Which of the following is an identity?
(1) $2^x + 3^x = 5^x$
(2) $2 \cdot 2^x = 2^{x+1}$
(3) $\sqrt{x^2 + 1} = x + 1$
(4) $\sin \theta \cdot \sec \theta = 1$

13-2 BASIC TRIGONOMETRIC IDENTITIES

In Chapter 8, the definition of the trigonometric functions established basic identities derived from the relationships among these functions. These basic identities are summarized in the following chart.

Reciprocal Identities	Quotient Identities	Pythagorean Identities
$\sec \theta = \dfrac{1}{\cos \theta}$	$\tan \theta = \dfrac{\sin \theta}{\cos \theta}$	$\cos^2 \theta + \sin^2 \theta = 1$
$\csc \theta = \dfrac{1}{\sin \theta}$	$\cot \theta = \dfrac{\cos \theta}{\sin \theta}$	$1 + \tan^2 \theta = \sec^2 \theta$
$\cot \theta = \dfrac{1}{\tan \theta}$		$\cot^2 \theta + 1 = \csc^2 \theta$

Some useful alternate forms of these eight basic identities are shown in the following chart.

Basic Identity	Alternate Forms
$\cos^2 \theta + \sin^2 \theta = 1 \longrightarrow \cos^2 \theta = 1 - \sin^2 \theta$ OR $\sin^2 \theta = 1 - \cos^2 \theta$	
$\sec \theta = \dfrac{1}{\cos \theta} \longrightarrow \sec \theta \cos \theta = 1$ OR $\cos \theta = \dfrac{1}{\sec \theta}$	
$\csc \theta = \dfrac{1}{\sin \theta} \longrightarrow \csc \theta \sin \theta = 1$ OR $\sin \theta = \dfrac{1}{\csc \theta}$	
$\cot \theta = \dfrac{1}{\tan \theta} \longrightarrow \cot \theta \tan \theta = 1$ OR $\tan \theta = \dfrac{1}{\cot \theta}$	

By using the eight basic identities or their alternate forms, we can change an expression involving trigonometric functions into an equivalent expression.

For example: $\dfrac{\sin^2 \theta}{\cos^2 \theta} = \left(\dfrac{\sin \theta}{\cos \theta} \right)^2 = \tan^2 \theta$

Furthermore, the methods we learned to add, subtract, multiply, and divide algebraic expressions can be applied to trigonometric ones.

Algebraic Examples

1. $x(x + 1) = x^2 + x$

2. $x^2 - 1 = (x - 1)(x + 1)$

3. $\dfrac{1 - \dfrac{1}{b}}{\dfrac{a}{b}} = \left(\dfrac{1 - \dfrac{1}{b}}{\dfrac{a}{b}} \right) \cdot \dfrac{b}{b}$

$= \dfrac{b - 1}{a}$

Trigonometric Examples

1. $\cos \theta (\cos \theta + 1) = \cos^2 \theta + \cos \theta$

2. $1 - \sin^2 \theta = (1 - \sin \theta)(1 + \sin \theta)$

3. $\dfrac{1 - \dfrac{1}{\cos \theta}}{\dfrac{\sin \theta}{\cos \theta}} = \left(\dfrac{1 - \dfrac{1}{\cos \theta}}{\dfrac{\sin \theta}{\cos \theta}} \right) \cdot \dfrac{\cos \theta}{\cos \theta}$

$= \dfrac{\cos \theta - 1}{\sin \theta}$

| MODEL PROBLEMS |

1. Express $\sec \theta \cot \theta$ as a single function.

Solution

1. Use basic identities to express $\sec \theta$ and $\cot \theta$ in terms of $\sin \theta$ and $\cos \theta$.

$\sec \theta \cot \theta = \dfrac{1}{\cos \theta} \cdot \dfrac{\cos \theta}{\sin \theta}$

2. Divide numerator and denominator by the common factor, $\cos \theta$.

3. Use a basic identity to express $\dfrac{1}{\sin \theta}$ as $\csc \theta$.

$$\sec \theta \cot \theta = \frac{1}{\cancel{\cos \theta}} \cdot \frac{\overset{1}{\cancel{\cos \theta}}}{\sin \theta}$$

$$= \frac{1}{\sin \theta}$$

$$= \csc \theta \quad Ans.$$

2. Show that $(1 - \cos \theta)(1 + \cos \theta) = \sin^2 \theta$.

Solution

1. Find the product of the binomials.

2. Using $\cos^2 \theta + \sin^2 \theta = 1$, *either* substitute $\cos^2 \theta + \sin^2 \theta$ for 1 and simplify (method 1) *or* use an alternate form of the identity, namely, $\sin^2 \theta = 1 - \cos^2 \theta$ (method 2).

Method 1

$$(1 - \cos \theta)(1 + \cos \theta) = 1 - \cos^2 \theta$$
$$= (\cos^2 \theta + \sin^2 \theta) - \cos^2 \theta$$
$$= \sin^2 \theta$$

Method 2

$$(1 - \cos \theta)(1 + \cos \theta) = 1 - \cos^2 \theta$$
$$= \sin^2 \theta$$

EXERCISES

In 1–21, write the given expression as a monomial containing a single function or a constant.

1. $1 - \cos^2 \theta$
2. $1 - \sin^2 \theta$
3. $\tan^2 \theta + 1$
4. $\cot^2 \theta + 1$
5. $\sin \theta \cot \theta$
6. $\cos \theta \tan \theta$
7. $\sin \theta \sec \theta$
8. $\cos \theta \csc \theta$
9. $\sec \theta \cot \theta \sin \theta$
10. $\cos \theta \tan \theta \csc \theta$
11. $\sin \theta \cot \theta \tan \theta$
12. $\tan \theta \cot \theta \cos \theta$
13. $\sec \theta \sin \theta \csc \theta$
14. $\sec \theta \csc \theta \cos \theta$
15. $\csc \theta (1 - \cos^2 \theta)$
16. $\sec \theta (1 - \sin^2 \theta)$
17. $\sin \theta (\cot^2 \theta + 1)$
18. $\cos \theta (\tan^2 \theta + 1)$
19. $\sec \theta \cos \theta - \cos^2 \theta$
20. $\csc \theta \sin \theta - \sin^2 \theta$
21. $\tan \theta (\cot \theta + \tan \theta)$

In 22–33, write the expression as a single fraction.

22. $\dfrac{a}{c} + \dfrac{b}{c}$

23. $\dfrac{1}{\cos \theta} + \dfrac{\sin \theta}{\cos \theta}$

24. $\dfrac{1 + a}{b} - \dfrac{a}{b}$

25. $\dfrac{1 + \cos \theta}{\sin \theta} - \dfrac{\cos \theta}{\sin \theta}$

26. $1 - \dfrac{a}{b}$

27. $1 - \dfrac{\sin \theta}{\cos \theta}$

28. $\dfrac{a}{1 - a^2} - \dfrac{1}{1 - a}$

29. $\dfrac{\cos \theta}{1 - \cos^2 \theta} - \dfrac{1}{1 - \cos \theta}$

30. $\dfrac{a}{b} - \dfrac{b}{a}$

31. $\dfrac{\sin \theta}{\cos \theta} - \dfrac{\cos \theta}{\sin \theta}$

32. $\dfrac{1}{a} + \dfrac{1}{a^2}$

33. $\dfrac{1}{\sin \theta} + \dfrac{1}{\sin^2 \theta}$

In 34–39, simplify the complex fraction.

34. $\dfrac{\dfrac{1}{\cos \theta}}{1 - \dfrac{1}{\cos \theta}}$

35. $\dfrac{\dfrac{1}{\cos^2 \theta} - 1}{\dfrac{\sin^2 \theta}{\cos^2 \theta}}$

36. $\dfrac{\dfrac{\sin \theta}{\cos \theta} + \dfrac{\cos \theta}{\sin \theta}}{\dfrac{1}{\sin \theta}}$

37. $\dfrac{\dfrac{1}{\sin \theta} - \dfrac{1}{\cos \theta}}{\dfrac{\cos \theta}{\sin \theta} - \dfrac{\sin \theta}{\cos \theta}}$

38. $\dfrac{\dfrac{1}{\sin \theta \cos \theta}}{\dfrac{1}{\sin \theta} + \dfrac{1}{\cos \theta}}$

39. $\dfrac{1 - \dfrac{1}{\cos \theta}}{1 - \dfrac{1}{\cos^2 \theta}}$

In 40–42, select the numeral preceding the expression that best completes the sentence.

40. The expression $\cot \theta \sec \theta$ is equivalent to:
    (1) $\sin \theta$ (2) $\cos \theta$ (3) $\tan \theta$ (4) $\csc \theta$

41. The expression $(\cos^2 \theta - 1)$ is equivalent to:
    (1) $\sin^2 \theta$ (2) $\cos^2 \theta$ (3) $-\sin^2 \theta$ (4) $-\cos^2 \theta$

42. The expression $\sin^2 \theta - \cos^2 \theta$ is equivalent to:
    (1) 1 (2) $(\sin \theta - \cos \theta)^2$ (3) $(1 - \cos^2 \theta)(\sin^2 \theta + 1)$
    (4) $(\sin \theta + \cos \theta)(\sin \theta - \cos \theta)$

13-3 PROVING TRIGONOMETRIC IDENTITIES

To prove that an equality is an identity, we need to show that both members of the equality can be written in identical form. To do this, we use valid substitutions and operations that allow us either:

1. to transform the more complicated member into the form of the simpler member

OR

2. to transform each member separately into some common form.

Simple Substitution

Some identities can be proved by making a substitution based on one or more of the eight basic identities.

☐ EXAMPLE 1: Prove that $\tan^2 \theta + \sin^2 \theta + \cos^2 \theta = \sec^2 \theta$.

1. Simplify the more complicated left-hand member. $\tan^2 \theta + \sin^2 \theta + \cos^2 \theta = \sec^2 \theta$

2. Replace $\cos^2 \theta + \sin^2 \theta$ by 1. $\tan^2 \theta + 1 = \sec^2 \theta$

3. Replace $\tan^2 \theta + 1$ by $\sec^2 \theta$. $\sec^2 \theta = \sec^2 \theta$

Since a proof should proceed from a statement that is known to be true to the one that is to be proved, the steps of this proof should really progress in reverse order to prove that the given statement is true. However, the form in which the proof is written is the generally accepted form. Note that this identity is undefined for the odd multiples of $\frac{\pi}{2}$.

Factors and Products

We can change the form of a member of an identity by performing an indicated multiplication or by factoring.

☐ EXAMPLE 2: Prove the identity $\cos \theta (\sec \theta - \cos \theta) = \sin^2 \theta$.

1. Simplify the more complicated left-hand member by first performing the indicated multiplication.

$$\cos \theta (\sec \theta - \cos \theta) = \sin^2 \theta$$
$$\cos \theta \sec \theta - \cos^2 \theta =$$

2. Use the reciprocal identity and simplify.

$$\cos \theta \left(\frac{1}{\cos \theta}\right) - \cos^2 \theta =$$
$$1 - \cos^2 \theta =$$

3. Use the alternate form of the Pythagorean identity.

$$\sin^2 \theta = \sin^2 \theta$$

Factors often enable us to simplify fractions.

☐ EXAMPLE 3: Prove the identity $\dfrac{\sin^2 \theta}{1 - \cos \theta} = 1 + \cos \theta$.

1. Replace $\sin^2 \theta$ in the more complicated left-hand member by an expression in terms of $\cos \theta$.

$$\dfrac{\sin^2 \theta}{1 - \cos \theta} = 1 + \cos \theta$$

$$\dfrac{1 - \cos^2 \theta}{1 - \cos \theta} =$$

2. Factor the numerator of the fraction.

$$\dfrac{(1 + \cos \theta)(1 - \cos \theta)}{1 - \cos \theta} =$$

3. Cancel common factors in the numerator and the denominator.

$$\dfrac{(1 + \cos \theta)\overset{1}{\cancel{(1 - \cos \theta)}}}{\underset{1}{\cancel{1 - \cos \theta}}} =$$

$$1 + \cos \theta = 1 + \cos \theta$$

Note: It is incorrect to add terms to both members or to multiply or divide both members in order to prove the identity. For example, it is *not* correct to use the rule that the product of the means equals the product of the extremes in the preceding example, writing $\sin^2 \theta = (1 + \cos \theta)(1 - \cos \theta)$, or $\sin^2 \theta = 1 - \cos^2 \theta$, or $\sin^2 \theta = \sin^2 \theta$. In doing this, we are changing the equation and are no longer working with the given equality. We work with an identity as we have learned to work with a *check* rather than as we have learned to solve an equation.

Adding Fractions

The quotient and reciprocal identities often transform identities into forms that include fractions. To add fractions, we need a common denominator.

☐ EXAMPLE 4: Prove the identity $1 + \sec \theta = \dfrac{\cos \theta + 1}{\cos \theta}$.

Each member can be transformed to prove the identity in either of two ways.

Method 1	*Method 2*

Method 1

$$1 + \sec \theta = \dfrac{\cos \theta + 1}{\cos \theta}$$

$$1 + \dfrac{1}{\cos \theta} =$$

$$\dfrac{\cos \theta}{\cos \theta} + \dfrac{1}{\cos \theta} =$$

$$\dfrac{\cos \theta + 1}{\cos \theta} = \dfrac{\cos \theta + 1}{\cos \theta}$$

Method 2

$$1 + \sec \theta = \dfrac{\cos \theta + 1}{\cos \theta}$$

$$= \dfrac{\cos \theta}{\cos \theta} + \dfrac{1}{\cos \theta}$$

$$1 + \sec \theta = 1 + \sec \theta$$

☐ EXAMPLE 5: Prove the identity $\tan \theta + \cot \theta = \sec \theta \csc \theta$.

In proving this identity, we will work with each member separately in order to express each one as a common form.

$$\tan \theta + \cot \theta \qquad\qquad = \qquad\qquad \sec \theta \csc \theta$$

$$\frac{\sin \theta}{\cos \theta} + \frac{\cos \theta}{\sin \theta} \qquad\qquad\qquad \frac{1}{\cos \theta} \cdot \frac{1}{\sin \theta}$$

$$\frac{\sin \theta}{\sin \theta} \cdot \frac{\sin \theta}{\cos \theta} + \frac{\cos \theta}{\sin \theta} \cdot \frac{\cos \theta}{\cos \theta} \qquad\qquad \frac{1}{\cos \theta \sin \theta}$$

$$\frac{\sin^2 \theta}{\sin \theta \cos \theta} + \frac{\cos^2 \theta}{\sin \theta \cos \theta}$$

$$\frac{\sin^2 \theta + \cos^2 \theta}{\sin \theta \cos \theta}$$

$$\frac{1}{\sin \theta \cos \theta} \qquad = \qquad \frac{1}{\sin \theta \cos \theta}$$

Note: By showing that each member of the equality reduces to a common form, we have proved the identity. If we wish to show that the left-hand member of the equality can be transformed into the form given on the right, however, we can now simply reverse the steps taken on the right and place them at the left. In other words, we add these steps to the identity just shown:

$$\frac{1}{\sin \theta \cos \theta}$$

$$\frac{1}{\cos \theta} \cdot \frac{1}{\sin \theta}$$

$$\sec \theta \csc \theta \qquad = \qquad \sec \theta \csc \theta$$

Complex Fractions

A complex fraction can be simplified in various ways, as illustrated in the following example.

☐ EXAMPLE 6: Prove the identity $\dfrac{\sec \theta}{\csc \theta} = \tan \theta$.

Method 1

Multiply the complex fraction by a form of the identity element 1. In this example, use the form:

$$\frac{\sin \theta}{\sin \theta}, \text{ or } \frac{\sin \theta \cos \theta}{\sin \theta \cos \theta}$$

Method 2

If the denominator of the complex fraction is a monomial, express the fraction as the product of the numerator times the reciprocal of the denominator.

Solution for Method 1	*Solution for Method 2*

Solution for Method 1

$$\frac{\sec \theta}{\csc \theta} = \tan \theta$$

$$\frac{\dfrac{1}{\cos \theta}}{\dfrac{1}{\sin \theta}} =$$

$$\frac{\dfrac{1}{\cos \theta}}{\dfrac{1}{\sin \theta}} \cdot \frac{\sin \theta \cos \theta}{\sin \theta \cos \theta} =$$

$$\frac{\sin \theta}{\cos \theta} =$$

$$\tan \theta = \tan \theta$$

Solution for Method 2

$$\frac{\sec \theta}{\csc \theta} = \tan \theta$$

$$\sec \theta \cdot \frac{1}{\csc \theta} =$$

$$\frac{1}{\cos \theta} \cdot \sin \theta =$$

$$\frac{\sin \theta}{\cos \theta} =$$

$$\tan \theta = \tan \theta$$

There are no rules that apply to proving all identities, but the following basic principles can aid in finding a proof.

1. Start with the more complicated member, and write it more simply. If both members are complicated, work on one member of the equality until no further step is evident. Then, work on the other member of the equation.

2. Transform different functions into the same function. It is often useful to express each member in terms of sines and cosines.

3. Simplify complex fractions. Look for a common factor in the numerator and denominator of a fraction in order to reduce the fraction to lowest terms.

| **MODEL PROBLEM** |

a. Prove the identity $\dfrac{\sin \theta}{1 + \cos \theta} = \csc \theta - \cot \theta$.

b. For what values of θ is the identity in part a undefined?

Solution

a. 1. Express the right-hand member in terms of $\sin \theta$ and $\cos \theta$.

$$\frac{\sin \theta}{1 + \cos \theta} = \csc \theta - \cot \theta$$

$$= \frac{1}{\sin \theta} - \frac{\cos \theta}{\sin \theta}$$

2. Add the fractions.

$$\frac{\sin \theta}{1 + \cos \theta} = \frac{1 - \cos \theta}{\sin \theta}$$

3. The factor $(1 - \cos \theta)$ in the numerator and $(1 + \cos \theta)$ in the denominator of the form we want to reach suggests that $(1 - \cos \theta)(1 + \cos \theta) = 1 - \cos^2 \theta = \sin^2 \theta$ will be useful. Therefore, multiply by $\frac{1 + \cos \theta}{1 + \cos \theta}$.

$$= \frac{1 - \cos \theta}{\sin \theta} \cdot \frac{1 + \cos \theta}{1 + \cos \theta}$$

$$= \frac{1 - \cos^2 \theta}{\sin \theta \, (1 + \cos \theta)}$$

4. Then, substitute $\sin^2 \theta$ for $1 - \cos^2 \theta$, and simplify.

$$= \frac{\sin^2 \theta}{\sin \theta \, (1 + \cos \theta)}$$

$$\frac{\sin \theta}{1 + \cos \theta} = \frac{\sin \theta}{1 + \cos \theta}$$

b. In the left-hand member, $1 + \cos \theta = 0$ when $\cos \theta = -1$. Therefore, the left-hand member is undefined for $\theta = \pi + 2k\pi$, where k is any integer. In the right-hand member, $\csc \theta = \frac{1}{\sin \theta}$ and $\cot \theta = \frac{\cos \theta}{\sin \theta}$ are undefined when $\sin \theta = 0$, that is, when $\theta = k\pi$, where k is any integer.

Answer: The identity is undefined for all multiples of π.

| **EXERCISES** |

In 1–36, prove that the equation is an identity.

1. $\sin \theta \cot \theta = \cos \theta$
2. $\cos \theta \tan \theta = \sin \theta$
3. $\sec \theta \cot \theta = \csc \theta$
4. $\csc \theta \tan \theta = \sec \theta$
5. $\cos \theta \, (\sec \theta - \cos \theta) = \sin^2 \theta$
6. $\sin \theta \, (\csc \theta - \sin \theta) = \cos^2 \theta$
7. $\sec \theta \, (\sec \theta - \cos \theta) = \tan^2 \theta$
8. $\csc \theta \, (\csc \theta - \sin \theta) = \cot^2 \theta$
9. $\tan^2 \theta \, (1 + \cot^2 \theta) = \sec^2 \theta$
10. $\sin^2 \theta \, (\csc^2 \theta - 1) = \cos^2 \theta$
11. $\cos^2 \theta \, (\sec^2 \theta - 1) = \sin^2 \theta$
12. $\sec^2 \theta \, (1 - \cos^2 \theta) = \tan^2 \theta$
13. $\sec \theta \sin \theta + \csc \theta \cos \theta = \tan \theta + \cot \theta$
14. $(\sin \theta + \cos \theta)^2 = 1 + 2 \sin \theta \cos \theta$
15. $\dfrac{\sin \theta - \cos \theta}{\sin \theta} = 1 - \cot \theta$
16. $\dfrac{\sin \theta - \cos \theta}{\cos \theta} = \tan \theta - 1$
17. $\dfrac{\cos^2 \theta - \cos \theta}{\cos \theta} = \cos \theta - 1$
18. $\dfrac{\cos \theta - 1}{\cos^2 \theta} = \sec \theta - \sec^2 \theta$

19. $\dfrac{\sec \theta}{\tan \theta} = \csc \theta$

20. $\dfrac{\csc \theta}{\cot \theta} = \sec \theta$

21. $\dfrac{\tan \theta}{\cot \theta} + 1 = \sec^2 \theta$

22. $\dfrac{\cot \theta}{\tan \theta} + 1 = \csc^2 \theta$

23. $\dfrac{\sin \theta}{\cot \theta} + \cos \theta = \sec \theta$

24. $\dfrac{\sin^2 \theta}{1 + \cos \theta} = 1 - \cos \theta$

25. $\dfrac{\sin \theta}{1 + \cos \theta} = \dfrac{1 - \cos \theta}{\sin \theta}$

26. $\dfrac{1 - \sec^2 \theta}{1 + \sec \theta} = \dfrac{\cos \theta - 1}{\cos \theta}$

27. $\dfrac{1 - \csc^2 \theta}{1 - \csc \theta} = \dfrac{1 + \sin \theta}{\sin \theta}$

28. $\dfrac{1 + \sec \theta}{1 - \sec^2 \theta} = \dfrac{\cos \theta}{\cos \theta - 1}$

29. $\sin^4 \theta - \cos^4 \theta = \sin^2 \theta - \cos^2 \theta$

30. $\sec^4 \theta - \tan^4 \theta = \sec^2 \theta + \tan^2 \theta$

31. $\dfrac{1 + 2 \sin \theta \cos \theta}{(\sin \theta + \cos \theta)^2} = 1$

32. $\dfrac{(1 + \sin \theta)^2}{\cos^2 \theta} = \dfrac{1 + \sin \theta}{1 - \sin \theta}$

33. $\sin^2 \theta + \sin^2 \theta \tan^2 \theta = \tan^2 \theta$

34. $\sec^2 \theta + \csc^2 \theta = \sec^2 \theta \csc^2 \theta$

35. $\dfrac{1}{1 + \cos \theta} + \dfrac{1}{1 - \cos \theta} = 2 \csc^2 \theta$

36. $\dfrac{1}{1 + \sin \theta} + \dfrac{1}{1 - \sin \theta} = 2 \sec^2 \theta$

37. a. Prove the identity $\dfrac{(1 - \cos x)(1 + \cos x)}{\sin x} = \sin x$

    b. For what values of x is the identity proved in part a undefined?

38. a. Prove the identity $(1 - \cos \theta)(1 + \sec \theta) = \sin \theta \tan \theta$.

    b. For what values of θ is the identity proved in part a undefined?

13-4 COSINE OF THE DIFFERENCE OF TWO ANGLE MEASURES

Which of the following equalities, if any, is an identity? Let us first test each equality by selecting replacements for the variables A and B.

☐ EXAMPLE 1: $\cos (A - B) \overset{?}{=} \cos A - \cos B$

$$\cos (60° - 0°) \overset{?}{=} \cos 60° - \cos 0°$$

$$\cos 60° \overset{?}{=} \cos 60° - \cos 0°$$

$$\dfrac{1}{2} \overset{?}{=} \dfrac{1}{2} - 1$$

$$\dfrac{1}{2} \neq -\dfrac{1}{2}$$

(Not an identity)

☐ EXAMPLE 2: $\cos (A - B) \overset{?}{=} \cos A \cos B + \sin A \sin B$

$\cos (60° - 0°) \overset{?}{=} \cos 60° \cos 0° + \sin 60° \sin 0°$

$\cos 60° \overset{?}{=} \cos 60° \cos 0° + \sin 60° \sin 0°$

$$\frac{1}{2} \overset{?}{=} \frac{1}{2}(1) + \frac{\sqrt{3}}{2}(0)$$

$$\frac{1}{2} \overset{?}{=} \frac{1}{2} + 0$$

$$\frac{1}{2} = \frac{1}{2} \text{ (True)}$$

It is clear that the equality in example 1 is not an identity because there are replacements for the variables for which the equality is not true. In example 2, however, we are still not certain if the equality is an identity. While the replacements chosen result in a true statement, other replacements for the variables might show us that the equality is not true.

Let us apply some mathematical principles and definitions to prove that the following equality is an identity:

Proof: $\cos (A - B) = \cos A \cos B + \sin A \sin B$

1. Consider a unit circle whose center is at the origin. Let A and B be the measures of two angles in standard position whose terminal rays intersect the unit circle at points P and Q, respectively. Therefore, the coordinates of point P are $(\cos A, \sin A)$, and the coordinates of point Q are $(\cos B, \sin B)$. The measure of $\angle QOP$ is $(A - B)$.

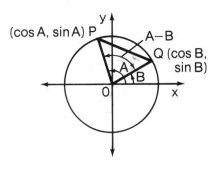

2. Express $(PQ)^2$, the square of the distance from P ($\cos A$, $\sin A$) to Q ($\cos B$, $\sin B$), by using the *distance formula*. After substituting coordinate values in the formula, square each binomial and group terms as follows:

$d^2 = (x_2 - x_1)^2 + (y_2 - y_1)^2$

$(PQ)^2 = (\cos A - \cos B)^2 + (\sin A - \sin B)^2$

$(PQ)^2 = \cos^2 A - 2 \cos A \cos B + \cos^2 B + \sin^2 A - 2 \sin A \sin B + \sin^2 B$

$(PQ)^2 = (\cos^2 A + \sin^2 A) + (\cos^2 B + \sin^2 B) - 2 \cos A \cos B - 2 \sin A \sin B$

$(PQ)^2 = 1 + 1 - 2 (\cos A \cos B + \sin A \sin B)$

3. Consider $\overline{PQ}$ as the side of $\triangle POQ$ that is opposite the angle whose measure is $(A - B)$. Since sides $\overline{OP}$ and $\overline{OQ}$ of $\triangle POQ$ are also radii of a unit circle, $OP = 1$ and $OQ = 1$. Express $(PQ)^2$ by using the *Law of Cosines*.

$$(PQ)^2 = (OP)^2 + (OQ)^2 - 2(OP)(OQ) \cos (A - B)$$
$$(PQ)^2 = 1^2 + 1^2 - 2(1)(1) \cos (A - B)$$
$$(PQ)^2 = 1 + 1 - 2 \cos (A - B)$$

4. From steps 2 and 3, there are two expressions for $(PQ)^2$, each in terms of the function values of A and B. These two expressions are equal.

$$(PQ)^2 = (PQ)^2$$
$$1 + 1 - 2 \cos (A - B) = 1 + 1 - 2 (\cos A \cos B + \sin A \sin B)$$
$$-2 \cos (A - B) = -2 (\cos A \cos B + \sin A \sin B)$$
$$\cos (A - B) = \cos A \cos B + \sin A \sin B$$

This equation, which expresses the cosine of the difference of two angle measures, $(A - B)$, in terms of the sines and cosines of the individual angle measures, A and B, is an identity since it is true for all replacements of the variables.

We can illustrate the truth of this identity by using familiar values of A and B.

☐ EXAMPLE 1: Show that $\cos 30° = \dfrac{\sqrt{3}}{2}$ by finding $\cos (A - B)$ when $A = 90°$ and $B = 60°$.

Solution

1. Write the identity. $\cos (A - B) = \cos A \cos B + \sin A \sin B$

2. Substitute the given values. $\cos (90° - 60°) = \cos 90° \cos 60° + \sin 90° \sin 60°$

3. Substitute function values in the right-hand member, and simplify both members.

$$\cos 30° = 0 \left(\frac{1}{2}\right) + 1 \left(\frac{\sqrt{3}}{2}\right)$$
$$\cos 30° = 0 + \frac{\sqrt{3}}{2}$$
$$\cos 30° = \frac{\sqrt{3}}{2}$$

☐ EXAMPLE 2: Find the exact value of cos 15° by finding cos $(A - B)$ when $A = 45°$ and $B = 30°$.

Solution

$$\cos (A - B) = \cos A \cos B + \sin A \sin B$$
$$\cos (45° - 30°) = \cos 45° \cos 30° + \sin 45° \sin 30°$$
$$\cos 15° = \frac{\sqrt{2}}{2} \cdot \frac{\sqrt{3}}{2} + \frac{\sqrt{2}}{2} \cdot \frac{1}{2}$$
$$\cos 15° = \frac{\sqrt{6}}{4} + \frac{\sqrt{2}}{4} = \frac{\sqrt{6} + \sqrt{2}}{4} \quad Ans.$$

If we substitute approximate values for $\sqrt{6}$ and $\sqrt{2}$, we find:

$$\cos 15° = \frac{2.4495 + 1.4142}{4} = \frac{3.8637}{4} = .9659$$

This is the approximate value that is given in the table of trigonometric function values.

| MODEL PROBLEMS |

1. Use the identity for the cosine of the difference of two angle measures to prove that $\cos (180° - x) = -\cos x$.

Solution

1. Write the identity. $\cos (A - B) = \cos A \cos B + \sin A \sin B$

2. Substitute 180° for A $\cos (180° - x) = \cos 180° \cos x$
   and x for B. $+ \sin 180° \sin x$

3. Substitute the values of $\cos (180° - x) = -1 \cdot \cos x + 0 \cdot \sin x$
   sin 180° and cos 180°, $\cos (180° - x) = -\cos x + 0$
   and simplify. $\cos (180° - x) = -\cos x$

2. If sin A is $\frac{3}{5}$ and $\angle A$ is in quadrant II and cos $B = \frac{5}{13}$ and $\angle B$ is in quadrant I, find cos $(A - B)$.

Solution

1. In order to use the identity for cos $(A - B)$, we must know the sine and cosine values of both A and B. We will use basic identities to find the required values.

$\cos^2 A = 1 - \sin^2 A$	$\sin^2 B = 1 - \cos^2 B$
$\cos^2 A = 1 - (\frac{3}{5})^2$	$\sin^2 B = 1 - (\frac{5}{13})^2$
$\cos^2 A = 1 - \frac{9}{25}$	$\sin^2 B = 1 - \frac{25}{169}$
$\cos^2 A = \frac{16}{25}$	$\sin^2 B = \frac{144}{169}$
$\cos A = -\frac{4}{5}$	$\sin B = \frac{12}{13}$

Since $\angle A$ is in quadrant II, $\cos A$ is negative. | Since $\angle B$ is in quadrant I, $\sin B$ is positive.

2. Write the identity for $\cos (A - B)$. $\quad \cos (A - B) = \cos A \cos B + \sin A \sin B$

3. Substitute known values, and simplify.
$\cos (A - B) = -\frac{4}{5} \cdot \frac{5}{13} + \frac{3}{5} \cdot \frac{12}{13}$
$\cos (A - B) = -\frac{20}{65} + \frac{36}{65}$
$\cos (A - B) = \frac{16}{65}$ *Ans.*

Note: There is an *alternate method for step 1* of the solution just given. To find the values of $\cos A$ and $\sin B$, we may use: right triangles in a unit circle and the Pythagorean Theorem (see Fig. 1), or dilations of these right triangles and the Pythagorean Theorem (see Fig. 2). In quadrant II, x is negative.

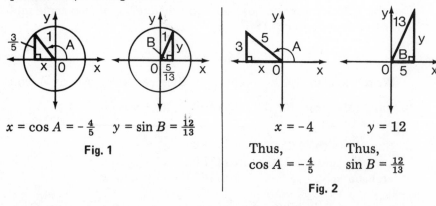

$x = \cos A = -\frac{4}{5}$ $y = \sin B = \frac{12}{13}$ $x = -4$ $y = 12$

Fig. 1   Thus,   Thus,

                                         $\cos A = -\frac{4}{5}$ $\sin B = \frac{12}{13}$

Fig. 2

Then, using the values found, perform steps 2 and 3 of the solution.

EXERCISES

1. Complete the identity: $\cos (x - y) = $ _____
2. Complete the identity: $\cos (\theta - \phi) = $ _____
3. Find $\cos 105°$ by using $\cos (135° - 30°)$.
4. Find $\cos 75°$ by using $\cos (120° - 45°)$.
5. Find $\cos 165°$ by using $\cos (225° - 60°)$.
6. Find $\cos (-15°)$ by using $\cos (30° - 45°)$.

In 7–10, use the identity for cos $(A - B)$ to prove the given statement.

7. cos $(270° - x) = -\sin x$ **8.** cos $(90° - x) = \sin x$

9. cos $(360° - x) = \cos x$

10. cos $(45° - x) = \dfrac{\sqrt{2}}{2}(\cos x + \sin x)$

11. If sin $A = \frac{4}{5}$, sin $B = \frac{4}{5}$, $\angle A$ is in quadrant II, and $\angle B$ is in quadrant I, find the value of cos $(A - B)$.

12. If sin $A = -\frac{12}{13}$, $\angle A$ is in quadrant III, sin $B = \frac{4}{5}$, and $\angle B$ is in quadrant II, find: **a.** cos $(A - B)$ **b.** cos $(B - A)$

13. If sin $A = -\frac{1}{2}$, cos $B = -\frac{1}{4}$, and both $\angle A$ and $\angle B$ are in quadrant III, find: **a.** cos $(A - B)$ **b.** cos $(B - A)$

14. If x is the measure of a positive acute angle and cos $x = \frac{3}{5}$, find the value of cos $(180° - x)$.

15. If x is the measure of a positive acute angle and sin $x = 0.8$, find the value of cos $\left(\dfrac{\pi}{2} - x\right)$.

In 16–19, select the numeral preceding the expression that best completes the sentence.

16. The expression cos $30°$ cos $12°$ + sin $30°$ sin $12°$ is equivalent to:
(1) cos $42°$ (2) cos $18°$
(3) cos $42°$ + sin $42°$ (4) cos^2 $42°$ + sin^2 $42°$

17. The expression cos $(\pi - x)$ is equivalent to:
(1) sin x (2) $-\sin x$ (3) cos x (4) $-\cos x$

18. If cos $(A - 30°) = \frac{1}{2}$, then the measure of angle A may be:
(1) $30°$ (2) $60°$ (3) $90°$ (4) $120°$

19. The value of cos $45°$ cos $15°$ + sin $45°$ sin $15°$ is:

(1) 1 (2) $\dfrac{1}{2}$ (3) $\dfrac{\sqrt{3}}{2}$ (4) 0

13-5 COSINE OF THE SUM OF TWO ANGLE MEASURES

The identity cos $(A - B) = \cos A \cos B + \sin A \sin B$ makes it possible for us to derive many other useful identities, some of which are already familiar. Since an identity is true for all replacements of the variables for which the terms are defined, we can assign special values to A or B, or both.

Proof: $\cos (90° - B) = \sin B$

Use the identity: $\cos (A - B) = \cos A \cos B + \sin A \sin B$

Let $A = 90°$. Then: $\cos (90° - B) = \cos 90° \cos B + \sin 90° \sin B$
$$\cos (90° - B) = 0 \cdot \cos B + 1 \cdot \sin B$$
$$\cos (90° - B) = 0 + \sin B$$
$$\cos (90° - B) = \sin B$$

Proof: $\cos A = \sin (90° - A)$

Use the identity: $\cos (90° - B) = \sin B$

Let $B = 90° - A$. Then: $\cos (90° - (90° - A)) = \sin (90° - A)$
$$\cos (90° - 90° + A) = \sin (90° - A)$$
$$\cos A = \sin (90° - A)$$

In Chapter 8, these two identities were derived when A and B were the measures of acute angles. These identities were the basis of the definition of cofunctions. The statements are true for all replacements of the variables, however, not just for values that are the measures of acute angles.

For example: $\cos (90° - B) = \sin B$

Let $B = 120°$. Then: $\cos (90° - 120°) = \sin 120°$
$$\cos (-30°) = \sin 120°$$
$$\frac{\sqrt{3}}{2} = \frac{\sqrt{3}}{2}$$

Proof: $\cos (-\theta) = \cos \theta$

Use the identity: $\cos (A - B) = \cos A \cos B + \sin A \sin B$

Let $A = 0°$ and $B = \theta$. Then: $\cos (0° - \theta) = \cos 0° \cos \theta + \sin 0° \sin \theta$
$$\cos (-\theta) = 1 \cdot \cos \theta + 0 \cdot \sin \theta$$
$$\cos (-\theta) = \cos \theta + 0$$
$$\cos (-\theta) = \cos \theta$$

Proof: $\sin (-\theta) = -\sin \theta$

Use the identity: $\sin B = \cos (90° - B)$

Let $B = -\theta$. Then: $\sin (-\theta) = \cos (90° - (-\theta))$
$$\sin (-\theta) = \cos (90° + \theta)$$
$$\sin (-\theta) = \cos (\theta + 90°)$$
$$\sin (-\theta) = \cos (\theta - (-90°))$$

$$\sin(-\theta) = \cos\theta \cos(-90°) + \sin\theta \sin(-90°)$$
$$\sin(-\theta) = \cos\theta \cdot (0) + \sin\theta \cdot (-1)$$
$$\sin(-\theta) = 0 - \sin\theta$$
$$\sin(-\theta) = -\sin\theta$$

We can see the results of the last two derivations in the following diagrams, each of which consists of a unit circle.

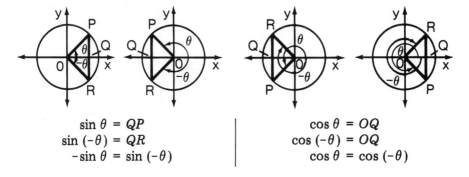

$$\sin\theta = QP$$
$$\sin(-\theta) = QR$$
$$-\sin\theta = \sin(-\theta)$$

$$\cos\theta = OQ$$
$$\cos(-\theta) = OQ$$
$$\cos\theta = \cos(-\theta)$$

We can use the identity for the cosine of the difference of two angle measures to find an identity for the sum of two angle measures.

Proof: $\cos(A + B) = \cos A \cos B - \sin A \sin B$

$$\cos(A + B) = \cos(A - (-B))$$
$$\cos(A + B) = \cos A \cos(-B) + \sin A \sin(-B)$$
$$\cos(A + B) = \cos A \cos B + \sin A (-\sin B)$$
$$\cos(A + B) = \cos A \cos B - \sin A \sin B$$

We can illustrate the truth of this identity by using familiar function values.

☐ EXAMPLE : Show that $\cos 90° = 0$ by using $\cos(60° + 30°)$.

Solution

$$\cos(60° + 30°) = \cos 60° \cos 30° - \sin 60° \sin 30°$$
$$\cos 90° = \frac{1}{2} \cdot \frac{\sqrt{3}}{2} - \frac{\sqrt{3}}{2} \cdot \frac{1}{2}$$
$$\cos 90° = \frac{\sqrt{3}}{4} - \frac{\sqrt{3}}{4} = 0$$

| MODEL PROBLEMS |

Find the exact value of cos 75° by using cos (45° + 30°).

Solution

1. Write the iden-
   tity in terms
   of the sines
   and cosines of
   45° and 30°.

   $\cos (45° + 30°) = \cos 45° \cos 30° - \sin 45° \sin 30°$

   $\cos 75° = \dfrac{\sqrt{2}}{2} \cdot \dfrac{\sqrt{3}}{2} - \dfrac{\sqrt{2}}{2} \cdot \dfrac{1}{2}$

2. Substitute
   the function
   values, and
   simplify.

   $\cos 75° = \dfrac{\sqrt{6}}{4} - \dfrac{\sqrt{2}}{4} = \dfrac{\sqrt{6} - \sqrt{2}}{4}$ *Ans.*

| EXERCISES |

1. Complete the identity: $\cos (\theta + \phi) = $ ____
2. Complete the identity: $\cos (x + y) = $ ____
3. Find $\cos 105°$ by using $\cos (45° + 60°)$.
4. Find $\cos 255°$ by using $\cos (210° + 45°)$.
5. Find $\cos 195°$ by using $\cos (135° + 60°)$.

In 6–9, use the identity for cos $(A + B)$ to prove the given statement.

6. $\cos (\pi + x) = -\cos x$

7. $\cos \left(\dfrac{\pi}{2} + x \right) = -\sin x$

8. $\cos \left(\dfrac{\pi}{4} + x \right) = \dfrac{\sqrt{2}}{2} (\cos x - \sin x)$

9. $\cos \left(\dfrac{3\pi}{2} + x \right) = \sin x$

10. If $\sin x = \frac{3}{5}$ and x is the measure of a positive acute angle, find the value of $\cos (x + 180°)$.

11. If $\sin A = \frac{5}{13}$, $\sin B = \frac{5}{13}$, $\angle A$ is in quadrant II, and $\angle B$ is in quadrant I, find the value of $\cos (A + B)$.

In 12–16, select the numeral preceding the expression that best completes the sentence.

12. If $\cos \theta = -.6$, then $\cos (-\theta)$ is equal to:
    (1) .6   (2) −.6   (3) .8   (4) −.8

13. If $\sin (90° - x) = \frac{1}{2}$, then $\cos x$ is equal to:

    (1) $\dfrac{1}{2}$ (2) $-\dfrac{1}{2}$ (3) $\dfrac{\sqrt{3}}{2}$ (4) $-\dfrac{\sqrt{3}}{2}$

14. If $\sin (-A) = \frac{3}{5}$, then $\sin A$ is equal to:

    (1) $\frac{4}{5}$ (2) $-\frac{4}{5}$ (3) $\frac{3}{5}$ (4) $-\frac{3}{5}$

15. The expression $\cos (90° + \theta)$ is equivalent to:

    (1) $\sin \theta$ (2) $-\sin \theta$ (3) $\cos \theta$ (4) $-\cos \theta$

16. The value of $(\cos 67°30')(\cos 22°30') - (\sin 67°30')(\sin 22°30')$ is:

    (1) 1   (2) $\dfrac{\sqrt{2}}{2}$ (3) $-\dfrac{\sqrt{2}}{2}$ (4) 0

17. If $\sin A = \frac{3}{5}$, $\angle A$ is in quadrant I, $\cos B = -\frac{5}{13}$, and $\angle B$ is in quadrant II, find: **a.** $\cos (A + B)$ **b.** $\cos (-B)$ **c.** $\sin (90° - A)$ **d.** $\sin (-A)$

18. If $\sin x = -\frac{1}{3}$, x is the measure of an angle in quadrant III, $\cos y = -\frac{1}{5}$, and y is the measure of an angle in quadrant II, find:
    a. $\cos (x + y)$ **b.** $\cos (-x)$ **c.** $\cos (90° - x)$

13-6 SINE OF THE SUM OR DIFFERENCE OF ANGLE MEASURES

We can combine the identities that we learned in the previous section to find identities for the sine of the sum of two angle measures and for the sine of the difference of two angle measures.

Proof: $\sin (A + B) = \sin A \cos B + \cos A \sin B$

Use the identity: $\sin \theta = \cos (90° - \theta)$

Let $\theta = A + B$.

Then: $\sin (A + B) = \cos (90° - (A + B))$

     $\sin (A + B) = \cos (90° - A - B)$

     $\sin (A + B) = \cos ((90° - A) - B)$

     $\sin (A + B) = \cos (90° - A) \cos B + \sin (90° - A) \sin B$

     $\sin (A + B) = \sin A \cos B + \cos A \sin B$

Proof: $\sin (A - B) = \sin A \cos B - \cos A \sin B$

     $\sin (A - B) = \sin (A + (-B))$

     $\sin (A - B) = \sin A \cos (-B) + \cos A \sin (-B)$

     $\sin (A - B) = \sin A \cos B + \cos A (-\sin B)$

     $\sin (A - B) = \sin A \cos B - \cos A \sin B$

We can verify these identities by using familiar function values.

☐ Example 1: Show that $\sin 90° = 1$ by using $\sin (60° + 30°)$.

Solution

$$\sin (A + B) = \sin A \cos B + \cos A \sin B$$
$$\sin (60° + 30°) = \sin 60° \cos 30° + \cos 60° \sin 30°$$
$$\sin 90° = \frac{\sqrt{3}}{2} \cdot \frac{\sqrt{3}}{2} + \frac{1}{2} \cdot \frac{1}{2}$$
$$\sin 90° = \frac{3}{4} + \frac{1}{4} = \frac{4}{4} = 1$$

☐ Example 2: Show that $\sin 120° = \frac{\sqrt{3}}{2}$ by using $\sin (180° - 60°)$.

Solution

$$\sin (A - B) = \sin A \cos B - \cos A \sin B$$
$$\sin (180° - 60°) = \sin 180° \cos 60° - \cos 180° \sin 60°$$
$$\sin 120° = 0 \cdot \frac{1}{2} - (-1) \cdot \frac{\sqrt{3}}{2}$$
$$\sin 120° = 0 + \frac{\sqrt{3}}{2} = \frac{\sqrt{3}}{2}$$

MODEL PROBLEMS

1. Use the identity for the sine of the sum of two angle measures to show that $\sin (180° + x) = -\sin x$.

Solution

1. Write the identity for the sum of two angle measures.

$$\sin (A + B) = \sin A \cos B + \cos A \sin B$$

2. Substitute $180°$ for A and x for B.

$$\sin (180° + x) = \sin 180° \cos x + \cos 180° \sin x$$

3. Substitute values for $\sin 180°$ and $\cos 180°$.

$$\sin (180° + x) = 0 \cdot \cos x + (-1) \cdot \sin x$$
$$\sin (180° + x) = 0 - \sin x$$

4. Simplify the $\sin (180° + x) = -\sin x$
   right-hand
   member.

2. Find the exact value of sin 15° by using sin (45° - 30°).

Solution

$$\sin (45° - 30°) = \sin 45° \cos 30° - \cos 45° \sin 30°$$

$$\sin 15° = \frac{\sqrt{2}}{2} \cdot \frac{\sqrt{3}}{2} - \frac{\sqrt{2}}{2} \cdot \frac{1}{2}$$

$$\sin 15° = \frac{\sqrt{6}}{4} - \frac{\sqrt{2}}{4} = \frac{\sqrt{6} - \sqrt{2}}{4} \quad Ans.$$

EXERCISES

1. Complete the identity: $\sin (x + y) = $ ____
2. Complete the identity: $\sin (x - y) = $ ____

In 3-6, use the identity for sin $(A + B)$ or sin $(A - B)$ to verify the given statement.

3. $\sin (90° + x) = \cos x$ 4. $\sin (90° - x) = \cos x$
5. $\sin (270° + x) = -\cos x$ 6. $\sin (180° - x) = \sin x$

In 7-12, express the given sine value in terms of sin θ, or cos θ, or both.

7. $\sin \left(\dfrac{\pi}{2} + \theta \right)$ 8. $\sin (\theta - \pi)$ 9. $\sin \left(\dfrac{\pi}{4} + \theta \right)$

10. $\sin \left(\theta - \dfrac{\pi}{2} \right)$ 11. $\sin \left(\dfrac{\pi}{4} - \theta \right)$ 12. $\sin (\pi + \theta)$

13. If $\sin x = \frac{4}{5}$, $\cos y = \frac{4}{5}$, and x and y are measures of angles in the first quadrant, find the value of $\sin (x + y)$.
14. If $\angle B$ is acute and $\sin B = \frac{12}{13}$, find the value of $\sin (90° - B)$.

In 15-20, use the identity for the sine of the sum or difference of two angle measures to find the exact value of the given function.

15. $\sin 150° = \sin (60° + 90°)$ 16. $\sin 90° = \sin (135° - 45°)$
17. $\sin 75° = \sin (45° + 30°)$ 18. $\sin 75° = \sin (120° - 45°)$
19. $\sin 105° = \sin (135° - 30°)$ 20. $\sin 105° = \sin (60° + 45°)$

In 21–24, select the numeral preceding the expression that best completes the sentence.

21. The expression $\sin 40° \cos 15° + \cos 40° \sin 15°$ is equivalent to:
    (1) $\sin 55°$ (2) $\sin 25°$ (3) $\cos 55°$ (4) $\cos 25°$

22. The expression $\sin\left(\dfrac{\pi}{6} - x\right)$ is equivalent to:

    (1) $\dfrac{1}{2} - \sin x$ (2) $\dfrac{\sqrt{3}}{2} - \sin x$

    (3) $\dfrac{\sqrt{3}}{2} \cos x - \dfrac{1}{2} \sin x$ (4) $\dfrac{1}{2} \cos x - \dfrac{\sqrt{3}}{2} \sin x$

23. If $\sin(A - 30°) = \cos 60°$, the number of degrees in the measure of angle A is: (1) 30   (2) 60   (3) 90   (4) 120

24. If x and y are the measures of positive acute angles, $\sin x = \frac{1}{2}$, and $\sin y = \frac{4}{5}$, then $\sin(x + y)$ equals:

    (1) $\dfrac{3 + 4\sqrt{3}}{10}$ (2) $\dfrac{3 - 4\sqrt{3}}{10}$ (3) $\dfrac{\sqrt{3}}{4} + \dfrac{12}{25}$ (4) $\dfrac{\sqrt{3}}{4} - \dfrac{12}{25}$

25. If $\sin x = -\dfrac{1}{3}$, $\sin y = -\dfrac{\sqrt{5}}{3}$, and x and y are the measures of angles in the third quadrant, find: a. $\sin(x + y)$ b. $\sin(x - y)$ c. $\sin(y - x)$

26. If $\sin A = \frac{3}{5}$, $\angle A$ is in quadrant I, $\cos B = -\frac{5}{13}$, and $\angle B$ is in quadrant II, find: a. $\sin(A + B)$ b. $\sin(A - B)$ c. $\sin(B - A)$ d. $\cos(A + B)$ e. $\cos(A - B)$ f. $\cos(B - A)$

13-7 TANGENT OF THE SUM OR DIFFERENCE OF ANGLE MEASURES

Since $\tan\theta = \dfrac{\sin\theta}{\cos\theta}$, we can use the identities for $\sin(A + B)$ and $\cos(A + B)$ to derive an identity for $\tan(A + B)$.

Proof: $\tan(A + B) = \dfrac{\tan A + \tan B}{1 - \tan A \tan B}$

Use the identity: $\tan(A + B) = \dfrac{\sin(A + B)}{\cos(A + B)}$

$$\tan(A + B) = \dfrac{\sin A \cos B + \cos A \sin B}{\cos A \cos B - \sin A \sin B}$$

This equality is an identity for tan $(A + B)$, but since it is convenient to write the identity in terms of tan A and tan B, we will divide the preceding ratio by $\dfrac{\cos A \cos B}{\cos A \cos B}$.

$$\tan (A + B) = \frac{\sin A \cos B + \cos A \sin B}{\cos A \cos B - \sin A \sin B} \div \frac{\cos A \cos B}{\cos A \cos B}$$

$$\tan (A + B) = \frac{\dfrac{\sin A \cancel{\cos B}}{\cos A \cancel{\cos B}} \overset{1}{} + \dfrac{\cancel{\cos A} \sin B}{\cancel{\cos A} \cos B} \overset{1}{}}{\dfrac{\cancel{\cos A \cos B}}{\cancel{\cos A \cos B}} - \dfrac{\sin A \sin B}{\cos A \cos B}}$$

$$\tan (A + B) = \frac{\tan A \cdot 1 + 1 \cdot \tan B}{1 \cdot 1 - \tan A \tan B}$$

$$\tan (A + B) = \frac{\tan A + \tan B}{1 - \tan A \tan B}$$

In a similar manner, using the identities for sin $(A - B)$ and cos $(A - B)$, we can derive the identity:

$$\tan (A - B) = \frac{\tan A - \tan B}{1 + \tan A \tan B}$$

These two identities are true for all replacements of the variables for which tan $(A \pm B)$, tan A, and tan B are defined.

☐ EXAMPLE 1: Show that tan $120° = -\sqrt{3}$ by using tan $(60° + 60°)$.

Solution

$$\tan (60° + 60°) = \frac{\tan 60° + \tan 60°}{1 - \tan 60° \tan 60°}$$

$$\tan 120° = \frac{\sqrt{3} + \sqrt{3}}{1 - \sqrt{3} \cdot \sqrt{3}} = \frac{2\sqrt{3}}{1 - 3} = \frac{2\sqrt{3}}{-2} = -\sqrt{3}$$

☐ EXAMPLE 2: Show that tan $90°$ is undefined by using tan $(60° + 30°)$.

Solution

$$\tan (60° + 30°) = \frac{\tan 60° + \tan 30°}{1 - \tan 60° \tan 30°}$$

$$\tan 90° = \frac{\sqrt{3} + \dfrac{\sqrt{3}}{3}}{1 - \sqrt{3} \cdot \dfrac{\sqrt{3}}{3}}$$

$$\tan 90° = \frac{\dfrac{3\sqrt{3}}{3} + \dfrac{\sqrt{3}}{3}}{1 - \dfrac{3}{3}} = \frac{\dfrac{4\sqrt{3}}{3}}{1 - 1} = \frac{\dfrac{4\sqrt{3}}{3}}{0}$$

Division by 0 is undefined.

MODEL PROBLEMS

1. Use the identity for the tangent of the sum of two angle measures to show that $(\tan 180° + x) = \tan x$.

Solution

1. Write the identity.
$$\tan (A + B) = \frac{\tan A + \tan B}{1 - \tan A \tan B}$$

2. Substitute $180°$ for A and x for B.
$$\tan (180° + x) = \frac{\tan 180° + \tan x}{1 - \tan 180° \tan x}$$

3. Substitute the value of $\tan 180°$.
$$\tan (180° + x) = \frac{0 + \tan x}{1 - 0 \cdot \tan x}$$

$$\tan (180° + x) = \frac{\tan x}{1 - 0} = \tan x$$

2. If $\tan A = \dfrac{5}{4}$ and $\tan B = \dfrac{1}{5}$, find $\tan (A - B)$.

Solution

$$\tan (A - B) = \frac{\tan A - \tan B}{1 + \tan A \tan B}$$

$$\tan (A - B) = \frac{\dfrac{5}{4} - \dfrac{1}{5}}{1 + \dfrac{5}{4} \cdot \dfrac{1}{5}} = \frac{\dfrac{5}{4} - \dfrac{1}{5}}{1 + \dfrac{1}{4}} \cdot \frac{20}{20}$$

$$\tan (A - B) = \frac{25 - 4}{20 + 5} = \frac{21}{25} \quad Ans.$$

EXERCISES

1. Complete the identity: $\tan (\theta + \phi)$ = ____

2. Complete the identity: $\tan (x - y)$ = ____

In 3-6, use the identity for $\tan (A + B)$ or $\tan (A - B)$ to verify the given statement.

3. $\tan (360° + x) = \tan x$

4. $\tan (180° - x) = -\tan x$

5. $\tan (45° + x) = \dfrac{1 + \tan x}{1 - \tan x}$

6. $\tan (315° + x) = \dfrac{-1 + \tan x}{1 + \tan x}$

In 7-9, express the given tangent value in terms of $\tan \theta$.

7. $\tan (\pi + \theta)$

8. $\tan (2\pi - \theta)$

9. $\tan \left(\dfrac{3\pi}{4} - \theta \right)$

10. Prove that $\tan (-x) = -\tan x$. (*Hint:* Start with the identity for $\tan (A - B)$, and let $A = 0°$ and $B = x$.)

11. If $\tan A = 4$ and $\tan B = 3$, find: a. $\tan (A + B)$ b. $\tan (A - B)$

12. If $\tan \theta = -6$ and $\tan \phi = \frac{1}{2}$, find: a. $\tan (\theta + \phi)$ b. $\tan (\theta - \phi)$

13. If $\tan x = -\frac{2}{3}$ and $\tan y = \frac{9}{4}$, find: a. $\tan (x + y)$ b. $\tan (x - y)$

14. Let $\tan A = 2$ and $\tan B = 3$. a. Find the value of $\tan (A + B)$.
b. One possible measure of the angle $(A + B)$ is:
(1) $0°$ (2) $45°$ (3) $90°$ (4) $135°$

In 15-18, use the identity for $\tan (A + B)$ or for $\tan (A - B)$ to find the exact value of the given function. Rationalize the denominator of the answer.

15. $\tan 75° = \tan (30° + 45°)$

16. $\tan 15° = \tan (45° - 30°)$

17. $\tan 105° = \tan (60° + 45°)$

18. $\tan 195° = \tan (135° + 60°)$

In 19-22, select the numeral preceding the expression that best completes the sentence.

19. The expression $\dfrac{\tan 40° + \tan 30°}{1 - \tan 40° \tan 30°}$ is equivalent to:

(1) $\tan 70°$ (2) $\tan 10°$ (3) $\dfrac{\tan 40°}{1 - \tan 40°}$ (4) $\dfrac{\tan 70°}{1 - \tan 70°}$

20. The expression $\tan (x + y)$ is undefined when:
(1) $\tan x \tan y = 0$ (2) $\tan x \tan y = 1$
(3) $\tan x \tan y = -1$ (4) $\tan x + \tan y = 0$

21. The expression tan $(A - B)$ is undefined when tan $A = \frac{1}{2}$ and tan B equals: (1) 1   (2) 2   (3) –2   (4) 0

22. Since tan $165° =$ tan $(135° + 30°)$, the exact value of tan 165 can be found to be:

    (1) $-2 + \sqrt{3}$ (2) $-2 - \sqrt{3}$ (3) $2 + \sqrt{3}$ (4) $2 - \sqrt{3}$

13-8   FUNCTION VALUES OF DOUBLE ANGLES

Since $A + A = 2A$, we can derive an identity for the function value of an angle whose measure is twice that of a given angle. To do this, we start with the identity for the function value of the sum of two angle measures.

Sine of a Double Angle

Proof: sin $2A = 2$ sin A cos A

Use the identity: sin $(A + B) =$ sin A cos $B +$ cos A sin B

Let $B = A$. Then: sin $(A + A) =$ sin A cos $A +$ cos A sin A

$$\text{sin } 2A = \text{sin } A \text{ cos } A + \text{sin } A \text{ cos } A$$
$$\text{sin } 2A = 2 \text{ sin } A \text{ cos } A$$

☐ EXAMPLE 1:   Show that sin $90° = 1$ by using sin $2(45°)$.

Solution

$$\text{sin } 2A = 2 \text{ sin } A \text{ cos } A$$
$$\text{sin } 2(45°) = 2 \text{ sin } 45° \text{ cos } 45°$$
$$\text{sin } 90° = 2 \cdot \frac{\sqrt{2}}{2} \cdot \frac{\sqrt{2}}{2} = \frac{4}{4} = 1$$

Cosine of a Double Angle

Proof: cos $2A = \cos^2 A - \sin^2 A$

Use the identity: cos $(A + B) =$ cos A cos $B -$ sin A sin B

Let $B = A$. Then: cos $(A + A) =$ cos A cos $A -$ sin A sin A

$$\text{cos } 2A = \cos^2 A - \sin^2 A$$

By using the basic identity $\sin^2 A = 1 - \cos^2 A$, we can write this identity in terms of cos A only.

$$\text{cos } 2A = \cos^2 A - \sin^2 A$$
$$\text{cos } 2A = \cos^2 A - (1 - \cos^2 A)$$
$$\text{cos } 2A = \cos^2 A - 1 + \cos^2 A$$
$$\text{cos } 2A = 2 \cos^2 A - 1$$

Using a similar replacement, $\cos^2 A = 1 - \sin^2 A$, we can express $\cos 2A$ in terms of $\sin A$ only.

$$\cos 2A = \cos^2 A - \sin^2 A$$
$$\cos 2A = (1 - \sin^2 A) - \sin^2 A$$
$$\cos 2A = 1 - 2 \sin^2 A$$

☐ EXAMPLE 2: Show that $\cos 60° = \dfrac{1}{2}$ by using $\cos 2(30°)$.

Solution

$$\cos 2A = \cos^2 A - \sin^2 A$$
$$\cos 2(30°) = \cos^2 30° - \sin^2 30°$$
$$\cos 60° = \left(\frac{\sqrt{3}}{2}\right)^2 - \left(\frac{1}{2}\right)^2$$
$$\cos 60° = \frac{3}{4} - \frac{1}{4} = \frac{2}{4} = \frac{1}{2}$$

Tangent of a Double Angle

Proof: $\tan 2A = \dfrac{2 \tan A}{1 - \tan^2 A}$

Use the identity: $\tan (A + B) = \dfrac{\tan A + \tan B}{1 - \tan A \tan B}$

Let $B = A$. Then: $\tan (A + A) = \dfrac{\tan A + \tan A}{1 - \tan A \tan A}$

$$\tan 2A = \frac{2 \tan A}{1 - \tan^2 A}$$

☐ EXAMPLE 3: Show that $\tan 120° = -\sqrt{3}$ by using $\tan 2(60°)$.

Solution

$$\tan 2A = \frac{2 \tan A}{1 - \tan^2 A}$$
$$\tan 2(60°) = \frac{2 \tan 60°}{1 - \tan^2 60°}$$
$$\tan 120° = \frac{2(\sqrt{3})}{1 - (\sqrt{3})^2} = \frac{2\sqrt{3}}{1 - 3} = \frac{2\sqrt{3}}{-2} = -\sqrt{3}$$

Note that the identity is true for all values of A for which $\tan A$ and $\tan 2A$ are defined. If $A = 90°$, $\tan 2A = \tan 180° = 0$ but this value cannot be found by using the expression $\dfrac{2 \tan A}{1 - \tan^2 A}$.

| MODEL PROBLEM |

Cos $A = -\frac{5}{13}$, and A is the measure of an angle in quadrant II. **a.** Find sin $2A$. **b.** Find cos $2A$. **c.** Find tan $2A$. **d.** Determine the quadrant in which the angle whose measure is $2A$ lies.

Solution

a. 1. Find sin A. Since A is the measure of an angle in quadrant II, sin A is positive.

$$\sin^2 A = 1 - \cos^2 A$$
$$\sin^2 A = 1 - \left(-\frac{5}{13}\right)^2$$
$$\sin^2 A = 1 - \frac{25}{169} = \frac{169}{169} - \frac{25}{169} = \frac{144}{169}$$
$$\sin A = \frac{12}{13}$$

2. Write the identity for sin $2A$.

$$\sin 2A = 2 \sin A \cos A$$

3. Substitute the values of sin A and cos A, and simplify.

$$\sin 2A = 2\left(\frac{12}{13}\right)\left(-\frac{5}{13}\right)$$
$$\sin 2A = -\frac{120}{169} \quad Ans.$$

b. 1. Write the identity for cos $2A$.

$$\cos 2A = \cos^2 A - \sin^2 A$$

2. Substitute the values for sin A and cos A, and simplify.

$$\cos 2A = \left(-\frac{5}{13}\right)^2 - \left(\frac{12}{13}\right)^2$$
$$\cos 2A = \frac{25}{169} - \frac{144}{169}$$
$$\cos 2A = -\frac{119}{169} \quad Ans.$$

c. 1. Find tan A.

$$\tan A = \frac{\sin A}{\cos A} = \frac{\frac{12}{13}}{-\frac{5}{13}} = \frac{\frac{12}{13}}{-\frac{5}{13}} \cdot \frac{13}{13} = -\frac{12}{5}$$

2. Write the identity for tan $2A$.

$$\tan 2A = \frac{2 \tan A}{1 - \tan^2 A}$$

3. Substitute the value of tan A, and simplify.

$$\tan 2A = \frac{2\left(-\dfrac{12}{5}\right)}{1 - \left(-\dfrac{12}{5}\right)^2}$$

$$\tan 2A = \frac{-\dfrac{24}{5}}{1 - \dfrac{144}{25}}$$

$$\tan 2A = \frac{-\dfrac{24}{5}}{-\dfrac{119}{25}} = +\frac{\dfrac{24}{5}}{\dfrac{119}{25}} \cdot \frac{25}{25} = \frac{120}{119} \quad Ans.$$

Note: Tan $2A$ could have been found by using the answers to parts a and b:

$$\tan 2A = \frac{\sin 2A}{\cos 2A} = \frac{-\dfrac{120}{169}}{-\dfrac{119}{169}} = +\frac{\dfrac{120}{169}}{\dfrac{119}{169}} \cdot \frac{169}{169} = \frac{120}{119}$$

d. Since sin $2A$ and cos $2A$ are both negative, $2A$ must be the measure of a third-quadrant angle. *Ans.*

| EXERCISES |

1. Complete the identity: sin $2x$ = ____
2. Complete the identity: tan $2x$ = ____
3. Write the identity for cos $2x$ in terms of sin x and cos x.
4. Write the identity for cos $2x$ in terms of cos x.
5. Write the identity for cos $2x$ in terms of sin x.
6. If x is the measure of a positive acute angle and sin $x = \frac{4}{5}$, find sin $2x$.
7. If sin $A = -0.8$, what is the value of cos $2A$?
8. If tan $A = \frac{1}{4}$, find tan $2A$.
9. Show that cos $90° = 0$ by using cos $2(45°)$.
10. Show that sin $180° = 0$ by using sin $2(90°)$.
11. Show that tan $360° = 0$ by using tan $2(180°)$.
12. Show that sin $270° = -1$ by using sin $2(135°)$.
13. If sin $\theta = b$, express cos 2θ in terms of b.
14. If cos $\theta = a$, express cos 2θ in terms of a.
15. If tan $\theta = c$, express tan 2θ in terms of c.

16. If $\sin A = -\frac{3}{5}$ and $\angle A$ is in quadrant III, find: **a.** $\sin 2A$
    b. $\cos 2A$ **c.** $\tan 2A$ **d.** the quadrant in which $\angle A$ terminates
17. If $\cos A = \frac{1}{3}$ and $\angle A$ is acute, find: **a.** $\sin 2A$ **b.** $\cos 2A$
    c. $\tan 2A$
18. If $\cos \theta = -.6$ and θ is the measure of an angle in quadrant II, find:
    a. $\cos 2\theta$ **b.** $\sin 2\theta$ **c.** $\tan 2\theta$ **d.** the quadrant in which the
    angle whose measure is 2θ terminates

In 19-21, select the numeral preceding the expression that best completes the sentence.

19. If $\cos \theta = \sin \theta$, then $\cos 2\theta$ is equal to:
    (1) 1   (2) 0   (3) $2 \cos^2 \theta$ (4) $2 \sin^2 \theta$
20. The expression $(\sin x - \cos x)^2$ is equivalent to:
    (1) 1   (2) $\sin^2 x - \cos^2 x$ (3) $1 - \cos 2x$ (4) $1 - \sin 2x$
21. If $\tan A = -1$, then $\tan 2A$:
    (1) equals 1   (2) equals 2   (3) equals -2 (4) is undefined

13-9 FUNCTION VALUES OF HALF ANGLES

Since the angle measure θ is half of the angle measure 2θ, we can use the identities that were developed in the last section to derive identities for the function values of an angle whose measure is half that of a given angle.

Sine of a Half Angle

Proof: $\sin \frac{1}{2}A = \pm\sqrt{\dfrac{1 - \cos A}{2}}$

1. Use the identity for $\cos 2\theta$ in $\cos 2\theta = 1 - 2 \sin^2 \theta$
   terms of $\sin \theta$. $2 \sin^2 \theta = 1 - \cos 2\theta$

2. Solve for $\sin \theta$. $\sin^2 \theta = \dfrac{1 - \cos 2\theta}{2}$
   (*Note:* Since $\cos 2\theta \le 1$, then
   $-\cos 2\theta \ge -1$ or $1 - \cos 2\theta \ge 0$.
   Thus, the right-hand member is $\sin \theta = \pm\sqrt{\dfrac{1 - \cos 2\theta}{2}}$
   a real number.)

3. Let $\theta = \dfrac{1}{2}A$. $\sin \frac{1}{2}A = \pm\sqrt{\dfrac{1 - \cos A}{2}}$

   Then, $2\theta = 2\left(\dfrac{1}{2}A\right) = A$.

☐ Example 1: Show that $\sin 30° = \dfrac{1}{2}$ by using $\sin \dfrac{1}{2}(60°)$.

Solution

$$\sin \frac{1}{2}A = \pm\sqrt{\frac{1 - \cos A}{2}}$$

$$\sin \frac{1}{2}(60°) = +\sqrt{\frac{1 - \cos 60°}{2}}$$

$$\sin 30° = \sqrt{\frac{1 - \frac{1}{2}}{2}} = \sqrt{\frac{\frac{1}{2}}{2}} = \sqrt{\frac{1}{2} \cdot \frac{2}{2}} = \sqrt{\frac{1}{4}} = \frac{1}{2}$$

Since $\frac{1}{2}(60°) = 30°$ is an acute-angle measure, we chose the positive value for the sine.

Cosine of a Half Angle

In a similar way, we will derive an identity for $\cos \frac{1}{2}A$ by starting with the identity for $\cos 2\theta$ in terms of $\cos \theta$:

Proof: $\cos \dfrac{1}{2}A = \pm\sqrt{\dfrac{1 + \cos A}{2}}$

1. Use the identity for $\cos 2\theta$ in terms of $\cos \theta$.

$$\cos 2\theta = 2 \cos^2 \theta - 1$$
OR
$$2 \cos^2 \theta - 1 = \cos 2\theta$$

2. Solve for $\cos \theta$.
   (*Note:* Since $\cos 2\theta \geq -1$, then $1 + \cos 2\theta \geq 0$. Thus, the right-hand member is a real number.)

$$2 \cos^2 \theta = 1 + \cos 2\theta$$

$$\cos^2 \theta = \frac{1 + \cos 2\theta}{2}$$

$$\cos \theta = \pm\sqrt{\frac{1 + \cos 2\theta}{2}}$$

3. Let $\theta = \dfrac{1}{2}A$.

Then, $2\theta = 2\left(\dfrac{1}{2}A\right) = A$.

$$\cos \frac{1}{2}A = \pm\sqrt{\frac{1 + \cos A}{2}}$$

☐ EXAMPLE 2: Show that $\cos 45° = \dfrac{\sqrt{2}}{2}$ by using $\cos \dfrac{1}{2}(90°)$.

Solution

$$\cos \frac{1}{2}A = \pm\sqrt{\frac{1 + \cos A}{2}}$$

$$\cos \frac{1}{2}(90°) = +\sqrt{\frac{1 + \cos 90°}{2}}$$

$$\cos 45° = \sqrt{\frac{1 + 0}{2}} = \sqrt{\frac{1}{2}} = \frac{\sqrt{1}}{\sqrt{2}} = \frac{1}{\sqrt{2}} \cdot \frac{\sqrt{2}}{\sqrt{2}} = \frac{\sqrt{2}}{2}$$

Here, again, we chose the positive value for the function of an acute-angle measure.

Tangent of a Half Angle

We will use the identities for $\sin \frac{1}{2}A$ and $\cos \frac{1}{2}A$ to derive an identity for $\tan \frac{1}{2}A$.

Proof: $\tan \dfrac{1}{2}A = \pm\sqrt{\dfrac{1 - \cos A}{1 + \cos A}}$

Use the identity: $\tan \dfrac{1}{2}A = \dfrac{\sin \frac{1}{2}A}{\cos \frac{1}{2}A}$

$$\tan \frac{1}{2}A = \pm\frac{\sqrt{\dfrac{1 - \cos A}{2}}}{\sqrt{\dfrac{1 + \cos A}{2}}}$$

$$\tan \frac{1}{2}A = \pm\sqrt{\frac{1 - \cos A}{\cancel{2}}} \cdot \sqrt{\frac{\cancel{2}}{1 + \cos A}}$$

$$\tan \frac{1}{2}A = \pm\sqrt{\frac{1 - \cos A}{1 + \cos A}}$$

☐ EXAMPLE 3: Show that $\tan 120° = -\sqrt{3}$ by using $\tan \dfrac{1}{2}(240°)$.

Solution

$$\tan \frac{1}{2}A = \pm\sqrt{\frac{1 - \cos A}{1 + \cos A}}$$

$$\tan \frac{1}{2}(240°) = -\sqrt{\frac{1 - \cos 240°}{1 + \cos 240°}}$$

$$\tan 120° = -\sqrt{\frac{1 - \left(-\frac{1}{2}\right)}{1 + \left(-\frac{1}{2}\right)}} = -\sqrt{\frac{\frac{3}{2}}{\frac{1}{2}}} = -\sqrt{\frac{\frac{3}{2}}{\frac{1}{2}} \cdot \frac{2}{2}} = -\sqrt{3}$$

Here we chose the negative value of the tangent of a second-quadrant angle.

In using each of the identities just derived for a function value of a half angle, we must be careful to choose the correct sign. For example, if $180° < x < 270°$, then $90° < \frac{1}{2}x < 135°$ and $\frac{1}{2}x$ is the measure of a second-quadrant angle. Therefore, $\sin \frac{1}{2}x$ would be positive, and $\cos \frac{1}{2}x$ and $\tan \frac{1}{2}x$ would be negative. If $360° < x < 540°$, then $180° < \frac{1}{2}x < 270°$, and $\frac{1}{2}x$ is the measure of a third-quadrant angle whose tangent value is positive and whose sine and cosine values are negative.

| MODEL PROBLEMS |

1. If $\cos x = \frac{1}{9}$, what is the positive value of $\sin \frac{1}{2}x$?

Solution

1. Write the identity for $\sin \frac{1}{2}x$. Use the positive value.

$$\sin \frac{1}{2}x = \sqrt{\frac{1 - \cos x}{2}}$$

2. Substitute the given value for $\cos x$, and simplify.

$$\sin \frac{1}{2}x = \sqrt{\frac{1 - \frac{1}{9}}{2}}$$

$$\sin \frac{1}{2}x = \sqrt{\frac{\frac{8}{9}}{2} \cdot \frac{\frac{1}{2}}{\frac{1}{2}}} = \sqrt{\frac{4}{9}}$$

$$\sin \frac{1}{2}x = \frac{2}{3} \quad Ans.$$

2. If $A = \text{Arc cos } \frac{1}{5}$, what is the value of $\tan \frac{A}{2}$?

Solution

1. Write the identity for $\tan \frac{A}{2}$.

$$\tan \frac{A}{2} = \pm\sqrt{\frac{1 - \cos A}{1 + \cos A}}$$

2. Since A is the principal value and $\cos A$ is positive, $0° < A < 90°$.

Therefore, $0° < \frac{A}{2} < 45°$, and

$\tan \frac{A}{2}$ is positive. Substitute the given value for $\cos A$.

$$= \sqrt{\frac{1 - \frac{1}{5}}{1 + \frac{1}{5}}}$$

$$= \sqrt{\frac{\frac{4}{5}}{\frac{6}{5}}} = \sqrt{\frac{\cancel{4}^{2}}{\cancel{5}} \cdot \frac{\cancel{5}^{1}}{\cancel{6}_{3}}}$$

$$= \sqrt{\frac{2}{3} \cdot \frac{3}{3}}$$

$$\tan \frac{A}{2} = \frac{\sqrt{6}}{3} \quad Ans.$$

| **EXERCISES** |

In 1–3, complete the identity.

1. $\sin \frac{1}{2}x = $ ___   2. $\cos \frac{1}{2}x = $ ___   3. $\tan \frac{1}{2}x = $ ___

4. If x is the measure of a positive acute angle and $\cos x = \frac{7}{32}$, find the value of $\sin \frac{1}{2}x$.

5. If $\cos A = \frac{1}{8}$ and angle A is positive and acute, find $\cos \frac{1}{2}A$.

6. If $A = \text{Arc cos } \frac{1}{2}$, find the value of $\sin \frac{1}{2}A$.

7. If $B = \text{Arc cos } \frac{24}{25}$, find the value of $\tan \frac{B}{2}$.

8. If $\cos x = -\frac{7}{18}$ and $180° < x < 270°$, find the value of $\sin \frac{1}{2}x$.

9. If $\cos y = -\frac{1}{2}$ and $180° < y < 270°$, find the value of $\cos \frac{y}{2}$.

10. If $\cos \theta = -\frac{15}{17}$ and $180° < \theta < 270°$, find the value of $\tan \frac{1}{2}\theta$.

11. Find $\cos \frac{1}{2}y$ if $y = \text{arc cos } .28$ and $270° < y < 360°$.

12. Find $\sin \frac{1}{2}\theta$ if $\theta = \text{arc cos } .68$ and $360° < \theta < 450°$.

13. Find $\tan \frac{1}{2}x$ if $x = \text{arc sin } \frac{5}{13}$ and $90° < x < 180°$.

14. Find $\sin \frac{1}{2}B$ if $\tan B = \frac{3}{4}$ and $\angle B$ is acute.

15. Find the exact value of $\sin 60°$ by using $\sin \frac{1}{2}(120°)$.

16. Find the exact value of $\cos 225°$ by using $\cos \frac{1}{2}(450°)$.

17. Find the exact value of $\tan 135°$ by using $\tan \frac{1}{2}(270°)$.

18. If $\cos A = \frac{7}{25}$ and $\angle A$ is a positive acute angle, find: **a.** $\sin \frac{1}{2}A$ **b.** $\cos \frac{1}{2}A$ **c.** $\tan \frac{1}{2}A$

19. If $\cos B = \frac{5}{9}$ and $\angle B$ is a positive acute angle, find: **a.** $\sin \frac{1}{2}B$ **b.** $\cos \frac{1}{2}B$ **c.** $\tan \frac{1}{2}B$

20. If $\cos \theta = \frac{7}{8}$ and $0° < \theta < 90°$, find: **a.** $\sin \dfrac{\theta}{2}$ **b.** $\cos \dfrac{\theta}{2}$

c. $\tan \dfrac{\theta}{2}$

21. If $\sin A = 0.6$ and $\angle A$ is a positive acute angle, find: **a.** $\sin \frac{1}{2}A$ **b.** $\cos \frac{1}{2}A$ **c.** $\tan \frac{1}{2}A$

In 22-27, select the numeral preceding the expression that best completes the sentence or answers the question.

22. Which of the following is not an identity?

(1) $\sin \frac{1}{2}x = \pm\sqrt{\dfrac{1 - \cos x}{2}}$ (2) $\cos^2 \frac{1}{2}x = \dfrac{1 + \cos x}{2}$

(3) $\tan 2x = \pm\sqrt{\dfrac{1 - \cos 4x}{1 + \cos 4x}}$ (4) $\sin \frac{1}{2}x = \frac{1}{2}\sin x$

23. The expression $\sqrt{\dfrac{1 - \cos 80°}{2}}$ is equivalent to:

(1) $\cos 40°$ (2) $\sin 40°$ (3) $\frac{1}{2} - \cos 40°$ (4) $\frac{1}{2}\sin 80°$

24. The expression $\dfrac{1 - \cos 100°}{1 + \cos 100°}$ is equivalent to:

(1) $\tan 50°$ (2) $-\cos 100°$ (3) $\sqrt{\cos 50°}$ (4) $\tan^2 50°$

25. The expression $2 \sin^2 \frac{1}{2}\theta$ is equivalent to:

(1) $1 - \cos \theta$ (2) $1 + \cos \theta$ (3) $\sin^2 \theta$ (4) $2 - 2 \cos \theta$

26. If $\cos \frac{1}{2}\theta = \frac{1}{3}$, then $\cos \theta$ equals:

(1) $\frac{2}{3}$ (2) $\frac{1}{6}$ (3) $-\frac{7}{9}$ (4) $\frac{7}{9}$

27. Using $\tan \frac{1}{2}(30°)$, we can find the exact value of $\tan 15°$ to be:

(1) $2 + \sqrt{3}$ (2) $2 - \sqrt{3}$ (3) $\dfrac{2 + \sqrt{3}}{4}$ (4) $\dfrac{2 - \sqrt{3}}{4}$

13-10 SUMMARY OF IDENTITIES

Sum of Angle Measures	*Difference of Angle Measures*
$\sin (A + B) = \sin A \cos B + \cos A \sin B$	$\sin (A - B) = \sin A \cos B - \cos A \sin B$
$\cos (A + B) = \cos A \cos B - \sin A \sin B$	$\cos (A - B) = \cos A \cos B + \sin A \sin B$
$\tan (A + B) = \dfrac{\tan A + \tan B}{1 - \tan A \tan B}$	$\tan (A - B) = \dfrac{\tan A - \tan B}{1 + \tan A \tan B}$

Double-Angle Measures	*Half-Angle Measures*
$\sin 2A = 2 \sin A \cos A$	$\sin \dfrac{1}{2}A = \pm \sqrt{\dfrac{1 - \cos A}{2}}$
$\cos 2A = \cos^2 A - \sin^2 A$	
$\cos 2A = 2 \cos^2 A - 1$	$\cos \dfrac{1}{2}A = \pm \sqrt{\dfrac{1 + \cos A}{2}}$
$\cos 2A = 1 - 2 \sin^2 A$	
$\tan 2A = \dfrac{2 \tan A}{1 - \tan^2 A}$	$\tan \dfrac{1}{2}A = \pm \sqrt{\dfrac{1 - \cos A}{1 + \cos A}}$

These fundamental identities can be used to prove other identities. When proving an identity, we should express functions of a double angle, a half angle, or the sum or difference of angle measures in terms of functions of the same angle. Then, we employ the same techniques that were used at the beginning of this chapter.

MODEL PROBLEM

Prove the identity: $\dfrac{1 - \cos 2\theta}{\sin 2\theta} = \tan \theta$

Solution

1. Write the identity.

$$\frac{1 - \cos 2\theta}{\sin 2\theta} = \tan \theta$$

2. Replace cos 2θ and sin 2θ by expressions using sin θ and cos θ. Note that other replacements can also be made.

$$\frac{1 - (1 - 2 \sin^2 \theta)}{2 \sin \theta \cos \theta} =$$

3. Simplify the fraction in the left-hand member.

$$\frac{1 - 1 + 2 \sin^2 \theta}{2 \sin \theta \cos \theta} =$$

$$\frac{2 \sin^2 \theta}{2 \sin \theta \cos \theta} =$$

$$\frac{\sin \theta}{\cos \theta} =$$

$$\tan \theta = \tan \theta$$

| EXERCISES |

In 1-26, prove the identity.

1. $\sin 2\theta \sec \theta = 2 \sin \theta$

2. $\sin 2\theta \csc \theta = 2 \cos \theta$

3. $\sin \theta = \dfrac{\sin 2\theta}{2 \cos \theta}$

4. $\dfrac{2 \cos \theta}{\sin 2\theta} = \csc \theta$

5. $\dfrac{\cos 2\theta}{\sin \theta} + \sin \theta = \dfrac{\cot \theta}{\sec \theta}$

6. $\tan \theta + \cot \theta = \dfrac{2}{\sin 2\theta}$

7. $\sin 2\theta = \dfrac{2 \tan \theta}{1 + \tan^2 \theta}$

8. $\cos 2\theta = \dfrac{1 - \tan^2 \theta}{1 + \tan^2 \theta}$

9. $\sin 2\theta \sec^2 \theta = 2 \tan \theta$

10. $2 - \sec^2 \theta = \cos 2\theta \sec^2 \theta$

11. $(\cos \theta - \sin \theta)^2 = 1 - \sin 2\theta$

12. $(\cos \theta + \sin \theta)^2 = 1 + \sin 2\theta$

13. $\dfrac{2 \sin^2 \theta}{\sin 2\theta} + \cot \theta = \sec \theta \csc \theta$

14. $\cos 2\theta = \dfrac{\cot^2 \theta - 1}{\cot^2 \theta + 1}$

15. $\dfrac{\cos 2\theta}{\sin \theta} + \sin \theta = \csc \theta - \sin \theta$

16. $\dfrac{2 \tan \theta - \sin 2\theta}{2 \sin^2 \theta} = \tan \theta$

17. $2 \cos \theta - \dfrac{\cos 2\theta}{\cos \theta} = \sec \theta$

18. $\dfrac{\cos 2\theta + \cos \theta + 1}{\sin 2\theta + \sin \theta} = \cot \theta$

19. $\dfrac{\sin \theta + \sin 2\theta}{\sec \theta + 2} = \sin \theta \cos \theta$

20. $\dfrac{1 + \cos 2\theta}{1 - \cos 2\theta} = \cot^2 \theta$

21. $\sin \left(\dfrac{\pi}{6} - \theta \right) + \sin \left(\dfrac{\pi}{6} + \theta \right) = \cos \theta$

22. $\cos \left(\dfrac{\pi}{4} - \theta \right) - \cos \left(\dfrac{\pi}{4} + \theta \right) = \sqrt{2} \sin \theta$

23. $\sin \left(\dfrac{\pi}{2} + \theta \right) - \sin \left(\dfrac{\pi}{2} - \theta \right) = 0$

24. $\sin \left(\dfrac{\pi}{4} + \theta \right) + \cos \left(\dfrac{\pi}{4} + \theta \right) = \sqrt{2} \cos \theta$

25. $\dfrac{\tan \left(\dfrac{\pi}{4} + \theta \right)}{\tan \left(\dfrac{3\pi}{4} - \theta \right)} = -1$

26. $\dfrac{\sin \left(\dfrac{\pi}{4} + \theta \right)}{\cos \left(\dfrac{\pi}{4} - \theta \right)} = 1$

27. a. Prove the identity: $\dfrac{\sin 2A}{1 + \cos 2A} = \tan A$

b. Use the identity in part a to write an identity for $\tan \frac{1}{2}\theta$ in terms of $\sin \theta$ and $\cos \theta$. (*Hint:* Let $A = \frac{1}{2}\theta$.)

c. Prove the identity: $\pm \sqrt{\dfrac{1 - \cos \theta}{1 + \cos \theta}} = \dfrac{\sin \theta}{1 + \cos \theta}$

13-11   FIRST-DEGREE TRIGONOMETRIC EQUATIONS

A *trigonometric equation* is an equation in which the variable is expressed in terms of a trigonometric function value.

Algebraic Equation	*Trigonometric Equation*
$2x - 1 = 0$	$2 \cos \theta - 1 = 0$

To find the values of θ that make $2 \cos \theta - 1 = 0$ true, we first find a value or values for $\cos \theta$. To do this, we use the same procedures that we used to solve algebraic equations.

Solving an Algebraic Equation	*Solving a Trigonometric Equation*
$2x - 1 = 0$	$2 \cos \theta - 1 = 0$
$\underline{+ 1 \quad +1}$	$\underline{+ 1 \quad +1}$
$2x = 1$	$2 \cos \theta = 1$
$\dfrac{2x}{2} = \dfrac{1}{2}$	$\dfrac{2 \cos \theta}{2} = \dfrac{1}{2}$
$x = \dfrac{1}{2}$	$\cos \theta = \dfrac{1}{2}$

There is one value of x that makes $2x - 1 = 0$ true, namely, $x = \frac{1}{2}$. There is one value of $\cos \theta$ that makes $2 \cos \theta - 1 = 0$ true, namely, $\cos \theta = \frac{1}{2}$. When $\cos \theta = \frac{1}{2}$, however, there are infinitely many values of θ that make the equation $2 \cos \theta - 1 = 0$ true.

In terms of degree measure, if $\cos \theta = \frac{1}{2}$ and $0° \leq \theta \leq 360°$, then $\theta = 60°$ or $\theta = 300°$. In terms of radian measure, if $\cos \theta = \frac{1}{2}$ and $0 \leq \theta \leq 2\pi$, then $\theta = \frac{\pi}{3}$ and $\theta = \frac{5\pi}{3}$. If $\cos \theta = \frac{1}{2}$ and θ is any real number, then any value of θ that differs from $\frac{\pi}{3}$ or $\frac{5\pi}{3}$ by a multiple of 2π is also a solution. The general solution of $2 \cos \theta - 1 = 0$ in radians is:

$$\theta = \frac{\pi}{3} + 2\pi k \text{ or } \theta = \frac{5\pi}{3} + 2\pi k \text{ for } k \text{ an integer}$$

When we solve an algebraic equation, the replacement set is usually the set of all real numbers. When we solve a trigonometric equation, the replacement set for $\sin \theta$ or $\cos \theta$ is the set of numbers between -1 and 1 inclusive, and the replacement set for $\tan \theta$ is the set of all real numbers.

$3x - 5 = 4$	$3 \sin \theta - 5 = 4$	$3 \tan \theta - 5 = 4$
$3x = 9$	$3 \sin \theta = 9$	$3 \tan \theta = 9$
$x = 3$	$\sin \theta = 3$	$\tan \theta = 3$
	(There is no solution.)	If $0° \leq \theta < 360°$:

In quadrant I, $\theta \approx 71°30'$.
In quadrant III,
$\theta \approx 180° + 71°30'$, or
$\theta \approx 251°30'$.

MODEL PROBLEMS

1. **a.** Solve for $\cos \theta$: $\cos \theta = 3 \cos \theta + 1$. **b.** Find all values of θ in the interval $0 \leq \theta < 2\pi$ that satisfy the equation in part **a.**

Solution

a.
$$\cos \theta = 3 \cos \theta + 1$$
$$\underline{-3 \cos \theta \qquad -3 \cos \theta \qquad\quad}$$
$$-2 \cos \theta = \qquad\qquad 1$$
$$\frac{-2 \cos \theta}{-2} = \frac{1}{-2}$$
$$\cos \theta = -\frac{1}{2}$$

b. 1. Since $\cos \frac{\pi}{3} = \frac{1}{2}$, the reference angle is an angle of $\frac{\pi}{3}$ radians.

2. In quadrants II and III, $\cos \theta$ is negative.

3. In quadrant II, $\theta = \pi - \frac{\pi}{3} = \frac{2\pi}{3}$.

In quadrant III, $\theta = \pi + \frac{\pi}{3} = \frac{4\pi}{3}$.

Answer: **a.** $\cos \theta = -\frac{1}{2}$ **b.** $\theta = \frac{2\pi}{3}$ or $\theta = \frac{4\pi}{3}$

2. Find to the nearest degree the measure of the positive acute angle that satisfies the equation $3(\csc \theta - 1) = \csc \theta + 2$.

Solution

1. Solve the equation for $\csc \theta$.

$$3(\csc \theta - 1) = \csc \theta + 2$$
$$3 \csc \theta - 3 = \csc \theta + 2$$
$$2 \csc \theta = 5$$
$$\csc \theta = \frac{5}{2}$$

2. Write the reciprocal of each member of the equation, and use the identity $\dfrac{1}{\csc \theta} = \sin \theta$.

$$\dfrac{1}{\csc \theta} = \dfrac{2}{5}$$

$$\sin \theta = \dfrac{2}{5}$$

3. Write the value of $\sin \theta$ as a decimal and use the table of trigonometric function values to find θ to the nearest degree.

$$\sin \theta = .4$$
$$\theta = 24° \quad Ans.$$

EXERCISES

In 1–6, solve for θ in the interval $0° \leq \theta \leq 360°$.

1. $2 \sin \theta - 1 = 0$ 2. $2 \sin \theta - \sqrt{3} = 0$ 3. $3 \cos \theta + 1 = 1$
4. $\sqrt{3} \tan \theta - 3 = 0$ 5. $8 \sin \theta + 1 = -3$ 6. $3 \tan \theta - 2 = \tan \theta$

In 7–12, solve for θ in the interval $0 \leq \theta \leq 2\pi$.

7. $4(\cos \theta + 1) = 0$ 8. $3 \cos \theta - \sqrt{3} = \cos \theta$
9. $2(\sin \theta + \sqrt{2}) = \sqrt{2}$ 10. $3 \cos \theta - 1 = 1 - \cos \theta$
11. $5 \tan \theta + 2 = 3 \tan \theta$ 12. $2(\sin \theta + 1) = \sin \theta + 3$

In 13–18, find θ to the nearest degree in the interval $0° \leq \theta \leq 360°$.

13. $5 \cos \theta - 1 = 0$ 14. $2 \tan \theta + 3 = 7$
15. $6 \sin \theta + 2 = \sin \theta$ 16. $10(\cos \theta + 1) = 6$
17. $\csc \theta + 8 = 3 \csc \theta$ 18. $3 \tan \theta + 2 = 7 - \tan \theta$

19. Find in radian measure five values of θ that satisfy the equation $4 \cos \theta - \sqrt{2} = \sqrt{2}$.
20. Find to the nearest degree five values of θ that satisfy the equation $3(\sin \theta - 2) = 1 - 5 \sin \theta$.

In 21–24, select the numeral preceding the expression that best completes the sentence.

21. One root of the equation $2 \cos \theta + \sqrt{3} = 0$ is:

   (1) $\dfrac{2\pi}{3}$ (2) $\dfrac{\pi}{3}$ (3) $\dfrac{5\pi}{6}$ (4) $\dfrac{\pi}{6}$

22. One root of the equation $\tan x - 1 = 2 \tan x$ is:

   (1) $\dfrac{\pi}{4}$ (2) π (3) $\dfrac{5\pi}{4}$ (4) $\dfrac{7\pi}{4}$

23. If $0° \le \theta \le 360°$, the solution set of $3 \tan \theta + \sqrt{3} = 2\sqrt{3}$ is:
    (1) $\{60°, 120°\}$ (2) $\{60°, 240°\}$
    (3) $\{30°, 210°\}$ (4) $\{30°, 150°\}$

24. If $0° \le \theta \le 360°$, the solution set of $\sin \theta + 1 = 3$ is:
    (1) $\{90°, 270°\}$ (2) $\{30°, 150°\}$ (3) $\{0°, 180°\}$ (4) $\{\ \}$

13-12 SECOND-DEGREE TRIGONOMETRIC EQUATIONS

An equation such as $\tan^2 \theta - 3 \tan \theta - 4 = 0$ is a second-degree equation in terms of $\tan \theta$. To find the values of θ that make this equation true, we first solve the equation for $\tan \theta$, using the same method we would use to solve $x^2 - 3x - 4 = 0$.

Quadratic equations with rational roots can be solved by factoring.

□ EXAMPLE 1:

Solve for x:

$$x^2 - 3x - 4 = 0$$
$$(x + 1)(x - 4) = 0$$
$$x + 1 = 0 \quad | \quad x - 4 = 0$$
$$x = -1 \quad | \quad x = 4$$

□ EXAMPLE 2:

Solve for $\tan \theta$:

$$\tan^2 \theta - 3 \tan \theta - 4 = 0$$
$$(\tan \theta + 1)(\tan \theta - 4) = 0$$
$$\tan \theta + 1 = 0 \quad | \quad \tan \theta - 4 = 0$$
$$\tan \theta = -1 \quad | \quad \tan \theta = 4$$

To determine the values of θ that satisfy $\tan^2 \theta - 3 \tan \theta - 4 = 0$, we can use the values of $\tan \theta$ that we just found. If the replacement set for θ is the set of numbers in the interval $0° \le \theta \le 360°$ and if θ is to be found to the nearest degree:

$\tan \theta = -1$
Since $\tan 45° = 1$, then:

$\tan \theta = 4$

In quadrant II, $\theta = 180° - 45° = 135°$.
In quadrant IV, $\theta = 360° - 45° = 315°$.

In quadrant I, $\theta = 76°$.
In quadrant III,
$\theta = 180° + 76° = 256°$.

Therefore, if $\tan^2 \theta - 3 \tan \theta - 4 = 0$ and the domain is $0° \le \theta \le 360°$, the solution set of all measures of θ to the nearest degree is:

$$\{76°, 135°, 256°, 315°\}$$

If the replacement set for θ is the set of all real numbers, then, to the nearest degree, θ can be any measure that differs from one of the measures just found by a multiple of $360°$, that is, for k an integer:

$$\theta = 135° + 360°k \qquad \theta = 76° + 360°k$$
$$\theta = 315° + 360°k \qquad \theta = 256° + 360°k$$

Any quadratic equation can be solved by using the quadratic formula. Compare the solutions that follow:

☐ EXAMPLE 3:   Find x to the nearest tenth if $3x^2 - 5x - 4 = 0$.

$$a = 3, b = -5, c = -4$$

$$x = \frac{-b \pm \sqrt{b^2 - 4ac}}{2a}$$

$$x = \frac{-(-5) \pm \sqrt{(-5)^2 - 4(3)(-4)}}{2(3)}$$

$$x = \frac{5 \pm \sqrt{25 + 48}}{6}$$

$$x = \frac{5 \pm \sqrt{73}}{6}$$

$$x = \frac{5 + 8.54}{6} \qquad x = \frac{5 - 8.54}{6}$$

$$x = \frac{13.54}{6} \qquad x = \frac{-3.54}{6}$$

$$x = 2.25 \qquad x = -.59$$

$$x = 2.3 \qquad x = -.6$$

Answer: $x = 2.3$ or $x = -.6$ OR   solution set $= \{2.3, -.6\}$

☐ EXAMPLE 4:   Find θ to the nearest degree if $0° \leq \theta \leq 360°$.

$$3 \cos^2 \theta - 5 \cos \theta - 4 = 0$$
$$a = 3, b = -5, c = -4$$

$$\cos \theta = \frac{-b \pm \sqrt{b^2 - 4ac}}{2a}$$

$$\cos \theta = \frac{-(-5) \pm \sqrt{(-5)^2 - 4(3)(-4)}}{2(3)}$$

$$\cos \theta = \frac{5 \pm \sqrt{25 + 48}}{6}$$

$$\cos \theta = \frac{5 \pm \sqrt{73}}{6}$$

$$\cos \theta = \frac{5 + 8.54}{6}$$

$$\cos \theta = \frac{13.54}{6}$$

$$\cos \theta = 2.25$$

(There is no solution for this value of $\cos \theta$.)

$$\cos \theta = \frac{5 - 8.54}{6}$$

$$\cos \theta = \frac{-3.54}{6}$$

$$\cos \theta = -.59$$

Since $\cos 54° \approx .59$, then:
In quadrant II, $\theta = 180° - 54° = 126°$.
In quadrant III, $\theta = 180° + 54° = 234°$.

Answer: $\theta = 126°$ or $\theta = 234°$ OR   solution set = $\{126°, 234°\}$

A quadratic equation in which there is no first-degree term can be solved by any of the methods shown in example 5.

☐ EXAMPLE 5: Solve $\sin^2 \theta - 1 = 0$ for $\sin \theta$.

Method 1

Use square root.

$$\sin^2 \theta - 1 = 0$$
$$\sin^2 \theta = 1$$
$$\sin \theta = \pm 1$$

Method 2

Use factoring, and set each factor equal to 0.

$$\sin^2 \theta - 1 = 0$$
$$(\sin \theta - 1)(\sin \theta + 1) = 0$$

$\sin \theta - 1 = 0$ | $\sin \theta + 1 = 0$
$\sin \theta = 1$ | $\sin \theta = -1$

Method 3

Use the quadratic formula.

$$\sin^2 \theta - 1 = 0$$
$$a = 1, b = 0, c = -1$$

$$\sin \theta = \frac{-b \pm \sqrt{b^2 - 4ac}}{2a}$$

$$\sin \theta = \frac{0 \pm \sqrt{0^2 - 4(1)(-1)}}{2(1)}$$

$$\sin \theta = \frac{0 \pm \sqrt{0 + 4}}{2} = \frac{\pm\sqrt{4}}{2}$$

$$\sin \theta = \pm \frac{2}{2} = \pm 1$$

| MODEL PROBLEMS |

1. Solve the equation $2 \cos^2 \theta = \cos \theta$ for all values of θ in the interval $0° \leq \theta < 360°$.

Solution

1. Write the qua-
    dratic equation in
    standard form.

$$2 \cos^2 \theta = \cos \theta$$
$$2 \cos^2 \theta - \cos \theta = 0$$

2. Factor the left-
    hand member of
    the equation.

$$\cos \theta (2 \cos \theta - 1) = 0$$

3. Let each factor
    equal 0, and solve
    for $\cos \theta$.

$\cos \theta = 0$ | $2 \cos \theta - 1 = 0$
$2 \cos \theta = 1$
$\cos \theta = \frac{1}{2}$

4. For each value of
    $\cos \theta$, find the
    value of θ in the
    given interval.

$\theta = 90°$ | $\theta = 60°$
$\theta = 270°$ | $\theta = 360° - 60° = 300°$

Answer: $60°, 90°, 270°, 300°$

2. In the interval $0° \leq \theta < 360°$, find to the nearest degree all values of θ that satisfy the equation $3 \sin \theta + 4 = \dfrac{1}{\sin \theta}$.

Solution

1. Write the equa-
    tion as an equiv-
    alent equation
    without fractions
    by multiplying
    each member by
    $\sin \theta$.

$$3 \sin \theta + 4 = \frac{1}{\sin \theta}$$
$$3 \sin^2 \theta + 4 \sin \theta = 1$$

2. Write the equa-
    tion in standard
    form, and solve
    by using the
    quadratic for-
    mula.

$$3 \sin^2 \theta + 4 \sin \theta - 1 = 0$$
$$a = 3, b = 4, c = -1$$
$$\sin \theta = \frac{-b \pm \sqrt{b^2 - 4ac}}{2a}$$
$$\sin \theta = \frac{-(4) \pm \sqrt{4^2 - 4(3)(-1)}}{2(3)}$$

$$\sin \theta = \frac{-4 \pm \sqrt{16 + 12}}{6}$$

$$\sin \theta = \frac{-4 \pm \sqrt{28}}{6}$$

3. Use logarithms or the square-root algorithm to find the approximate value of $\sqrt{28}$ to the nearest hundredth.

$\sin \theta = \dfrac{-4 + 5.29}{6}$	$\sin \theta = \dfrac{-4 - 5.29}{6}$
$\sin \theta = \dfrac{1.29}{6}$	$\sin \theta = \dfrac{-9.29}{6}$
$\sin \theta = .21$	$\sin \theta = -1.54$

In quadrant I, $\theta = 12°$.
In quadrant II, $\theta = 180° - 12°$
$= 168°$.

(There is no solution for this value of $\sin \theta$.)

Answer: $\theta = 12°$ or $\theta = 168°$ OR   solution set = $\{12°, 168°\}$

| **EXERCISES** |

In 1–8, find all values of θ in the interval $0° \le \theta < 360°$ that satisfy the given equation.

1. $\sin^2 \theta - \sin \theta = 0$
2. $\tan^2 \theta - 3 = 0$
3. $2 \cos^2 \theta - 1 = 0$
4. $\cos^2 \theta - \cos \theta = 2$
5. $2 \sin^2 \theta - \sin \theta = 1$
6. $3 \sec^2 \theta + 5 \sec \theta = 2$

7. $2 \sin \theta + 1 = \dfrac{1}{\sin \theta}$
8. $\tan \theta = \dfrac{1}{\tan \theta}$

9. Find the measure of the smallest positive acute angle for which $2 \sin^2 \theta - 3 \sin \theta + 1 = 0$.
10. Find a value of x, where $0° \le x < 360°$ and $\sin^2 x - 3 \sin x - 4 = 0$.
11. How many solutions to the equation $2 \sin^2 \theta - 3 \sin \theta + 1 = 0$ are in the interval $0° \le \theta \le 90°$?
12. How many solutions to the equation $2 \cos^2 \theta + 3 \cos \theta + 1 = 0$ are in the interval $0° \le \theta \le 90°$?
13. How many solutions to the equation $4 \cos^2 x + 3 \cos x - 1 = 0$ are in the interval $0° \le x < 180°$?
14. How many solutions to the equation $9 \sin^2 x - 6 \sin x + 1 = 0$ are in the interval $0° \le x < 360°$?

In 15–19, select the numeral preceding the expression that best completes the sentence.

15. The solution set of $\cos^2 \theta - \cos \theta - 2 = 0$ for $0° \le \theta < 360°$ is:
    (1) $\{0°\}$ (2) $\{180°\}$ (3) $\{0°, 180°\}$ (4) $\{60°, 180°, 300°\}$

16. The smallest positive measure for which $2 \cos^2 \theta = \cos \theta$ is:
    (1) $0°$ (2) $30°$ (3) $60°$ (4) $90°$

17. A value of θ that is *not* a solution of $\cos^2 \theta - \cos \theta = 0$ is:

    (1) 0 (2) $\dfrac{\pi}{2}$ (3) π (4) $\dfrac{3\pi}{2}$

18. A value of θ that is a solution of the equation $\sec \theta = \dfrac{1}{\sec \theta}$ is:

    (1) 0 (2) $\dfrac{\pi}{4}$ (3) $\dfrac{\pi}{2}$ (4) $\dfrac{3\pi}{2}$

19. A value of θ that is a solution of the equation $2 \cos^2 \theta + \cos \theta = 0$ is:
    (1) $60°$ (2) $-60°$ (3) $0°$ (4) $240°$

In 20–27, find to the nearest degree all values of θ in the interval $0° \le \theta < 360°$ that satisfy the given equation.

20. $\tan^2 \theta - 5 \tan \theta + 6 = 0$ 21. $4 \sin^2 \theta - 3 \sin \theta - 1 = 0$
22. $3 \cos^2 \theta - 7 \cos \theta + 2 = 0$ 23. $5 \sin^2 \theta + 4 \sin \theta = 0$
24. $25 \sin^2 \theta - 1 = 0$ 25. $\tan^2 \theta = 16$

26. $2 \tan \theta - 7 = \dfrac{4}{\tan \theta}$ 27. $\dfrac{2}{\cos \theta} = 5 \cos \theta + 3$

In 28–35, use the quadratic formula to find to the nearest degree all values of θ in the interval $0° \le \theta < 360°$ that satisfy the given equation.

28. $4 \sin^2 \theta - 2 \sin \theta - 3 = 0$ 29. $2 \tan^2 \theta - \tan \theta - 2 = 0$
30. $9 \cos^2 \theta - 6 \cos \theta = 2$ 31. $3 \sin^2 \theta - 1 = \sin \theta$
32. $5 \cos^2 \theta - 2 = 4 \cos \theta$ 33. $8(\sin^2 \theta - \sin \theta) = 1$
34. $1 - 4 \cos \theta = 2 \cos^2 \theta$ 35. $\tan^2 \theta = 8 \tan \theta - 5$

36. If $0° \le x < 360°$, find all values of x for which $4 \sin^3 x - \sin x = 0$.

13-13 EQUATIONS INVOLVING MORE THAN ONE FUNCTION

To solve a trigonometric equation that contains two or more functions of a variable, such as $2 \cos^2 \theta - \sin \theta = 1$, we will find it useful to express each variable term as the same function of the same variable. To do this in the given equation, we would need to express $\cos^2 \theta$ in terms of $\sin \theta$ or $\sin \theta$ in terms of $\cos \theta$.

Since $\cos^2 \theta = 1 - \sin^2 \theta$ and $\sin^2 \theta = 1 - \cos^2 \theta$ or $\sin \theta = \pm\sqrt{1 - \cos^2 \theta}$, it is simpler to express $\cos^2 \theta$ in terms of $\sin \theta$ than to use the radical form necessary to express $\sin \theta$ in terms of $\cos \theta$.

☐ EXAMPLE 1: Find all values of θ in the interval $0° \leq \theta < 360°$ for which $2 \cos^2 \theta - \sin \theta = 1$.

$$2 \cos^2 \theta - \sin \theta = 1$$
$$2(1 - \sin^2 \theta) - \sin \theta = 1$$
$$2 - 2 \sin^2 \theta - \sin \theta = 1$$
$$-2 \sin^2 \theta - \sin \theta + 1 = 0$$
$$2 \sin^2 \theta + \sin \theta - 1 = 0$$
$$(2 \sin \theta - 1)(\sin \theta + 1) = 0$$

$2 \sin \theta - 1 = 0$	$\sin \theta + 1 = 0$
$2 \sin \theta = 1$	$\sin \theta = -1$
$\sin \theta = \frac{1}{2}$	$\theta = 270°$
$\theta = 30°$ or $150°$	

The solution set = $\{30°, 150°, 270°\}$.

In a similar way, if the equation contains function values of two different but related angle measures, such as θ and 2θ, write the equation in terms of a single function of a single variable.

☐ EXAMPLE 2: Find to the nearest degree the measure of the positive acute angle that satisfies the equation $\cos 2\theta - 2 \sin \theta + 2 = 0$.

There are three identities that express $\cos 2\theta$ in terms of function values of θ. Since the equation also has a term in $\sin \theta$, we will use the identity for $\cos 2\theta$ that uses $\sin \theta$, that is, $\cos 2\theta = 1 - 2 \sin^2 \theta$.

Solution

$$\cos 2\theta - 2 \sin \theta + 2 = 0$$
$$1 - 2 \sin^2 \theta - 2 \sin \theta + 2 = 0$$
$$-2 \sin^2 \theta - 2 \sin \theta + 3 = 0$$
$$a = -2, b = -2, c = 3$$

$$\sin \theta = \frac{-b \pm \sqrt{b^2 - 4ac}}{2a}$$

$$\sin \theta = \frac{-(-2) \pm \sqrt{(-2)^2 - 4(-2)(3)}}{2(-2)}$$

$$\sin \theta = \frac{2 \pm \sqrt{28}}{-4}$$

$$\sin \theta = \frac{2 + 5.29}{-4}$$

$$\sin \theta = \frac{7.29}{-4}$$

$$\sin \theta = -1.82$$

(There is no solution for this value of $\sin \theta$.)

$$\sin \theta = \frac{2 - 5.29}{-4}$$

$$\sin \theta = \frac{-3.29}{-4}$$

$$\sin \theta = 0.82$$

$$\theta = 55° \quad Ans.$$

Equations that contain two different functions can sometimes be solved without substitution if the two functions can be separated by factoring.

☐ EXAMPLE 3: If $0 \leq \theta \leq \frac{\pi}{2}$, find θ when $2 \cos \theta \sin \theta - \cos \theta = 0$.

Solution

1. Factor the left-hand member.

$$2 \cos \theta \sin \theta - \cos \theta = 0$$
$$\cos \theta (2 \sin \theta - 1) = 0$$

2. Set each factor equal to 0. Notice that each factor contains only one function.

$$\cos \theta = 0 \quad \bigg| \quad 2 \sin \theta - 1 = 0$$

$$\theta = \frac{\pi}{2} \quad \bigg| \quad 2 \sin \theta = 1$$

$$\sin \theta = \frac{1}{2}$$

$$\theta = \frac{\pi}{6}$$

Answer: $\frac{\pi}{6}, \frac{\pi}{2}$

KEEP IN MIND

To solve trigonometric equations:

1. If the equation involves different but related angle measures, use identities to write each function in terms of the same angle measure.

2. If the equation involves different functions, separate the functions by factoring if possible, or use identities to express different functions in terms of the same function.

MODEL PROBLEMS

1. Find all values of θ in the interval $0° \le \theta < 360°$ that satisfy the equation $2(\sin \theta + \csc \theta) = 5$.

Solution

1. Replace $\csc \theta$ by $\dfrac{1}{\sin \theta}$.

$$2(\sin \theta + \csc \theta) = 5$$
$$2\left(\sin \theta + \frac{1}{\sin \theta}\right) = 5$$

2. Simplify the left-hand member by multiplication.

$$2 \sin \theta + \frac{2}{\sin \theta} = 5$$

3. Clear fractions by multiplying both members by $\sin \theta$.

$$2 \sin^2 \theta + 2 = 5 \sin \theta$$

4. Solve the quadratic equation for $\sin \theta$. (Note that the quadratic formula could have been used to solve for $\sin \theta$.)

$$2 \sin^2 \theta - 5 \sin \theta + 2 = 0$$
$$(2 \sin \theta - 1)(\sin \theta - 2) = 0$$

$2 \sin \theta - 1 = 0$	$\sin \theta - 2 = 0$
$2 \sin \theta = 1$	$\sin \theta = 2$
$\sin \theta = \frac{1}{2}$	

5. Find the values of θ in the given interval.

$\theta = 30°$	(No solution is possible for this value of $\sin \theta$.)
$\theta = 150°$	

Answer: $\theta = 30°$ or $\theta = 150°$ OR   solution set = $\{30°, 150°\}$

2. Solve for θ in the interval $-\dfrac{\pi}{2} \le \theta \le \dfrac{\pi}{2}$ when $2 \sin \theta = \tan \theta$.

Solution

1. Write the equation in terms of sines and cosines.

$$2 \sin \theta = \tan \theta$$
$$2 \sin \theta = \frac{\sin \theta}{\cos \theta}$$

2. Write an equivalent equation with one member equal to 0.

$$2 \sin \theta - \frac{\sin \theta}{\cos \theta} = 0$$

3. Factor the left-hand member.

$$\sin \theta \left(2 - \frac{1}{\cos \theta}\right) = 0$$

4. Set each factor equal to 0, and solve the resulting equations.

$$\sin \theta = 0 \qquad 2 - \frac{1}{\cos \theta} = 0$$

$$\theta = 0 \qquad 2 = \frac{1}{\cos \theta}$$

$$2 \cos \theta = 1$$

$$\cos \theta = \frac{1}{2}$$

$$\theta = \frac{\pi}{3}$$

$$\theta = -\frac{\pi}{3}$$

Answer: $\left\{ -\dfrac{\pi}{3}, \ 0, \ \dfrac{\pi}{3} \right\}$

| EXERCISES |

In 1–10, find all values of θ in the interval $0° \leq \theta \leq 360°$ that satisfy the given equation.

1. $4 \sin^2 \theta + 4 \cos \theta = 5$
2. $2 \cos^2 \theta + 3 \sin \theta - 3 = 0$
3. $\sec^2 \theta - \tan \theta - 1 = 0$
4. $2 \cos \theta + 1 = \sec \theta$
5. $2 \sin^2 \theta + 3 \cos \theta = 0$
6. $\cos^2 \theta + \sin \theta = 1$
7. $\tan \theta = 3 \cot \theta$
8. $2 \sin \theta = \csc \theta$
9. $\cos \theta = \sec \theta$
10. $\sin 2\theta = \tan \theta$

In 11–14, find all values of θ in the interval $0 \leq \theta < 2\pi$ that satisfy the given equation.

11. $\cos 2\theta + \cos \theta + 1 = 0$
12. $2 \sin^2 \theta - \cos 2\theta = 0$
13. $\cos 2\theta + \sin \theta = 0$
14. $\cos 2\theta + 3 \cos \theta + 2 = 0$

In 15–30, find to the nearest degree the values of θ in the interval $0° \leq \theta < 360°$ that satisfy the given equation.

15. $5 \sin^2 \theta + 3 \cos \theta = 3$
16. $5 \cos^2 \theta - 4 \sin \theta - 4 = 0$
17. $\sec^2 \theta + \tan \theta - 7 = 0$
18. $2 \sec^2 \theta + 7 \tan \theta - 6 = 0$
19. $5 \cos \theta + 4 = \sec \theta$
20. $7 \sin \theta + 1 = 6 \csc \theta$
21. $\tan \theta + 5 = 6 \cot \theta$
22. $\cot \theta - 2 \tan \theta - 1 = 0$
23. $3 \sin^2 \theta + 5 \cos \theta - 4 = 0$
24. $\cos^2 \theta + \sin \theta = 0$
25. $3 \cos^2 \theta + 2 \sin \theta - 1 = 0$
26. $2 \sec^2 \theta - 3 \tan \theta - 1 = 0$
27. $\cos \theta + 1 = \sec \theta$
28. $3 \cos 2\theta + 8 \sin \theta + 5 = 0$
29. $3 \cos 2\theta + 5 \cos \theta + 2 = 0$
30. $2 \cos 2\theta + \cos \theta = 0$

In 31–36: **a.** Express θ in inverse trigonometric form. **b.** Find the principal values of θ.

31. $2 \sin \theta \cos \theta + \sin \theta = 0$ **32.** $\sec \theta \tan \theta - 2 \tan \theta = 0$

33. $\cot \theta + \cot \theta \cos \theta = 0$ **34.** $2 \sin \theta \cos \theta + \sqrt{2} \cos \theta = 0$

35. $\sin 2\theta + \cos \theta = 0$ **36.** $\sin \theta - \sin 2\theta = 0$

37. Solve the equation $1 + \sin x = 2 \cos^2 x$ for the measure of a positive acute angle.

In 38 and 39, select the numeral preceding the expression that best completes the sentence.

38. For values of θ in the interval $0° \leq \theta < 360°$, the solution set of the equation $\dfrac{2 \sin 2\theta}{\cos \theta} - \dfrac{1}{\sin \theta} = 0$ is:

(1) $\{60°, 300°\}$ (2) $\{30°, 150°\}$

(3) $\{60°, 120°\}$ (4) $\{30°, 150°, 210°, 330°\}$

39. One value of x that satisfies the equation $\sin \theta = \cos 2\theta$ is:

(1) $0°$ (2) $30°$ (3) $90°$ (4) $210°$

13-14   REVIEW EXERCISES

1. If x and y are the measures of acute angles and $\sin x = \frac{5}{13}$ and $\cos y = \frac{3}{5}$, find: **a.** $\sin (x + y)$ **b.** $\cos (x - y)$ **c.** $\tan 2x$

d. $\cos \frac{1}{2} y$ **e.** $\sin \left(\frac{\pi}{2} + x \right)$ **f.** $\tan (-x)$

2. If $\sin \frac{1}{2} \theta = \dfrac{\sqrt{5}}{3}$, find $\cos \theta$.

In 3–6, prove the identity.

3. $\dfrac{1 + \cos 2\theta}{\sin 2\theta} = \cot \theta$ **4.** $\dfrac{2 \tan \theta - \sin 2\theta}{2 \tan \theta} = \sin^2 \theta$

5. $\sin 2\theta = \dfrac{2}{\tan \theta + \cot \theta}$ **6.** $\dfrac{1}{1 + \sin \theta} + \dfrac{1}{1 - \sin \theta} = 2 \sec^2 \theta$

In 7–10, solve the given equation for θ in the interval $0° \leq \theta \leq 360°$.

7. $3 - 3 \sin \theta - 2 \cos^2 \theta = 0$ **8.** $\sin 2\theta + 2 \cos \theta = 0$

9. $\sec^2 \theta - \tan \theta - 1 = 0$ **10.** $\cos 2\theta + \sin^2 \theta - 1 = 0$

In 11–14, find to the nearest degree the values of θ in the interval $0° \leq \theta \leq 360°$ that satisfy the given equation.

11. $3 \cos 2\theta + \cos \theta + 2 = 0$ **12.** $7 \cos^2 \theta - 4 \sin \theta = 4$

13. $2 \tan \theta - 2 \cot \theta - 3 = 0$ **14.** $9 \sin^2 \theta + 6 \cos \theta - 8 = 0$

15. Find the three factors of $9 \cos^3 \theta - \cos \theta$.
16. Find the three factors of $\cos \theta - \cos \theta \sin^2 \theta$.
17. Find the smallest positive value of θ for which $3 \tan^2 \theta - 1 = 0$.
18. Find the measure θ of the smallest acute angle that is a solution of the equation $2 \cos^2 \theta - 5 \cos \theta + 2 = 0$.
19. If x is the measure of an acute angle and $\cos x = \frac{1}{9}$, find $\sin \frac{1}{2}x$.
20. If θ is the measure of an obtuse angle and $\cos \theta = -\frac{8}{17}$, find $\tan \frac{1}{2}\theta$.
21. If x is the measure of an acute angle and $\sin x = .6$, find the value of $\sin (\pi + x)$.
22. If θ is the measure of an acute angle and $\sin \theta = \frac{1}{3}$, find $\cos 2\theta$.
23. If x is the measure of an angle in quadrant III and $\cos x = -\frac{3}{5}$, find $\sin 2x$.

In 24–37, select the numeral preceding the expression that best completes the sentence or answers the question.

24. The expression $\sin 2\theta \csc \theta$ is equivalent to:
    (1) $\sin^2 2\theta$ (2) $\sin^2 \theta$ (3) $2 \sin \theta$ (4) $2 \cos \theta$
25. The expression $\cos^2 2\theta + \sin^2 2\theta$ is equivalent to:
    (1) 1   (2) 2   (3) $\cos \theta$ (4) $\cos 4\theta$
26. Which of the following is *not* an identity?
    (1) $\sin (-x) = -\sin x$ (2) $\cos (-x) = -\cos x$
    (3) $\tan (-x) = -\tan x$ (4) $\cot (-x) = -\cot x$
27. If $\tan x = -\frac{1}{4}$ and $\tan y = 2$, then the value of $\tan (x + y)$ is:
    (1) $\frac{9}{2}$ (2) $\frac{7}{2}$ (3) $\frac{9}{6}$ (4) $\frac{7}{6}$
28. The expression $\cos 40° \cos 30° - \sin 40° \sin 30°$ is equivalent to:
    (1) $\sin 70°$ (2) $\sin 10°$ (3) $\cos 70°$ (4) $\cos 10°$
29. The value of $\sin 10° \cos 20° + \cos 10° \sin 20°$ is:
    (1) $\dfrac{\sqrt{3}}{2}$ (2) $\dfrac{1}{2}$ (3) approximately equal to .1736
    (4) approximately equal to .9848
30. The expression $\cos (x - 90°)$ is equivalent to:
    (1) $-\cos x$ (2) $\cos x$ (3) $-\sin x$ (4) $\sin x$
31. The expression $\sin \left(x - \dfrac{3\pi}{2} \right)$ is equivalent to:
    (1) $-\cos x$ (2) $\cos x$ (3) $-\sin x$ (4) $\sin x$
32. How many solutions to the equation $2 \cos^2 \theta + 3 \cos \theta + 1 = 0$ are there in the interval $0° \leq \theta \leq 360°$?
    (1) 0   (2) 2   (3) 3   (4) 4

33. How many solutions to the equation $\cos^2 \theta - 5 \cos \theta + 6 = 0$ are there in the interval $0° \le \theta \le 360°$?
    (1) 1   (2) 2   (3) 3   (4) 0

34. If $\cos (\theta - 40°) = \sin 50°$, then the value of θ in degrees is:
    (1) 40   (2) 50   (3) 80   (4) 90

35. If $\tan x = \frac{1}{2}$ and $\tan y = \frac{1}{3}$, then $\tan (x + y)$ equals:
    (1) 1   (2) $\frac{5}{6}$ (3) $\frac{6}{5}$ (4) $\frac{5}{7}$

36. If $\cos \theta = \frac{1}{8}$ and $270° < \theta < 360°$, then $\cos \frac{1}{2}\theta$ equals:
    (1) $\dfrac{3}{4}$ (2) $\dfrac{\sqrt{7}}{4}$ (3) $-\dfrac{3}{4}$ (4) $-\dfrac{\sqrt{7}}{4}$

37. If $A = \text{Arc cos } \dfrac{5}{13}$, what is the value of $\tan \dfrac{A}{2}$?
    (1) $\frac{2}{3}$ (2) $\frac{6}{5}$ (3) $\frac{3}{2}$ (4) $\frac{12}{5}$

38. Diagonal $\overline{BD}$ of quadrilateral $ABCD$ is perpendicular to $\overline{AB}$ and to $\overline{DC}$, $AD = BC$, and $m\angle A = 2\ m\angle DBC$.
    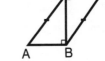
    a. Find $m\angle A$ and $m\angle DBC$.
    b. Prove: $ABCD$ is a parallelogram.

    (*Hint:* Express BD as function values of $\angle A$ and $\angle DCB$.)

Chapter 14

The Complex Numbers

14-1 IMAGINARY NUMBERS

In Chapter 4, we learned how to solve quadratic equations having real roots. For example:

Solve: $x^2 - 4 = 0$ *Solve:* $x^2 - 3 = 0$

$$x^2 = 4 \qquad\qquad x^2 = 3$$

$$x = \pm 2 \qquad\qquad x = \pm\sqrt{3}$$

Answer: $x = 2$ or $x = -2$ *Answer:* $x = \sqrt{3}$ or $x = -\sqrt{3}$

There are quadratic equations, however, that have no roots in the set of real numbers. For example, to solve $x^2 + 1 = 0$, we write the equivalent equation $x^2 = -1$. But the square of a real number is either positive or zero; that is, the square of a real number cannot be negative.

The equation $x^2 + 1 = 0$ will have two roots only if we agree to *extend our number system beyond the set of real numbers* to include numbers such as $\sqrt{-1}$ and $-\sqrt{-1}$.

Solve: $x^2 + 1 = 0$

$$x^2 = -1$$

$$x = \pm\sqrt{-1}$$

Answer: $x = \sqrt{-1}$ or $x = -\sqrt{-1}$

The numbers $\sqrt{-1}$ and $-\sqrt{-1}$ are *not* real numbers. Early mathematicians called these numbers "imaginary" because the numbers seemed to have little value and they certainly were not "real." In today's world, these numbers are of great use in solving many practical problems, especially in the field of electricity.

To simplify notation, we use i to represent the square root of negative one. Therefore:

$$i = \sqrt{-1} \qquad \text{AND} \qquad -i = -\sqrt{-1}$$

The number i, or the number $\sqrt{-1}$, is called the *imaginary unit* because it is the basis upon which a new set of numbers, called "imaginary numbers," is built.

■ **DEFINITION.** A *pure imaginary number* is any number that can be expressed in the form bi, where b is a real number such that $b \neq 0$, and i is the imaginary unit $\sqrt{-1}$.

Examples of pure imaginary numbers include $\sqrt{-25}, \sqrt{-7}, -2\sqrt{-9}$, and $\sqrt{-12}$ because:

1. $\sqrt{-25} = \sqrt{25}\sqrt{-1} = 5\sqrt{-1} = 5i$. (Although $5\sqrt{-1}$ and $5i$ indicate the same number, we will write the number in the form $5i$ in this book.)

2. $\sqrt{-7} = \sqrt{7}\sqrt{-1} = \sqrt{7}\,i = i\sqrt{7}$. (By writing $\sqrt{-7}$ as $i\sqrt{7}$, we make it clear that i is not a term under the radical sign.)

3. $-2\sqrt{-9} = -2\sqrt{9}\sqrt{-1} = -2 \cdot 3i = -6i$.

4. $\sqrt{-12} = \sqrt{12}\sqrt{-1} = \sqrt{4}\sqrt{3}\sqrt{-1} = 2\sqrt{3}\,i = 2i\sqrt{3}$.

■ **In general, for any real number b where $b > 0$:**

$$\sqrt{-b^2} = \sqrt{b^2}\sqrt{-1} = bi$$

Note: By the zero property of multiplication, $0i = 0 \cdot i = 0$, a real number.

Powers of i

Since the solution of $x^2 = -1$ is $x = \sqrt{-1}$ or $x = -\sqrt{-1}$, it follows that:

$$(\sqrt{-1})^2 = -1 \quad \text{AND} \quad (-\sqrt{-1})^2 = -1$$

However, $i = \sqrt{-1}$. Therefore, the solution of $x^2 = -1$ can be written as $x = i$ or $x = -i$. By substitution:

$$(i)^2 = -1 \quad \text{AND} \quad (-i)^2 = -1$$

Using $i = \sqrt{-1}$ and $i^2 = -1$, we can build a table of the powers of i.

$$i^3 = i \cdot i \cdot i = i^2 \cdot i = -1\sqrt{-1} = -\sqrt{-1}, \text{ or simply } -i.$$
$$i^4 = i \cdot i \cdot i \cdot i = i^2 \cdot i^2 = (-1)(-1) = 1.$$

Watch what happens now: $i^5 = i^4 \cdot i = 1 \cdot i = i$. Then, $i^6 = i^4 \cdot i^2 = 1(-1) = -1$, and $i^7 = i^4 \cdot i^3 = 1(-i) = -i$. Similarly, $i^8 = i^4 \cdot i^4 = 1 \cdot 1 = 1$. In the table, the powers of i repeat in a definite cycle, as seen below. Notice that $i^0 = 1$ by definition.

$i^0 = 1$	$i^1 = i$	$i^2 = -1$	$i^3 = -i$
$i^4 = 1$	$i^5 = i$	$i^6 = -1$	$i^7 = -i$
$i^8 = 1$	$i^9 = i$	$i^{10} = -1$	$i^{11} = -i$

The powers of i that fall into each of these four columns can be described by a rule.

■ **In general, for any whole number k:**

$$i^{4k} = 1 \qquad i^{4k+1} = i \qquad i^{4k+2} = -1 \qquad i^{4k+3} = -i$$

By moving in a clockwise direction within the circle at the right, we see that the powers of i behave exactly like a clock 4 system. In fact, when a whole number exponent is divided by 4, its remainder must be 0, 1, 2, or 3 (the same numbers on a clock 4). Powers of i can be simplified by using the rules stated above or the remainders 0, 1, 2, 3.

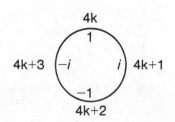

□ EXAMPLE: Write i^{82} in simplest terms.

Method 1: The Rules	*Method 2: Remainders*
Since $i^{82} = i^{4(20)+2}$ and $\quad i^{4k+2} = -1$, then	Since $82 \div 4 = 20$ with a *remainder of* 2, then i^{82} is equivalent to i^2. Thus: $\quad \dfrac{20}{4)\overline{82}}$ R.2
$i^{82} = i^{4(20)+2} = -1 \quad$ *Ans.*	$i^{82} = i^2 = -1 \quad$ *Ans.*

Properties and Operations

Many familiar properties are true for the set of imaginary numbers, including the *commutative* and *associative* properties of both addition and multiplication, and the *distributive* property of multiplication over addition. These properties are used in performing operations with imaginary numbers. In each of the following examples, notice that numbers are first expressed in terms of i, and then operations are performed.

Addition: $\sqrt{-16} + \sqrt{-9} = \sqrt{16}\sqrt{-1} + \sqrt{9}\sqrt{-1}$
$$= 4i + 3i = (4 + 3)i = 7i \quad Ans.$$

Subtraction: $\sqrt{-16} - \sqrt{-16} = \sqrt{16}\sqrt{-1} - \sqrt{16}\sqrt{-1}$
$$= 4i - 4i = (4 - 4)i = 0i = 0 \quad Ans.$$

Multiplication: $\sqrt{-16}\sqrt{-9} = \sqrt{16}\sqrt{-1} \cdot \sqrt{9}\sqrt{-1}$
$$= 4i \cdot 3i$$
$$= (4 \cdot 3)(i \cdot i)$$
$$= 12i^2 = 12(-1) = -12 \quad Ans.$$

Note: It is incorrect to use the rule $\sqrt{a}\sqrt{b} = \sqrt{ab}$ with pure imaginary numbers since this rule is true only when $a \geq 0$ and $b \geq 0$.

Division: $\sqrt{-16} \div \sqrt{-9} = \dfrac{\sqrt{-16}}{\sqrt{-9}} = \dfrac{\sqrt{16}\sqrt{-1}}{\sqrt{9}\sqrt{-1}} = \dfrac{4i}{3i} = \dfrac{4}{3}$ *Ans.*

The examples just studied illustrate that:

1. The sum, or the difference, of two pure imaginary numbers is a *pure imaginary number*, except when the result is $0i = 0$.
2. The product, or the quotient, of two pure imaginary numbers is always a *real number*.

| **MODEL PROBLEM** |

Express in terms of i the sum $4\sqrt{-18} + \sqrt{-50}$.

How to Proceed	*Solution*
1. Write each number in simplest radical form and in terms of i.	$4\sqrt{-18} + \sqrt{-50}$ $= 4\sqrt{9}\sqrt{2}\sqrt{-1} + \sqrt{25}\sqrt{2}\sqrt{-1}$ $= 4 \cdot 3\sqrt{2} \cdot i + 5\sqrt{2} \cdot i$ $= 12i\sqrt{2} + 5i\sqrt{2}$
2. Use the distributive property to simplify the expression.	$= (12 + 5)i\sqrt{2}$ $= 17i\sqrt{2}$ *Ans.*

| **EXERCISES** |

In 1–24, express each number in terms of i, and simplify.

1. $\sqrt{-36}$ 2. $\sqrt{-100}$ 3. $-\sqrt{-81}$ 4. $2\sqrt{-49}$

5. $\frac{1}{8}\sqrt{-64}$ 6. $-\frac{2}{3}\sqrt{-9}$ 7. $\frac{3}{4}\sqrt{-144}$ 8. $\frac{1}{3}\sqrt{-25}$

9. $\sqrt{-\frac{1}{4}}$ 10. $\sqrt{-\frac{16}{25}}$ 11. $4\sqrt{-\frac{49}{64}}$ 12. $\frac{3}{5}\sqrt{-\frac{100}{9}}$

13. $\sqrt{-3}$ 14. $\sqrt{-29}$ 15. $3\sqrt{-11}$ 16. $-\sqrt{-10}$

17. $\sqrt{-20}$ 18. $-\sqrt{-28}$ 19. $2\sqrt{-75}$ 20. $5\sqrt{-8}$

21. $\frac{2}{3}\sqrt{-72}$ 22. $-\frac{1}{2}\sqrt{-300}$ 23. $-\sqrt{-\frac{1}{3}}$ 24. $4\sqrt{-\frac{1}{8}}$

In 25–34, write the given power of i in simplest terms as 1, i, -1, or $-i$.

25. i^{12} 26. i^{7} 27. i^{49} 28. i^{72} 29. i^{54}

30. i^{99} 31. i^{300} 32. i^{246} 33. i^{91} 34. i^{2001}

In 35–56, write each number in terms of i, perform the indicated operation, and write the answer in simplest terms.

35. $\sqrt{-64} + \sqrt{-36}$

36. $3\sqrt{-4} + \sqrt{-121}$

37. $\sqrt{-100} - \sqrt{-9}$

38. $\sqrt{-16} - 2\sqrt{-4}$

39. $\sqrt{-45} + \sqrt{-5}$

40. $8\sqrt{-3} - \sqrt{-12}$

41. $\frac{1}{2}\sqrt{-200} - \sqrt{-32}$

42. $-2\sqrt{-18} - \frac{1}{5}\sqrt{-50}$

43. $\sqrt{-196} - \sqrt{-225}$

44. $\sqrt{-289} + \sqrt{-169}$

45. $\sqrt{-49} \cdot \sqrt{-1}$

46. $\sqrt{-81} \cdot \sqrt{-25}$

47. $\sqrt{-2} \cdot \sqrt{-18}$

48. $\sqrt{-5} \cdot \sqrt{-80}$

49. $-4\sqrt{-3} \cdot \sqrt{-3}$

50. $-3\sqrt{-10} \cdot 2\sqrt{-10}$

51. $6\sqrt{-6} \cdot \frac{2}{3}\sqrt{-6}$

52. $\frac{2}{3}\sqrt{-7} \cdot \sqrt{-63}$

53. $\dfrac{\sqrt{-400}}{\sqrt{-25}}$

54. $\dfrac{\sqrt{-98}}{\sqrt{-2}}$

55. $\dfrac{-\sqrt{-48}}{4\sqrt{-3}}$

56. $\dfrac{-\sqrt{-20}}{\sqrt{-180}}$

57. Express in terms of i the sum $\sqrt{-81} + 3\sqrt{-25} + \sqrt{-4}$.

58. Express in terms of i the sum $\sqrt{-72} + \sqrt{-32} + 3\sqrt{-8}$.

In 59–66, select the numeral preceding the expression that best completes the sentence or answers the question.

59. The expression $\sqrt{-192}$ is equivalent to:
(1) $-8\sqrt{3}$ (2) $8\sqrt{3}$ (3) $8i\sqrt{3}$ (4) $-8i\sqrt{3}$

60. If $\sqrt{-60}$ is subtracted from $\sqrt{-135}$, the difference is:
(1) $\sqrt{-75}$ (2) $i\sqrt{15}$ (3) $-i\sqrt{15}$ (4) i

61. The product $i^8 \cdot i^9 \cdot i^{10}$ equals:
(1) 1 (2) i (3) -1 (4) $-i$

62. The product of $2i^2 \cdot 3i^3$ is:
(1) 6 (2) $6i$ (3) -6 (4) $-6i$

63. The value of $(3i^3)^2$ is:
(1) 6 (2) -6 (3) 9 (4) -9

64. The sum of $5i^5$ and i^9 is:
(1) $6i$ (2) $5i$ (3) -5 (4) -6

65. The solution set of $x^2 + 9 = 0$ is:
(1) $\{-3\}$ (2) $\{3, -3\}$ (3) $\{3i, -3i\}$ (4) $\{i, -i\}$

66. Which expression is equal to *zero*?
(1) $i^2 \cdot i^2$ (2) $i^2 + i^2$ (3) $i^4 \cdot i^2$ (4) $i^4 + i^2$

In 67–69, solve the equation, and express roots in terms of i.

67. $x^2 + 64 = 0$ **68.** $x^2 = 2x^2 + 16$ **69.** $2x^2 + 10 = 0$

70. **a.** Is the set of pure imaginary numbers closed under addition? Explain why. **b.** Is there an identity element for addition in the set of pure imaginary numbers? Explain why.
71. Give as many reasons as possible to indicate why the set of pure imaginary numbers is not a group under multiplication.

14-2 COMPLEX NUMBERS

The discovery of the set of pure imaginary numbers enables us to define still another set of numbers, called the complex numbers.

■ **DEFINITION.** A *complex number* is any number that can be expressed in the form $a + bi$, where a and b are real numbers and i is the imaginary unit.

Examples of complex numbers include $2 + 5i$, $-3 + 0i$, $0 + 2i$, and $0 + 0i$.

In Fig. 1, we see the intersection of two number lines. The *real number line* is drawn horizontally, and the *pure imaginary number line* is drawn vertically. Since $0i = 0$, it is natural that these number lines intersect at a point that represents 0 on the real number line and $0i$ on the imaginary number line. Therefore, this point of intersection represents the complex number $0 + 0i$.

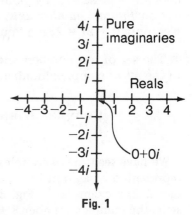

Fig. 1

In Fig. 2, the complex number plane is formed by using the two axes we have just studied. The real number axis is called the x-axis, and the pure imaginary number axis is called the yi-axis.

The complex number plane is similar to the rectangular coordinate system studied earlier. In the same way that we located the point $(x, y) = (2, 5)$, we now locate the point for the complex number $x + yi = 2 + 5i$. That is, we use the rectangular grid to locate the point of intersection of the real component 2 and the imaginary component $5i$.

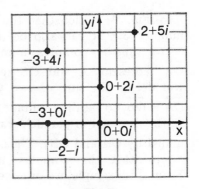

Fig. 2

By studying other points on the complex number plane, we observe:

1. *Any complex number a + bi where b = 0 is a real number.*

For example, $-3 + 0i = -3$, represented by a point on the real number axis. Conversely, every real number can be expressed as a complex number, as in $5 = 5 + 0i$.

2. *Any complex number a + bi where a = 0 and b ≠ 0 is a pure imaginary number.*

For example, $0 + 2i = 2i$, represented by a point on the pure imaginary axis. Conversely, every pure imaginary number can be expressed as a complex number, as in $-4i = 0 - 4i$.

3. *Any complex number a + bi where b ≠ 0 is an imaginary number.*

For example, the complex numbers $-3 + 4i$, $-2 - i$, $2 + 5i$, and $0 + 2i$ can simply be called imaginary numbers. Notice that these numbers are represented by points in the complex number plane that are *not on the real number axis*. Of the imaginary numbers cited as examples, only $0 + 2i = 2i$ is a "pure imaginary" number.

■ **The set of real numbers and the set of imaginary numbers are subsets of the set of complex numbers.**

Complex Numbers, Points, and Vectors

We have seen that a complex number can be represented by a *point* in the complex number plane. For example, in Fig. 3, point C represents the complex number $3 + 2i$.

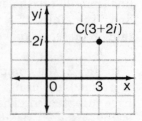

Fig. 3

In Fig. 4, notice that the same complex number $3 + 2i$ can also be represented as a *vector* in the complex number plane. Let $\overrightarrow{OA}$ represent 3, and $\overrightarrow{OB}$ represent $2i$. We will define the sum of two vectors, such as $\overrightarrow{OA} + \overrightarrow{OB}$, to be the resultant vector $\overrightarrow{OC}$ (that is, the diagonal in parallelogram $OACB$ determined by $\overrightarrow{OA}$ and $\overrightarrow{OB}$). Thus, $\overrightarrow{OC}$ is a vector representing the complex number $3 + 2i$. From this example, we observe:

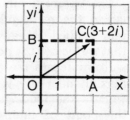

Fig. 4

■ **Every complex number can be represented as a point or as a vector in the complex number plane.**

Equality of Complex Numbers

Two complex numbers are equal if and only if their real components are equal and their imaginary components are equal. In symbols, $a + bi = c + di$ if and only if $a = c$ and $bi = di$.

For example, $x + 3i = -5 + yi$ if and only if $x = -5$ and $3i = yi$ (or $y = 3$).

EXERCISES

1. Write the complex number that is represented by each of the vectors drawn in the accompanying diagram:

   a. $\overrightarrow{OA}$ b. $\overrightarrow{OB}$ c. $\overrightarrow{OC}$

   d. $\overrightarrow{OD}$ e. $\overrightarrow{OE}$ f. $\overrightarrow{OF}$

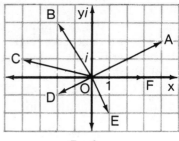

Ex. 1

In 2-7, find the real numbers a and b that will make the equation true.

2. $a + bi = 7 + 2i$ 3. $a - 6i = 4 + bi$

4. $a + bi = 12 + i - 3$ 5. $3i = a + bi$

6. $a + \sqrt{-16} = 16 + bi$ 7. $-\sqrt{25} = a + bi$

In 8-11: **a.** Tell whether the statement is *true* or *false*. **b.** If the statement is false, explain why.

8. The set of real numbers is a subset of the set of complex numbers.
9. If $b = 0$, then the number $a + bi$ is a real number.
10. If $a = 0$, then the number $a + bi$ is a pure imaginary number.
11. Every point on the complex number plane that does not represent a real number must represent an imaginary number.

14-3 ADDITION AND SUBTRACTION OF COMPLEX NUMBERS

It seems natural to treat the sum of two complex numbers as the sum of two binomials. By adding like terms, we find the sum of the real components and the sum of the pure imaginary components. At the right, an addition is performed in a vertical format. This addition may also be written in a horizontal format:

Add:
$2 + 3i$
$5 + \ i$
$7 + 4i$

$$(2 + 3i) + (5 + i) = (2 + 5) + (3i + i) = 7 + 4i$$

We can use vector addition to add the same two complex numbers. In the diagram at the right, let $\overrightarrow{OA}$ represent the complex number $2 + 3i$, and let $\overrightarrow{OB}$ represent $5 + i$. The sum of the vectors $\overrightarrow{OA}$ and $\overrightarrow{OB}$ is the resultant vector $\overrightarrow{OC}$ (that is, the diagonal in parallelogram $OACB$ determined by $\overrightarrow{OA}$ and $\overrightarrow{OB}$). Thus, $\overrightarrow{OC}$ represents the complex number $7 + 4i$, and this geometric demonstration verifies the sum found earlier by algebraic methods.

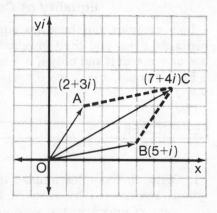

In general, whether we treat complex numbers as binomials or as vectors, we define the addition of complex numbers to be:

$$(a + bi) + (c + di) = (a + c) + (b + d)i$$

Properties

The set of complex numbers under addition, or (Complex numbers, +), can be shown to have the following properties:

1. *Closure.* By the definition of addition, the sum of two complex numbers is a complex number. Thus, (Complex numbers, +) is closed.

2. *Associativity.* (Complex numbers, +) is associative. In symbols:

$$((a + bi) + (c + di)) + (e + fi) = (a + bi) + ((c + di) + (e + fi))$$

3. *Identity.* The identity element for the addition of real numbers is 0. However, $0 = 0 + 0i$. By this extension, we state that the identity element for (Complex numbers, +) is $0 + 0i$. This is true because:

$$(a + bi) + (0 + 0i) = a + bi \quad \text{AND} \quad (0 + 0i) + (a + bi) = a + bi$$

4. *Inverses.* The additive inverse of the complex number $a + bi$ is $-a - bi$ because:

$$(a + bi) + (-a - bi) = 0 + 0i \quad \text{AND} \quad (-a - bi) + (a + bi) = 0 + 0i$$

5. *Commutativity.* (Complex numbers, +) is commutative. In symbols:

$$(a + bi) + (c + di) = (c + di) + (a + bi)$$

Since these five properties are true, it follows that:

■ **(Complex numbers, +) is a commutative group.**

Subtraction

Subtraction has been defined as the addition of an additive inverse. That is, $x - y = x + (-y)$. In the following example, $3 + 2i$ is subtracted from $1 + 3i$ by adding $-3 - 2i$ (the additive inverse of $3 + 2i$) to $1 + 3i$.

$$(1 + 3i) - (3 + 2i)$$
$$= (1 + 3i) + (-3 - 2i) = (1 - 3) + (3i - 2i) = -2 + i \quad Ans.$$

Let us use vectors to demonstrate this subtraction. In the diagram at the right, let $\overrightarrow{OA}$ represent $1 + 3i$, and let $\overrightarrow{OB}$ represent $3 + 2i$. Since $-(3 + 2i) = -3 - 2i$, the additive inverse of $\overrightarrow{OB}$ is $\overrightarrow{OB'}$ or $-3 - 2i$.

Therefore, the subtraction of vectors, $\overrightarrow{OA} - \overrightarrow{OB}$, is treated as an addition of vectors, $\overrightarrow{OA} + \overrightarrow{OB'}$, where $\overrightarrow{OB'}$ is the additive inverse of $\overrightarrow{OB}$. The resultant $\overrightarrow{OC}$ represents $-2 + i$, which is the difference $\overrightarrow{OA} - \overrightarrow{OB}$, or $(1 + 3i) - (3 + 2i)$.

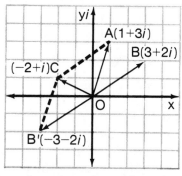

$$\overrightarrow{OA} - \overrightarrow{OB} = \overrightarrow{OA} + \overrightarrow{OB'} = \overrightarrow{OC}$$

In general, subtraction of complex numbers is defined as follows:

$$(a + bi) - (c + di) = (a - c) + (b - d)i$$

Note that the additive inverse of a complex number presented graphically is equivalent to moving the complex number through a point reflection in the origin, or a rotation of 180° about the origin.

| MODEL PROBLEMS |

1. Express the sum of $(5 + \sqrt{-36})$ and $(3 - \sqrt{-16})$ in the form $a + bi$.

   Solution

   $$(5 + \sqrt{-36}) + (3 - \sqrt{-16})$$
   $$= (5 + 6i) + (3 - 4i)$$
   $$= (5 + 3) + (6i - 4i)$$
   $$= 8 + 2i \quad Ans.$$

2. Subtract $6 - 2i\sqrt{3}$ from $5 - 3i\sqrt{3}$.

   Solution

   $$(5 - 3i\sqrt{3}) - (6 - 2i\sqrt{3})$$
   $$= (5 - 3i\sqrt{3}) + (-6 + 2i\sqrt{3})$$
   $$= (5 - 6) + (-3i\sqrt{3} + 2i\sqrt{3})$$
   $$= -1 - i\sqrt{3} \quad Ans.$$

| EXERCISES |

In 1–14, perform the operation and express the result in the form $a + bi$.

1. $(10 + 3i) + (5 + 8i)$
2. $(7 - 2i) + (3 - 6i)$
3. $(4 - 2i) + (-3 + 2i)$
4. $(6 + 3i) - (2 + i)$
5. $(-8 + 5i) - (5 - 7i)$
6. $(9 - 2i) - (9 - 5i)$
7. $(1.3 + 4i) + (2.9 - 1.7i)$
8. $(3.1 - 0.6i) - (4.8 - 0.4i)$
9. $\left(\dfrac{2}{3} - \dfrac{i}{4}\right) + \left(\dfrac{1}{6} - \dfrac{i}{2}\right)$
10. $\left(\dfrac{4}{5} + \dfrac{3}{8}i\right) - \left(\dfrac{3}{10} + \dfrac{3}{4}i\right)$
11. $(8 + \sqrt{-9}) - (10 + \sqrt{-4})$
12. $(-2 + \sqrt{-12}) + (8 + \sqrt{-27})$
13. $(-1 - \sqrt{-80}) - (3 + \sqrt{-20})$
14. $(5 - \sqrt{-128}) + (-5 - \sqrt{-98})$

15. Add: $5 + i, 7 - 3i, 12 + 6i, -10 - 8i,$ and $-14 + 4i$.
16. Subtract $2 - 13i$ from $7 - 5i$.
17. From the sum of $3 - i$ and $-2 - 2i$, subtract $4 - 5i$.
18. Express the sum of $9 + \sqrt{-9}$ and $5 - \sqrt{-16}$ in the form $a + bi$.
19. Express the difference, $(5 - \sqrt{-50}) - (-2 + \sqrt{-162})$, in the form $a + bi$.

In 20–25: a. Express the indicated sum or difference in the form $a + bi$. b. Demonstrate how this sum or difference is found using *vectors* in a complex number plane. (*Hint:* The parallelogram rule does not apply in exercises 24 and 25.)

20. $(2 + 3i) + (3 - 2i)$
21. $(-3 - i) - (1 + 4i)$
22. $(-2 - i) - (-2 + 4i)$
23. $(5 - 4i) - (2 + 0i)$
24. $(3 + i) + (6 + 2i)$
25. $(6 - 4i) - (3 - 2i)$

26. *True* or *False:* The sum of any complex number and its additive inverse is a real number. Explain why.

14-4 MULTIPLICATION OF COMPLEX NUMBERS

The product of two complex numbers can be treated as the product of two binomials. Notice, however, in the following example that i^2 is replaced by -1. In this way, the product is then written in the form of a complex number.

☐ Example: Find the product of $(3 + 2i)$ and $(2 + i)$.

Method 1: The Distributive Property

$$(3 + 2i)(2 + i) = 3(2 + i) + 2i(2 + i)$$
$$= 6 + 3i + 4i + 2i^2$$
$$= 6 + 3i + 4i + 2(-1)$$
$$= 6 + 3i + 4i - 2$$
$$= 4 + 7i \quad Ans.$$

Method 2: Mental Arithmetic

$$(3 + 2i)(2 + i) = 6 + 7i + 2i^2$$
$$= 6 + 7i + 2(-1)$$
$$= 6 + 7i - 2$$
$$= 4 + 7i \quad Ans.$$

We may use either method to prove the following statement:

■ **The product of two complex numbers is a complex number.**

Proof:

$$(a + bi)(c + di) = a(c + di) + bi(c + di)$$
$$= ac + adi + bci + bdi^2$$
$$= ac + adi + bci + bd(-1)$$
$$= ac + adi + bci - bd$$
$$= (ac - bd) + (adi + bci)$$

Thus, $(a + bi)(c + di) = (ac - bd) + (ad + bc)i$

In this proof, since $(a + bi)$ and $(c + di)$ are complex numbers, then a, b, c, and d are reals. By the closure properties of addition and multiplication, both $(ac - bd)$ and $(ad + bc)$ are real numbers. Therefore, the product $(ac - bd) + (ad + bc)i$ is a complex number by definition.

Multiplication, Transformations, and Vectors

To understand the multiplication of complex numbers from a graphic point of view, we will multiply $(3 + 2i)(2 + i)$ in a step-by-step manner. Observe how transformations are used in these steps. Since $(3 + 2i)(2 + i) = 3(2 + i) + 2i(2 + i)$, let $\overrightarrow{OA}$ represent $(2 + i)$ in each step.

Step 1:

$$3(2 + i) = 6 + 3i$$
$$3 \cdot \overrightarrow{OA} = \overrightarrow{OA'}$$

Multiplication by 3 is equivalent to D_3, a *dilation of 3* with the origin as the center of dilation.

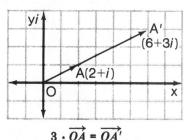

$$3 \cdot \overrightarrow{OA} = \overrightarrow{OA'}$$

Step 2: $i(2 + i) = 2i + i^2$
$= 2i + (-1)$
$= -1 + 2i$
$i \cdot \overrightarrow{OA} = \overrightarrow{OB}$

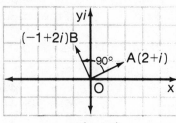

$i \cdot \overrightarrow{OA} = \overrightarrow{OB}$

Multiplication by i is equivalent to $R_{90°}$, a counterclockwise rotation of 90° about the origin.

Step 3: $2i(2 + i) = 4i + 2i^2$
$= 4i - 2$
$= -2 + 4i$
$2i \cdot \overrightarrow{OA} = \overrightarrow{OB'}$, or
$2(\overrightarrow{OB}) = \overrightarrow{OB'}$

Since multiplication by 2 is equivalent to D_2 and multiplication by i is equivalent to $R_{90°}$, *multiplication by 2i is equivalent to the composition of $D_2 \circ R_{90°}$.*

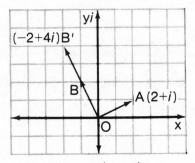

$2i \cdot \overrightarrow{OA} = \overrightarrow{OB'}$
OR, $2(\overrightarrow{OB}) = \overrightarrow{OB'}$

Step 4:   Use vector addition to find the resultant $\overrightarrow{OC}$. This single graph includes steps 1, 2, and 3.

$(3 + 2i)(2 + i)$
$= 3(2 + i) + 2i(2 + i)$
$= 3 \cdot \overrightarrow{OA} + 2i \cdot \overrightarrow{OA}$
$= \overrightarrow{OA'} + \overrightarrow{OB'}$
$= \overrightarrow{OC}$
$= 4 + 7i$

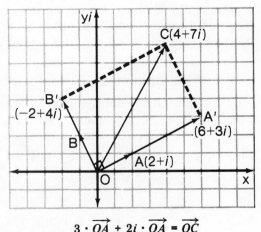

$3 \cdot \overrightarrow{OA} + 2i \cdot \overrightarrow{OA} = \overrightarrow{OC}$
OR, $\overrightarrow{OA'} + \overrightarrow{OB'} = \overrightarrow{OC}$

Conjugates

The *conjugate* of the complex number $a + bi$ is the complex number $a - bi$. For example, the conjugate of $5 + 2i$ is $5 - 2i$. Similarly, $3 - i$ and $3 + i$ are conjugates of each other. Consider the following products:

$$(5 + 2i)(5 - 2i) = 25 + 10i - 10i - 4i^2$$
$$= 25 + 10i - 10i + 4$$
$$= 29 + 0i$$
$$= 29$$

$$(3 - i)(3 + i) = 9 - 3i + 3i - i^2$$
$$= 9 - 3i + 3i + 1$$
$$= 10 + 0i$$
$$= 10$$

It is true that the product of two complex numbers is a complex number. However, when we multiply complex numbers that are conjugates, the imaginary component in the product is $0i$, which equals 0. We may now say:

■ **The product of two complex numbers that are conjugates is a real number.**

Proof:

$$(a + bi)(a - bi) = a^2 + abi - abi - b^2i^2$$
$$= a^2 + abi - abi + b^2$$
$$= (a^2 + b^2) + (abi - abi)$$
$$= (a^2 + b^2) + (ab - ab)i$$
$$= (a^2 + b^2) + 0i$$

Thus, $(a + bi)(a - bi) = a^2 + b^2$

Since $(a + bi)$ and $(a - bi)$ are complex numbers, then a and b are real. By the closure properties of multiplication and addition, the product $a^2 + b^2$ must be a real number.

In the next section, we will study properties involving the multiplication of complex numbers. For now, let us practice working with the operation.

| **MODEL PROBLEMS** |

1. Express the product of $(3 + 7i)$ and $(1 - 2i)$ in the form $a + bi$.

   Solution: Use either method. Let $i^2 = -1$, and simplify.

   Method 1

   $$(3 + 7i)(1 - 2i) = 3(1 - 2i) + 7i(1 - 2i)$$
   $$= 3 - 6i + 7i - 14i^2$$
   $$= 3 - 6i + 7i - 14(-1)$$
   $$= 3 - 6i + 7i + 14$$
   $$= 17 + 1i$$

   Method 2

   $$(3 + 7i)(1 - 2i) = 3 + 1i - 14i^2$$
   $$= 3 + i - 14(-1)$$
   $$= 3 + i + 14$$
   $$= 17 + i$$

   Answer: $17 + 1i$ or $17 + i$

2. Express the number $(4 - i)^2 - 8i^3$ in *simplest terms*.

How to Proceed

Perform the operations. Reduce terms so that i^1 or i is the highest power of i. (*Note:* In simplest form, $a + 0i = a$.)

Solution

$$(4 - i)^2 - 8i^3 = (4 - i)(4 - i) - 8i^3$$
$$= 16 - 8i + i^2 - 8i^2 \cdot i$$
$$= 16 - 8i + (-1) - 8(-1)i$$
$$= 16 - 8i - 1 + 8i$$
$$= 15 + 0i$$
$$= 15 \quad Ans.$$

| EXERCISES |

In 1–13, write the product of the given numbers in the form $a + bi$.

1. $(3 + i)(4 + i)$ **2.** $(5 - i)(3 + i)$ **3.** $(1 + 3i)(2 - i)$
4. $(4 - 5i)(2 + i)$ **5.** $(7 - i)(1 - 2i)$ **6.** $(8 - 5i)(3 - i)$
7. $(1 - 3i)(5 - 3i)$ **8.** $(6 - i)(6 + i)$ **9.** $(2 - i)(4 + 2i)$
10. $(8 + \sqrt{-25})(2 + \sqrt{-1})$ **11.** $(3 - \sqrt{-49})(2 + \sqrt{-16})$
12. $(2 + \sqrt{-9})(3 - \sqrt{-4})$ **13.** $(5 - \sqrt{-36})(2 - \sqrt{-100})$

In 14–19, multiply the conjugates, and write the product as a real number.

14. $(4 + i)(4 - i)$ **15.** $(9 - i)(9 + i)$ **16.** $(1 + 5i)(1 - 5i)$
17. $(7 - 2i)(7 + 2i)$ **18.** $(\sqrt{5} + i)(\sqrt{5} - i)$ **19.** $(\sqrt{6} - 3i)(\sqrt{6} + 3i)$

20. a. Express the product of $(2 + i)$ and $(3 + i)$ in the form $a + bi$.
    b. Find the product $(2 + i)(3 + i)$ using graphic methods. (*Hint:* Let $\overrightarrow{OA}$ represent $(3 + i)$. Use $2(3 + i) + i(3 + i)$ to find the product.)

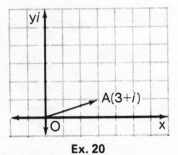

Ex. 20

21. a. Express the product $(1 + 2i)(2 + 5i)$ in the form $a + bi$.
    b. Find the product $(1 + 2i)(2 + 5i)$ by using graphic methods.

In 22-27, raise the complex number to the indicated power, and write the answer in the form $a + bi$.

22. $(6 + i)^2$ 23. $(8 - i)^2$ 24. $(3 - 2i)^2$
25. $(2 + 5i)^2$ 26. $(1 + i)^3$ 27. $(2 - 2i)^3$

In 28-48, perform the indicated operations, and write the answer in *simplest terms*. That is, whenever possible, write the complex number answer as a real number or a pure imaginary number.

28. $(10 + i)(10 - i)$ 29. $(3 + 5i)(3 - 5i)$ 30. $(1 + 6i)(6 + i)$
31. $(2 + 5i)(5 + 2i)$ 32. $(9 + 3i)(6 - 2i)$ 33. $(4 - 3i)(3 - 4i)$
34. $(7 - 10i)(1 + 0i)$ 35. $(8 + 9i)(1 + 0i)$ 36. $(2 + 0i)(8 - 5i)$
37. $i(4 + i) - 4i$ 38. $i(9 - 2i) - 2$ 39. $i(1 + i) + i^3$
40. $5(2 - i) - 5i^3$ 41. $(8 + i)(8 - i)$ 42. $8 + i(8 - i)$
43. $3i^3 - 3(2 - i)$ 44. $6i^3 + 2(7 + 3i)$ 45. $i^3 + i(10 - i)$
46. $(2 + i)^2 - 3$ 47. $(3 - 4i)^2 + 7$ 48. $(1 + i)^2 - 2i$

49. Express the product of $2 + i\sqrt{5}$ and its conjugate in simplest terms.

In 50 and 51, select the numeral preceding the expression that best completes the sentence.

50. The product $(3 + 3i)(3 - 3i)$ equals:
    (1) 0   (2) 9   (3) 18   (4) $9 - 9i$
51. The complex number $i^3 - i(5 - i)$ is equivalent to:
    (1) $1 - 6i$ (2) $-1 - 6i$ (3) $-1 - 4i$ (4) $-5 + i$

14-5 MULTIPLICATIVE INVERSES AND DIVISION OF COMPLEX NUMBERS

As we define the multiplicative identity and inverses in the complex number system, let us recall these elements in the real number system.

Real Numbers	Complex Numbers
The identity element for multiplication is 1 because, for any real number n: $$n \cdot 1 = n \quad \text{AND} \quad 1 \cdot n = n$$	The *identity element for multiplication* is $1 + 0i$ because, for any complex number $a + bi$: $$(a + bi)(1 + 0i) = a + bi$$ AND $$(1 + 0i)(a + bi) = a + bi$$

Since $1 + 0i = 1$, we will often use this simpler form of the identity when working with complex numbers. Let us now consider inverses.

Real Numbers	Complex Numbers
The multiplicative inverse of a real number n is $\dfrac{1}{n}$, where $n \neq 0$, because: $$n \cdot \frac{1}{n} = 1 \quad \text{AND} \quad \frac{1}{n} \cdot n = 1$$	The **multiplicative inverse** of a complex number $a + bi$ is $\dfrac{1}{a + bi}$, where $a + bi \neq 0 + 0i$, because: $$(a + bi) \cdot \frac{1}{a + bi} = 1$$ AND $$\frac{1}{a + bi} \cdot (a + bi) = 1$$

There is a problem, however, in stating that $\dfrac{1}{a + bi}$ is the multiplicative inverse of $a + bi$. For example, the multiplicative inverse of $3 - i$ is $\dfrac{1}{3 - i}$. But $\dfrac{1}{3 - i}$ is not written in the form of a complex number.

To find the complex number that is equivalent to $\dfrac{1}{3 - i}$, we must *rationalize the denominator*. That is, we multiply both numerator and denominator by the *conjugate* $3 + i$, which, in effect, multiplies the fraction by the identity element 1. Therefore:

$$\frac{1}{3 - i} = \frac{1}{(3 - i)} \cdot \frac{(3 + i)}{(3 + i)} = \frac{3 + i}{9 - i^2} = \frac{3 + i}{9 + 1} = \frac{3 + i}{10}, \text{ or } \frac{3}{10} + \frac{1}{10}i$$

It can be shown that $(3 - i)$ and $(\frac{3}{10} + \frac{1}{10}i)$ are multiplicative inverses because their product is the identity element $1 + 0i$, or 1.

$$\begin{aligned} \textit{Check:} \quad (3 - i)(\tfrac{3}{10} + \tfrac{1}{10}i) &= \tfrac{9}{10} + \tfrac{3}{10}i - \tfrac{3}{10}i - \tfrac{1}{10}i^2 \\ &= \tfrac{9}{10} + \tfrac{3}{10}i - \tfrac{3}{10}i + \tfrac{1}{10} \\ &= (\tfrac{9}{10} + \tfrac{1}{10}) + (\tfrac{3}{10}i - \tfrac{3}{10}i) \\ &= \tfrac{10}{10} + 0i = 1 + 0i = 1 \end{aligned}$$

Properties

The identity and inverses are only two properties of the complex number system under multiplication. Let us list some important properties for (Complex numbers, $\cdot$).

1. *Closure.* Recall that $(a + bi)(c + di) = (ac - bd) + (ad + bc)i$. Since the product of two complex numbers is a complex number, it follows that (Complex numbers, ·) is closed.

2. *Associativity.* (Complex numbers, ·) is associative. In symbols:

$$((a + bi) + (c + di)) + (e + fi) = (a + bi) + ((c + di) + (e + fi))$$

3. *Identity.* The identity element in (Complex numbers, ·) is $1 + 0i$, which may be written in simplest terms as 1.

4. *Inverses.* To find the multiplicative inverse of a complex number $a + bi$, where $a + bi \neq 0 + 0i$, we write a fraction and rationalize its denominator:

$$\frac{1}{a + bi} = \frac{1}{(a + bi)} \cdot \frac{(a - bi)}{(a - bi)} = \frac{a - bi}{a^2 - b^2 i^2} = \frac{a - bi}{a^2 + b^2} = \frac{a}{a^2 + b^2} - \frac{bi}{a^2 + b^2}$$

In the final form of the inverse, the denominator $a^2 + b^2$ cannot equal 0. However, there is only one complex number, namely, $0 + 0i$, where $a^2 + b^2 = 0^2 + 0^2 = 0$. Therefore, every complex number except $0 + 0i$ has an inverse under multiplication.

■ The multiplicative inverse of $a + bi$, where $a + bi \neq 0 + 0i$, is:

$$\frac{a - bi}{a^2 + b^2} \qquad \text{OR} \qquad \frac{a}{a^2 + b^2} - \frac{bi}{a^2 + b^2}$$

5. *Commutativity.* (Complex numbers, ·) is commutative. In symbols:

$$(a + bi)(c + di) = (c + di)(a + bi)$$

By excluding the element $0 + 0i$, we can show that the remaining complex numbers under multiplication are still closed, associative, and commutative. The set of "complex numbers less $0 + 0i$" contains the identity element $1 + 0i$ and inverses under multiplication. Therefore:

■ (Complex numbers/$\{0 + 0i\}$, ·) is a commutative group.

Division

Division by a number is equivalent to multiplication by the reciprocal (or multiplicative inverse) of that number. That is, $x \div y = x \cdot \frac{1}{y}$, or $\frac{x}{y} = x \cdot \frac{1}{y}$. Consider the problem "6 divided by 2."

$$6 \div 2 = 6 \cdot \tfrac{1}{2} = 3 \qquad \text{OR} \qquad \tfrac{6}{2} = 6 \cdot \tfrac{1}{2} = 3$$

We may extend this definition of division to the set of complex numbers. As seen in the following example, however, it now seems natural to rationalize the denominator once the division is written in a fractional form.

☐ EXAMPLE: Divide $8 + i$ by $2 - i$.

How to Proceed	*Solution*

Write the division problem in fractional form. To rationalize the denominator, multiply the fraction by a form of the identity element 1. That is, multiply by $\dfrac{2 + i}{2 + i}$, where $2 + i$ is the conjugate of the denominator $2 - i$. Then, simplify.

$$(8 + i) \div (2 - i) = \frac{8 + i}{2 - i} = \frac{(8 + i)}{(2 - i)} \cdot \frac{(2 + i)}{(2 + i)}$$

$$= \frac{16 + 10i + i^2}{4 - i^2}$$

$$= \frac{16 + 10i - 1}{4 + 1}$$

$$= \frac{15 + 10i}{5} = \frac{15}{5} + \frac{10i}{5}$$

$$= 3 + 2i \quad Ans.$$

A check is performed for a division problem by using multiplication. For example, to check that $6 \div 2 = 3$ or $\frac{6}{2} = 3$, we multiply $3 \cdot 2 = 6$. The product of the quotient 3 and the divisor 2 is equal to the dividend 6. Let us use this process to check the division problem just performed with complex numbers. That is, if $\dfrac{8 + i}{2 - i} = 3 + 2i$, it should be true that $(3 + 2i)(2 - i) = 8 + i$.

$$\begin{aligned} Check: \quad (3 + 2i)(2 - i) &= 6 + 4i - 3i - 2i^2 \\ &= 6 + 4i - 3i + 2 = 8 + i \quad \text{(True)} \end{aligned}$$

MODEL PROBLEMS

1. Write the multiplicative inverse of $2 + 4i$ in the form $a + bi$. Simplify, if possible.

Solution

1. The multiplicative inverse of $2 + 4i$ is $\dfrac{1}{2 + 4i}$.

2. Rationalize the denominator. Since $2 - 4i$ is the conjugate of the denominator $2 + 4i$, multiply the fraction by $\dfrac{2 - 4i}{2 - 4i}$.

$$\frac{1}{2+4i} = \frac{1}{(2+4i)} \cdot \frac{(2-4i)}{(2-4i)} = \frac{2-4i}{4-16i^2} = \frac{2-4i}{4+16} = \frac{2-4i}{20}$$

3. This inverse may be simplified by various methods.

$$\frac{2-4i}{20} = \frac{\overset{1}{\cancel{2}}(1-2i)}{\underset{10}{\cancel{20}}} = \frac{1-2i}{10} \quad \text{OR} \quad \frac{2-4i}{20} = \frac{\overset{1}{\cancel{2}}}{\underset{10}{\cancel{20}}} - \frac{\overset{1}{\cancel{4i}}}{\underset{5}{\cancel{20}}} = \frac{1}{10} - \frac{i}{5}$$

Answer: $\dfrac{1-2i}{10}$ OR $\dfrac{1}{10} - \dfrac{i}{5}$ OR $\dfrac{1}{10} - \dfrac{1}{5}i$

2. Divide and check: $(3 + 12i) \div (4 - i)$

Solution	*Check*
$(3+12i) \div (4-i) = \dfrac{3+12i}{4-i} = \dfrac{(3+12i)}{(4-i)} \cdot \dfrac{(4+i)}{(4+i)}$	$3i(4-i) \overset{?}{=} 3+12i$
	$12i - 3i^2 \overset{?}{=} 3+12i$
$= \dfrac{12+51i+12i^2}{16-i^2} = \dfrac{12+51i-12}{16+1}$	$12i + 3 \overset{?}{=} 3+12i$
	$3+12i = 3+12i$
$= \dfrac{0+51i}{17} = \dfrac{0}{17} + \dfrac{51i}{17} = 0+3i$	(True)
$= 3i \quad Ans.$	

EXERCISES

In 1–12, write the multiplicative inverse of the given complex number in the form $a + bi$. Simplify, if possible.

1. $3 + i$ **2.** $6 - i$ **3.** $1 + 5i$ **4.** $4 - 3i$

5. $9 - 2i$ **6.** $7 + 5i$ **7.** $4 - 4i$ **8.** $5 + 5i$

9. $3 + 6i$ **10.** $8 - 4i$ **11.** $\sqrt{5} + i$ **12.** $2 - i\sqrt{3}$

In 13–19, perform the indicated division, and check your answer.

13. $(4 - 2i) \div (1 + i)$ **14.** $(3 - i) \div (2 + i)$ **15.** $(5 + 5i) \div (3 - i)$

16. $\dfrac{5 - 3i}{1 - i}$ **17.** $\dfrac{7 - 4i}{1 - 2i}$ **18.** $\dfrac{21 + i}{5 - i}$ **19.** $\dfrac{18 + i}{2 + 3i}$

In 20–27, perform the given division.

20. $\dfrac{4 + i}{2 - 3i}$ **21.** $\dfrac{6 + i}{6 - i}$ **22.** $\dfrac{1 - 3i}{2 - 7i}$ **23.** $\dfrac{5 + i}{5 - i}$

24. $\dfrac{7 - i}{7 + i}$ **25.** $\dfrac{1 + 3i}{2 + 4i}$ **26.** $\dfrac{3 - 5i}{i}$ **27.** $\dfrac{7 + 2i}{4i}$

28. Match the descriptions of complex numbers in Column 1 with their corresponding values in Column 2.

Column 1	*Column 2*
1. The additive inverse of 2 - i	*a.* 2 + i
2. The conjugate of 2 - i	*b.* 2 - i
3. The multiplicative inverse of 2 - i	*c.* 1 + 0i
4. The additive inverse of -2 + i	*d.* 0 + 0i
5. The conjugate of -2 + i	*e.* -2 + i
6. The multiplicative inverse of 2 + i	*f.* -2 - i
7. The identity for (Complex numbers, +)	*g.* $\frac{2}{5} + \frac{1}{5}i$
8. The identity for (Complex numbers/{0 + 0i}, ·)	*h.* $\frac{2}{5} - \frac{1}{5}i$

In 29 and 30, select the numeral preceding the expression that best completes the sentence.

29. The multiplicative inverse of $\frac{1}{2} - \frac{1}{2}i$ is:

(1) $\frac{1}{2} + \frac{1}{2}i$ (2) $\frac{1}{2} - \frac{1}{2}i$ (3) 1 + i (4) 2 + 2i

30. The expression $\dfrac{1}{7 - 3i}$ is equivalent to:

(1) $\dfrac{7 + 3i}{40}$ (2) $\dfrac{7 + 3i}{58}$ (3) $\dfrac{7 - 3i}{40}$ (4) $\dfrac{7 - 3i}{58}$

31. **a.** What is the conjugate of (.6 - .8i)? **b.** Express the product of (.6 - .8i) and its conjugate in the form of a complex number. **c.** What is the multiplicative inverse of (.6 - .8i)? (*Hint:* See part *b*.)

In 32-35, tell whether the statement is *true* or *false*. If it is false, explain why.

32. The conjugate of $(\frac{5}{13} + \frac{12}{13}i)$ is $(\frac{5}{13} - \frac{12}{13}i)$, and $(\frac{5}{13} + \frac{12}{13}i)(\frac{5}{13} - \frac{12}{13}i) =$ 1 + 0i.

33. The conjugate of $\left(\dfrac{\sqrt{3}}{2} - \dfrac{i}{2}\right)$ is $\left(\dfrac{\sqrt{3}}{2} + \dfrac{i}{2}\right)$, and $\left(\dfrac{\sqrt{3}}{2} - \dfrac{i}{2}\right)\left(\dfrac{\sqrt{3}}{2} + \dfrac{i}{2}\right) = 1 + 0i$.

34. The conjugate of $(\frac{1}{3} - \frac{2}{3}i)$ is $(\frac{1}{3} + \frac{2}{3}i)$, and $(\frac{1}{3} - \frac{2}{3}i)(\frac{1}{3} + \frac{2}{3}i) =$ 1 + 0i.

35. For any complex number, its conjugate is equal to its multiplicative inverse.

36. For what value of $a^2 + b^2$ does the conjugate of $a + bi$ equal the multiplicative inverse of $a + bi$?

14-6 THE FIELD OF COMPLEX NUMBERS

A system that consists of a set of elements and two operations, usually addition and multiplication, is a *field* if eleven properties are true. We have studied ten field properties involving complex numbers earlier in this chapter; the last property, which is the distributive property, is also true. Therefore:

■ (Complex numbers, +, ·) is a field because:

1. (Complex numbers, +) is a commutative group.
   These *five* properties were studied in section 3 of this chapter.
2. (Complex numbers/{0 + 0i}, ·) is a commutative group.
   These *five* properties were studied in section 5 of this chapter.
3. Multiplication is distributive over addition. In symbols:

$$(a + bi)((c + di) + (e + fi)) = (a + bi)(c + di) + (a + bi)(e + fi)$$

In this statement of the distributive property, it can be proved that both numbers of the equation equal the same complex number, namely:

$$(ac + ae - bd - bf) + (ad + af + bc + be)i$$

Complex Numbers Cannot Be Ordered

In an *ordered field*, there are fifteen properties: the eleven properties of a field and four other properties that relate to order. One of the properties of an ordered field is the *Trichotomy Property*, namely: For any elements a and b, one and only one of the following statements is true:

$$a > b \quad \text{OR} \quad a = b \quad \text{OR} \quad a < b$$

☐ For example, in the ordered field of (*Rational numbers*, +, ·, <), consider -4 and 3. By the Trichotomy Property:

$$-4 > 3 \text{ (False)} \quad \text{OR} \quad -4 = 3 \text{ (False)} \quad \text{OR} \quad -4 < 3 \text{ (True)}$$

☐ In the ordered field of (*Real numbers*, +, ·, <), consider $\sqrt{5}$ and 2. By the Trichotomy Property:

$$\sqrt{5} > 2 \text{ (True)} \quad \text{OR} \quad \sqrt{5} = 2 \text{ (False)} \quad \text{OR} \quad \sqrt{5} < 2 \text{ (False)}$$

Any complex number of the form $x + 0i$ equals the real number x. Since the set of real numbers can be ordered, complex numbers of the form $x + 0i$ can be ordered. However, it is true that:

■ Complex numbers that are *not* real *cannot* be ordered.

For example, consider the complex numbers $2 + 3i$ and $3 + 2i$, represented by vectors $\overrightarrow{OA}$ and $\overrightarrow{OB}$, respectively, in the diagram at the right. Although the vectors are equal in length, they have different directions; thus, the vectors are not equal. Also, by the definition of equality of complex numbers, we know that $2 + 3i \neq 3 + 2i$. However, it is *not* possible to say that one of these numbers is greater than the other.

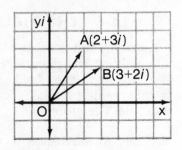

Let us study some convincing arguments about elements in an ordered field that will lead us to conclusions about complex numbers.

Argument 1: In an ordered field, the square of any number is greater than or equal to zero. That is:

If n is an element of an ordered field, then $n^2 \geq 0$.

Consider the complex number $0 + i$, which is simply i. Since $i^2 = -1$, and $-1 < 0$, it follows that $i^2 < 0$. Therefore, $i^2 \ngeq 0$. Similarly, $(3i)^2 = -9$, and $(3i)^2 \ngeq 0$. By the contrapositive of the original statement, it is true that:

*If $n^2 \ngeq 0$, then n is **not** an element of an ordered field.*

Since the squares of many complex numbers are neither greater than nor equal to zero, we conclude:

■ **(Complex numbers, $+$, $\cdot$) is *not* an ordered field.**

Argument 2: In an ordered field, if a number is positive ($n > 0$) or if a number is negative ($n < 0$), then its square is greater than zero. That is:

If $n > 0$ or $n < 0$, then $n^2 > 0$.

But we have seen that $i^2 = -1$. Thus, $i^2 \ngtr 0$. By the contrapositive of the original statement, it is true that:

If $n^2 \ngtr 0$, then $n \ngtr 0$ and $n \nless 0$.

Since $i^2 \ngtr 0$, it follows that $i \ngtr 0$ and $i \nless 0$. In other words:

■ *i* **is not a positive number, and** *i* **is not a negative number.**

In the same way, any complex number of the form $a + bi$, where $b \neq 0$, is neither a positive number nor a negative number.

EXERCISES

1. List the eleven field properties of (Complex numbers, +, ·).
2. a. What complex number is equal to $(4 - i)((3 + 2i) + (1 - 4i))$?
   b. What complex number is equal to $(4 - i)(3 + 2i) + (4 - i)(1 - 4i)$?
   c. If the answers to parts a and b are equal, what field property is illustrated by this equality?

In 3-10, identify the property of the complex numbers that is illustrated by the given statement.

3. $(7 - 4i) + (0 + 0i) = 7 - 4i$ 4. $(2 + i)(8 - i) = (8 - i)(2 + i)$
5. $(2 + i)(\frac{2}{5} - \frac{1}{5}i) = 1 + 0i$ 6. $(-3 + i) + (3 - i) = 0 + 0i$
7. $(17 - 9i)(1 + 0i) = 17 - 9i$
8. $(6 - i) + (8 + 5i) = (8 + 5i) + (6 - i)$
9. $(10 + 4i) + ((3 + 7i) + (2 - 6i)) = ((10 + 4i) + (3 + 7i)) + (2 - 6i)$
10. $(3 + 2i)((3 + 7i) + (-2 - 6i)) = (3 + 2i)(3 + 7i) + (3 + 2i)(-2 - 6i)$

In 11-17, tell whether the statement is *true* or *false*. If the statement is false, explain why.

11. (Complex numbers, +, ·) is a field.
12. (Complex numbers, +, ·, <) is an ordered field.
13. In the set of complex numbers, $(8 + 6i) > (3 + 2i)$.
14. For the complex numbers i and $(1 - i)$, it is true that $i > (1 - i)$ or $i = (1 - i)$ or $i < (1 - i)$.
15. Every complex number has an additive inverse.
16. Every complex number has a multiplicative inverse.
17. The square of a complex number is greater than or equal to zero.

18. In an ordered field, the following conditional statement is true:

    $$If\ x > y,\ then\ x - y > 0.$$

    a. Write the contrapositive of the given conditional statement.
    b. Let $x = 5i$, $y = 4i$, and $x - y = 5i - 4i = i$. Since we have proved that $i \not> 0$ or that $x - y \not> 0$, what conclusion must be true?
       (1) $5i > 4i$ (2) $5i \not> 4i$ (3) $i < 0$ (4) $i = 0$

14-7 SOLVING QUADRATIC EQUATIONS WITH IMAGINARY ROOTS

In Chapter 4, we solved quadratic equations with real roots. Recall that a *quadratic equation* is an equation of the form $ax^2 + bx + c = 0$, where a, b, and c are real numbers, and $a \neq 0$.

To find the roots of a quadratic equation, we may use the *quadratic formula* (derived on page 193):

$$x = \frac{-b \pm \sqrt{b^2 - 4ac}}{2a}$$

In Chapter 4, we learned that when the *discriminant* $b^2 - 4ac$ is a negative number, the roots of the quadratic equation are not real. Now that we have extended the number system from the real numbers to the complex numbers, let us attempt to solve a quadratic equation whose discriminant is negative. We will let the domain be the set of complex numbers.

□ EXAMPLE: Solve the equation: $x^2 - 8x + 17 = 0$

How to Proceed	*Solution*
1. Compare the equation to $ax^2 + bx + c = 0$ to determine a, b, and c.	$x^2 - 8x + 17 = 0$ $a = 1$, $b = -8$, $c = 17$
2. Substitute the values of a, b, and c in the quadratic formula, and simplify.	$x = \dfrac{-b \pm \sqrt{b^2 - 4ac}}{2a}$ $x = \dfrac{-(-8) \pm \sqrt{(-8)^2 - 4(1)(17)}}{2(1)}$ $x = \dfrac{8 \pm \sqrt{64 - 68}}{2}$
(*Note:* If $b^2 - 4ac$ is negative, then $\sqrt{b^2 - 4ac}$ is an imaginary number.)	$x = \dfrac{8 \pm \sqrt{-4}}{2} = \dfrac{8 \pm \sqrt{4}\sqrt{-1}}{2}$ $x = \dfrac{8 \pm 2i}{2} = \dfrac{8}{2} \pm \dfrac{2i}{2}$
3. Check both roots in the original equation. (The check is left as a student exercise.)	$x = 4 \pm i$

Answer: $x = 4 \pm i$ OR   solution set $= \{4 + i, 4 - i\}$

It is true that $4 + i$ and $4 - i$ are complex numbers. However, any number of the form $a + bi$ where $b \neq 0$ can also be called an *imaginary number*. Therefore, the roots $4 \pm i$ are imaginary.

In the next section, we will discuss the nature of the roots of any quadratic equation. For now, however, we will restrict the equations in the exercises that follow to those having imaginary roots only.

| MODEL PROBLEM |

a. Solve the equation $\dfrac{x^2}{2} = x - 5$, and express its roots in the form $a + bi$.

b. Check the roots of the equation.

a.

| *How to Proceed* | *Solution* |

1. Write the equation. Then multiply by the L.C.D. to clear all fractions.

$$\frac{x^2}{2} = x - 5$$

$$\cancel{2} \cdot \frac{x^2}{\cancel{2}} = 2(x - 5)$$

$$x^2 = 2x - 10$$

2. Transform the equation so that one side is 0.

$$x^2 - 2x + 10 = 0$$

3. Compare the equation to $ax^2 + bx + c = 0$ to determine a, b, and c.

$$a = 1, \; b = -2, \; c = 10$$

4. Substitute the values of a, b, and c in the quadratic formula, and simplify. Remember, the domain is now the set of complex numbers.

$$x = \frac{-b \pm \sqrt{b^2 - 4ac}}{2a}$$

$$x = \frac{-(-2) \pm \sqrt{(-2)^2 - 4(1)(10)}}{2(1)}$$

$$x = \frac{2 \pm \sqrt{4 - 40}}{2}$$

$$x = \frac{2 \pm \sqrt{-36}}{2} = \frac{2 \pm \sqrt{36}\sqrt{-1}}{2}$$

$$x = \frac{2 \pm 6i}{2} = \frac{2}{2} \pm \frac{6i}{2}$$

$$x = 1 \pm 3i$$

b. Check both roots in the original equation only.

Check for x = 1 + 3i	*Check for x = 1 - 3i*
$\dfrac{x^2}{2} = x - 5$	$\dfrac{x^2}{2} = x - 5$
$\dfrac{(1 + 3i)(1 + 3i)}{2} \overset{?}{=} (1 + 3i) - 5$	$\dfrac{(1 - 3i)(1 - 3i)}{2} \overset{?}{=} (1 - 3i) - 5$
$\dfrac{1 + 6i + 9i^2}{2} \overset{?}{=} 1 + 3i - 5$	$\dfrac{1 - 6i + 9i^2}{2} \overset{?}{=} 1 - 3i - 5$
$\dfrac{1 + 6i - 9}{2} \overset{?}{=} -4 + 3i$	$\dfrac{1 - 6i - 9}{2} \overset{?}{=} -4 - 3i$
$\dfrac{-8 + 6i}{2} \overset{?}{=} -4 + 3i$	$\dfrac{-8 - 6i}{2} \overset{?}{=} -4 - 3i$
$-4 + 3i = -4 + 3i$ (True)	$-4 - 3i = -4 - 3i$ (True)

Answer: **a.** $x = 1 \pm 3i$ OR   solution set = $\{1 + 3i, 1 - 3i\}$
b. See checks.

| **EXERCISES** |

In 1–12: **a.** Solve the given equation, and express the roots in the form $a + bi$. **b.** Check the roots of the given equation.

1. $x^2 - 4x + 5 = 0$ 2. $x^2 - 6x + 25 = 0$ 3. $x^2 + 26 = 2x$
4. $x^2 + 29 = 10x$ 5. $x^2 = 8x - 25$ 6. $3x^2 = 6(x - 1)$

7. $\dfrac{x^2}{3} = 4x - 15$ 8. $\dfrac{x^2}{2} + 3x + 5 = 0$ 9. $\dfrac{x^2 + 21}{4} = 1 - 2x$

10. $2x^2 + 72 = 0$ 11. $\dfrac{x^2}{5} + 5 = 0$ 12. $\dfrac{x^2}{8} + 10 = 2$

In 13–21, find in $a + bi$ form the roots of the given equation.

13. $9x^2 - 6x + 2 = 0$ 14. $2x^2 + 17 = 6x$ 15. $4x(x - 2) + 5 = 0$
16. $4x(x + 5) + 29 = 0$ 17. $x^2 + 3 = 2x$ 18. $x^2 + 20x = 2x - 86$

19. $\dfrac{25x}{3} + \dfrac{3}{x} = 0$ 20. $\dfrac{x - 4}{10} = \dfrac{x - 5}{x}$ 21. $\dfrac{x - 2}{x} = \dfrac{x + 27}{17}$

In 22–24, select the numeral preceding the expression that best completes the sentence.

22. The roots of $x^2 - x + 5 = 0$ are:

(1) $\dfrac{1 \pm i\sqrt{19}}{2}$ (2) $\dfrac{1 \pm i\sqrt{21}}{2}$ (3) $\dfrac{1 \pm 19i}{2}$ (4) $\dfrac{1 \pm 21i}{2}$

23. The roots of $2x^2 + x + 6 = 0$ are:

(1) $\dfrac{1 \pm i\sqrt{47}}{2}$ (2) $\dfrac{-1 \pm i\sqrt{47}}{2}$ (3) $\dfrac{1 \pm i\sqrt{47}}{4}$ (4) $\dfrac{-1 \pm i\sqrt{47}}{4}$

24. The roots of $3x^2 + 2x + 1 = 0$ are:

(1) $-\dfrac{1}{3} \pm \dfrac{1}{3}i$ (2) $-\dfrac{1}{3} \pm \dfrac{2}{3}i$ (3) $-\dfrac{1}{3} \pm \dfrac{i\sqrt{2}}{3}$ (4) $-1 \pm i\sqrt{2}$

14-8 THE NATURE OF THE ROOTS OF ANY QUADRATIC EQUATION

Quadratic Equations and Graphs

We have learned that the graph of any quadratic equation of the form $ax^2 + bx + c = y$, where a, b, and c are real numbers and $a \neq 0$, is a *parabola*. A parabola contains an infinite number of points, and each point represents some ordered pair (x, y) in the coordinate plane. Do not confuse the *coordinate plane* in which each point represents a pair of real numbers with the *complex plane* in which each point represents one complex number.

If the coefficient a is positive $(a > 0)$, then the parabola contains a minimum turning point and opens upward.

If the coefficient a is negative $(a < 0)$, then the parabola contains a maximum turning point and opens downward.

Parabola: $ax^2 + bx + c = y$

$(a > 0)$ $(a < 0)$

Imagine that a parabola crosses the x-axis at two points. We can use substitution to find the equation of this intersection, as follows:

Equation of the parabola:	$ax^2 + bx + c = y$
Equation of the x-axis:	$y = 0$
Equation of the intersection of the parabola and the x-axis:	$ax^2 + bx + c = 0$

Therefore, a *quadratic equation* is an equation of the form $ax^2 + bx + c = 0$, where a, b, and c are real numbers and $a \neq 0$. When the parabola crosses the x-axis at two points, the roots of the equation $ax^2 + bx + c = 0$ are the x-coordinates of the two points of intersection.

Quadratic Equation: $ax^2 + bx + c = 0$

(Intersection of parabola and x-axis)

Roots of $ax^2 + bx + c = 0$ are x_1 and x_2.

If these points of intersection represent the ordered pairs $(x_1, 0)$ and $(x_2, 0)$, then the roots of the equation $ax^2 + bx + c = 0$ are x_1 and x_2. As we will see in the examples that follow, the parabola may be tangent to the x-axis and, in some cases, the parabola may not intersect the x-axis at all.

The Discriminant

By the quadratic formula, the two roots of $ax^2 + bx + c = 0$ are:

$$x_1 = \frac{-b + \sqrt{b^2 - 4ac}}{2a} \quad \text{AND} \quad x_2 = \frac{-b - \sqrt{b^2 - 4ac}}{2a}$$

The *discriminant* $b^2 - 4ac$, which is the expression under the radical sign, determines the nature of the roots of a quadratic equation when a, b, and c are rational numbers. Let us study possible values for the discriminant and, at the same time, examine the quadratic equation graphically.

Case 1. If $b^2 - 4ac > 0$ and $b^2 - 4ac$ is a perfect square, then the roots of the equation $ax^2 + bx + c = 0$ are *real*, *rational*, and *unequal*.

Solve: $x^2 - 2x - 3 = 0$

$$a = 1, b = -2, c = -3$$

$$x = \frac{-b \pm \sqrt{b^2 - 4ac}}{2a}$$

$$x = \frac{-(-2) \pm \sqrt{(-2)^2 - 4(1)(-3)}}{2(1)}$$

$$x = \frac{2 \pm \sqrt{4 + 12}}{2}$$

$$x = \frac{2 \pm \sqrt{16}}{2} = \frac{2 \pm 4}{2}$$

$$x_1 = \frac{2 + 4}{2} = \frac{6}{2} \quad \bigg| \quad x_2 = \frac{2 - 4}{2} = \frac{-2}{2}$$

$$x_1 = 3 \quad \bigg| \quad x_2 = -1$$

Graph: $x^2 - 2x - 3 = y$

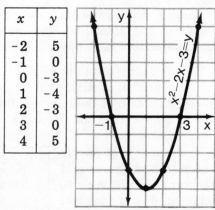

x	y
-2	5
-1	0
0	-3
1	-4
2	-3
3	0
4	5

The roots of $x^2 - 2x - 3 = 0$ are -1 and 3. These values can be read directly from the graph.

Observation: The discriminant $b^2 - 4ac = 16$, a positive number that is a perfect square. The parabola intersects the x-axis at -1 and 3. The roots, -1 and 3, of $x^2 - 2x - 3 = 0$ are real, rational, and unequal.

Case 2. If $b^2 - 4ac > 0$ and $b^2 - 4ac$ is *not* a perfect square, then the roots of the equation $ax^2 + bx + c = 0$ are *real*, *irrational*, and *unequal*.

Solve: $x^2 - 6x + 7 = 0$

$$a = 1, \ b = -6, \ c = 7$$

$$x = \frac{-b \pm \sqrt{b^2 - 4ac}}{2a}$$

$$x = \frac{-(-6) \pm \sqrt{(-6)^2 - 4(1)(7)}}{2(1)}$$

$$x = \frac{6 \pm \sqrt{36 - 28}}{2} = \frac{6 \pm \sqrt{8}}{2}$$

$$x = \frac{6 \pm 2\sqrt{2}}{2} = \frac{\cancel{2}(3 \pm \sqrt{2})}{\cancel{2}}$$

$$x_1 = 3 + \sqrt{2} \qquad x_2 = 3 - \sqrt{2}$$

Graph: $x^2 - 6x + 7 = y$

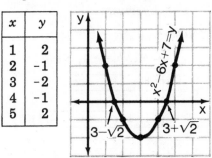

x	y
1	2
2	-1
3	-2
4	-1
5	2

While the roots $3 \pm \sqrt{2}$ cannot be read exactly from the graph, use $\sqrt{2} \approx 1.4$ to approximate their values:

$$3 + \sqrt{2} \approx 3 + 1.4 = 4.4$$
$$3 - \sqrt{2} \approx 3 - 1.4 = 1.6$$

Observation: The discriminant $b^2 - 4ac = 8$, a positive number that is not a perfect square. The parabola intersects the x-axis at $3 + \sqrt{2}$ and $3 - \sqrt{2}$. The roots, $3 \pm \sqrt{2}$, of $x^2 - 6x + 7 = 0$ are real, irrational, and unequal.

Case 3. If $b^2 - 4ac = 0$, then the roots of the equation $ax^2 + bx + c = 0$ are *real*, *rational*, and *equal*.

Solve: $x^2 - 4x + 4 = 0$

$$a = 1, \ b = -4, \ c = 4$$

$$x = \frac{-b \pm \sqrt{b^2 - 4ac}}{2a}$$

$$x = \frac{-(-4) \pm \sqrt{(-4)^2 - 4(1)(4)}}{2(1)}$$

$$x = \frac{4 \pm \sqrt{16 - 16}}{2} = \frac{4 \pm \sqrt{0}}{2}$$

$$x_1 = \frac{4 + 0}{2} = \frac{4}{2} \qquad x_2 = \frac{4 - 0}{2} = \frac{4}{2}$$

$$x_1 = 2 \qquad x_2 = 2$$

Graph: $x^2 - 4x + 4 = y$

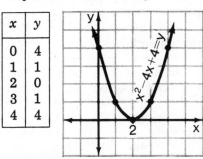

x	y
0	4
1	1
2	0
3	1
4	4

The parabola is tangent to the x-axis at $x = 2$. The two roots of $x^2 - 4x + 4 = 0$ are each equal to 2.

Observation: The discriminant $b^2 - 4ac = 0$. The parabola is tangent to the x-axis at 2. The roots of $x^2 - 4x + 4 = 0$ are 2 and 2. That is, the roots are real, rational, and equal.

Case 4. If $b^2 - 4ac < 0$, then the roots of $ax^2 + bx + c = 0$ are **imaginary**.

Solve: $x^2 - 4x + 5 = 0$

$$a = 1, b = -4, c = 5$$

$$x = \frac{-b \pm \sqrt{b^2 - 4ac}}{2a}$$

$$x = \frac{-(-4) \pm \sqrt{(-4)^2 - 4(1)(5)}}{2(1)}$$

$$x = \frac{4 \pm \sqrt{16 - 20}}{2} = \frac{4 \pm \sqrt{-4}}{2}$$

$$x = \frac{4 \pm 2i}{2} = \frac{\cancel{2}(2 \pm i)}{\cancel{2}}$$

$$x_1 = 2 + i \qquad x_2 = 2 - i$$

Graph: $x^2 - 4x + 5 = 0$

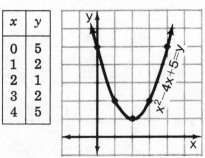

x	y
0	5
1	2
2	1
3	2
4	5

The minimum point of the parabola occurs at $(2, 1)$, and the parabola opens upward. Thus, the parabola does not intersect the x-axis at all.

Observation: The discriminant $b^2 - 4ac = -4$, a negative number. The parabola does not intersect the x-axis. Therefore, the equation $x^2 - 4x + 5 = 0$ has no real roots. By the quadratic formula, however, it is shown that the roots of $x^2 - 4x + 5 = 0$ are $2 + i$ and $2 - i$. That is, the roots are imaginary.

Let us summarize our findings in a table:

Value of Discriminant	Nature of Roots of $ax^2 + bx + c = 0$
$b^2 - 4ac > 0$, and $b^2 - 4ac$ is a perfect square	real, rational, unequal
$b^2 - 4ac > 0$, and $b^2 - 4ac$ is not a perfect square	real, irrational, unequal
$b^2 - 4ac = 0$	real, rational, equal
$b^2 - 4ac < 0$	imaginary

Note: If $b^2 - 4ac \geq 0$, the roots of $ax^2 + bx + c = 0$ are real numbers.

| MODEL PROBLEMS |

1. The roots of the equation $2x^2 + 6x + 3 = 0$ are:
   (1) real, rational, and unequal   (2) real, rational, and equal
   (3) real, irrational, and unequal   (4) imaginary

Solution

1. Compare $2x^2 + 6x + 3 = 0$ to $ax^2 + bx + c = 0$ to find: $a = 2$, $b = 6, c = 3$.
2. The discriminant $b^2 - 4ac = (6)^2 - 4(2)(3) = 36 - 24 = 12$. Since $b^2 - 4ac > 0$ and $b^2 - 4ac$ is *not* a perfect square, the roots of the equation are irrational.

Answer: (3) real, irrational, and unequal

2. The graph of $y = -x^2 - 6$:
   (1) is tangent to the x-axis
   (2) intersects the x-axis at 2 points
   (3) lies entirely above the x-axis
   (4) lies entirely below the x-axis

Solution

1. Compare $-x^2 - 6$ to $ax^2 + bx + c$ to find: $a = -1, b = 0, c = -6$

2. The discriminant $b^2 - 4ac = (0)^2 - 4(-1)(-6) = 0 - 24 = -24$. Since $b^2 - 4ac < 0$, the roots of $-x^2 - 6 = 0$ are imaginary, and the parabola $y = -x^2 - 6$ does not intersect the x-axis.

3. Since the coefficient $a = -1$ (or $a < 0$), the parabola opens downward. Select choice (4).

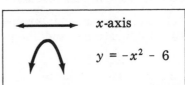

Alternate Method: Graph the parabola.

Answer: (4) The graph of $y = -x^2 - 6$ lies entirely below the x-axis.

3. Find the largest integral value of k for which the roots of the equation $2x^2 + 7x + k = 0$ are real.

Solution

1. For the equation $2x^2 + 7x + k = 0$: $a = 2, b = 7, c = k$

2. If $b^2 - 4ac \geq 0$, the roots of the quadratic equation are real. Substitute the given values in $b^2 - 4ac \geq 0$, and simplify.

Method 1	*Method 2*
$b^2 - 4ac \geq 0$	$b^2 - 4ac \geq 0$
$(7)^2 - 4(2)k \geq 0$	$(7)^2 - 4(2)k \geq 0$
$49 - 8k \geq 0$	$49 - 8k \geq 0$
$49 \geq 8k$	$-8k \geq -49$
$\dfrac{49}{8} \geq \dfrac{8k}{8}$	$\dfrac{-8k}{-8} \leq \dfrac{-49}{-8}$
$6\dfrac{1}{8} \geq k$	$k \leq 6\dfrac{1}{8}$

3. The largest integral value of k means the largest *integer* for which $6\frac{1}{8} \geq k$, or $k \leq 6\frac{1}{8}$. This integer is 6.

Answer: 6

Check both $k = 6$ and $k = 7$ (the next-larger integer) to see:

If $k = 6$, $b^2 - 4ac = 1$, and the roots of $2x^2 + 7x + 6 = 0$ are real. If $k = 7$, $b^2 - 4ac = -7$, and the roots of $2x^2 + 7x + 7 = 0$ are imaginary.

EXERCISES

In 1–12: a. For the given equation, evaluate the discriminant $b^2 - 4ac$. b. Using the value of $b^2 - 4ac$, select the choice below that describes the nature of the roots of the equation:

(1) real, rational, and unequal   (2) real, rational, and equal
(3) real, irrational, and unequal   (4) imaginary

1. $x^2 + 3x - 5 = 0$ 2. $x^2 + 3x - 10 = 0$ 3. $x^2 + 3x + 4 = 0$
4. $x^2 - 10x + 25 = 0$ 5. $x^2 + 8x + 17 = 0$ 6. $x^2 + 7x - 30 = 0$
7. $x^2 = 7x - 6$ 8. $2x^2 + 3x = 4$ 9. $4x^2 + 9 = 12x$
10. $x^2 + 9 = 2x^2 + x$ 11. $x - 8 = x^2 - 2x$ 12. $x^2 + 6 = 3x^2 + x$

In 13–21, select the choice below that describes the graph of the given parabola.

(1) It is tangent to the x-axis.
(2) It intersects the x-axis at 2 points.
(3) It lies entirely above the x-axis.
(4) It lies entirely below the x-axis.

13. $y = x^2 - 2x - 8$ 14. $y = x^2 - 2x + 1$ 15. $y = x^2 + 3$
16. $y = x^2 - 6$ 17. $y = 2x^2 - 5x + 4$ 18. $y = -x^2 - 10$
19. $y = 10x^2$ 20. $y = -x^2 + 3x - 7$ 21. $y = 12 - x^2$

In 22–33, solve the quadratic equation, and check both roots.

22. $x^2 - 6x + 5 = 0$ 23. $x^2 - 6x + 13 = 0$ 24. $x^2 = 2x + 6$
25. $x^2 = 2x - 10$ 26. $x^2 + 20x = 6x - 49$ 27. $x^2 + 4x = x$
28. $9x^2 + 4 = 0$ 29. $3x^2 + 11x = 4$ 30. $x^2 + 10x + 26 = 0$

31. $x - x^2 = \dfrac{1}{4}$ 32. $\dfrac{x^2}{2} = x - 2$ 33. $\dfrac{x-2}{2} = \dfrac{x-1}{x}$

In 34–41, select the numeral preceding the expression that best completes the sentence.

34. The roots of $x^2 + 2x + k = 0$ are equal when k is:
    (1) 1   (2) 2   (3) 3   (4) 4
35. The roots of $x^2 + kx + 3 = 0$ are real when k is:
    (1) 1   (2) 2   (3) 3   (4) 4
36. The roots of $x^2 + bx + 8 = 0$ are imaginary when b equals:
    (1) 8   (2) -7   (3) 6   (4) -5
37. The roots of $ax^2 + 6x + 4 = 0$ are imaginary if a equals:
    (1) 1   (2) 2   (3) 3   (4) -1
38. If the graph of $y = ax^2 + bx + c$ is tangent to the x-axis, then the roots of $ax^2 + bx + c = 0$ are:
    (1) rational and unequal   (2) rational and equal
    (3) irrational and unequal   (4) imaginary
39. The graph of $y = ax^2 + bx + c$ where $a \neq 0$ lies entirely above the x-axis. The roots of $ax^2 + bx + c = 0$ are:
    (1) rational and unequal   (2) rational and equal
    (3) irrational   (4) imaginary
40. The graph of $y = x^2$ is symmetric with respect to:
    (1) the x-axis   (2) the y-axis
    (3) the line $y = x$ (4) the point $(0, 0)$
41. If the roots of $x^2 + bx + 16 = 0$ are equal, then b is:
    (1) 8, only   (2) -8, only   (3) 8 or -8   (4) neither 8 nor -8

In 42–44, find the largest integral value of k such that the roots of the given equation are real.

42. $x^2 + 6x + k = 0$ 43. $2x^2 + 5x + k = 0$ 44. $kx^2 - 7x + 3 = 0$

In 45–47, find the smallest integral value of k such that the given equation has imaginary roots.

45. $x^2 - 3x + k = 0$ 46. $kx^2 - 2x + 5 = 0$ 47. $x^2 + 4x + k = 0$

48. If $a \neq 0$, $c = a$, and $b = 2a$, describe the nature of the roots of the equation $ax^2 + bx + c = 0$.
49. If $a \neq 0$, $b = a$, and $c = a$, describe the nature of the roots of the equation $ax^2 + bx + c = 0$.

14-9 USING GIVEN CONDITIONS TO WRITE A QUADRATIC EQUATION

Certain quadratic equations contain polynomial expressions that are factorable. If we factor such an expression and set each factor equal to zero, we can find the roots of the equation. For example:

$$\textit{Solve:} \qquad x^2 + 3x - 10 = 0$$
$$(x - 2)(x + 5) = 0$$
$$x - 2 = 0 \mid x + 5 = 0$$
$$x = 2 \mid \qquad x = -5 \quad \textit{Solution:} \ \{2, -5\}$$

If we are given the roots of a quadratic equation, we can reverse the process just described to find the equation. This new procedure, which we will call the *"Reverse Factoring Technique,"* is explained in the following problem.

☐ EXAMPLE 1: Write a quadratic equation whose roots are 3 and -7.

How to Proceed	*Solution*
1. Write the roots.	$x = 3 \quad\mid\quad x = -7$
2. Transform each equation so that one side is zero.	$x - 3 = 0 \mid x + 7 = 0$
3. Multiply the binomials, showing that their product is zero.	$(x - 3)(x + 7) = 0$ $x^2 + 4x - 21 = 0 \quad Ans.$

To check that the roots of $x^2 + 4x - 21 = 0$ are 3 and -7, substitute these values in the equation. (This exercise is left to the student.)

Let us extend the Reverse Factoring Technique to the general case.

☐ EXAMPLE 2: Write a quadratic equation whose roots are r_1 and r_2.

$$\textit{Solution:} \qquad x = r_1 \ \mid \ x = r_2$$
$$x - r_1 = 0 \mid x - r_2 = 0$$
$$(x - r_1)(x - r_2) = 0$$
$$x^2 - r_1 x - r_2 x + r_1 r_2 = 0$$

$$\textit{Answer:} \ \ x^2 - (r_1 + r_2)x + r_1 r_2 = 0$$

Observe these terms: $(r_1 + r_2)$ is the sum of the roots $r_1 r_2$ is the product of the roots

The Sum and the Product of the Roots
of a Quadratic Equation

Let us discover some rules involving the sum and the product of the roots of a quadratic equation. Two proofs are offered here.

Proof I:

1. Given the general quadratic equation $ax^2 + bx + c = 0$ where $a \neq 0$, multiply each member of the equation by $\frac{1}{a}$ so that the coefficient of x^2 is 1.

$$\frac{1}{a}(ax^2 + bx + c) = \frac{1}{a}(0) \longrightarrow x^2 + \frac{b}{a}x + \frac{c}{a} = 0$$

2. Using the preceding Example 2, write the quadratic equation whose roots are r_1 and r_2.

$$\longrightarrow x^2 - (r_1 + r_2)x + r_1r_2 = 0$$

3. Compare the coefficients of the corresponding terms of the equations written in steps 1 and 2.

$$-(r_1 + r_2) = \frac{b}{a} \quad \text{OR} \quad (r_1 + r_2) = \frac{-b}{a} \qquad \qquad r_1r_2 = \frac{c}{a}$$

The sum of the roots $= \frac{-b}{a}$. | **The product of the roots** $= \frac{c}{a}$.

Proof II (An Alternate Approach):

1. If r_1 and r_2 are the roots of the quadratic equation $ax^2 + bx + c = 0$ where $a \neq 0$, it follows by the quadratic formula that:

$$r_1 = \frac{-b + \sqrt{b^2 - 4ac}}{2a} \quad \text{AND} \quad r_2 = \frac{-b - \sqrt{b^2 - 4ac}}{2a}$$

2. Add: $r_1 + r_2 = \frac{-b + \sqrt{b^2 - 4ac} - b - \sqrt{b^2 - 4ac}}{2a} = \frac{-2b}{2a} = \frac{-b}{a}$

The sum of the roots $= \frac{-b}{a}$.

3. Multiply: $r_1r_2 = \left(\frac{-b + \sqrt{b^2 - 4ac}}{2a}\right)\left(\frac{-b - \sqrt{b^2 - 4ac}}{2a}\right)$

Notice that the numerators are conjugates.

$$r_1r_2 = \frac{+b^2 - (b^2 - 4ac)}{4a^2} = \frac{b^2 - b^2 + 4ac}{4a^2} = \frac{4ac}{4a \cdot a} = \frac{c}{a}$$

The product of the roots $= \frac{c}{a}$.

The rules for the sum and the product of the roots can be used to help us find a quadratic equation to fit certain given information. In fact, if we are given roots that are irrational or imaginary and we are asked to find the quadratic equation that contains these roots, these rules are especially useful.

☐ EXAMPLE 3:   Write a quadratic equation whose roots are $3 + \sqrt{5}$ and $3 - \sqrt{5}$.

Solution

1. The sum of the roots = $(3 + \sqrt{5}) + (3 - \sqrt{5}) = 6$. Thus, $\dfrac{-b}{a} = 6$.

2. The product of the roots = $(3 + \sqrt{5})(3 - \sqrt{5}) = 9 - 5 = 4$. Thus, $\dfrac{c}{a} = 4$.

3. Let $a = 1$. | Then, $\dfrac{-b}{a} = 6$, or $\dfrac{-b}{1} = 6$ | Also, $\dfrac{c}{a} = 4$, or $\dfrac{c}{1} = 4$
$$-b = 6$$
$$b = -6$$
$$c = 4$$

4. Substitute $a = 1$, $b = -6$, and $c = 4$ in the equation $ax^2 + bx + c = 0$.

Answer: $x^2 - 6x + 4 = 0$

To *check* that $x^2 - 6x + 4 = 0$ is a correct equation, apply the quadratic formula and find the roots of the equation.

$$x = \frac{-b \pm \sqrt{b^2 - 4ac}}{2a} = \frac{-(-6) \pm \sqrt{(-6)^2 - 4(1)(4)}}{2(1)} = \frac{6 \pm \sqrt{36 - 16}}{2}$$

$$x = \frac{6 \pm \sqrt{20}}{2} = \frac{6 \pm \sqrt{4}\sqrt{5}}{2} = \frac{6 \pm 2\sqrt{5}}{2} = \frac{6}{2} \pm \frac{2\sqrt{5}}{2} = 3 \pm \sqrt{5}$$

(True)

─── **KEEP IN MIND** ───

For any quadratic equation $ax^2 + bx + c = 0$, $a \neq 0$:

The sum of the roots = $-\dfrac{b}{a}$

The product of the roots = $\dfrac{c}{a}$

| MODEL PROBLEMS |

1. For the quadratic equation $2x^2 + 5x + 8 = 0$, find:
   a. the sum of its roots
   b. the product of its roots

Solution

For the quadratic equation $2x^2 + 5x + 8 = 0$: $a = 2$, $b = 5$, $c = 8$

a. The sum of the roots $= \dfrac{-b}{a} = \dfrac{-5}{2}$ *Ans.*

b. The product of the roots $= \dfrac{c}{a} = \dfrac{8}{2} = 4$ *Ans.*

2. Write a quadratic equation whose roots are $\frac{1}{2}$ and -2.

Solution

Method 1: Use the Reverse Factoring Technique. To clear fractions in $x = \frac{1}{2}$, multiply each member of the equation by 2.

$$x = \tfrac{1}{2} \quad\bigg|\quad x = -2$$
$$2x = 1$$
$$2x - 1 = 0 \quad\big|\quad x + 2 = 0$$
$$(2x - 1)(x + 2) = 0$$
$$2x^2 + 3x - 2 = 0 \quad Ans.$$

Method 2: Use the rules for the sum and the product of the roots. In step 3, notice that a may take on different values.

1. The sum of the roots $= \dfrac{1}{2} + (-2) = \dfrac{1}{2} + \left(-\dfrac{4}{2}\right) = -\dfrac{3}{2}$. Thus,

   $-\dfrac{b}{a} = -\dfrac{3}{2}$.

2. The product of the roots $= \left(\dfrac{1}{2}\right)(-2) = -1$. Thus, $\dfrac{c}{a} = -1$.

3. Let $a = 1$.

   $-\dfrac{b}{a} = -\dfrac{3}{2}$, or $-\dfrac{b}{1} = -\dfrac{3}{2}$

   $b = \dfrac{3}{2}$

   $\dfrac{c}{a} = -1$, or $\dfrac{c}{1} = -1$

   $c = -1$

4. Thus, $ax^2 + bx + c = 0$ is

   $x^2 + \dfrac{3}{2}x - 1 = 0$.

3. Alternate: Let $a = 2$.

   $-\dfrac{b}{a} = -\dfrac{3}{2}$, or $-\dfrac{b}{2} = -\dfrac{3}{2}$

   $-b = -3$

   $b = 3$

   $\dfrac{c}{a} = -1$, or $\dfrac{c}{2} = -1$

   $c = -2$

4. Thus, $ax^2 + bx + c = 0$ is

   $2x^2 + 3x - 2 = 0$.

Note: If the members of $x^2 + \frac{3}{2}x - 1 = 0$ are multiplied by 2, the equivalent equation $2x^2 + 3x - 2 = 0$ is formed. Both equations are correct answers because they are *equivalent* equations; that is, they have the same roots.

Answer: $2x^2 + 3x - 2 = 0$, or any equivalent equation

3. Write a quadratic equation whose roots are $5i$ and $-5i$.

Solution

Method 1: Use the Reverse Factoring Technique, extended to include the set of imaginary numbers.

$$
\begin{array}{c|c}
x = 5i & x = -5i \\
x - 5i = 0 & x + 5i = 0
\end{array}
$$

$$(x - 5i)(x + 5i) = 0$$
$$x^2 - 25i^2 = 0$$
$$x^2 - 25(-1) = 0$$
$$x^2 + 25 = 0 \quad Ans.$$

Method 2: Use the rules for the sum and the product of the roots of a quadratic equation.

1. The sum of the roots = $5i + (-5i) = 0$.   Thus, $-\dfrac{b}{a} = 0$.

2. The product of the roots = $(5i)(-5i) = -25i^2 = 25$. Thus, $\dfrac{c}{a} = 25$.

3. Let $a = 1$. $\quad \left| \quad -\dfrac{b}{a} = 0, \text{ or } -\dfrac{b}{1} = 0 \quad \right| \quad \dfrac{c}{a} = 25, \text{ or } \dfrac{c}{1} = 25$

$$-b = 0 \qquad\qquad c = 25$$
$$b = 0$$

4. Substitute $a = 1$, $b = 0$, and $c = 25$ in $ax^2 + bx + c = 0$ to form the equation $1x^2 + 0x + 25 = 0$, or $x^2 + 25 = 0$.

Answer: $x^2 + 25 = 0$

4. a. If one root of a quadratic equation is $3 + 2i$, what is its other root? b. Write a quadratic equation having these roots.

Solution

a. If a quadratic equation has imaginary roots, the roots are conjugates. Thus, if one root is $3 + 2i$, the other root is $3 - 2i$.

b. 1. The sum of the roots = $(3 + 2i) + (3 - 2i) = 6$. Thus, $-\dfrac{b}{a} = 6$.

2. The product of the roots $= (3 + 2i)(3 - 2i) = 9 - 4i^2 =$ $9 + 4 = 13$. Thus, $\frac{c}{a} = 13$.

3. Let $a = 1$.

$-\frac{b}{a} = 6$, or $-\frac{b}{1} = 6$ $\qquad$ $\frac{c}{a} = 13$, or $\frac{c}{1} = 13$

$\qquad\qquad\qquad -b = 6$ $\qquad\qquad\qquad\qquad c = 13$

$\qquad\qquad\qquad\quad b = -6$

4. By substitution, $ax^2 + bx + c = 0$ is $x^2 - 6x + 13 = 0$.

Answer: a. $3 - 2i$ $\qquad$ b. $x^2 - 6x + 13 = 0$

5. If one root of $x^2 - 6x + k = 0$ is 4, find the other root.

Method 1

1. In $x^2 - 6x + k = 0$, $a = 1, b = -6$.

2. Let the root $r_1 = 4$.

3. $r_1 + r_2 = -\frac{b}{a}$

$\qquad 4 + r_2 = \frac{-(-6)}{1}$

$\qquad 4 + r_2 = 6$

$\qquad\quad r_2 = 2$

Method 2

1. Substitute 4 for x: $\quad x^2 - 6x + k = 0$

$\qquad\qquad\qquad\qquad\qquad 4^2 - 6(4) + k = 0$

$\qquad\qquad\qquad\qquad\qquad 16 - 24 + k = 0$

$\qquad\qquad\qquad\qquad\qquad\qquad -8 + k = 0$

$\qquad\qquad\qquad\qquad\qquad\qquad\qquad k = 8$

2. Solve the equation: $\quad x^2 - 6x + 8 = 0$

$\qquad\qquad\qquad\qquad (x - 2)(x - 4) = 0$

$\qquad\qquad x - 2 = 0 \quad | \quad x - 4 = 0$

$\qquad\qquad\quad x = 2 \quad | \quad\quad x = 4$

Answer: The second root of the equation is 2.

EXERCISES

In 1-12, for the given quadratic equation, find: a. the sum of its roots b. the product of its roots

1. $x^2 - 2x - 15 = 0$ $\qquad$ 2. $x^2 + 9x + 5 = 0$ $\qquad$ 3. $2x^2 - 7x + 3 = 0$
4. $4x^2 + x - 3 = 0$ $\qquad$ 5. $x^2 + 6x = 16$ $\qquad$ 6. $3x^2 + 9x = 2$
7. $2x^2 = 3x - 6$ $\qquad$ 8. $x^2 + 9 = 0$ $\qquad$ 9. $4x^2 - 8x = 0$
10. $2m^2 + 2 = 5m$ $\qquad$ 11. $5k - 10 = 2k^2$ $\qquad$ 12. $y^2 + 2y = 2$

In 13-18, write a quadratic equation whose roots have the indicated sum and product.

13. sum $= 4$, product $= 3$ $\qquad\qquad$ 14. sum $= 16$, product $= -80$

15. sum = -3, product = -10 16. sum = -6, product = 8

17. $r_1 + r_2 = 8, r_1 r_2 = 25$ 18. $r_1 + r_2 = -\frac{5}{2}, r_1 r_2 = 1$

In 19–37, write a quadratic equation whose roots are given below.

19. $-3, 5$ 20. $2, 10$ 21. $-4, -6$

22. $-8, 8$ 23. $7, 0$ 24. $\sqrt{3}, -\sqrt{3}$

25. $\frac{1}{2}, 4$ 26. $-2, \frac{3}{2}$ 27. $-\frac{1}{3}, -\frac{2}{3}$

28. $i, -i$ 29. $4i, -4i$ 30. $2 + i, 2 - i$

31. $6 - i, 6 + i$ 32. $2 + \sqrt{3}, 2 - \sqrt{3}$ 33. $1 - \sqrt{7}, 1 + \sqrt{7}$

34. $5 + 3i, 5 - 3i$ 35. $-4 + 5i, -4 - 5i$

36. $-2 + i\sqrt{5}, -2 - i\sqrt{5}$ 37. $6 + 3i\sqrt{2}, 6 - 3i\sqrt{2}$

In 38–45, for the given quadratic equation, one root r_1 is given. Find the second root r_2 of the equation.

38. $x^2 - 11x + k = 0; r_1 = 5$ 39. $x^2 - x + k = 0; r_1 = -4$

40. $x^2 + 9x + k = 0; r_1 = -2$ 41. $x^2 + kx + 18 = 0; r_1 = 6$

42. $x^2 + kx - 16 = 0; r_1 = -8$ 43. $x^2 + kx + 4 = 0; r_1 = 12$

44. $2x^2 + kx - 12 = 0; r_1 = \frac{3}{2}$ 45. $3x^2 - x + k = 0; r_1 = -\frac{5}{3}$

46. If the roots of the equation $x^2 - 4x + c = 0$ are $2 + 3i$ and $2 - 3i$, what is the value of c?

47. If the roots of the equation $x^2 + kx + 6 = 0$ are $4 + \sqrt{10}$ and $4 - \sqrt{10}$, what is the value of k?

48. a. If one root of a quadratic equation is $4 - i$, what is the other root?

     b. Write a quadratic equation having these roots.

49. a. If one root of a quadratic equation with rational coefficients is $3 + \sqrt{7}$, what is the other root?

     b. Write a quadratic equation having these roots.

50. a. Find the roots of the equation $x^2 - 6x + 34 = 0$.

     b. Demonstrate that the sum of these roots is 6.

     c. Demonstrate that the product of these roots is 34.

In 51 and 52, select the numeral preceding the expression that best completes the sentence.

51. A quadratic equation having the roots 0 and -2 is:

     (1) $x^2 - 2 = 0$ (2) $x^2 - 2x = 0$

     (3) $x^2 + 2x = 0$ (4) $x^2 + 2 = 0$

52. A quadratic equation with roots $7 + i$ and $7 - i$ is:
(1) $x^2 - 14x + 50 = 0$ (2) $x^2 - 14x + 48 = 0$
(3) $x^2 + 14x + 50 = 0$ (4) $x^2 + 14x + 48 = 0$

53. Write, in simplest form, a quadratic equation whose roots are k and $-k$.

54. Write, in simplest form, a quadratic equation whose roots are $e + fi$ and $e - fi$.

14-10 SOLUTION OF SYSTEMS OF EQUATIONS

In *Course I*, you learned how to solve a system of linear equations by algebraic and graphic methods. If the equations are consistent, the graphs will be intersecting lines and there will be one solution.

☐ EXAMPLE: Solve $x - 2y = 4$
$y = 3 - 2x$

Algebraic Solution

$$x - 2y = 4$$
$$y = 3 - 2x$$

$$x - 2(3 - 2x) = 4$$
$$x - 6 + 4x = 4$$
$$5x = 10$$
$$x = 2$$

$$y = 3 - 2x$$
$$y = 3 - 2(2)$$
$$y = 3 - 4$$
$$y = -1$$

Answer: $(2, -1)$

Graphic Solution

On the same set of axes, graph both equations. The coordinates of the point of intersection are the solution of the system.

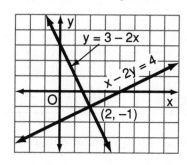

Answer: $(2, -1)$

These same procedures are used to solve a quadratic-linear system. In the graphic solution, the graphs may intersect in 2, 1, or 0 points. Since the points in the coordinate plane correspond to the set of ordered pairs of real numbers, these intersections indicate 2, 1, or 0 solutions that are pairs of real numbers.

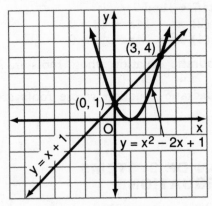

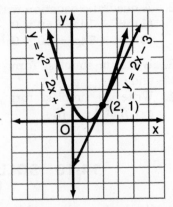

Two real solutions

One real solution

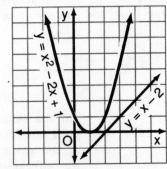

No real solution

When a quadratic-linear system has no solutions in the set of real numbers, two complex roots can often be found by solving algebraically, as shown in the following model problem.

MODEL PROBLEMS

Solve the system of equations algebraically and check.

$$y = x^2 - 2x + 3$$
$$2x - y = 2$$

How to Proceed   *Solution*

1. Solve the linear equation for one of the variables.

$$2x - y = 2$$
$$2x - 2 = y$$

2. Substitute the value of y, $2x - 2$, for y in the quadratic equation.

$$y = x^2 - 2x + 3$$
$$2x - 2 = x^2 - 2x + 3$$

3. Write the resulting equation in standard form and solve by using the quadratic formula.

$$0 = x^2 - 4x + 5$$

$$x = \frac{4 \pm \sqrt{16 - 20}}{2}$$

$$x = \frac{4 \pm 2i}{2}$$

$$x = 2 \pm i$$

4. Substitute each value of x in the linear equation of the system to find y.

$x = 2 + i$	$x = 2 - i$
$y = 2x - 2$	$y = 2x - 2$
$y = 2(2 + i) - 2$	$y = 2(2 - i) - 2$
$y = 4 + 2i - 2$	$y = 4 - 2i - 2$
$y = 2 + 2i$	$y = 2 - 2i$

5. Write the solution.

$(2 + i, 2 + 2i)$, $(2 - i, 2 - 2i)$

Note that the graphs of the equations have no points of intersection. There are no solutions in the set of ordered pairs of real numbers.

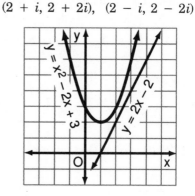

6. Check each ordered pair in both equations.

$$(2 + i, 2 + 2i)$$

$y = x^2 - 2x + 3$
$2 + 2i \stackrel{?}{=} (2 + i)^2 - 2(2 + i) + 3$
$2 + 2i \stackrel{?}{=} 4 + 4i + i^2 - 4 - 2i + 3$
$2 + 2i \stackrel{?}{=} 4 + 4i - 1 - 4 - 2i + 3$
$2 + 2i = 2 + 2i$ (True)

$2x - y = 2$
$2(2 + i) - (2 + 2i) \stackrel{?}{=} 2$
$4 + 2i - 2 - 2i \stackrel{?}{=} 2$
$2 = 2$
(True)

$$(2 - i, 2 - 2i)$$

$y = x^2 - 2x + 3$
$2 - 2i \stackrel{?}{=} (2 - i)^2 - 2(2 - i) + 3$
$2 - 2i \stackrel{?}{=} 4 - 4i + i^2 - 4 + 2i + 3$
$2 - 2i \stackrel{?}{=} 4 - 4i - 1 - 4 + 2i + 3$
$2 - 2i = 2 - 2i$ (True)

$2x - y = 2$
$2(2 - i) - (2 - 2i) \stackrel{?}{=} 2$
$4 - 2i - 2 + 2i \stackrel{?}{=} 2$
$2 = 2$
(True)

Answer: $(2 + i, 2 + 2i)$, $(2 - i, 2 - 2i)$

EXERCISES

1. Name the maximum number of points of intersection for:
   a. two lines   **b.** a line and a parabola
   c. a line and a circle   **d.** a line and an ellipse
   e. a line and a hyperbola

2. What does it mean if the two graphs of a system do not intersect?

3. **a.** Draw the graph of $y = x^2 - 4x + 5$ for $-1 \le x \le 6$.
   b. On the same set of axes, draw the graph of $y = 5 - x$.
   c. Determine from the graphs drawn in parts **a** and **b** the solution of the system. Check in both equations.

In 4–9, solve the system of equations graphically and check.

4. $2x - y = 5$
   $3x + 2y = 4$

5. $x^2 + y^2 = 10$
   $y = 3x$

6. $x^2 + 25y^2 = 25$
   $x = -5$

7. $y = x^2 - 4x + 3$
   $y = x - 1$

8. $x^2 - y^2 = 9$
   $y = 4$

9. $xy = 8$
   $y = x + 2$

10. $y = -x^2 + 4$
    $y = x + 2$

11. $9x^2 + y^2 = 9$
    $3x - y = 3$

12. $x^2 + y^2 = 8$
    $x + y = 4$

In 13–21, solve the system of equations algebraically and check.

13. $x^2 + 4y^2 = 4$
    $x = 2y - 2$

14. $y = x^2 - 3$
    $x + y = -1$

15. $y = 2 - x^2$
    $y = 2x + 4$

16. $x^2 + y^2 = 16$
    $x - y = 4$

17. $xy = -6$
    $x + 3y = 3$

18. $y = x^2 + x - 4$
    $y = 2x - 2$

19. $9x^2 - 4y^2 = 36$
    $y = 3x - 6$

20. $x^2 + y^2 = 18$
    $x + y = 6$

21. $x^2 - 2y^2 = 11$
    $y = x + 1$

In 22–24: **a.** Solve graphically. **b.** Solve algebraically. **c.** Check.

22. $x^2 + y^2 = 25$
    $y = x - 1$

23. $xy = 12$
    $y = 2x + 2$

24. $y = x^2 + 4x + 1$
    $y = 2x + 1$

25. **a.** Draw the graphs of $25x^2 + 4y^2 = 100$ and $y = 2x + 5$ on the same set of axes.
    b. How many solutions does this system have?
    c. Solve the system algebraically to obtain the exact values of the roots.
    d. Use a calculator to check your solution.

14-11 QUADRATIC INEQUALITIES

We have learned that certain quadratic equations can be solved by graphic methods. For example, in Fig. 1, the graph of the parabola $y = x^2 - 1$ crosses the x-axis at -1 and 1. Recall that the equation of the x-axis is $y = 0$. The intersection of $y = x^2 - 1$ and $y = 0$ is named by the equation $x^2 - 1 = 0$. Thus, as shown in the graph of the number line to the right, the solution set of $x^2 - 1 = 0$ is $x = -1$ or $x = 1$.

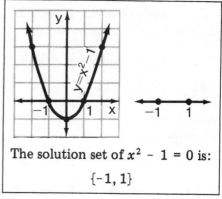

The solution set of $x^2 - 1 = 0$ is:

$$\{-1, 1\}$$

Fig. 1

The set of points that form the parabola $y = x^2 - 1$ separates the plane into two regions, or *open half-planes*, named by the inequalities:

$$y > x^2 - 1 \quad \text{AND} \quad y < x^2 - 1$$

In Fig. 2 and Fig. 3 below, each shaded region is an open half-plane. The points of the parabola $y = x^2 - 1$ are drawn as a dashed line to show that they are not points in the open half-plane. Notice that the intersection of each open half-plane and the x-axis (that is, $y = 0$) is the graph of the solution set of a quadratic inequality on a number line.

In Fig. 2, the intersection of $y > x^2 - 1$ and $y = 0$ is:

$$0 > x^2 - 1$$
or
$$x^2 - 1 < 0$$

In Fig. 3, the intersection of $y < x^2 - 1$ and $y = 0$ is:

$$0 < x^2 - 1$$
or
$$x^2 - 1 > 0$$

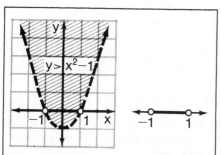

The solution set of $x^2 - 1 < 0$ is:

$$\{x \mid -1 < x < 1\}$$

Fig. 2

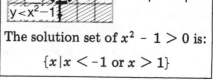

The solution set of $x^2 - 1 > 0$ is:

$$\{x \mid x < -1 \text{ or } x > 1\}$$

Fig. 3

Let us examine these quadratic relationships from an algebraic point of view. Since $x^2 - 1$ is factorable, a principle involving factors is applied to each situation. Notice the use of the logical *"and"* and *"or"* in the various principles stated.

☐ EXAMPLE 1: Solve the quadratic equation: $x^2 - 1 = 0$

Since the product of two factors is zero, the first factor is zero *or* the second factor is zero. Thus, we use the principle:

If $ab = 0$, then $a = 0$ *or* $b = 0$.

Solution

$$x^2 - 1 = 0$$
$$(x + 1)(x - 1) = 0$$
$$x + 1 = 0 \text{ or } x - 1 = 0$$
$$x = -1 \text{ or } x = 1 \quad Ans.$$

☐ EXAMPLE 2: Solve the quadratic inequality: $x^2 - 1 < 0$

Here, the product of two factors is less than zero; that is, the product is negative. Therefore, the first factor is negative *and* the second factor is positive, *or* the first factor is positive *and* the second factor is negative. This principle is stated as follows:

If $ab < 0$, then $a < 0$ *and* $b > 0$, or $a > 0$ *and* $b < 0$.

Solution

$$x^2 - 1 < 0$$
$$(x + 1)(x - 1) < 0$$

$x + 1 < 0$ *and* $x - 1 > 0$	OR	$x + 1 > 0$ *and* $x - 1 < 0$
$x < -1$ *and* $x > 1$		$x > -1$ *and* $x < 1$

There are no numbers that are less than -1 *and* greater than 1 at the same time. Since this intersection is empty, *disregard* this case.

The numbers that are greater than -1 *and* less than 1 are in the intersection:

$$\{x | -1 < x < 1\} \quad Ans.$$

Compare this answer to Fig. 2, seen earlier.

☐ EXAMPLE 3: Solve the quadratic inequality: $x^2 - 1 > 0$

Here, the product of two factors is greater than zero; that is, the product is positive. Therefore, both factors are negative *or* both factors are positive. This principle is stated as follows:

If $ab > 0$, then $a < 0$ *and* $b < 0$, or $a > 0$ *and* $b > 0$.

Solution

$$x^2 - 1 > 0$$
$$(x + 1)(x - 1) > 0$$

$x + 1 < 0$ *and* $x - 1 < 0$	OR	$x + 1 > 0$ *and* $x - 1 > 0$
$x < -1$ *and* $x < 1$		$x > -1$ *and* $x > 1$

The word *"and"* indicates an intersection of sets. The numbers that are less than -1 *and* also less than 1 form the intersection:	The word *"and"* indicates an intersection of sets. The numbers that are greater than -1 *and* also greater than 1 form the intersection:
$x < -1$ OR	$x > 1$

Answer: $\{x \mid x < -1 \text{ or } x > 1\}$

Compare this answer to Fig. 3, seen earlier.

Caution: In section 6 of this chapter, we learned that imaginary numbers cannot be ordered. That is, the relations ">" and "<" are meaningless for imaginary numbers. Therefore, quadratic inequalities cannot have solutions that are imaginary numbers. In other words:

■ **Solution sets of quadratic inequalities are restricted to sets of real numbers only.**

Summary

The three examples we have just studied serve as illustrations of the following general cases where roots r_1 and r_2 are real numbers and $r_1 \neq r_2$, and where $a > 0$:

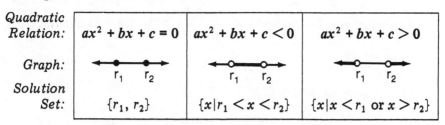

Quadratic Relation:	$ax^2 + bx + c = 0$	$ax^2 + bx + c < 0$	$ax^2 + bx + c > 0$
Graph:	$r_1 \quad r_2$	$r_1 \quad r_2$	$r_1 \quad r_2$
Solution Set:	$\{r_1, r_2\}$	$\{x \mid r_1 < x < r_2\}$	$\{x \mid x < r_1 \text{ or } x > r_2\}$

When an equality is combined with an inequality, other quadratic relations can be formed, as shown on the next page.

Relation:	$ax^2 + bx + c \leq 0$	$ax^2 + bx + c \geq 0$
Graph:		
	$r_1 \qquad r_2$	$r_1 \qquad r_2$
Solution Set:	$\{x \mid r_1 \leq x \leq r_2\}$	$\{x \mid x \leq r_1 \text{ or } x \geq r_2\}$

If the roots of $ax^2 + bx + c = 0$ are not distinct real numbers, that is, if $r_1 = r_2$ or if r_1 and r_2 are imaginary, then the parabola $y = ax^2 + bx + c$ intersects the x-axis in 1 or 0 points respectively. In these cases, the solution set of the inequality $ax^2 + bx + c > 0$ or $ax^2 + bx + c < 0$ is either the set of all real numbers or the empty set.

MODEL PROBLEMS

1. Which is the graph of the solution set of $x^2 + 2x - 8 > 0$?

   (1)   (2)   (3)   (4)
      $-4 \quad 2$ $-4 \quad 2$ $-2 \quad 4$ $-2 \quad 4$

Solution

Use the principle:
    If $ab > 0$, then $a < 0$ and $b < 0$, or $a > 0$ and $b > 0$.

$$x^2 + 2x - 8 > 0$$
$$(x + 4)(x - 2) > 0$$

$x + 4 < 0$ and $x - 2 < 0$ OR $x + 4 > 0$ and $x - 2 > 0$
   $x < -4$ and $x < 2$ $x > -4$ and $x > 2$

This intersection is: This intersection is:

       $x < -4$ $x > 2$

The solution set is $\{x \mid x < -4$ OR $x > 2\}$, graphed in choice (2).

Answer: (2)

2. Graph the solution set of $x^2 - 4x - 5 \leq 0$.

Solution

Use the principle:
    If $ab \leq 0$, then $a \leq 0$ and $b \geq 0$, or $a \geq 0$ and $b \leq 0$.

$$x^2 - 4x - 5 \leq 0$$
$$(x + 1)(x - 5) \leq 0$$

$$x + 1 \leq 0 \quad \text{and} \quad x - 5 \geq 0 \quad \text{or} \quad x + 1 \geq 0 \quad \text{and} \quad x - 5 \leq 0$$
$$x \leq -1 \quad \text{and} \qquad x \geq 5 \qquad\qquad x \geq -1 \quad \text{and} \qquad x \leq 5$$

Since this intersection is empty, disregard this set.

The intersection is:

$$\{x | -1 \leq x \leq 5\}$$

Answer: ◄––●––––●––►
$\quad\quad\quad\,$ −1 $\quad$ 5

EXERCISES

In 1-7, select the numeral preceding the expression that best completes the sentence or answers the question.

1. The solution set of $x^2 - 4 < 0$ is:
   (1) $-2 < x < 2$ $\qquad\qquad$ (2) $-4 < x < 4$
   (3) $x < -2$ or $x > 2$ $\qquad$ (4) $x < -4$ or $x > 4$
2. The solution set of $x^2 - 9 > 0$ is:
   (1) $-3 < x < 3$ $\qquad\qquad$ (2) $-9 < x < 9$
   (3) $x < -3$ or $x > 3$ $\qquad$ (4) $x < -9$ or $x > 9$
3. Which is the graph of the solution set of $x^2 + 3x - 4 > 0$?

   (1) ◄––○––○––► $\quad$ (2) ◄––○––○––► $\quad$ (3) ◄––○––○––► $\quad$ (4) ◄––○––○––►
   $\quad\quad$ −4 $\,$ 1 $\qquad\qquad$ −1 $\,$ 4 $\qquad\qquad$ −4 $\,$ 1 $\qquad\qquad$ −1 $\,$ 4

4. Which is the graph of the solution set of $x^2 - 8x + 12 < 0$?

   (1) ◄––○––○––► $\quad$ (2) ◄––○––○––► $\quad$ (3) ◄––○––○––► $\quad$ (4) ◄––○––○––►
   $\quad\quad$ 2 $\,$ 6 $\qquad\qquad$ −6 $\,$ −2 $\qquad\qquad$ 2 $\,$ 6 $\qquad\qquad$ −6 $\,$ −2

5. If $x^2 - 3x \leq 0$, which is the graph of its solution set?

   (1) ◄––○––○––► $\quad$ (2) ◄––●––●––► $\quad$ (3) ◄––○––○––► $\quad$ (4) ◄––●––●––►
   $\quad\quad$ 0 $\,$ 3 $\qquad\qquad$ 0 $\,$ 3 $\qquad\qquad$ −3 $\,$ 0 $\qquad\qquad$ −3 $\,$ 0

6. The graph of the solution set of $x^2 - 4x \geq 0$ is:

   (1) ◄––○––○––► $\quad$ (2) ◄––●––●––► $\quad$ (3) ◄––○––○––► $\quad$ (4) ◄––●––●––►
   $\quad\quad$ 0 $\,$ 4 $\qquad\qquad$ 0 $\,$ 4 $\qquad\qquad$ −4 $\,$ 0 $\qquad\qquad$ −4 $\,$ 0

7. Which of the following is a quadratic inequality whose solution is graphed at the right?
   (1) $x^2 - 3 < 0$ $\qquad\qquad$ (2) $x^2 - 3 > 0$
   (3) $x^2 - 3 \leq 0$ $\qquad\qquad$ (4) $x^2 - 3 \geq 0$

   ◄––○––○––►
   $-\sqrt{3}$ $\sqrt{3}$
   Ex. 7

In 8-22: **a.** Write the solution set for the given quadratic inequality. **b.** Graph the solution set.

8. $x^2 - 16 < 0$ $\qquad$ 9. $x^2 - 25 > 0$ $\qquad$ 10. $2x^2 - 8 > 0$
11. $x^2 + 7x \leq 0$ $\qquad$ 12. $x^2 - 4x \geq 0$ $\qquad$ 13. $x^2 - 7x + 10 < 0$
14. $x^2 - x - 6 > 0$ $\qquad$ 15. $x^2 - 5x - 24 \leq 0$ $\qquad$ 16. $x^2 + 8x + 7 \geq 0$
17. $x^2 + 2x < 15$ $\qquad$ 18. $x^2 > 8x + 20$ $\qquad$ 19. $x^2 + 27 < 12x$
20. $4x^2 - 9 \geq 0$ $\qquad$ 21. $2x^2 - 11x + 5 \geq 0$ $\qquad$ 22. $3x^2 + 10x \leq 8$

23. a. The quadratic inequality $x^2 - 14x + 49 \leq 0$ has a solution set consisting of only one number. Find the solution set. b. Explain why the solution set is limited to one number.
24. Explain why the solution set of $x^2 + 16 < 0$ is empty.
25. Explain why the solution set of $x^2 + 9 > 0$ is the set of all real numbers.

14-12 REVIEW EXERCISES

In 1–4, write each number in terms of i, perform the indicated operation, and write the answer in simplest terms.

1. $\sqrt{-49} + 3\sqrt{-25}$
2. $\sqrt{-18} + \sqrt{-2} - \sqrt{-32}$
3. $5\sqrt{-5} \cdot \sqrt{-20}$
4. $\sqrt{-300} - 2\sqrt{-75} + \sqrt{-48}$

In 5–10, perform the operation, and express the result in the form $a + bi$.

5. $(7 - 3i) + (5 + 8i)$
6. $(4 - 10i) - (23 - 6i)$
7. $(2 + 5i)(3 - 7i)$
8. $(4 - 3i)^2$
9. $(5 + \sqrt{-16})(2 - \sqrt{-9})$
10. $(6 + 7i) \div (2 - i)$

In 11–18, 31–33, and 40, select the numeral preceding the expression that best completes the sentence.

11. The complex number $5i^3 - 2i^2$ is equivalent to:
    (1) $-2 + 5i$ (2) $-2 - 5i$ (3) $2 - 5i$ (4) $2 + 5i$
12. The product $4i^3 \cdot 2i^2$ is:
    (1) 8 (2) $8i$ (3) -8 (4) $-8i$
13. The number $3 + i(3 + i)$ is equivalent to:
    (1) 8 (2) $2 + 3i$ (3) $8 + 6i$ (4) $4 + 3i$
14. The expression $\sqrt{-192}$ equals:
    (1) $8\sqrt{3}$ (2) $3\sqrt{8}$ (3) $8i\sqrt{3}$ (4) $3i\sqrt{8}$
15. The product of $5 - 2i$ and its conjugate is:
    (1) 21 (2) 29 (3) $21 - 20i$ (4) $29 - 20i$
16. The multiplicative identity for the set of complex numbers is:
    (1) $1 + 0i$ (2) $1 + 1i$ (3) $0 + 1i$ (4) $0 + 0i$
17. The multiplicative inverse of $3 - i$ is:
    (1) $3 + i$ (2) $\dfrac{3 + i}{10}$ (3) $\dfrac{3 + i}{8}$ (4) $\dfrac{3 - i}{10}$
18. If $x + yi = 1 + 2i + 3i^2$, then:
    (1) $x = 1, y = 2$ (2) $x = -2, y = 2$
    (3) $x = 2, y = 2$ (4) $x = 4, y = 2$

In 19-22, perform the indicated operations and write the answer in *simplest* terms.

19. $(5 - i)(10 + 2i)$

20. $3i(2 - i) - 3(i + 1)$

21. $(1 - 3i)^2 + 6i$

22. $(9 - i\sqrt{2})(9 + i\sqrt{2})$

23. *True* or *False:* (Complex numbers, +, ·) is a field.

24. *True* or *False:* Every complex number has a multiplicative inverse.

In 25-30, find in $a + bi$ form the roots of the given equation.

25. $x^2 - 6x + 10 = 0$ **26.** $x^2 + 10x + 26 = 0$ **27.** $x^2 + 29 = 4x$

28. $\dfrac{x^2}{4} = 3x - 10$ **29.** $2x^2 = 6x - 5$ **30.** $\dfrac{16x}{7} + \dfrac{7}{x} = 0$

31. The roots of $x^2 - 16x + 61 = 0$ are:
   (1) real, rational, and equal   (2) real, rational, and unequal
   (3) real, irrational, and unequal   (4) imaginary

32. The graph of the parabola $y = x^2 + 8$:
   (1) is tangent to the x-axis
   (2) intersects the x-axis at 2 points
   (3) lies entirely above the x-axis
   (4) lies entirely below the x-axis

33. The roots of $2x^2 + bx + 1 = 0$ are imaginary when b equals:
   (1) -4 (2) -2 (3) 3   (4) 4

34. Find the largest integral value of k such that the roots of $2x^2 + 7x + k = 0$ are real.

35. What is the sum of the roots of $3x^2 - 12x + 10 = 0$?

36. If the roots of $x^2 - 6x + k = 0$ are $3 + \sqrt{7}$ and $3 - \sqrt{7}$, find k.

In 37-39, write a quadratic equation whose roots are given.

37. $6i, -6i$ **38.** $2 + 4i, 2 - 4i$ **39.** $-8, 3$

40. The graph of the solution set of $x^2 - 9x + 14 < 0$ is:

   (1) ⬅—○——○—➡
         2 7   (2) ⬅—○——○—➡
           2 7   (3) ⬅—○——○—➡
        −7 −2   (4) ⬅—○——○—➡
         −7 −2

In 41–43: **a.** Solve graphically. **b.** Solve algebraically. **c.** Check.

41. $x^2 - 2x = 2 - y$ **42.** $xy = -4$ **43.** $x^2 + 4y^2 = 4$
    $y = 2x + 1$ $x + 2y = 2$ $x = 6$

Chapter **15**

Statistics

In order to understand and compare events in the daily lives of individuals and society, we study statistics. *Statistics* is the science that deals with the collection, organization, and interpretation of related pieces of numerical information called *data*.

15-1 THE SUMMATION SYMBOL

In statistics, we often work with sums. The Greek capital letter Σ, pronounced "sigma," is used to indicate *the sum of* related terms. For example, $\sum_{i=1}^{6} i$ means "the sum of the integers designated by i, from $i = 1$ to $i = 6$." Therefore:

$$\sum_{i=1}^{6} i = 1 + 2 + 3 + 4 + 5 + 6 = 21$$

In the preceding example, the letter i is called the **index**; and i is replaced by a series of consecutive integers, starting with the lower limit of summation and ending with the upper limit of summation.

The *lower limit of summation* is the value of the index placed below the summation symbol, and the *upper limit of summation* is the value of the index placed above the summation symbol. In $\sum_{i=1}^{6} i$, the lower limit is 1, and the upper limit is 6.

Any letter can be used as an index; but i, j, k, and n are the most frequently used letters. Since an index acts as a placeholder, an index of i is *not* equal to the imaginary unit, $\sqrt{-1}$.

$\square$ EXAMPLE 1: $\sum_{k=0}^{5} 3^k = 3^0 + 3^1 + 3^2 + 3^3 + 3^4 + 3^5$
$= 1 + 3 + 9 + 27 + 81 + 243 = 364$

684

☐ EXAMPLE 2: $\displaystyle\sum_{j=3}^{6} j^2 = 3^2 + 4^2 + 5^2 + 6^2$
$$= 9 + 16 + 25 + 36 = 86$$

If a variable quantity is denoted by x, then successive values of that variable can be indicated by using a *subscript*, as in x_1, x_2, x_3 (read as "x sub-one, x sub-two, x sub-three"). Notice how sigma notation and subscripted variables are used in the next example.

☐ EXAMPLE 3: Mr. Cook teaches five classes. The recorded number of students who are absent from his classes today are 3, 1, 2, 0, and 1.
a. If x represents the number of students absent from one of Mr. Cook's classes, use subscripted variables to show the number of absences in each of his classes.
b. Write an expression in sigma notation to find the total number of students absent from Mr. Cook's classes, and find the total.

Solution

a. Using subscripted variables: $x_1 = 3$ $x_2 = 1$ $x_3 = 2$ $x_4 = 0$ $x_5 = 1$
b. $\displaystyle\sum_{i=1}^{5} x_i = x_1 + x_2 + x_3 + x_4 + x_5$
$$= 3 + 1 + 2 + 0 + 1 = 7$$

When the summation symbol is used without an index and without upper and lower limits of summation, then $\sum$ designates the sum of *all* values of the given variable under consideration. Summation symbols without an index are found in the following example.

☐ EXAMPLE 4: Mrs. Gallagher, a science teacher, has assigned 5 labs to her class. In the frequency distribution at the right, she has recorded the number of students who have completed 0, 1, 2, 3, 4, or 5 lab reports. Using the table, find:
a. the number of students in the class.
b. the total number of lab reports completed.

Reports x	Frequency f	xf
5	1	5
4	5	20
3	8	24
2	7	14
1	4	4
0	2	0

Solution

a. $\sum f = 1 + 5 + 8 + 7 + 4 + 2 = 27$
b. $\sum xf = 5 + 20 + 24 + 14 + 4 + 0 = 67$

Answer: a. 27 students   b. 67 lab reports

MODEL PROBLEMS

1. a. Evaluate $\sum_{k=1}^{5} 3k$.   b. Evaluate $3 \sum_{k=1}^{5} k$.

Solution

a. $\sum_{k=1}^{5} 3k = 3(1) + 3(2) + 3(3) + 3(4) + 3(5)$
$= 3 + 6 + 9 + 12 + 15 = 45$ *Ans.*

b. $3 \sum_{k=1}^{5} k = 3(1 + 2 + 3 + 4 + 5) = 3(15) = 45$ *Ans.*

Note: $\sum_{k=1}^{5} 3k = 3 \sum_{k=1}^{5} k$. This example illustrates a general statement that is proved in the next model problem.

2. If c is a constant, show that $\sum_{j=1}^{n} cj = c \sum_{j=1}^{n} j$.

Solution

$\sum_{j=1}^{n} cj = c(1) + c(2) + c(3) + \cdots + c(n)$
$= c(1 + 2 + 3 + \cdots + n)$ by the distributive property.
$= c \sum_{j=1}^{n} j$

3. Using the summation symbol, write an expression that equals $1 + \frac{1}{4} + \frac{1}{9} + \frac{1}{16}$.

Solution

$$1 + \frac{1}{4} + \frac{1}{9} + \frac{1}{16} = \frac{1}{1^2} + \frac{1}{2^2} + \frac{1}{3^2} + \frac{1}{4^2} = \sum_{i=1}^{4} \frac{1}{i^2} \quad Ans.$$

Note: Other answers also exist. For example, $1 + \frac{1}{4} + \frac{1}{9} + \frac{1}{16} = \sum_{n=0}^{3} \frac{1}{(n+1)^2}$.

| EXERCISES |

In 1-15, find the value indicated by the summation symbol.

1. $\sum_{k=0}^{6} 2k$ 　　　　 2. $\sum_{i=1}^{5} (i + 1)$ 　　　　 3. $\sum_{j=0}^{4} j^2$

4. $\sum_{n=2}^{5} (n - 2)^2$ 　　 5. $\sum_{j=1}^{3} j^3$ 　　　　 6. $\sum_{n=0}^{3} 2^n$

7. $\sum_{n=1}^{4} \frac{1}{n}$ 　　　 8. $\sum_{k=0}^{5} (10 - k)$ 　　 9. $\sum_{i=3}^{7} (2 - i)^2$

10. $5 \sum_{n=1}^{4} (n - 1)$ 　 11. $\sum_{k=1}^{3} 3k^2$ 　　　 12. $\sum_{k=1}^{3} k^{k-1}$

13. $\sum_{n=0}^{2} \cos(n\pi)$ 　 14. $\sum_{n=1}^{3} \sin\left(\frac{n\pi}{2}\right)$ 　 15. $\sum_{k=0}^{4} \cos\left(\frac{k\pi}{2}\right)$

In 16-21, use the summation symbol to write an expression to indicate the sum.

16. $2(1) + 2(2) + 2(3) + 2(4) + 2(5)$ 　　　 17. $0 + 1 + 4 + 9 + 16$

18. $\frac{1}{3} + \frac{1}{6} + \frac{1}{9} + \frac{1}{12} + \frac{1}{15}$ 　　　 19. $\frac{1}{2} + \frac{2}{3} + \frac{3}{4} + \frac{4}{5}$

20. $1^1 + 2^2 + 3^3 + 4^4 + 5^5 + 6^6$ 　　　 21. $2(1) + 3(2) + 4(3)$

In 22-25, find the value of the indicated sum when $x_1 = 12, x_2 = 5$, $x_3 = 4, x_4 = 8$, and $x_5 = 7$.

22. $\sum_{i=1}^{5} x_i$ 　 23. $\sum_{k=3}^{5} x_k$ 　 24. $3 \sum_{j=1}^{3} x_j$ 　 25. $\sum_{n=2}^{4} 5x_n$

In 26-29, select the numeral preceding the expression that best completes the sentence.

26. The value of $3 \sum_{k=1}^{5} (k - 1)$ is:

　　 (1) 10 　　　 (2) 15 　　　 (3) 30 　　　 (4) 45

27. The value of $\sum_{n=1}^{3} \tan \frac{(2n - 1)\pi}{4}$ is:

　　 (1) 1 　　　 (2) 2 　　　 (3) 3 　　　 (4) -1

28. $\sum_{n=0}^{2} \left(\text{Arc sin } \dfrac{n}{2} \right)$ equals:

    (1) $60°$ (2) $90°$ (3) $120°$ (4) $150°$

29. The sum $1 + 8 + 27 + 64$ is *not* equal to:

    (1) $\sum_{i=1}^{4} i^3$ (2) $\sum_{k=0}^{4} (k + 1)^3$

    (3) $\sum_{n=0}^{3} (n + 1)^3$ (4) $\sum_{j=2}^{5} (j - 1)^3$

30. Represent the sum $5 + 10 + 15 + 20 + 25 + 30$ by *three* different expressions, each involving the summation symbol.

31. In **a–d**, evaluate the expression:

    a. $\sum_{n=4}^{7} (3n + 2)$ b. $3 \sum_{n=4}^{7} (n + 2)$

    c. $\sum_{n=4}^{7} (3n + 6)$ d. $3 \sum_{n=4}^{7} (n + 6)$

    e. Which expressions in this exercise, if any, represent the same sum?

32. Show that $\sum_{k=1}^{n} 7k = 7 \sum_{k=1}^{n} k$.

33. If b is a constant, show that $\sum_{n=1}^{6} bn = b \sum_{n=1}^{6} n$.

34. Show that $\sum_{k=1}^{n} (k - 1) = \sum_{k=0}^{n-1} k$.

35. Evaluate $\sum_{n=1}^{100} n$. (*Hint:* Add $1 + 100$, add $2 + 99$, etc.)

15-2 MEASURES OF CENTRAL TENDENCY

In analyzing data, we often find it useful to represent all the collected measures by a single number called a *measure of central tendency*. Each measure is a number, like an average, that "tends" to fall somewhere in the "center" of a set of organized data. We have studied three measures of central tendency called the *mean*, the *median*, and the *mode*.

The Mean

The *mean*, or *arithmetic mean*, of a set of n numbers is the sum of the numbers divided by n. The number n is sometimes written as a capital letter N. The symbol for the mean is $\bar{x}$, read as "x-bar." Therefore, for a sample of scores, $x_1, x_2, x_3, \ldots, x_n$:

$$\text{The mean} = \bar{x} = \frac{\sum\limits_{i=1}^{n} x_i}{n} = \frac{x_1 + x_2 + x_3 + \cdots + x_n}{n}$$

☐ EXAMPLE 1: Cindy Woodhouse, a cabdriver, collected the following fares one afternoon: $4.50, $3.00, $6.50, $5.75, $11.00, $7.50, $3.25, $5.75, $6.25, and $8.00. What was the mean fare?

Solution

$$\bar{x} = \frac{\sum\limits_{i=1}^{10} x_i}{10} = \frac{\$61.50}{10} = \$6.15 \quad Ans.$$

Most people use the word "average" to describe the mean. Thus, if Ms. Woodhouse had collected $6.15 from each of her 10 fares that afternoon, she would have collected the same total amount of money, that is, $61.50.

The Median

When a set of data is arranged in numerical order, the "middle" score is called the *median*. Thus, in an ordered set, the number of values that precede the median is equal to the number of values that follow the median.

☐ EXAMPLE 2: Kurt's grades on his report card are 88, 81, 91, 83, and 86. What is his median grade?

How to Proceed	*Solution*
1. Arrange the grades in numerical order.	81, 83, 86, 88, 91
2. Find the middle number by counting from the top or bottom grade.	81, 83, 86, 88, 91
	median = 86 *Ans.*

Alternate Method. If N represents an *odd* number of scores arranged in numerical order, the median is the score that is $\dfrac{N+1}{2}$ from the top or the bottom. Since $N = 5$ grades, the median grade is the one that is

$\dfrac{N+1}{2} = \dfrac{5+1}{2} = \dfrac{6}{2} = 3$ from the top or the bottom. That is, the median grade is 86.

☐ EXAMPLE 3: Sarah's grades on her report card are 88, 90, 85, 84, 84, and 88. What is her median grade?

How to Proceed	*Solution*
1. Arrange the grades in numerical order.	90, 88, 88, 85, 84, 84
2. There are *two* middle numbers.	⌊90, 88,⌉88, 85,⌊84, 84⌋
3. Find the mean, or arithmetic average, of these two middle scores.	$\dfrac{88+85}{2} = \dfrac{173}{2} = 86.5$
	median = 86.5 *Ans.*

Alternate Method. If N represents an *even* number of scores arranged in numerical order, the median is the mean (average) of the two scores that are $\dfrac{N}{2}$ and $\dfrac{N+2}{2}$ from the top or the bottom. Here, $N = 6$ grades.

Thus, $\dfrac{N}{2} = \dfrac{6}{2} = 3$, and $\dfrac{N+2}{2} = \dfrac{6+2}{2} = \dfrac{8}{2} = 4$. The median is the mean (average) of the 3rd and 4th scores, counting from the top or the bottom. That is, the median is 86.5.

The Mode

The *mode* is the score that appears most often in a set of data.

☐ EXAMPLE 4: The chart at the right shows the number of children in each of 20 families answering a survey. What is the mode?

Number of children	Frequency
6	1
5	0
4	2
3	4
2	8
1	5

Solution

The mode for this set of data is 2, the entry that has the highest frequency.

Answer: mode = 2

For some sets of data, there may be more than one mode and, in some cases, no mode whatsoever.

1. The ages of employees at a fast-food restaurant are 17, 17, 17, 18, 18, 19, 19, 19, 21, 23, and 37. This set of data contains *two* modes: 17 and 19. When two modes appear, the data is **bimodal**, and both modes are reported. We do *not* take an average of these modes, since a mode tells us where most scores occur.

2. Mrs. Mangold found the following number of spelling errors on compositions that she graded: 0, 0, 0, 1, 2, 2, 3, 4, 4, 4, 5, 5, 5, 6, 7, 7, 8, 9, 9, and 12. This data has *three* modes: 0, 4, and 5.

3. On his last five trips to Sound Beach, Mr. Fernandes caught the following number of fish: 3, 1, 5, 0, 2. Since no number appears more often than others, this data has *no* mode.

| **MODEL PROBLEMS** |

---

1. Mrs. Taggart bought a new car. She kept a record of the number of miles that she drove per gallon of gas for each of the first six times she filled the tank. Her record showed the following number of miles per gallon: 29, 32, 32, 33, 35, 37. **a.** Find the mean. **b.** Find the median.   **c.** Find the mode.

Solution

a. $\bar{x} = \dfrac{\sum\limits_{i=1}^{n} x_i}{n} = \dfrac{29 + 32 + 32 + 33 + 35 + 37}{6} = \dfrac{198}{6} = 33$ *Ans.*

b. There are six entries in this set of data. The median is the mean (average) of the entries that fall into positions 3 and 4 from the top or the bottom.

$29, 32, | 32, 33, | 35, 37$

$\text{median} = \dfrac{32 + 33}{2}$

$= 32.5$ *Ans.*

c. The mode is the entry that occurs most often: 32   *Ans.*

2. The ages of 25 students in a senior high school mathematics class are recorded in the frequency distribution table at the right. For these ages, find: **a.** the mean   **b.** the median   **c.** the mode.

Age in years x	Frequency f
18	2
17	11
16	12

Solution

a. (1) For each interval, find *fx*.
(Note that 18 + 18 = 36,
and 2 · 18 = 36. Similarly,
11 · 17 = 187, the same
number obtained by add-
ing eleven values, each of
which is 17.)

x	f	fx
18	2	36
17	11	187
16	12	192
—	$N = 25$	$\sum fx = 415$

(2) Find totals: $N = \sum f = 25$,
and $\sum fx = 415$.

(3) The mean = $\bar{x} = \dfrac{\sum fx}{N} = \dfrac{415}{25} = 16.6.$ *Ans.*

b.

x	f
18	2
17	11
16	12

The median is the middle score. For $N = 25$, the
median falls in a position that is $\dfrac{N+1}{2} = \dfrac{25+1}{2} = \dfrac{26}{2} = 13$ scores from the top or the bottom. Thus,
the median lies in the interval, 17. The median =
17. *Ans.*

c. The age appearing most often is 16, because this interval has the
greatest frequency, 12. Therefore, the mode = 16.   *Ans.*

EXERCISES

1. Find the mean grade for each of the following students. If necessary, round off the mean to the nearest tenth.
   a. Peter: 90, 70, 88, 82, 70   b. Maria: 80, 82, 93, 91, 94
   c. Elizabeth: 82, 75, 100, 83   d. Thomas: 92, 91, 75, 93, 98
   e. Al: 80, 70, 92, 78, 78, 98   f. Joanna: 90, 90, 61, 90
2. a. Mr. Katzel will give a grade of *A* on the report card to any student whose mean average is 90.0 or higher. Which students in
exercise 1, if any, will receive a grade of *A*?
   b. If Mr. Katzel leaves out the lowest test grade for each student
listed in exercise 1, which students will then receive a grade of *A*?
3. For each set of student grades in exercise 1, find the median grade.
4. In a–h, find the mode for each distribution. If no mode exists,
write "none."
   a. 3, 3, 4, 5, 9   b. 4, 4, 5, 9, 9   c. 4, 4, 6, 6, 6
   d. 3, 4, 7, 8, 9   e. 3, 8, 3, 8, 3   f. 5, 2, 2, 5, 2, 5
   g. 1, 7, 4, 3, 2, 4, 3, 1, 7, 1   h. 5, 2, 7, 2, 8, 5, 7, 9, 3
5. What is the median for the digits 0, 1, 2, . . . , 9?

6. The median age of 4 children is 9.5 years. If Jeanne is 11, Debbie is 8, and Jimmy is 5 years old, then Kathy's age *cannot* be:
   (1) 10   (2) 11   (3) 13   (4) 16

7. The set of data 6, 8, 9, x, 9, 8 is given. In each part, find all possible values of x such that: (a) There is no mode because all scores appear an equal number of times. (b) There is only one mode. (c) There are two modes.

8. Four tests have been given to students in a class, and their test scores are recorded below. Find the score needed by each student on a fifth test so that the mean average of all 5 tests is exactly 80, or explain why such an average is not possible.
   a. Edna: 80, 75, 92, 85   b. Rosemary: 77, 81, 76, 83
   c. Joe: 78, 72, 70, 75   d. Jerry: 68, 82, 79, 71

9. Alice Garr typed a seven-page report. She made the following number of typing errors, reported by successive pages of the report: 2, 0, 2, 1, 4, 5, 7. For the number of errors per page, find:
   a. the mean   b. the median   c. the mode

10. David enters bicycle races. His times to complete the last 6 races, each covering a distance of 20 miles, were recorded to the nearest minute as follows: 55, 58, 53, 50, 52, 50. For these recorded times, find: a. the mean   b. the median   c. the mode

11. In the last 8 times that Mary ran the 100-yard dash, her times to the nearest tenth of a second were: 13.5, 13.3, 13.1, 13.3, 13.2, 13.0, 12.8, and 13.0. The mean time for Mary to run the 100-yard dash is:
    (1) 13.1   (2) 13.15   (3) 13.2   (4) 13.25

12. The ages of 30 students enrolled in a health class are shown in the accompanying table.
    a. Find the mean age to the *nearest tenth*.
    b. For this data, which statement is true?
       (1) median $>$ mode
       (2) median $=$ mode
       (3) median $<$ mode
       (4) median $>$ mean

Ages x_i	Frequency f_i
18	2
17	12
16	14
15	2

In 13–17, for the given frequency distribution, find: a. the mean b. the median c. the mode.

13.

Measure (x_i)	Frequency (f_i)
10	3
11	5
12	2

14.

Index i	Measure x_i	Frequency f_i
1	100	3
2	90	3
3	80	4

15.

x_i	f_i
10	1
20	5
30	2
40	4

16.

x_i	f_i
5	5
4	5
3	2
2	3

17.

x_i	f_i
15	12
10	4
5	4
0	4

18. Eddie Dunn scored the following number of goals in his last 7 hockey games: 3, 4, 0, 2, 3, 1, and 2. What is the least number of goals that Eddie must score in his next game to claim that his mean average number of goals per game is greater than 2?

(1) 1   (2) 2   (3) 3   (4) 4

15-3 MEASURES OF DISPERSION: THE RANGE AND THE MEAN ABSOLUTE DEVIATION

The mean is the measure of central tendency most frequently used to describe statistical data. The mean, however, does not give us sufficient information about the data to draw conclusions.

Below are frequency diagrams for four sets of data. The pictures clearly show us that these sets of data are very different.

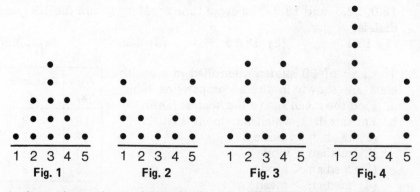

And yet, for each set of data displayed here, *the mean is 3.*

For example, using the data in Fig. 4, we can complete the frequency table at the right and find sums. It follows that:

$$\text{the mean} = \bar{x} = \frac{\sum f_i x_i}{n} = \frac{39}{13} = 3$$

x_i	f_i	$f_i x_i$
1	0	0
2	8	16
3	1	3
4	0	0
5	4	20
—	$n = 13$	$\sum f_i x_i = 39$

Since each of these sets of data has a mean of 3, we need another measure to show how these sets are different. The new measure should indicate how individual scores are scattered, or distributed, about the mean. A number that indicates the spread, or variation, of scores about the mean is called a *measure of dispersion*.

One such measure of dispersion might have a value of 1.0 for the data in Fig. 1 and a value of 1.7 for the data in Fig. 2. The lower number 1.0 tells us that scores are close to the mean of 3 in Fig. 1, while the higher number 1.7 tells us that scores are distributed further away from the mean of 3 in Fig. 2.

In this section, we will study two measures of dispersion: the *range* and the *mean absolute deviation*. In the next section, we will study a third measure of dispersion, called the *standard deviation*.

The Range

The simplest of the measures of dispersion is the range. The *range* is the difference between the highest score and the lowest score in a set of data.

For example, if Eve Lucano's grades for this marking period are 97, 94, 92, 89, and 87, then the range of Eve's grades is 97 − 87 or 10.

Since the range is dependent on only the highest and lowest scores in a distribution, the range is *often unreliable* as a measure of dispersion. Consider the following cases in which we compare data.

Comparison 1. For each set of data in Fig. 5 and Fig. 6, the mean is 6 and the range = 10 − 2 = 8.

Although these frequency diagrams are very different, the range does not help us in comparing the distribution of data in these sets.

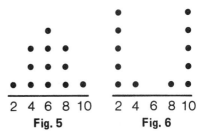

Fig. 5   Fig. 6

Comparison 2. The test scores for two students are shown below.

Heidi: 75, 88, 91, 92, 95, 99   range = 99 − 75 = 24
Eric: 87, 88, 91, 92, 95, 99   range = 99 − 87 = 12

While the range of Heidi's grades is twice the range of Eric's grades, their grades are exactly the same, except for one extreme score. In fact, both students have an average in the low nineties. Here, the sets of data are very much alike, but we have been led to believe that they are different because of differences in the range.

The Mean Absolute Deviation

Let us study a more useful measure of dispersion, based upon the information in the accompanying table.

x_i	$\bar{x}$	$x_i - \bar{x}$	$\lvert x_i - \bar{x} \rvert$
93	86	7	7
90	86	4	4
89	86	3	3
87	86	1	1
85	86	-1	1
72	86	-14	14
$\sum x_i =$ 516	—	$\sum (x_i - \bar{x}) =$ 0	$\sum \lvert x_i - \bar{x} \rvert =$ 30

1. The data in column 1 is used to find the mean:

$$\bar{x} = \frac{\sum x_i}{n} = \frac{516}{6} = 86$$

2. The mean of 86 is entered in every row of column 2.

3. The difference between each entry x_i in the sample and the mean $\bar{x}$ is recorded in column 3. Notice that the sum of these differences is zero; that is, $\sum (x_i - \bar{x}) = 0$. If the differences had been reversed, it would also be true that $\sum (\bar{x} - x_i) = 0$. This example illustrates the fact that *the sum of the differences between each entry in a sample and the mean of that sample is always equal to zero*.

4. In column 4, however, by taking the absolute value of the deviation of each score from the mean, we see that the sum of these absolute values is usually a number other than zero. Here, $\sum \lvert x_i - \bar{x} \rvert = 30$.

5. By definition, the *mean absolute deviation* $= \dfrac{\displaystyle\sum_{i=1}^{n} \lvert x_i - \bar{x} \rvert}{n} = \dfrac{30}{6} = 5$.

■ **DEFINITION.** If $\bar{x}$ is the mean of a set of numbers denoted by x_i, then the *mean absolute deviation*, or simply the *mean deviation*, is:

$$\frac{\displaystyle\sum_{i=1}^{n} \lvert x_i - \bar{x} \rvert}{n} \qquad \text{OR} \qquad \frac{1}{n} \sum_{i=1}^{n} \lvert x_i - \bar{x} \rvert$$

Since $\lvert a - b \rvert = \lvert b - a \rvert$, it is also correct to write the formula for the mean absolute deviation as:

$$\frac{\displaystyle\sum_{i=1}^{n} \lvert \bar{x} - x_i \rvert}{n} \qquad \text{OR} \qquad \frac{1}{n} \sum_{i=1}^{n} \lvert \bar{x} - x_i \rvert$$

| MODEL PROBLEMS |

1. George Goldstein is a business student. His last six grades on tests in accounting were 87, 76, 93, 83, 84, and 81. For these grades, find:
   a. the range   b. the mean deviation

Solution

a. The range is the difference between the highest and the lowest scores. Here, the range = 93 − 76 = 17.

b. 1. Organize the data in column 1 of a chart.

   2. Find the mean:

   $$\bar{x} = \frac{\sum x_i}{n} = \frac{504}{6} = 84$$

   3. After the mean is entered in every row of column 2, find the values of $x_i - \bar{x}$ in column 3 and $|x_i - \bar{x}|$ in column 4.

   4. Find the mean deviation:

x_i	$\bar{x}$	$x_i - \bar{x}$	$|x_i - \bar{x}|$
93	84	9	9
87	84	3	3
84	84	0	0
83	84	−1	1
81	84	−3	3
76	84	−8	8
$\sum x_i =$ 504	—	—	$\sum |x_i - \bar{x}| =$ 24

$$\frac{\sum |x_i - \bar{x}|}{n} = \frac{24}{6} = 4$$

Answer: a. range = 17   b. mean deviation = 4

2. Two students in a computer course have the same average, based on grades received by each student for six computer programs.

   Thomas: 90, 70, 85, 100, 80, 85   (mean = 85)
   Robert: 100, 90, 65, 90, 65, 100   (mean = 85)

   a. For each set of grades, find the mean absolute deviation to the nearest tenth.
   b. Which student has more widely dispersed grades? Explain why.

Solution

a. For each student, organize the data in a table, find $\sum|x_i - \bar{x}|$, and find the mean absolute deviation.

Thomas

x_i	$\bar{x}$	$|x_i - \bar{x}|$
100	85	15
90	85	5
85	85	0
85	85	0
80	85	5
70	85	15

$$\sum|x_i - \bar{x}| = 40$$

$$\text{mean deviation} = \frac{\sum|x_i - \bar{x}|}{n}$$

$$= \frac{40}{6} = 6.66$$

$$= 6.7 \ \textit{Ans.}$$

Robert

x_i	$\bar{x}$	$|x_i - \bar{x}|$
100	85	15
100	85	15
90	85	5
90	85	5
65	85	20
65	85	20

$$\sum|x_i - \bar{x}| = 80$$

$$\text{mean deviation} = \frac{\sum|x_i - \bar{x}|}{n}$$

$$= \frac{80}{6} = 13.33$$

$$= 13.3 \ \textit{Ans.}$$

b. Robert has more widely dispersed grades because the mean deviation of his grades, 13.3, is about twice as much as the mean deviation, 6.7, for the other student.

EXERCISES

1. For each set of student grades, find the range.
   a. Ann: 83, 87, 92, 92, 95
   b. Barbara: 94, 90, 86, 86, 85
   c. Bill: 78, 97, 82, 86, 90
   d. Cathy: 88, 81, 90, 74, 72
   e. Stephen: 91, 65, 92, 94, 98
   f. Tom: 90, 90, 90, 90, 90
2. If the students in Mr. Pedersen's class are either 16 or 17 years old, what is the range of student ages in the class?
3. The most expensive item in Dale Singer's shopping basket is meat at $5.60, and the least expensive item is fruit at $.39. What is the range of the prices of items in the shopping basket?
4. a. Anna has fourteen grandchildren. If the oldest is 18 and the youngest is 3, what is the range of ages of the grandchildren?

b. If Andrea, who is Anna's fifteenth grandchild, is born today, what is now the range of ages of Anna's grandchildren?

5. For the data given in the accompanying table:
   a. Find the mean.
   b. Copy and complete the table.
   c. Find the mean absolute deviation.

x_i	$\bar{x}$	$x_i - \bar{x}$	$|x_i - \bar{x}|$
27			
26			
25			
22			
20			

6. "Curveball" Klopfer is a baseball pitcher. In his last 6 games, he struck out the following number of batters: 16, 20, 14, 13, 21, 12. For this set of data, find: **(a)** the range **(b)** the mean **(c)** the mean deviation.

7. *True* or *False*: For any set of data, $\dfrac{1}{n}\sum|x_i - \bar{x}| = \dfrac{1}{n}\sum|\bar{x} - x_i|$.

   Explain why.

8. Frances owns and manages a printing business. Over the last 6 days, she has processed the following number of jobs: 5, 8, 12, 7, 3, 4. For the number of jobs processed: **a.** Find the mean. (*Hint:* The mean is not an integer.) **b.** Find the mean absolute deviation.

9. For the last week, Florence kept a log of the number of hours that she watched television each day. The times, recorded to the nearest half-hour, are 5, 4, 3.5, 3, 2, 1.5, and 0 hours. For these times: **a.** Find the mean to the nearest tenth. **b.** Using the mean from part a, find the mean deviation to the nearest hundredth.

10. Three sets of data (Set I, Set II, and Set III) are displayed in the following frequency diagrams. For each set of data, find: **(a)** the range **(b)** the mean **(c)** the mean deviation. Then, by using the mean deviations found, tell: **(d)** which set of data is most closely grouped about the mean and **(e)** which set of data is most widely dispersed.

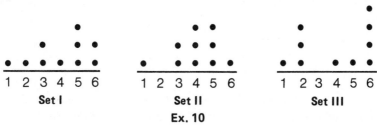

Set I   Set II   Set III

Ex. 10

11. Two students in a mathematics class are comparing their grades.

    Mary Murray: 87, 98, 82, 96, 99, 84
    Thea Olmstead: 95, 92, 79, 94, 90, 96

    a. For Mary's grades, find (1) the range (2) the mean (3) the mean deviation to the nearest tenth.
    b. For Thea's grades, find (1) the range (2) the mean (3) the mean deviation to the nearest tenth.
    c. Which student has more widely dispersed grades? Explain why.
12. The highest daily temperatures were recorded in Celsius for two weeks in the summer. Each week had the same mean daily reading.

    Week 1: 37, 35, 34, 30, 32, 36, 34   (mean = 34)
    Week 2: 37, 36, 40, 33, 31, 30, 31   (mean = 34)

    a. For each week's data, find the mean deviation to the nearest tenth.
    b. Using part a, tell which week had the more consistent readings.
13. The set of data 3, 4, 4, 5, 5, 5, 6, 6, 7 contains 9 numbers and has a range of 4 and a mean of 5. Write three more sets of data where $n = 9$, the range = 4, and the mean = 5.

15-4 THE STANDARD DEVIATION

When we were working with mean deviation, the use of absolute value allowed us to change all non-zero differences to positive values. Non-zero differences can also be changed to positive values by squaring the differences, as in $(5 - 7)^2 = (-2)^2 = 4$ and $(1 - 7)^2 = (-6)^2 = 36$. In fact, the use of squaring leads to a measure of dispersion that gives greatest weight to those scores that are farthest from the mean.

■ DEFINITION.   The *variance*, v, of a set of data is the average of the squares of the deviations from the mean. In symbols:

$$v = \frac{\sum_{i=1}^{n} (x_i - \bar{x})^2}{n} \qquad \text{OR} \qquad v = \frac{1}{n} \sum_{i=1}^{n} (x_i - \bar{x})^2$$

☐ EXAMPLE 1: On five mathematics tests taken this month, Fred earned the following scores: 92, 86, 95, 84, and 78. Calculate the variance.

Solution

1. Organize the data in a table. See column 1.
2. Find the mean:

$$\bar{x} = \frac{\sum x_i}{n} = \frac{435}{5} = 87$$

x_i	$\bar{x}$	$x_i - \bar{x}$	$(x_i - \bar{x})^2$
95	87	8	64
92	87	5	25
86	87	-1	1
84	87	-3	9
78	87	-9	81
435 $\sum x_i$	—	—	180 $\sum (x_i - \bar{x})^2$

3. Subtract the mean in column 2 from each entry in column 1, and write this set of deviations from the mean in column 3.
4. In column 4, square each deviation from the mean.

5. Therefore, variance $v = \dfrac{\displaystyle\sum_{i=1}^{5} (x_i - \bar{x})^2}{5} = \dfrac{180}{5} = 36$ *Ans.*

Some people object to variance as a measure of dispersion because it contains a distortion that results from squaring the differences. To overcome this objection, we can find the square root of the variance and thus obtain a measure of dispersion that has the same units as the given data. This measure, called the standard deviation, is the most important and widely used measure of dispersion in the world today.

■ DEFINITION. The *standard deviation*, *s*, of a set of data is equal to the square root of the variance. In symbols:

$$s = \sqrt{v} = \sqrt{\frac{\displaystyle\sum_{i=1}^{n} (x_i - \bar{x})^2}{n}} \qquad \text{OR} \qquad s = \sqrt{\frac{1}{n} \sum_{i=1}^{n} (x_i - \bar{x})^2}$$

Other commonly used symbols for standard deviation are S.D. and σ (the lower-case Greek letter, sigma).

☐ EXAMPLE 2: On five mathematics tests, Fred earned the following scores: 92, 86, 95, 84, and 78. Calculate the standard deviation.

Solution

This is the same set of data as given in the previous Example 1. After following steps 1, 2, 3, and 4, as outlined in Example 1, we write:

$$\text{Standard deviation} = s = \sqrt{\frac{\sum_{i=1}^{n} (x_i - \bar{x})^2}{n}} = \sqrt{\frac{180}{5}} = \sqrt{36} = 6 \quad Ans.$$

Note: Let us compare the various measures of dispersion we have learned for the set of data just studied: 92, 86, 95, 84, 78.

$$\begin{array}{ll}
\text{Range} = 95 - 78 = 17 & \text{(unreliable)} \\
\text{Mean Deviation} = 5.2 & \text{(acceptable, but rarely used today)} \\
\text{Variance} = 36 & \text{(contains a distortion)} \\
\text{Standard Deviation} = 6 & \text{(the most commonly used measure)}
\end{array}$$

Later in this chapter, we will see how standard deviation is used to help us study frequency distributions in greater detail.

MODEL PROBLEM

The time in minutes required by each of five students to complete a test was as follows: 35, 27, 30, 25, and 38. For this data, find the standard deviation to the nearest tenth.

Solution

1. Organize the data, as shown in column 1 of the table.
2. Find the mean:

$$\bar{x} = \frac{\sum x_i}{n} = \frac{155}{5} = 31$$

3. Write the mean in column 2, the deviations from the mean in column 3, and the squares of the deviations in column 4.

x_i	$\bar{x}$	$x_i - \bar{x}$	$(x_i - \bar{x})^2$
38	31	7	49
35	31	4	16
30	31	-1	1
27	31	-4	16
25	31	-6	36
155 $\sum x_i$	—	—	118 $\sum (x_i - \bar{x})^2$

4. Standard deviation $= \sqrt{\dfrac{1}{n} \sum\limits_{i=1}^{n} (x_i - \bar{x})^2}$

$= \sqrt{\tfrac{1}{5} (118)}$

$= \sqrt{23.6} = 4.85$

$= 4.9$ (nearest tenth)

Answer: 4.9

```
                    4. 8  5
              √23.60 00
                 16
           88│  7 60
            8│  7 04
          965│  56 00
            5│  48 25
```

EXERCISES

1. For the data given in the accompanying table:
   a. Find the mean.
   b. Copy and complete the table.
   c. Find the standard deviation to the nearest tenth.

x_i	$\bar{x}$	$x_i - \bar{x}$	$(x_i - \bar{x})^2$
30			
29			
26			
23			
12			

2. The scores of six students on an IQ test are listed below.

   Fred: 130   Toni: 127   Lee: 125
   Paul: 122   Lynn: 128   John: 118

   For these scores, find: a. the mean   b. the standard deviation.

3. The highest number of points that a student can score in a mathematics competition is five. In the last six competitions, Jennifer's scores were 2, 1, 4, 2, 4, 5. For these scores, calculate: a. the mean   b. the standard deviation to the nearest tenth.

4. On his last five fishing trips, Jim caught the following number of fish: 6, 5, 12, 3, 9. For this data, find: a. the mean   b. the standard deviation to the nearest tenth.

5. On a test, five students received scores of 63, 60, 59, 57, and 56. For these scores, find: a. the mean   b. the standard deviation to the nearest tenth.

6. The heights in centimeters of five players on the basketball team are listed at the right. For these heights, find: a. the mean b. the standard deviation to the nearest tenth.

Player	Height (cm)
R. Melendy	192
C. Cronin	189
D. Schmeling	187
M. Natale	184
R. Weinrich	183

7. Two students were comparing their grades on their report cards.

Sean O'Brien: 83, 92, 79, 65, 82, 85
John Parisi: 83, 75, 78, 86, 77, 87

a. Which student, if either, has the higher mean average? b. For Sean's grades, calculate the standard deviation to the nearest tenth. c. For John's grades, calculate the standard deviation to the nearest tenth. d. Which student has the greater dispersion in grades?

8. During a 5-day work week, Helen worked the following number of hours per day: 8, 10, 9, 11, 10. For this data, calculate: a. the mean b. the standard deviation to the nearest tenth.

9. Dave worked the following number of hours at a fast-food restaurant over a period of 7 days: 5, 4, 0, 9, 7, 0, 3. Using $n = 7$, find: a. the mean number of hours worked per day b. the standard deviation to the nearest tenth.

10. Over the last 7 days, Mr. Kavanagh spent the following number of hours reading a novel: 3, 1.5, 2, 1.5, 4, 2.5, and 3. For these times, find: a. the mean b. the standard deviation to the nearest tenth.

11. In a statistical study, if the variance is 64, then the standard deviation is:
    (1) 6.4   (2) 16   (3) 8   (4) 64

12. If the standard deviation for a set of data is 2.5, find the value of the variance.

13. In a statistical study, variance and standard deviation are usually not equal. For what two numerical values would these measures be equal?

15-5 NORMAL DISTRIBUTION

In most statistical studies, the number of scores to be considered is large, and these scores are organized into groups. For example, 10 coins are tossed, and the number of heads obtained is recorded. The table that follows shows the result after 100 trials of tossing 10 coins.

Number of heads	0	1	2	3	4	5	6	7	8	9	10
Frequency	1	2	4	11	20	24	20	11	4	2	1

The *histogram* for the distribution of the number of heads is the bar graph, shown on the next page. The *frequency polygon* is the line graph, determined by connecting the midpoints of the upper parts of each bar, shown in the same diagram.

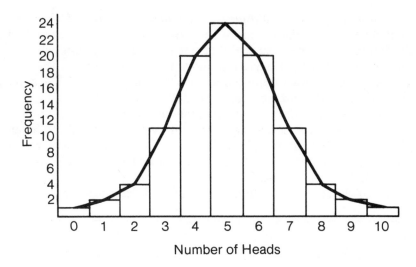

Number of Heads

If we were to draw a smooth curve through the points that determine the frequency polygon, the curve would be shaped somewhat like a bell. In fact, by increasing the number of trials as well as the number of coins tossed in each trial, the frequency polygon would more closely resemble a *bell-shaped curve* called the **normal curve**.

In a normal curve, the greatest frequency occurs at a score in the center of the distribution. This center score is the *mean*, $\bar{x}$. A vertical line drawn through the mean serves as an axis of symmetry for the normal curve. Since half the scores in the distribution lie below the mean and half lie above it,

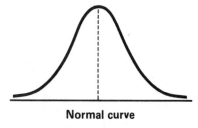

Normal curve

this center score is also the *median*. Since this center score has the greatest frequency in the distribution, it is also the *mode*.

As we move further to the right or to the left of the mean in a normal curve, the frequencies decrease. The normal curve is essentially a theoretical model that is used to study sets of data. For example, a frequency polygon of the heights of thousands of high-school juniors would closely approximate the normal curve. So too would data involving weights, clothing sizes, numerical test grades, and the like.

Standard Deviation and the Normal Curve

The distribution of data represented by the normal curve is called a **normal distribution**. In a normal distribution, the pattern of scores on the next page will occur, correct to the nearest one-half of one percent.

1. **68%** *of the scores lie within one standard deviation of the mean.* That is, 68%, or slightly more than two-thirds of the scores, lie in the interval from $\bar{x} - s$ to $\bar{x} + s$, or one standard deviation below the mean to one standard deviation above the mean. Because the curve is symmetrical, 34% of the scores lie in the interval from $\bar{x} - s$ to $\bar{x}$, and 34% lie in the interval from $\bar{x}$ to $\bar{x} + s$.

2. **95%** *of the scores lie within two standard deviations of the mean.* Thus, 47.5% of the scores lie in the interval from $\bar{x} - 2s$ to $\bar{x}$, and 47.5% lie in the interval from $\bar{x}$ to $\bar{x} + 2s$. By arithmetic, we can show that each interval, $\bar{x} - 2s$ to $\bar{x} - s$ and $\bar{x} + s$ to $\bar{x} + 2s$, contains 13.5% of the scores.

3. **99.5%** *of the scores lie within three standard deviations of the mean.*

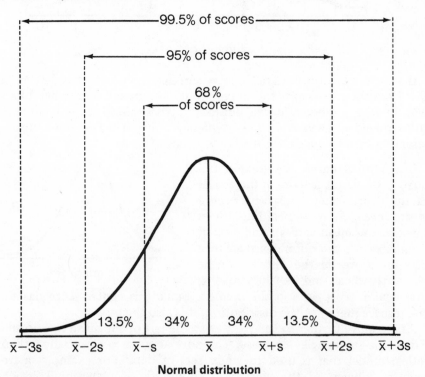

Normal distribution

Let us see how these concepts are used in a statistical study.

☐ EXAMPLE: In a normal distribution, the mean height of 10-year-old children is 138 cm and the standard deviation is 5 cm. Find those heights that are exactly one standard deviation above and below the mean, and two standard deviations above and below the mean.

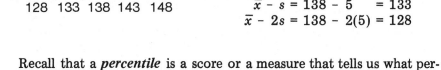

Solution

The normal curve is sketched here as an aid.

Let $\bar{x} = 138$ and $s = 5$.

Answer

$\bar{x} + 2s = 138 + 2(5) = 148$
$\bar{x} + s = 138 + 5 \quad = 143$
$\bar{x} - s = 138 - 5 \quad = 133$
$\bar{x} - 2s = 138 - 2(5) = 128$

Recall that a *percentile* is a score or a measure that tells us what percent of the total frequency scored "at or below that measure."

By using the percents observed earlier, we can list the percentiles for five important scores in a normal distribution (see the diagram at the right). For example, the mean $\bar{x}$ is at the 50th percentile because 50% of the scores lie at or below the mean. Similarly, the score that lies one standard deviation above the mean, $\bar{x} + s$, is at the 84th percentile because 84% of the scores lie at or below $\bar{x} + s$.

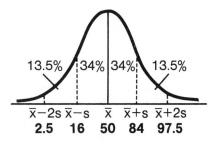

**Percentiles in a
normal distribution**

Let us combine the scores from the statistical study involving heights of 10-year-old children with the percents and percentiles of a normal distribution. Many interesting observations can now be made, only a few of which are listed on the next page.

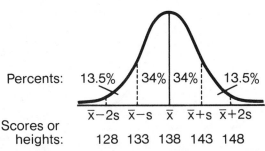

1. 68% of the children are between 133 cm and 143 cm tall.
2. 95% of the children are between 128 cm and 148 cm tall.
3. 34% of the children are between 138 cm and 143 cm tall.
4. A 10-year-old child who is 133 cm tall is at the 16th percentile. That is, 16% of 10-year-old children are 133 cm or shorter in height.
5. A 10-year-old child who is 148 cm tall is at the 97.5 percentile. That is, only 2.5% of 10-year-old children are taller than 148 cm.
6. If the height of a 10-year-old child is at the 90th percentile, then the child's height is somewhere between 143 cm and 148 cm.
7. A height of 137 cm is above the 16th percentile but less than the 50th percentile for this population.
8. Heights that would occur *less than* 5% of the time for these children are heights less than 128 cm and heights greater than 148 cm.

In courses of higher mathematics, statistical tables are provided that allow us to be very precise when assigning percentiles to scores. For now, let us simply remember the percents and percentiles that occur in a normal distribution for these key scores: $\bar{x}, \bar{x} \pm s,$ and $\bar{x} \pm 2s$.

MODEL PROBLEMS

1. Scores on the Preliminary Scholastic Aptitude Test (PSAT) range from 20 to 80. For a certain population of students, the mean is 52 and the standard deviation is 9.
   a. A score at the 65th percentile might be: (1) 49   (2) 56   (3) 64 (4) 65.
   b. Which of the following scores can be expected to occur *less than* 3% of the time?   (1) 39   (2) 47   (3) 65   (4) 71

Solution

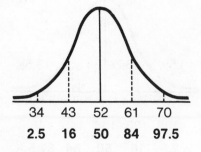

Scores:	34	43	52	61	70
Percentiles:	**2.5**	**16**	**50**	**84**	**97.5**

Let $\bar{x} = 52$ and $s = 9$. Then:
$\bar{x} + 2s = 52 + 2(9) = 70$
$\bar{x} + s = 52 + 9$ $= 61$
$\bar{x} - s = 52 - 9$ $= 43$
$\bar{x} - 2s = 52 - 2(9) = 34$

Sketch a normal curve. Place appropriate scores and percentiles at $\bar{x}$, $\bar{x} \pm s$, and $\bar{x} \pm 2s$.

a. The 65th percentile lies in the interval from the 50th percentile to the 84th percentile, that is, from $\bar{x}$ to $\bar{x} + s$. The scores in this interval range from 52 to 61. Of the four choices given, only choice (2), a score of 56, lies within the interval.

b. The score of 70 lies at the 97.5 percentile. Since 71 is greater than 70, then 71 is above the 97.5 percentile, and this score should occur less than 3% of the time.

Answer: **a.** (2) 56   **b.** (4) 71

2. In the diagram, the shaded area represents approximately 68% of the scores in a normal distribution. If the scores range from 12 to 40 in this interval, find the standard deviation.

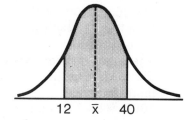

12 $\bar{x}$ 40

Solution

Since 68% of the scores in a normal distribution lie in the interval from $\bar{x} - s$ to $\bar{x} + s$, the interval contains exactly 2 standard deviations. Thus, 40 - 12 = 28 = 2s, and 28 ÷ 2 = 14 = s.

Answer: 14

| EXERCISES |

In 1-4, write the number that makes the sentence true.

1. In a normal distribution, the mean is at the ____ percentile.
2. Approximately ____% of the scores in a normal distribution lie within one standard deviation of the mean.
3. A score that is one standard deviation below the mean in a normal distribution is at the ____ percentile.
4. In a normal distribution, about ____% of the scores lie in the interval from $\bar{x} - 2s$ to $\bar{x} + 2s$.

5. On a standardized test, the mean is 82 and the standard deviation is 6. Find the scores for: **a.** $\bar{x} + 2s$ **b.** $\bar{x} + s$ **c.** $\bar{x} - s$ **d.** $\bar{x} - 2s$

6. On a standardized test, the mean is 70 and the standard deviation is 10. If Rita Keane scored 90 on this test, then her score is at the: (1) 84th percentile   (2) 90th percentile   (3) 95th percentile (4) 97.5 percentile

7. If the mean score on an IQ test is 104 and the standard deviation is 7, which score will occur less than 5% of the time? (1) 93 (2) 97   (3) 116   (4) 119

8. Mr. Noren bowls regularly and has an average of 182. His bowling scores approximate a normal distribution with a standard deviation of 8.5. He can expect to bowl a score of 200 or better: (1) about 16% of the time   (2) between 10% and 15% of the time   (3) between 5% and 10% of the time   (4) less than 3% of the time

9. The times of telephone calls approximate a normal distribution. If the mean time is 4.3 minutes, with a standard deviation of 1.4 minutes, which length call should occur less than 5 times out of 100? (1) 2.1 minutes   (2) 7.4 minutes   (3) 3 minutes (4) 6 minutes

10. For a certain population, the mean score on the Scholastic Aptitude Test (SAT) is 430, with a standard deviation of 108.
    a. Which of the following scores should occur the greatest number of times? (1) 301   (2) 335   (3) 539   (4) 550
    b. Which of the following scores might represent the 70th percentile? (1) 505   (2) 540   (3) 590   (4) 700
    c. A student with a score of 322 does as well as or better than what percent of the population? (1) 2.5% (2) 16% (3) 30% (4) 32.2%

11. Ms. Surber owns a real-estate business. Sale prices of homes in her area approximate a normal distribution, with a mean of $72,000 and a standard deviation of $7,600. A home that sells for $87,600 would rank: (1) below the 75th percentile (2) between the 75th and 85th percentiles (3) between the 85th and 95th percentiles (4) above the 95th percentile

12. In a large school district, the years of service for the teaching staff approximate a normal distribution. The mean is 16 years of teaching, and the standard deviation is 7.4 years. Four teachers, and the number of years of service for each, are listed below.

    Robert Novak: 16 years   Mabel Oestrich: 35 years
    Samuel Backer: 21 years   Nancy Garbowski: 12 years

    According to years of service, name the teacher or teachers of those listed (if any) who might possibly rank at the:
    a. 10th percentile   b. 35th percentile   c. 50th percentile
    d. 70th percentile   e. 90th percentile   f. 99th percentile

13. In a contest, the mean time for pianists to play the "Minute Waltz" was 56 seconds, with a standard deviation of 3 seconds. The times approximated a normal distribution. If Al Cavallaro played the "Minute Waltz" in 50 seconds, his time was faster than what percent of the population? (1) 2.5% (2) 16% (3) 84% (4) 97.5%

14. In the given diagram, about 95% of the scores fall in the shaded area. If the distribution is normal and these scores range from 25 to 51, find: a. the standard deviation b. the mean

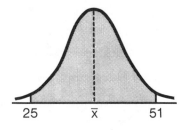

25    $\bar{x}$ 51

Ex. 14

In 15–18, assume a normal distribution. Using the given information, find: a. the standard deviation b. the mean

15. $\bar{x} - s = 75$ and $\bar{x} + s = 109$
16. $\bar{x} - s = 4.5$ and $\bar{x} + 2s = 25.5$
17. Scores within one standard deviation of the mean range from 9.2 to 14.
18. About 2.5% of the scores are at or below 45, and 2.5% are above 77.
19. In a normal distribution, $\bar{x} = 27.9$, and 16% of the scores are at or below 25.2. Find: a. s b. $\bar{x} + 2s$

15-6 GROUPED DATA

In this section, we will work with data that is organized into groups. The measures of dispersion found here are based on the rules learned earlier, but we must take into account the *frequency* for each of the intervals of grouped data.

☐ EXAMPLE 1: Mrs. Fowles, an English teacher, recorded the number of misspelled words in 25 book reports as follows:

12, 7, 7, 7, 6, 6, 5, 5, 5, 5, 5, 4, 4, 4, 3, 3, 3, 2, 2, 2, 2, 1, 0, 0, 0

For this data, calculate the *standard deviation* to the nearest tenth.

Solution

1. Organize the data in a table as shown below. In columns 1 and 2, let x_i represent the number of misspelled words, and let f_i represent the frequency for each measure x_i. (Note that $\sum f_i = n$.)
2. In column 3, find each product, $f_i x_i$, and the sum, $\sum f_i x_i$.
3. Find the mean: $\bar{x} = \dfrac{\sum f_i x_i}{n} = \dfrac{100}{25} = 4$.

Col. 1	Col. 2	Col. 3	Col. 4	Col. 5	Col. 6
Measure x_i	Frequency f_i	$f_i x_i$	$x_i - \bar{x}$	$(x_i - \bar{x})^2$	$f_i (x_i - \bar{x})^2$
12	1	12	8	64	$1 \cdot 64 = 64$
7	3	21	3	9	$3 \cdot 9 = 27$
6	2	12	2	4	$2 \cdot 4 = 8$
5	5	25	1	1	$5 \cdot 1 = 5$
4	3	12	0	0	$3 \cdot 0 = 0$
3	3	9	-1	1	$3 \cdot 1 = 3$
2	4	8	-2	4	$4 \cdot 4 = 16$
1	1	1	-3	9	$1 \cdot 9 = 9$
0	3	0	-4	16	$3 \cdot 16 = 48$
—	25 $\sum f_i = n$	100 $\sum f_i x_i$	—	—	180 $\sum f_i (x_i - \bar{x})^2$

4. In column 4, find the values of $x_i - \bar{x}$ by subtracting the mean of 4 from each measure x_i. Then, in column 5, square these deviations from the mean.
5. Since the frequency changes from one interval to another, multiply each value of $(x_i - \bar{x})^2$ by its corresponding frequency f_i before finding the sum, $\sum f_i (x_i - \bar{x})^2$, in column 6.
6. The rule for standard deviation s with grouped data includes the frequencies f_i seen in the table:

$$s = \sqrt{\frac{\sum f_i (x_i - \bar{x})^2}{n}} \quad \text{OR} \quad s = \sqrt{\frac{1}{n} \sum f_i (x_i - \bar{x})^2}$$

Here, $s = \sqrt{\dfrac{180}{25}} = \sqrt{7.2} \approx 2.68 = 2.7$ (nearest tenth)

Answer: 2.7

It is interesting to see how closely this data approximates a normal distribution. Here, the mean number of misspelled words is 4, and the standard deviation is 2.7. Look at the frequency distribution in the table, and notice the following diagram:

1. There are 17 students who have 2, 3, 4, 5, or 6 misspelled words. Thus, from 1.3 to 6.7, or within one standard deviation of the mean, exactly $\frac{17}{25}$ or 68% of the population is found.
2. Every student except one is found in the interval from -1.4 to 9.4. Thus, $\frac{24}{25}$ or 96% of the population lies within two standard deviations of the mean.

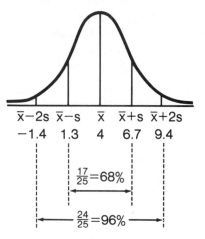

3. The one student with 12 spelling errors is found outside of the interval from -1.4 to 9.4. This is $\frac{1}{25}$ or 4% of the population. We expect scores in a normal distribution to fall outside of the interval from $\bar{x} - 2s$ to $\bar{x} + 2s$ less than 5% of the time.

Although the number, $n = 25$, is relatively small in this study, the percents that we have just found indicate that this data is close to that of a normal distribution.

□ EXAMPLE 2: Using the data given in Example 1, calculate the *mean absolute deviation* to the nearest tenth.

Solution

1. A table is constructed with the headings:

x_i	f_i	$f_i x_i$	$x_i - \bar{x}$	$\lvert x_i - \bar{x} \rvert$	$f_i \lvert x_i - \bar{x} \rvert$

The first four columns are the same as those in Example 1, and the mean $\bar{x} = 4$. We will concentrate on columns 5 and 6 in the limited table that follows.

2. In column 5, the absolute value of each deviation from the mean, $|x_i - \bar{x}|$, is noted. Then, in column 6, each deviation is multiplied by its corresponding frequency, f_i.

3. The rule for mean absolute deviation with grouped data is:

$$\frac{\sum f_i |x_i - \bar{x}|}{n}$$

Here, the mean deviation = $\frac{52}{25}$ = 2.08 = 2.1 (nearest tenth)

Answer: 2.1

| | | Col. 5 | Col. 6 |
x_i	f_i	$\|x_i - \bar{x}\|$	$f_i\|x_i - \bar{x}\|$
12	1	8	8
7	3	3	9
6	2	2	4
5	5	1	5
4	3	0	0
3	3	1	3
2	4	2	8
1	1	3	3
0	3	4	12
—	25	—	52
	n		$\sum f_i\|x_i - \bar{x}\|$

Remember that the percents studied earlier apply to standard deviation only, and not to mean absolute deviation.

Intervals Other Than Length One

If the range of a set of data is large, it is useful to group the data into intervals larger than length one. The *length* of an interval is found by subtracting the starting points of two consecutive intervals. For example, the intervals 75 to 79, 80 to 84, 85 to 89, and so on, have a length of 5 each because 80 – 75 = 5 and 85 – 80 = 5.

To work with data grouped into such intervals, we must be sure that the intervals are of equal length. Then, let x_i represent the *midpoint of each interval*. For example, the six scores 89, 88, 88, 86, 85, 85 lie in the interval "85–89," whose midpoint is 87. By addition, 89 + 88 + 88 + 86 + 85 + 85 = 521. We can obtain a number that is not significantly different from this sum by multiplying the midpoint, $x_i = 87$, by the frequency, $f_i = 6$, in the stated interval.

Interval	Measure (x_i)	Frequency (f_i)	$f_i x_i$
85–89	87	6	6 · 87 = 522

These procedures are used in the model problem that follows.

─────── **KEEP IN MIND** ───────

For grouped data involving intervals with given frequencies f_i:

1. Mean absolute deviation $= \dfrac{\sum f_i |x_i - \bar{x}|}{n}$

2. Standard deviation $= s = \sqrt{\dfrac{\sum f_i (x_i - \bar{x})^2}{n}}$

MODEL PROBLEM

The test scores of 50 students are grouped into intervals as shown in the accompanying table.

For this set of grouped data, find: a. the mean b. the standard deviation to the nearest tenth

Interval	Frequency
96–100	2
91–95	4
86–90	10
81–85	14
76–80	10
71–75	6
66–70	4

Solution

a. 1. In the table at the right, let each measure x_i represent the midpoint of the interval.

2. Find the products, $f_i x_i$, and their sum, $\sum f_i x_i$.

3. Determine the mean:

$$\bar{x} = \frac{\sum f_i x_i}{n} = \frac{4100}{50} = 82$$

Interval	Measure x_i	Frequency f_i	$f_i x_i$
96–100	98	2	196
91–95	93	4	372
86–90	88	10	880
81–85	83	14	1162
76–80	78	10	780
71–75	73	6	438
66–70	68	4	272
—	—	50 n	4100 $\sum f_i x_i$

Interval	Measure x_i	Frequency f_i	$f_i x_i$	$x_i - \bar{x}$	$(x_i - \bar{x})^2$	$f_i(x_i - \bar{x})^2$
96–100	98	2	196	16	256	512
91–95	93	4	372	11	121	484
86–90	88	10	880	6	36	360
81–85	83	14	1162	1	1	14
76–80	78	10	780	-4	16	160
71–75	73	6	438	-9	81	486
66–70	68	4	272	-14	196	784
		50	4100			2800
—	—	n	$\sum f_i x_i$	—	—	$\sum f_i(x_i - \bar{x})^2$

b. 1. To obtain the numbers in the column headed "$x_i - \bar{x}$," subtract the mean of 82 from each measure x_i. Then, in the next column, square these deviations from the mean. For example, $98 - 82 = 16$, and $(16)^2 = 256$.

2. In the last column, multiply the square of each deviation by its corresponding frequency. For example, $2 \cdot 256 = 512$.

3. Add the entries in the last column, and find the standard deviation.

$$s = \sqrt{\frac{1}{n} \sum f_i (x_i - \bar{x})^2} = \sqrt{\frac{1}{50}(2800)} = \sqrt{56} \approx 7.48$$

$s = 7.5$ (nearest tenth)

Answer: **a.** $\bar{x} = 82$ **b.** $s = 7.5$

EXERCISES

1. The set of data 12, 15, 12, 10, 9, 12, 9, 10, 12, 9 has been organized into a table that follows. For this grouped data: a. Find the mean. b. Copy and complete the table. c. Find the standard deviation to the nearest tenth.

Measure (x_i)	Frequency (f_i)	$f_i x_i$	$x_i - \bar{x}$	$(x_i - \bar{x})^2$	$f_i(x_i - \bar{x})^2$
15	1				
12	4				
10	2				
9	3				

In 2-4, for the grouped data, find: **a.** the mean   **b.** the standard deviation to the nearest tenth

2.

Measure x_i	Frequency f_i
31	1
25	5
22	2
20	2

3.

Measure x_i	Frequency f_i
40	4
30	10
20	4
10	2

4.

Measure x_i	Frequency f_i
10	4
8	2
7	2
5	7

5. In her last 10 mathematics tests, Karla Adasse scored 90, 100, 80, 100, 100, 70, 90, 80, 90, and 100. Organize these test grades into a table of grouped data, and find: **a.** the mean grade   **b.** the standard deviation

6. Joan kept a record of the number of miles that she drove each day during the last week: 20, 20, 15, 20, 10, 40, and 15. Organize the record of her mileage into a table of grouped data, and find: **a.** the mean   **b.** the standard deviation to the nearest mile

7. Ten high-school juniors were chosen at random. Their non-verbal scores on the PSAT are: 68, 65, 70, 70, 68, 58, 65, 68, 70, and 68. Find the standard deviation of these scores to the nearest tenth.

8. Mr. McEntee is a guidance counsellor. During the last 8 school days, he saw the following number of students: 12, 8, 10, 12, 15, 10, 10, and 3. For this data, find the standard deviation to the nearest tenth.

9. Mrs. Cerulli drives to work. On her last 10 trips: 5 trips took 20 minutes each, 4 trips took 18 minutes each, and 1 trip took 38 minutes. For these times, find: **a.** the range   **b.** the median **c.** the mode   **d.** the mean   **e.** the standard deviation to the nearest tenth

10. Ten Speedo cars were tested at random. These cars were found to average the following number of miles per gallon: 30, 31, 31, 29, 32, 30, 23, 31, 32, and 31. **a.** Find the mean for this data. **b.** Find the standard deviation to the nearest tenth. **c.** Mr. Pappas owns the car that averages 23 mpg. Does his car have a rate that lies within two standard deviations of the mean?

11. The test scores of 50 students are grouped into intervals as shown in the table at the right. For this set of grouped data, find:
a. the mean   b. the standard deviation
(*Hint:* Let x_i represent the midpoint of each interval, as shown in the model problem in this section.)

Interval	Frequency
96-100	2
91-95	8
86-90	8
81-85	5
76-80	11
71-75	10
66-70	5
61-65	1

Interval	Midpoint x_i	Frequency f_i
94-100	97	2
87-93	90	5
80-86	83	15
73-79	76	7
66-72	69	1

12. Mrs. Barash analyzed a test given to 30 students in her class. Using the accompanying table, find: a. the mean score   b. the standard deviation to the nearest tenth

13. Tony Spartalis organized the grades he earned on 20 lab reports into the accompanying table. For this data:
a. Find the mean grade.
b. Find the standard deviation to the nearest tenth.
c. Expand the table to include columns for $|x_i - \bar{x}|$, and $f_i|x_i - \bar{x}|$. Use this information to find the mean absolute deviation to the nearest tenth.

x_i	f_i
100	8
90	9
80	2
70	1

14. The ages of 10 teachers in the mathematics department are: 38, 41, 34, 28, 41, 41, 34, 38, 41, and 34. For this data, find: a. the mean   b. the standard deviation to the nearest tenth   c. the mean absolute deviation to the nearest tenth

Interval	Midpoint x_i	Frequency f_i
45-49		1
40-44		0
35-39		3
30-34		10
25-29		6

15. The Vromans taught a group of children how to grow stringbeans. In one week, 20 plants were harvested and the number of stringbeans on each plant was recorded. For the grouped data in the accompanying table: a. Find the mean. b. Find the standard deviation to the nearest tenth.
c. Joanna grew the plant that had the greatest number of stringbeans, and Zack grew the plant having the least. Whose plant, if any, lies outside the interval that is two standard deviations from the mean?

15-7 REVIEW EXERCISES

In 1–4, find the value indicated by the summation symbol.

1. $\displaystyle\sum_{n=1}^{3} n^n$ **2.** $\displaystyle\sum_{i=2}^{4} (5 - i)^2$ **3.** $\displaystyle\sum_{k=0}^{4} 3(k + 1)$ **4.** $\displaystyle\sum_{n=0}^{2} \sin (n\pi)$

5. Which statistical term represents the score that occurs most often in a distribution? (1) mean (2) median (3) mode (4) standard deviation

6. If Sal's test grades are 73, 90, 87, 98, and 92, what is the range of his grades?

7. The hours that Carol worked over the last 8 weeks are recorded in the table at the right. For this data, find: **a.** the mean **b.** the median **c.** the mode

Measure x_i	Frequency f_i
34	1
32	4
30	1
28	2

Index i	Measure x_i	Frequency f_i
1	40	4
2	25	2
3	20	3
4	10	1

8. For the grouped data in the accompanying table, which is true?
   (1) The mean equals the median.
   (2) The mean exceeds the median by 3.
   (3) The median exceeds the mean by 3.
   (4) The mean equals the mode.

9. In a class, 15 students are 17 years old and 10 students are 16 years old. The mean age of these students is: (1) 16.3 (2) 16.5 (3) 16.6 (4) 16.8

10. Two players at a mini-golf course had the same average for the first nine holes of golf.

   Len: 3, 2, 5, 3, 3, 1, 5, 3, 2   (mean = 3)
   Stan: 3, 2, 5, 3, 3, 3, 2, 3, 3   (mean = 3)

   a. Find the mean absolute deviation for each player, expressed as a fraction. **b.** Which player has more widely dispersed scores?

11. The weights of a group of people approximate a normal distribution. If the mean weight is 64 kg and the standard deviation is 4.8, which weight is expected to occur less than 3% of the time? (1) 56 (2) 59.3 (3) 69.8 (4) 74

12. On a standardized exam, the mean is 83 and the standard deviation is 5.5. Which score might be assigned to a student at the 70th percentile? (1) 70 (2) 77.5 (3) 86 (4) 90

13. In the diagram, about 68% of the scores fall into the shaded area. If the distribution is normal and the scores range from 55 to 80, find: a. the standard deviation b. the mean c. the score at $\bar{x} + 2s$

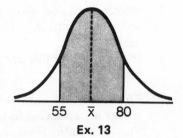

55 $\bar{x}$ 80

Ex. 13

14. Over the last 5 days, Mrs. Tountas received the following numbers of letters in the mail: 7, 4, 9, 2, 3. For this data, find: a. the mean b. the standard deviation to the nearest tenth

15. The scores earned by 5 students on the PSAT are 62, 57, 55, 51, and 50. Calculate the standard deviation of these scores to the nearest tenth.

16. For the grouped data in the accompanying table, find:
    a. the mode   b. the median
    c. the mean   d. the standard
                       deviation

Measure (x_i)	Frequency (f_i)
60	1
55	3
50	3
45	5

17. Doris scored the following grades on her last 10 science tests: 85, 100, 90, 90, 100, 85, 80, 90, 100, and 90. Organize these grades into a table of grouped data, and find: a. the mean b. the standard deviation to the nearest tenth

Chapter **16**

Probability and the Binomial Theorem

16-1 PERMUTATIONS AND COMBINATIONS

A bank contains 4 coins: a penny, a nickel, a dime, and a quarter, represented by P, N, D, and Q, respectively. If one coin is taken out of the bank at random, there are 4 possible outcomes. The *sample space*, or the set of all possible outcomes, is $\{P, N, D, Q\}$. If a coin is tossed, there are 2 possible outcomes, heads and tails, represented by H and T, respectively. This sample space is $\{H, T\}$. If a coin is taken out of the bank and tossed, the sample space is now:

$$\{(P, H), (P, T), (N, H), (N, T), (D, H), (D, T), (Q, H), (Q, T)\}$$

This sample space is represented on a graph at the right. We see that there are $4 \cdot 2$ or 8 possible outcomes.

■ **THE COUNTING PRINCIPLE.** If one activity can occur in any of m ways and, following this, a second activity can occur in any of n ways, then both activities can occur in the order given in $m \cdot n$ ways.

The 4 coins in the bank are to be drawn out one at a time without replacement. There are 4 possible outcomes for the first draw. Then, since 3 coins remain in the bank, there are 3 possible coins to be

Draw	Number of Coins in the Bank	Number of Possible Outcomes
1st	4	4
2nd	3	$4 \cdot 3$
3rd	2	$4 \cdot 3 \cdot 2$
4th	1	$4 \cdot 3 \cdot 2 \cdot 1$

drawn and $4 \cdot 3$ ways in which the first 2 coins could be removed. On the third draw there are 2 coins left to be selected and $4 \cdot 3 \cdot 2$ ways in which the first 3 coins could be removed. On the last draw there is one coin left to be selected and $4 \cdot 3 \cdot 2 \cdot 1$ or 24 possible orders in which the coins could have been drawn.

A *permutation* is an arrangement of objects in some specific order. The symbol $_4P_4$ is read "the number of permutations of 4 things taken 4 at a time."

$$_4P_4 = 4! = 4 \cdot 3 \cdot 2 \cdot 1 = 24$$

There are 24 permutations of the 4 coins.

■ In general, the number of permutations of n things taken n at a time is:

$$_nP_n = n! = n(n - 1)(n - 2) \ldots 2 \cdot 1$$

A bank contains 4 coins, and 2 are to be drawn. There are 4 ways in which the first coin can be selected and 3 ways in which the second coin can be selected. Therefore, there are $4 \cdot 3$ or 12 orders in which the coins can be selected. The symbol $_4P_2$ is read "the number of permutations of 4 things taken 2 at a time."

$$_4P_2 = 4 \cdot 3 = 12$$

■ In general, the number of permutations of n things taken r at a time is:

$$_nP_r = \underbrace{n(n - 1)(n - 2) \ldots}_{r \text{ factors}}$$

Permutations With Repetition

A bank contains 6 coins: a penny, a nickel, a dime, and 3 quarters. In how many different orders can the 6 coins be removed from the bank?

If the coins are considered to be all different, then the number of orders is:

$$_6P_6 = 6! = 6 \cdot 5 \cdot 4 \cdot 3 \cdot 2 \cdot 1 = 720$$

But within any given order, such as $P N Q D Q Q$, there are 3! or 6 permutations of the 3 quarters that produce this arrangement of coins. These permutations are indicated by using subscripts:

$$P N Q_1 D Q_2 Q_3 \qquad P N Q_2 D Q_1 Q_3 \qquad P N Q_3 D Q_1 Q_2$$
$$P N Q_1 D Q_3 Q_2 \qquad P N Q_2 D Q_3 Q_1 \qquad P N Q_3 D Q_2 Q_1$$

We can divide the 720 arrangements of coins into groups of 6 that are the same. Therefore, the number of different orders of the 6 coins, 3 of which are quarters, is:

$$\frac{6!}{3!} = \frac{6 \cdot 5 \cdot 4 \cdot 3 \cdot 2 \cdot 1}{3 \cdot 2 \cdot 1} = \frac{720}{6} = 120$$

■ The number of permutations of n things taken n at a time when r are identical is $\frac{n!}{r!}$.

Combinations

A bank contains 4 coins: a penny, a nickel, a dime, and a quarter. Two coins are to be drawn from the bank and the sum of the values noted. The number of ways in which 2 coins can be drawn from the bank, one after another, is a permutation: $_4P_2 = 4 \cdot 3 = 12$. Here, order is important, and the 12 permutations can be written as 12 ordered pairs.

Let us now find the number of possible *sums of the values* of the 2 coins selected. If the penny is drawn first and then the dime, the sum is 11 cents. If the dime is drawn first and then the penny, the sum is also 11 cents. Here, the order of the coins drawn is *not* important in finding the sum. A selection in which order is not important is called a *combination*.

For any 2 coins selected, such as the penny and the dime, there are $2! = 2 \cdot 1 = 2$ orders. Therefore, to find the number of combinations of 4 things taken 2 at a time, we divide the number of permutations of 4 things taken 2 at a time by the 2! orders.

Permutations		Combinations
$(P, D)(D, P)$	$\longrightarrow$	$\{P, D\}$
$(P, N)(N, P)$	$\longrightarrow$	$\{P, N\}$
$(P, Q)(Q, P)$	$\longrightarrow$	$\{P, Q\}$
$(N, D)(D, N)$	$\longrightarrow$	$\{N, D\}$
$(N, Q)(Q, N)$	$\longrightarrow$	$\{N, Q\}$
$(D, Q)(Q, D)$	$\longrightarrow$	$\{D, Q\}$

$$_4C_2 = \frac{_4P_2}{2!} = \frac{4 \cdot 3}{2 \cdot 1} = 6$$

While permutations are regarded as ordered elements, such as the ordered pairs (P, D) and (D, P), combinations may be regarded as *sets*, such as $\{P, D\}$, in which order is not important. Each of the 6 combinations listed in the accompanying box has a unique sum; these sums are 11¢, 6¢, 26¢, 15¢, 30¢, and 35¢.

■ In general, the number of combinations of n things taken r at a time is:

$$_nC_r = \frac{_nP_r}{r!} \qquad \text{OR} \qquad \binom{n}{r} = \frac{_nP_r}{r!}$$

Note: Do not confuse the symbol for the combination, $\binom{n}{r}$, with the fraction, $\left(\dfrac{n}{r}\right)$.

For example, from a class of 15 students, 5 are to represent the class at a mathematics contest. How many selections are possible?

Since the order in which the students are chosen is not important, this problem is a combination problem.

$$_{15}C_5 = \frac{_{15}P_5}{5!} = \frac{15 \cdot 14 \cdot 13 \cdot 12 \cdot 11}{5 \cdot 4 \cdot 3 \cdot 2 \cdot 1} = 3003$$

Certain relationships with combinations can be shown to be true:

1. $_3C_3 = \dfrac{_3P_3}{3!} = \dfrac{3 \cdot 2 \cdot 1}{3 \cdot 2 \cdot 1} = 1$ AND $\binom{4}{4} = \dfrac{_4P_4}{4!} = \dfrac{4 \cdot 3 \cdot 2 \cdot 1}{4 \cdot 3 \cdot 2 \cdot 1} = 1$

   In general, for any counting number n:

   $$_nC_n = 1 \qquad \text{OR} \qquad \binom{n}{n} = 1$$

2. There is only one way to take 0 objects from a set of n objects.

   In general, for any counting number n:

   $$_nC_0 = 1 \qquad \text{OR} \qquad \binom{n}{0} = 1$$

3. Since $_7C_2 = \dfrac{7 \cdot 6}{2 \cdot 1}$ and $_7C_5 = \dfrac{7 \cdot 6 \cdot \cancel{5} \cdot \cancel{4} \cdot \cancel{3}}{\cancel{5} \cdot \cancel{4} \cdot \cancel{3} \cdot 2 \cdot 1} = \dfrac{7 \cdot 6}{2 \cdot 1}$, then $_7C_2 = {}_7C_5$.

   In general, for any whole numbers n and r, when $r \leq n$:

   $$_nC_r = {}_nC_{n-r} \qquad \text{OR} \qquad \binom{n}{r} = \binom{n}{n-r}$$

MODEL PROBLEMS

1. In how many different orders can the program for a music recital be arranged if 7 students are to perform?

Solution

This is a permutation of 7 things taken 7 at a time.

$$_7P_7 = 7 \cdot 6 \cdot 5 \cdot 4 \cdot 3 \cdot 2 \cdot 1 = 5040 \quad Ans.$$

2. In how many ways can one junior and one senior be selected from a group of 8 juniors and 6 seniors?

Solution

Use the counting principle. The junior can be selected in 8 ways and the senior can be selected in 6 ways. Therefore, there are 8 · 6 = 48 possible selections. *Ans.*

3. How many 5-letter arrangements can be made of the letters of the word "books"?

Solution

This is a permutation of 5 things taken 5 at a time where 2 are identical.

$$\frac{_5P_5}{2!} = \frac{5!}{2!} = \frac{5 \cdot 4 \cdot 3 \cdot \cancel{2} \cdot \cancel{1}}{\cancel{2} \cdot \cancel{1}} = 60 \quad Ans.$$

4. A reading list gives the titles of 20 novels and 12 biographies from which each student is to choose 3 novels and 2 biographies to read. How many different combinations of titles could be chosen?

Solution

1. Find the number of ways in which 3 novels could be chosen.

$$\binom{20}{3} = \frac{20 \cdot 19 \cdot 18}{3 \cdot 2 \cdot 1}$$
$$= 1140$$

2. Find the number of ways in which 2 biographies could be chosen.

$$\binom{12}{2} = \frac{12 \cdot 11}{2 \cdot 1}$$
$$= 66$$

3. Use the counting principle to find the number of possible choices of novels and biographies.

$$1140 \cdot 66 = 75{,}240 \quad Ans.$$

EXERCISES

In 1–12, evaluate the given expression.

1. $_6P_6$ **2.** $_8P_4$ **3.** $_{12}P_2$ **4.** $_8C_5$ **5.** $_8C_3$ **6.** $_{12}C_{12}$

7. $_{12}C_0$ **8.** $4!$ **9.** $\binom{13}{4}$ **10.** $\binom{17}{15}$ **11.** $\frac{_5P_3}{3!}$ **12.** $\frac{_6P_2}{2!}$

In 13–18, find the number of "words" (arrangements of letters) that can be formed from the letters of the given word, using each letter.

13. axis
14. circle
15. identity
16. parabola
17. abscissa
18. minimum

19. In how many different ways can 6 runners be assigned to 6 lanes at the start of a race?

20. In how many different ways can a "hand" of 5 cards be dealt from a deck of 52 cards?

21. There are 7 students in a history club. **a.** In how many ways can a president, a vice-president, and a treasurer be elected from the members of the club? **b.** In how many ways can a committee of 3 of the club members be selected to plan a visit to a museum?

22. There are 5 red and 4 white marbles in an urn. A marble is drawn from the urn and not replaced. Then, a second marble is drawn. **a.** In how many ways can a red marble and a white marble be drawn in that order? **b.** In how many ways can a red marble and a white marble be drawn in either order?

23. From a standard deck of 52 cards, 2 cards are drawn without replacement. **a.** How many combinations of 2 hearts are possible? **b.** How many combinations of 2 kings are possible?

24. Each week, Mark does the dishes on 4 days and Lisa does them on the remaining 3 days. In how many different orders could they choose to do the dishes? (*Hint:* This is an arrangement of 7 things with repetition.)

25. Each week, Albert does the dishes on 3 days, Rita does them on 2 days, and Marie does them on the remaining 2 days. In how many different orders could they choose to do the dishes?

26. There are 4 boys and 5 girls who are members of a chess club.
    a. How many games must be played if each member is to play every other member once?
    b. In how many ways could one boy and one girl be selected to play a demonstration game?
    c. In how many ways could a group of 3 members be selected to represent the club at a regional meet?
    d. In how many ways could 2 boys and 2 girls be selected to attend the state tournament?

16-2 PROBABILITY

A sample space is the set of all possible outcomes or results of an activity. An event is a subset of a sample space. For example, if a die is rolled, the sample space is {1, 2, 3, 4, 5, 6}. The event of rolling a number less than 3 is {1, 2}. If the die is fair, or unbiased, each outcome is equally likely to occur. The probability of rolling a number less than 3 is the ratio of the number of elements in the set {1, 2} to the number of elements in the sample space {1, 2, 3, 4, 5, 6}.

$$P \text{ (rolling a number less than 3)} = \tfrac{2}{6} = \tfrac{1}{3}$$

■ **The theoretical probability of an event is the number of ways that an event can occur divided by the total number of possible outcomes when each outcome is equally likely to occur.**

If $P(E)$ represents the probability of an event E,
   $n(E)$ represents the number of ways that E can occur,
and $n(S)$ represents the number of possible outcomes in the sample space S, then:

$$P(E) = \frac{n(E)}{n(S)}$$

□ EXAMPLE 1: What is the probability of drawing a red queen from a standard deck of 52 cards?

Solution

1. Count the total number of outcomes in the sample space.

$$S = \{\text{cards in the deck}\} \qquad n(S) = 52$$

2. Count the number of outcomes in the event.

$$E = \{\text{queen of hearts, queen of diamonds}\} \qquad n(E) = 2$$

3. Substitute these values in the formula.

$$P(\text{red queen}) = \frac{n(E)}{n(S)} = \frac{2}{52} = \frac{1}{26} \quad Ans.$$

The number of outcomes in the sample space and in the event can often be determined by using permutations or combinations.

□ EXAMPLE 2: Two cards are to be drawn from a standard deck of 52 cards without replacement. What is the probability that both cards will be red?

Solution

1. The total number of outcomes is $_{52}C_2 = \dfrac{52 \cdot 51}{2 \cdot 1} = 26 \cdot 51$.

2. The number of favorable outcomes is $_{26}C_2 = \dfrac{26 \cdot 25}{2 \cdot 1} = 13 \cdot 25$.

3. Substitute these values in the formula.

$$P(\text{two red cards}) = \frac{n(E)}{n(S)} = \frac{\overset{1}{\cancel{13}} \cdot 25}{\underset{2}{\cancel{26}} \cdot 51} = \frac{25}{102} \quad Ans.$$

Recall that the probability of an event that is *certain* is 1. The probability of rolling a number less than 7 on one roll of a die is $\frac{6}{6}$ or 1. The probability of an *impossible* event is 0. The probability of rolling a number greater than 7 on one roll of a die is $\frac{0}{6} = 0$. The probability of any event is greater than or equal to 0 and less than or equal to 1.

$$0 \le P(E) \le 1$$

If $P(E) = p$, then $P(\text{not } E) = 1 - p$
Notice that $P(E) + P(\text{not } E) = 1$

□ EXAMPLE 3: If the probability that it will rain tomorrow is 40%, what is the probability that it will not rain?

Solution: If $P(\text{rain}) = .40$, then $P(\text{not rain}) = 1 - .40 = .60$. *Ans.*

| MODEL PROBLEMS |

1. A choral group is composed of 6 juniors and 8 seniors. If a junior and a senior are chosen at random to sing a duet at the spring concert, what is the probability that the choice is Emira, who is a junior, and Jean, who is a senior?

Solution

1. Since there are 6 ways of choosing the junior and 8 ways of choosing the senior, there are $6 \cdot 8 = 48$ possible choices.

2. There is one choice that includes Emira and Jean.

3. $P(\text{Emira and Jean}) = \frac{1}{48}$ *Ans.*

Alternate Solution

Since the choices are independent events, we may use the counting principle for probabilities.

1. The probability of choosing Emira is $\frac{1}{6}$.

2. The probability of choosing Jean is $\frac{1}{8}$.

3. The probability of choosing Emira and Jean is $\frac{1}{6} \cdot \frac{1}{8} = \frac{1}{48}$. *Ans.*

2. What is the probability that a 3-letter "word" formed from letters of the word "coinage" consists of all vowels?

Solution

1. Find $n(S)$, the number of 3-letter permutations of the letters in the word "coinage."

$$n(S) = {}_7P_3 = 7 \cdot 6 \cdot 5 = 210$$

2. Find $n(E)$, the number of 3-letter permutations of the 4 vowels (o, i, a, e) in the word.

$$n(E) = {}_4P_3 = 4 \cdot 3 \cdot 2 = 24$$

3. Find the probability of event E.

$$P(E) = \frac{n(E)}{n(S)} = \frac{24}{210} = \frac{4}{35} \quad Ans.$$

EXERCISES

1. What is the probability of getting a number less than 5 on one throw of a fair die?
2. If one of the letters of the word "element" is chosen at random, what is the probability that an "e" is chosen?
3. A bag contains only 5 red marbles and 3 blue marbles. If one marble is drawn at random from the bag, what is the probability that it is blue?
4. If the letters of the word "equal" are rearranged at random, what is the probability that the first letter of the new "word" is a vowel?
5. If 2 coins are tossed, what is the probability that both show heads?
6. If two coins are tossed, what is the probability that neither shows heads?
7. If a card is drawn from a standard deck of 52 cards, what is the probability that the card is a queen?
8. If 2 cards are drawn from a standard deck of 52 cards without replacement, what is the probability that both cards are queens?
9. A bank contains 4 coins: a penny, a nickel, a dime, and a quarter. One coin is removed at random and tossed. **a.** What is the probability that the dime is removed? **b.** What is the probability that the coin shows heads? **c.** What is the probability that the coin removed is a quarter that shows heads? **d.** What is the probability that the coin has a value less than 20 cents and shows heads?
10. The weather report gives the probability of rain on Saturday as 20% and the probability of rain on Sunday as 10%. **a.** What is the probability that it will not rain on Saturday? **b.** What is the

probability that it will not rain on Sunday? **c.** What is the probability that it will not rain either day?

11. A seed company advertises that if its geranium seed is properly planted, the probability that the seed will grow is 90%. **a.** What is the probability that a geranium seed that has been properly planted will fail to grow? **b.** If 5 geranium seeds are properly planted, what is the probability that all will fail to grow?

12. Of the 5 sandwiches that Mrs. Muth made for her children's lunches, 2 contain tuna fish and 3 contain peanut butter and jelly. Her son, Tim, took 2 sandwiches at random. What is the probability that he took 2 sandwiches containing peanut butter and jelly?

13. Mrs. Gillis' small son, Brian, tore all the labels off the soup cans on the kitchen shelf. If Mrs. Gillis knows that she bought 4 cans of tomato soup and 2 cans of vegetable soup, what is the probability that the first 2 cans of soup she opens are both tomato?

14. Of the 15 students in Mrs. Barney's mathematics class, 10 take Spanish. If 2 students are absent from Mrs. Barney's class on Monday, what is the probability that they are both students who take Spanish?

15. At a card party, 2 door prizes are to be awarded by drawing 2 names at random from a box. The box contains the names of 40 persons, including Patricia Sullivan and Joe Ramirez. **a.** What is the probability that Joe's name is not drawn for either prize? **b.** What is the probability that Patricia does win one of the prizes?

16-3 PROBABILITY WITH TWO OUTCOMES

In the spinner at the right, the arrow can land on one of 3 equally likely regions, numbered 1, 2, and 3. If the arrow lands on a line, the spin is not counted, and the arrow is spun again. Let us define an experiment with 2 outcomes for this spinner: obtaining an odd number or obtaining an even number. Therefore:

$$P(\text{odd}) = P(O) = \tfrac{2}{3} \quad \text{and} \quad P(\text{even}) = P(E) = \tfrac{1}{3}$$

When the arrow is spun several times, the result of each spin is independent of the result of the other spins. To find the probability of getting an odd number each time in several spins, we can use the *counting principle with probabilities*.

■ If E and F are independent events, and if the probability of E is m ($0 \le m \le 1$) and the probability of F is n ($0 \le n \le 1$), then the probability of E and F occurring jointly is $m \cdot n$ ($0 \le m \cdot n \le 1$).

If the arrow is spun twice:

$$P(2 \text{ odd numbers in 2 spins}) = \tfrac{2}{3} \cdot \tfrac{2}{3} = \tfrac{4}{9}$$

If the arrow is spun 3 times:

$$P(3 \text{ odd numbers in 3 spins}) = \tfrac{2}{3} \cdot \tfrac{2}{3} \cdot \tfrac{2}{3} = \tfrac{8}{27}$$

If the arrow is spun 4 times:

$$P(4 \text{ odd numbers in 4 spins}) = \tfrac{2}{3} \cdot \tfrac{2}{3} \cdot \tfrac{2}{3} \cdot \tfrac{2}{3} = \tfrac{16}{81}$$

We can also apply the counting principle with probabilities to situations where the arrow is spun n times and an odd number is obtained *less than n* times.

□ EXAMPLE 1: Find the probability of obtaining exactly one odd number on 4 spins of the arrow.

Solution

Let us begin by considering the case of getting an odd number on the first spin and an even number on each of the other 3 spins.

$$P(\text{odd on first spin only}) = \tfrac{2}{3} \cdot \tfrac{1}{3} \cdot \tfrac{1}{3} \cdot \tfrac{1}{3} = \tfrac{2}{81}$$

We could also get exactly one odd number, however, by getting an odd number on the second spin, or on the third spin, or on the fourth spin, and an even number on the other spins in each case.

$$P(\text{odd on the second spin only}) = \tfrac{1}{3} \cdot \tfrac{2}{3} \cdot \tfrac{1}{3} \cdot \tfrac{1}{3} = \tfrac{2}{81}$$

$$P(\text{odd on the third spin only}) = \tfrac{1}{3} \cdot \tfrac{1}{3} \cdot \tfrac{2}{3} \cdot \tfrac{1}{3} = \tfrac{2}{81}$$

$$P(\text{odd on the fourth spin only}) = \tfrac{1}{3} \cdot \tfrac{1}{3} \cdot \tfrac{1}{3} \cdot \tfrac{2}{3} = \tfrac{2}{81}$$

Notice that the probability of each possible way of getting exactly one odd number on 4 spins of the arrow is $(\tfrac{2}{3})^1(\tfrac{1}{3})^3$ or $\tfrac{2}{81}$. There are 4 possible ways of doing this, as shown in the diagram at the right. Therefore:

O	E	E	E
E	O	E	E
E	E	O	E
E	E	E	O

$$P(\text{exactly one odd number on 4 spins}) = 4(\tfrac{2}{3})^1(\tfrac{1}{3})^3 = \tfrac{8}{81} \quad Ans.$$

□ EXAMPLE 2: Find the probability of obtaining exactly 2 odd numbers on 4 spins of the arrow.

Solution

One possible way to obtain exactly 2 odd numbers is on the first 2 spins of the arrow. Thus, an even number is obtained on the last 2 spins.

$$P(\text{odd on first 2 spins only}) = \tfrac{2}{3} \cdot \tfrac{2}{3} \cdot \tfrac{1}{3} \cdot \tfrac{1}{3} = (\tfrac{2}{3})^2(\tfrac{1}{3})^2 = \tfrac{4}{81}$$

O	O	E	E
O	E	O	E
O	E	E	O
E	O	O	E
E	O	E	O
E	E	O	O

The probability $\frac{4}{81}$ indicates only one possible way of obtaining exactly 2 odd numbers on 4 spins of the arrow. The diagram at the right shows that there are 6 possible combinations of 2 odd numbers out of 4 spins, obtained by the formula $_4C_2 = \dfrac{4 \cdot 3}{2 \cdot 1} = 6$. Thus:

$$P(\text{exactly 2 odds on 4 spins}) = {}_4C_2 \cdot (\tfrac{2}{3})^2 (\tfrac{1}{3})^2$$

$$= 6 \cdot \tfrac{4}{9} \cdot \tfrac{1}{9}$$

$$= \tfrac{24}{81} \text{ or } \tfrac{8}{27} \ Ans.$$

☐ EXAMPLE 3: Find the probability of obtaining exactly 3 odd numbers on 4 spins of the arrow.

Solution

1. Find the probability of one way of obtaining exactly 3 odd numbers, for example, getting odd numbers on the first 3 spins and an even number on the fourth spin.

$$P(\text{an odd number on only the first 3 spins}) = \tfrac{2}{3} \cdot \tfrac{2}{3} \cdot \tfrac{2}{3} \cdot \tfrac{1}{3}$$

$$= (\tfrac{2}{3})^3 (\tfrac{1}{3})^1 = \tfrac{8}{81}$$

2. Find the number of possible ways of obtaining exactly 3 odd numbers on 4 spins of the arrow.

$$_4C_3 = \frac{4 \cdot \cancel{3} \cdot \cancel{2}}{\cancel{3} \cdot \cancel{2} \cdot 1} = 4$$

3. $P(\text{exactly 3 odds on 4 spins}) = 4(\tfrac{2}{3})^3(\tfrac{1}{3})^1 = 4 \cdot \tfrac{8}{81} = \tfrac{32}{81} \ Ans.$

The steps that we have been using enable us to determine the probability of exactly r successes in n independent trials of an experiment with 2 outcomes, called a *Bernoulli experiment*. An experiment such as tossing a coin has 2 outcomes: heads and tails. An experiment such as tossing a die can be thought of as having two outcomes, for example, rolling a one and not rolling a one. If an event E, such as rolling a one on a die, is to occur exactly r times in n trials, then the event *not E*, rolling a number that is not one, must occur $n - r$ times in n trials.

■ In general, for a given experiment, if the probability of success is p and the probability of failure is $1 - p = q$, then the probability of exactly r successes in n independent trials is:

$$_nC_r p^r q^{n-r}$$

| MODEL PROBLEMS |

1. If a fair coin is tossed 10 times, what is the probability that it falls tails exactly 6 times?

Solution

In one toss of a fair coin, the probability of success $p = P(\text{tails}) = \frac{1}{2}$ and the probability of failure $q = P(\text{heads}) = \frac{1}{2}$. Here, the number of trials, n, is 10, and the number of successes, r, is 6. Use the formula:

$$_nC_r \, p^r q^{n-r}$$

$$_{10}C_6 \left(\frac{1}{2}\right)^6 \left(\frac{1}{2}\right)^{10-6} = \frac{10 \cdot 9 \cdot 8 \cdot 7 \cdot 6 \cdot 5}{6 \cdot 5 \cdot 4 \cdot 3 \cdot 2 \cdot 1} \left(\frac{1}{2}\right)^6 \left(\frac{1}{2}\right)^4$$

$$= 210 \cdot \frac{1}{64} \cdot \frac{1}{16} = \frac{210}{1024} = \frac{105}{512} \quad Ans.$$

2. If 5 fair dice are tossed, what is the probability that they show exactly 3 fours?

Solution

In this problem:

$p = P(4) = \dfrac{1}{6}$ Use the formula: $_nC_r \, p^r q^{n-r}$

$q = P(\text{not } 4) = \dfrac{5}{6}$ $P(3 \text{ fours in 5 trials}) = {}_5C_3 \left(\dfrac{1}{6}\right)^3 \left(\dfrac{5}{6}\right)^2$

$r = 3$ successes $= \dfrac{5 \cdot 4 \cdot 3}{3 \cdot 2 \cdot 1} \cdot \dfrac{1}{216} \cdot \dfrac{25}{36}$

$n = 5$ trials $= \dfrac{125}{3888} \quad Ans.$

| EXERCISES |

1. A fair coin is tossed 5 times. Find the probability of tossing:
   a. exactly 2 heads   b. exactly 3 heads
   c. exactly 4 heads   d. exactly 5 heads
2. A fair coin is tossed 4 times. Find the probability of tossing:
   a. exactly 2 tails   b. exactly 3 tails
3. If 4 fair dice are tossed, find the probability of getting:
   a. exactly 3 fives   b. exactly 4 fives

4. If 4 fair dice are tossed, find the probability of getting:
   a. exactly 3 even numbers b. exactly 2 odd numbers
5. Jan's record shows that her probability of success on a basketball free throw is $\frac{3}{5}$. Find the probability that Jan will be successful on 2 out of 3 shots.
6. The probability that the Wings will win a baseball game is $\frac{2}{3}$. What is the probability that they will win:
   a. exactly 2 out of 4 games? b. exactly 3 out of 4 games?

In 7 and 8, select the numeral preceding the expression that best answers the question.

7. What is the probability of getting a number less than 3 on 6 out of 10 tosses of a fair die?
   (1) $6(\frac{1}{3})^6$ (2) $210(\frac{1}{3})^6$ (3) $6(\frac{1}{3})^6(\frac{2}{3})^4$ (4) $210(\frac{1}{3})^6(\frac{2}{3})^4$
8. A coin is loaded so that the probability of heads is $\frac{1}{4}$. What is the probability of getting exactly 3 heads on 8 tosses of the coin?
   (1) $_8C_3(\frac{1}{4})^3$ (2) $_8C_3(\frac{1}{4})^3(\frac{3}{4})^5$
   (3) $_8C_3(\frac{1}{4})^3(\frac{1}{4})^5$ (4) $3(\frac{1}{4})^3(\frac{3}{4})^5$

In 9–13, answers may be expressed in exponential form, as in the choices in exercise 8.

9. The probability that a flashbulb is defective is found to be $\frac{1}{20}$. What is the probability that a package of 6 flashbulbs has only one defective bulb?
10. A multiple-choice test gives 5 possible choices for each answer, of which one is correct. The probability of selecting the correct answer by guessing is $\frac{1}{5}$. a. What is the probability of getting 5 out of 10 questions correct by guessing? b. What is the probability of getting only 1 out of 10 questions correct by guessing? c. What is the probability of getting 9 correct answers out of 10 by guessing?
11. Mrs. Shusda gave a true-false test of 10 questions. The probability of selecting the correct answer by guessing is $\frac{1}{2}$. a. What is the probability that Fred, who guessed at every answer, will get 9 out of 10 correct? b. What is the probability that Fred will get 10 out of 10 correct? c. What is the probability that Fred will get either 9 or 10 out of 10 correct?
12. In a group of 100 persons who were born in June, what is the probability that exactly 2 were born on June 1? (Assume that a person born in June is equally likely to have been born on any one of the 30 days.)

13. In a box there are 4 red marbles and 5 white marbles. Marbles are drawn one at a time and replaced after each drawing. What is the probability of drawing:
    a. exactly 2 red marbles when 3 marbles are drawn?
    b. exactly 3 white marbles when 5 marbles are drawn?
    c. exactly 7 red marbles when 12 marbles are drawn?

16-4 AT LEAST AND AT MOST

When we are anticipating success on repeated trials of an experiment, we often require *at least* a given number. For example, in a game, David rolls 5 dice. To win, *at least* 3 of the 5 dice that David rolls must be "ones." Therefore, David will win if he rolls 3, 4, or 5 "ones." In general:

■ At least r successes in n trials means $r, r + 1, r + 2, \ldots, n$ successes.

If a manufacturer considers *at most* two defective parts in a lot of one hundred parts to be an acceptable standard, then there can be 2, 1, or 0 defective parts in every 100 parts.

■ At most r successes in n trials means $r, r - 1, r - 2, \ldots, 0$ successes.

The model problems that follow show how these concepts are applied to probability.

| MODEL PROBLEMS |
---

1. Rose is the last person to compete in a basketball free-throw contest. In order to win, Rose must be successful in at least 4 out of 5 throws. If the probability that Rose will be successful on any one throw is $\frac{3}{4}$, what is the probability that Rose will win the contest?

Solution

To be successful in at least 4 out of 5 throws means to be successful in 4 or in 5 throws.

On one throw, $P(\text{success}) = \frac{3}{4}$ and $P(\text{failure}) = 1 - \frac{3}{4} = \frac{1}{4}$.

1. $P(4 \text{ out of 5 successes}) = {}_5C_4 \left(\frac{3}{4}\right)^4 \left(\frac{1}{4}\right)^1$

$$= \frac{5 \cdot \cancel{4} \cdot \cancel{3} \cdot \cancel{2}}{\cancel{4} \cdot \cancel{3} \cdot \cancel{2} \cdot 1} \cdot \frac{81}{4^4} \cdot \frac{1}{4^1} = \frac{405}{4^5}$$

2. $P(\text{5 out of 5 successes}) = {}_5C_5 \left(\dfrac{3}{4}\right)^5 \left(\dfrac{1}{4}\right)^0 = 1 \cdot \dfrac{243}{4^5} \cdot 1 = \dfrac{243}{4^5}$

3. $P(\text{at least 4 out of 5 successes})$:

$$= P(\text{4 out of 5 successes}) + P(\text{5 out of 5 successes})$$

$$= \dfrac{405}{4^5} + \dfrac{243}{4^5}$$

$$= \dfrac{648}{4^5} = \dfrac{648}{1024} = \dfrac{81}{128} \quad Ans.$$

2. What is the probability of at most 2 boys in a family of 5 children if the family has been chosen at random?

Solution

To have at most 2 boys means to have 0, 1, or 2 boys. Let us assume that $P(\text{boy}) = \frac{1}{2}$ and $P(\text{girl}) = \frac{1}{2}$.

1. $P(\text{0 boys in 5 children}) = {}_5C_0 \left(\dfrac{1}{2}\right)^5 \left(\dfrac{1}{2}\right)^0 = 1 \cdot \left(\dfrac{1}{2}\right)^5 \cdot 1 = \dfrac{1}{32}$

2. $P(\text{1 boy in 5 children}) = {}_5C_1 \left(\dfrac{1}{2}\right)^1 \left(\dfrac{1}{2}\right)^4 = 5 \cdot \dfrac{1}{2} \cdot \dfrac{1}{16} = \dfrac{5}{32}$

3. $P(\text{2 boys in 5 children}) = {}_5C_2 \left(\dfrac{1}{2}\right)^2 \left(\dfrac{1}{2}\right)^3 = \dfrac{5 \cdot 4}{2 \cdot 1} \cdot \dfrac{1}{4} \cdot \dfrac{1}{8} = \dfrac{10}{32}$

4. $P(\text{at most 2 boys in 5 children})$:

$$= P(\text{0 boys}) + P(\text{1 boy}) + P(\text{2 boys})$$

$$= \dfrac{1}{32} + \dfrac{5}{32} + \dfrac{10}{32}$$

$$= \dfrac{16}{32} = \dfrac{1}{2} \quad Ans.$$

3. A coin is loaded so that the probability of heads is 4 times the probability of tails.
    a. What is the probability of heads on a single throw?
    b. What is the probability of at least one tail in 5 throws?

Solution

a. Let $P(\text{tails}) = q$ and $P(\text{heads}) = 4q$.

$$q + 4q = 1$$
$$5q = 1$$
$$q = \tfrac{1}{5}, \text{ or } P(\text{tails}) = \tfrac{1}{5}$$

Since $4q = 4 \cdot \frac{1}{5} = \frac{4}{5}$, then $P(\text{heads}) = \frac{4}{5}$. *Ans.*

b. P(at least one tail):

$$= P(1T) + P(2T) + P(3T) + P(4T) + P(5T)$$

$$= {}_5C_1\left(\frac{1}{5}\right)^1\left(\frac{4}{5}\right)^4 + {}_5C_2\left(\frac{1}{5}\right)^2\left(\frac{4}{5}\right)^3 + {}_5C_3\left(\frac{1}{5}\right)^3\left(\frac{4}{5}\right)^2$$

$$+ {}_5C_4\left(\frac{1}{5}\right)^4\left(\frac{4}{5}\right)^1 + {}_5C_5\left(\frac{1}{5}\right)^5$$

$$= 5 \cdot \frac{256}{5^5} + 10 \cdot \frac{64}{5^5} + 10 \cdot \frac{16}{5^5} + 5 \cdot \frac{4}{5^5} + 1 \cdot \frac{1}{5^5}$$

$$= \frac{2101}{5^5} = \frac{2101}{3125} \quad Ans.$$

Alternate Solution

There is one way to fail to get at least one tail in 5 throws, that is, to get 5 heads in 5 throws.

$$P(\text{failure}) = P(5 \text{ heads}) = {}_5C_0\left(\frac{4}{5}\right)^5 = 1 \cdot \frac{1024}{5^5} = \frac{1024}{3125}$$

$$P(\text{success}) = 1 - P(\text{failure}) = 1 - \frac{1024}{3125} = \frac{2101}{3125} \quad Ans.$$

EXERCISES

1. A fair coin is tossed 4 times. Find the probability of tossing:
   - a. exactly 3 heads
   - b. exactly 4 heads
   - c. at least 3 heads
   - d. at most 1 head
2. If a fair coin is tossed 6 times, find the probability of obtaining at most 2 heads.
3. If 5 fair coins are tossed, what is the probability that at least 2 are tails?
4. A fair die is rolled 3 times. Find the probability of rolling:
   - a. exactly 1 five
   - b. no fives
   - c. at most 1 five
   - d. at least 2 sixes
   - e. at most 2 fours
   - f. exactly 1 even number
   - g. at most 1 even
   - h. at most 1 number less than three
   - i. at least 1 even
   - j. at least 1 number greater than four
5. If a fair die is tossed 5 times, find the probability of obtaining at most 2 ones.
6. If 4 fair dice are rolled, what is the probability that at least 2 are fives?

In 7–9, an arrow can land on one of 5 equally likely regions on a spinner, numbered 1, 2, 3, 4, and 5. If the arrow lands on a line, the spin is not counted, and the arrow is spun again.

Ex. 7 to 9

7. For the spinner described, find:
   a. P(even number)
   b. P(odd number)
   c. P(both even on 2 spins)
   d. P(both odd on 2 spins)
   e. P(at least 1 even on 2 spins)
   f. P(at least 1 odd on 2 spins)
   g. P(exactly 2 evens on 3 spins)
   h. P(at least 2 evens on 3 spins)
   i. P(exactly 2 odds on 4 spins)
   j. P(at most 2 odds on 4 spins)

8. *True* or *False*: If the arrow is spun on the spinner as described, then P(at least 3 evens on 5 spins) = P(at most 2 odds on 5 spins). Explain why.

9. For the spinner described: a. Find $P(2)$. b. If the arrow is spun 5 times, will a "2" appear exactly once? Support your answer by finding P(exactly one "2" on 5 spins).

10. A coin is weighted so that the probability of heads is $\frac{4}{5}$.
    a. What is the probability of getting at least 3 heads when the weighted coin is tossed 4 times?
    b. What is the probability of getting at most 2 tails when the weighted coin is tossed 5 times?

11. Each evening the members of the Sanchez family are equally likely to watch the news on any one of 3 possible TV channels: 5, 8, and 13. a. What is the probability that they watch the news on channel 8 on at least 3 out of 5 evenings? b. What is the probability that they watch the news on channel 13 on at most 2 out of 5 evenings?

12. In a game, the probability of winning is $\frac{1}{5}$ and the probability of losing is $\frac{4}{5}$. If 3 games are played, what is the probability of winning at least 2 games?

13. In each game that the school basketball team plays, the probability that the team will win is $\frac{2}{3}$. What is the probability that the team will win at least 3 of the next 4 games?

14. A die is loaded so that the probability of rolling a one is $\frac{3}{4}$. What is the probability of rolling at least 2 ones when the die is tossed 3 times?

15. If a family of 4 children is selected at random, what is the probability that at most 3 of the children are boys?

In 16 and 17, an electronic game contains 9 keys. As shown at the right, 5 keys have numbers, and 4 keys have colors. Each key is equally likely to be pressed on each move.

Red	1	Blue
2	3	4
Green	5	Yellow

Ex. 16 and 17

16. If one key is pressed at random in the electronic game, find:
    a. P(number key) b. P(color key)
17. If 3 keys are pressed at random in the electronic game, find the probability of selecting:
    a. exactly 1 color key b. at least 2 color keys
    c. at most 1 color key d. all color keys
    e. exactly 1 number key f. at least 1 number key

In 18–22, answers may be expressed in exponential form.

18. Of last year's graduates, 3 out of 5 are enrolled in college. If the names of 10 of last year's graduates are chosen at random, what is the probability that at least 8 out of 10 are in college?
19. A coin is loaded so that the probability of heads on a single throw is 3 times the probability of tails. a. What is the probability of heads and the probability of tails on a single throw? b. What is the probability of at most 3 heads when the coin is tossed 6 times?
20. A license number consists of 5 letters of the alphabet selected at random. Each letter can be selected any number of times. a. What is the probability that the license number has at most 2 Q's? b. What is the probability that the license number has at least 4 X's?
21. A manufacturer tests her product and finds that the probability of a defective part is .02. What is the probability that out of 5 parts selected at random at most 1 will be defective?
22. A seed company advertises that if their seeds are properly planted, 95% of them will germinate. What is the probability that when 20 seeds are properly planted, at least 15 will germinate?

16-5 THE BINOMIAL THEOREM

Any binomial may be represented by $(x + y)$ so that x represents the first term and y represents the second term. If a binomial is raised to a positive integral power, the result is a polynomial called the *expansion of the binomial*. Observe the expansion of the first 4 powers of $(x + y)$.

$(x + y)^0 = 1$

$(x + y)^1 = 1x + 1y$

$(x + y)^2 = (x + y)(x + y) = 1x^2 + 2xy + 1y^2$

$(x + y)^3 = (x + y)(x + y)^2 = (x + y)(x^2 + 2xy + y^2)$
$$= x^3 + 2x^2y + xy^2 + x^2y + 2xy^2 + y^3$$
$$= 1x^3 + 3x^2y + 3xy^2 + 1y^3$$

$(x + y)^4 = (x + y)(x + y)^3 = (x + y)(x^3 + 3x^2y + 3xy^2 + y^3)$
$$= x^4 + 3x^3y + 3x^2y^2 + xy^3 + x^3y + 3x^2y^2 + 3xy^3 + y^4$$
$$= 1x^4 + 4x^3y + 6x^2y^2 + 4xy^3 + 1y^4$$

Notice the pattern that is developing in each expansion of the binomial, shown below on the left.

A display that contains only the coefficients of the terms in the expansions is called *Pascal's Triangle*, shown below on the right.

Binomial Expansions	Pascal's Triangle
$(x + y)^0 =$ *1*	
$(x + y)^1 =$ *1x + 1y*	
$(x + y)^2 =$ *1x² + 2xy + 1y²*	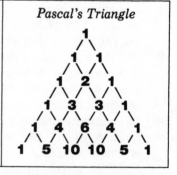
$(x + y)^3 =$ *1x³ + 3x²y + 3xy² + 1y³*	
$(x + y)^4 = 1x⁴ + 4x³y + 6x²y² + 4xy³ + 1y⁴*	
$(x + y)^5 =$ (to be answered)	

In Pascal's Triangle, notice that elements inside the triangle can also be obtained by adding a pair of adjacent entries from the row above.

To write the expansion of $(x + y)^5$, we can use the entries from the sixth row of Pascal's Triangle as the coefficients. Each coefficient is multiplied by factors of x and y, starting with x^5y^0, or simply x^5. In each successive term, we decrease the number of factors of x by 1 and increase the number of factors of y by 1. Therefore:

$$(x + y)^5 = 1x^5 + 5x^4y + 10x^3y^2 + 10x^2y^3 + 5xy^4 + 1y^5$$

Since $(x + y)^5 = (x + y)(x + y)(x + y)(x + y)(x + y)$, we can think of the expansion of $(x + y)^5$ as the result of choosing all possible combinations of either x or y from each of 5 factors.

1. We can choose x from 5 factors and y from 0 factors: $_5C_0x^5y^0$
   The number of ways of doing this is $_5C_0$.
2. We can choose x from 4 factors and y from 1 factor: $_5C_1x^4y^1$
   The number of ways of doing this is $_5C_1$.
3. We can choose x from 3 factors and y from 2 factors: $_5C_2x^3y^2$
   The number of ways of doing this is $_5C_2$.
4. We can choose x from 2 factors and y from 3 factors: $_5C_3x^2y^3$
   The number of ways of doing this is $_5C_3$.
5. We can choose x from 1 factor and y from 4 factors: $_5C_4x^1y^4$
   The number of ways of doing this is $_5C_4$.
6. We can choose x from 0 factors and y from 5 factors: $_5C_5x^0y^5$
   The number of ways of doing this is $_5C_5$.

Therefore:

$$(x + y)^5 = {_5C_0}x^5y^0 + {_5C_1}x^4y^1 + {_5C_2}x^3y^2 + {_5C_3}x^2y^3 + {_5C_4}x^1y^4 + {_5C_5}x^0y^5$$

OR

$$(x + y)^5 = \binom{5}{0}x^5y^0 + \binom{5}{1}x^4y^1 + \binom{5}{2}x^3y^2 + \binom{5}{3}x^2y^3 + \binom{5}{4}x^1y^4 + \binom{5}{5}x^0y^5$$

OR SIMPLY

$$(x + y)^5 = 1x^5 + 5x^4y + 10x^3y^2 + 10x^2y^3 + 5xy^4 + 1y^5$$

■ In general:

$$(x + y)^n = {_nC_0}x^ny^0 + {_nC_1}x^{n-1}y^1 + {_nC_2}x^{n-2}y^2 + \ldots + {_nC_{n-1}}x^1y^{n-1} + {_nC_n}x^0y^n$$

OR

$$(x + y)^n = \binom{n}{0}x^ny^0 + \binom{n}{1}x^{n-1}y^1 + \binom{n}{2}x^{n-2}y^2 + \ldots + \binom{n}{n-1}x^1y^{n-1} + \binom{n}{n}x^0y^n$$

For example, to write the expansion of $(2a - 1)^5$, substitute $2a$ for x and -1 for y in the expansion of $(x + y)^5$.

$$(2a - 1)^5 = 1(2a)^5 + 5(2a)^4(-1)^1 + 10(2a)^3(-1)^2 + 10(2a)^2(-1)^3 +$$
$$5(2a)(-1)^4 + 1(-1)^5$$
$$= 32a^5 - 80a^4 + 80a^3 - 40a^2 + 10a - 1$$

It can also be shown that:

1. For any binomial expansion $(x + y)^n$, there are $n + 1$ terms.
2. In general, the rth term of the expansion is $_nC_{r-1}x^{n-r+1}y^{r-1}$.

MODEL PROBLEMS

1. Write the expansion of $(b^2 - 3)^3$.

Solution

$$(x + y)^3 = {}_3C_0x^3y^0 + {}_3C_1x^2y^1 + {}_3C_2x^1y^2 + {}_3C_3x^0y^3$$

Let $x = b^2$, and $y = -3$.

$$(b^2 - 3)^3 = 1 \cdot (b^2)^3(-3)^0 + \frac{3}{1} \cdot (b^2)^2(-3)^1 + \frac{3 \cdot 2}{2 \cdot 1}(b^2)^1(-3)^2 + \frac{3 \cdot 2 \cdot 1}{3 \cdot 2 \cdot 1}(b^2)^0(-3)^3$$

$$(b^2 - 3)^3 = 1 \cdot b^6 \cdot 1 \quad + 3 \cdot b^4 \cdot (-3) \quad + 3 \cdot b^2 \cdot 9 \quad + 1 \cdot 1 \cdot (-27)$$

$$(b^2 - 3)^3 = b^6 \quad\quad\quad - 9b^4 \quad\quad\quad + 27b^2 \quad\quad\quad - 27$$

Answer: $b^6 - 9b^4 + 27b^2 - 27$

2. Write the 11th term of the expansion of $(2a - 1)^{12}$.

Solution

1. The rth term of a binomial expansion is: $\quad {}_nC_{r-1}x^{n-r+1}y^{r-1}$

2. To find the 11th term of $(2a - 1)^{12}$, let $n = 12$, $r = 11$, $x = 2a$, and $y = -1$.

$$= {}_{12}C_{11-1}(2a)^{12-11+1}(-1)^{11-1}$$
$$= {}_{12}C_{10}(2a)^2(-1)^{10}$$

Substitute in the formula, and simplify.

$$= 66 \cdot 4a^2 \cdot 1$$
$$= 264a^2 \quad Ans.$$

(*Note:* ${}_{12}C_{10} = {}_{12}C_2 = \dfrac{12 \cdot 11}{2 \cdot 1} = 66.$)

EXERCISES

In 1–6, write the expansion of the given binomial.

1. $(x + y)^6$
2. $(x + y)^7$
3. $(x + y)^{10}$
4. $(3a - 1)^3$
5. $(x - 2)^4$
6. $(1 - b^3)^5$

In 7–12, write in simplest form the *third term* of the expansion.

7. $(a + b)^4$
8. $(k + 2)^5$
9. $(x - 3y)^5$
10. $(a - 7)^6$
11. $(2 + y)^7$
12. $(2x - 1)^6$

In 13–15, write in simplest form the *fourth term* of the expansion.

13. $(5 - b)^4$
14. $(2a - 3)^5$
15. $(k^2 + 1)^7$

16. Write in expanded form the volume of a cube if the measure of each edge is represented by $(2x - 3)$.
17. Find $(1.01)^5$ by using the expansion of $(1 + .01)^5$.

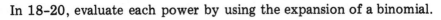

In 18–20, evaluate each power by using the expansion of a binomial.

18. $(1.2)^3$ **19.** $(1.02)^4$ **20.** $(1.05)^3$

In 21–24, select the numeral preceding the expression that best completes the sentence.

21. The third term of the expansion of $(a + 2b)^4$ is:
(1) $2a^2b^2$ (2) $4a^2b^2$ (3) $12a^2b^2$ (4) $24a^2b^2$

22. The eighth term of the expansion of $(2r - 1)^8$ is:
(1) $16r$ (2) $-16r$ (3) 1 (4) -1

23. The middle term of the expansion of $(x - 2y)^4$ is:
(1) $12x^2y^2$ (2) $-12x^2y^2$ (3) $24x^2y^2$ (4) $-24x^2y^2$

24. The last term in the expansion of $(6 - y)^9$ is:
(1) $54y^9$ (2) $-54y^9$ (3) y^9 (4) $-y^9$

16-6   REVIEW EXERCISES

In 1–8, evaluate the expression.

1. $_5P_5$ **2.** $_6P_3$ **3.** $_6C_3$ **4.** $_7C_0$

5. $_{20}C_{20}$ **6.** $\binom{12}{5}$ **7.** $\dfrac{_4P_4}{2!}$ **8.** $\dfrac{7!}{6!}$

9. In how many ways can the letters of the following words be rearranged? a. chef b. career c. bookkeeper

10. In how many ways can first, second, and third prize be awarded in an art contest if 6 of the entries are being considered for the prizes?

11. In how many ways can 3 honorable-mention awards be given in an art contest if 9 of the entries are being considered for the awards?

12. How many different sums of money can be obtained by taking 2 coins from a purse containing a half dollar, a quarter, a dime, a nickel, and a penny?

13. From a standard deck of 52 cards, 4 cards are drawn without replacement.
a. How many combinations of 4 cards are possible?
b. How many combinations of 4 hearts are possible?
c. What is the probability that the 4 cards are all hearts?

14. If 6 fair coins are tossed, find the probability of obtaining:
a. exactly 5 heads   b. at least 5 heads
c. at most 3 heads   d. at least 3 heads

15. In a basketball free-throw contest, the probability that Eric will be successful on each shot is $\frac{4}{5}$. Find the probability that Eric will be successful in:
    a. exactly 2 out of 3 shots
    b. at least 2 out of 3 shots
    c. at most 1 out of 3 shots
    d. exactly 2 out of 4 shots
16. A calculator has 10 number keys and 5 operation keys.
    a. If one key is pressed at random, what is the probability that the key pressed is (i) a number key? (ii) an operation key?
    b. When 5 keys are pressed at random, what is the probability that (i) exactly 3 are operation keys? (ii) at least 3 are operation keys? (iii) at most 2 are number keys?

In 17 and 18, 27 and 28, select the numeral preceding the expression that best completes the sentence or answers the question.

17. When a certain machine makes parts, the probability that one is defective is $\frac{1}{20}$. In a sample of 50 parts, the probability that exactly 5 are defective is:
    (1) $(\frac{1}{20})^5(\frac{19}{20})^{45}$
    (2) $_{45}C_5(\frac{1}{20})^5(\frac{19}{20})^{45}$
    (3) $_{50}C_5(\frac{1}{20})^5(\frac{19}{20})^{50}$
    (4) $_{50}C_5(\frac{1}{20})^5(\frac{19}{20})^{45}$
18. What is the probability of getting exactly 7 fours when 10 dice are rolled?
    (1) $(\frac{1}{6})^7(\frac{5}{6})^{10}$
    (2) $(\frac{1}{6})^7(\frac{5}{6})^3$
    (3) $_{10}C_3(\frac{1}{6})^7(\frac{5}{6})^3$
    (4) $_7C_3(\frac{1}{6})^7(\frac{5}{6})^3$

In 19–22, write the expansion of the binomial.

19. $(x + y)^8$
20. $(1 + 3x)^4$
21. $(5 - b)^3$
22. $\left(a - \frac{1}{a}\right)^5$

In 23–26, write in simplest form the fifth term of the expansion.

23. $(x + y)^{10}$
24. $(5 - y^2)^5$
25. $(3 - 2a)^6$
26. $(x^2 + 1)^9$

27. The fourth term of the expansion of $(1 - y^3)^7$ is:
    (1) $35y^6$
    (2) $35y^9$
    (3) $-35y^6$
    (4) $-35y^9$
28. The middle term of the expansion of $(a + b)^8$ is:
    (1) a^4b^4
    (2) $56a^4b^4$
    (3) $70a^4b^4$
    (4) $56a^5b^3$

Appendix

Squares and Square Roots

No.	Square	Square Root	No.	Square	Square Root	No.	Square	Square Root
1	1	1.000	51	2,601	7.141	101	10,201	10.050
2	4	1.414	52	2,704	7.211	102	10,404	10.100
3	9	1.732	53	2,809	7.280	103	10,609	10.149
4	16	2.000	54	2,916	7.348	104	10,816	10.198
5	25	2.236	55	3,025	7.416	105	11,025	10.247
6	36	2.449	56	3,136	7.483	106	11,236	10.296
7	49	2.646	57	3,249	7.550	107	11,449	10.344
8	64	2.828	58	3,364	7.616	108	11,664	10.392
9	81	3.000	59	3,481	7.681	109	11,881	10.440
10	100	3.162	60	3,600	7.746	110	12,100	10.488
11	121	3.317	61	3,721	7.810	111	12,321	10.536
12	144	3.464	62	3,844	7.874	112	12,544	10.583
13	169	3.606	63	3,969	7.937	113	12,769	10.630
14	196	3.742	64	4,096	8.000	114	12,996	10.677
15	225	3.873	65	4,225	8.062	115	13,225	10.724
16	256	4.000	66	4,356	8.124	116	13,456	10.770
17	289	4.123	67	4,489	8.185	117	13,689	10.817
18	324	4.243	68	4,624	8.246	118	13,924	10.863
19	361	4.359	69	4,761	8.307	119	14,161	10.909
20	400	4.472	70	4,900	8.367	120	14,400	10.954
21	441	4.583	71	5,041	8.426	121	14,641	11.000
22	484	4.690	72	5,184	8.485	122	14,884	11.045
23	529	4.796	73	5,329	8.544	123	15,129	11.091
24	576	4.899	74	5,476	8.602	124	15,376	11.136
25	625	5.000	75	5,625	8.660	125	15,625	11.180
26	676	5.099	76	5,776	8.718	126	15,876	11.225
27	729	5.196	77	5,929	8.775	127	16,129	11.269
28	784	5.292	78	6,084	8.832	128	16,384	11.314
29	841	5.385	79	6,241	8.888	129	16,641	11.358
30	900	5.477	80	6,400	8.944	130	16,900	11.402
31	961	5.568	81	6,561	9.000	131	17,161	11.446
32	1,024	5.657	82	6,724	9.055	132	17,424	11.489
33	1,089	5.745	83	6,889	9.110	133	17,689	11.533
34	1,156	5.831	84	7,056	9.165	134	17,956	11.576
35	1,225	5.916	85	7,225	9.220	135	18,225	11.619
36	1,296	6.000	86	7,396	9.274	136	18,496	11.662
37	1,369	6.083	87	7,569	9.327	137	18,769	11.705
38	1,444	6.164	88	7,744	9.381	138	19,044	11.747
39	1,521	6.245	89	7,921	9.434	139	19,321	11.790
40	1,600	6.325	90	8,100	9.487	140	19,600	11.832
41	1,681	6.403	91	8,281	9.539	141	19,881	11.874
42	1,764	6.481	92	8,464	9.592	142	20,164	11.916
43	1,849	6.557	93	8,649	9.644	143	20,449	11.958
44	1,936	6.633	94	8,836	9.695	144	20,736	12.000
45	2,025	6.708	95	9,025	9.747	145	21,025	12.042
46	2,116	6.782	96	9,216	9.798	146	21,316	12.083
47	2,209	6.856	97	9,409	9.849	147	21,609	12.124
48	2,304	6.928	98	9,604	9.899	148	21,904	12.166
49	2,401	7.000	99	9,801	9.950	149	22,201	12.207
50	2,500	7.071	100	10,000	10.000	150	22,500	12.247

Formulas

Pythagorean and Quotient Identities

$$\sin^2 A + \cos^2 A = 1$$
$$\tan^2 A + 1 = \sec^2 A$$
$$\cot^2 A + 1 = \csc^2 A$$

$$\tan A = \frac{\sin A}{\cos A}$$

$$\cot A = \frac{\cos A}{\sin A}$$

Functions of the Sum of Two Angles

$$\sin (A + B) = \sin A \cos B + \cos A \sin B$$
$$\cos (A + B) = \cos A \cos B - \sin A \sin B$$
$$\tan (A + B) = \frac{\tan A + \tan B}{1 - \tan A \tan B}$$

Functions of the Difference of Two Angles

$$\sin (A - B) = \sin A \cos B - \cos A \sin B$$
$$\cos (A - B) = \cos A \cos B + \sin A \sin B$$
$$\tan (A - B) = \frac{\tan A - \tan B}{1 + \tan A \tan B}$$

Functions of the Double Angle

$$\sin 2A = 2 \sin A \cos A$$
$$\cos 2A = \cos^2 A - \sin^2 A$$
$$\cos 2A = 2 \cos^2 A - 1$$
$$\cos 2A = 1 - 2 \sin^2 A$$
$$\tan 2A = \frac{2 \tan A}{1 - \tan^2 A}$$

Functions of the Half Angle

$$\sin \tfrac{1}{2} A = \pm \sqrt{\frac{1 - \cos A}{2}}$$

$$\cos \tfrac{1}{2} A = \pm \sqrt{\frac{1 + \cos A}{2}}$$

$$\tan \tfrac{1}{2} A = \pm \sqrt{\frac{1 - \cos A}{1 + \cos A}}$$

Law of Sines

$$\frac{a}{\sin A} = \frac{b}{\sin B} = \frac{c}{\sin C}$$

Law of Cosines

$$a^2 = b^2 + c^2 - 2bc \cos A$$

Area of Triangle

$$K = \tfrac{1}{2} ab \sin C$$

Standard Deviation

$$\text{S.D.} = \sqrt{\frac{1}{n} \sum_{i=1}^{n} (\bar{x} - x_i)^2}$$

Common Logarithms of Numbers*

N	0	1	2	3	4	5	6	7	8	9
10	0000	0043	0086	0128	0170	0212	0253	0294	0334	0374
11	0414	0453	0492	0531	0569	0607	0645	0682	0719	0755
12	0792	0828	0864	0899	0934	0969	1004	1038	1072	1106
13	1139	1173	1206	1239	1271	1303	1335	1367	1399	1430
14	1461	1492	1523	1553	1584	1614	1644	1673	1703	1732
15	1761	1790	1818	1847	1875	1903	1931	1959	1987	2014
16	2041	2068	2095	2122	2148	2175	2201	2227	2253	2279
17	2304	2330	2355	2380	2405	2430	2455	2480	2504	2529
18	2553	2577	2601	2625	2648	2672	2695	2718	2742	2765
19	2788	2810	2833	2856	2878	2900	2923	2945	2967	2989
20	3010	3032	3054	3075	3096	3118	3139	3160	3181	3201
21	3222	3243	3263	3284	3304	3324	3345	3365	3385	3404
22	3424	3444	3464	3483	3502	3522	3541	3560	3579	3598
23	3617	3636	3655	3674	3692	3711	3729	3747	3766	3784
24	3802	3820	3838	3856	3874	3892	3909	3927	3945	3962
25	3979	3997	4014	4031	4048	4065	4082	4099	4116	4133
26	4150	4166	4183	4200	4216	4232	4249	4265	4281	4298
27	4314	4330	4346	4362	4378	4393	4409	4425	4440	4456
28	4472	4487	4502	4518	4533	4548	4564	4579	4594	4609
29	4624	4639	4654	4669	4683	4698	4713	4728	4742	4757
30	4771	4786	4800	4814	4829	4843	4857	4871	4886	4900
31	4914	4928	4942	4955	4969	4983	4997	5011	5024	5038
32	5051	5065	5079	5092	5105	5119	5132	5145	5159	5172
33	5185	5198	5211	5224	5237	5250	5263	5276	5289	5302
34	5315	5328	5340	5353	5366	5378	5391	5403	5416	5428
35	5441	5453	5465	5478	5490	5502	5514	5527	5539	5551
36	5563	5575	5587	5599	5611	5623	5635	5647	5658	5670
37	5682	5694	5705	5717	5729	5740	5752	5763	5775	5786
38	5798	5809	5821	5832	5843	5855	5866	5877	5888	5899
39	5911	5922	5933	5944	5955	5966	5977	5988	5999	6010
40	6021	6031	6042	6053	6064	6075	6085	6096	6107	6117
41	6128	6138	6149	6160	6170	6180	6191	6201	6212	6222
42	6232	6243	6253	6263	6274	6284	6294	6304	6314	6325
43	6335	6345	6355	6365	6375	6385	6395	6405	6415	6425
44	6435	6444	6454	6464	6474	6484	6493	6503	6513	6522
45	6532	6542	6551	6561	6571	6580	6590	6599	6609	6618
46	6628	6637	6646	6656	6665	6675	6684	6693	6702	6712
47	6721	6730	6739	6749	6758	6767	6776	6785	6794	6803
48	6812	6821	6830	6839	6848	6857	6866	6875	6884	6893
49	6902	6911	6920	6928	6937	6946	6955	6964	6972	6981
50	6990	6998	7007	7016	7024	7033	7042	7050	7059	7067
51	7076	7084	7093	7101	7110	7118	7126	7135	7143	7152
52	7160	7168	7177	7185	7193	7202	7210	7218	7226	7235
53	7243	7251	7259	7267	7275	7284	7292	7300	7308	7316
54	7324	7332	7340	7348	7356	7364	7372	7380	7388	7396
N	0	1	2	3	4	5	6	7	8	9

* This table gives the mantissas of numbers with the decimal point omitted in each case. Characteristics are determined by inspection from the numbers.

Common Logarithms of Numbers*

N	0	1	2	3	4	5	6	7	8	9
55	7404	7412	7419	7427	7435	7443	7451	7459	7466	7474
56	7482	7490	7497	7505	7513	7520	7528	7536	7543	7551
57	7559	7566	7574	7582	7589	7597	7604	7612	7619	7627
58	7634	7642	7649	7657	7664	7672	7679	7686	7694	7701
59	7709	7716	7723	7731	7738	7745	7752	7760	7767	7774
60	7782	7789	7796	7803	7810	7818	7825	7832	7839	7846
61	7853	7860	7868	7875	7882	7889	7896	7903	7910	7917
62	7924	7931	7938	7945	7952	7959	7966	7973	7980	7987
63	7993	8000	8007	8014	8021	8028	8035	8041	8048	8055
64	8062	8069	8075	8082	8089	8096	8102	8109	8116	8122
65	8129	8136	8142	8149	8156	8162	8169	8176	8182	8189
66	8195	8202	8209	8215	8222	8228	8235	8241	8248	8254
67	8261	8267	8274	8280	8287	8293	8299	8306	8312	8319
68	8325	8331	8338	8344	8351	8357	8363	8370	8376	8382
69	8388	8395	8401	8407	8414	8420	8426	8432	8439	8445
70	8451	8457	8463	8470	8476	8482	8488	8494	8500	8506
71	8513	8519	8525	8531	8537	8543	8549	8555	8561	8567
72	8573	8579	8585	8591	8597	8603	8609	8615	8621	8627
73	8633	8639	8645	8651	8657	8663	8669	8675	8681	8686
74	8692	8698	8704	8710	8716	8722	8727	8733	8739	8745
75	8751	8756	8762	8768	8774	8779	8785	8791	8797	8802
76	8808	8814	8820	8825	8831	8837	8842	8848	8854	8859
77	8865	8871	8876	8882	8887	8893	8899	8904	8910	8915
78	8921	8927	8932	8938	8943	8949	8954	8960	8965	8971
79	8976	8982	8987	8993	8998	9004	9009	9015	9020	9025
80	9031	9036	9042	9047	9053	9058	9063	9069	9074	9079
81	9085	9090	9096	9101	9106	9112	9117	9122	9128	9133
82	9138	9143	9149	9154	9159	9165	9170	9175	9180	9186
83	9191	9196	9201	9206	9212	9217	9222	9227	9232	9238
84	9243	9248	9253	9258	9263	9269	9274	9279	9284	9289
85	9294	9299	9304	9309	9315	9320	9325	9330	9335	9340
86	9345	9350	9355	9360	9365	9370	9375	9380	9385	9390
87	9395	9400	9405	9410	9415	9420	9425	9430	9435	9440
88	9445	9450	9455	9460	9465	9469	9474	9479	9484	9489
89	9494	9499	9504	9509	9513	9518	9523	9528	9533	9538
90	9542	9547	9552	9557	9562	9566	9571	9576	9581	9586
91	9590	9595	9600	9605	9609	9614	9619	9624	9628	9633
92	9638	9643	9647	9652	9657	9661	9666	9671	9675	9680
93	9685	9689	9694	9699	9703	9708	9713	9717	9722	9727
94	9731	9736	9741	9745	9750	9754	9759	9763	9768	9773
95	9777	9782	9786	9791	9795	9800	9805	9809	9814	9818
96	9823	9827	9832	9836	9841	9845	9850	9854	9859	9863
97	9868	9872	9877	9881	9886	9890	9894	9899	9903	9908
98	9912	9917	9921	9926	9930	9934	9939	9943	9948	9952
99	9956	9961	9965	9969	9974	9978	9983	9987	9991	9996
N	0	1	2	3	4	5	6	7	8	9

* This table gives the mantissas of numbers with the decimal point omitted in each case. Characteristics are determined by inspection from the numbers.

Values of Trigonometric Functions

Angle		Sin	Cos	Tan	Cot		
0°	00′	.0000	1.0000	.0000	—	90°	00′
	10	.0029	1.0000	.0029	343.77		50
	20	.0058	1.0000	.0058	171.89		40
	30	.0087	1.0000	.0087	114.59		30
	40	.0116	.9999	.0116	85.940		20
	50	.0145	.9999	.0145	68.750		10
1°	00′	.0175	.9998	.0175	57.290	89°	00′
	10	.0204	.9998	.0204	49.104		50
	20	.0233	.9997	.0233	42.964		40
	30	.0262	.9997	.0262	38.188		30
	40	.0291	.9996	.0291	34.368		20
	50	.0320	.9995	.0320	31.242		10
2°	00′	.0349	.9994	.0349	28.636	88°	00′
	10	.0378	.9993	.0378	26.432		50
	20	.0407	.9992	.0407	24.542		40
	30	.0436	.9990	.0437	22.904		30
	40	.0465	.9989	.0466	21.470		20
	50	.0494	.9988	.0495	20.206		10
3°	00′	.0523	.9986	.0524	19.081	87°	00′
	10	.0552	.9985	.0553	18.075		50
	20	.0581	.9983	.0582	17.169		40
	30	.0610	.9981	.0612	16.350		30
	40	.0640	.9980	.0641	15.605		20
	50	.0669	.9978	.0670	14.924		10
4°	00′	.0698	.9976	.0699	14.301	86°	00′
	10	.0727	.9974	.0729	13.727		50
	20	.0756	.9971	.0758	13.197		40
	30	.0785	.9969	.0787	12.706		30
	40	.0814	.9967	.0816	12.251		20
	50	.0843	.9964	.0846	11.826		10
5°	00′	.0872	.9962	.0875	11.430	85°	00′
	10	.0901	.9959	.0904	11.059		50
	20	.0929	.9957	.0934	10.712		40
	30	.0958	.9954	.0963	10.385		30
	40	.0987	.9951	.0992	10.078		20
	50	.1016	.9948	.1022	9.7882		10
6°	00′	.1045	.9945	.1051	9.5144	84°	00′
	10	.1074	.9942	.1080	9.2553		50
	20	.1103	.9939	.1110	9.0098		40
	30	.1132	.9936	.1139	8.7769		30
	40	.1161	.9932	.1169	8.5555		20
	50	.1190	.9929	.1198	8.3450		10
7°	00′	.1219	.9925	.1228	8.1443	83°	00′
	10	.1248	.9922	.1257	7.9530		50
	20	.1276	.9918	.1287	7.7704		40
	30	.1305	.9914	.1317	7.5958		30
	40	.1334	.9911	.1346	7.4287		20
	50	.1363	.9907	.1376	7.2687		10
8°	00′	.1392	.9903	.1405	7.1154	82°	00′
	10	.1421	.9899	.1435	6.9682		50
	20	.1449	.9894	.1465	6.8269		40
	30	.1478	.9890	.1495	6.6912		30
	40	.1507	.9886	.1524	6.5606		20
	50	.1536	.9881	.1554	6.4348		10
9°	00′	.1564	.9877	.1584	6.3138	81°	00′
		Cos	Sin	Cot	Tan	Angle	

Values of Trigonometric Functions

Angle		Sin	Cos	Tan	Cot		
9°	00'	.1564	.9877	.1584	6.3138	81°	00'
	10	.1593	.9872	.1614	6.1970		50
	20	.1622	.9868	.1644	6.0844		40
	30	.1650	.9863	.1673	5.9758		30
	40	.1679	.9858	.1703	5.8708		20
	50	.1708	.9853	.1733	5.7694		10
10°	00'	.1736	.9848	.1763	5.6713	80°	00'
	10	.1765	.9843	.1793	5.5764		50
	20	.1794	.9838	.1823	5.4845		40
	30	.1822	.9833	.1853	5.3955		30
	40	.1851	.9827	.1883	5.3093		20
	50	.1880	.9822	.1914	5.2257		10
11°	00'	.1908	.9816	.1944	5.1446	79°	00'
	10	.1937	.9811	.1974	5.0658		50
	20	.1965	.9805	.2004	4.9894		40
	30	.1994	.9799	.2035	4.9152		30
	40	.2022	.9793	.2065	4.8430		20
	50	.2051	.9787	.2095	4.7729		10
12°	00'	.2079	.9781	.2126	4.7046	78°	00'
	10	.2108	.9775	.2156	4.6382		50
	20	.2136	.9769	.2186	4.5736		40
	30	.2164	.9763	.2217	4.5107		30
	40	.2193	.9757	.2247	4.4494		20
	50	.2221	.9750	.2278	4.3897		10
13°	00'	.2250	.9744	.2309	4.3315	77°	00'
	10	.2278	.9737	.2339	4.2747		50
	20	.2306	.9730	.2370	4.2193		40
	30	.2334	.9724	.2401	4.1653		30
	40	.2363	.9717	.2432	4.1126		20
	50	.2391	.9710	.2462	4.0611		10
14°	00'	.2419	.9703	.2493	4.0108	76°	00'
	10	.2447	.9696	.2524	3.9617		50
	20	.2476	.9689	.2555	3.9136		40
	30	.2504	.9681	.2586	3.8667		30
	40	.2532	.9674	.2617	3.8208		20
	50	.2560	.9667	.2648	3.7760		10
15°	00'	.2588	.9659	.2679	3.7321	75°	00'
	10	.2616	.9652	.2711	3.6891		50
	20	.2644	.9644	.2742	3.6470		40
	30	.2672	.9636	.2773	3.6059		30
	40	.2700	.9628	.2805	3.5656		20
	50	.2728	.9621	.2836	3.5261		10
16°	00'	.2756	.9613	.2867	3.4874	74°	00'
	10	.2784	.9605	.2899	3.4495		50
	20	.2812	.9596	.2931	3.4124		40
	30	.2840	.9588	.2962	3.3759		30
	40	.2868	.9580	.2994	3.3402		20
	50	.2896	.9572	.3026	3.3052		10
17°	00'	.2924	.9563	.3057	3.2709	73°	00'
	10	.2952	.9555	.3089	3.2371		50
	20	.2979	.9546	.3121	3.2041		40
	30	.3007	.9537	.3153	3.1716		30
	40	.3035	.9528	.3185	3.1397		20
	50	.3062	.9520	.3217	3.1084		10
18°	00'	.3090	.9511	.3249	3.0777	72°	00'
		Cos	Sin	Cot	Tan	Angle	

Values of Trigonometric Functions

Angle		Sin	Cos	Tan	Cot		
18°	00′	.3090	.9511	.3249	3.0777	72°	00′
	10	.3118	.9502	.3281	3.0475		50
	20	.3145	.9492	.3314	3.0178		40
	30	.3173	.9483	.3346	2.9887		30
	40	.3201	.9474	.3378	2.9600		20
	50	.3228	.9465	.3411	2.9319		10
19°	00′	.3256	.9455	.3443	2.9042	71°	00′
	10	.3283	.9446	.3476	2.8770		50
	20	.3311	.9436	.3508	2.8502		40
	30	.3338	.9426	.3541	2.8239		30
	40	.3365	.9417	.3574	2.7980		20
	50	.3393	.9407	.3607	2.7725		10
20°	00′	.3420	.9397	.3640	2.7475	70°	00′
	10	.3448	.9387	.3673	2.7228		50
	20	.3475	.9377	.3706	2.6985		40
	30	.3502	.9367	.3739	2.6746		30
	40	.3529	.9356	.3772	2.6511		20
	50	.3557	.9346	.3805	2.6279		10
21°	00′	.3584	.9336	.3839	2.6051	69°	00′
	10	.3611	.9325	.3872	2.5826		50
	20	.3638	.9315	.3906	2.5605		40
	30	.3665	.9304	.3939	2.5386		30
	40	.3692	.9293	.3973	2.5172		20
	50	.3719	.9283	.4006	2.4960		10
22°	00′	.3746	.9272	.4040	2.4751	68°	00′
	10	.3773	.9261	.4074	2.4545		50
	20	.3800	.9250	.4108	2.4342		40
	30	.3827	.9239	.4142	2.4142		30
	40	.3854	.9228	.4176	2.3945		20
	50	.3881	.9216	.4210	2.3750		10
23°	00′	.3907	.9205	.4245	2.3559	67°	00′
	10	.3934	.9194	.4279	2.3369		50
	20	.3961	.9182	.4314	2.3183		40
	30	.3987	.9171	.4348	2.2998		30
	40	.4014	.9159	.4383	2.2817		20
	50	.4041	.9147	.4417	2.2637		10
24°	00′	.4067	.9135	.4452	2.2460	66°	00′
	10	.4094	.9124	.4487	2.2286		50
	20	.4120	.9112	.4522	2.2113		40
	30	.4147	.9100	.4557	2.1943		30
	40	.4173	.9088	.4592	2.1775		20
	50	.4200	.9075	.4628	2.1609		10
25°	00′	.4226	.9063	.4663	2.1445	65°	00′
	10	.4253	.9051	.4699	2.1283		50
	20	.4279	.9038	.4734	2.1123		40
	30	.4305	.9026	.4770	2.0965		30
	40	.4331	.9013	.4806	2.0809		20
	50	.4358	.9001	.4841	2.0655		10
26°	00′	.4384	.8988	.4877	2.0503	64°	00′
	10	.4410	.8975	.4913	2.0353		50
	20	.4436	.8962	.4950	2.0204		40
	30	.4462	.8949	.4986	2.0057		30
	40	.4488	.8936	.5022	1.9912		20
	50	.4514	.8923	.5059	1.9768		10
27°	00′	.4540	.8910	.5095	1.9626	63°	00′
		Cos	Sin	Cot	Tan	Angle	

Values of Trigonometric Functions

Angle		Sin	Cos	Tan	Cot		
27°	**00′**	.4540	.8910	.5095	1.9626	**63°**	**00′**
	10	.4566	.8897	.5132	1.9486		50
	20	.4592	.8884	.5169	1.9347		40
	30	.4617	.8870	.5206	1.9210		30
	40	.4643	.8857	.5243	1.9074		20
	50	.4669	.8843	.5280	1.8940		10
28°	**00′**	.4695	.8829	.5317	1.8807	**62°**	**00′**
	10	.4720	.8816	.5354	1.8676		50
	20	.4746	.8802	.5392	1.8546		40
	30	.4772	.8788	.5430	1.8418		30
	40	.4797	.8774	.5467	1.8291		20
	50	.4823	.8760	.5505	1.8165		10
29°	**00′**	.4848	.8746	.5543	1.8040	**61°**	**00′**
	10	.4874	.8732	.5581	1.7917		50
	20	.4899	.8718	.5619	1.7796		40
	30	.4924	.8704	.5658	1.7675		30
	40	.4950	.8689	.5696	1.7556		20
	50	.4975	.8675	.5735	1.7437		10
30°	**00′**	.5000	.8660	.5774	1.7321	**60°**	**00′**
	10	.5025	.8646	.5812	1.7205		50
	20	.5050	.8631	.5851	1.7090		40
	30	.5075	.8616	.5890	1.6977		30
	40	.5100	.8601	.5930	1.6864		20
	50	.5125	.8587	.5969	1.6753		10
31°	**00′**	.5150	.8572	.6009	1.6643	**59°**	**00′**
	10	.5175	.8557	.6048	1.6534		50
	20	.5200	.8542	.6088	1.6426		40
	30	.5225	.8526	.6128	1.6319		30
	40	.5250	.8511	.6168	1.6212		20
	50	.5275	.8496	.6208	1.6107		10
32°	**00′**	.5299	.8480	.6249	1.6003	**58°**	**00′**
	10	.5324	.8465	.6289	1.5900		50
	20	.5348	.8450	.6330	1.5798		40
	30	.5373	.8434	.6371	1.5697		30
	40	.5398	.8418	.6412	1.5597		20
	50	.5422	.8403	.6453	1.5497		10
33°	**00′**	.5446	.8387	.6494	1.5399	**57°**	**00′**
	10	.5471	.8371	.6536	1.5301		50
	20	.5495	.8355	.6577	1.5204		40
	30	.5519	.8339	.6619	1.5108		30
	40	.5544	.8323	.6661	1.5013		20
	50	.5568	.8307	.6703	1.4919		10
34°	**00′**	.5592	.8290	.6745	1.4826	**56°**	**00′**
	10	.5616	.8274	.6787	1.4733		50
	20	.5640	.8258	.6830	1.4641		40
	30	.5664	.8241	.6873	1.4550		30
	40	.5688	.8225	.6916	1.4460		20
	50	.5712	.8208	.6959	1.4370		10
35°	**00′**	.5736	.8192	.7002	1.4281	**55°**	**00′**
	10	.5760	.8175	.7046	1.4193		50
	20	.5783	.8158	.7089	1.4106		40
	30	.5807	.8141	.7133	1.4019		30
	40	.5831	.8124	.7177	1.3934		20
	50	.5854	.8107	.7221	1.3848		10
36°	**00′**	.5878	.8090	.7265	1.3764	**54°**	**00′**
		Cos	Sin	Cot	Tan	**Angle**	

Values of Trigonometric Functions

Angle		Sin	Cos	Tan	Cot		
36°	**00'**	.5878	.8090	.7265	1.3764	**54°**	**00'**
	10	.5901	.8073	.7310	1.3680		50
	20	.5925	.8056	.7355	1.3597		40
	30	.5948	.8039	.7400	1.3514		30
	40	.5972	.8021	.7445	1.3432		20
	50	.5995	.8004	.7490	1.3351		10
37°	**00'**	.6018	.7986	.7536	1.3270	**53°**	**00'**
	10	.6041	.7969	.7581	1.3190		50
	20	.6065	.7951	.7627	1.3111		40
	30	.6088	.7934	.7673	1.3032		30
	40	.6111	.7916	.7720	1.2954		20
	50	.6134	.7898	.7766	1.2876		10
38°	**00'**	.6157	.7880	.7813	1.2799	**52°**	**00'**
	10	.6180	.7862	.7860	1.2723		50
	20	.6202	.7844	.7907	1.2647		40
	30	.6225	.7826	.7954	1.2572		30
	40	.6248	.7808	.8002	1.2497		20
	50	.6271	.7790	.8050	1.2423		10
39°	**00'**	.6293	.7771	.8098	1.2349	**51°**	**00'**
	10	.6316	.7753	.8146	1.2276		50
	20	.6338	.7735	.8195	1.2203		40
	30	.6361	.7716	.8243	1.2131		30
	40	.6383	.7698	.8292	1.2059		20
	50	.6406	.7679	.8342	1.1988		10
40°	**00'**	.6428	.7660	.8391	1.1918	**50°**	**00'**
	10	.6450	.7642	.8441	1.1847		50
	20	.6472	.7623	.8491	1.1778		40
	30	.6494	.7604	.8541	1.1708		30
	40	.6517	.7585	.8591	1.1640		20
	50	.6539	.7566	.8642	1.1571		10
41°	**00'**	.6561	.7547	.8693	1.1504	**49°**	**00'**
	10	.6583	.7528	.8744	1.1436		50
	20	.6604	.7509	.8796	1.1369		40
	30	.6626	.7490	.8847	1.1303		30
	40	.6648	.7470	.8899	1.1237		20
	50	.6670	.7451	.8952	1.1171		10
42°	**00'**	.6691	.7431	.9004	1.1106	**48°**	**00'**
	10	.6713	.7412	.9057	1.1041		50
	20	.6734	.7392	.9110	1.0977		40
	30	.6756	.7373	.9163	1.0913		30
	40	.6777	.7353	.9217	1.0850		20
	50	.6799	.7333	.9271	1.0786		10
43°	**00'**	.6820	.7314	.9325	1.0724	**47°**	**00'**
	10	.6841	.7294	.9380	1.0661		50
	20	.6862	.7274	.9435	1.0599		40
	30	.6884	.7254	.9490	1.0538		30
	40	.6905	.7234	.9545	1.0477		20
	50	.6926	.7214	.9601	1.0416		10
44°	**00'**	.6947	.7193	.9657	1.0355	**46°**	**00'**
	10	.6967	.7173	.9713	1.0295		50
	20	.6988	.7153	.9770	1.0235		40
	30	.7009	.7133	.9827	1.0176		30
	40	.7030	.7112	.9884	1.0117		20
	50	.7050	.7092	.9942	1.0058		10
45°	**00'**	.7071	.7071	1.0000	1.0000	**45°**	**00'**
		Cos	Sin	Cot	Tan	Angle	

Logarithms of Trigonometric Functions
(Subtract 10 from each logarithm)

Angle		L Sin	L Cos	L Tan	L Cot		
0°	00′	—	10.0000	—	12.5363	90°	00′
	10	7.4637	10.0000	7.4637	12.2352		50
	20	7.7648	10.0000	7.7648	12.2352		40
	30	7.9408	10.0000	7.9409	12.0591		30
	40	8.0658	10.0000	8.0658	11.9342		20
	50	8.1627	10.0000	8.1627	11.8373		10
1°	00′	8.2419	9.9999	8.2419	11.7581	89°	00′
	10	8.3088	9.9999	8.3089	11.6911		50
	20	8.3668	9.9999	8.3669	11.6331		40
	30	8.4179	9.9999	8.4181	11.5819		30
	40	8.4637	9.9998	8.4638	11.5362		20
	50	8.5050	9.9998	8.5053	11.4947		10
2°	00′	8.5428	9.9997	8.5431	11.4569	88°	00′
	10	8.5776	9.9997	8.5779	11.4221		50
	20	8.6097	9.9996	8.6101	11.3899		40
	30	8.6397	9.9996	8.6401	11.3599		30
	40	8.6677	9.9995	8.6682	11.3318		20
	50	8.6940	9.9995	8.6945	11.3055		10
3°	00′	8.7188	9.9994	8.7194	11.2806	87°	00′
	10	8.7423	9.9993	8.7429	11.2571		50
	20	8.7645	9.9993	8.7652	11.2348		40
	30	8.7857	9.9992	8.7865	11.2135		30
	40	8.8059	9.9991	8.8067	11.1933		20
	50	8.8251	9.9990	8.8261	11.1739		10
4°	00′	8.8436	9.9989	8.8446	11.1554	86°	00′
	10	8.8613	9.9989	8.8624	11.1376		50
	20	8.8783	9.9988	8.8795	11.1205		40
	30	8.8946	9.9987	8.8960	11.1040		30
	40	8.9104	9.9986	8.9118	11.0882		20
	50	8.9256	9.9985	8.9272	11.0728		10
5°	00′	8.9403	9.9983	8.9420	11.0580	85°	00′
	10	8.9545	9.9982	8.9563	11.0437		50
	20	8.9682	9.9981	8.9701	11.0299		40
	30	8.9816	9.9980	8.9836	11.0164		30
	40	8.9945	9.9979	8.9966	11.0034		20
	50	9.0070	9.9977	9.0093	10.9907		10
6°	00′	9.0192	9.9976	9.0216	10.9784	84°	00′
	10	9.0311	9.9975	9.0336	10.9664		50
	20	9.0426	9.9973	9.0453	10.9547		40
	30	9.0539	9.9972	9.0567	10.9433		30
	40	9.0648	9.9971	9.0678	10.9322		20
	50	9.0755	9.9969	9.0786	10.9214		10
7°	00′	9.0859	9.9968	9.0891	10.9109	83°	00′
	10	9.0961	9.9966	9.0995	10.9005		50
	20	9.1060	9.9964	9.1096	10.8904		40
	30	9.1157	9.9963	9.1194	10.8806		30
	40	9.1252	9.9961	9.1291	10.8709		20
	50	9.1345	9.9959	9.1385	10.8615		10
8°	00′	9.1436	9.9958	9.1478	10.8522	82°	00′
	10	9.1525	9.9956	9.1569	10.8431		50
	20	9.1612	9.9954	9.1658	10.8342		40
	30	9.1697	9.9952	9.1745	10.8255		30
	40	9.1781	9.9950	9.1831	10.8169		20
	50	9.1863	9.9948	9.1915	10.8085		10
9°	00′	9.1943	9.9946	9.1997	10.8003	81°	00′
		L Cos	L Sin	L Cot	L Tan	Angle	

Logarithms of Trigonometric Functions
(Subtract 10 from each logarithm)

Angle		L Sin	L Cos	L Tan	L Cot		
9°	00'	9.1943	9.9946	9.1997	10.8003	81°	00'
	10	9.2022	9.9944	9.2078	10.7922		50
	20	9.2100	9.9942	9.2158	10.7842		40
	30	9.2176	9.9940	9.2236	10.7764		30
	40	9.2251	9.9938	9.2313	10.7687		20
	50	9.2324	9.9936	9.2389	10.7611		10
10°	00'	9.2397	9.9934	9.2463	10.7537	80°	00'
	10	9.2468	9.9931	9.2536	10.7464		50
	20	9.2538	9.9929	9.2609	10.7391		40
	30	9.2606	9.9927	9.2680	10.7320		30
	40	9.2674	9.9924	9.2750	10.7250		20
	50	9.2740	9.9922	9.2819	10.7181		10
11°	00'	9.2806	9.9919	9.2887	10.7113	79°	00'
	10	9.2870	9.9917	9.2953	10.7047		50
	20	9.2934	9.9914	9.3020	10.6980		40
	30	9.2997	9.9912	9.3085	10.6915		30
	40	9.3058	9.9909	9.3149	10.6851		20
	50	9.3119	9.9907	9.3212	10.6788		10
12°	00'	9.3179	9.9904	9.3275	10.6725	78°	00'
	10	9.3238	9.9901	9.3336	10.6664		50
	20	9.3296	9.9899	9.3397	10.6603		40
	30	9.3353	9.9896	9.3458	10.6542		30
	40	9.3410	9.9893	9.3517	10.6483		20
	50	9.3466	9.9890	9.3576	10.6424		10
13°	00'	9.3521	9.9887	9.3634	10.6366	77°	00'
	10	9.3575	9.9884	9.3691	10.6309		50
	20	9.3629	9.9881	9.3748	10.6252		40
	30	9.3682	9.9878	9.3804	10.6196		30
	40	9.3734	9.9875	9.3859	10.6141		20
	50	9.3786	9.9872	9.3914	10.6086		10
14°	00'	9.3837	9.9869	9.3968	10.6032	76°	00'
	10	9.3887	9.9866	9.4021	10.5979		50
	20	9.3937	9.9863	9.4074	10.5926		40
	30	9.3986	9.9859	9.4127	10.5873		30
	40	9.4035	9.9856	9.4178	10.5822		20
	50	9.4083	9.9853	9.4230	10.5770		10
15°	00'	9.4130	9.9849	9.4281	10.5719	75°	00'
	10	9.4177	9.9846	9.4331	10.5669		50
	20	9.4223	9.9843	9.4381	10.5619		40
	30	9.4269	9.9839	9.4430	10.5570		30
	40	9.4314	9.9836	9.4479	10.5521		20
	50	9.4359	9.9832	9.4527	10.5473		10
16°	00'	9.4403	9.9828	9.4575	10.5425	74°	00'
	10	9.4447	9.9825	9.4622	10.5378		50
	20	9.4491	9.9821	9.4669	10.5331		40
	30	9.4533	9.9817	9.4716	10.5284		30
	40	9.4576	9.9814	9.4762	10.5238		20
	50	9.4618	9.9810	9.4808	10.5192		10
17°	00'	9.4659	9.9806	9.4853	10.5147	73°	00'
	10	9.4700	9.9802	9.4898	10.5102		50
	20	9.4741	9.9798	9.4943	10.5057		40
	30	9.4781	9.9794	9.4987	10.5013		30
	40	9.4821	9.9790	9.5031	10.4969		20
	50	9.4861	9.9786	9.5075	10.4925		10
18°	00'	9.4900	9.9782	9.5118	10.4882	72°	00'
		L Cos	L Sin	L Cot	L Tan	Angle	

Logarithms of Trigonometric Functions
(Subtract 10 from each logarithm)

Angle		L Sin	L Cos	L Tan	L Cot		
18°	00′	9.4900	9.9782	9.5118	10.4882	72°	00′
	10	9.4939	9.9778	9.5161	10.4839		50
	20	9.4977	9.9774	9.5203	10.4797		40
	30	9.5015	9.9770	9.5245	10.4755		30
	40	9.5052	9.9765	9.5287	10.4713		20
	50	9.5090	9.9761	9.5329	10.4671		10
19°	00′	9.5126	9.9757	9.5370	10.4630	71°	00′
	10	9.5163	9.9752	9.5411	10.4589		50
	20	9.5199	9.9748	9.5451	10.4549		40
	30	9.5235	9.9743	9.5491	10.4509		30
	40	9.5270	9.9739	9.5531	10.4469		20
	50	9.5306	9.9734	9.5571	10.4429		10
20°	00′	9.5341	9.9730	9.5611	10.4389	70°	00′
	10	9.5375	9.9725	9.5650	10.4350		50
	20	9.5409	9.9721	9.5689	10.4311		40
	30	9.5443	9.9716	9.5727	10.4273		30
	40	9.5477	9.9711	9.5766	10.4234		20
	50	9.5510	9.9706	9.5804	10.4196		10
21°	00′	9.5543	9.9702	9.5842	10.4158	69°	00′
	10	9.5576	9.9697	9.5879	10.4121		50
	20	9.5609	9.9692	9.5917	10.4083		40
	30	9.5641	9.9687	9.5954	10.4046		30
	40	9.5673	9.9682	9.5991	10.4009		20
	50	9.5704	9.9677	9.6028	10.3972		10
22°	00′	9.5736	9.9672	9.6064	10.3936	68°	00′
	10	9.5767	9.9667	9.6100	10.3900		50
	20	9.5798	9.9661	9.6136	10.3864		40
	30	9.5828	9.9656	9.6172	10.3828		30
	40	9.5859	9.9651	9.6208	10.3792		20
	50	9.5889	9.9646	9.6243	10.3757		10
23°	00′	9.5919	9.9640	9.6279	10.3721	67°	00′
	10	9.5948	9.9635	9.6314	10.3686		50
	20	9.5978	9.9629	9.6348	10.3652		40
	30	9.6007	9.9624	9.6383	10.3617		30
	40	9.6036	9.9618	9.6417	10.3583		20
	50	9.6065	9.9613	9.6452	10.3548		10
24°	00′	9.6093	9.9607	9.6486	10.3514	66°	00′
	10	9.6121	9.9602	9.6520	10.3480		50
	20	9.6149	9.9596	9.6553	10.3447		40
	30	9.6177	9.9590	9.6587	10.3413		30
	40	9.6205	9.9584	9.6620	10.3380		20
	50	9.6232	9.9579	9.6654	10.3346		10
25°	00′	9.6259	9.9573	9.6687	10.3313	65°	00′
	10	9.6286	9.9567	9.6720	10.3280		50
	20	9.6313	9.9561	9.6752	10.3248		40
	30	9.6340	9.9555	9.6785	10.3215		30
	40	9.6366	9.9549	9.6817	10.3183		20
	50	9.6392	9.9543	9.6850	10.3150		10
26°	00′	9.6418	9.9537	9.6882	10.3118	64°	00′
	10	9.6444	9.9530	9.6914	10.3086		50
	20	9.6470	9.9524	9.6946	10.3054		40
	30	9.6495	9.9518	9.6977	10.3023		30
	40	9.6521	9.9512	9.7009	10.2991		20
	50	9.6546	9.9505	9.7040	10.2960		10
27°	00′	9.6570	9.9499	9.7072	10.2928	63°	00′
		L Cos	L Sin	L Cot	L Tan	Angle	

Logarithms of Trigonometric Functions
(Subtract 10 from each logarithm)

Angle		L Sin	L Cos	L Tan	L Cot		
27°	**00′**	9.6570	9.9499	9.7072	10.2928	**63°**	**00′**
	10	9.6595	9.9492	9.7103	10.2897		50
	20	9.6620	9.9486	9.7134	10.2866		40
	30	9.6644	9.9479	9.7165	10.2835		30
	40	9.6668	9.9473	9.7196	10.2804		20
	50	9.6692	9.9466	9.7226	10.2774		10
28°	**00′**	9.6716	9.9459	9.7257	10.2743	**62°**	**00′**
	10	9.6740	9.9453	9.7287	10.2713		50
	20	9.6763	9.9446	9.7317	10.2683		40
	30	9.6787	9.9439	9.7348	10.2652		30
	40	9.6810	9.9432	9.7378	10.2622		20
	50	9.6833	9.9425	9.7408	10.2592		10
29°	**00′**	9.6856	9.9418	9.7438	10.2562	**61°**	**00′**
	10	9.6878	9.9411	9.7467	10.2533		50
	20	9.6901	9.9404	9.7497	10.2503		40
	30	9.6923	9.9397	9.7526	10.2474		30
	40	9.6946	9.9390	9.7556	10.2444		20
	50	9.6968	9.9383	9.7585	10.2415		10
30°	**00′**	9.6990	9.9375	9.7614	10.2386	**60°**	**00′**
	10	9.7012	9.9368	9.7644	10.2356		50
	20	9.7033	9.9361	9.7673	10.2327		40
	30	9.7055	9.9353	9.7701	10.2299		30
	40	9.7076	9.9346	9.7730	10.2270		20
	50	9.7097	9.9338	9.7759	10.2241		10
31°	**00′**	9.7118	9.9331	9.7788	10.2212	**59°**	**00′**
	10	9.7139	9.9323	9.7816	10.2184		50
	20	9.7160	9.9315	9.7845	10.2155		40
	30	9.7181	9.9308	9.7873	10.2127		30
	40	9.7201	9.9300	9.7902	10.2098		20
	50	9.7222	9.9292	9.7930	10.2070		10
32°	**00′**	9.7242	9.9284	9.7958	10.2042	**58°**	**00′**
	10	9.7262	9.9276	9.7986	10.2014		50
	20	9.7282	9.9268	9.8014	10.1986		40
	30	9.7302	9.9260	9.8042	10.1958		30
	40	9.7322	9.9252	9.8070	10.1930		20
	50	9.7342	9.9244	9.8097	10.1903		10
33°	**00′**	9.7361	9.9236	9.8125	10.1875	**57°**	**00′**
	10	9.7380	9.9228	9.8153	10.1847		50
	20	9.7400	9.9219	9.8180	10.1820		40
	30	9.7419	9.9211	9.8208	10.1792		30
	40	9.7438	9.9203	9.8235	10.1765		20
	50	9.7457	9.9194	9.8263	10.1737		10
34°	**00′**	9.7476	9.9186	9.8290	10.1710	**56°**	**00′**
	10	9.7494	9.9177	9.8317	10.1683		50
	20	9.7513	9.9169	9.8344	10.1656		40
	30	9.7531	9.9160	9.8371	10.1629		30
	40	9.7550	9.9151	9.8398	10.1602		20
	50	9.7568	9.9142	9.8425	10.1575		10
35°	**00′**	9.7586	9.9134	9.8452	10.1548	**55°**	**00′**
	10	9.7604	9.9125	9.8479	10.1521		50
	20	9.7622	9.9116	9.8506	10.1494		40
	30	9.7640	9.9107	9.8533	10.1467		30
	40	9.7657	9.9098	9.8559	10.1441		20
	50	9.7675	9.9089	9.8586	10.1414		10
36°	**00′**	9.7692	9.9080	9.8613	10.1387	**54°**	**00′**
		L Cos	L Sin	L Cot	L Tan	Angle	

Logarithms of Trigonometric Functions
(Subtract 10 from each logarithm)

Angle		L Sin	L Cos	L Tan	L Cot		
36°	00′	9.7692	9.9080	9.8613	10.1387	54°	00′
	10	9.7710	9.9070	9.8639	10.1361		50
	20	9.7727	9.9061	9.8666	10.1334		40
	30	9.7744	9.9052	9.8692	10.1308		30
	40	9.7761	9.9042	9.8718	10.1282		20
	50	9.7778	9.9033	9.8745	10.1255		10
37°	00′	9.7795	9.9023	9.8771	10.1229	53°	00′
	10	9.7811	9.9014	9.8797	10.1203		50
	20	9.7828	9.9004	9.8824	10.1176		40
	30	9.7844	9.8995	9.8850	10.1150		30
	40	9.7861	9.8985	9.8876	10.1124		20
	50	9.7877	9.8975	9.8902	10.1098		10
38°	00′	9.7893	9.8965	9.8928	10.1072	52°	00′
	10	9.7910	9.8955	9.8954	10.1046		50
	20	9.7926	9.8945	9.8980	10.1020		40
	30	9.7941	9.8935	9.9006	10.0994		30
	40	9.7957	9.8925	9.9032	10.0968		20
	50	9.7973	9.8915	9.9058	10.0942		10
39°	00′	9.7989	9.8905	9.9084	10.0916	51°	00′
	10	9.8004	9.8895	9.9110	10.0890		50
	20	9.8020	9.8884	9.9135	10.0865		40
	30	9.8035	9.8874	9.9161	10.0839		30
	40	9.8050	9.8864	9.9187	10.0813		20
	50	9.8066	9.8853	9.9212	10.0788		10
40°	00′	9.8081	9.8843	9.9238	10.0762	50°	00′
	10	9.8096	9.8832	9.9264	10.0736		50
	20	9.8111	9.8821	9.9289	10.0711		40
	30	9.8125	9.8810	9.9315	10.0685		30
	40	9.8140	9.8800	9.9341	10.0659		20
	50	9.8155	9.8789	9.9366	10.0634		10
41°	00′	9.8169	9.8778	9.9392	10.0608	49°	00′
	10	9.8184	9.8767	9.9417	10.0583		50
	20	9.8198	9.8756	9.9443	10.0557		40
	30	9.8213	9.8745	9.9468	10.0532		30
	40	9.8227	9.8733	9.9494	10.0506		20
	50	9.8241	9.8722	9.9519	10.0481		10
42°	00′	9.8255	9.8711	9.9544	10.0456	48°	00′
	10	9.8269	9.8699	9.9570	10.0430		50
	20	9.8283	9.8688	9.9595	10.0405		40
	30	9.8297	9.8676	9.9621	10.0379		30
	40	9.8311	9.8665	9.9646	10.0354		20
	50	9.8324	9.8653	9.9671	10.0329		10
43°	00′	9.8338	9.8641	9.9697	10.0303	47°	00′
	10	9.8351	9.8629	9.9722	10.0278		50
	20	9.8365	9.8618	9.9747	10.0253		40
	30	9.8378	9.8606	9.9772	10.0228		30
	40	9.8391	9.8594	9.9798	10.0202		20
	50	9.8405	9.8582	9.9823	10.0177		10
44°	00′	9.8418	9.8569	9.9848	10.0152	46°	00′
	10	9.8431	9.8557	9.9874	10.0126		50
	20	9.8444	9.8545	9.9899	10.0101		40
	30	9.8457	9.8532	9.9924	10.0076		30
	40	9.8469	9.8520	9.9949	10.0051		20
	50	9.8482	9.8507	9.9975	10.0025		10
45°	00′	9.8495	9.8495	10.0000	10.0000	45°	00′
		L Cos	L Sin	L Cot	L Tan	Angle	

Index